T0419252

MATERIALS SCIENCE AND TECHNOLOGIES

ADVANCED ORGANIC-INORGANIC COMPOSITES

MATERIALS, DEVICES AND ALLIED APPLICATIONS

MATERIALS SCIENCE AND TECHNOLOGIES

Additional books in this series can be found on Nova's website
under the Series tab.

Additional E-books in this series can be found on Nova's website
under the E-book tab.

MATERIALS SCIENCE AND TECHNOLOGIES

ADVANCED ORGANIC-INORGANIC COMPOSITES

MATERIALS, DEVICES AND ALLIED APPLICATIONS

INAMUDDIN
EDITOR

Nova Science Publishers, Inc.
New York

For permission to use material from this book please contact us:
Telephone 631-231-7269; Fax 631-231-8175
Web Site: http://www.novapublishers.com

NOTICE TO THE READER

Additional color graphics may be available in the e-book version of this book.

Library of Congress Cataloging-in-Publication Data

Advanced organic-inorganic composites : materials, devices, and allied applications / editor, Inamuddin.
p. cm.
Includes bibliographical references and index.
ISBN 978-1-61324-264-3 (hardcover)
1. Composite materials. I. Inamuddin, 1980-
TA418.9.C6A2824 2011
620.1'18--dc22

201100905

Published by Nova Science Publishers, Inc. † New York

To my parents,
my wife Mrs. Khushbu Jahan and
my loving son Mohammad Uzair

CONTENTS

Preface **ix**

Chapter 1 Organic-Inorganic Composite Polymer Electrolyte Membranes for Fuel Cells **1**
Zhongqing Jiang, Zhong-Jie Jiang and Yuedong Meng

Chapter 2 Surface-Confined Inorganic-Organic Hybrid Materials for Sensor Functionalities **33**
Prakash Chandra Mondal, Bhawna Gera and Tarkeshwar Gupta

Chapter 3 Organic-Inorganic Composite Materials as Sorbents and Ion-exchangers Used in Environmental Protection **69**
Dorota Kołodyńska

Chapter 4 Organic-Inorganic Composites for Solar Cells **131**
Jing-Shun Huang, Chen-Yu Chou and Ching-Fuh Lin

Chapter 5 Organic-Inorganic Composite Coatings **169**
Eram Sharmin, Fahmina Zafar,Manawwar Alamand Sharif Ahmad

Chapter 6 Polymer Electrolyte Membranes and Electrodes Fabricated Using Plasma Method and their Applications in Proton Exchange Membrane Fuel Cells **191**
Zhongqing Jiang, Xinyao Yu, Zhong-Jie Jiang, and Yuedong Meng

Chapter 7 Polymer/Ceramic/Metal Composites for Automobiles **243**
Mukesh Kumar

Chapter 8 Light Emission from Organic-Inorganic Composite Materials **257**
Muhammd Jamil, Farzana Ahmad, Jin Woo Lee, and Young Jae Jeon

Chapter 9 Bimetal Composites for Application to a Cleaner Environment **291**
Ahmed Hamood Ahmed Dabwan, Satoshi Kaneco, Hideyuki Katsumata, Tohru Suzuki and Kiyohisa Ohta

Chapter 10	CeO2 Based Electro-Ceramic Composite Materials for Solid Oxide Fuel Cell Applications *Suddhasatwa Basu and Rajalekshmi Chockalingam*	**311**
Chapter 11	Composite Materials in Dental Applications *Geeta Rajput and Sarwat Husain Hashmi*	**353**
Chapter 12	Organic/Inorganic Composites for Energy Storage in Electrochemical Capacitors *Sadaf Zaidi and Inamuddin*	**369**
Chapter 13	Nanocomposite Coatings for Corrosion Control *Rana Sardar*	**403**
Chapter 14	Composite Materials for Orthopaedic Aids *Geeta Rajput, and Aftab Ahmad Iraqi*	**429**
Chapter 15	Biphasic Layer Materials in Thin Layer Chromatographic Analysis of Inorganic and Organic Matrices *Ali Mohammad, and Abdul Moheman*	**437**
Chapter 16	Spectroscopic, Morphological and Electrochemical Properties of Polyaniline: Fly Ash Nano Composites *Sipho Mavundla, Amir-Al-Ahmed, Leslie Petrik, Priscilla G. L. Baker and Emmanuel I. Iwuoha*	**465**
Chapter 17	Synthesis and Spectro-chemical Interrogation of Metal Oxide Modified Polypyrrole Hybrid Composites *Richard Akinyeye, Priscilla G. L. Baker and Emmanuel Iwuoha*	**477**
Chapter 18	Polymer Composites: Toxicity, Environment and Human Health *Abhinav Srivastava, Mayank Pandey, Pradeep K. Mathur and Vinod P. Sharma*	**491**
Chapter 19	Synthesis, Characterisation of Polyaniline-co-poly(2,2'-dithiodianiline) and its Sensor Application for Trace Metal Determination *Vernon Somerset, Emmanuel I. Iwuoha and Lucas Hernandez*	**499**
Chapter 20	Basic Properties of High Temperature Superconductors and its Applications *Intikhab Aalam Ansari*	**511**
Chapter 21	Polymers and Composites *Zeid A. AL-Othman and Mu. Naushad*	**529**
Acknowledgments		**539**
Index		**541**

PREFACE

Composite materials appeared in human technology since several hundred years before Christ has applied innovations to improve the quality of life. The wisdom of the dead civilization may be testifying as the Long Ancient Pharaohs made their slaves using bricks with straw to enhance the structural integrity of their buildings. Although, composites were known to mankind since prehistoric times, the concept and technology have undergone a sea change with better understanding of the basics like the bonding mechanism between the matrix and dispersoids, dispersoid size and distribution, morphological features etc. In fact, composites are now one of the most important classes of engineered materials, as they offer several outstanding properties as compared to conventional materials. These materials have found increasingly wider applications in the general areas of chemistry, physics, nanotechnology, material science and engineering. *Advanced Organic-Inorganic Composites: Materials, Devices and Allied Applications* have been compiled to broadly explore the developments in the field of composite materials.

This book is the outcome of remarkable contributions of 49 experts of interdisciplinary fields of science, with comprehensive, in-depth and up-to-date research works and reviews. The book edition thoroughly covers the latest development of composites especially organic-inorganic composite materials. This book is highly useful for various disciplines of science, engineering, biomedicine, dental medicine, nanotechnology, etc. This book is compiled using more than 3900 references, 200 figures, 20 tables and 25 equations. This edition hopefully will prove worthy to the scientist working in the fields of chemistry, polymer chemistry, electrochemistry, material science, polymer electrolyte membranes, fuel cells, environmental chemistry, sensors, coatings, automobiles parts, corrosion, biomedical, conducting polymers and solar cells. Based on thematic topics, the book edition contains the following 21 chapters:

Chapter 1: In this chapter, various types of inorganic and organic materials that are used to prepare organic-inorganic composite polymer electrolyte membranes have been reviewed. The properties and performances of organic-inorganic composite membranes applied in polymer electrolyte membrane fuel cells are also discussed.

Chapter 2: In this chapter, surface-confined inorganic-organic hybrid materials for sensor functionalities are reviewed.

Chapter 3: This chapter provides the review of the organic-inorganic composites used in the environmental protection processes.

Chapter 4: This chapter focuses on the development of highly efficient, air-stable, and low-cost ZnO nanostructure/organic composite solar cells.

Chapter 5: This chapter provides a brief overview of organic-inorganic hybrid coatings, metal-polymer, conducting polymer, UV curable and waterborne as well as some vegetable oil based composite coatings. A few characterization methods of the composite coatings have also been exemplified.

Chapter 6: In this chapter, polymer electrolyte membranes and electrodes fabricated using plasma method and their applications in proton exchange membrane fuel cells are reviewed.

Chapter 7: In this chapter, preparation methods and properties of different types of polymer/ceramic/metal composites for automobiles are discussed.

Chapter 8: This chapter reviews and addresses the light emission characteristics, working principle, the basic characteristics and the fabrication methods of organic, inorganic and organic-inorganic hybrid materials. Moreover, some of the applications of organic, inorganic and organic-inorganic hybrid composite materials are also presented.

Chapter 9: In this chapter, the single metal and bimetallic metal composites applied for the removal of trihalomethanes from the aqueous solution in the flow system to safeguard the environment are discussed.

Chapter 10: In this chapter, an attempt has been made to briefly review the composite mixed ionic-electronic conductors used as electrolytes and electrodes.

Chapter 11: In this chapter, properties, clinical aspects and applications of dental composites are reviewed.

Chapter 12: In this chapter, the fundamentals of electrochemical capacitors (ECs), their comparison with batteries, the storage principles and characteristics of electrode materials for ECs such as, carbon-based materials, transition metal oxides and conductive polymers, are briefly discussed. Special emphasis is laid on an up-to-date, state-of-the-art review of the organic-inorganic composite materials being developed for use as electrodes in ECs.

Chapter 13: In this chapter, a brief introduction to corrosion and application of nanocomposite coatings in the field of corrosion prevention of metals is reviewed. Recent research and developments in designing efficient coating materials for corrosion control are also discussed.

Chapter 14: In this chapter properties and application of composite materials used in orthopaedic aids are reviewed.

Chapter 15: In this chapter the use of biphasic layer materials during last thirty years (1980—2010) in the analysis of organic and inorganic compounds by thin layer chromatography (TLC) are reviewed.

Chapter 16: In this chapter, preparation, spectroscopic, morphological and electrochemical properties of polyaniline: fly ash nano composites are discussed.

Chapter 17: In this chapter, synthesis and spectro-chemical interrogation of metal oxide modified polypyrrole hybrid composites are discussed.

Chapter 18: In this chapter, toxicity, environment implication and human health related issues of polymer composites are discussed.

Chapter 19: In this chapter, synthesis, characterization of polyaniline-co-poly(2,2'-dithiodianiline) and its sensor application for trace metal determination are discussed.

Chapter 20: In this chapter, basic properties and applications of high temperature superconductors are discussed.

Chapter 21: In this chapter, the history, introduction and applications of polymer composite are discussed.

In: Advanced Organic-Inorganic Composites
Editor: Inamuddin

ISBN 978-1-61324-264-3

Chapter 1

ORGANIC-INORGANIC COMPOSITE POLYMER ELECTROLYTE MEMBRANES FOR FUEL CELLS

***Zhongqing Jiang*[a,b,*], *Zhong-Jie Jiang*[c,*] *and Yuedong Meng*[a]**

[a]Institute of Plasma Physics, Chinese Academy of Sciences,
Hefei 230031, Anhui, China

[b]Department of Chemical Engineering, Ningbo University of Technology,
Ningbo 315016, Zhejiang, China

[c]Department of Nature and Sciences, University of California,
Merced 95348, U. S.

ABSTRACT

Proton exchange membrane (PEM), which usually serves as a barrier for separating the reactants and as a medium for conducting protons, plays an important role in determining the performance of proton exchange membrane fuel cells (PEMFCs). Currently, commercially available PEMs, such as Nafion, do not meet the commercialization requirements for high power density direct methanol fuel cells (DMFCs), partly due to their relatively high membrane thickness, high cost, limited operation temperature (<80 °C) and high methanol permeability. Great efforts have therefore been devoted to the exploitation of new types of PEMs. Among various types of PEMs, organic-inorganic composite polymer electrolyte membranes have received particular attention. It is practically rational to prepare PEMs by the addition of a functional or nonfunctional inorganic into an organic material to obtain the properties and performances that cannot be achieved by their individual components alone. In this chapter, various types of inorganic and organic materials that are used to prepare organic-inorganic composite polymer electrolyte membranes have been reviewed. The properties and performances of the obtained organic-inorganic composite membranes with applications in PEMFCs are discussed.

* Corresponding author's e-mail: zhongqingjiang@hotmail.com

ABBREVIATIONS

ABPBI	Combination of phosphomolybdic acid $H_3PMo_{12}O_{40}$ (PMo_{12}) and poly (2,5-benzimidazole)
AMPS	2-acrylamido-2-methyl-1-propanesulfonic acid
ATRP	Atom transfer radical polymerization
BS	Bifunctional sulfonated phenethylsilica
CHP	Calciumhydroxyphosphate ($Ca_5(PO_4)_3OH$)
CO	Carbon monoxide
$CsHSO_4$	Hydrogen cesium sulfate
DDS	Diphenyldimethoxysilane
DI	Deionized
DMFC	Direct methanol fuel cell
DS	Degree of sulfonation
FTIR	Forier transform infrared
HPA	Heteropoly acid
IEC	Ion exchange capacity
OCV	Open circuit voltage
ORR	Oxygen reduction reaction
PAC	Polymer acid complexe
PAEEKK	Poly (arylene ether ether ketone ketone)
PAES	Poly (arylene ether sulfone)
PBI	Polybenzimidazole
PEEK	Polyether ether ketone
PEM	Proton exchange membrane
PEMFC	Proton exchange membrane fuel cell
PETMS	Phenethyltrimethoxysilane
PFI	Perfluorinated ionomer
PFSA	Perfluorosulfonic acid
PMA	Phosphomolybdic acid
POM	Polyoxometallate
PPEK	Poly (phthalazinone ether ketone)
PSSA	Poly (styrene sulfonic acid)
PTA	Phosphotungstic acid
PTFE	Polytetrafluoroethylene
PVA	Poly (vinyl alcohol)
PVDF	Polyvinylidenedifluoride
PVDF-*g*-PSSA	Poly (vinylidene fluoride) grafted polystyrene sulfonated acid
PVDF-HFP	PVDF copolymerized with hexafluoropropylene
PWA	Phosphotungstic acid
MEA	Membrane electrode assembly
MMT	Montmorillonite
NASTA	A Nafion solution and a HPA, silicotungstic acid
NBS	Nafion/Bifunctional sulphonated phenethylsilica
NP-PCM	Nanoporous proton-conducting membrane
N-Sdds	Nafion-sulfonated diphenyldimethoxysilane
RH	Relative humidity
SPEEKS	Sulfonated poly (ether ether ketone sulfone)

SPSU-BP	Sulfonated poly (biphenyl ether sulfone)
STA	Silicotungstic acid
S-ZrO_2	Sulfated-ZrO_2
S-ZrO_2 (p)	Sulfated-ZrO_2 (p) (where p means that the powdered state $ZrO_2 \cdot nH_2O$ was used as the precursor.)
$Zr(HPO_4)(O_3P{-}C_6H_4SO_3H)$	Zirconium phosphate sulfophenylenphosphonate
ZrP	Zirconium phosphate
Zr(PBTC)	Zirconium tricarboxybutylphosphonate
ZrSPP	Zirconium sulphophenyl phosphate

1. INTRODUCTION

Proton exchange membrane fuel cells (PEMFCs) with polymer membranes as electrolytes, originally termed as solid polymer electrolyte fuel cells, have been developed since 1950's [1]. Due to attractive features of high power density, rapid start-up, environmental benign and high efficiency, PEMFCs have stimulated great research interests among scientists and engineers of various disciplines in recent years [2-4]. Currently, great efforts have been put on the development of methanol fueled PEMFCs, e.g direct methanol fuel cells (DMFCs) [5-10]. In comparison to other PEMFCs, DMFC is especially attractive since it operates at room temperature, uses liquid methanol as the fuel, and eliminates the expensive setup of hydrogen reformers and the hydrogen storage problems associated with H_2 fueled PEMFCs. However, to become commercially viable, PEMFCs have to overcome the barrier of low activity and poor durability of electrocatalysts in the fuel cell electrode. Up to now, research in this area has received great progress [2,11-14]. For example, catalyst loading has been significantly reduced, power density has been greatly increased, and prototype PEMFCs vehicles have been successfully tested in many continents. Despite these progressive advances, there are still strong needs to further enhance the performance of current PEMFCs. Several technical obstacles hinder their widespread commercialization for transportation and stationary applications. These obstacles include inadequate water and heat management, intolerance to impurities such as carbon monoxide (CO), sluggish electrochemical cathode kinetics, high cost, and limitations in operating temperature ranging from 0 to 100 °C in order to prevent water, an important medium for proton transportation, from freezing and evaporating in the flow fields of the fuel cell. All these problems convergence to a point that the properties of proton exchange membranes (PEMs) used in fuel cells must be improved. As an indispensable component, PEM, which usually serves as a barrier for separating the reactants and as a medium for conducting protons, plays an important role in determining the performance of a PEMFC. The configuration of a PEM typically contains a network of hydrophilic nanopores embedded in a hydrophobic domain. The hydrophilic nanopores, which play an important role in transporting the protons, contain water and acidic moieties. The proton migrating through the hydrophilic nanopores usually associates with water. The mechanical reinforcement of membranes mainly depends on the hydrophobic phase of the membrane. It is generally believed the conduction of proton is through a mixture of the vehicle mechanism [15] in which protons are transferred through the medium as complexes of $H^+(H_2O)_n$, and the Grotthus mechanism [16], or proton hopping, in which the proton passes down the chain of water molecules, with successive formations and deformations of H_3O^+, between the sulfonic acid groups. Currently, fully hydrated sulfonic acid groups are known to have sufficient proton conductivity to be practicable in PEMFCs, such as perfluorosulfonic acid (PFSA) membranes. Due to the great strength of the C−F bond, the membranes with a perfluoro backbone usually possess excellent chemical and thermal stabilities in oxidative and reductive media. In the presence of water, the terminal sulfonic acid groups of PFSA are hydrated to form ion clusters, which give rise to high proton conductivity, e.g. $\sigma_{H^+} = \sim 10^{-2}$ S cm^{-1} for Nafion® 117, at room temperature. However, PFSA has the following major disadvantages in DMFCs applications [17]:

- Very high methanol crossover. Because of the good affinity of methanol molecules to water molecules, methanol can permeate from the anode to the cathode through the hydrophilic nanopores, and over 40% of the methanol will be wasted in DMFCs. Methanol crossover, which would lead to the loss of fuel and the reduction of cathode voltage and cell performance, is therefore one of the important problems that must be addressed in DMFCs application.
- Very high cost. Currently, for a 50 kW PEMFCs in automotive applications, Nafion® polymer membranes account for 20% of the total cost of Nafion® based membrane electrode assemblies (MEAs). The cost will be even higher when the thicker membranes are needed to reduce methanol crossover. Typically, Nafion® membranes for DMFCs applications have a price in the range of \$600–1200 m^{-2}, depending on the thickness [17].
- Marked dependence of the conductivity of PEM on its water content. Because the PEM functions only in a highly hydrated state, it is limited to be operated at temperatures <80 °C under ambient pressure in order to maintain a high water content in the membrane. Thus, to prevent dry-out of the membrane and keep the membrane at high conductive state, the reactant gases are usually humidified through an external humidification system before entering the fuel cells. This process, however, increases the weight and complexity of the fuel cell system, making PEMFCs unsuitable for portable application.

Challenges associated with the PEMFC power technology are mainly related to the water contents. It is well known that the proton conductivity of a PEM is strictly dependent upon the content of water inside. The content of water of the PEMs is usually affected by its operation temperature. In this sense, the design of a new membrane that can conduct protons at desired temperature is perhaps the greatest challenge in the science community. There are several compelling reasons for operation of DMFCs at a higher temperature [18, 19]:

i. At the higher temperature, electrochemical kinetics for both electrode reactions can be enhanced, and poisoning of the Pt catalyst caused by CO are thus reduced [20];
ii. Water management can be simplified because only a single phase of water needs to be considered;
iii. The cooling system is simplified due to the increased temperature gradient between the fuel cell stack and the coolant;
iv. Waste heat can be recovered as a practical energy source;

v. CO tolerance is dramatically increased thereby allowing fuel cells to use lower quality reformed fuel;
vi. Methanol crossover is reduced. Since the methanol is partially or entirely in the vapor phase over 80 °C, the solubility and molecular activity of methanol could be decreased to reduce the methanol crossover with increasing operating temperature in a DMFC [19,21].

However, PEMs which are operated at higher temperature (>100 °C) might suffer from the disadvantage of dehydration. A decrease in the water content of PEMs will result in an increase in fuel permeability and reduction in the proton conductivity. Recently, great efforts have been put on the syntheses of new PEMs with the ultimate goal of being operable at elevated temperature and dry gas environment [18]. These approaches mainly include:

1) Producing new polymeric ionomers by acidifying, usually sulfonating polymers with known thermal and chemical stability, or synthesizing new polymers using sulfonated and nonsulfonated monomers as either block or random polymers;
2) Fabricating composite membranes with inorganic proton conductors or potential proton conductors either in an inert matrix in which the inorganic moiety is the proton conductor and water reservoir or in a hybrid composite membrane with a polymeric ionomer. Furthermore, the methodology of introducing functional groups onto inorganic moieties has been successfully used to improve proton conductivity at higher temperature with low fuel flux and high mechanical and oxidation stability.

The latter category of organic-inorganic composite proton exchange membranes and their perspectives will be the subject of this chapter. The production of composite membranes by hybridization of organic and inorganic materials at a molecular level is an extensive and fascinating field of investigation. A major benefit of such research activities is linked to synergetic effects of organic and inorganic components with desired and improved properties which are unavailable for isolated materials [22].

2. Definition of Organic-Inorganic Composite Polymer Electrolyte Membranes

An organic-inorganic composite polymer electrolyte membrane is referred to as a composite membrane in which an inorganic moiety is combined with a polymeric material, including the extremes of an inorganic moiety being dissolved in a highly polar polymer in which the inorganic moiety has a strong interaction with the polymer and a mixture of two components in which there are no interactions. In general, a composite membrane used in fuel cells could be identified either as proton exchange membranes (PEMs) in which the components are anionic with the charges being balanced by protons, or as polymer acid complexes (PACs) in which a basic polymer is doped with an acidic component [23]. In this sense, two extremes of composite membranes can be envisaged, one in which an inorganic proton conductor is the only proton conducting element and is suspended in a non-interacting inert support such as polyvinylidenedifluoride (PVDF), or the other in which a super-acidic inorganic proton conductor is intimately mixed with a strongly interacting ionomer to create a hybrid composite. Non-interacting composite membranes are expected to have the proton conducting properties of the dispersed inorganic "filler" whilst hybrid membrane properties are dependent upon the interaction of the composite components and the components acting independently [20, 23, 24].

3. Commonly Used Organic and Inorganic Materials for Composite Proton Exchange Membranes (PEMs)

Commonly used organic materials for composite PEMs mainly include perfluorinated polymer, such as Nafion and polytetrafluoroethylene (PTFE), partial fluorinated polymer, such as PVDF and sulfonated trifluorostyrene, etc., and various nonfluorinated polymers as alternatives to perfluorinated materials including poly(arylene ether)s, poly(arylene ether sulfone)s, acid functionalized or doped polybenzimidazole (PBI), poly(vinyl alcohol) (PVA), poly(sulfides), poly(phynelene oxides), poly(phosphazenes), etc [24, 25].

Commonly used inorganic additives for composite PEMs mainly include: (a) Less proton conductive hygroscopic oxides, such as $(SiO_2)_n$, ZrO_2, TiO_2, Al_2O_3, zeolites, clays, etc. (b) Sulfated–ZrO_2, SO_3H–SiO_2, PO_3H_2–SiO_2, zirconium phosphate (ZrP), heteropoly acids (HPAs), etc., which have the ability of proton conductivity; (c) Miscellaneous inorganic materials, such as Hydrogen cesium phosphates or sulfates and other metal phosphates, etc.; (d) In some cases, the mineral acids, such as HCl, HNO_3, $HClO_4$ or H_3PO_4.

Based on literature reports, the proton conductivity of these organic and inorganic materials can be arranged in order of decrease as: the mineral acids; the solid super acids; PFSAs; HPAs; or sulfonated zirconium phosphonates; hydrogen cesium sulfate ($CsHSO_4$) and related compounds above the super-protonic phase change; tin oxides and certain uranium oxides; silica, titania, alumina, and other related hydroscopic oxides and ice [26].

4. Organic-Inorganic Composite Polymer Electrolyte Membranes

Research in the development of self-humidifying organic-inorganic composite polymer electrolyte membranes are mainly focused on the following directions: (1) incorporating hygroscopic oxides, such as $(SiO_2)_n$, ZrO_2, TiO_2, Al_2O_3, zeolites, clays, etc., into the ionomers which have the ability of proton conductivity, to adsorb water and accordingly improve the proton conductivity of the ionomers [6, 7, 27-43]; and/or incorporating a mixture of hygroscopic oxides and Pt into the ionomers to catalytically combine the permeated oxygen with hydrogen and/or methanol to produce water and subsequently humidify the membrane [41, 44-46]; (2) incorporating some proton-conductive particles, such as sulfated–ZrO_2, SO_3H–SiO_2, PO_3H_2–SiO_2, ZrP, HPAs, etc., which have the ability of proton conductivity into the non-interacting inert polymers, to provide these inert membranes with proton conductivity under dry operation condition [47-51]; (3) incorporating some proton-conductive particles, such as ZrP, HPAs, etc., which have the ability of proton conductivity, into the membranes which also have the ability of proton conductivity, to further improve the proton conductivity of the membranes through the interactions between them under dry operation condition [52-64].

4.1. Perfluorinated Polymer Based Composite Polymer Electrolyte Membranes (PEMs)

Due to strong bond energy of C–F, the presence of the perfluorinated ionomer moiety can enhance the strength of the polymer matrix. In this sense, the matrix used for PEMs may be composed of Nafion or other perfluorinated and partial fluorinated polymer membranes, such as PTFE, PVDF, sulfonated trifluorostyrene, and so on. Such matrixes usually have strong mechanical strength. As a matter of fact, apart from the improvement in mechanical strength of the polymer matrix, the proper combination of a polymer matrix and filler, usually inorganic materials, can also lead to improvement in solvent resistance, water retentivity, fuel antipermeability and other properties.

Nafion is a best-known perfluorinated ionomer membrane widely used as a solid proton conducting electrolyte in electrochemical technology. It has a copolymer molecular architecture and consists of a hydrophobic polytetrafluoroethylene backbone chain and regularly spaced shorter perfluorovinyl ether side-chains terminated by a strongly hydrophilic sulfonate ionic group. Figure 1 displays a molecular structure of the Nafion® membrane produced by DuPont Company. Although the exact structure of a Nafion based membrane is not known, it is usually modeled as a three zones with interconnecting channels within the polymer as shown in Figure 2, e.g. a low dielectric zone with the hydrophobic fluorocarbon polymer matrix, a high dielectric zone containing ion clusters with the sulfonate exchange sites, counterions, and sorbed water, and an interfacial zone containing the pendant side chains of the sulfonate groups and a small amount of water. The hydration water that is indispensable in the process of proton conduction is mainly adsorbed on the sulfonic acid group with three water molecules per sulfonic acid group. A small amount of water is located in the nanopores between the hydrated, sulfonated polymeric backbone [65]. As described above, in DMFC applications Nafion® membranes have disadvantages of high methanol crossover, high cost, low temperature limit (<80 °C), and high humidification requirements [30,33,66-70]. High methanol crossover which reduces the efficiency of the oxygen reduction reaction (ORR) by the known mixed potential effect is the major barrier inhibiting Nafion® membranes from being used commercially in DMFCs. In order to solve these problems, incorporation of inorganic fillers into the Nafion mambranes with aims to improve their water retentivity, control the sizes of the channels of the proton transport and produce winding pathways for methanol permeation are commonly adopted [71,72]. These inorganic fillers can be hygroscopic oxides, mineral acids, or miscellaneous inorganic materials [58,73,74].

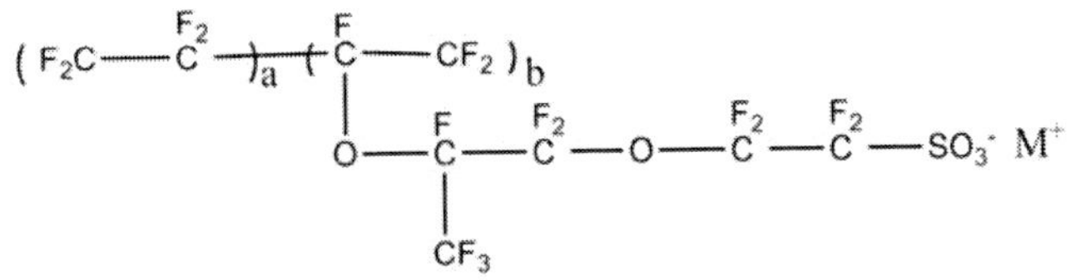

Figure 1. Molecular structure of Nafion® from the DuPont company.

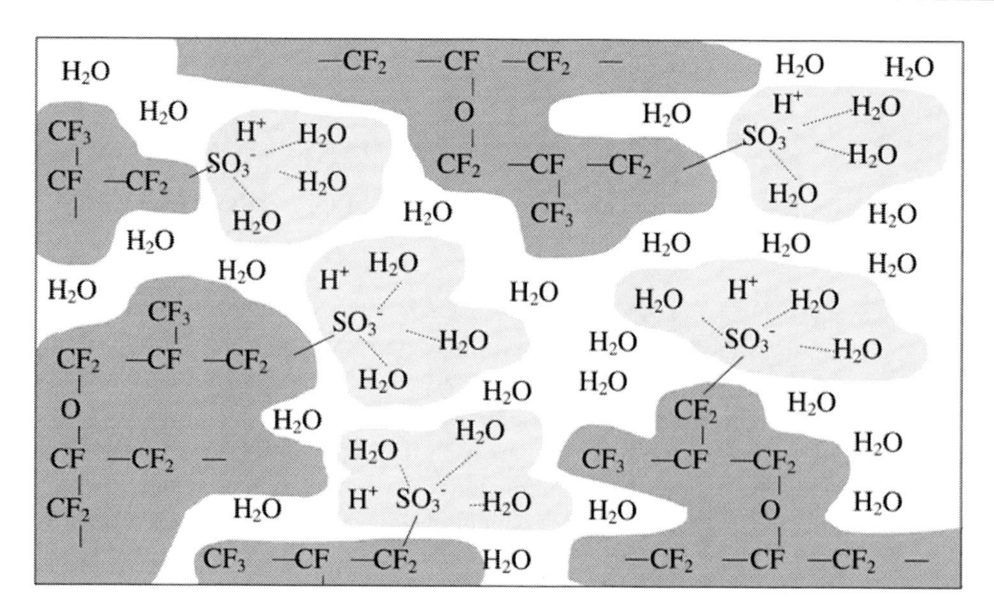

Figure 2. Schematic of the chemical make-up of a Nafion membrane. Reproduced with permission from Klein, L. C. et al. *Polymer* 2005, *46*, 4504-4509. Copyright © 2005, Elsevier Science Ltd.

4.1.1. Perfluorinated Polymer/Hygroscopic Oxides Composite Polymer Electrolyte Membranes (PEMs)

The hydroscopic oxides, such as $(SiO_2)_n$, ZrO_2, TiO_2, Al_2O_3, zeolites, and clays, etc., are commomly used additives for composite PEMs. These oxides intrinsically do not have proton conductivity, but are strongly hydroscopic, so they can be added to an ionomer to increase the water retentivity of composite PEMs under hotter and drier conditions and enhance the mechanical strength of membranes.

Crystalline $(SiO_2)_n$ involves a high degree of ordering with a three-dimensional structure [75]. The active surface of a crystalline $(SiO_2)_n$, which participates in chemical or physical interactions, is the external surface of the crystalline particles, and thus its active surface area is relative small. A main interest in $(SiO_2)_n$ is the porosity of its amorphous form. Porosity introduces a large surface area to the silica particles. As interphase processes require a large surface/mass ratio, amorphous silica is far more interesting for chemical and physical applications than its crystalline counterparts. It is commonly utilized in modifying the properties of the ionic polymers. Pioneering work for preparing ionic nanocomposites has been done *via* sol-gel techniques [33,76,77], where the inorganic $(SiO_2)_n$ was molecularly dispersed into the Nafion® matrix. The obtained Nafion/SiO_2 composites showed a higher water uptake at room temperature (~20%) compared with unfilled Nafion (~15%).

In comparison to SiO_2, ZrO_2 has a stronger ability to form several coordination states upon reaction with water due to a lower electronegativity of Zr. It usually exists in forms ranging from a gelatinous hydrate, $ZrO_2{\cdot}nH_2O$, of variable water content to an insoluble ZrO_2. Its use in the composite PEMs has been demonstrated in the work by Saccà et al. [34] where they prepared the recast Nafion composite membranes containing different percentages of commercial ZrO_2 by a Doctor-Blade casting technique. It was reported the introduction of the inorganic powder improved the mechanical characteristics and the water uptake of the composite PEMs. Moreover, power density values of 604 mW cm^{-2} and of 387 mW cm^{-2} were obtained at 0.6 V and at T = 110 °C (100% of relative humidity) (RH) and T = 130 °C (85% RH), respectively, for the composite Nafion/ZrO_2 membrane containing 10 wt% inorganic fillers in PEMFC application, even a decrease of the ion exchange capacity (IEC) with the increase of the ZrO_2 powder was observed. The decrease of IEC was caused by the decrease of the relative contents of proton exchange groups with the increase of the ZrO_2 powder, because that ZrO_2 is softly hydrophilic, but not proton conductive. So, the incorpotion of ZrO_2 could only increase the water retentivity of composite PEMs.

TiO_2 can exist either as the octahedral rutile, or the more distorted anatase or brookite. The solubility of TiO_2 depends greatly on its thermal and chemical history. These oxides are available as colloidal solutions or they may be formed in situ *via* sol-gel chemistry from the appropriate alkoxide, such as titanium (IV) tert-butoxide [35]. Nanocrystalline TiO_2 can be used as filler in the preparation of Nafion based composite membranes as polymer electrolytes in PEMFCs. Work by Licoccia and Traversa [78] showed that TiO_2, which was calcined at lower temperatures (The best calcination temperature was 500 °C), showed a larger content of oxygen species due to the presence of surface –OH groups. The lower calcination temperature led to the formation of TiO_2 in form of anatase with higher specific surface areas and smaller particle sizes. The large amount of surface –OH groups might be responsible for interactions with water and could facilitate the operation of the composite PEMs at temperatures >100 °C. The presence of TiO_2 promoted the inhibition of direct permeation of reaction gases by modifying the transport pathways. A maximum power density of 350 mW/cm^2 was achieved at a current density of about 1.1 A/cm^2 for DMFC with the membrane containing 5 wt% nanocrystalline TiO_2. A gradual increase in calcination temperatures (from 500 to 800 °C) resulted in the removal of hydroxyl groups attached on TiO_2, accompanied by the change of its crystallographic structure from anatase to rutile. Correspondingly, a decrease in the current and power density of DMFC was observed, due to the decrease of water retentivity caused by the loss of hydroxyl groups on TiO_2. Based on this result, the influence of the specific surface area of fillers by preparing composite membranes containing mesoporous TiO_2 on the power density and circuit current density of DMFC was investigated. The mesoporous anatase TiO_2 possesses higher surface/mass ratio and thus the content of oxygen species than the well crystallized rutile. A significant increase in the maximum power density and the circuit current density was observed when these mesoporous TiO_2 was used in DMFC [78].

Zeolites are the most important class of framework silicates and have been used as additives in composite PEMs for fuel cell application [75]. They are characterized as microporous aluminosilicates with a framework structure enclosing cavities that accommodate a wide variety of cations or water molecules. The framework consists of an open arrangement of corner sharing tetrahedral where SiO_4 are partially replaced by AlO_4 tetrahedral. Up to now, more than 175 unique zeolite frameworks have been identified, while those used in fuel cell membranes mainly include chabazite, clinoptilolite, and mordentite. Chabazite and clinoptilolite are stable in aqueous solution in the pH range 3–12. Clinoptilolite is zeolite with a high silica content (Si/Al ≥4), having a layered structure and a two-dimensional system of channels with mean diameters ranging from 0.35 to 0.50 nm. Composite PEMs with Nafion and chabazite or clinoptilolite have been reported by Tricoli et al. [39]. However, the obtained composite membranes with a zeolite content of 40 vol% exhibited the proton conductivities with about an order of magnitude lower than the pure

Nafion membrane at 60 °C, and a slightly higher activity energy of proton conductivity E_{H+}=17 kJ/mol (the E_{H+} for pure Nafion, 12 kJ/mol), although their MeOH permeabilities were comparable. Actually, the advantage of Nafion composite membranes with the inclusion of chabazite and clinoptilolite is that they can be used at the higher operation temperature in DMFCs. Work by Baglio et al. [79] showed these membranes had good characteristics for DMFC operated at high temperature. Maximum power densities of 350–370 mW/cm^2 and 200–210 mW/cm^2 were recorded at 140 °C with the 3 and 6 vol% zeolite-based membranes under oxygen and air operation, respectively. The high performances obtained were attributed to the enhancement of water retentivity of the composite membranes provided by the hygroscopic zoelite filler, as confirmed by cell resistance measurements.

Clays are naturally occurring and strong hydroscopic materials composed primarily of fine-grained minerals, and show a great promising for application in composite PEMs. They have sheet structures with the sheets bound together by the cations that lie between them. Variable amounts of water are trapped in the mineral structure by polar attraction. Clays may be dispersed in a polymer in three ways as schematically illustrated in Figure 3 (a) polymer are simply mixed with clays and the silicate layers of clays are not penetrated; (b) polymer is intercalated between the silicate sheets, and (c) polymers totally exfoliate the silicate layers of clays. In these dispersion ways, the exfoliation of the multi-layered silicate by polymers is most interested. It is reported that an addition of 1 wt% of organo-nanoclay montmorillonite (MMT) in this way could decrease the methanol permeability of DMFCs from 2.3×10^{-6} to 1.6×10^{-7} cm^2/s, but did not cause any apparent proton conductivity loss [40,80]. In this case, the tensile strength and elongation of PEMs at break were improved dramatically by the addition of nanodispersed clay particles [80].

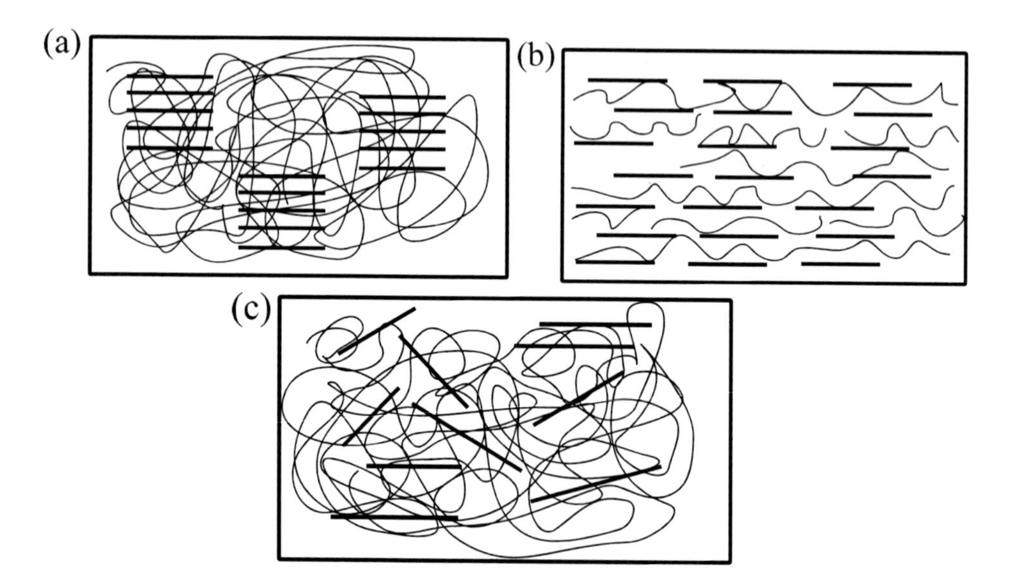

Figure 3. Typical dispersion states of a nanoclay within a polymer (a) no chain penetration, (b) chain intercalation, and (c) clay exfoliation. Reproduced with permission from Thomassin, J. M. et al. *J. Membr. Sci.* 2006, *270*, 50-56. Copyright © 2006, Elsevier Science Ltd.

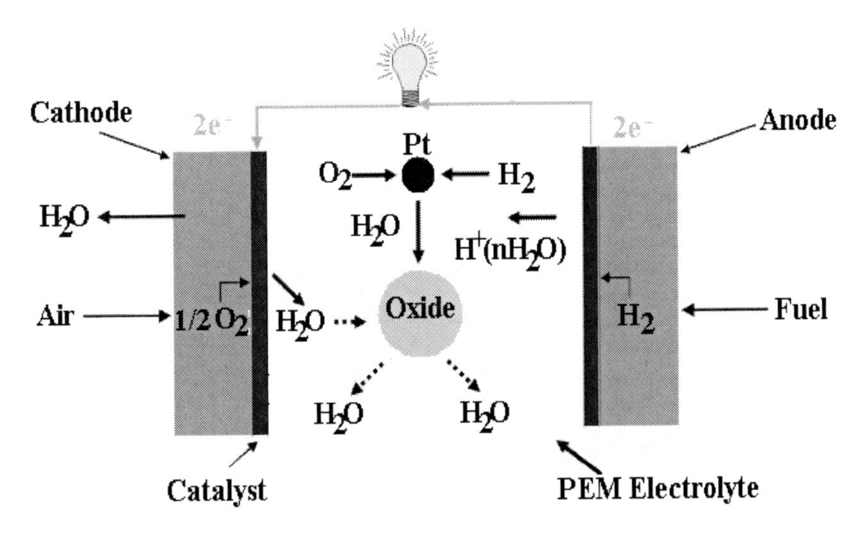

Figure 4. Schematic operation concept of a new PEMFC using self-humidifying Pt-oxide-PEM.

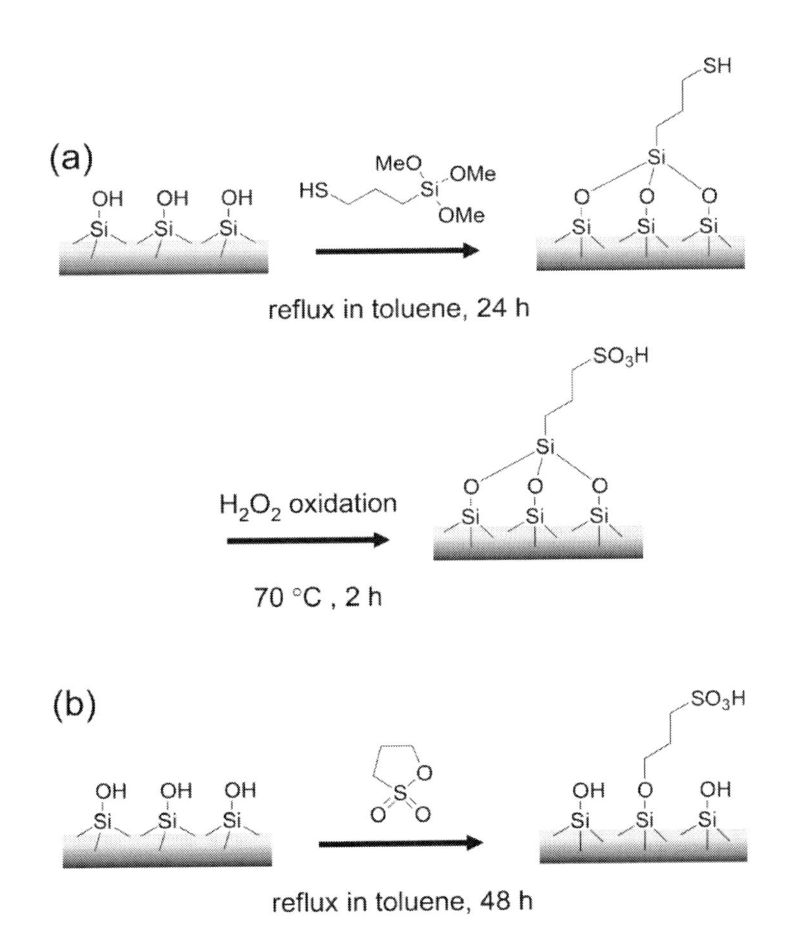

Figure 5. Outline of surface sulfonation procedures by SH oxidation method (a) and direct reaction method (b). Reproduced with permission from Munakata, H. et al. *Solid State Ionics* 2005, *176*, 2445-2450. Copyright © 2005, Elsevier Science Ltd.

Although the hygroscopic oxides can improve the water retentivity of composite PEMs, in some cases, noble metal nanoparticles, like Pt, are deposited onto these hygroscopic oxides to further improve the water retentivity of composite PEMs. The deposited Pt nanoparticles works as a catalyst to combine the permeated oxygen with hydrogen and/or methanol to produce water and subsequently humidify the membrane. These electron-insulated catalysts could not only avoid electron circuit in the whole membrane, but improve the humidity of membranes by in-situ adsorbing water produced at Pt particles on the surface of hygroscopic supports. Figure 4 displays a simple but effective schematical operation concept that explains how the Pt-oxide-PEM works to self-humidify in a PEMFC. It shows the dispersed Pt catalyze the production of H_2O from the H_2 and O_2 permeating the membranes. In this case, losses due to fuel crossover can be eliminated and the produced H_2O are adsorbed by the hygroscopic oxides, allowing the membranes to operate under hotter and drier conditions. This plot has been realized by Watanabe et al. [44,45] where they prepared a PEM which was comprised of 50 μm thick Nafion membrane containing 0.07 mg/cm^2 of Pt catalyst particles (d = 1 to 2 nm) and a few weight percent of a hygroscopic material such as 3 wt% colloidal TiO_2 (5 nm). It was found that by using these novel PEMs (Pt-PEM, TiO_2-PEM, Pt-TiO_2-PEM) the cell performances were dramatically improved in the voltages and the current densities for PEMFC application. In particular, the combination of Pt and TiO_2 showed superiority in the maximum output, i.e., 0.7 W/cm^2 at 1.6 A/cm^2 and about 0.50 V. More recently, the same authors [41] have formed TiO_2 nanoparticles in a commercial Nafion 112 membrane with highly dispersed Pt nanoparticles *via* in situ sol-gel reactions, resulting in a transparent membrane with uniform distribution of TiO_2 in the PEM. Water adsorbability increased more than two times by dispersing only 2 wt% TiO_2 in the PEM.

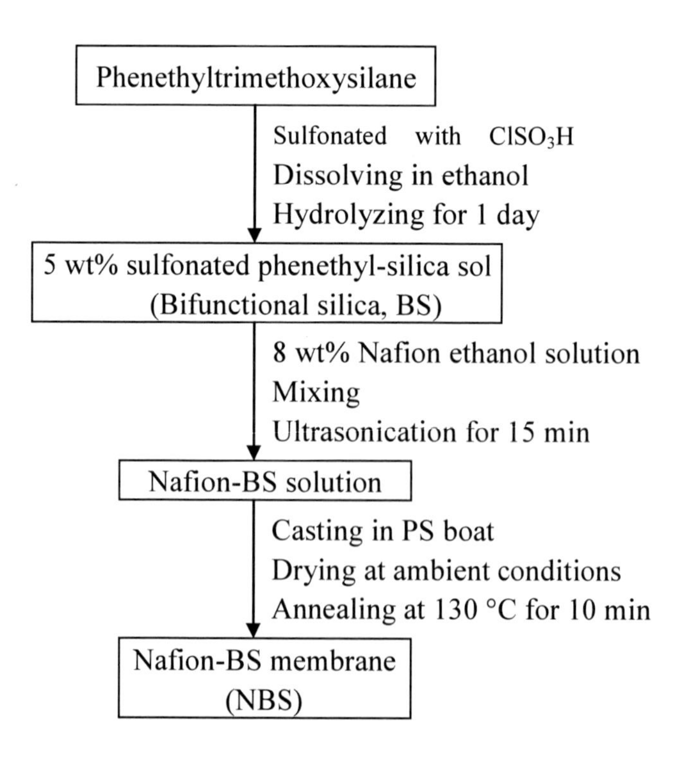

Figure 6. Preparation protocol of NBS composite membranes.

4.1.2. Perfluorinated Polymer/Hygroscopic Oxides Superacidic Composite Ionic Polymer Electrolyte Membranes (PEMs)

The oxides described above do not have any great intrinsic proton conductivity. The reason that the obtained composite PEMs exhibit higher performances is that the incorporated oxides are hydroscopic, which ensure them higher proton conductivities, even at temperature signifantly higher than 100 °C. In addition, these oxides can also act as filler to change the methanol crossover pathways and thus decrease the crossover but without improving, in some case even decreasing, the proton conductivity of the composite PEMs. Due to their relative inertness of oxide surface, the interactions between the oxides and ionomers are weak. However, the situation, deficiency in interactions between the inorganic filler and the ionomer, can be changed by modifications to the inorganic filler, e.g. functionalization of the inorganic filler with acid groups to promote interactions between the ionomer and inorganic materials and improve the performance of the composite PEMs. All of these oxides mentioned above can be sulfonated to give acidic materials, e.g. SiO_2, ZrO_2 and TiO_2 superacidic materials, which exhibit higher proton conductivities [81,82].

It is well known that silanol groups (Si-OH) can easily react with various organosilane compounds, so one can introduce desired properties on glass surfaces. If acidic groups, especially, sulfonic acid groups are introduced into the porous silica matrix, the proton conductivity of the porous silica matrix is enhanced by this surface treatment. The surface sulfonation of the porous silica membrane can be conducted by mercapto groups (–SH) oxidation method or direct reaction method as shown in Figure 5 [83]. In both cases, sulfonic acid groups are introduced on the silica surface *via* covalent bonds. However, the prepared porous silica membrane usually hold few silanol groups which were important reactive points to introduce sulfonic acid groups on the silica surface, due to high temperature treatments. In order to address this problem, a hydrothermal treatment was adopted at 170 °C to increase silanol groups. In the case of SH oxidation method, the porous silica membrane was firstly refluxed in 2.6 wt% 3-mercaptopropyltrimethoxysilane toluene solution in order to introduce mercapto groups (–SH) on the silica surface as precursors. Then, the mercapto groups were converted to sulfonic acid groups ($–SO_3H$) by an oxidation in a H_2O_2 solution. In the direct reaction method, the introduction of sulfonic acid groups onto the porous silica can be done with one step by refluxing the porous silica in toluene solution containing 1,3-propanesultone. Munakata et al. [83] prepared the sulfonated silica matrix using SH oxidation method, which exhibited high proton conductivity of 6.0×10^{-3} S/cm at 60 °C under 90% RH, a value that was about 400 times higher than that of unmodified silica matrix. The proton conductivity of the composite membrane filled by a proton-conducting gel polymer, 2-acrylamido-2-methyl-1-propanesulfonic acid (AMPS), was considerably enhanced by using the sulfonated silica matrix.

Lin et al. [10] introduced sulfonic acid groups on the mesoporous silica surface using the direct reaction method. The proton conductivity (σ) of the composite membrane containing sulfated mesoporous silica increased from 0.10 to 0.12 S/cm as the modified mesoporous silica content increased from 0 to 3 wt%. The methanol permeability of the composite membrane declined as the sulfonic mesoporous silica content increased. The reasons that the Nafion®/M–SiO_2–SO_3H composite PEMs exhibited superious performance are: (i) the functionalized mesoporous silica has more hydrophilic regions than pristine Nafion®, in the form of sulfonic acid groups; and (ii) the mesoporous silica effectively blocks the passage of methanol.

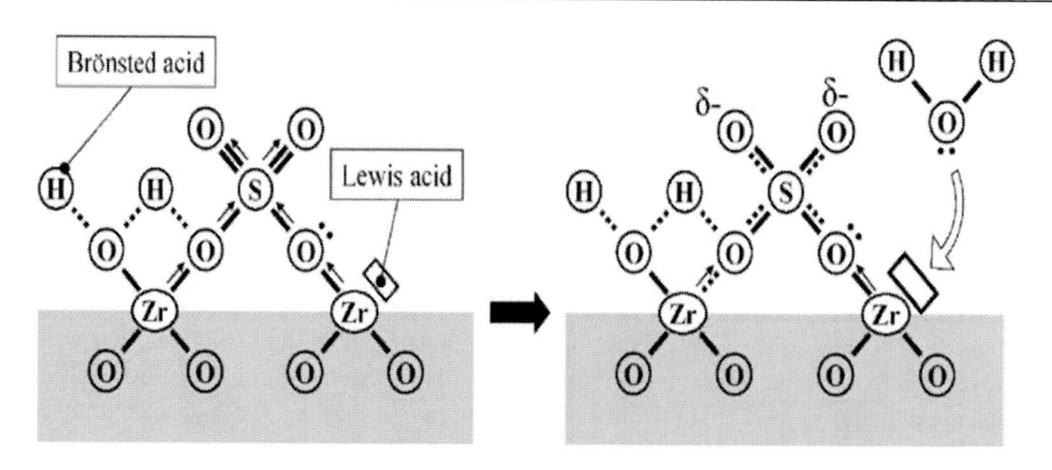

Figure 7. Surface model of S-ZrO_2(p) (620 °C). Reproduced with permission from Hara, S. et al. *Solid State Ionics* 2004, *168*, 111-116. Copyright © 2004, Elsevier Science Ltd.

Another method to introduce sulfonic acid groups into the porous silica matrix was also obtained by sulfonating the porous silica matrix with chlorosulfonic acid. For example, Wang et al. [84] made attempts to develop bifunctional sulfonated phenethylsilica (BS) nanoparticles using this synthetic protocol followed by complexing with Nafion to improve its proton conductivity and water uptake characteristics of composite PEMs. In this synthetic protocol, silica precursor, i.e. phenethyltrimethoxysilane (PETMS), sulfonated by chlorosulfonic acid. After sulfonation, measured amounts of absolute ethanol and deionized (DI) water were added for the hydrolysis condensation of sulfonated PETMS to obtain sulfated silica sol. The obtained sol was mixed to appropriate amount of Nafion solution in ethanol to give composite membranes with specific composition. The preparation protocol of Nafion/BS (NBS) composite membranes is shown in Figure 6. It was found that the water uptake as well as the IEC of the composite PEMs increased with the amount of silica incorporated. The proton conductivity of the composite membranes at 80 °C and 100% RH with 2.5-5.0 wt% SiO_2 is 2.7-5.8 times higher than that of the pure Nafion®117 membrane. By the similar preparation protocol, Li et al. [85] prepared Nafion-sulfonated diphenyldimethoxysilane (N-sDDS) nano-composite membranes. The performances of fuel cell with the N-sDDS membrane were evaluated and compared with those with the unsulfonated diphenyldimethoxysilane (DDS) composite membrane and Nafion®117. It showed that the nano-composite membrane with the thickness of 125 μm had a lower methanol permeability and better single cell performance than the unsulfonated DDS composite membrane, where the sol-gel derived sulfonated diphenyldimethoxysilane (sDDS) with hydrophilic -SO_3H functional groups were considered to be able to work as additive to reduce the methanol permeability of Nafion.

Sulfated-ZrO_2 (S-ZrO_2)/Nafion composite PEMs were investigated by Datta et al. [87]. Higher current density in comparison to pure Nafion membranes at 110 °C and 0.5 V was reported when they were used in PEMFC. This approach opened the window for acid modification of ZrO_2 to improve fuel cell performance. The reason that the Sulfated-ZrO_2(S-ZrO_2)/Nafion composite PEMs exhibited higher fuel cell performance was the formation of acid sites on S-ZrO_2. S-ZrO_2, a solid state super-acid, exhibiting a Hammett acid strength H_0 of −16.03, is recognized as the strongest solid acid [82]. Hara and Miyayama [86] studied the proton conductivity of S-ZrO_2 and found that it had a high conductivity of 2.3×10^{-1} S/cm at

105-135 °C. The electronic polarization induced by SO_x was assumed to improve the Lewis and Brønsted acidities on Zr, thus leading to higher proton conductivities. As shown in Figure 7, it is assumed that the increased partial charge on oxygen reduces the electron density of Zr. As a result, the Lewis acidity of S-ZrO_2 (p) (where p means that the powdered state $ZrO_2{\cdot}nH_2O$ was used as the precursor.) (620 °C) becomes larger than that of S-ZrO_2(p) (520 °C). The Lewis acid points can easily convert to Brønsted acid points in the presence of water, and the Brønsted acid points work as proton donors. Therefore, it is thought that only S-ZrO_2(p) (620 °C) shows higher proton conductivity. Zhang et al. [88] synthesized fine particles (less 0.5 μm) of S-ZrO_2 by an ameliorated method based on the work reported by Hara et al. [86]. It was reported that the IEC of composite membrane increased with the content of S-ZrO_2. S-ZrO_2 was well compatible with the Nafion matrix and the incorporation of the S-ZrO_2 could increase the crystallinity and also improve the initial degradation temperature of the composite membrane. The performance of the single cell with optimized S-ZrO_2 was far more than that of the Nafion at the same condition (e.g. 1.28 W/cm^2 at 80 °C, 0.75 W/cm^2 at 120 °C). It shown the 15 wt% S-ZrO_2/Nafion composite membrane had lower fuel cell internal resistance than Nafion membranes at high temperature and low relative humidity (RH) [88], and 15 wt% is the optimum content, considering the water uptake and IEC simultaneously.

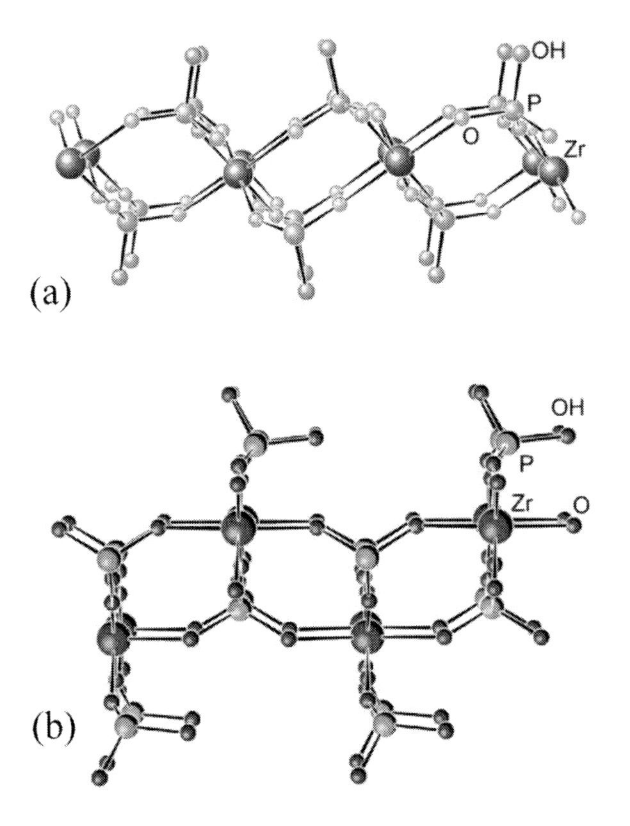

Figure 8. Ball and stick structure of a single lamella of α−ZrP (a) and γ−ZrP (b). Reproduced with permission from Herring, A. M. et al. *Polym. Rev.* 2006, *46*, 245-296. Copyright © 2006, Taylor & Francis.

The investigation of a nanocomposite PEM in combination of sulfated titanate nanosheets and Nafion was done by Lee et al. [89]. The titanate is a layer-structured material with a large surface area of over 500 m^2/g, and has a proton conductivity of 2.66×10^{-6} S/cm at 573 K [90]. The functionalization of titanate with organic sulfonic acid (HSO_3^-) was conducted by grafting various thiol and sultone groups onto the surface of its nanosheets [89]. The composite PEMs containing surface-sulfonated titanates showed higher proton conductivity than those containing untreated TiO_2 (P25, Degussa, mixed-phase titania) particles, and better mechanical and thermal stability than Nafion alone. The methanol permeability of such composite membranes decreased with increasing the content of the sulfonated titanate in the nanocomposite membranes. The membrane electrode assembly using Nafion/sulfonated titanate nanocomposite membranes exhibited up to 57% higher power density than that containing a pristine Nafion membrane.

The composite membrane from sulfonic acid modified clay, e.g. laponite and Nafion composite, was studied by Bébin et al. [4]. The clay powder was firstly activated by helium plasma. and then immersed in a solution of p-styrene sulfonate in DMF and refluxed to get sulfated. The composite PEMs were prepared by a recasting Nafion solution mixed with modified laponite particles. The measured proton conductivity of the composite membrane was higher than that of the commercial Nafion in the temperature ranges 20-95 °C with a more significant difference at low relative humidity. The optimized composite membrane showed 20% improvement in power density. Under the condition of higher temperature (120 °C) with dilute cathode gas and lower air pressure, it will induce a drastic dehydration effect in the cell. The sensitivity of the membrane to humidity is easily observed with these cell conditions. Nafion membrane could only deliver a current of 390 mA/cm^2 at 0.6 V, while the composite PEMs with sulfated clay can reach a current of 500 mA/cm^2, a 30% improvement in power density.

4.1.3. Perfluorinated Polymer/Solid Proton Conductors Particles Composite Ionic Polymer Electrolyte Membranes (PEMs)

As described above, the incorporation of sulfonated hygroscopic oxides could effectively improve the performance of the obtained composite PEMs in fuel cell. However, in some case, the introduction of sulfonate groups is relatively difficult due to the deficiency of active points. Morever, sulfonate grafting would increase the complexity and the cost of PEMs. In this sense, the incorporation of the inorganic fillers that are intrinsic proton conductors into PEMs is particularly desired. It is generally expected the inorganic fillers could be proton conductors as good as the ionomer, or even better. For this reason, a class of solid proton conductors such as ZrP, HPAs, are developed and incorporated into the ionomers to improve the proton conductivity of the composite membranes under dry operation condition.

Zirconium phosphate is the most extensively studied inorganic material for PEMFCs. As a desirable additive, ZrP has the following attributes [91]:

(i) Moderate proton conductivity when it is humidified ($\sigma \sim 10^{-3}$ S/cm);
(ii) Brønsted acidity with the ability to donate protons;
(iii) Higher thermal stability up to temperatures >180 °C;
(iv) Hygroscopic and hydrophilic character;
(v) Easily fabricated in a manner which is compatible with the chemical and physical limits of the polymer membrane.

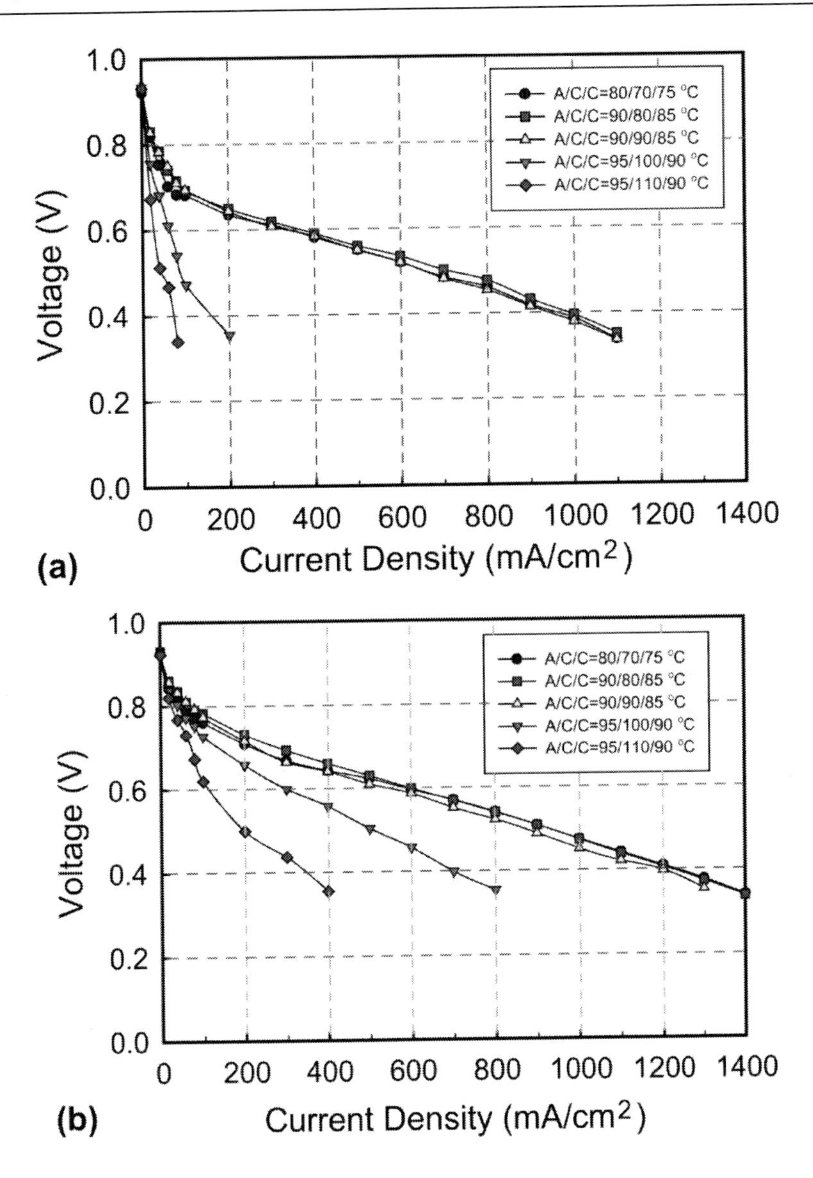

Figure 9. Unit cell performance of recast Nafion (a) and Nafion/ZrSPP (b) 20 wt% at different temperatures. Reproduced with permission from Kim, Y. T. et al. *Curr. Appl. Phys.* 2006, *6*, 612-615. Copyright © 2006, Elsevier Science Ltd.

The ZrP phase may exist in amorphous to crystalline forms with the proton conductivity of ZrP varying from 10^{-7} to 10^{-3} S/cm depending on the phase composition, structure, and hydration state. Amorphous zirconium phosphate (PO_4:Zr = 2) can be obtained by mixing a soluble zirconium salt with an excess of phosphoric acid or soluble phosphate. Crystalline ZrP was synthesized by Clearfield and Stynes [92] for the first time in 1964. It was obtained by the refluxtion of ZrP gels in the presence of phosphoric acid. Usually, there are two types of crystalline ZrP with layered structures, e. g. α−and γ−ZrP. Ball and stick structure of a single lamella of α−ZrP and γ−ZrP is shown in Figure 8. In α−ZrP, the layers consist of ZrO_6 octahedra in which zirconium atoms are located in same plane and bonded with one another by tetrahedral phosphate groups (HPO_4^{2-}) lying above and below the plane composed of zirconium atoms. While in γ–ZrP, ZrO_6 is coordinated with two kinds of tetrahedron, $H_2PO_4^-$ and PO_4^{3-} [93]. Both α and γ structures of ZrP are good for proton transporting due to the presence of pendent –OH groups. Generally, α–ZrP has higher chemical reactivity, ion-exchange ability and intercalation property, than the γ form due to its abundance of hydroxyl group, which extends into interlayer region and forms hydrogen bonded network with water. Ion-exchange of the phosphate hydrogens (e.g. Na^+, K^+) or intercalation of other molecules can be lead to an increase in the interlayer distance of ZrP to accommodate the guest ions or molecules. The ZrP particles/ionomer composite membranes can be prepared by incorporating the zirconium phosphate into the membrane using the procedure described by Grot and Rajendra [94] or by recasting from a Nafion solution with zirconium ions as described by Savadogo [21]. A comparison of the physiochemical properties between Nafion 115 and a composite Nafion 115/zirconium phosphate (~25 wt%) membranes have been done by Benziger et al. [91]. It showed the composite membrane exhibited higher water uptake ability than Nafion at the same water activity. However, the proton conductivity of the composite membrane was slightly less than that of Nafion 115. Small angle X-ray scattering showed that the hydrophilic phase domains in the composite membrane were spaced further apart than in Nafion 115, and the composite membrane exhibited less restructuring with water uptake. Despite its lower proton conductivity, the composite membranes displayed a better fuel cell performance than Nafion 115 when operated at reduced humidity conditions. It was suggested that a greater rigidity of the composite membrane accounted for its improved fuel cell performance. Modification has been done to ZrP by functionalization with organic moieties or Brønsted bases. Significant improvement in conductivity was achieved by intercalation of strong acidic functional groups, $-SO_3H$, into the interlayer region. Both α and γ–ZrP may be derivatized with an organic group to form the phosphonate with general formula of $Zr(IV)(O_3P-G)_{2-x}(O_3P-RX)_x$, where G may be inorganic or organic group, R is an organic group, X is an acidic functionality, and x can vary from 0 to 1.5. A large variety of structures of increasing acidity in the series R = $(CH_2)_n$, X = Me, OH, COOH, R = Ph, X = SO_3H have been reported [95,96]. The composite membranes synthesized with these functionalized ZrP exhibit a great performance when used in fuel cells. For example, Rhee et al. [52] made efforts to make composites of these materials with the Nafion. An increase in the conductivity at high temperatures >100 °C was found. A comparison of cell performance of Nafion/zirconium sulphophenyl phosphate (ZrSPP) composite membranes with that of recast Nafion is shown in Figure 9, where A/C/C denotes the temperature of anode humidifier, cell and cathode humidifier, respectively. The composite membrane showed 700 mA/cm^2 current density in PEMFC application which was four times better than recasted Nafion at 100 °C.

Besides zirconium phosphate, various other metal phosphates have also been used as inorganic additives in polymeric composite PEMs for fuel cells. These phosphates include tin phosphate, boron phosphates, calcium hydrogen phosphate, and phosphatoantimonic acids. Layered phosphoantimonic acids ($H_nSb_nP_2O_{(3n+5)}\cdot xH_2O$, with n = 1 or 3) are super acids and exhibit similar swelling to smectite clays [97]. Calciumhydroxyphosphate ($Ca_5(PO_4)_3OH$), (CHP) shows proton conductivity, good compatibility with various polymers, high thermal stability and high crystallinity. As crystalline filler, CHP can act as a nucleating agent for crystallization of various polymers, in which high crystallinity of CHP may affect the crystalline kinetics of Nafion and change crystallinity of the composite membranes. Work by Park and Yamazaki [53] demonstrated that crystallinity of the composite PEMs was increased with CHP content compared to the recast Nafion as indicated by the formation of new crystalline peaks in the original amorphous region of Nafion. The methanol crossover through the Nafion membrane was substantially reduced in these crystallized Nafion. In addition, as the content of CHP filler in the composite was increased, a decrease in water uptake, methanol diffusivity, and methanol crossover in terms of permeability was identified. The possible reason for reduced methanol crossover suggetsted by the authors was the increased higher crystallinity of the composite membrane, a concept that can be explained based on the general fact that water and methanol molecules can not infiltrate into the crystalline region due to the compact carbon-fluoro structure even though consist in the amorphous region due to the large free volume [53].

Heteropoly acid is a subset of a large class of inorganic oxides called the polyoxometallates (POMs), made up of a particular combination of hydrogen and oxygen with certain metals such as tungsten, molybdenum or vanadium and an element from the p-block of the periodic table, such as silicon, phosphorous or arsenic and acidic hydrogen atoms. Among a wide variety of HPAs, most stable and more easily available are those with the Keggin structure, which consists of a central tetrahedron XO_4 surrounded by 12 WO or MoO octahedra arranged in a tetrahedral fashion as four groups of three octahedral (M_3O_{13}), where three octahedral share edges and have a common oxygen atom shared with the central tetrahedron XO_4, and are represented by the formula $XM_{12}O_{40}{}^{x-8}$, where X is the central atom (Si^{4+}, P^{5+}, etc.), x is its oxidation state, and M is the metal ion (Mo^{6+} or W^{6+}), as shown in Figure 10. The HPAs are known to have high proton conductivity at room temperature. Therein the Keggin structure based HPAs are reported to have some of the highest reported acidity and solid state proton conductivities at room temperature, and thus become an increasingly popular proton conducting additive in proton exchange composite membranes for use under dry and/or elevated temperature conditions of fuel cell. The studies on the Nafion/HPAs composite membranes have attracted tremendous attentions. Their strong acidity and proton conductivity of HPAs is caused by two main factors: (1) dispersion of the negative charge over many atoms of the polyanion and (2) the fact that the negative charge is less distributed over the outer surface of the polyanion owing to the double-bond character of the $M = O_t$ bond, which polarises the negative charge of O_t to M. For example, hydrated phosphotungstic acid ($H_3PW_{12}O_{40}{\cdot}29H_2O$) has a conductivity of 1.9×10^{-1} S/cm at room temperature. However, the HPAs are water-soluble materials. Water produced in the electrochemical processes may dissolve the HPAs and the acids may leak out through the gas outlet holes, which would ultimately result in a decline in cell performance. Consequently, research to fix the HPAs in stable structures and maintain its high proton conductivity has been the focus of this issue. Embedment of HPAs into a hydrophilic polymer matrix might be one of the best choices. The obtained composite membranes usually exhibit high proton conductivity, and the mechanical properties of the polymer membranes are improved while the depletion of HPAs caused by water dissolution also decreased. For example, Tazi et al. [99] synthesized several new cation exchange membranes of different thicknesses (15-500 μm) based on a Nafion solution and a HPA, silicotungstic acid (NASTA) by a simple chemical route for PEMFC applications. It was found that the water uptake (60%) and the ionic conductivity (10.10×10^{-2} S/cm) of the NASTA membrane was significantly higher than those of Nafion 117, which were 27% and 1.23×10^{-2} S/cm measured at same condition, respectively. The membranes fabricated with Nafion and silicotungstic acid exhibited good mechanical strength and stability even if it was dipped in an acid or a basic medium for more than 10 months. Additionally, they [58] studied the properties of PEMs based on a Nafion solution and HPAs such as silicotungstic acid (STA), phosphotungstic acid (PTA) and phosphomolybdic acid (PMA) for H_2/O_2 fuel cells. It was reported that the water uptake of the various membranes in the increasing order is: Nafion 117 (27%) < NASTA (60%) < NAPTA (70%) < NAPMA (95%); the ionic conductivity is: Nafion 117 (1.3×10^{-2} S/cm) < NAPMA (1.5×10^{-2} S/cm) < NAPTA (2.5×10^{-2} S/cm) < NASTA (9.5×10^{-2} S/cm); The current density at 0.600 V of the PEMFC based on the various membranes varing in the increasing order is: Nafion 117 (640 mA/cm^2) < NASTA (695 mA/cm^2) < NAPTA (810 mA/cm^2) < NAPMA (940 mA/cm^2). Shul et al. [54] prepared a nanocomposite membrane of Nafion containing sulfonic functionalized heteropoly acid–silica nanofiller, which was then applied in DMFCs. The DMFC performance in terms of open circuit voltage (OCV) and power density of the nanocomposites membranes increased with the operating temperature from 80 to 200 °C. In this work, the function of the sulfonic-functionalized heteropolyacid–SiO_2 nanoparticles was to provide a proton carrier and act as a water reservoir in the composite membrane at elevated temperature. The power densities at different temperatures were 33 mW/cm^2 at 80 °C, 39 mW/cm^2 at 160 °C and 44 mWcm^{-2} at 200 °C, respectively. However, the performance of the DMFC with the composite membrane was still lower than the virgin Nafion membrane.

4.1.4. Non-interacting Fluorinated Inert Polymer Based Composite Membranes

High cost is one of main obstacles for Nafion membrane used in DMFC. Thus, composite membranes with non-interacting polymers usually comprise of inert polymers, such as PTFE and PVDF, and inorganic materials with high proton conductivity have been developed. It is reasonable to assume in this case that the proton conduction observed are attributable only to conduction pathways from the inorganic moieties, but does not benefit from the chemically inert hydrophobic supporting polymer. As forementioned, many of the inorganic species, such as sulfonated SiO_2, ZrP and HPAs, usually exhibit high proton conductivities. The disadvantage of these inorganic materials, however, is that they are often crystalline and are not suitable for fuel cell applications as films of these materials are fragile. The way to resolve this problem is to suspend the inorganic proton conductor in an inert polymer such as PVDF or more often PVDF copolymerized with hexafluoropropylene (PVDF-HFP), a polymer that is moderately soluble in polar solvents. PVDF-HFP is developed for the battery industry as it is more processable than PVDF. PVDF-HFP films of the composite membranes can be obtained by casting directly from the polar solution with PVDF-HFP and inorganic proton conductors. A new nanoporous proton-conducting membrane (NP-PCM), in combination of PVDF and sulfated nanosize silica powder, was developed by Peled [49] and applied to a DMFC. The use of the NP-PCM in the DMFC effectively decreased membrane cost and methanol crossover. In this work, the sulfonic acid groups functionalized nanosize silica particles worked as water retainers in the membrane and thus the obtained membranes could be operated at elevated temperatures. It was demonstrated that the membrane had a thermal stablilty up to 250 °C. Preliminary tests on fuel cell showed cell maximum power density of 32 mW/cm^2 at 70 °C. The crossover-current density for a 100 μm thick membrane, measured in 1 M methanol at 80 °C, was 110 mA/cm^2, a value showing that methanol crossover was reduced.

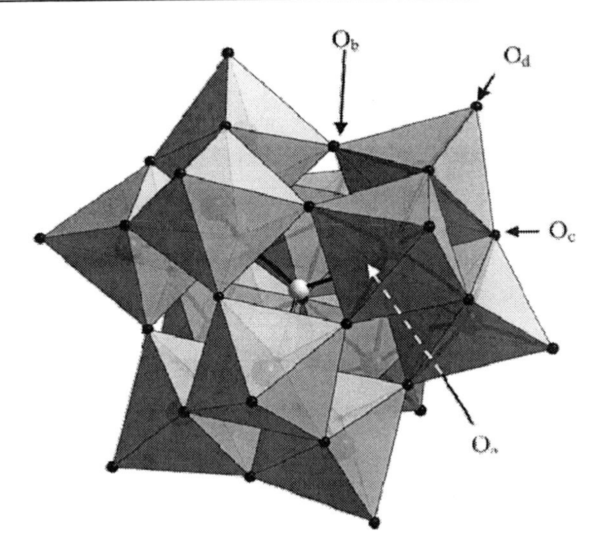

Figure 10. Keggin structure of the $XM_{12}O_{40}^{x-8}$ anion. O_a: oxygen atoms connected with MO_4 and XO_4 units; O_b, O_c: bridge oxygen atoms linking MO_4 octahedra from the same (O_c) or different (O_b) M_3O_{13} sets. O_d: external (corner) oxygen atoms. Reproduced with permission from Checkiewicz, K. et al. *Chem. Mater.* 2001, *13,* 379-384. Copyright © 2001, American Chemical Society.

The preparation of composite PEMs by the incorporation of amorphous and crystallized layered zirconium phosphate sulfophenylenphosphonate into PVDF also has been reported [50]. The obtained composite membranes exhibited higher relative humidity values, which kept at almost constant even when the temperature increases up to 120 °C. The maximum conductivity of these membranes appeared to be 2×10^{-3} S/cm at 120 °C with ~9% of zirconium phosphate sulfophenylenphosphonate ($Zr(HPO_4)(O_3P-C_6H_4SO_3H)$). The proton conductivity of PVDF/$\alpha-Zr(HPO_4)(O_3P-C_6H_4SO_3H)$ could be improved by use of a $\alpha-Zr(HPO_4)(O_3P-C_6H_4SO_3H)$ precursor solution in DMF. It showed that transparent and flexible composite membranes of PVDF and 40% $\alpha-Zr(HPO_4)(O_3P-C_6H_4SO_3H)$ had a conductivity of 5×10^{-3} S/cm at 100 °C and 95% RH.

Novel proton conducting gels formed by combination of HPAs with a polar aprotic DMF solution containing PVDF was reported by Checkiewicz. [51] It was found that the conductivity of the obtained membrane at room temperature could be up to as 5×10^{-3} S/cm. Actually, the HPA based fuel cells have higher proton conductivities and can create impressive currents at room temperature even without external humidification. Malers et al. [47] synthesized composite membranes from PVDF-HFP and different HPA, such as $H_3PW_{12}O_{40}$, $H_3P_2W_{18}O_{62}$, $H_6P_2W_{21}O_{71}$, and $H_6As_2W_{21}O_{69}$, with weight ratio of 1:1. In this work, it was firstly confirmed that HPAs were the only proton conducting component in the PEMs. Due to high conductivity and thermal stability, such PEMs hold great promising for solid acid fuel cell applications. However, all of the materials studied were somewhat porous and the open circuit potentials measured were somewhat low. The lifetime of a HPA fuel cell could be shorted by the reduction of the HPA to a heteropoly blue under some circumstances. In addition, HPAs themselves do not have sufficient protons mobility for fuel cell operated at temperature <100 °C, although they may have surprising application in high temperature PEMs at >200 °C. In this regard, an optimization of HPA based fuel cells must be done before used widely.

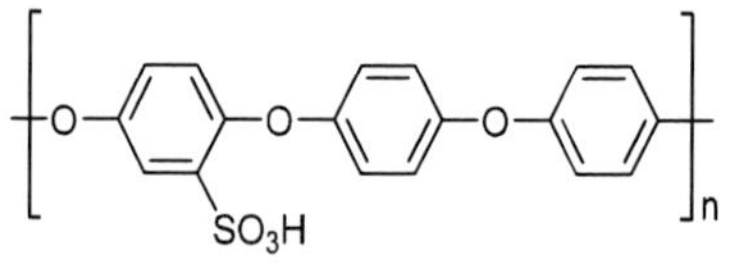

SPEEK

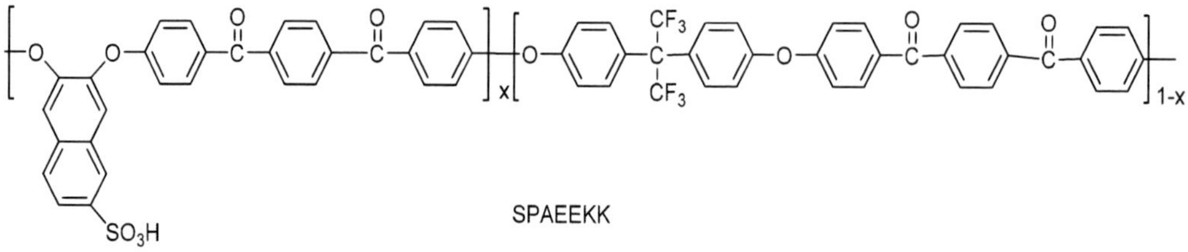

SPAEEKK

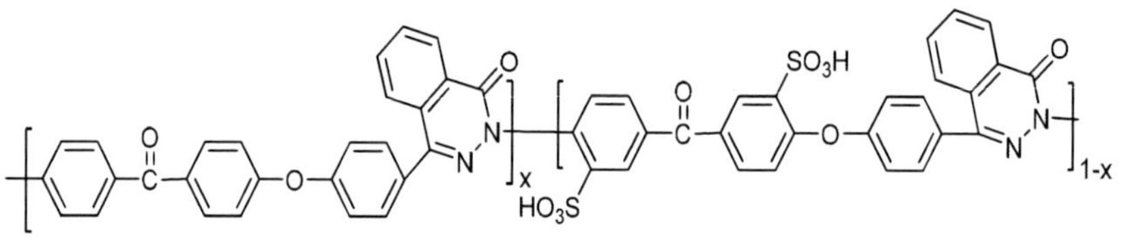

SPPEK

Figure 11. Chemical structures of sulfated PEEK, PAEEKK and PPEK.

In addtion, these inert polymers can also be changed as proton conductive polymers by grafting several organic groups like hydroxyl, amine, carboxylate, sulfonate, and quaternary ammonium, etc. Lehtinen et al. [100] prepared poly(vinylidene fluoride) grafted polystyrene sulfonated acid (PVDF-*g*-PSSA) proton exchange membranes based on a radiation-grafting technique, and they found that PVDF-*g*-PSSA membranes had a higher proton conductivity, water-uptake and lower oxygen solution ability compared to Nafion membranes. The hydroscopic oxides, such as Al_2O_3 and zeolite, etc., can be added to these membranes to increase water retentivity under hotter and drier conditions of operation and enhance the mechanical strength of membranes. For example, PVDF-*g*-PSSA membranes doped with different amount of Al_2O_3 (PVDF/Al_2O_3-*g*-PSSA) have been prepared by a solution-grafting technique [101]. The results showed that the obtained composite PEM with 10% Al_2O_3 had a lower methanol permeability of 6.6×10^{-8} cm^2/s (a value that is almost one-fortieth of that of Nafion 117) and a moderate proton conductivity of 4.5×10^{-2} S/cm. A test on fuel cells showed that a DMFC based on such composite membrane had a better performance than that based on Nafion 117. Although Al_2O_3 has some negative influence on the stability of the membrane, it can still be used in DMFCs in the moderate temperature.

As mentioned above, zeolite is an important material in the composite PEMs. Kim and Min et al. [102] synthesized a proton conducting comb copolymer consisting of poly(vinylidene fluoride-co-chlorotrifluoroethylene) backbone and poly(styrene sulfonic acid) (PSSA) side chains, i.e. (PVDF-co-CTFE)-g-PSSA (graft copolymer), with 47 wt% of PSSA and zeolite using a atom transfer radical polymerization (ATRP) method. The forier transform infrared (FTIR) spectra showed that the zeolite particles strongly interacted with the sulfonic acid groups of PSSA chains. The weight percent of zeolite in membranes had a great effect on the proton conductivity. The proton conductivity of the composite PEM at room temperature was reduced to 0.011 S/cm when the weight percent of zeolite 5A was increased to 7%. The reduction in the proton conductivity could be explained by the decrease in water uptake. The increase of the weight percent of zeolite resulted in the decrease in the number of available water absorption sites because of the hydrogen bonding interactions between the zeolite particles and the graft copolymer matrix. It was demonstrated that the water uptake of the composite membranes decreased from 234 to 125% with an increase of the zeolite 5A weight percent from 0% to 10 wt%.

4.2. Poly(arylene Ether)s Based Composite Membranes

The necessity to reduce the cost of PEM and improve the performance of PEMs stimulates a large body of research, with aims to rationalize the remarkably efficient combination of properties of the perfluorinated ionomers (PFI) or develop new polymers with similar properties by a less expensive chemistry. In recent years, great efforts have been devoted to synthesizing PEM based on fluorine free hydrocarbon ionomer as alternatives to the perfluorinated ionomers [5,103-107]. Among them membranes based on aromatic polyether ether ketone (PEEK), poly(arylene ether ether ketone ketone) (PAEEKK), poly(phthalazinone ether ketone) (PPEK), etc., show great promising in this area, since they possess good mechanical properties, thermal stability, toughness and some conductivity depending on degree of sulfonation (DS). To meet the requirements for applications in fuel cells, these poly(arylene ether)s are usually needed to be sulfonated. The chemical structures of these sulfonated poly(arylene ether)s are shown in Figure 11. After sulfonation, however, their mechanical properties are progressively deteriorated. Morever, these sulfonated membranes show a strong swelling when the DS is high enough to achieve high proton conductivity. One strategy to address these problems is to incorporate inorganic fillers into their polymer network. As discussed above, the conduction mechanism of sulfonated polymers is, in fact, water accociated, and their hydration state is an important factor in determining electrochemical performance of membranes. Therefore, the strong water retentivity of the inorganic filler surfaces could lead to an improved performance in the hybrid materials where the two components are held together by weak non-covalent interactions. It is, however, hard to achieve homogeneity at molecular level. The control at atomic level of the proper dosage and dispersion of the inorganic component can be reached *via* the preparation of hybrid membranes, where the organic and inorganic moieties are linked through covalent bonds.

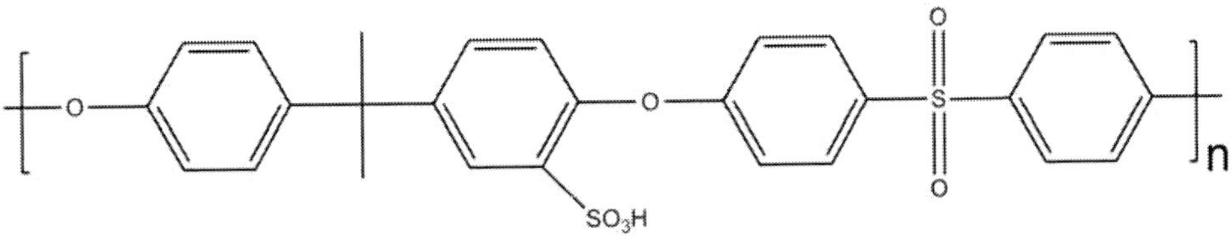

Figure 12. Chemical structure of a sulfated poly(arylene ether sulfone).

The incorporation of SiO_2, TiO_2, laponite and montmorillonite (MMT), etc., into the sulfonated poly(arylene ether)s have been reported [37,38,71,108-112]. In these cases, the dispersities of the inorganic fillers in the polymer matrix were not well controlled and they tended to aggregate into relatively hydrophobic domain, resulting in separated hydrophilic domains, due to a weakly non-covalent interaction between polymer matrix and the inorganic fillers. The conduction of the protons through membranes was mainly *via* Grotthus mechanism, and the obtained composite membranes usually exhibited relatively lower proton conductivity, although their methanol permeabilities of membranes have been reduced largely. Improvement of the hybrid characteristics was achieved by adding a larger amount of sulfation functionalized or intrinsically acidic inorganic fillers, such as SO_3H-functionalized SiO_2 nanoparticles (silica-SO_3H), zirconium phosphate (ZrP), boron phosphate (BPO_4) etc. [8,9,46,62,63,105,106,113-117]. The presence of organic proton conductive groups increased the interactions between polymer matrix and the inorganic fillers, which prevented the aggregation of inorganic filler molecules, resulting in smaller and better dispersed hydrophilic domains than in the sulfonated poly(arylene ether)s/inorganic filler without organic proton conductive groups composite membrane system. The significant improvements of the performance for these composite membranes have been demonstrated, as indicated by the high IEC value, high proton conductivity, low methanol cross-over and high bound water degree. In these cases, the protons within sulfonated poly(arylene ether)s/inorganic filler with organic proton conductive groups composite membrane were transported through both Grotthus and vehicle mechanisms.

4.3. Poly(arylene Ether Sulfone)s Based Composite Membranes

Due to excellent thermal and chemical stabilities and high glass transition temperatures, poly(arylene ether sulfone)s (PAES) have been identified as a promising class of polymers for use in PEMs. High proton conductivities of PAES are usually achieved through sulfonation. The chemical structure of a sulfated PAES is shown in Figure 12. In most cases, however, the sulfonation of PAES will result in a very high water sorption or even water solubility. Introduction of an inorganic component into PAES materials based membranes can improve the properties by potentially decreasing the water uptake and methanol permeability, while increasing the modulus and mechanical strength of the membrane.

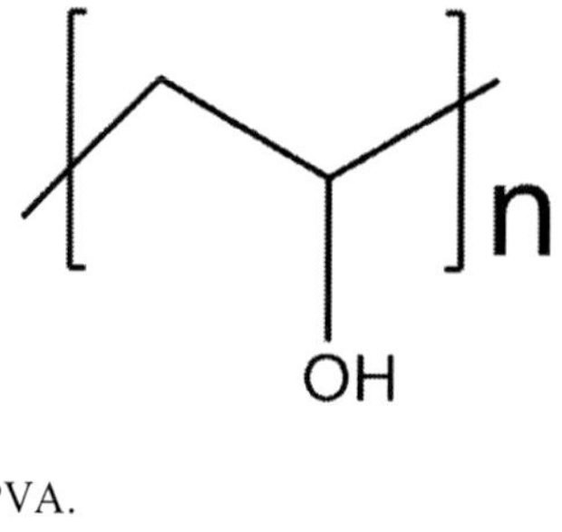

Figure 13. Chemical structure of a PVA.

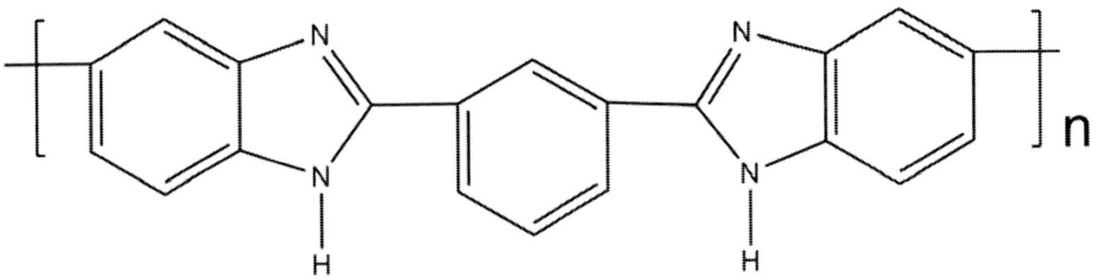

Figure 14. Schematic of a basic chemical structure of polybenzimidzole (PBI).

As memtioned above, the presence of organic proton conductive groups on the surface of inorganic filler can enhance the interactions between polymer matrix and the inorganic fillers which allows for the preparartion of membranes with high degree of homogeneity by preventing aggregation of inorganic fillers in the membranes. The strong water retentivity of the inorganic fillers could lead to an improved performance in the hybrid materials where the organic and inorganic moieties are linked through covalent bonds. Hill et al. [55] prepared a series of zirconium hydrogen phosphate/disulfonated poly(arylene ether sulfone) random copolymer composite membranes for PEMFC applications. Fully hydrated membranes (40 mol% disulfonation) with 38 wt% zirconium hydrogen phosphate had a conductivity of 0.06 S/cm at room temperature and linearly increased up to 0.13 S/cm in water vapor at 130 °C, whereas the pure copolymer only had a conductivity of 0.07 S/cm at room temperature and 0.09 S/cm at 130 °C. Wang et al. [56] investigated the sulfonated poly(ether ether ketone sulfone) (SPEEKS)/heteropolyacid (HPA) composite membranes for high temperature PEMFCs. Its proton conductivity linearly increased from 0.068 S/cm at 25 °C to 0.095 S/cm at 120 °C, which was significantly higher than that of Nafion 117.

However, the chemical stability of PAES membranes is always questionable owing to oxidative degradation of nonfluorinated backbone. The membranes prepared with PAESs are susceptible to degradation caused by the attack of hydroxyl and hydroperoxyl radicals. These hydroxyl and hydroperoxyl radicals are generated either in the anode or in the cathode after a 2-electron reduction reaction between H_2 and O_2, which across over the PEM to opposite electrode during the electrochemical prossess.

In a perfluorinated PEM, the membrane has a good oxidation resistance against degradation, due to the high energy of C–F bond (485 kJ/mol) and the close wrap of the C–C main chain by the fluorine atoms. Therefore, the attack from oxidative free radicals is effectively avoided. In a non-fluoride PEMs, however, the C–H bond dissociation enthalpy is relatively lower and the oxidative degradation easily occurrs. For the non-fluoride proton exchange membrane, oxidative degradation problem has become more prominent. In order to reduce the degradation of membrane caused by hydroxyl and hydroperoxyl radicals, Xing et al. [118] fabricated a Ag-SiO_2/sulfonated poly(biphenyl ether sulfone) (SPSU-BP) composite membrane. In this work, the Ag/SiO_2 in the composite membranes could work as a catalyst to decompose hydroperoxyl radicals. The durability of PEMFC with the Ag-SiO_2/SPSU-BP composite membrane was improved.

4.4. Poly(vinyl Alcohol) Based COMPOSITE Membranes

PVA is a water-soluble synthetic polymer. Bcause of high tensile strength, flexibility and humidity, and excellent methanol antipermeability, it has been employed as very attractive material for preparing composite PEMs. The chemical structure of PVA is shown in Figure 13. Due to proton inertness of PVA itself, various inorganic proton conductive particles, such as sulfated–ZrO_2, SO_3H–SiO_2, PO_3H_2–SiO_2, ZrP, HPAs, etc., are usually incorporated to impart it proton conductivity when combined to prepare PEMs [42,103,119][9]Nguyen and P. Schaetzel, Novel hydrophilic membrane materials: sulfonated polyethersulfone Cardo, *J. Membr. Sci. 186* (2001), p. 267. The main problem associated with these PVA based PEMs is the dissolution of PVA caused by water produced during the electrochemical process. For this reason, an approach that is used to create cross-linked network is usually employed. Chemical cross-linking is a highly versatile method to create and modify polymers, in particular, to limit the swelling, since the cross-linked polymers are ususlly water insoluble due to the presence of highly networked structures. In addition, the methanol antipermeability and proton conductivity of PEMs can also be improved by adjusting the cross-linking density of the membranes prepared. Therefore, for PEMFCs, in particular, for DMFC applications at low operating temperatures with a liquid methanol feed system, cross-linking techniques have attracted attention owing to their low cost and easy processility [120,121]. Tripathi et al. [122] and Ogoshi et al. [123] prepared a phosphonic acid grafted cross-linked PVA composite PEM. Cross-linking between organic and inorganic segments was carried out by gluteraldehyde in order to increase the hydrophobic nature of resultant composite for achieving polyelectrolyte membranes impervious to methanol. Phosphonic acid groups were introduced at the inorganic segment, because it shows high intrinsic proton conductivity due to high mobility of protonic charge carriers and degree of self-dissociation [124,125].

4.5. Polybenzimidazole Based Composite Membranes

Polybenzimidazole (PBI) is an aromatic heterocyclic basic polymer ($pK_a = 5.5$) which can be complexed with strong acids or very strong bases. The chemical structure of a PBI is shown in Figure 14. Due to its excellent thermo-chemical and mechanical stability properties, extremely high melting point and lower permeability for hydrogen than Nafion®, it has been widely used in the preparation of composite PEMs applicable at higher temperature PEMFCs. Althoug pure PBI is an electronic and ionic insulator, its ionic conductivity can be conferred by doping with the concentrated mineral acids, such as HCl, H_2SO_4, H_3PO_4, etc., and/or inorganic proton conductive particles, such as ZrP, HPAs, etc., in the proper conditions, while its electronic insulativity remains. For example, Savadogo et al. [126] systemically studied the protonic conductivity of PBI in various electrolytes. It was shown that blank PBI is a protonic insulator. After doped with H_2SO_4, HCl, HNO_3, $HClO_4$ or H_3PO_4, it exhibited high proton conductivity. The highest conductivity obtained (i.e. 0.0601 S/cm for PBI doped with 16 mol/L H_2SO_4) was as good as that of Nafion 117. The conductivity changing in the order of decrease is: $H_2SO_4 > H_3PO_4 > HClO_4 > HNO_3 > HCl$. The major disadvantage of these membranes is leaching of the low molecular weight acid in hot methanol solutions. The problems could be solved by the addition of high molecular weight ZrP, HPAs, etc. The proton conductivity of zirconium tricarboxybutylphosphonate [Zr(PBTC)]/PBI nanocomposite membranes has been studied by Yamazaki and coworkers [127,128]. Composite membranes with 50 wt% Zr(PBTC) exhibited uniform distribution of the particles throughout the thickness. The conductivities measured for compacted Zr(PBTC) powder and PBI-based membranes with 50 wt% Zr(PBTC) were 6.74×10^{-2} and 3.82×10^{-3} S/cm, respectively, at 200 °C for $PH_2O/Ps = 1$. Functionalization of the PBI with phosphoric acid groups or sulfonic acid groups further improved the proton conductivity of 50 wt% Zr(PBTC) membranes to 5.24×10^{-3} and 8.14×10^{-3} S/cm, repectively, under the same measurement conditions. In these cases, acid additives in PBI membranes were relatively stable due to the increased interaction between them after combination.

HPA doped PBI membranes have been investigated by Romero et al. [129]. They reported a novel hybrid organic–inorganic material formed by combination of phosphomolybdic acid $H_3PMo_{12}O_{40}$ (PMo_{12}) and poly(2,5-benzimidazole) (ABPBI). A enhancement of proton conducity was achieved by impregnation with phosphoric acid. Upon impregnation with phosphoric acid, the hybrid membranes presented higher conductivity than the best ABPBI polymer membranes measured in the same conditions. These PEMs had a thermal stability up to 200 °C, and a proton conductivity of 3×10^{-2} S/cm at 185 °C without humidification. These properties make them very good candidates as membranes for PEMFC at temperatures of 100–200 °C.

Staiti et al. [130,131] prepared composite membranes with combination of phosphotungstic acid (PWA), SiO_2 and PBI. The obtained membranes exhibited a high tensile strength and were chemically stable in boiling water with thermal stability in air up to 400 °C. It showed the proton conductivity of membranes were influenced by the temperature range 30–100 °C, relative humidity and PWA loading in the membrane. Maximum conductivity of 3.0×10^{-3} S/cm was obtained at 100% relative humidity and 100 °C with membrane containing 60 wt% PWA/SiO_2 in PBI. Conductivity measurements performed at higher temperatures, in the range from 90-150 °C, gave almost stable values of $1.4–1.5 \times 10^{-3}$ S/cm at 100% relative humidity.

Conclusions and Outlook

Fuel cell, as one of the important sources of non-conventional energies, delivering electrical energy through electrochemical oxidation of fuels, has been widely applied in military, portable devices, family house, transportation systems, etc. As an important component of PEMFCs, PEMs are compulsively required to have high proton conductivity and low methanol/water crossover. Evidently, the addition of inorganic materials effectively improves the thermal, mechanical and chemical stability, ionic conductivity and greatly decreases methanol/water crossover of PEMs at desired operation temperature, which results from the combined effect of organic and inorganic phases. Based on advances and progress obtained recently, it is reasonable to believe that more problems associated with membranes of PEMFCs could be addressed by proper incorporation of inorganic fillers into polymer electrolyte membranes. Presently, a great deal of work is underway in the area of composite membranes for PEMFCs. Unquestionably, there are a great many of issues needing to be accomplished, such as, what is the exact mechanism of proton conduction in the PEMs, how to operate H_2 fueled PEMFCs at elevated temperatures and dry conditions, how to further improve the tolarance of catalysts to poisonous matters produced during the electrochemical process and the fuel antipermeability of membranes, etc. Most important is to better understand the interaction between components in composite membranes. It is believed that a greater understanding of the interaction between components in composite membranes will help the rational design of new PEMs with desired properties.

REFERENCES

[1] Kordesch, K. V. *J. Electrochem. Soc.* 1978, *125*, C77-C91.
[2] Devanathan, R. *Energy Environ. Sci.* 2008, *1*, 101-119.
[3] Sattler, G. *J. Power Sources* 2000, *86*, 61-67.
[4] Bebin, P.; Caravanier, M.; Galiano, H. *J. Membr. Sci.* 2006, *278*, 35-42.
[5] Fu, R. Q.; Hong, L.; Lee, J. Y. *Fuel Cells* 2008, *8*, 52-61.
[6] Gribov, E. N.; Parkhomchuk, E. V.; Krivobokov, I. M.; Darr, J. A.; Okunev, A. G. *J. Membr. Sci.* 2007, *297*, 1-4.
[7] Kalappa, P.; Lee, J. H. *Polym. Int.* 2007, *56*, 371-375.
[8] Su, Y. H.; Liu, Y. L.; Sun, Y. M.; Lai, J. Y.; Wang, D. M.; Gao, Y.; Liu, B. J.; Guiver, M. D. *J. Membr. Sci.* 2007, *296*, 21-28.
[9] Zhao, C. J.; Lin, H. D.; Cui, Z. M.; Li, X. F.; Na, H.; Xing, W. *J. Power Sources* 2009, *194*, 168-174.
[10] Lin, Y. F.; Yen, C. Y.; Hung, C. H.; Hsiao, Y. H.; Ma, C. *J. Power Sources* 2007, *168*, 162-166.
[11] Lavorgna, M.; Gilbert, M.; Mascia, L.; Mensitieri, G.; Scherillo, G.; Ercolano, G. *J. Membr. Sci.* 2009, *330*, 214-216.
[12] Miyatake, K.; Tombe, T.; Chikashige, Y.; Uchida, H.; Watanabe, M. *Angew. Chem. Int. Ed.* 2007, *46*, 6646-6649.
[13] Nagarale, R. K.; Gohil, G. S.; Shahi, V. K. *Adv. Colloid Interface Sci.* 2006, *119*, 97-130.
[14] Kim, Y.; Lee, J. S.; Rhee, C. H.; Kim, H. K.; Chang, H. *J. Power Sources* 2006, *162*, 180-185.
[15] Kreuer, K. D. *J. Membr. Sci.* 2001, *185*, 29-39.
[16] Eigen, M. *Angew. Chem. Int. Ed.* 1964, *3*, 1-19.
[17] Neburchilov, V.; Martin, J.; Wang, H. J.; Zhang, J. J. *J. Power Sources* 2007, *169*, 221-238.
[18] Li, Q. F.; He, R. H.; Jensen, J. O.; Bjerrum, N. J. *Chem. Mater.* 2003, *15*, 4896-4915.
[19] Zhang, J. L.; Xie, Z.; Zhang, J. J.; Tanga, Y. H.; Song, C. J.; Navessin, T.; Shi, Z. Q.; Song, D. T.; Wang, H. J.; Wilkinson, D. P.; Liu, Z. S.; Holdcroft, S. *J. Power Sources* 2006, *160*, 872-891.
[20] Kundu, P. P.; Sharma, V.; Shul, Y. G. *Crit. Rev. Solid State Mater. Sci.* 2007, *32*, 51-66.
[21] Savadogo, O. *J. Power Sources* 2004, *127*, 135-161.
[22] Sanchez, C.; Julian, B.; Belleville, P.; Popall, M. *J. Mater. Chem.* 2005, *15*, 3559-3592.
[23] Herring, A. M. *Polym. Rev.* 2006, *46*, 245-296.
[24] Nagarale, R. K.; Shin, W.; Singh, P. K. *Polym. Chem.* 2010, 1, 388-408.

[25] Roziere, J.; Jones, D. J. *Annu. Rev. Mater. Sci.* 2003, *33*, 503-555.

[26] Kreuer, K. D. *Chem. Mater.* 1996, *8*, 610-641.

[27] Lee, C. H.; Min, K. A.; Park, H. B.; Hong, Y. T.; Jung, B. O.; Lee, Y. M. *J. Membr. Sci.* 2007, *303*, 258-266.

[28] Lavorgna, M.; Mascia, L.; Mensitieri, G.; Gilbert, M.; Scherillo, G.; Palomba, B. *J. Membr. Sci.* 2007, *294*, 159-168.

[29] Ladewig, B. P.; Knott, R. B.; Martin, D. J.; Da, Costa J.; Lu, G. Q. *Electrochem. Commun.* 2007, *9*, 781-786.

[30] Ren, S. Z.; Sun, G. Q.; Li, C. N.; Song, S. Q.; Xin, Q.; Yang, X. F. *J. Power Sources* 2006, *157*, 724-726.

[31] Ren, S. Z.; Sun, G. Q.; Li, C. N.; Liang, Z. X.; Wu, Z. M.; Jin, W.; Qin, X.; Yang, X. F. *J. Membr. Sci.* 2006, *282*, 450-455.

[32] Kim, J. D.; Mori, T.; Honma, I. *J. Membr. Sci.* 2006, *281*, 735-740.

[33] Daiko, Y.; Klein, L. C.; Kasuga, T.; Nogami, M. *J. Membr. Sci.* 2006, *281*, 619-625.

[34] Sacca, A.; Gatto, I.; Carbone, A.; Pedicini, R.; Passalacqua, E. *J. Power Sources* 2006, *163*, 47-51.

[35] Jalani, N. H.; Dunn, K.; Datta, R. *Electrochim. Acta* 2005, *51*, 553-560.

[36] Thomassin, J. M.; Pagnoulle, C.; Caldarella, G.; Germain, A.; Jerome, R. *J. Membr. Sci.* 2006, *270*, 50-56.

[37] Di, V. M. L.; Ahmed, Z.; Bellitto, S.; Lenci, A.; Traversa, E.; Licoccia, S. *J. Membr. Sci.* 2007, *296*, 156-161.

[38] Zhang, G. W.; Zhou, Z. T. *J. Membr. Sci.* 2005, *261*, 107-113.

[39] Tricoli, V.; Nannetti, F. *Electrochim. Acta* 2003, *48*, 2625-2633.

[40] Jung, D. H.; Cho, S. Y.; Peck, D. H.; Shin, D. R.; Kim, J. S. *J. Power Sources* 2003, *118*, 205-211.

[41] Uchida, H.; Ueno, Y.; Hagihara, H.; Watanabe, M. *J. Electrochem. Soc.* 2003, *150*, A57-A62.

[42] Kim, D. S.; Park, H. B.; Rhim, J. W.; Lee, Y. M. *J. Membr. Sci.* 2004, *240*, 37-48.

[43] Chang, H. Y.; Lin, C. W. *J. Membr. Sci.* 2003, *218*, 295-306.

[44] Watanabe, M.; Uchida, H.; Emori, M. *J. Phys. Chem. B* 1998, *102*, 3129-3137.

[45] Watanabe, M.; Uchida, H.; Seki, Y.; Emori, M.; Stonehart, P. *J. Electrochem. Soc.* 1996, *143*, 3847-3852.

[46] Zhang, Y.; Zhang, H. M.; Bi, C.; Zhu, X. B. *Electrochim. Acta* 2008, *53*, 4096-4103.

[47] Malers, J. L.; Sweikart, M. A.; Horan, J. L.; Turner, J. A.; Herring, A. M. *J. Power Sources* 2007, *172*, 83-88.

[48] Li, L.; Xu, L.; Wang, Y. X. *Mater. Lett.* 2003, *57*, 1406-1410.

[49] Duvdevani, T.; Philosoph, M.; Rakhman, M.; Golodnitsky, D.; Peled, E. *J. Power Sources* 2006, *161*, 1069-1075.

[50] Casciola, M.; Alberti, G.; Ciarletta, A.; Cruccolini, A.; Piaggio, P.; Pica, M. *Solid State Ionics* 2005, *176*, 2985-2989.

[51] Checkiewicz, K.; Zukowska, G.; Wieczorek, W. *Chem. Mater.* 2001, *13*, 379-384.

[52] Kim, Y. T.; Kim, K. H.; Song, M. K.; Rhee, H. W. *Curr. Appl. Phys.* 2006, *6*, 612-615.

[53] Park, Y. S.; Yamazaki, Y. *Eur. Polym. J.* 2006, *42*, 375-387.

[54] Kim, H. J.; Shul, Y. G.; Han, H. *J. Power Sources* 2006, *158*, 137-142.

[55] Hill, M. L.; Kim, Y. S.; Einsla, B. R.; Mcgrath, J. E. *J. Membr. Sci.* 2006, *283*, 102-108.

[56] Wang, Z.; Ni, H. Z.; Zhao, C. J.; Li, X. F.; Fu, T. Z.; Na, H. *J. Polym. Sci., Part B: Polym. Phys.* 2006, *44*, 1967-1978.
[57] Tan, A. R.; De, Carvalho, L. M.; Gomes, A. D. *Macromol. Symp.* 2005, *229*, 168-178.
[58] Tazi, B.; Savadogo, O. *J. New Mater. Electrochem. Syst.* 2001, *4*, 187-196.
[59] Yang, C.; Srinivasan, S.; Arico, A. S.; Creti, P.; Baglio, V.; Antonucci, V. *Electrochem. Solid-State Lett.* 2001, *4*, A31-A34.
[60] Alberti, G.; Costantino, U.; Casciola, M.; Ferroni, S.; Massinelli, L.; Staiti, P. *Solid State Ionics* 2001, *145*, 249-255.
[61] Costamagna, P.; Yang, C.; Bocarsly, A. B.; Srinivasan, S. *Electrochim. Acta* 2002, *47*, 1023-1033.
[62] Silva, V. S.; Weisshaar, S.; Reissner, R.; Ruffmann, B.; Vetter, S.; Mendes, A.; Madeira, L. M.; Nunes, S. *J. Power Sources* 2005, *145*, 485-494.
[63] Zaidi, S.; Mikhailenko, S. D.; Robertson, G. P.; Guiver, M. D.; Kaliaguine, S. *J. Membr. Sci.* 2000, *173*, 17-34.
[64] Smitha, B.; Sridhar, S.; Khan, A. A. *J. Polym. Sci., Part B: Polym. Phys.* 2005, *43*, 1538-1547.
[65] Klein, L. C.; Daiko, Y.; Aparicio, M.; Damay, F. *Polymer* 2005, *46*, 4504-4509.
[66] Mahreni, A.; Mohamad, A. B.; Kadhum, A.; Daud, W.; Iyuke, S. E. *J. Membr. Sci.* 2009, *327*, 32-40.
[67] Choi, P.; Jalani, N. H.; Datta, R. *J. Electrochem. Soc.* 2005, *152*, E84-E89.
[68] Choi, P.; Jalani, N. H.; Datta, R. *J. Electrochem. Soc.* 2005, *152*, E123-E130.
[69] Choi, P.; Jalani, N. H.; Datta, R. *J. Electrochem. Soc.* 2005, *152*, A1548-A1554.
[70] Kim, Y. J.; Choi, W. C.; Woo, S. I.; Hong, W. H. *J. Membr. Sci.* 2004, *238*, 213-222.
[71] Mccabe, C.; Glotzer, S. C.; Kieffer, J.; Neurock, M.; Cummings, P. T. *J. Comput. Theor. Nanosci.* 2004, *1*, 265-279.
[72] Won, J.; Park, H. H.; Kim, Y. J.; Choi, S. W.; Ha, H. Y.; Oh, I. H.; Kim, H. S.; Kang, Y. S.; Ihn, K. J. *Macromolecules* 2003, *36*, 3228-3234.
[73] Young, S. K.; Jarrett, W. L.; Mauritz, K. A. *Polymer* 2002, *43*, 2311-2320.
[74] Li, T.; Yang, Y. *J. Power Sources* 2009, *187*, 332-340.
[75] Cotton, F. A.; Wilkinson, G.; Murillo, C. A.; Bochmann, M. Advanced Inorganic Chemistry; John Wiley & Sons: New York, 1999.
[76] Mauritz, K. A. *Mater. Sci. Eng., C* 1998, *6*, 121-133.
[77] Klein, L. C.; Daiko, Y.; Aparicio, M.; Damay, F. *Polymer* 2005, *46*, 4504-4509.
[78] Licoccia, S.; Traversa, E. *J. Power Sources* 2006, *159*, 12-20.
[79] Baglio, V.; Di, B. A.; Arico, A. S.; Antonucci, V.; Antonucci, P. L.; Nannetti, F.; Tricoli, V. *Electrochim. Acta* 2005, *50*, 5181-5188.
[80] Song, M. K.; Park, S. B.; Kim, Y. T.; Kim, K. H.; Min, S. K.; Rhee, H. W. *Electrochim. Acta* 2004, *50*, 639-643.
[81] Lonyi, F.; Valyon, J.; Engelhardt, J.; Mizukami, F. *J. Catal.* 1996, *160*, 279-289.
[82] Yadav, G. D.; Nair, J. J. *Microporous Mesoporous Mater.* 1999, *33*, 1-48.
[83] Munakata, H.; Chiba, H.; Kanamura, K. *Solid State Ionics* 2005, *176*, 2445-2450.
[84] Wang, H. T.; Holmberg, B. A.; Huang, L. M.; Wang, Z. B.; Mitra, A.; Norbeck, J. M.; Yan, Y. S. *J. Mater. Chem.* 2002, *12*, 834-837.
[85] Li, C. N.; Sun, G. Q.; Ren, S. Z.; Liu, J.; Wang, Q.; Wu, Z. M.; Sun, H.; Jin, W. *J. Membr. Sci.* 2006, *272*, 50-57.
[86] Hara, S.; Miyayama, M. *Solid State Ionics* 2004, *168*, 111-116.

[87] Jalani, N. H.; Dunn, K.; Datta, R. *Electrochim. Acta* 2005, *51*, 553-560.
[88] Zhai, Y. F.; Zhang, H. M.; Hu, J. W.; Yi, B. L. *J. Membr. Sci.* 2006, *280*, 148-155.
[89] Rhee, C. H.; Kim, Y.; Lee, J. S.; Kim, H. K.; Chang, H. *J. Power Sources* 2006, *159*, 1015-1024.
[90] Corcoran, D.; Tunstall, D. P.; Irvine, J. *Solid State Ionics* 2000, *136*, 297-303.
[91] Yang, C.; Srinivasan, S.; Bocarsly, A. B.; Tulyani, S.; Benziger, J. B. *J. Membr. Sci.* 2004, *237*, 145-161.
[92] Clearfield, A.; Stynes, J. A. *J. Inorg. Nucl. Chem.* 1964, *26*, 117-129.
[93] Clearfield, A. *Annu. Rev. Mater. Sci.* 1984, *14*, 205-229.
[94] Grot, W. G.; Rajendran, G. *US Patent*, US 5 919 583, 1999.
[95] Alberti, G.; Casciola, M. *Solid State Ionics* 1997, *97*, 177-186.
[96] Clearfield, A.; Wang, J. D.; Tian, Y.; Stein, E.; Bhardwaj, C. *J. Solid State Chem.* 1995, *117*, 275-289.
[97] Alberti, G.; Casciola, M. *Annu. Rev. Mater. Res.* 2003, *33*, 129-154.
[98] Checkiewicz, K.; Zukowska, G.; Wieczorek, W. *Chem. Mater.* 2001, *13,* 379-384.
[99] Tazi, B.; Savadogo, O. *Electrochim. Acta* 2000, *45*, 4329-4339.
[100] Lehtinen, T.; Sundholm, G.; Holmberg, S.; Sundholm, F.; Bjornbom, P.; Bursell, M. *Electrochim. Acta* 1998, *43*, 1881-1890.
[101] Shen, Y.; Qiu, X. P.; Shen, J.; Xi, J. Y.; Zhu, W. T. *J. Power Sources* 2006, *161*, 54-60.
[102] Patel, R.; Park, J. T.; Lee, W. S.; Kim, J. H.; Min, B. R. *Polym. Adv. Technol.* 2009, *20*, 1146-1151.
[103] Tripathi, B. P.; Shahi V. K. *ACS Appl. Mater. Interfaces* 2009, *1*, 1002-1012.
[104] Hickner, M. A.; Ghassemi, H.; Kim, Y. S.; Einsla, B. R.; Mcgrath, J. E. *Chem. Rev.* 2004, *104*, 4587-4611.
[105] Silva, V. S.; Ruffmann, B.; Vetter, S.; Mendes, A.; Madeira, L. M.; Nunes, S. P. *Catal. Today* 2005, *104*, 205-212.
[106] Ponce, M. L.; Prado, L.; Silva, V.; Nunes, S. P. *Desalination* 2004, *162*, 383-391.
[107] Jung, D. H.; Myoung, Y. B.; Cho, S. Y.; Shin, D. R.; Peck, D. H. *Int. J. Hydrogen Energy* 2001, *26*, 1263-1269.
[108] Di, V. M. L.; Marani, D.; D'Epifanio, A.; Traversa, E.; Trombetta, M.; Licoccia, S. *Polymer* 2005, *46*, 1754-1758.
[109] Yen, Y. C.; Ye, Y. S.; Cheng, C. C.; Lu, C. H.; Tsai, L. D.; Huang, J. M.; Chang, F. C. *Polymer* 2010, *51*, 84-91.
[110] Kim, D. S.; Liu, B.; Guiver, M. D. *Polymer* 2006, *47*, 7871-7880.
[111] Nunes, S. P.; Ruffmann, B.; Rikowski, E.; Vetter, S.; Richau, K. *J. Membr. Sci.* 2002, *203*, 215-225.
[112] Chang, J. H.; Park, J. H.; Park, G. G.; Kim, C. S.; Park, O. O. *J. Power Sources* 2003, *124*, 18-25.
[113] Bonnet, B.; Jones, D. J.; Roziere, J.; Tchicaya, L.; Alberti, G.; Casciola, M.; Massinelli, L.; Bauer, B.; Peraio, A.; Ramunni, E. *J. New Mater. Electrochem. Syst.* 2000, *3*, 87-92.
[114] Tripathi, B. P.; Kumar, M.; Shahi, V. K. *J. Membr. Sci.* 2009, *327*, 145-154.
[115] Ahmad, M. I.; Zaidi, S.; Rahman, S. U. *Desalination* 2006, *193*, 387-397.
[116] Krishnan, P.; Park, J. S.; Kim, C. S. *J. Membr. Sci.* 2006, *279*, 220-229.
[117] Ponce, M. L.; Prado, L.; Ruffmann, B.; Richau, K.; Mohr, R.; Nunes, S. P. *J. Membr. Sci.* 2003, *217*, 5-15.

[118] Xing, D. M.; Zhang, H. M.; Wang, L.; Zhai, Y. F.; Yi, B. L. *J. Membr. Sci.* 2007, *296*, 9-14.
[119] Chang, Y. W.; Wang, E.; Shin, G.; Han, J. E.; Mather, P. T. *Polym. Adv. Technol.* 2007, *18*, 535-543.
[120] Fujimoto, Y.; Heishi, M.; Shimojima, A.; Kuroda, K. *J. Mater. Chem.* 2005, *15*, 5151-5157.
[121] Shea, K. J.; Loy, D. A. *Chem. Mater.* 2001, *13*, 3306-3319.
[122] Tripathi, B. P.; Saxena, A.; Shahi, V. K. *J. Membr. Sci.* 2008, *318*, 288-297.
[123] Ogoshi, T.; Chujo, Y. *J. Mater. Chem.* 2005, *15*, 315-322.
[124] Steininger, H.; Schuster, M.; Kreuer, K. D.; Maier, J. *Solid State Ionics* 2006, *177*, 2457-2462.
[125] Schuster, M.; Rager, T.; Noda, A.; Kreuer, K. D.; Maier, J. *Fuel Cells* 2005, *5*, 355-365.
[126] Xing, B. Z.; Savadogo, O. *J. New Mater. Electrochem. Syst.* 1999, *2*, 95-101.
[127] Jang, M. Y.; Yamazaki, Y. *J. Power Sources* 2005, *139*, 2-8.
[128] Jang, M. Y.; Yamazaki, Y. *Solid State Ionics* 2004, *167*, 107-112.
[129] Gomez-Romero, P.; Asensio, J. A.; Borros, S. *Electrochim. Acta* 2005, *50*, 4715-4720.
[130] Staiti, P. *Mater. Lett.* 2001, *47*, 241-246.
[131] Staiti, P.; Minutoli, M.; Hocevar, S. *J. Power Sources* 2000, *90*, 231-235.

In: Advanced Organic-Inorganic Composites
Editor: Inamuddin
ISBN 978-1-61324-264-3

Chapter 2

SURFACE-CONFINED INORGANIC-ORGANIC HYBRID MATERIALS FOR SENSOR FUNCTIONALITIES

Prakash Chandra Mondal, Bhawna Gera and Tarkeshwar Gupta*
Department of Chemistry, University of Delhi,
Delhi 110 007, India

ABSTRACT

The combination of inorganic or organic surfaces and molecular chemistry has led to the development of inorganic-organic hybrid materials with advanced functionalities. These ideas provide a means of bridging the gap between molecular chemistry, materials sciences, nanotechnology and surface chemistry. Recently, three major research areas transform our vision of molecular chemistry and material sciences: supramolecular chemistry, metal-organic frameworks (reticular chemistry) and surface chemistry (thin film chemistry). Each area or combination of these at nanoscale level provides a technologically exciting ground to nanochemistry or nanotechnology. Moreover, on functional aspect these combined areas allow the fine-tuning of optical and electronic properties of the materials which offers new perspective for the development of electronic and optical sensors with enhanced molecular properties.

ABBREVIATIONS

AAO	Anodized aluminium oxide
AES	Auger electron spectroscopy
AFM	Atomic force microscope
ALD	Atomic layer deposition
BTC	Benzene tricarboxylic acid
CA	Contact angle

* Corresponding author's e-mail: tgupta@chemistry.du.ac.in

CPs	Coordination polymers
CV	Cyclic voltammetry
DEF	N,N-diethylformamide
EIS	Electrochemical impedance spectroscopy
GaAs	Gallium arsenide
HKUST	Hong Kong University of Science and Technology
IEPs	Isoelectric points
IOHMs	Inorganic-organic hybrid materials
ISS	Ion scattering spectroscopy
ITO	Indium tin oxide
LB	Langmuir-Blodgett
LBL	Layer-by-layer
LPE	Liquid phase epitaxy
MIL	Materials of Institute Lavoisier
MOFs	Metal-organic frameworks
MITD	Microwave-induced thermal deposition
NACS	Nitroaromatic compounds
Pc	Pthalocyanine
PL	Photoluminescence
PXRD	Powder X-ray diffraction
QCM	Quartz crystal microbalance
RCA	Radio-Corporation of America
SAMs	Self assembled monolayers
SAWs	Surface acoustic waves
SCPs	Surface coordination polymers
SEM	Scanning electron microscope
SIMS	Secondary ion mass spectroscopy
SPR	Surface plasmon resonance
STM	Scanning tunneling microscope
SURIOHMs	Surface confined inorganic-organic hybrid materials
SURMOFs	Surface confined MOFs
TTF	Tetrathiafulvalene
UVOCS	Ultra violet ozone cleaning system
VOCs	Volatile organic compounds
XPS	X-ray photoelectron spectroscopy
XRD	X-ray diffraction
XRR	X-ray reflectivity

1. Introduction: Linking Molecular Chemistry with Solid Surfaces

Inorganic-organic hybrid materials [1,2] (IOHMs) are substances prepared by chemical reaction between organic and inorganic building blocks. The development of such materials, which have already found numerous applications in the field of optics [3-7], electronics [8], opto-electronics [9, 10], mechanics [11], catalysts [12], sensors [13-16], biology [17,18], and nanotechnology [19-22], is one of the big achievements for the material scientists. The initial plan was to create smart materials with new combinations of structures-properties and functions by combining inorganic and organic building blocks on a molecular level, which recently appeared as a creative alternative for obtaining new high-technology nanomaterials [23-25] with unusual features. For instance, the combination of designed inorganic and organic molecular building blocks yields fascinating structures with various dimensionalities in the form of supramolecular structures [26-28], coordination polymers (CPs) [29, 30] or metal-organic frameworks (MOFs) [31-33].

On the basis of the nature of chemical interactions at the inorganic-organic interface, the IOHMs have been classified into two classes [2]. Class I, where the organic and inorganic components interact weakly (i.e., hydrogen bonds, halogen bonds, π-π interactions, hydrophobic interactions, Van der Waals interactions etc.), such materials were consequently designated as supramolecular network [26-28]. Class II, where the organic and inorganic components linked through strong chemical bonds (i.e., covalent bond, coordinate bond etc.), such materials were advanced as metal-organic frameworks [31-33] or coordination polymers [29,30]. Both classes of IOHMs are independently developing rapidly in chemical science. Recently, class II IOHMs has received enormous interest with huge potentialities because of its gas adsorption [34-36] properties in its nanopores which encouraged chemist, physicist, and material scientist to fascinate advanced research in this field.

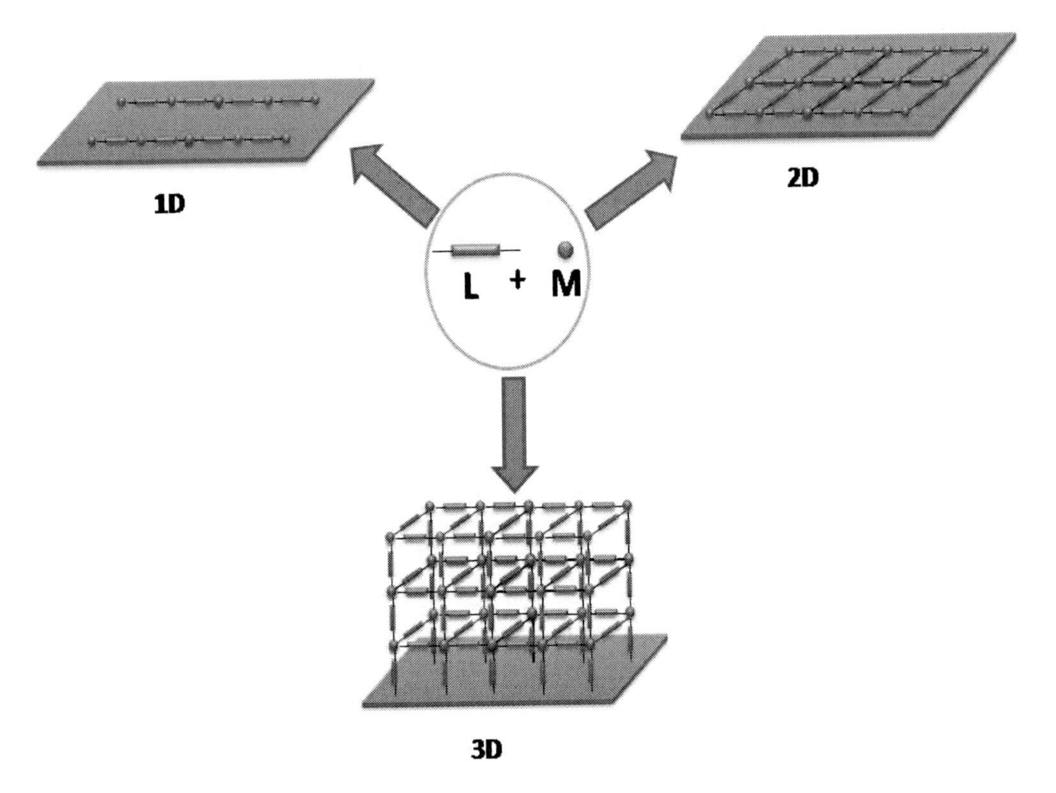

Figure 1. Schematic representation for the fabrication of 1D, 2D and 3D structure on solid support. M = inorganic unit (metal ions), L = organic unit (ligand/coupling molecules). Solid support is either inorganic surface or organic surface (inorganic surface functionalized with organic molecules).

On elementary level, inorganic and organic materials individually or in combination show a large variety of physical properties such as conductivity, magnetism, optics, which render them attractive for new generation of optical and electronic devices, but this requires confinement of these materials onto the solid surface. Therefore, modification of substrate surfaces especially inorganic scaffold with various organic/inorganic molecules is developing at an extraordinary pace, with an exponential growth in the number of research papers [37-41] and reviews [42-46] appearing in the chemical literature. Basically, this area of research, which is also nominated as "molecular-based thin film chemistry" [47] combine organic/inorganic molecules with solid surfaces to produce technology-generating properties in a way that is extraordinarily experienced in science. Interestingly, the molecular-based thin film chemistry in particular monolayer/multilayer assembly with inorganic-organic materials is a fast emerging field. For instance, a large flexible design of ordered porous and non-porous inorganic-organic hybrid material on a variety of surfaces has been reported by Zacher et al. [45], Bekermann et al. [48], Fischer et al. [50], Shekhah et al. [37,49,51,53], Manuera et al. [52], and others [54-58] in order to establish structure-properties relationships. In addition, they have shown that apart from amorphous solid material on surface, a controlled growth of IOHMs on various surfaces offers a new route to design crystalline material with a variety of functionalities and dimensionalities *viz*, one dimensional (1D), two dimensional (2D), and three dimensional (3D) as shown in Figure 1.

The major challenges for fabrication of surface-confined inorganic-organic hybrid materials (SURIOHMs) are to transfer the properties of the target inorganic/organic molecules from the solution or gas phase to the solid substrate and maintain/enhance the desired molecular properties/functions of these materials at the solid-state interface. Such a transfer needs a fine control over molecular orientation, interactions, pore size, topology, order, thickness and properties in functional nanoscale thin films. Among these, most of the issues have been resolved during past few years and some of the system has been studied extensively for their application and device fabrication.

There are varieties of potential application of SURIOHMs reported in the literature, nevertheless, probably the opto-electronic properties and (bio-)chemical sensors are the most studied one. This is due to the fact that SURIOHMs-based sensor system works more efficiently than the solution-based sensor system because all the receptors/active sites are in direct contact (pre-organized receptors), with analyte medium which results in very short response time, easy reproducibility, amplification of sensing. The preorganized receptors can be obtained *via* fine control on organization and packing of individual molecules on platform such as transparent and/or conducting glass slides, metal surfaces or other supports such as nanoparticles and polymers. The chemical sensing events on SURIOHMs depend mainly on interaction between the immobilized receptors and the analytes in solution or gas phase. Thus, the sensing process can be monitored at solid surface/solution or surface/gas interfaces. In this chapter we will discuss the various methods for the fabrication of thin films of IOHMs on a variety of substrates and chemical sensing functionalities based on the recent reports. We will also discuss an overview of the analytical techniques used for characterization of SURIOHMs, since these techniques can also be used to study the interaction of the analytes at functionalized surfaces.

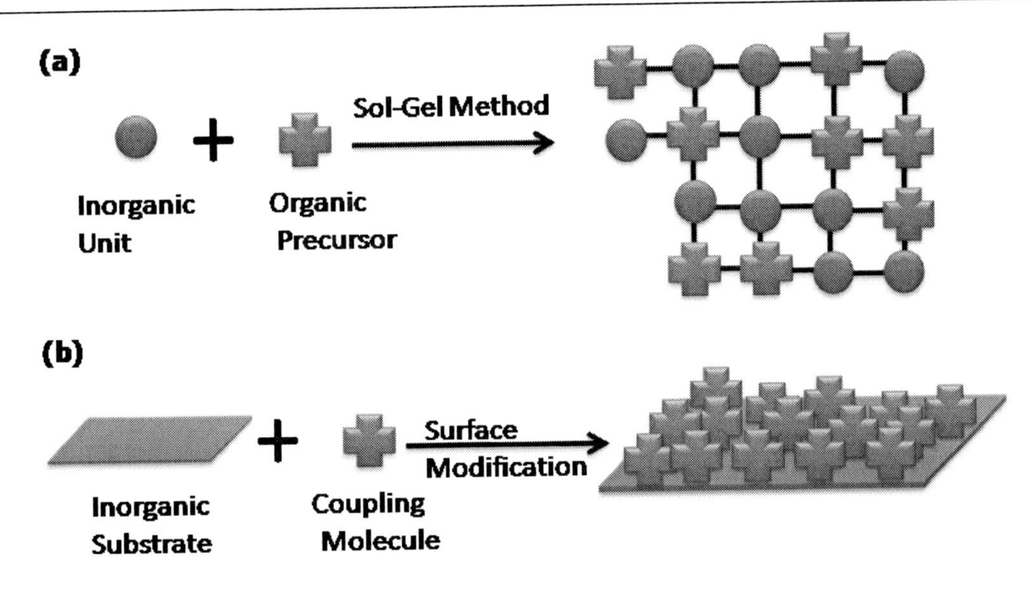

Figure 2. Schematic representation of inorganic-organic hybrid materials prepared by sol-gel process (a) and by surface modification process (b).

2. Immobilization of Inorganic/Organic Hybrid Materials

The solution or bulk phase studies of materials lends to a wide variety of analytical techniques, it is perhaps, not the most beneficial environment for the operation of advanced materials because it generally lacks coherence, prohibiting large-scale uniformity and coordination in the operations of functional materials. However, immobilizing molecules on surface, impart a very high degree of coherence in their orientation, alignment, and packing provided that well defined uniform adlayers of the absorbates can be obtained. Moreover, solid support provide a mean of integrating molecular adlayer with well known and thoroughly studied material substrates (e.g. SiO_2, ITO, gold etc.) in a conventional device setting. Therefore, assembling inorganic-organic compounds on solid surface (organic or inorganic) represents a vitally important step towards developing nanoscopic material properties.

There are mainly two approaches adopted by the material scientists to study the surface-confined materials at nanoscopic level a) top-down approach which included a famous sentence by Richard P. Feynman i.e. "*there is plenty of room at the bottom*" [59] and b) bottom-up approach which included another famous reply from Jean-Marie Lehn i.e., "*there is even more room at the top*" [60]. Top-down approaches seek to create nanoscale devices by using larger, externally-controlled ones to direct their assembly such as micro patterning, electron-beam, soft and chemical/photo lithography, however, most of these techniques have limitations below 50-100 nm^2. Alternatively, bottom-up approaches seek to have smaller (usually molecular) components built up into more complex assemblies such as molecular self assembly, molecular layer-by-layer assembly. Following a bottom-up strategy, the tools of material chemistry provides the guideline how to design molecular building blocks in order to achieve a balance between molecule-molecule, molecule-substrate, molecule-solvent and solvent-substrate interaction leading to targeted functional patterns. Molecules can be assembled onto surfaces by a variety of methods (vide infra). Here we will discuss briefly the important techniques for the deposition of organic/inorganic materials on various surfaces.

2.1. Deposition of Organic Materials on Inorganic Surfaces

The coupling of organic and inorganic moiety is most often implemented by sol-gel process [61] (anchoring of organic group to inorganic network) as a bulk hybrid material. However, coupling of organic moiety with inorganic surface is further developed through surface modification (anchoring of organic group to inorganic surface) process [62] as a nanoscale hybrid thin film (Figure 2). The organic moiety is preferably named as "precursors" or "modifiers" by sol-gel community and "coupling agent" or "coupling molecules" by surface community. Here we will highlight the fabrication of inorganic-organic hybrid materials by surface modification process while sol-gel process is out of scope of this chapter.

The surface modification with organic molecules, in other words adsorption or immobilization of organic molecules on inorganic substrates, can be achieved by two ways: a) physical adsorption (weak interaction: class I SURIOHMs) and/or b) chemical adsorption (strong interaction: class II SURIOHMs). In class I, generally molecules adhere to the surface through non-covalent interactions. However, in class II, the molecules have chemical functionality (or head group) with specific affinity for the solid surface by which it forms covalent or coordinate bonds. Both types of adsorption can be formed by spontaneous immobilization either from the liquid or vapor phase. There exist number of methods [63,64] as illustrated in Figure 3, for the deposition of organic coupling molecules on inorganic surfaces such as: a) Langmuir-Blodgett (LB) films, b) Self-assembled monolayers (SAMs), c) Drop casting, d) Spin coating, e) Stamping, f) Electrochemical deposition etc..

In surface modification process, selection of substrates (insulating, metallic, transparent etc.), is of primary importance because of their functional contributions to the film-substrate hetero-structures and also because they should enable the preparation of well defined interfaces. Various inorganic substrates such as Si(100), Si(111), SiO_2, ITO, TiO_2, Au, Ag, Ni, Al_2O_3, mica, alkali halide, glass, quartz, metal dichalcogenides etc. have been used to deposit organic molecules such as siloxane, thiols, phosphonates, acenes, alkenes, alkynes etc., in order to fabricate surface-confined hybrid materials.

The old but elegant Langmuir-Blodgett (LB) technique [65] is one of the ways to arrange molecules, especially amphiphilic molecules onto various solid supports in the form of organized assemblies. It consists of one or more monolayers of an organic material, deposited from the surface of a liquid onto a solid by immersing (or emersing) the solid substrate into (or from) the liquid. A monolayer is adsorbed homogeneously with each immersion or emersion step, thus films with very accurate thickness can be formed. This thickness is accurate because the thickness of each monolayer is known and can therefore be added to find the total thickness of a Langmuir-Blodgett film. Langmuir–Blodgett films are named after Irving Langmuir and Katharine B. Blodgett, who invented this technique while working in Research and Development for General Electric Co.

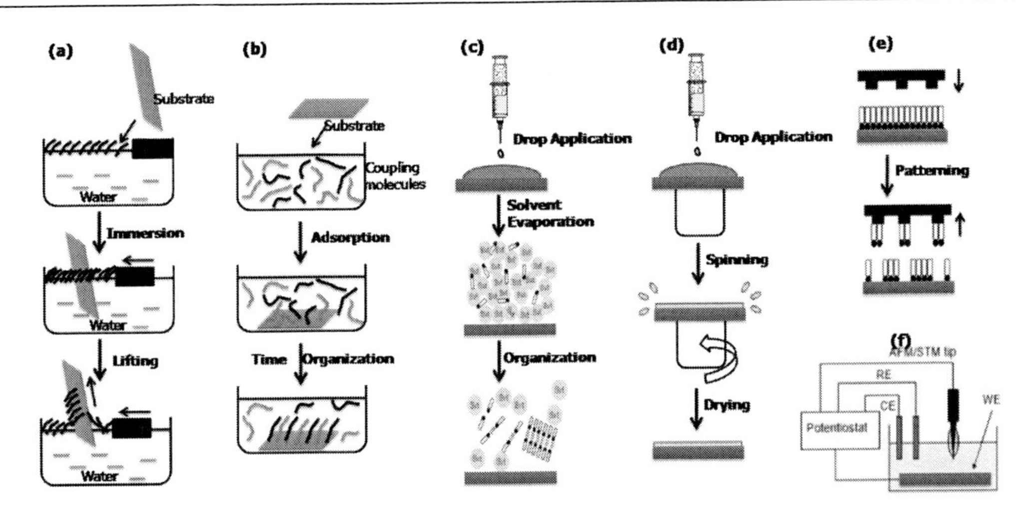

Figure 3. Schematic representation of the various routes for the deposition of organic molecules onto inorganic substrates or inorganic/organic molecules onto any substrates. Formation of Langmuir-Blodgett films (a), Formation of self-assembled monolayers (b), Drop casting method (c), Spin coating method (d), Stamping or patterning of materials (e) and Electrochemical deposition (f) (RE: reference electrode; CE: counter electrode; WE: working electrode).

An alternative technique for creating single molecular layers on surfaces is that of self-assembled monolayers [66,67]. SAMs can be created by the chemisorption of "head groups" onto a substrate either from the vapor or liquid phase followed by a slow two-dimensional organization of "tail groups" as shown in Figure 3 b.

For instance, chemical adsorption of various organosulfur, organoselenium molecules such as thiols, thioether, disulfides or thioctic ester derivatives, onto soft metal substrates (eg. Au, Ag) can be achieved by this technique. Self-assembly of organosulfur compounds [68] is probably the most studied system among all molecular-based self-assembly on surface. Organosulfur compounds coordinate strongly to the soft metal surfaces such as gold, silver, copper, platinum, mercury, iron etc.. However, most investigations have been done on gold surface. The crystalline nature of gold substrate are the main attraction in this area. It was also suggested that gold does not have a stable surface oxide; therefore, its surface can be cleaned simply by removing the physically and chemically adsorbed contaminants.

Techniques based on a controlled solvent evaporation such as drop casting [69] and spin coating [70] (Figure 3 c and d) are other approach to deposit organic material onto inorganic substrates from solution phase. In these techniques a careful choice of different parameters such as polarity and volatility of solvent, concentration and size of the coupling molecules and nature of surface involved. Drop casting consists of letting a small volume of solution of coupling molecules fall onto the substrate surface and leaving it to dry either in air or in solvent atmosphere to ensure slower evaporation.

Another similar method is spin coating where the solution of functional molecules is applied on the substrate which is then rotated on a spinning wheel at high speed so the centrifugal forces push the excess solution over the edge of the substrate and a residue on the substrate remains thanks to surface tension.

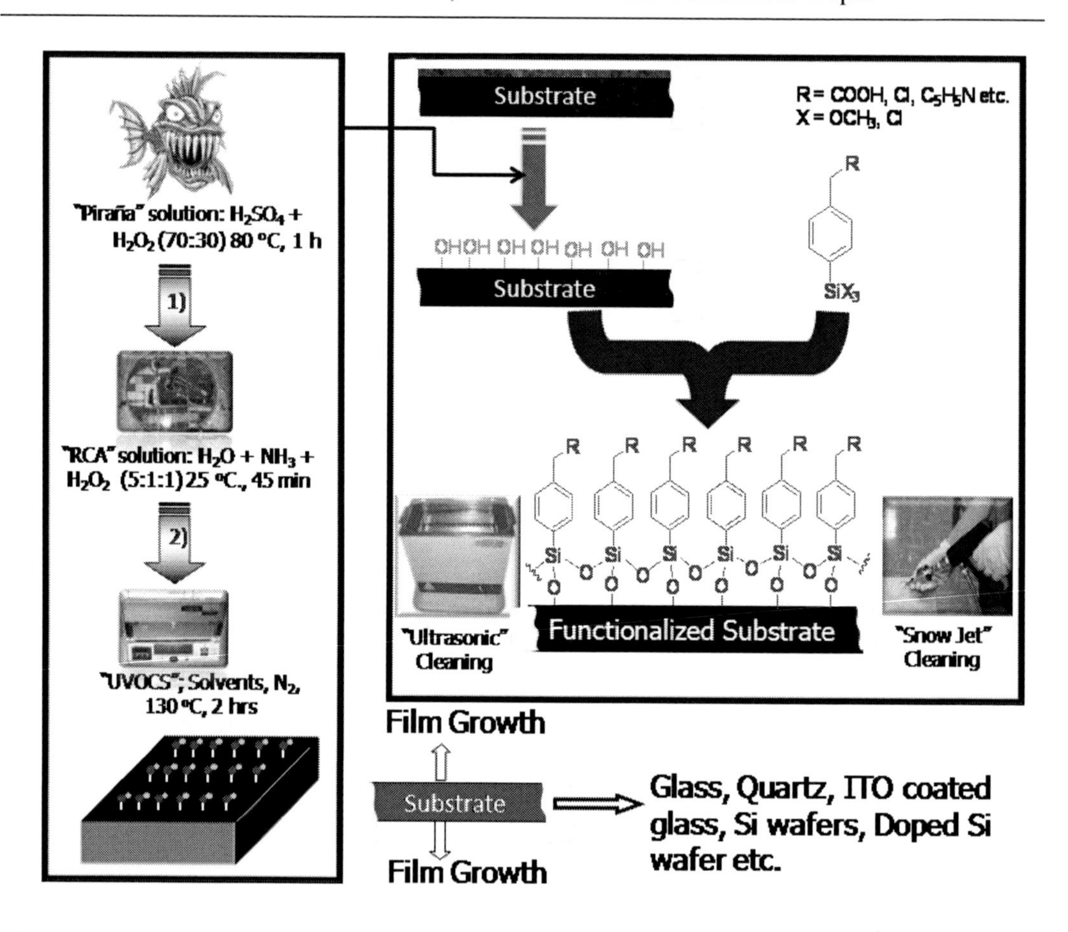

Figure 4. Methods for cleaning and modification/functionalization of inorganic substrates (SiOx) with organic coupling molecules having different functionalities.

The use of functional molecules as ink on a rubber stamp [71] (Figure 3 e) is a conceptually simple and attractive way to deposit organic molecules onto substrates. Basically, the stamp is pressed into the solution of organic molecules and is then removed and pressed onto substrate surface. This method has very wide variety of appearance within the field of lithography, printing and patterning. Another stylish way to deposit organic material on electrode substrate is electrochemical deposition [72,73] (Figure 3 f). In this technique the ability exists to produce films accurately in regard to nucleation and growth, by controlling the potential and current flow of the system. The electrodeposition process is carried out in a three electrode electrochemical cell consist of a working electrode (conducting substrate surface), a reference electrode and a counter electrode. The resulting structure can be studied in situ at nano-scale if the cell is integrated in an atomic force microscopy (AFM) or scanning tunnelling microscopy (STM).

Another elegant way to assembled organic coupling molecules such as organoalkoxysilanes, organochlorosilanes or organophosphorous acids and their derivatives (salt, ester) onto SiO_x-based surfaces such as silica, silicate glasses is covalent assembly of various coupling molecules, which results in the formation of M-O-Si bond between the inorganic phase and organosilane/organophosphate [62,66,74,75]. Provided that the substrate has been cleaned properly with piranha [74], Radio Corporation of America (RCA) solutions followed by ultraviolet ozone cleaning system (UVOCS) [76] in order to generate hydroxyl-terminated surface (Figure 4). In principle, it simply has to be dipped into the corresponding solution of organic molecules for a period of time determined by the solvent and the reactivity of the head group. These are well suited to the design of hybrid materials based on SiO_x-OH terminated inorganic substrates, although they have also been used for the coupling of organic groups to many other inorganic matrices and supports. In this kind of surface modification, presence of water appears necessary for the formation of complete monolayers [77-80]. However, homocondensation increases as the water content increases and the risk is the formation of multilayers by polymerization of the multifunctional organosilicon molecules [81]. As a consequence, the quality of the surface modification is very sensitive to the amount of water in the solvent or adsorbed on the surface of the oxide. Phosphorous derivatives are much less sensitive to nucleophilic substitution than the silicon parent derivative. The organophosphorous coupling molecules cannot homocondense (forming P-O-P bonds), they form only monolayers, whatever the water content is. Therefore, recently surface modification using organophosphorous coupling molecules has been applied to a large variety of metal oxide supports [82-85], including the native oxide layer at the surface of metals such as titanium, aluminium, iron, copper and silver. Immobilization of functional organic molecules with variety of open receptor sites [86-90] or coordination sites is a huge interest for material chemists.

Similarly, assembly of organic molecules such as alkenes and alkynes onto silicon surfaces [both Si(100)-H and Si(111)-H] by covalent assembly method can be achieved [91-93]. First step is to clean silicon surface (removal of silicon oxide) by treating with HF, followed by reaction of the hydrogen terminated silicon surface with alkene or alkyne which results in the formation of robust monolayer with stable covalent Si-C bond. After complete formation of the monolayers an appropriate rinsing/sonication procedures followed by CO_2 snow jet treatment [94] has usually to be performed to remove unreacted material. The high stability of these monolayers and the relatively mild conditions require for their formation allow the construction of functionalized silicon surfaces.

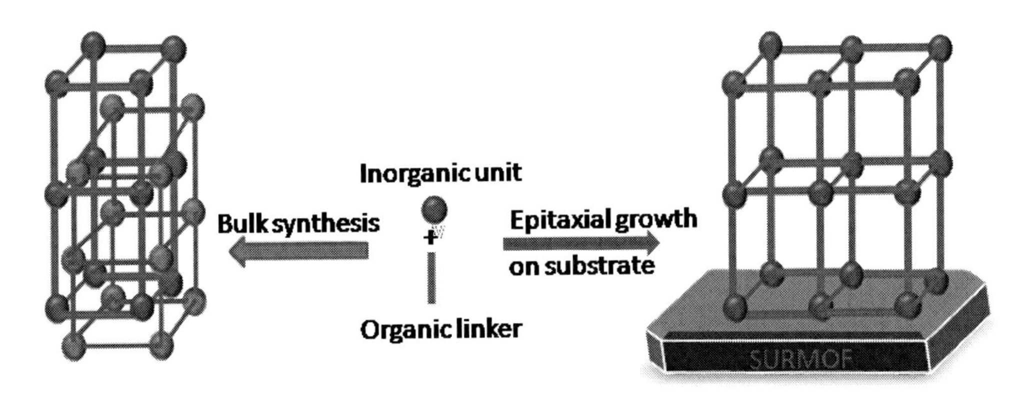

Figure 5. Schematic representation of MOF synthesis. Using conventional synthesis, often two equivalent networks (coloured red and blue) are formed at the same time. Using liquid-phase epitaxy, the equivalence of these two networks is lifted by the presence of the substrate and the formation of interpenetrated networks is suppressed, yielding SURMOFs containing only one network. Red and blue colours are guide to the eyes.

2.2. Deposition of Inorganic/Organic Materials on Inorganic/Organic Surfaces

MOFs and/or PCPs represent a class of advanced inorganic-organic hybrid crystalline, nanoporous materials with well-defined coordination between metal ions and organic linkers. A wide variety of one-, two-, and three-dimensional structure results through the choice of metal ions and linkers. Recently, Yaghi and co-workers [95-97] reported the growth of MOFs in bulk which shows exceptionally high surface areas, exceeding those of zeolites (up to 6,000 m^2/g), thermal stability in air as high as 400°C, and gas sorption behavior. Most of the studies [34,98-101] have shown the potential of MOFs mainly in gas storage (especially H_2), adsorption and separation applications. Beyond this use as bulk active composite materials, the controlled growth of MOF materials on surfaces [45] offers a route to new types of functional materials. For example, recent limited reports by Yoo et al. [54] and Hermes et al. [102-103] presented the surface crystallization of MOF-5 [$Zn_4O(bdc)_3$, bdc = 1,4-benzenedicarboxylate], HKUST-1 [$Cu_3(btc)_2$, btc = 1,3,5-benzene-tricarboxylates] [53], [$Cu_2(pzdc)_2(pyz)$, pzdc = pyrazine-2,3-dicarboxylate; pyz = pyrazine] [102], [Mn(HCOO)] [57], MIL-53 [Fe(OH)(bdc)] and MIL-88B [$Fe_3O(bdc)_3(Ac)$, Ac = CH_3COO^-] [58] on various substrates and support materials.

Figure 6. The deposition of MOF-5, HKUST-1 and [$Zn_2(bdc)_2$(dabco)] on silica, ALD-Al_2O_3 (ALD; Atomic layer deposition) and COOH/CF_3-modified surfaces. Reproduced with permission from Zacher, D. et al. *J. Mater. Chem.* 2007, *17,* 2785-2792. Copyright © 2007, Royal Society of Chemistry.

The ultimate goal in surface-confined MOFs (SURMOFs) materials is the ability to control the arrangements of channels of porous modules for various applications. Thus there are two challenges: the growth of MOF thin film on substrates, ideally in dense, homogeneous and oriented fashion, and the preparation of size-, shape-, and surface-functionalized MOF nanocrystals. One major problem, in MOFs is the phenomenon of 'interpenetration' [103-105]: as the size of the pores is increased by extending the length of the linkers, new sub-lattices synthesized within the existing frameworks (Figure 5). The cages essentially become enmeshed, reducing the effective size of the pores. Now a team led by Christof Wöll of Ruhr University Bochum in Germany believes it might have cracked the problem. Traditionally, MOFs are made by reacting the linkers and coupling units together in a solvent at high temperatures. Wöll's team [49], however, has used an approach based on surface chemistry to lay down the individual components sequentially onto a geometrically precise template, consisting of an organic self-assembling monolayer. Because of the highly ordered nature of the surface, the newly assembled frameworks can align in only one plane. In turn, the growth of subsequent layers of frameworks is also constrained to a specific position on the underlying layer, regardless of the size of the pore, preventing interpenetration (Figure 5).

There are three different approach adopted for MOF thin film fabrication: a) direct deposition of MOF on inorganic substrates from solvothermal method, b) deposition of MOF on functionalized substrates and c) the stepwise layer-by-layer growth onto substrate.

2.2.1. Deposition of MOFs on Alumina, Silica and Carbon (Inorganic Substrates)

Several research groups have been successful to deposit the MOFs microcrystal on various inorganic substrates. Specifically, the MOF-5 has been the most extensively studied among the various MOFs materials due to the great interest to control the microstructure of MOF membranes such as the grain size, orientation, thickness, grain boundary structures and their potential applications. The deposition of MOF-5 on the sapphire (c-plane) substrate has been reported by Hermes et al. [102,103] using cooled solvothermal mother liquor made from N, N-diethylformamide (DEF) solvent. The amorphous Al_2O_3 type buffer was deposited onto the sapphire (c plane) to grow the MOF-5 crystal. They observed the deposition of MOF-5 on the silica substrate is not possible under the identical conditions which indicate the deposition of MOFs is extremely substrate selective. D. Zacher et al. [45,108] also observed similar result for the deposition of HKUST-1 directly from the solvothermal mother liquor. Under the solvothermal condition at higher temperature, the deposition of $[Zn_2bdc_2(dabco)]$ never showed any substrate selectivity (Figure 6). The difference of the activity between SiO_2 and Al_2O_3 substrates are mainly due to their different isoelectric points (IEPs) and they should be electrostatically well-matched surface for anchoring the MOFs.

The silica (SiO_2) and alumina (Al_2O_3) substrates are acidic and basic in nature respectively. As a result, the pillared MOFs having the bifunctionality (both acidic and basic) in nature due to bdc linker and dabco pillar shows the strong acid base interactions through the acidic and basic substrates. The common advantage is that both silica and alumina substrate provides the intense coating towards the growth of MOFs. It is a great challenging job to the chemist to fabricate uniform and defect-free films. Among other issues, one main difficulty to deposit the MOF films on various substrates is the organic linkers in MOFs, which make the materials unique and extremely versatile.

Figure 7. SEM images of MOF-5 grown on a bare AAO (a), on an amorphous carbon-coated AAO (b) and on a graphite-coated AAO (c). Reproduced with permission from Yoo, Y. et al. *Chem. Commun.* 2008, 2441-2443. Copyright © 2008, Royal Society of Chemistry.

Recently, Yoo et al. [54] reported the fabrication of MOF-5 films on porous support under microwave-induced thermal deposition (MITD) in a microwave oven with 500 W for 5–30 seconds. Figure 7 displays the scanning electron microscope (SEM) of MOF-5 crystals grown on three different substrates: a) bare anodized aluminium oxide (AAO), b) an amorphous carbon-coated AAO, and c) a graphite-coated AAO. From the SEM image it is clear that the MOF/G-AAO (graphite coated) shows smaller and more densely-packed MOF-5 crystals as compared to MOF/C-AAO (carbon coated) and MOF/AAO. The above observation can be explained due to the rapid formation of more nuclei on the graphite substrate under the identical conditions (microwaves irradiation) over the amorphous carbon and the bare alumina substrate. The deposition of the microcrystal on bare AAO is very little due to the lack of conductive surface, but when it is coated, sufficient deposition is noticed.

The larger and rapidly nucleation on graphite coated AAO is due to the better absorption of the microwaves. Another important observation that the MOF-5 film grown on G-AAO shows an out-of-plane preferred orientation direction perpendicular to the substrate, which is confirmed by the powder X-ray diffraction (PXRD) (data not shown here). In addition important thing is that the thin layers of the conductive materials like carbon, graphite, and other metals such as Au significantly enhance the kinetics of heterogeneous nucleation, growth of the MOF-5 microcrystal. It has been observed that about 80% of the coating remained after sonication over one hour. The advantages of this process are that it is very rapid; growth of nucleation is also high but the problem is that the microwave process needs a conductive surface for the growth of MOF. Although the adsorption and porosity properties of the MOF-5 was not reported, however, recently Yoo et. al. described the activation, porosity as well as adsorption property of the MOF-5 microcrystal [54].

It is difficult to grow the cracks and fractures free MOF materials. However, Guerrero et al. [109] have developed a new seeding method, known as "thermal seeding" for growing defect free HKUST 1. They reported the synthesis of cracks and fractures free HKUST-1 on the porous alumina supports by the help of secondary growth method. At first they synthesized HKUST-1 seed crystals using $Cu(NO_3)_2$, 1,3,5-benzenetricarboxylic acid (BTC), under hydrothermal condition, then the autoclave was slowly cooled down to 60°C at the rate of 1°C/min. The secondary growth is selected over in situ growth due to the controlled expansion of the materials which includes control over thickness, grain size, and nucleation. The mother solution was seeded onto the alumina supports. It was observed that low binding interaction of the HKUST-1 seed crystals with the porous alumina surface. Experimentally it is found that about 30% of the seed crystals are tightly attached to the surface even after 1 h of continuous sonication. This observation suggests the HKUST-1 seed crystals loosely bounded on the supports. Whereas the zeolite seed crystals show the strong binding interaction with the supports. This is due to the fact that the HKUST-1 seed crystals are non-covalently bounded, but the zeolite seed crystals are covalently bounded with the supports.

2.2.2. Deposition of MOFs on Self-assembled Organic Monolayers (Organic Surfaces)

The term organic surfaces means self-assembled monolayers on gold or silica surfaces i.e., the inorganic substrate is functionalized with organic molecules for further use. The advantage of such surfaces is to control the highly oriented growth of IOHMs and to monitor the phase selective deposition. For instance, recently growth of MOF thin films or crystalline IOHMs on functionalized substrates has been reported by several researchers for smart membrane, catalytic coating, chemical sensors and related nanodevices. The influence of surface functionality of the chosen substrate and the use of MOF precursors on the nucleation, orientation as well as on the adhesion of the MOF films, are important factors to be addressed. In particular, SAMs have been used to direct the nucleation, orientation and structure of the immobilized MOFs.

Hermes et al. [103] reported the selective deposition of well-shaped MOF-5 cubes at the COOH-terminated sites of the patterned, mixed COOH/CF_3-terminated SAM surface on Au at room temperature. On the other hand very soon Biemmi et al. [55] had chosen HKUST-1 as candidate for a study aiming at the oriented growth of HKUST-1 on differently terminated SAMs. The growth of HKUST-1 microcrystals on Au substrates modified with thiol-based SAMs using 11-mercapto-undecanoicacid [$HS(CH_2)_{10}COOH$], 11-mercapto-undecanol [$HS(CH_2)_{10}CH_2OH$] and 11-mercapto-undecane [$HS(CH_2)_{10}CH_3$] was monitored by SEM over a period of 16 to 100 h. The key finding was a highly oriented growth of individual crystals over time. Under the conditions of the experiment, the COOH-terminated SAM favored orientation along the [110] direction, which resulted in the formation of pyramids. The OH-terminated SAM favoured the [111] direction, which led to the formation of octahedral crystals resting on one triangular face (Figure 8).

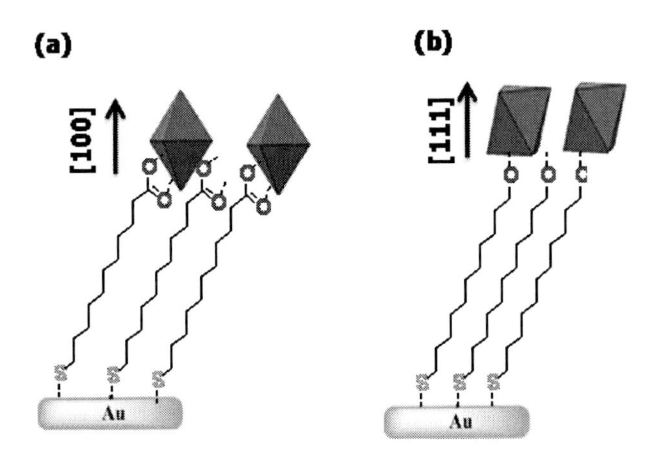

Figure 8. Schematic illustrations of the oriented growth of HKUST-1 nanocrystals [103] controlled *via* surface fictionalization; on an 11-mercapto-undecanoic acid SAM (a) and on 11-mercapto-undecanol-modified gold surfaces (b). The alkanethiol self-assembled monolayers are represented with a tilt of *ca.* 30° from the surface normal as reported in the literature.

Interestingly, the authors observed the growth on CH_3-terminated SAMs as well. An even faster growth process was reported than on the other two polar and coordinatively more strongly interacting surfaces. However, a strongly preferred orientation was absent. The growth on CH_3-termination is a somewhat surprising result; however, dispersive forces between the supposedly organic-terminated crystal faces and the alkyl-terminated SAM may be dominant in that case. It is noteworthy that in the case of CF_3-terminated thiol-based SAMs on Au, as well as silane-based SAMs on SiO_2/Si substrates (*vide infra*), neither MOF-5 nor HKUST-1 could be grown [106-108]. Nevertheless, the protocol to achieve the highly selective growth developed by Biemmi et al. [55] appears a bit complicated. The authors argue that the thermal pre-treatment of the synthesis solution (8 days at 75°C) induces the crystallization process. After filtration and removal of the deposited solid product (see also Gascon et al. [32,56]), they suggested the existence of colloidal nanocrystals or small molecular building blocks of HKUST-1 in the possibly rather diluted solution.

These frameworks are quite flexible and the exact cell parameters strongly depend on the interaction with guest molecules, which makes these materials particularly interesting candidates for sensors and smart membranes. The obvious and common drawback of all the above mentioned pioneering reports on the deposition of MOFs on SAMs from somehow pre-treated mother solutions is that in fact smooth and dense MOF thin films were not obtained at all. Rather scattered, more or less isolated crystals or island of crystals, or rough coatings with many cracks were deposited, even after reaction times of several days.

2.2.3. Deposition of Inorganic-organic Materials by Layer-by-layer (LBL) Methods

The LBL assembly [37,110-113] is probably one of the simplest, versatile, and significantly inexpensive approach by which nanocomponents of different groups can be combined to coat both macroscopically flat and non-planar surfaces. Further, LBL approach is not only useful for assembly of inorganic-organic hybrid materials but also can be used to combine a wide variety of species including nanoparticles, nanosheets and nanowires with polymers thus merging the properties of each type of materials. This versatility has led to recent exceptional growth in the use of LBL-generated nanocomposites.

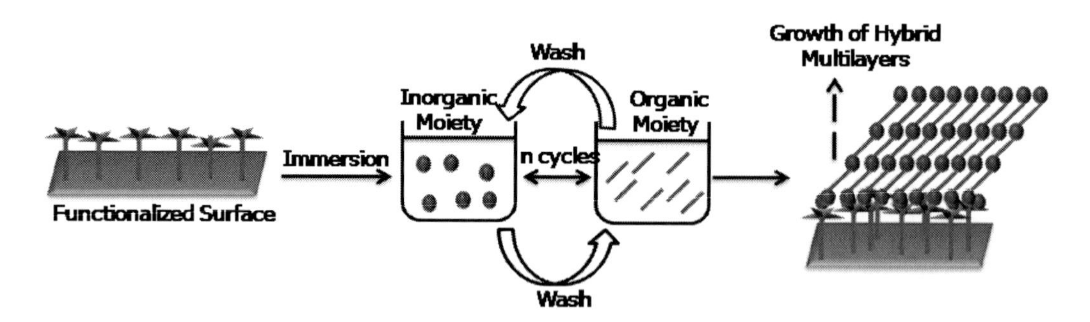

Figure 9. Schematic representation for layer-by-layer immobilization of inorganic-organic hybrid materials on functionalized surface. The approach is based on the sequential immersion of functionalized organic surfaces into solution of the molecular building blocks of the MOFs or CPs, i.e., the organic ligand and inorganic unit, followed by washing after each immersion step.

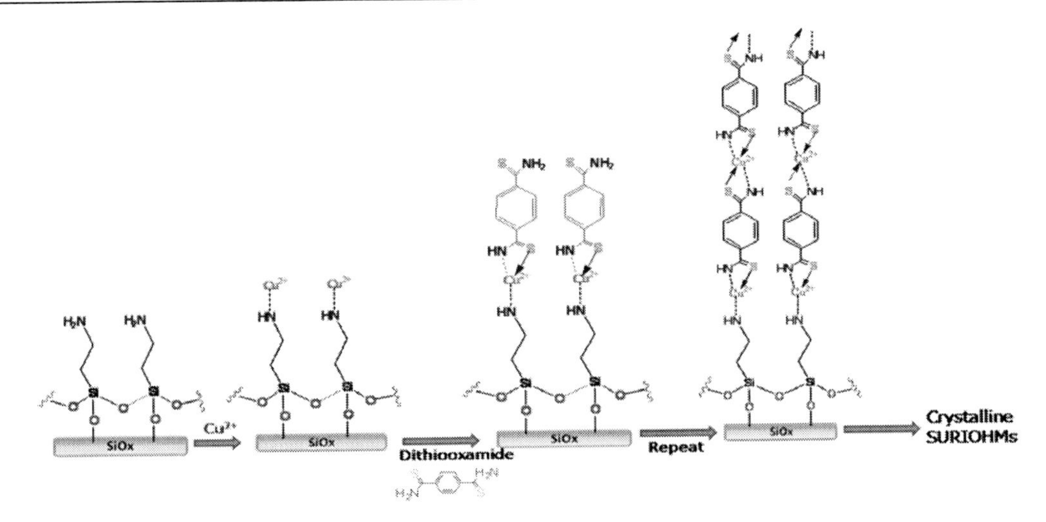

Figure 10. Fabrication of crystalline coordination polymers on SiO_x substrate using LBL technique [117].

The main advantage of this kind of assembly is that the chemical properties of the materials can be programmed through smart design of the building blocks; also the mechanistic and kinetic study for the MOF formation can be performed in more details from a new perspective. Further, the layer-by-layer method offers the unique opportunity to grow controlled 3D-ordered MOF like structures, which may not be possible to obtain by established solvothermal routes. For example, the deposition of MOFs with alternating layers (heterostructures), possibly with non-periodic combinations of different metal ions and/or different linkers should be practicable and useful for many applications including gas storage, gases/metals sensing etc.

The initial reports using LBL methods focused on the use of synthetic, charged polymers (polyelectrolytes) for the assembly of the molecular/polymer films [114-116]. The LBL method has also been expanded to include a host of different materials, including metal and inorganic nanoparticles, dyes, peptides, organonucleotides, proteins and enzymes. This demonstrates the utility of approach for preparing films with diverse chemical and biological functions. Further, the multilayer films have been constructed on the basis of different non-electrostatic interactions [110] such as hydrogen bonding, sequential chemical reactions, metal-ligand complexation and hydrophobic interactions. As a result, the LBL technique became a versatile approach for the construction of thin films of inorganic-organic hybrid materials which offers a number of advantages when compared to other methods of surface modification. An important feature is the precision with which the layer thickness can be controlled. This control can be achieved by varying a) the specific materials being used b) the number of layer assembled and/or c) the specific adsorption condition used.

The principle of this technique is well adapted for the construction of inorganic-organic hybrid materials on surface (Figure 9). The ideal substrate to start with in such a layer-by-layer deposition of organic molecules and metal coupling units, is the inorganic substrate exposed by organic SAMs. For example, Shekhah et al. [53] demonstrated the growth of HKUST-1 on COOH-terminated organic surface on gold. The two components, cupric acetate and 1,3,5-benzenetricorboxylic acid were separately dissolved in ethanol and the substrate was immersed into each solution in a cyclic way as represented in Figure 9, while each step was followed by rinsing with ethanol. The X-ray diffraction (XRD) data (not shown here) revealed that the orientation of the growth is (100) direction for COOH-terminated SAMS on the other hand they observed (111) direction for OH-terminated SAMs. The gas-loading properties of deposited HKUST-1 were also studied *via* NH_3/H_2O exchange experiments and the data showed that loading of NH_3 is similar to bulk HKUST-1 including a substantial irreversibility. Interestingly when Cu^{2+} is replaced by Zn^{2+} then non-porous polymer was observed in contrast to solvothermal method, which leads to MOFs. The overall kinetics of layer-by-layer MOFs growth were found to be strictly linear, however, nonlinear growth modes for the related formation of nanoarchitechtures at surfaces were also observed in case of polyelectrolytes and other inorganic-organic hybrid system on silane surfaces (vide infra).

Another particular advantage of the layer-by-layer method is the possibility to monitor the growth of material using spectral techniques and also to divert the growth process by the doping of new species at any stage of the nucleation process. Further development for this method is to investigate the deposition parameter(s), kinetics and selective growth of MOF films on predefined area by patterning of SAMs. In contrast to solvothermal method, the layer-by-layer method yields extraordinary homogeneous films of >100 nm thickness and a roughness in the order of only one elementary cell in the range of several μm^2. As a result, the layer-by-layer approach may also be suited for the fabrication of thicker layers with a very homogeneous, flat surface, e.g. in sensor applications and for fabricating membranes.

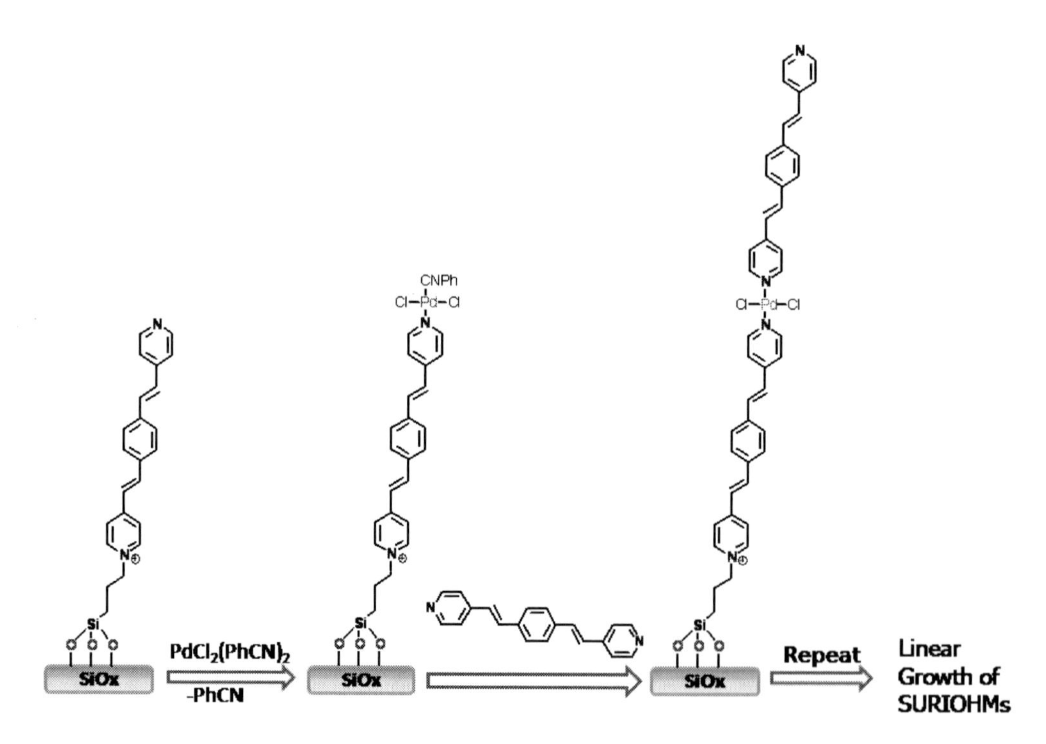

Figure 11. Schematic description of the LBL assembly of two palladium coordination-based multilayers on silicon and glass substrates [120].

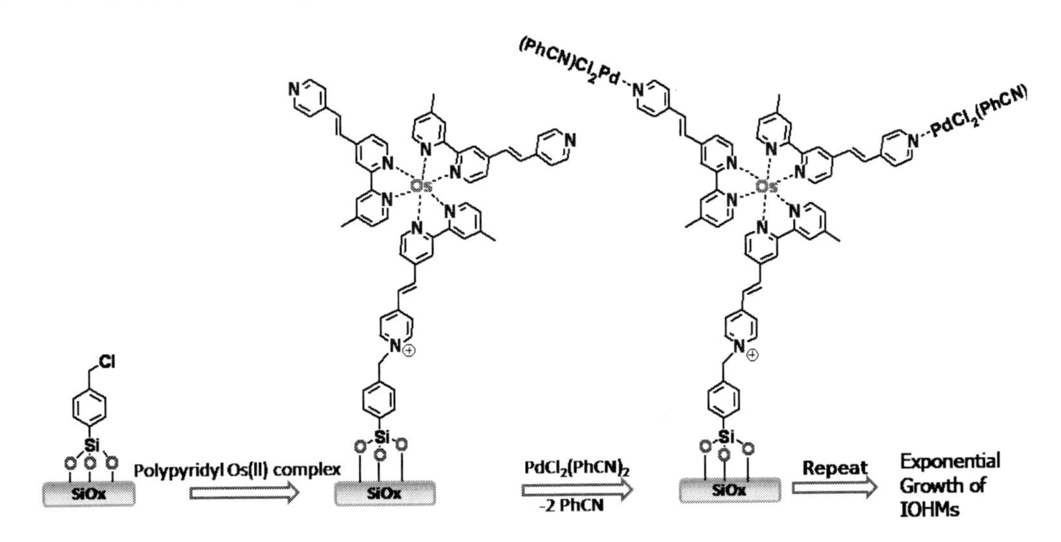

Figure 12. Schematic description of the LBL assembly of two palladium coordination-based nonlinear multilayer films on silicon and glass substrates [113].

Kanaizuka et al. [117] have effectively fabricated the surface coordination polymers (SCPs) which are perfectly crystalline and highly oriented using the liquid phase epitaxy (LPE) technique. In this process Cu^{2+} and dithiooxamide units were used on amino group terminated alkyl silane SAMs on sapphire and glass plates (Figure 10). The multilayer formation process was monitored using quartz crystal microbalance (QCM). The absorbance of the film is increased linearly with the number of the immersion cycles.

Recently, Altman et al. [118-120] have reported the growth of well defined surface coordination polymers (SCPs) using the LBL technique as shown in Figure 11. The bottom-up fabrication of the inorganic-organic materials on SiO_2 surface are composed by rigid-rod like organic chromophores and Pd(II) salt/colloid. Both colloidal palladium and bis(benzonitrile) dichloropalladium(II) gives ordered multilayers with a linear increase in the average chromophore density for each additional molecular layer. Another interesting report by Motiei et al. [113] demonstrates the nonlinear growth of multilayer assembly using polypyridyl Os(II) chromophore and Pd(II) salt as shown in Figure 12. The result shows the self-propagative growth of multilayer films using osmium polypyridyl complexes and Pd salt as starting component on functionalized surface.

3. Common Surface Characterization Techniques

Along with the discovery of the surface-confined inorganic-organic hybrid materials on various substrates in early 1980s, analytical techniques have been developed with sufficient sensitivity to allow the detailed characterization of SURIOHMs. The amount of IOHMs confined on surfaces is in the order of micro to pico mol/cm^2. Such a small quantities make it rather difficult to characterize with single analytical technique. Therefore, it is essential to use a combination of characterization techniques in order to ascertain the important properties of the film e.g., surface morphology, molecular orientation, intralayer spacing, thickness, chemical composition, packing density, order etc. A wide range of physical and chemical techniques are available; most of them involve electrons, photons (light), x-rays, neutral species, or ions as a probe beam striking the material to be analyzed. The beam interacts with the material in some way. In some techniques the changes induced by beam (energy, intensity) are monitored after the interaction and analytical information is derived from observing these changes. In other techniques the information used for analysis comes from electrons, photons, x-rays or ions that are ejected from the specimen under the stimulation of the probe beam. In addition, some techniques involve mechanical contact between probe and surface. Some of the important techniques for the characterization of SURIOHMs are listed in Table 1 and some of them are discussed below briefly.

Table 1. Analytical techniques for characterization of SURIOHMs

Analytical techniques		Structural Information
General	Contact Angle [121]	Wetability, hydrophobicity, order
	Quartz Crystal Microbalance/ Surface Acoustic Waves [122]	Change in mass upon deposition
Optical	ATRIR Spectroscopy [123]	Functional group, orientation
	UV-Vis Spectroscopy [124]	Packing density, foot print
	Fluorescence Spectroscopy [125]	Packing density, foot print
	Ellipsometry [126]	Film thickness
	Surface Plasmon Resonance [127]	Film thickness
Vacuum	X-ray Photoelectron Spectroscopy [124]	Elemental composition
	Auger Electron Spectroscopy [128]	Elemental composition
	Ion Scattering Spectroscopy [129]	Elemental composition
	X-Ray Reflectivity [87]	Thickness, roughness
	Near Edge X-ray Absorption Fine Structure [130]	Bond angle, Bond length
	Secondary Ion Mass Spectrometry [131]	Molecular mass, fragments
Microscopy	Atomic Force Microscopy [132]	Morphology, packing
	Scanning Tunneling Microscopy [133]	Morphology, packing, structure
	Scanning Electron Microscopy [134-136]	Imaging, Grain structure
	Transmission Electron Microscopy [137]	Atomic structure, chemical bonding
	X-Ray Diffraction [58,138]	Crystalline phase, strain, depth
	Low Energy Electron Diffraction [139,140]	Microstructure, order, imaging, periodicity
Electrochemical	Cyclic Voltammetry [87,141,142]	Surface coverage, defect, thickness
	Impedance Spectroscopy [143,144]	Surface coverage, defect, thickness

Contact angle (CA) is used to determine the wettability of a surface after applying a drop of liquid. When a drop of liquid is placed on a solid surface, an angle (θ) is formed. This angle is the result of the balance between the surface interfacial as described by Young's equation, $\theta = \arccos\left[(\gamma_{SV} - \gamma_{SL})/\gamma_{LV}\right]$, where γ is the surface interfacial tension and LV, SV, and SL, refer to the liquid/vapor, solid/vapor, and solid/liquid interfaces respectively.

Piezoelectric oscillators, for example, the quartz crystal microbalance (QCM) and surface acoustic waves (SAWs) are capable of detecting mass changes in the range of nanograms. They have been used to study the kinetics of material adsorption on surface. When the liquid is water, CA provides information about the hydrophobicity and hydrophilicity of the surface.

Attenuated total reflection infra-red (ATRIR) spectroscopy is a powerful tool for characterization of SURIOHMs. The incoming infra-red light is reflected under a large angle of incidence to maximize the absorbance by the film. As a consequence, besides the usual structural data, the spectra also provide the average orientation of the molecules on metallic surface.

When a UV-Vis light passes through a sample its intensity is reduced due to the absorbance of the species in its pathway. The decrease in absorbance is expressed by the Lambert Beer law $A = -\log_{10}(I/I_0) = \varepsilon \cdot c \cdot L$ where A is the measured absorbance, I_0 is the intensity of the incident light at a given wavelength, I is the transmitted intensity, L the path length through the sample and c the concentration of the absorbing species. For each species and wavelength, ε is a constant known as the extinction coefficient. While the change in intensity provides concentration in solution or thickness of the films, optical spectroscopy peak position identifies specific electron transitions in the species being measured. In multilayer chemistry, the change in spectrum with each additional layer provides information about the structure of the system. The information concerning the packing of IOHMs can also be obtained from fluorescence spectroscopy. Many aromatic hydrocarbons exhibit concentration-dependant emission.

Ellipsometry is a measurement technique that uses polarized light to characterize thin films, surfaces and material microstructure. When plane-polarized light strikes the surface of a sample at some angle, it is resolved into its parallel (s) and perpendicular (p) components. The amplitude and phase of both components are changed as they are reflected from a surface. The interaction of the light with the sample causes a polarization change in the light from linear to elliptical polarization. The polarization change or a change in the shape of the polarization is then measured by analyzing the light reflected from the sample. The ellipsometer measures two parameters: Δ, the phase shift induced by reflection and Ψ, a measure of the ratio of the intensities of the s-type and p-type waves. Variable-angle spectroscopic ellipsometry acquires this data as a function of both wavelength and angle of incidence. For complex systems, in order to obtain thickness and optical constants from Δ and Ψ, a model must be postulated. Expected values of Δ and Ψ, as calculated by a dispersion relationship such as the Cauchy approximation are compared with the measured quantities. The fit is then improved by regression analysis. Due to limitations of uniqueness for complex or novel films independent thickness measurements are required in order to determine the optical constants of the film.

A related optical method, surface plasmon resonance (SPR) is based on the angle-dependent reflection of a polarized laser beam at metal surfaces (Ag or Au). At a certain angle of the incident light, the energy of the laser light excites the surface plasmon into resonance, which results in a minimum in the reflectance. The incident angle is strongly affected by change in the refractive index of the contacting medium in close proximity to the metal surface and can be used to determine the thickness of the film.

X-Ray photoelectron spectroscopy (XPS) is a form of electron spectroscopy where photoelectrons emitted from the top few atomic layers of a sample are sorted according to their kinetic energy by a hemispherical analyzer. Electrons emitted from the sample are into a hemispherical analyzer which sorts out the electrons according to their kinetic energy. The XPS spectra are displayed as a plot of intensity versus binding energy in eV. Binding energy is a measure of how tightly the photoelectron was originally held in the atom. It is measured as the difference between the kinetic energy of the X-ray photon, the kinetic energy of the emitted electron and the work function of the spectrometer. XPS can measure the oxidation state of the elements on the surface of a sample, detect all elements except H and He and can measure the electronic properties of valence electrons in complex molecular structures. Like XPS, auger electron spectroscopy (AES) and ion scattering spectroscopy (ISS) allows the selective detection of different elements present on the surface.

X-ray reflectivity (XRR) measures the intensity of an x-ray beam reflected by a sample at grazing angles. The reflectivity recorded is function of the electron density profile $\rho(z)$ perpendicular to the sample surface. For the monolayer and multilayer films, a model comprising a silicon substrate (with native layer of silicon oxide) and organic layer (SAM) is used:

$$\frac{R(q_z)}{R_F(q_z)} = \left| \sum_{i=0}^{N} \frac{(\rho_i - \rho_{i+1})}{\rho_{Si}} e^{-iq_z D_i} e^{-q_z^2 \sigma_{i+1}^2 / 2} \right|^2$$

where, N is the number of layers of different electron densities, ρ_i, with Gaussian broadened interfaces, σ_i, ρ_{Si} is the electron density of the substrate; $D_i = \sum_{j=1}^{i} T_j$ is the distance from the substrate surface to the *ith* interface, and T_i is the thickness of the ith layer. The reflectivity data are then fitted to obtain the thickness of each layer, the electron density of each layer, and the root-mean-square width of each interface. The density of electrons per unit of substrate area for the monolayer film, ρ_{exp}, is calculated using the obtained electron density profiles according to $\rho_{\exp} = \int \rho_z(z)dz$, where integration is taken over the entire film. The molecular "footprint" can then be calculated from: N_e/ρ_{exp}, where N_e is total density of electrons in a single chromophore.

Secondary ion mass spectroscopy (SIMS) uses a high energy ion beam (usually Ar^+ or Xe^+) to irradiate the sample. The primary ions penetrate the surface where they cause a cascade of collisions which results loss in energy, and secondary ions are ejected only from the outermost surface and accelerate into mass spectrometer.

The AFM and STM head uses beam deflection to monitor the cantilever displacement. This scheme permits registration of both normal deflection of the cantilever with sub-angstrom resolution and its twisting angle, so normal and lateral force can be measured simultaneously. A laser beam is focused onto the back surface of cantilever close to tip position, and reflected beam falls onto the quadrant photodiode. Cantilever deflection causes displacement of the reflected beam over sections of the photodiode. An amplified differential signal from the quadrant photodiode permits measurement of cantilever deflection of the order of 0.05 nm. Semicontact mode is a modulation technique for non-destructive imaging of samples. It measures topography by tapping the surface with an oscillating probe tip and allow the characterization of SURIOHMs with molecular resolution.

Cyclic voltammetry (CV) is an electrochemical experiment in which the potential continuously changes as a linear function of time. At the end of the first scan, the direction of the potential is reversed. The technique is used to determine the redox potential and to determine the electron transfer kinetics of both species in solution and of thin films. For a reversible system, the redox system stays in equilibrium throughout the scan. In impedance spectroscopy very small sinusoidal potential sweep (5 mV) with frequencies varing from 100 mHz to 10 kHz are applied to the electrode, which makes it a less disturbing technique than CV. From the impedance spectra, values for the film capacitance and resistance are obtained.

4. Molecular Interactions with Functionalized Surfaces

Interfacial reactions are becoming an increasingly important subject for the understanding of device fabrication process [41,145-147]. The ability to control the chemical and structural properties of surface is crucial for advancements in selective and environmental friendly chemical sensing application. Specifically, a combination of SURIOHMs and molecular recognition has received a vast attention in recent years because modification of surfaces with organic/inorganic functional groups allows the exact tuning of surface properties. Such devices/materials are urgently sought after in fields as diverse as analytical chemistry, environment protection, process control, and security applications. Here we will focus on the subject related to the study of molecular recognition events at SURIOHMs.

The pioneer work of Pedersen, Cram and Lehn have led to an intensive study of supramolecular host-guest chemistry [148-157] which revealed a route for designing sensor system based on host-guest interactions. They have designed a number of receptor molecules such as crown ethers, cryptands and spherands for the recognition of cations, anions or neutral molecules. Such molecules have been extensively studied for molecular recognition in solution phase and open the area for improving the studies of sensor functionalities and organization of such molecules on surface.

Several strategies such as direct grafting approach or post grafting approach are reported for developing chemical sensors that can be used to control the recognition and signaling abilities of analyte species. Direct grafting approach enables the design of chemosensors without pore blockage by grafting any coupling agent which is used to tune the surface properties. In post grafting approach thiol or silane coupling agents are used to enhance the surface properties and to create highly tuned and functional nanostructured surfaces. This method is used to overcome the difficulties in the direct immobilization of probes onto the surface. Further the enhancement of sensing events through the influence of the surface has been reported by preorganization of the receptor site on the surface.

Over last few years several researcher immobilized receptor molecules on metal and semiconductor nanoparticles in order to create solution-based sensors [24]. In these nanoparticle systems, molecular recognition is achieved by covalent attachment of receptors on the nanoparticles coupled with noncovalent interactions to target analytes. For instance, Labande et al. [158] designed mixed-monolayer-protected clusters featuring ferrocene units on the surface, which convey the interdependence between molecular recognition and redox events to selectively bind oxo anions. Similarly various receptor sites such as crown ether [159], pseudorotaxanes [160,161], porphyrins [162] etc. has been immobilized on nanoparticle and their molecular recognition events has been addressed for cations, anions and neutral moleculaes.

Although there are several flexible SURIOHMs designed for the detection of several physical and chemical components including temperature, humidity, pH, metal ion, anion, small organic/inorganic molecules, gases and biomolecules sensing system presented in the literature. Probably among them the most studied sensor system is metal ions-, gases- and bio- sensors. Here with particular interest we will discuss SURIOHMs-based sensor systems for selected metal ions and gases while others analytes (vide supra) are beyond the scope of this chapter and readers are referred to other specialized reviews [163-168].

4.1. Recognition of Metal Ions

SURIOHMs play a leading role for the development of optical and/or electrochemical sensors that are highly suitable for sensing applications for a wide range of metal ions. Moreover, sensing of metal ions from environmental sample is of critical importance due to their toxicity, longevity and other environmental and biological cause. Several strategies have been adopted for the selective recognition of metal ion in solution viz host-guest interaction, ionic interaction, electron transfer, and other weak interactions such as hydrogen bonding, hydrophobic interactions or Van der Wall interactions between receptors and analytes. The common techniques used for the recognition is optical or electrochemical.

Metal ion sensors based on Langmuir-Blodgett films have been reviewed by Lednev et al. [169,170]. They have presented a nice collection of reports stating the deposition of amphiphilic macrocyclic ionophores, chromoionophore, valimycine and cyclodextrins on various substrates *via* LB method for the detection of various cations such as Ag^{2+}, Pb^{2+}, Ba^{2+}, Hg^{2+}, K^{+} etc. Although many LB layers have demonstrated usefull properties such as reversibility, high sensitivity and specificity, however, further work is needed to address the chemical and thermal stabilities of LB layers and quantitative determination of analytes over wide range of concentration.

The design of well defined molecular receptors such as carboxylic acids, pyridines, amino acids, peptides, crown ethers, metal complexes and calixarenes for surface immobilization and study of their metal ion sensing capability have gained intensive interest. These receptors can be preorganized on the surface for the detection of both redox-active and redox-inactive metal ions. As illustrated in Figure 13, several inorganic/organic receptors have been immobilized on inorganic substrates for detection of different metal ions. For instance, Turyan et al. [171,172] reported a method for the electrochemical determination of 4×10^{-12} M of Cd^{2+} in an acetate buffer solution using ω-mercaptocarboxylic acid monolayer on gold electrode (Figure 13 a). This method was successful in determining ultra low level of cadmium in seawater. The authors also developed another highly sensitive electrode modified by 4-(2-mercaptoethyl)-pyridinium to recognize 0.02 nM level of Cr^{6+} in fluoride buffer solutions at pH 7.8 (Figure 13 b). Thereafter, a series of sensor platform have been developed for detection of several redox active metal ions by electrochemical means. For instance, Ruberstein and coworkers [173,174] immobilized mixed organic molecules on gold electrode which is capable to sense 10^{-7} M solution of Cu^{2+} or Pd^{2+} in presence of Fe^{2+} in 0.1 M aqueous H_2SO_4 (Figure 13 c). The selective bonding of these metals has been confirmed by electrochemical, ellipsometry and contact angle measurements.

Similarly, the redox-active monolayer has been developed for the detection of redox inactive metal ions also. For example, Moore et al. [175,176] have immobilized a redox active species, tetrathiafulvalene (TTF) on gold surface using thiol groups (Figure 13 d). Upon exposure of this system with redox inactive metal ions such as Li^+, K^+, Na^+ and Ba^{2+}, an anodic shift of the first oxidation process of TTF was observed i.e., a shift of 10-20 mV for Li^+ and K^+ and 45-55 mV for Na^+ and Ba^{2+}. The author attributed these shifts to the inductive effect on the polarizability of TTF imposed by the bound metal ions. This system has been optimized by several other groups especially taking account of stability and selectivity issues. The redox inactive metal ions can also be recognized by other mean such as fluorescent molecules (Figure 13 i and 13 j) has been immobilized on glass surfaces for the selective detection of Na^+, Pb^{2+}, Hg^{2+} and Cu^{2+}. The detection mechanism is based on the quenching/enhancing the emission intensity of the fluorophores after interaction with metal ions.

Electrochemical impedance spectroscopy (EIS) has proven to be an effective tool in the understanding of interfacial ion recognition phenomena when both the analyte and sensor are electrochemically inactive. The main advantage of this technique is its capability to detect any changes in the surface charge without disturbing the structure of the monolayer/multilayer and its ability to detect in aqueous media which brings sensor system closer to real world performance. This technique monitors binding events involving charged species, the change of capacitance and resistance. For instance, Flink et al. [177,178] showed binding of immobilized redox inactive crown ether with redox inactive alkali metal ions (Na^+, K^+) in water using impedance spectroscopy (Figure 13 e and 13 f). The complexation of metal with crown ether causes increase in charge transfer resistance which allows impedimetric sensing of these metals. They have corroborated their findings with XPS and AFM. Further, Bandyopadhyay et al. [179,180] showed monolayers on gold containing hexaethylene glycol receptors are selective for K^+ while those containing pentaethylene glycol receptors are selective for Na^+. In this study, impedance spectroscopy played a crucial role in obtaining these results. In another example, they have immobilized calix[4]crown-6 derivative in 1,3-alternate conformation on gold (Figure 13 h) and reported its efficiency for the selective detection of Cs^+ using impedance spectroscopy and cyclic voltammetry [181]. These monolayers are also capable to differentiate between alkaline earth metal ions from alkali metal ions in aqueous medium.

Optical detection of metal ions is advantageous because the sensing platform/device need not to be wired with large electronics. Recently, the electron transfer-based selective optical sensing of Fe^{3+} and Cr^{6+} in presence of different alkali, alkaline and transition metal ions by osmium polypyridyl complex-based monolayers on glass (Figure 13 k) has been introduced by Gupta and coworkers [182,183]. The interaction of Fe^{3+} with osmium polypyridyl complex immobilized on glass surface is physical which results in transfer of one electron from analyte to monolayer with change in the optical properties. Interestingly, the sensor can be used in aqueous media and also quantify these metals at part per million levels. This sensor system can be used for the detection of Fe^{3+} as well as Cr^{6+} as a function of pH change. The sensor performance, including reversibility, response time, reproducibility, selectivity, stability, on/off ratio, and detection limit has been demonstrated. In addition, the mechanism underlying the surface-confined redox chemistry has been explored.

Immobilization of redox- and optically-active polydentate ligands which shows geometrical isomerism has been the subject of extensive studies due to their strong affinity towards metal ions. For example, Wang et al. [184] immobilized (pyridylazo)phenol derivative of 8-[4-(2-pyridylazo) phenoxy]octyl disulfide on gold-coated optical waveguides (Figure 14 g). They have shown the monolayer can be transform into cis (inactive receptor) and trans (active receptor) isomeric forms using waveguided UV and visible light, respectively. The monolayer in the trans isomeric form provides a bidentate ligand that can chelate Ni^{2+} in the range 9.6×10^{-5}-2.44×10^{-3} M which can be detected by UV-Vis spectrophotometer. The monolayer shows no metal chelation with the cis isomeric form.

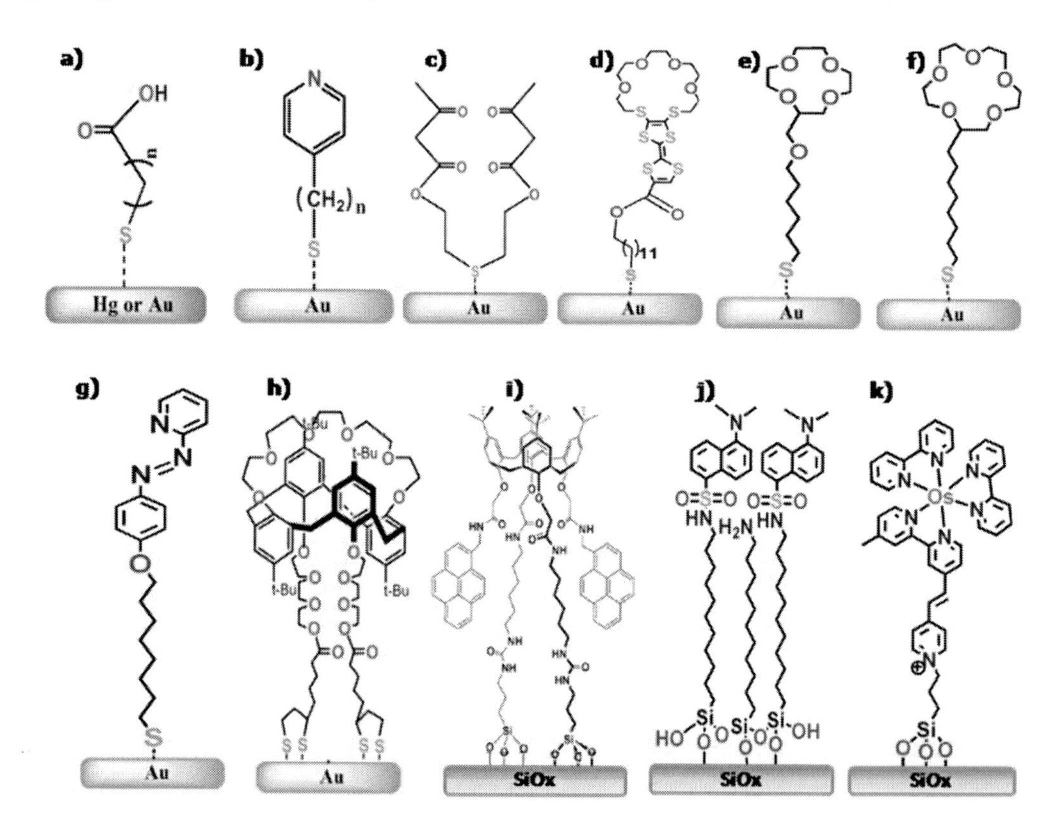

Figure 13. Representative examples of well-defined surface-confined inorganic/organic materials for the detection of various metal ions including Cd^{2+} (a) [171], Cr^{6+} (b) [172], Cu^{2+} (c) [173,174], Ba^{2+}/Ag^{+} (d) [175,176], Na^{+} (e) [160, 177], K^{+} (f) [160,177], Ni^{2+} (g) [184], Cs^{+} (h) [185], Na^{+} (i) [186], Cu^{2+}/Hg^{2+} (j) [187], Fe^{3+}/Cr^{6+} (k) [183,188].

4.2. Recognition of Gases

The design of efficient SURIOHMs-based gas sensors, which show characteristics of high selectivity, sensitivity and a good response, is an intriguing challenge in material science, as no common platform is available that is applicable to the detection of all gases. Nevertheless, detection of particular relevant gases such as O_2, CO_2, N_2, NO_x by SURIOHMs have attracted considerable attention during last two decades. Recent advances have also been achieved in the detection of some volatile organic compounds (VOCs) by SURIOHMs. For example, Zhang et al. [185] fabricated a class of fluorescent films in which pyrene was immobilized on glass surface (Figure 14 b) for detection of nitroaromatic compounds (NACs) in vapor phase. This design strategy offers several advantages for thin film fluorescent sensitive materials. They showed the fluorescent quenching of the films upon exposure to NACs vapors depends on several factors, including the evaporation rate of the NAC detected, spacer length, the substrate surface and density of the films.

The search for new luminescent oxygen-sensing materials has attracted considerable interest in the past few decades because luminescence-based detection methods show higher sensitivity, do not consume O_2, and can be combined with optical fibers for remote sensing. For example, Gulino et al. [89,190] grafted 5,10,15-tri-{*p*-dodecanoxyphenyl}-20-(*p*-hydroxyphenyl) porphyrin and 61-(*p*-hydroxyphenylmethano)fullerene [C60] molecules on functionalized silica substrates (Figure 14 a), which is proven to be suited for molecular oxygen optical recognition. Photoluminescence (PL) and UV-Vis measurements at room temperature indicate that the system is able to reveal low O_2 concentrations (0.2%) in N_2. Interestingly, exposing the sensor for only a few seconds to N_2 is sufficient to fully reset the system at room temperature. In another example, Chu et al. [191] described oxygen-sensing elements containing single-layered structures of luminescent indicators of ruthenium(II) bipyridyl complexes on glass surfaces (Figure 14 c) prepared by covalent attachment and LBL deposition. They are capable of detecting gaseous oxygen concentration by luminescence quenching of the indicator with reproducible and large quenching efficiencies that are comparable to the best quenching efficiencies obtained by other ruthenium(II) polypyridine based complexes immobilized in matrixes. It is noteworthy that, the single-layered sensing systems are advantageous over the matrix-based sensing materials in terms of signal response, material consumption, and process control. The prepared films achieved large quenching efficiencies and good reversibility. High quenching efficiency and the almost linear relationship of the quenching response for both films imply that the probe complexes are effectively quenched by oxygen, which is probably due to the thin single-layered structures with large surface-to-area ratio and short distance between the probe complexes and the quencher, resulting in better oxygen delivery to the complexes and making them suitable candidates as oxygen sensors. Further, several other examples are known for oxygen sensing such as monolayer assemblies of "hinged" iron-porphyrins on semiconductor surfaces [188] and $[Ru(dpp)_2Phen]^{2+}$ (dpp = 4, 7-diphenyl-1,10-phenanthroline), (Phen = 1,10-phenanthroline) covalently grafted to the ordered functionalized mesoporous SBA-15 and MCM-41 backbones [189].

Highly corrosive gases such as NO_2 and NO_x are governed by environmental factors, since it cause various environmental problems such as smog and acid rain. Therefore, selective monitoring of such gases at low concentration is challenging task. Recently, Gulino et al. [190] immobilized osmium polypyridyl complex on glass substrate for selective detection of NO_x in air. Optical detection of parts-per-million (ppm) levels of NO_2 (1–10 ppm in N_2) and NO_x (800–2550 ppm in air) by this platform has been demonstrated. Interestingly, this platform exhibits selectivity based on redox properties. Indeed, the monolayer displays high selectivity for NO_x since only minor changes were observed by in-situ UV/Vis spectroscopy upon exposure to 1 atm of pure CO, CO_2, H_2, N_2, O_2, N_2O, air, CH_4 and CH_2CH_2 for at least 5 min. In addition, the system is stable in concentrated HCl (36%, v/v) for 5 min, and in air for at least 6 months. The device's performance and characteristics, including selectivity, gain, drift, stability, reversibility and detection range were explored. The redox chemistry is fully reversible and is straightforward to control because water can be used to reset the system within 1 min. Furthermore, Gulino et al. [90] has shown optical NO_2 recognition capability by covalently-assembled monolayer of 5,10,15-tri-{*p*-dodecanoxyphenyl}-20-(*p*-hydroxyphenyl) porphyrin molecules on silica substrates (Figure 14 d). The monolayer is highly sensitive to 1 ppm of NO_2 in both anhydrous and humid conditions, selective towards several other gases and can be recovered by heating at 80 °C under vacuum. In another example, Simpson et al. [196] grafted phthalocyanine (Pc) molecules onto gold-coated optical waveguides (Figure 14 h). The thiol moiety is connected to the Pc macrocycle by either a $(CH_2)_{11}$ or $(CH_2)_3$ hydrocarbon chain. Fluorescence emission spectra from both of the Pc SAMs were obtained by exciting the monolayer *via* laser-induced evanescent wave stimulation. The use of the longer mercaptoalkyl connecting chain appears to inhibit quenching of the electronically excited state through energy transfer to the metal layer. The Pc SAM can detect concentrations of NO_2 down to 10 ppm with no interferent effect from other environmentally relevant gases such as carbon monoxide and carbon dioxide.

On the other hand, nitric oxide (NO) is important physiologically as a gaseous messenger in the central nervous system; its level is also associated with several disease states. The concentration of NO in these cases is in the nanomolar range; therefore accurate and direct detection of NO at these levels is important. Schwartz et al. [197] have shown that direct detection of NO can be achieved by XPS as a surface-attached iron(III) porphyrin complex on the SiO_2 surface (Figure 14 e). Further, Wu et al. [198] also showed that iron(III) porphyrin immobilized on gallium arsenide (GaAs) surface are much more sensitive to NO than O_2 and CO gases.

Carbon monoxide (CO) is a toxic gas, but being colorless, odorless, tasteless and non-irritating, it is very difficult for people to detect. CO often produced in domestic or industrial settings by motor vehicles and other gasoline-powered tools, and cooking equipment on daily basis. Therefore, CO detection is a crucial point for industry and domestic purposes since exposures at 100 ppm or greater can be dangerous to human health. Recently, many CO sensors have been developed. For instance, Gulino et al. [195] build up a well organized monolayer of the bimetallic–rhodium complex on glass (Figure 14 f) which exhibited excellent selectivity towards CO in the presence of a series of other gases and air. We have shown straightforward sensor regeneration by heating or purging the system with air, Ar or N_2. The demonstrated response time coupled with nearly immediate optical read-out is fast (1 min). However, the large sensitivity indicates that even shorter exposure times are sufficient. Interestingly, the CO detection range can be controlled and expanded as a function of the sensor temperature. Further, Ye et al. [200] have reported a unique dual electrochemical sensor system composed of trinuclear ruthenium complexes on gold surface (Figure 14 i) which can recognize CO as well as NO at low concentration. They demonstrated a precise control of the electronic environment of the Ru_3 ion is crucial to the CO/NO exchange ability of the system. First, the CO ligand can be reversibly introduced into the Ru_3 unit of the monolayer under electrochemical control. Second, and most importantly, the NO molecule is efficiently introduced into the monolayer *via* a $1e^-$ electrochemical oxidation of the Ru_3 cluster. They are able to repeatedly switch the CO, NO or solvent on the electrode surface as needed by electrochemically tuning the electronic state of the Ru site.

Ashwell et al. [201] prepared 4-{2-[N-(10-Thiodecyl)quinolinium-4-yl]vinyl}phenolate monolayer on gold (Figure 14 g) for detection of ammonia. The films undergo a change from purple (merocyanine form) to yellow (protonated form) and, by monitoring changes in the reflectance, used as sensors with a detection limit of 1 ppm for NH_3 in a carrier gas. Langmuir–Blodgett (LB) films of the N-octadecyl analogue show similar behavior but, for sensing applications, are disadvantaged because the phenolate group is adjacent to the substrate. The protonated form is transparent at the red end of the spectrum (cut-off, $\lambda = 600$ nm) whereas the merocyanine form is absorbing. The SPR response of LB films is also affected by the ammonium salt, which becomes trapped and results in the film becoming cloudy after a few cycles. However, this does not apply to SAM structures where the phenolate group is at the surface. These have a detection limit of ca. 1 ppm NH_3 in nitrogen.

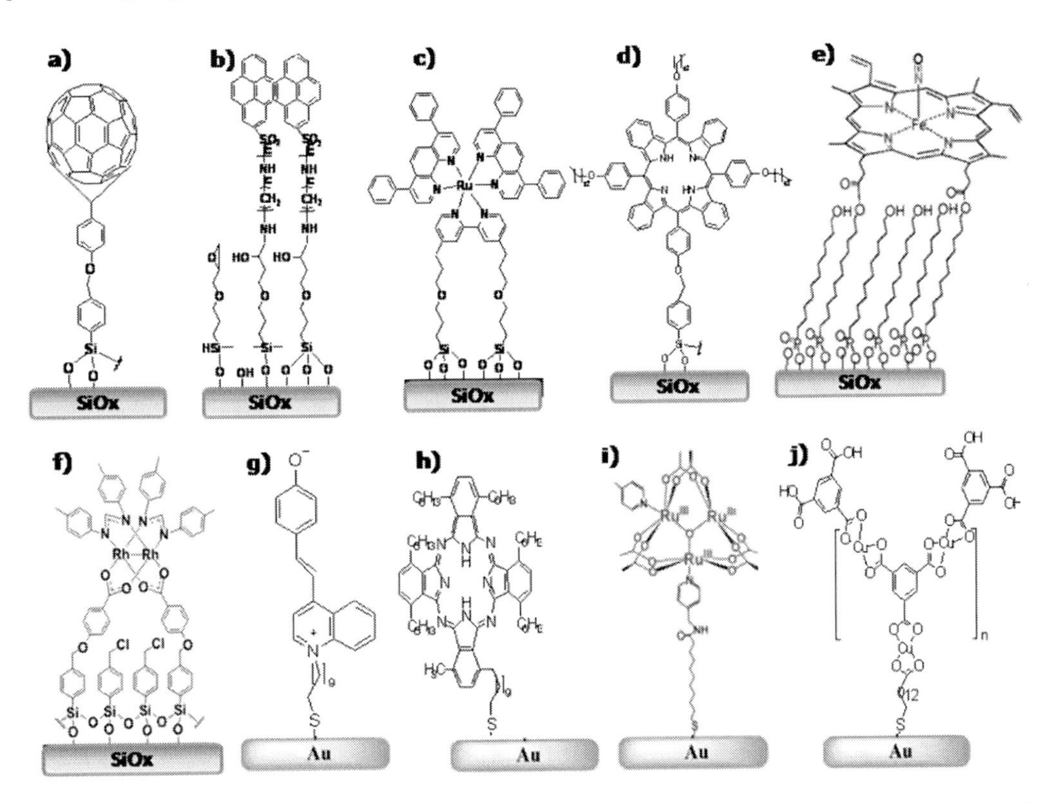

Figure 14. Representative examples of well-defined surface-confined inorganic/organic materials used for the detection of various gaseous analytes including O_2 (a) [186], NACs (b) [185], N_2/O_2 (c) [187], NO_2 (d) [90,191], NO (e) [193], CO (f) [195], NH_3 (g) [197], NO_2 (h) [192], CO/NO (i) [196] and VOCs (j) [198].

Allendorf et al. [202] communicated the concept of stress-induce chemical sensing using MOFs by integrating a thin film of the HKUST-1 with a microcantilever surface (Figure 14 j). Their result shows that the energy of molecular adsorption, which causes slight distortions in the MOF crystal structure, is converted to mechanical energy to create a highly responsive, reversible and selective sensor. This sensor responds to VOCs including water, methanol and ethanol vapours, but yields no response to either N_2 or O_2. The magnitude of the signal, which is measured by a build-in piezoresistor, is correlated with the concentration and can be fitted to a Langmuir isotherm.

CONCLUSION

The surface-confined inorganic-organic hybrid materials provide a powerful route to nanoscale systems with numerous technological applications including sensors, switches, memory elements, logic architectures, electro-chromic and photonic materials. The fabrication of SURIOHMs in modern technology requires controlled growth techniques for growing thin films. The organization and properties of IOHMs on surfaces can be influenced greatly using the different techniques for the immobilization of the molecules. Key factors for the formation and properties of these materials are chemical design and balance of chemical interaction which direct the assembly of the molecules onto the surfaces (molecule-molecule, molecule-solvent, molecule-substrate and solvent-substrate). In the last few years, several research groups have devoted considerable effort towards the preparation and characterization of high quality SURIOHMs. Moreover, the SURIOHMs also evolved towards the joint development of molecular and crystal engineering (which involve relation between properties and structure) for the preparation of complex hybrid multilayer structure.

However, most of the SURIOHMs which are appeared in the literature have been grown by immersion of selected substrates (silica, alumina, graphite etc.) into specifically pre-treated solvothermal mother liquors of particular inorganic/organic precursors. Nevertheless, stepwise layer-by-layer (LBL) assembly of multilayer, which is a spontaneous formation of well–defined structures from a given set of components, becomes more and more significant as an essential step in the fabrication of new SURIOHMs which offers new prospects to study the kinetics and mechanism of thin film formation itself in more details from a new perspective.

One of the most vital potential of SURIOHMs is chemical sensing. The advantages of SURIOHMs-based sensors include (i) only a small amount of compound is needed to generate a large active surface, (ii) no sensing material is consumed and (iii) there are no diffusion limitations because the surface-confined compounds are in direct contact with their environment. Most of the reported sensors based on SURIOHMs suited for device fabrication, but extensive research is still needed before these are viable. Selectivity, reversibility, on/off ratio, response time and more importantly real world operational conditions are the issue to be optimized. Such optimization could be achieved by utilizing microarrays of individual sensing groups. To enhance the sensitivity and selectivity, novel sensing arrays incorporating metal or semiconductor nanoparticles, dendritic structures and nanotubes are expected to extend two-dimensional (2D) sensing arrays on surface to three-dimensional (3D) recognition systems. The majority of available sensor systems rely on inorganic materials primarily because of their durability and suitability for operation at elevated temperatures. Inorganic materials will certainly be an important part of next-generation sensors, but organic materials will play an expanding role because of the vast menu of physical and chemical properties they provide. For example, organic materials lend themselves to synthetic flexibility, which implies that they can be tailored to exhibit a high level of chemical independence and structural order. Thus, the designing steps of the materials for the use as sensors should be as follow: First, the material should be synthetically flexible so that they can be tailored to be chemical independence. Second, they should be cheap, durable and easy to immobilize on transducer surface. Third, they should be configured on the surface such that they respond sufficiently and quickly to an analyte in order to meet the need of the application.

ACKNOWLEDGMENTS

The authors gratefully acknowledge Department of Science and Technology (DST), University Grant Commissions (UGC), India and University of Delhi, Delhi, India for financial support. TG thanks Alexander von Humboldt (AvH) foundation for the Humboldt fellowship of experienced researcher. PCM thanks to council of scientific and industrial research for junior research fellowship.

REFERENCES

[1] Sharp, K. G. *Adv. Mater.* 1998, *10*, 1243-1248.
[2] Judeinstein, P.; Sanchez, C. *J. Mater. Chem.* 1996, *6*, 511-525.
[3] Sanchez, C.; Lebeau, B.; Chaput, F.; Boilot, J. P. *Adv. Mater.* 2003, *15*, 1969-1994.
[4] Cariati, E.; Pizzotti, M.; Roberto, D.; Tessore, F.; Ugo, R. *Coord. Chem. Rev.* 2006, *250*, 1210-1233.
[5] Kang, D. J.; Bae, B. S. *Acc. Chem. Res.* 2007, *40*, 903-912.
[6] Innocenzi, P.; Lebeau, B. *J. Mater. Chem.* 2005, *15*, 3821-3831.
[7] Carlos, L. A. D.; Ferreira, R. A. S.; de Zea Bermudez, V. *Hyb. Mater.* 2007, 337-400.
[8] Mitzi, D. B. *Funct. Hybrid Mater.* 2004, 347-386.
[9] Holder, E.; Tessler, N.; Rogach, A. L. *J. Mater. Chem.* 2008, *18*, 1064-1078.
[10] Mirkin, C. A.; Ratner, M. A. *Annu. Rev. Phys. Chem.* 1992, *43*, 719-754.
[11] Lacroix, P. G. *Chem. Mater.* 2001, *13*, 3495-3506.
[12] Wight, A. P.; Davis, M. E. *Chem. Rev.* 2002, *102*, 3589-3613.
[13] Itoh, T. *Chem. Sens.* 2008, *24*, 31-35.
[14] Barbe, J. M.; Canard, G.; Brandes, S.; Guilard, R. *Chem. Eur. J.* 2007, *13*, 2118-2129.
[15] Matsubara, I.; Murayama, N.; Shin, W.; Izu, N. *Chem. Sens.* 2004, *20*, 106-114.
[16] Walcarius, A. *Chem. Mater.* 2001, *13*, 3351-3372.
[17] Hinks, N. J.; McKinlay, A. C.; Xiao, B.; Wheatley, P. S.; Morris, R. E.*Microporous Mesoporous Mater.* 2009, *129*, 330-334.
[18] Nilsson, K. P. R.; Hammarstroem, P. *Adv. Mater.* 2008, *20*, 2639-2645.
[19] Eckert, H.; Ward, M. *Chem. Mater.* 2001, *13*, 3059-3060.
[20] Sanchez, C.; de Soler-Illia, G. J.; Ribot, F.; Lalot, T.; Mayer, C. R.; Cabuil, V. *Chem. Mater.* 2001, *13*, 3061-3083.
[21] Sanchez, C.; Julian, B.; Belleville, P.; Popall, M. *J. Mater. Chem.* 2005, *15*, 3559-3592.
[22] Karg, M.; Hellweg, T. *J. Mater. Chem.* 2009, *19*, 8714-8727.
[23] Han, W. S.; Lee, H. Y.; Jung, S. H.; Lee, S. J.; Jung, J. H. *Chem. Soc. Rev.* 2009, *38*, 1904-1915.
[24] Drechsler, U.; Erdogan, B.; Rotello, V. M. *Chem. Eur. J.* 2004, *10*, 5570-5579.
[25] Hoffmann, F.; Cornelius, M.; Morell, J.; Froeba, M. *Angew. Chem. Int. Ed.* 2006, *45*, 3216-3251.
[26] Descalzo, A. B.; Martinez-Manez, R.; Sancenon, F.; Hoffmann, K.; Rurack, K. *Angew. Chem. Int. Ed.* 2006, *45*, 5924-5948.
[27] Videnova-Adrabinska, V. *Coord. Chem. Rev.* 2007, *251*, 1987-2016.
[28] Fischer, R. A.; Wöll, C. *Angew. Chem. Int. Ed.* 2008, *47*, 8164-8168.

[29] Hagrman, P. J.; Hagrman, D.; Zubieta, J. *Angew. Chem. Int. Ed.* 1999, *38*, 2639-2684.
[30] Champness, N. R. *Angew. Chem. Int. Ed.* 2009, *48*, 2274-2275.
[31] Parnham, E. R.; Morris, R. E. *Acc. Chem. Res.* 2007, *40*, 1005-1013.
[32] Gascon, J.; Kapteijn, F. *Angew. Chem. Int. Ed.* 2010, *49*, 1530-1532.
[33] Prakash, M. J.; Lah, M. S. *Chem. Commun.* 2009, 3326-3341.
[34] Li, J. R.; Kuppler, R. J.; Zhou, H. C. *Chem. Soc. Rev.* 2009, *38*, 1477-1504.
[35] Murray, L. J.; Dinca, M.; Long, J. R. *Chem. Soc. Rev.* 2009, *38*, 1294-1314.
[36] Ma, S. *Pure Appl. Chem.* 2009, *81*, 2235-2251.
[37] Shekhah, O. *Materials* 2010, *3*, 1302-1315.
[38] Ameloot, R.; Pandey, L.; Van der Auweraer, M.; Alaerts, L.; Sels, B. F.; De Vos, D. E. *Chem.Commun.* 2010, *46*, 3735-3737.
[39] Horcajada, P.; Serre, C.; Grosso, D.; Boissiere, C.; Perruchas, S.; Sanchez, C.; Ferey, G. *Adv. Mater.* 2009, *21*, 1931-1935.
[40] Yao, H. B.; Fang, H. Y.; Tan, Z. H.; Wu, L. H.; Yu, S. H. *Angew. Chem. Int. Ed.* 2010, *49*, 2140-2145.
[41] Gupta, T.; Tartakovsky, E.; Iron, M. A.; van der Boom, M. E. *ACS Appl. Mater. Interfaces* 2010, *2*, 7-10.
[42] Mitzi, D. B. *Chem. Mater.* 2001, *13*, 3283-3298.
[43] Fraxedas, J. *Adv. Mater.* 2002, *14*, 1603-1614.
[44] Kudernac, T.; Lei, S.; Elemans, J. A. A. W.; De Feyter, S. *Chem. Soc. Rev.* 2009, *38*, 402- 421.
[45] Zacher, D.; Shekhah, O.; Wöll, C.; Fischer, R. A. *Chem. Soc. Rev.* 2009, *38*, 1418-1429.
[46] Gomar-Nadal, E.; Puigmarti-Luis, J.; Amabilino, D. B. *Chem. Soc. Rev.* 2008, *37*, 490-504.
[47] Valade, L.; de Caro, D.; Basso-Bert, M.; Malfant, I.; Faulmann, C.; Garreau de Bonneval, B.; Legros, J. P. *Coord. Chem. Rev.* 2005, *249*, 1986-1996.
[48] Bekermann, D.; Rogalla, D.; Becker, H. W.; Winter, M.; Fischer, R. A.; Devi, A. *Eur. J. Inorg. Chem.* 2010, 1366-1372.
[49] Shekhah, O.; Wang, H.; Paradinas, M.; Ocal, C.; Schupbach, B.; Terfort, A.; Zacher, D.; Fischer R., A.; Wöll, C. *Nat. Mater.* 2009, *8*, 481-484.
[50] Fischer, R. A.; Wöll, C. *Angew. Chem. Int. Ed.* 2009, *48*, 6205-6208.
[51] Shekhah, O.; Wang, H.; Zacher, D.; Fischer, R. A.; Wöll, C. *Angew. Chem. Int. Ed.* 2009, *48*, 5038-5041.
[52] Munuera, C.; Shekhah, O.; Wang, H.; Wöll, C.; Ocal, C. *Phys. Chem. Chem. Phys.* 2008, *10*, 7257-7261.
[53] Shekhah, O.; Wang, H.; Kowarik, S.; Schreiber, F.; Paulus, M.; Tolan, M.; Sternemann, C.; Evers, F.; Zacher, D.; Fischer, R. A.; Wöll, C. *J. Am. Chem. Soc.* 2007, *129*, 15118-15119.
[54] Yoo, Y.; Jeong, H. K. *Chem. Commun.* 2008, 2441-2443; Ebid *Microporous Mesoporous Mater.* 2009, *123*, 100-106.
[55] Biemmi, E.; Scherb, C.; Bein, T. *J. Am. Chem. Soc.* 2007, *129*, 8054-8055.
[56] Gascon, J.; Aguado, S.; Kapteijn, F. *Microporous Mesoporous Mater.* 2008, *113*, 132-138.
[57] Arnold, M.; Kortunov, P.; Jones, D. J.; Nedellec, Y.; Kaerger, J.; Caro, J. *Eur. J. Inorg. Chem.* 2007, 60-64.

[58] Scherb, C.; Schoedel, A.; Bein, T. *Angew. Chem. Int. Ed.* 2008, *47*, 5777-5779.
[59] Feynman, R. P. *Eng. Sci.* 1960, *23*, 22-36.
[60] Lehn, J. M. *Supramolecular Chemistry: Concepts and perspectives, VCH, Weinheim* 1995.
[61] Wen, J.; Wilkes, G. L. *Chem. Mater.* 1996, *8*, 1667-1681.
[62] Mutin, P. H.; Guerrero, G.; Vioux, A. *J. Mater. Chem.* 2005, *15*, 3761-3768.
[63] Schreiber, F. *Prog. Surf. Sci.* 2000, *65*, 151-256.
[64] Petty, M. C. *Nanosci. Nanotech.* 2004, *8*, 295-304.
[65] Talham, D. R. *Chem. Rev.* 2004, *104*, 5479-5501.
[66] Ulman, A. *Chem. Rev.* 1996, *96*, 1533-1554.
[67] Schwartz, D. K. *Annu. Rev. Phys. Chem.* 2001, *52*, 107-137.
[68] Chechik, V.; Crooks, R. M.; Stirling, C. J. M. *Advan. Mater.* 2000, *12*, 1161-1171.
[69] Park, J.; Lee, S.; Lee, H. H. *Org. Electron.* 2006, *7*, 256-260.
[70] Larson, R. G.; Rehg, T. J. *Liq. Film Coat.* 1997, 709-734.
[71] Kawase, T.; Shimoda, T.; Newsome, C.; Sirringhaus, H.; Friend, R. H. *Thin Solid Films* 2003, *438-439*, 279-287.
[72] Lee, J.; Choi, J. *Thin Solid Films* 2008, 121-170.
[73] Yang, R.; Qian, Z.; Deng, J. *J. Electrochem. Soc.* 1998, *145*, 2231-2236.
[74] Gupta, T.; Altman, M.; Shukla, A. D.; Freeman, D.; Leitus, G.; Van der Boom, M. E. *Chem. Mater.* 2006, *18*, 1379-1382.
[75] Haensch, C.; Hoeppener, S.; Schubert, U. S. *Chem. Soc. Rev.* 2010, *39*, 2323-2334.
[76] Lippert, G.; Osten, H. J. *J. Cryst. Growth* 1993, *127*, 476-478.
[77] Angst, D. L.; Simmons, G. W. *Langmuir* 1991, *7*, 2236-2242.
[78] Le Grange, J. D.; Markham, J. L.; Kurkjian, C. R. *Langmuir* 1993, *9*, 1749-1743.
[79] McGovern, M. E.; Kallury, K. M. R.; Thompson, M. *Langmuir* 1994, *10*, 3607-3614.
[80] Glaser, A.; Foisner, J.; Hoffmann, H.; Friedbacher, G. *Langmuir* 2004, *20*, 5599-5604.
[81] Fadeev, A. Y.; McCarthy, T. J. *Langmuir* 2000, *16*, 7268-7274.
[82] Hanson, E. L.; Schwartz, J.; Nickel, B.; Koch, N.; Danisman, M. F. *J. Am. Chem. Soc.* 2003, *125*, 16074-16080.
[83] Jahne, E.; Henke, A.; Adler, H. J. P. *Coating* 2000, *33*, 218-222.
[84] Frantz, R.; Durand, J. O.; Granier, M.; Lanneau, G. F. *Tetrahedron Lett.* 2004, *45*, 2935-2937.
[85] Viornery, C.; Chevolot, Y.; Leonard, D.; Aronsson, B. O.; Pechy, P.; Mathieu, H. J.; Descouts, P.; Graetzel, M. *Langmuir* 2002, *18*, 2582-2589.
[86] Flink, S.; van Veggel, F. C. J. M.; Reinhoudt, D. N. *Adv. Mater.* 2000, *12*, 1315-1328.
[87] Gupta, T.; Cohen, R.; Evmenenko, G.; Dutta, P.; Van der Boom, M. E. *J. Phys. Chem. C* 2007, *111*, 4655-4660.
[88] Gupta, T.; Van der Boom, M. E. *J. Am. Chem. Soc.* 2006, *128*, 8400-8401.
[89] Gulino, A.; Giuffrida, S.; Mineo, P.; Purrazzo, M.; Scamporrino, E.; Ventimiglia, G.; van der Boom, M. E.; Fragala, I. *J. Phys. Chem. B* 2006, *110*, 16781-16786.
[90] Gulino, A.; Bazzano, S.; Mineo, P.; Scamporrino, E.; Vitalini, D.; Fragala, I. *Chem. Mater.* 2005, *17*, 521-526.
[91] Rohde, R. D.; Agnew, H. D.; Yeo, W. S.; Bailey, R. C.; Heath, J. R. *J. Am. Chem. Soc.* 2006, *128*, 9518-9525.
[92] Sieval, A. B.; Linke, R.; Zuilhof, H.; Sudholter, E. J. R. *Adv. Mater.* 2000, *12*, 1457-1460.

[93] Sieval, A. B.; Opitz, R.; Maas, H. P. A.; Schoeman, M. G.; Meijer, G.; Vergeldt, F. J.; Zuilhof, H.; Sudhoelter, E. J. R. *Langmuir* 2000, *16*, 10359-10368.

[94] Chow, B. Y.; Mosley, D. W.; Jacobson, J. M. *Langmuir* 2005, *21*, 4782-4785.

[95] Banerjee, R.; Phan, A.; Wang, B.; Knobler, C.; Furukawa, H.; O'Keeffe, M.; Yaghi, O. M. *Science* 2008, *319*, 939-943.

[96] Yaghi, O. M.; O'Keeffe, M.; Ockwig, N. W.; Chae, H. K.; Eddaoudi, M.; Kim, J. *Nature* 2003, *423*, 705-714.

[97] Deng, H.; Doonan, C. J.; Furukawa, H.; Ferreira, R. B.; Towne, J.; Knobler, C. B.; Wang, B.; Yaghi, O. M. *Science* 2010, *327*, 846-850.

[98] Rosi, N. L.; Eckert, J.; Eddaoudi, M.; Vodak, D. T.; Kim, J.; O'Keeffe, M.; Yaghi, O. M. *Science* 2003, *300*, 1127-1130.

[99] Britt, D.; Furukawa, H.; Wang, B.; Glover, T. G.; Yaghi, O. M. *Proc. Natl. Acad. Sci. U.S.A.* 2009, 20637-20640.

[100] Demessence, A.; Long, J. R. *Chem. Eur. J.* 2010, *16*, 5902-5908.

[101] Ma, S.; Zhou, H. C. *Chem. Commun.* 2010, *46*, 44-53.

[102] Hermes, S.; Zacher, D.; Baunemann, A.; Wöll, C.; Fischer, R. A. *Chem. Mater.* 2007, *19*, 2168-2173.

[103] Hermes, S.; Schroeder, F.; Chelmowski, R.; Wöll, C.; Fischer, R. A. *J. Am. Chem. Soc.* 2005, *127*, 13744-13745.

[104] Kubo, M.; Chaikittisilp, W.; Okubo, T. *Chem. Mater.* 2008, *20*, 2887-2889.

[105] Han, Z. B.; Zhang, G. X. *Cryst. Eng. Comm.* 2010, *12*, 348-351.

[106] Han, L.; Zhou, Y.; Zhao, W. N.; Li, X.; Liang, Y. X. *Cryst. Growth Des.* 2009, *9*, 660-662.

[107] Mir, M. H.; Kitagawa, S.; Vittal, J. J. *Inorg. Chem.* 2008, *47*, 7728-7733.

[108] Zacher, D.; Baunemann, A.; Hermes, S.; Fischer, R. A. *J. Mater. Chem.* 2007, *17*, 2785-2792.

[109] Guerrero, V. V.; Yoo, Y.; McCarthy, M. C.; Jeong, H. K. *J. Mater.Chem.* 2010, *20*, 3938-3943.

[110] Quinn, J. F.; Johnston, A. P. R.; Such, G. K.; Zelikin, A. N.; Caruso, F. *Chem. Soc. Rev.* 2007, *36*, 707-718.

[111] George, S. M.; Yoon, B.; Dameron, A. A. *Acc. Chem. Res.* 2009, *42*, 498-508.

[112] Srivastava, S.; Kotov, N. A. *Acc. Chem. Res.* 2008, *41*, 1831-1841.

[113] Motiei, L.; Altman, M.; Gupta, T.; Lupo, F.; Gulino, A.; Evmenenko, G.; Dutta, P.; van der Boom, M. E. *J. Am. Chem. Soc.* 2008, *130*, 8913-8915.

[114] DeLongchamp, D. M.; Kastantin, M.; Hammond, P. T. *Chem. Mater.* 2003, *15*, 1575-1586.

[115] Ji, J.; Fu, J.; Shen, J. *Adv. Mater.* 2006, *18*, 1441-1444.

[116] Kujawa, P.; Moraille, P.; Sanchez, J.; Badia, A.; Winnik, F. M. *J. Am. Chem. Soc.* 2005, *127*, 9224-9234.

[117] Kanaizuka, K.; Haruki, R.; Sakata, O.; Yoshimoto, M.; Akita, Y.; Kitagawa, H. *J. Am. Chem. Soc.* 2008, *130*, 15778-15779.

[118] Altman, M.; Zenkina, O.; Evmenenko, G.; Dutta, P.; van der Boom, M. E. *J. Am. Chem. Soc.* 2008, *130*, 5040-5041.

[119] Altman, M.; Gupta, T.; Zubkov, T.; Cohen, R.; van der Boom, M. E. *PMSE Prepr.* 2007, *96*, 890-892.

[120] Altman, M.; Shukla, A. D.; Zubkov, T.; Evmenenko, G.; Dutta, P.; van der Boom, M. E. *J. Am. Chem. Soc.* 2006, *128*, 7374-7382.
[121] Ulman, A. *in An Introduction to Ultra Thin Organic Films: From Langamuir-Blogett to Self Assembly, San Diago, CA* 1991.
[122] Dickert, F. L.; Haunschild, A. *Adv. Mater.* 1993, *5*, 887-895.
[123] Parikh, A. N.; Allara, D. L. *J. Chem. Phys.* 1992, *96*, 927-945.
[124] Zhou, M.; Laux, J. M.; Edwards, K. D.; Hemminger, J. C.; Hong, B. *Chem. Commun.* 1997, 1977-1978.
[125] Mathauer, K.; Frank, C. W. *Langmuir* 1993, *9*, 3002-3008.
[126] Azzam, R. M. A.; Bashara, N. M. *Ellipsometry and Polarized Light, North-Holland, Amsterdam* 1977.
[127] Peterlinz, K. A.; Georgiadis, R. *Langmuir* 1996, *12*, 4731-4740.
[128] Vickerman, J. C. *In Surface Analysis, The Principal Techniques, Wiley, Chichester* 1997.
[129] Rabalais, J. W. *Principles and Applications of Ion Scattering Spectrometry: Surface Chemical and Structural Analysis.* John Wiley & Sons, Inc 2003.
[130] J. Stöhr *NEXAFS Spectroscopy, Springer* 1992.
[131] McCarley, T. D.; McCarley, R. L. *Anal. Chem.* 1997, *69*, 130-136.
[132] Gewirth, A. A.; Niece, B. K. *Chem. Rev.* 1997, *97*, 1129-1162.
[133] Gewirth, A. A.; Niece, B. K. *Electrochem. Nanotech.* 1998, 113-124.
[134] Gracias, D. H.; Chen, Z.; Shen, Y. R.; Somorjai, G. A. *Acc. Chem. Res.* 1999, *32*, 930-940.
[135] Hansma, P. K.; Elings, V. B.; Marti, O.; Bracker, C. E. *Science* 1988, *242*, 209-216.
[136] Hu, X.; Cubillas, P.; Higgins, S. R. *Langmuir* 2010, *26*, 4769-4775.
[137] Boury, F.; Gulik, A.; Dedieu, J. C.; Proust, J. E. *Langmuir* 1994, *10*, 1654-1656.
[138] Haas, H.; Caetano, W.; Borissevitch, G. P.; Tabak, M.; Mosquera Sanchez, M. I.; Oliveira, O. N.; Scalas, E.; Goldmann, M. *Chem. Phys. Lett.* 2001, *335*, 510-516.
[139] Pendry *Low-Energy Electron Diffraction. Academic Press Inc. (London) LTD* 1974, 1-75.
[140] Van Hove, M. A.; Weinberg, W. H.; Chan, C. M. *Low-Energy Electron Diffraction. Springer-Verlag, Berlin Heidelberg New York* 1986.
[141] Brennan, J. L.; Keyes, T. E.; Forster, R. J. *Langmuir* 2006, *22*, 10754-10761.
[142] A. J. Bard, L. R. F. *Electrochemical Methods, Wiley, New York* 1980.
[143] Macdonald, J. R. *Impedance Spectroscopy: Emphasizing Solid Materials and Systems, Wiley, New York*,1987.
[144] E. Barsoukov, J. R. M. *Impedance spectroscopy: theory, experiment, and applications, Wiley-Interscience,* 2005.
[145] Gupta, T.; van der Boom, M. E. *Angew. Chem. Int. Ed.* 2008, *47*, 5322-5326.
[146] Gupta, T.; van der Boom, M. E. *Angew. Chem. Int. Ed.* 2008, *47*, 2260-2262.
[147] Choi, I. S.; Chi, Y. S. *Angew. Chem. Int. Ed.* 2006, *45*, 4894-4897.
[148] Lehn, J. M. *Angew. Chem. Int. Ed.* 1990, 29, 1304-1319.
[149] Lehn, J. M. *Science* 1985, *227*, 849-856.
[150] Lehn, J. M. *Angew. Chem.* 1988, *100*, 91-116.
[151] Lehn, J. M. *Science* 1993, *260*, 1762-1763.
[152] Pedersen, C. J. *Angew. Chem.* 1988, *100*, 1053-1059.
[153] Pedersen, C. J. *Science* 1988, *241*, 536-540.

[154] Cram, D. J. *Science* 1983, *219*, 1177-1183.
[155] Cram, D. J. *Angew. Chem.* 1988, *100*, 1041-1052.
[156] Cram, D. J. *Nature* 1992, *356*, 29-36.
[157] Cram, D. J.; Cram, J. M. *Science* 1974, *183*, 803-809.
[158] Labande, A.; Ruiz, J.; Astruc, D. *J. Am. Chem. Soc.* 2002, *124*, 1782-1789.
[159] Lin, S. Y.; Liu, S. W.; Lin, C. M.; Chen, C. H. *Anal. Chem.* 2002, *74*, 330-335.
[160] itzmaurice, D.; Rao, S. N.; Preece, J. A.; Stoddart, J. F.; Wenger, S.; Zaccheroni, N. *Angew. Chem. Int. Ed.* 1999, *38*, 1147-1150.
[161] Ryan, D.; Rao, S. N.; Rensmo, H.; Fitzmaurice, D.; Preece, J. A.; Wenger, S.; Stoddart, J. F.; Zaccheroni, N. *J. Am. Chem. Soc.* 2000, *122*, 6252-6257.
[162] Fantuzzi, G.; Pengo, P.; Gomila, R.; Ballester, P.; Hunter, C. A.; Pasquato, L.; Scrimin, P. *Chem. Commun.* 2003, 1004-1005.
[163] Clechet, P.; Jaffrezic-Renault, N. *Adv. Mater.* 1990, *2*, 293-298.
[164] Gooding, J. J.; King, G. C. *J. Mater. Chem.* 2005, *15*, 4876-4880.
[165] Tully, D. C.; Frechet, J. M. J. *Chem. Commun.* 2001, 1229-1239.
[166] Joo, S.; Brown, R. B. *Chem. Rev.* 2008, *108*, 638-651.
[167] Potyrailo, R. A.; Mirsky, V. M. *Chem. Rev.* 2008, *108*, 770-813.
[168] Willner, I.; Zayats, M. *Angew. Chem. Int. Ed.* 2007, *46*, 6408-6418.
[169] Lednev, I. K.; Petty, M. C. *Adv. Mater.* 1996, *8*, 615-630.
[170] Lednev, I. K.; Petty, M. C. *Thin Solid Films* 1996, *284-285*, 683-686.
[171] Turyan, I.; Mandler, D. *Anal. Chem.* 1994, *66*, 58-63.
[172] Turyan, I.; Mandler, D. *Anal. Chem.* 1997, *69*, 894-897.
[173] Rubinstein, I.; Steinberg, S.; Tor, Y.; Shanzer, A.; Sagiv, J. *Nature* 1988, *332*, 426-429.
[174] Steinberg, S.; Tor, Y.; Sabatani, E.; Rubinstein, I. *J. Am. Chem. Soc.* 1991, *113*, 5176-5182.
[175] Moore, A. J.; Goldenberg, L. M.; Bryce, M. R.; Petty, M. C.; Monkman, A. P.; Marenco, C.; Yarwood, J.; Joyce, M. J.; Port, S. N. *Adv. Mater.* 1998, *10*, 395-398.
[176] Moore, A. J.; Goldenberg, L. M.; Bryce, M. R.; Petty, M. C.; Moloney, J.; Howard, J. A. K.; Joyce, M. J.; Port, S. N. *J. Org. Chem.* 2000, *65*, 8269-8276.
[177] Flink, S.; Boukamp, B. A.; van den Berg, A.; van Veggel, F. C. J. M.; Reinhoudt, D. N. *J. Am. Chem. Soc.* 1998, *120*, 4652-4657.
[178] Flink, S.; Van Veggel, F. C. J. M.; Reinhoudt, D. N. *J. Phys. Chem. B* 1999, *103*, 6515-6520.
[179] Bandyopadhyay, K.; Liu, H.; Liu, S. G.; Echegoyen, L. *Chem.Commun.* 2000, 141-142.
[180] Bandyopadhyay, K.; Liu, S. G.; Liu, H.; Echegoyen, L. *Chemistry* 2000, *6*, 4385-4392.
[181] Zhang, S.; Echegoyen, L. *Tetrahedron Lett.* 2003, *44*, 9079-9082.
[182] Gupta, T.; van der Boom Milko, E. *J. Am. Chem. Soc.* 2007, *129*, 12296-12303.
[183] de Ruiter, G.; Gupta, T.; van der Boom, M. E. *J. Am. Chem. Soc.* 2008, *130*, 2744-2745.
[184] Wang, Z.; Cook, M. J.; Nygrd, A. M.; Russell, D. A. *Langmuir* 2003, *19*, 3779-3784.
[185] Zhang, S.; Echegoyen, L. *Abstracts, 55th Southeast Regional Meeting of the American Chemical Society, Atlanta, GA, United States, November 16-19,* 2003, 496.
[186] van der Veen, N. J.; Flink, S.; Deij, M. A.; Egberink, R. J. M.; van Veggel, F. C. J. M.; Reinhoudt, D. N. *J. Am. Chem. Soc.* 2000, *122*, 6112-6113.
[187] Zimmerman, R.; Basabe-Desmonts, L.; van der Baan, F.; Reinhoudt, D. N.; Crego-Calama, M. *J. Mater. Chem.* 2005, *15*, 2772-2777.

[188] Gupta, T.; Van der Boom, M. E. *J. Am. Chem. Soc.* 2007, *129*, 12296-12303.
[189] Zhang, S.; Lue, F.; Gao, L.; Ding, L.; Fang, Y. *Langmuir* 2007, *23*, 1584-1590.
[190] Gulino, A.; Bazzano, S.; Condorelli, G. G.; Giuffrida, S.; Mineo, P.; Satriano, C.; Scamporrino, E.; Ventimiglia, G.; Vitalini, D.; Fragala, I. *Chem. Mater.* 2005, *17*, 1079-1084.
[191] Chu, B. W. K.; Yam, V. W. W. *Langmuir* 2006, *22*, 7437-7443.
[192] Ashkenasy, G.; Ivanisevic, A.; Cohen, R.; Felder, C. E.; Cahen, D.; Ellis, A. B.; Shanzer, A. *J. Am. Chem. Soc.* 2000, *122*, 1116-1122.
[193] Lei, B.; Li, B.; Zhang, H.; Zhang, L.; Li, W. *J. Phys. Chem. C* 2007, *111*, 11291-11301.
[194] Gulino, A.; Gupta, T.; Mineo, P. G.; van der Boom, M. E. *Chem. Commun.* 2007, 4878-4880.
[195] Gulino, A.; Mineo, P.; Scamporrino, E.; Vitalini, D.; Fragala, I. *Chem. Mater.* 2004, *16*, 1838-1840.
[196] [Simpson, T. R. E.; Revell, D. J.; Cook, M. J.; Russell, D. A. *Langmuir* 1997, *13*, 460-464.
[197] Dubey, M.; Bernasek, S. L.; Schwartz, J. *J. Am. Chem. Soc.* 2007, *129*, 6980-6981.
[198] Wu, D. G.; Ashkenasy, G.; Shvarts, D.; Ussyshkin, R. V.; Naaman, R.; Shanzer, A.; Cahen, D. *Angew. Chem. Int. Ed.* 2000, *39*, 4496-4500.
[199] Gulino, A.; Gupta, T.; Altman, M.; Lo Schiavo, S.; Mineo, P. G.; Fragala, I. L.; Evmenenko, G.; Dutta, P.; van der Boom, M. E. *Chem. Commun.* 2008, 2900-2902.
[200] Ye, S.; Zhou, W.; Abe, M.; Nishida, T.; Cui, L.; Uosaki, K.; Osawa, M.; Sasaki, Y. *J. Am. Chem. Soc.* 2004, *126*, 7434-7435.
[201] Ashwell, G. J.; Paxton, G. A. N.; Whittam, A. J.; Tyrrel, W. D.; Berry, M.; Zhou, D. *J. Mater. Chem.* 2002, *12*, 1631-1635.
[202] Allendorf, M. D.; Houk, R. J. T.; Andruszkiewicz, L.; Talin, A. A.; Pikarsky, J.; Choudhury, A.; Gall, K. A.; Hesketh, P. J. *J. Am. Chem. Soc.* 2008, *130*, 14404-14405.

In: Advanced Organic-Inorganic Composites
Editor: Inamuddin

ISBN 978-1-61324-264-3

Chapter 3

ORGANIC-INORGANIC COMPOSITE MATERIALS AS SORBENTS AND ION-EXCHANGERS USED IN ENVIRONMENTAL PROTECTION

Dorota Kołodyńska*

Department of Inorganic Chemistry, Faculty of Chemistry,
Maria Curie-Sklodowska University,
Maria Curie-Sklodowska Sq.2. 20-031 Lublin, Poland

ABSTRACT

The recent environmental concerns and demands for new protection technologies and/or improvement of those already existing affects development of sorption methods particularly of those suitable for the solution of those problems. The challenges involve new, more efficient and selective adsorbents. To accomplish these needs, new trends point towards the development of adsorbents of a combined or hybrid nature. Organic-inorganic composite materials have found numerous applications in the areas of chemistry, biochemistry, engineering, and material science. The combination of organic and inorganic precursors yields hybrid materials of mechanical properties not present in the pure materials. Therefore, organic-inorganic composite ion exchange materials show the improvement in their properties that makes them more suitable for the application in environmental protection. This chapter provides the review of the organic-inorganic composites used in the environmental protection processes. Among others, the removal of toxic heavy metal ions from waters and wastewaters, radionuclides, gaseous organic and inorganic contaminants, carbon dioxide should be mentioned.

ABBREVIATIONS

AA	Acetic acid
ABB	Acidic brilliant blue 6B

* Corresponding author's e-mail: kolodyn@poczta.onet.pl

AC	Activated carbon
AC-Mg	Active carbons with magnesium
AC-OG	Outgassed active carbon
ALG-PUCF	Alginate-polyurethane composite foams
AM3	Absorptionsmittel 3
APRB	Weak acidic pink red B
APTES	***3- a***minopropyltriethoxysilane
ATSDR	Agency for toxic substances and disease registry
AWWA	American water works association
BCF	Bead cellulose loaded with iron oxyhydroxide
Bip	Biphenyl
BPA	Biphenol A
BPR	Basic pink red B
BV	Bed volume
CB-1	Poly(allylamine)-silica-5-chloro-8-hydroxy quinoline composite
CEMT	2-chloroethoxymethyl thiirane
CNTs	Carbon nanotubes
COD	Chemical oxygen demand
CPTCS	Chloropropyltrichlorosilane
CS	Chitosan
CS-Nb	Chitosan-niobium(V) oxide composite
CSH	Cellulose silica hybrid
CT	Chitin
CTMABr	Cetyltrimethylammonium bromide
CV	Crystal violet
DDT	Dichlorodiphenyltrichloroethane
DMF	Dimethylformamide
DMSO	Dimethylsulfoxide
DTPA	Diethylenetriaminepentaacetic acid
DVB	Divinylbenzene
DWEL	Drinking water equivalent level
EDA	Ethylenediamine
EDTA	Ethylenediaminetetraacetic acid
EGDMA	Ethylene glycol dimethacrylate
EPA	Environmental protection agency
EtSH	Ethanethiol
EV	Ethyl violet
F108	Nonionic difunctional block polymer
FA	Formic acid
FDA	Food and drug administration
GAC	Granulated activative carbon
GIH	Granulated iron hydroxide
GMA	Glycidyl methacrylate resin
GMA-MBA	Glycidyl methacrylate-methylenebis-acrylamide resin

GMA-MBA-A	Glycidyl methacrylate-methylenebis-acrylamide resin with amine functional groups
GMA-MBA-SH	Glycidyl methacrylate-methylenebis-acrylamide resin with thiol functional groups
HA	Humic acids
HAIX-F	Hybrid anion exchange fiber
HAP	Hydroxyapatite
HAP-CT	Hydroxyapatite with chitin
HDL	High-density lipoprotein
HEMA	Hydroxyethylmethacrylate
HFO	Hydrated Fe(III) oxide
HGMS	High gradient magnetic separation
HIX	Hybrid ion exchangers
HMDA	Hexamethylenediamine
HMO	Hydrous metal oxides
IARC	International agency for research on cancer
ICS	Isocyanurate
ICSF	Iron hydroxide-chitosan flakes
ICSG	Iron hydroxide-chitosan granules
IDA	Iminodiacetic acid
IMC	Iron containing mesoporous carbon
IMI	Imidazol
LHMM	Luminescent hybrid microporous material
MAPP	Magnetically active polymeric particle
MB	Methylene blue
MB9	Mordant blue 9
MBA	Methylenebis-acrylamide
MCL	Maximum contaminant level
MCM-48	***C***ubic structural mesoporous molecular sieves
MG	Malachite green
M-GMA-DVB-IDA	Magnetic-glycidyl methacrylate-divinylbenzene resin with iminodiacetate functionality
M-GMA-DVB-TEP	Magnetic-glycidyl methacrylate- divinylbenzene resin with tetraethylenepentamine
M-GMA-MBA	Magnetic-glycidyl methacrylate-methylenebis-acrylamide resin
M-GMA-MBA-A	Magnetic-glycidyl methacrylate-methylenebis-acrylamide resin with amine functional groups
M-GMA-MBA-SH	Magnetic-glycidyl methacrylate-methylenebis-acrylamide resin with thiol functional groups
M-GMA-MBA-TEP	Magnetic-glycidyl methacrylate-methylenebis-acrylamide resin with tetraethylenepentamine
MISGM	Molecularly imprinted sol gel material
MMA	Methylmethacrylate
MMA-GMA-IDA	Methylmethacrylate-glycidyl methacrylate resin with iminodiacetate functionality

MO	Methyl orange
MOF	Metal-organic framework
MPY	Mercaptopyridine
M-PMMA-EDA	Magnetic polymethyl methacrylate-ethylenediamine
MPTS	Mercaptopropyl trimethoxysilane
M-PVAC	Magnetic-polyvinyl acetate
M-PVAC-IDA	Magnetic-polyvinyl acetate-iminodiacetate
M-PVOH	Magnetic-polyvinyl alcohol
M-PVEP	Magnetic-polyvinyl propenepoxide
MTCS	Methyltrichlorosilane
NIP	Nanoscale inorganic particle
NOC	Nonionic organic contaminant
OMS	Ordered mesoporous silica
OSHA	Occupational safety and health administration
PAH	Polyacrylic hydrazid
PAHC	Polycyclic aromatic hydrocarbon
PAM	Polyacrylamide
PAM-B-A	Polyacrylamide-bentonite composite with amine functionality
PAM-B-A-HC	Polyacrylamide-bentonite composite with amine functionality after immobilization of humic acid
PAM-TMS	Phenylaminomethyl trimethoxysilane
PAN	Polyacrylonitrile
PEG	Polyethylene glycol
PEI	Poly(ethyleneimine)
PEO-b-PMMA	Poly(ethylene oxide)-b-poly(methyl methacrylate)
PEO-PPO	Poly(ethylene oxide)-poly(propylene oxide)
PEO-PPO-PEO	Poly(ethylene oxide)-poly(propylene oxide)-poly(ethylene oxide)
PGMA	Poly(glycidyl methacrylate)
Phe	Phenanthrene
PHEMA	Poly(2-hydroxyethyl methacrylate)
PHEMA-CS	Poly(2-hydroxyethyl methacrylate)-chitosan
Pluronic F127	Triblock copolymer
Pluronic P123	Triblock copolymer
PMDA	Pyromellitic acid dianhydride
PMOs	Periodic mesoporous organosilicas
Poly(MMA-DVB-GMA)]	Poly(methylmethacrylate-divinylbenzene-glycidylmethacrylate)
POP	Persistent organic pollutant
Ppy	Polypyrrole
PS-DVB	Polystyrene-divinylbenzene
PSF	Phenolsulfonic formaldehyde
PSNs	Polymer supported nanoparticles
PS-P4VP	Polystyrene-block-poly(4-vinylpyridine)
PVAC	Polyvinyl acetate

PVB	Polyvinylbutyral
PVOH	Polyvinyl alcohol
RDA	Recommended dietary allowance
QA-PVOH	Quaternized poly(vinyl alcohol)
HAC-CS	2-hydroxypropyltrimethyl ammonium chloride chitosan
RF	Resorcinol with formaldehyde
SB-1	Poly(allylamine)-silica-5-sulfonic acid -8-hydroxy quinoline composite
SCM	Surface complexation model
SDBS	Sodium dodecyl benzene sulfonate
SDS	Sodium dodecyl sulfate
SilicaEDTA	Silica gel modified with EDTA
SIR	Solvent impregnated resin
SOD	Superoxide dismutase
SPC	Silica polyamine composite
S-PS-DVB	Sulfonated polystyrene-divinylbenzene
SSP	Spraying suspension polymerization
TBOS	Tetrabutyl orthosilicate
TCA	Tichloroacetic acid
TEOA	Triethyl orthoacetate
TEOS	Tetraethyl orthosilicate
TEP	Tetraethylenepentamine
TM	Transition metalS
TNT	Titanium dioxide nanotubes
TPED	(Trimethoxysilyl)-propylethylenediamine
US EPA	United States environmental protection agency
VAC	Vinyl acetate
WAG	Weak acidic green
WHO	World health organization
WP-1	Poly(ethyleneimine) (PEI)-silica composite
WP-2	Poly(ethyleneimine) (PEI)-silica-iminodiacetate composite
WP-3	Poly(ethyleneimine) (PEI)-silica-ethanethiol composite
WP-4	Poly(allylamine)-silica-8-hydroxyquinoline composite
ZrP	Zirconium phosphate
ZrWP	Zirconium(IV) tungstophosphate

1. Heavy Metals as a Source of Chemical Pollution

Toxic metals are one of the oldest environmental problems. Industrial activity during the past years has caused a sharp increase in the concentration of many metals in air, soil, and waters. Pollution of the environment and human exposure to metallic elements may occur naturally by erosion of surface deposits of metal minerals, as well as from human anthrophogenic activities, such as mining of coal, natural gas, smelting, fossil fuel combustion, and industrial application of metals, among others, in chemical industry, production of paper, plastics, plating and the manufacture of lubricants and chlor-alkali industries [1,2].

Some heavy metals - such as cobalt, copper, iron, manganese, molybdenium, vanadium, strontium and zinc are essential to health in trace amounts. The others are non-essential and can be harmful to health in excessive amounts. These include cadmium, antimonium, chromium, mercury, lead, and arsenic - the last three being the most common in the cases of heavy metal toxicity. The toxicity of heavy metals depends on a number of factors. Specific symptomatology varies according to the metal in question, the total dose absorbed, the age of living organism (for example, young children are more susceptible to the effects of lead exposure because they absorb several times the percent ingested compared with adults) and whether the exposure was acute or chronic. The form in which metal ions occur is also important. The toxicology of metal ions ranges from comparatively simple ionic salts to organometallic compounds. All heavy metal ions exist in surface waters in colloidal, particulate, and dissolved phases, although dissolved concentrations are generally low. The colloidal and particulate metal ions may be found as hydroxides, oxides, silicates, or sulfides or adsorbed to clay, silica, or organic matter. The soluble forms are generally ions or unionized organometallic chelates or complexes. The solubility of trace metals in surface waters is predominately controlled by the water pH, the type and concentration of ligands on which the metal could adsorb, and the oxidation state of the mineral components as well as the redox environment of the system. Drinking water is also a significant route of exposure particularly if it includes arsenic, aluminium, iron, manganese, cadmium and lead. The common forms of the metal ions in water depending on the pH are presented in Table 1.

As far the penetration way is concerned exposure to metals may take place by inhalation, ingestion, or skin contact. Inhalation is usually the most important occupational exposure route in the case of elemental mercury which is relatively inert in the gastrointestinal tract and also poorly absorbed through intact skin, whereas inhaled or injected elemental mercury may have disastrous effects. In this group the inhalation of cigarette smoke, which contains cadmium, nickel, arsenic, and lead can not be neglected.

It should be also noticed that the exposure levels vary greatly, depending on geographical, cultural, and other circumstances. Urban environments are typically more polluted than rural areas, and industrialization increases exposures directly through emissions and indirectly through products.

Among the toxic metal species on the priority list of the Environmental Protection Agency (EPA) the following are of great importance: aluminium, antimony, arsenic, beryllium, cadmium, chromium, cobalt, copper, lead, manganese, mercury, nickel, selenium, vanadium and zinc.

Aluminium is not a heavy metal but its environmental exposure is frequent, leading to concerns about accumulative effects and possible connection with Alzheimer's disease.

Antimony is metalloid, occurring with sulfur, lead and copper. Antiomony and its compounds are toxic. Clinically, antimony poisoning is very similar to arsenic poisoning. While levels observed for bottled water are below drinking water guidelines, fruit juice concentrates produced in the United Kingdom were found to contain up to 44.7 μg/l of antimony, well above the European Union limits for tap water of 5 μg/l. The maximum contaminant level (MCL) promoted in 2009 by the US EPA (the United States Environmental Protection Agency) for antimony is equal to 6 μg/l and 20 μg/l by the WHO (World Health Organization) [4].

Table 1. Basic inorganic forms of heavy metal ions in the aqueous medium [3]

Metal ion	pH 4		pH 7		pH 10	
	Oxidative conditions	Redox conditions	Oxidative conditions	Redox conditions	Oxidative conditions	Redox conditions
Cr	$HCrO_4^-$	$CrOH^{2+}$	$HCrO_4^-$, CrO_4^{2-}	$CrOH^{2+}$, $Cr(OH)_2^+$	CrO_4^{2-}	$Cr(OH)_4^-$
Mn	Mn^{2+}	Mn^{2+}	MnO_2, $MnCl^+$ (s.w.)	Mn^{2+}, $MnCl^+$ (s.w.), $MnSO_4$	MnO_2	$MnCO_3$
Fe	$FeOH^{2+}$, $Fe(OH)_2^+$	Fe^{2+}	$Fe(OH)_3$	Fe^{2+}, $FeCO_3$	$Fe(OH)_4^-$	$FeOH^+$, $Fe(OH)_2$
Co	Co^{2+}	Co^{2+}	Co^{2+}, $CoCO_3$	$CoCO_3$	Co_3O_4	$CoCO_3$
Ni	Ni^{2+}, $NiSO_4$	Ni^{2+}	Ni^{2+}, $NiHCO_3^+$, $NiCl^+$ (s.w.)	Ni^{2+}, $NiHCO_3^+$, $NiCl^+$ (s.w.)	$NiOH^+$, $Ni(OH)_2$, $NiCO_3$	$NiOH^+$, $Ni(OH)_2$, $NiCO_3$
Cu	Cu^{2+}	Cu^{2+}	Cu^{2+}, $CuOH^+$, $CuHCO_3^+$, $CuCl^+$ (s.w.)	Cu^{2+}, $CuOH^+$, $CuHCO_3^+$, $CuCl^+$ (s.w.)	$Cu(OH)_2$, $Cu(CO_3)_2^{2-}$	$Cu(OH)_2$, $Cu(CO_3)_2^{2-}$
Zn	Zn^{2+}	Zn^{2+}	Zn^{2+}, $Zn(OH)_2$, $ZnCl^+$ (s.w.)	Zn^{2+}, $Zn(OH)_2$, $ZnCl^+$ (s.w.)	$Zn(OH)_2$	$Zn(OH)_2$
Pb	Pb^{2+}, $PbSO_4$	Pb^{2+}	Pb^{2+}, $PbOH^+$, $PbHCO_3^+$, $PbCl^+$ (s.w.), $PbSO_4$ (s.w.)	Pb^{2+}, $PbOH^+$, $PbHCO_3^+$, $PbCl^+$ (s.w.),	$Pb(OH)_2$, $Pb(CO_3)_2^{2-}$, $PbCO_3$	$Pb(OH)_2$, $Pb(CO_3)_2^{2-}$, $PbCO_3$

s.w. - sea water.

Arsenic common sources of exposure are found near or in hazardous waste sites and areas with high levels naturally occurring arsenic in soil, rocks, and water. It was clasiffied by the IARC (International Agency for Research on Cancer) as a Class A human carcinogen. Many community water systems in North America and around the world contain arsenic concentrations exceeding the MCL of 10 μg/l promoted by US EPA, EU and WHO [4,5]. However, many countries have retained the earlier WHO guideline of 50 μg/l as their standard or as an interim target including Bangladesh and China [6].

Beryllium is naturally found in mineral rocks, coal, soil, and volcanic dust. Occupational exposure most often occurs in mining, extraction, and in the processing of alloy metals containing beryllium. The IARC has determined that beryllium and beryllium compounds are human carcinogens. In the United States, the average concentration of beryllium is 0.03 ng/m^3

of air but in cities, the average air concentration is higher, and its value is 0.2 ng/m^3. The amount of beryllium that has been measured in drinking water in different parts of the United States by EPA is generally less than 2 trillion of a gram for every liter of water. The maximum MCL for beryllium promoted in 2009 by the US EPA is equal to 4 μg/l.

Cadmium acute exposure generally occurs in manufacturing processes of batteries and color pigments used in paint and plastics, as well as in electroplating and galvanizing processes. Cadmium is also emitted by non-ferrous metal mining and refining, manufacture and application of phosphate fertilizers, fossil fuel combustion, as well as waste incineration and disposal. The drinking water guideline value recommended by WHO and AWWA (American Water Works Association) and FDA (Food and Drug Administration) is 0.005 mg Cd/l. OSHA (Occupational Safety and Health Administration) set a legal limit of 5 μg/m^3 cadmium averaged over an 8 hour work day.

Chromium is a naturally occurring element found in rocks, animals, plants, soil, and in volcanic dust and gases. It is present in the environment in several different forms. Chromium can be found in air, soil, and water after release from the manufacture, use, and disposal of chromium-based products, and during the manufacturing process. The EPA has determined the exposure limit of chromium in drinking water at concentrations of 1 mg/l/day. OSHA set a legal limit for chromium(VI) of 0.005 mg/m^3, for chromium(III) of 0.5 mg/m^3 and for chromium(0) of 1.0 mg/m^3 chromium in air averaged over an 8 hour work day.

Cobalt is a metal found in the earth crust, associated with zinc, lead, and copper ores. Radioactive cobalt is used for commercial and medical purposes. ^{60}Co is used for sterilizing medical equipment and consumer products, radiation therapy for treating cancer patients, manufacturing plastics, and irradiating food. OSHA regulates levels of nonradioactive cobalt in workplace air. The limit for an 8 hour workday for cobalt is an average of 0.1 mg/m^3.

Copper is extensively mined and processed. It is primarily used as the metal or alloy. Its compounds are most commonly used in agriculture to treat plant diseases, or for water treatment and as preservatives for wood, leather, and fabrics. EPA has determined that drinking water should not contain more than 1.3 mg/l of copper. According to the OSHA a limit of 0.1 mg/m^3 for copper fumes, 1.0 mg/m^3 for copper dusts during an 8 hour workday should be set. The RDAs (Recommended Dietary Allowances) are equal to 340 μg/day for children aged 1-3, 440 μg/day for children aged 4-8, 700 μg/day for children aged 9-13, 890 μg/day for children aged 14-18, and 900 μg/day for adults.

Lead occurs in many industrial effluents and/or wastewater discharged from the production processes of lead-acid batteries, paint, oil, and lead-petrol containing a considerable amount of lead. Leakage of Pb from these effluents can cause serious groundwater contamination (as high as 1.0 μM). EPA requires that the concentration of lead in air that the public breathes be no higher than 1.5 μg/m^3 averaged over 3 months. The OSHA regulations limit the concentration of lead in workroom air to 50 μg/m^3 for an 8 hour workday.

Manganese is used principally in steel production to improve hardness, stiffness, and strength. It is an essential trace element necessary for good health. The FDA has established that the manganese concentration in bottled drinking water should not exceed 0.05 mg/l. OSHA set a legal limit of 5 mg/m^3 manganese in air averaged over an 8 hour work day.

Mercury common sources are mining, production, and transportation of mercury, as well as mining and refining of gold and silver ores. High mercury exposure results in permanent nervous system and kidney damage. The EPA has set a limit of 2 μg/l of drinking water. FDA has set a maximum permissible level of 1 ppm of methylmercury seafood. OSHA has set limits of 0.1 mg/m^3 for organic mercury and 0.05 mg/m^3 of metallic mercury vapour for 8 hour workday.

Nickel combined with other elements occurs naturally in the earth's crust. It is released into the atmosphere during nickel mining and by industries that make or use nickel, its alloys, or compounds. OSHA has set an enforceable limit of 1.0 mg $nickel/m^3$ for metallic nickel and nickel compounds during an 8 hour shift over a 40 hour work week. EPA recommends that drinking water levels for nickel should not be more than 0.1 mg/l.

Selenium is used in the electronics, glass industry, as a component of pigments in plastics, paints, enamels, inks, and rubber as well as in pesticide formulations. It can be applied in the preparation of pharmaceuticals, or as a nutritional supplement. Radioactive selenium is used in diagnostic medicine. According to the EPA public water supplies are not allowed to exceed 50 ppb of total selenium. The FDA regulations allow a level of 50 ppb of selenium in bottled water. The OSHA limit for selenium compounds in air for an 8 hour period is 0.2 mg/m^3.

Zinc occurs in the nature as sulfide, carbonate, silicate and oxide. The main sources of zinc in wastewaters are discharging waste streams from metals, chemicals, pulp and paper manufacturing processes, steel works with galvanizing lines, zinc and brass metal works, zinc and brass plating, viscose rayon yarn and fiber production. The WHO recommended the maximum acceptable concentration of zinc in drinking water as 5.0 mg/l.

Among others, contaminants of the environement persistent organic pollutants (POPs), which are organic compounds resistant to environmental degradation through chemical, biological, and photolitic properties, should be mentioned. In this group aldrin, chlordane, dichlorodiphenyltrichloroethane (DDT), dieldrin, endrin, hexachlorobenzene, mirex, polychlorinated biphenyls, dibenzo-*p*-dioxins, dibenzofurans and toxaphene are listed.

The serious health problems are also connected with perchlorates. They have been released into the environment during the fabrication and demilitarization of weaponry. Perchlorates in groundwater have also been associated with natural sources. The exposure to perchlorates has been implicated in thyroid hormone and neuro development disruption. The established reference dose of 0.0007 mg/kg/day which translated into a Drinking Water Equivalent Level (DWEL) of 24.5 μg/l [5].

Industrial waste materials can be classified as toxic, ignitable, corrosive or reactive industrial wastes. According to the EPA classification, industrial wastes can be also classified into three categories:

- as a source-specific waste that is mainly generated from specific industries such as wood preserving, lead smelting and petroleum refining,
- as a common waste such as the waste from manufacturing and industrial processes including degreasing operations, spent solvents, landfills and ink formulation waste,
- commercial chemical products including some pesticides, creosote and other commercial chemicals [12,13].

Table 2. Possitive and negative effects of selected metals

Metal	Positive effect	Negative effect
Al	—	Chronic exposure may occur in the workplace, contact with aluminum in foods and in water; from cookware and soft drink cans; from consuming items with high levels of aluminum (e.g., antacids, buffered aspirin, or treated drinking water; or even by using nasal sprays, toothpaste, and antiperspirants) Citric acid (e.g., in orange juice) may increase aluminum levels by its leaching activity. Symptoms of aluminum toxicity include memory loss, learning difficulty, loss of coordination, disorientation, mental confusion, colic, heartburn, flatulence, and headaches.
As	Arsenic is one of microelements indispensable for functioning of living organisms taking part in the synthesis of proteins and hemoglobin [7-9]. Once ingested, arsenic is metabolized by methylation and excreted in urine. The dietary intake of inorganic arsenic has been estimated to be 1-20 μg/day. However, the knowledge about nutritional factors affecting individual variation in methylation is limited.	Arsenic compounds show poor genotoxicity and mutagenicity. However, a very consistent effect of arsenic compounds is being a comutagen, possibly by inhibiting the repair of DNA lesions produced by genotoxic agents such as UV. Symptoms of arsenic poisoning include nausea or vomiting, abdominal pain, garlic odour in breath, excessive salivation, headache, paralysis, kidney failure, progressive blindness, and mental impairment. The signs include mottled brown skin, hyperkeratosis of palms and soles, cutis edema, transverse striate leukonychia, perforation of nasal septum, eyelid edema, coryza, limb paralysis and reduced deep tendon reflexes. Mental symptoms include apathy, dementia and anorexia.
Be	—	Beryllium can damage lungs. Small amounts of beryllium can cause immune or inflammatory reaction, whereas long exposure causes inflammatory reaction called granulomas. Long periods of exposure to beryllium have been reported to cause cancer in laboratory animals.
Cd	Cadmium may interfere with the metallothionein ability to regulate zinc and copper concentrations in the body.	Cadmium and its compounds are carcinogenic for humans. In general, Cd compounds are weakly mutagenic. The mechanism by which Cd may be genotoxic is by indirectly inducing oxidative stress in cells as a result of inhibition of antioxidant enzymes and depletion of antioxidant molecules. The symptoms of acute cadmium exposure are nausea, vomiting, abdominal pain, and breathing difficulty. The symptoms of chronic exposure could include alopecia, anemia, arthritis, learning disorders, migraines, growth impairment, emphysema, osteoporosis,

Metal	Positive effect	Negative effect
		loss of taste and smell, poor appetite, and cardiovascular disease. Chronic exposure to cadmium can result in chronic obstructive lung disease, renal disease, and fragile bones
Co	Cobalt is beneficial for humans because it is part of vitamin B_{12}, which is essential to maintain human health. Cobalt (0.16-1.0 mg cobalt/kg of body weight) has also been used as a treatment for anemia. It also increases red blood cell production in healthy people, but only at very high exposure levels. Cobalt is also essential for the health of various animals (cattle, cows and sheep).	Serious effects on the lungs, including asthma, pneumonia, and wheezing, have been found in people exposed to 0.005 mg cobalt/m^3. People exposed to 0.007 mg cobalt/m^3 have also developed allergies to cobalt that resulted in asthma and skin rashes. In the case of radiation from radioactive cobalt the damage of cells is observed. The amount of damage depends on the amount of radiation, related to the amount of activity in the radioactive material and the length of exposition time.
Cr	Chromium, an essential trace element required for normal carbohydrate, protein and fat metabolism, may improve impaired glucose tolerance, decrease elevated blood lipid concentrations and result in weight loss and improved body composition. Chromium has been suggested to potentiate the action of insulin, possibly by increasing insulin binding, insulin binding receptor number, improving insulin internalization and increasing insulin sensitivity [10].	IARC has determined that chromium(VI) compounds are carcinogenic to humans. Water-soluble chromate compounds are responsible for mutagenicity, and chromosomal damage and were found to be very active. The reduction of Cr(VI) to Cr(III), and the binding of Cr(III) to DNA are thought to be involved in its mutagenesis and genotoxicity, can yield a wide variety of DNA lesions, DNA-DNA crosslinks, DNA-protein crosslinks. The other most common health problems involve the respiratory tract. These health effects include irritation of the lining of the nose, runny nose, and breathing problems (asthma, cough, shortness of breath, wheezing). Allergies to chromium compounds, which can cause breathing difficulties and skin rashes, are also common.
Cu	Copper is an essential metal for living systems and is found in an assortment of enzymes, including superoxide dismutase (SOD), ferroxidases, and cytochrome oxidase. Its transport is highly regulated, and various metals, including zinc, cadmium, and molybdenum, have been reported to interfere with its transport or availability in biological systems.	Exposure to higher doses of copper is harmful. Long-term exposure to copper dust can irritate nose, mouth, and eyes, and cause headaches, dizziness, nausea, and diarrhea. Drinking of water containing higher than normal levels of copper may cause nausea, vomiting, stomach cramps, or diarrhea. Intentionally high intakes of copper can cause liver and kidney damage and even death. It is not known whether copper can cause cancer in humans.

Table 2. Continued

Metal	Positive effect	Negative effect
Hg	—	Mercury toxicity has been linked to dental fillings. The symptoms include metallic taste in the mouth, excess salivation, gingivitis, tremors, as well as stomach and kidney troubles. Mental symptoms include shyness, irritability, apathy and depression, psychosis, mental deterioration, and anorexia. Mercury, at high levels, may damage the brain, kidneys, and developing fetus. Effects on brain functioning may result in irritability, shyness, tremors, changes in vision or hearing, and memory problems. Short-term exposure to high levels of metallic mercury vapors may cause effects including lung damage, nausea, vomiting, diarrhea, increase in blood pressure or heart rate, skin rashes, and eye irritation.
Mn	Manganese is a trace mineral that helps to promote the proper function of the pituitary gland as well as aiding in the proper absorption of carbohydrates and protein. While only a small amount of the mineral is required in order to maintain a healthy balance, the absence of this trace amount can have severe repercussions such as cuts and scratches.	The most common health problems in the case of exposition to high levels of manganese involve the nervous system. The inhalation of a large quantity of dust or fumes containing manganese may cause irritation of the lungs which could lead to pneumonia.
Ni	—	The most common harmful health effect of nickel in humans is an allergic reaction. Approximately 10-20% of the population is sensitive to nickel. The most serious harmful health effects from exposure to nickel are chronic bronchitis, reduced lung function, and cancer of the lung and nasal sinus. Exposure to high levels of nickel compounds may also result in cancer. Ni compounds induce oxidative stress, genomic instability, chromosome damage, cell transformation accompanied by global deregulation of gene expression
Pb	—	Because of size and charge similarities, lead can substitute for calcium contained in bones. Children are especially susceptible to lead because developing skeletal systems require high calcium levels. It may cause nephrotoxicity, neurotoxicity, and hypertension.

Metal	Positive effect	Negative effect
		The other symptoms include combinations of gastrointestinal complaints, fatigue, hemolytic anemia, abdominal pain, nausea, weight loss, headache, weakness, convulsions, irritability, impotence, loss of libido, depression etc. Mental symptoms include restlessness, insomnia, irritability, confusion, excitement, delusions, and disturbing dreams. Lead is toxic to humans by interaction with the sulfhydryl group of proteins, resulting in disruption of the metabolism and biological activities of many proteins
Se	Selenium is an essential trace element with antioxidative, antimutagenic, antiviral and anticarcinogenic properties. Deficiencies can lead to immune dysfunction and cardiomyopathy [11].	Exposure to selenium causes dizziness, fatigue, and irritation of mucous membranes. In extreme cases, collection of fluid in the lungs (pulmonary edema) and severe bronchitis occur. Ingestion of selenium over long periods causes brittle hair and deformed nails can develop.
Sb	—	In small doses, antimony causes headache, dizziness and depression. Larger doses cause violent and frequent vomiting and will lead to death in a few days.
Zn	Zinc is an essential component of many metalloenzymes and transcription factors (it was found in more than 300 of them) that are involved in various cellular processes such as gene expression, signal transduction, transcription, and replication. However, cadmium, lead, and arsenic have been shown to affect the proper regulation of zinc. The average daily zinc intake through the diet in this country ranges from 5.2 to 16.2 mg.	Inhaling large amounts of zinc (as zinc dust or fumes from smelting or welding) can cause a specific short-term disease called metal fume fever. Ingestion of large doses of zinc (10-15 times higher than those of the RDA) even for a short time can cause stomach cramps, nausea, and vomiting. Ingestion of high levels of zinc for several months may cause anemia, damage the pancreas, and decrease levels of high-density lipoprotein (HDL) cholesterol.

Removal and extraction of dissolved heavy metal species from the aqueous medium are commonly accomplished by several well known methods including precipitation, coagulation, floatation, reverse osmosis, membrane filtration, solvent extraction, ion exchange etc. However, most of these processes are unacceptable owing to their high cost, low efficiency, disposal of sludge, inapplicability to a wide range of pollutants. Similarly, different methods have been used to destroy and reduce the levels of organic pollutants including treatment with activated sludge, chemical oxidation, biological oxidation, thermal degradation, ozonization, and photooxidation with ultraviolet radiation. However, the efficiency of many of these methods is reduced in the presence of heavy metal ions due to the complex formation between the metal and the organic species.

On the other hand, adsorption is one of the most recommended physico-chemical treatment processes that is well recognized as one of the highly efficient methods for recovery and treatment of heavy metals from their matrices, samples and aqueous solutions [12,13]. Additionally, methods based on the adsorption are simple, do not corrosive or toxic by-products and do not require special technological conditions. They enable application of cheap sorption materials like zeolites, silicas, active carbons or synthetic materials and organic-inorganic composites.

2. Organic-inorganic Composite Materials Based on the Silica – General Information

The possibility of combining the properties of organic and inorganic components in a unique composite material is an old challenge starting with the beginning of the industrial era. However, the concept of organic-inorganic composites emerged with the birth of soft inorganic chemistry processes, where mild synthetic conditions allow versatile access to chemically designed hybrid organic-inorganic materials. Sol-gel synthesis of hybrid materials in which the organic and the inorganic parts are linked by stable chemical bonds implies the use of coupling molecules, able to bind in a hydrolytically stable way both the organic and the inorganic moieties. During the last twenty years there have been published many papers and review papers on the materials prepared using the sol-gel technique considering in detail all aspects referring to their properties and synthesis particularly with respect to their specific applications.

Organic-inorganic hybrid materials synthesized at room temperature by a sol-gel route are particularly attractive for fabricating micro-optical elements, optical coatings and sorbents as they combine some advantages of organic polymers (ease of processing, elasticity and organic functionalities) with characteristics of inorganic species (hardness, thermal and chemical stability, transparency and a large range of refractive index) [14,15].

The notion – hybrid material is widely used for various systems such as largely ordered crystalline coordination polymers, amorphous sol-gel materials and those with or without interactions between organic and inorganic components. The most frequently used definition of hybrid material is based on the statement that the hybrid is the material prepared by mixing two or more components of significantly different utilization properties on the molecular level. In this system at least one of the components, organic or inorganic, is present with a size scaling from tenths to tens of nanometers. Components making up the hybrids could be molecules, oligomers or polymers, aggregates and even particles. Therefore, they are considered as nanocomposites or even composites at the molecular scale.

According to the Sanchez [16] depending on the nature of the links and interactions existing at the hybrid interface the organic-inorganic composites may be divided into two classes. *Class I* includes the materials where there are no covalent or iono-covalent bonds between the organic and inorganic components and only van der Waals, hydrogen bonding or electrostatic forces are present. *Class II* includes the materials in which at least parts of the organic and inorganic components are linked through strong covalent or iono-covalent bonds. These systems often display multiple structural levels on the scale ranging from nanometers to millimeters.

Various approaches have been adapted to prepare organic-inorganic composites materials. Among others, in-situ polymerization of tetraethoxysilane in several organic polymers as well as the properties of these materials has been studied [17-21]. Inorganic-organic hybrids of *class I* have been synthesized by hydrolysis and condensation of organosilates with metal alkoxides $Si(OR)_4$, $Ti(OR)_4$, $Zr(OR)_4$, or $Al(OR)_4$, where R is the organofunctional group. A good mutual dispersion of the two phases is ensured by the presence of hydrogen bonds between silanols groups (Si-OH) of the silica network and the carbonyl and amides functions present in the polymer. This homogeneity can be further improved by functionalizing the organic polymer by $Si(OR)_3$ groups which after hydrolysis increase the chemical affinity between the organic and inorganic components through covalent or partly covalent bonds. The resulting hybrids belong to class II.

As mentioned earlier, the research on hybrid organic-inorganic materials and combined adsorbents has undergone an explosive growth since the 1990s. Nowadays, these are one of the most important scientific developments of the 20th and 21st centuries. Their properties are related not only to the chemical nature of the organic and inorganic components but they rely on their synergy. Therefore, the interface between organic and inorganic components is of great importance. The work on these materials includes improvement and optimalization of their preparation in order to ensure the most favourable sorption properties compared to traditional ion exchangers and sorbents. Nowadays, they are processed as gels, monoliths, thin films, fibers, particles or powders. Various structure properties, low extent of pores filling, homogeneity, large specific surface area, identical size of pores and chanels, good sorption capacity, high stability, reproducibility and selectivity give these materials a high potential of applications in different kinds of fields such as environmental remediation, wastewater treatment, water softening, separation and preconcentration of metal ions, hydrometallurgy, nuclear separations, ion exchange fibers, redox system, electrodialysis, ion-selective electrodes, chromatography, biomolecular separations and heterogenic catalysis of organic synthesis [22-30]. Moreover, they can be applied as stores of gases (carbon dioxide, hydrogen etc.) or molecular sieves enabling selective adsorption of gases. Their lumino properties decide if they can be also used in new optical and telecommunication materials [31-35]. They are also used to improve durability, mechanical, chemical and heat resistance and adhesion of composites, adhesives, paints, medical and dental products, and weatherability, gloss and colour retention of coatings.

Composite materials initially obtained by chemists from the sol-gel scientific community nowadays are obtained by scientists coming from a variety of disciplines – polymer chemists, solid state chemists, catalysis and material researchers etc. Each of these research groups prepares hybrids using their own tools, specific disciplinary methods and their own raw materials. Therefore, different names have been given to these materials: CERAMERs, POLYCERAMs, ORMOSILs or ORMOCERSs [36,37].

As the chemical composition is one of the most important parameters since its variation leads to hybrid materials with distinctive physico-chemical behaviours and insight into different properties. In this chapter these parameters were taken into account as the basis for classification of organic-inorganic composite materials applied as ion exchangers and sorbents in environmental protection.

2.1. Organic-inorganic Composites Based on Silica – Synthesis and Aplications

Appearance in 1992 of the papers on ordered mesoporous silicas (OMS) from the series of Mobile Crystalline Material i.e. MCM-41 and then Santa Barbara Amorphous i.e. SBA-15 triggered great interest in them [38,39]. These materials can be synthesized by two major strategies including direct co-condensation and post-grafting methods. The co-condensation approach is an especially attractive alternative to the grafting due to the possibility of incorporation of high loadings of organic groups, cost-effectiveness and simplicity of the process.

Siloxane based hybrids can be easily synthesized due to the Si-C bonds which are covalent and therefore they are not broken upon the hydrolysis. The precursor of these compounds are organo-substituted silicic acid esters of the general formula $R'_nSi(OR)_{4-n}$, where R' is the organofunctional group. If R' is a simple non-hydrolyzable group bonded to silicon through a Si-C bond, it will have a network modifying effect. In the case of R' being for example methacryl, vinyl, or epoxide group it can react with itself or additional components, thus it is determined as a network former. Network modifiers and network formers can also introduce other properties such as mechanical, hydrophobic, electrochemical and optical [14]. The synthesis of mesoporous silicas was also possible when cheap and convenient silica sources, such as sodium silicate or heksafluorosilicic acid are used.

The alternative method of synthesis by co-condensation of tetraethyl orthosilicate (TEOS) and an organotriethoxysilane $RSi(OEt)_3$ in the presence of a structure directing agent was also proposed [40-43]. Supramolecular templating using amphiphilic molecules was first demonstrated in the early nineties of the last century, and, generally, surfactant and surfactant like aggregates act as templates or structure directing agents for the assembly of mesoporous metal oxides with the pore sizes in the 2-50 nm range. One of the first reports dealt with the preparation of micrometer sized mesoporous silicate spheres by modification of Stober's emulsion process.

Oligomeric templates were found to be suitable for the preparation of mesoporous spherical silica particles of the size useful for chromatographic applications, silica thin films, and spherical nanoparticles of vesicular, hexagonal, or cubic structures. Moreover, oligomeric surfactants were shown to be facile templates for the synthesis of functionalized mesoporous silicas by co-condensation of tetraethylorthosilicate with organotriethoxysilanes [44]. In this process the neutral diblock copolymers (such as polystyrene-polybutadiene and polybutadiene-poly(ethylene oxide)), as well as cationic and anionic diblock copolymers in acidic media were successfully used to template mesoporous silicas. Poly(alkylene oxide) triblock copolymers, such as poly(ethylene oxide)-poly(propylene oxide)-poly(ethylene oxide) (PEO-PPO-PEO) are also suitable templates for the synthesis of disordered mesoporous silicas. PEO-PPO-PEO and some other triblock copolymers were found to self-assemble with silicate species in acidic media to form periodic mesoporous silicas (PMOs), including hexagonal structures (referred to as SBA-15) with the ordered pores as large as 30 nm and cubic structures. In the case of MCM-41 materials regularly hexagonal arrays of cylindrical mesopores and changeable pore diameter between 1.5-20 nm were found.

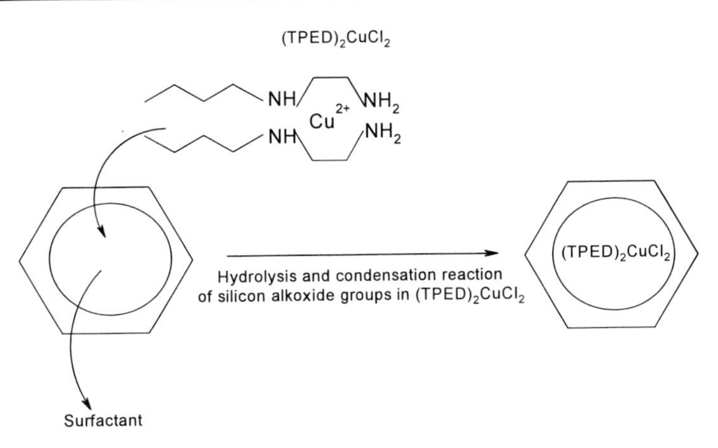

Figure 1. Removal of the surfactant template by the implantation of transition metal complexes [45]. Reproduced with permission from Lu, X. B. et al. *Chinese Chem. Lett.* 2002, *13*, 480-483. Copyright © 2002, Elsevier B.V.

The preparation of MCM-41 materials involves the process of high temperature calcination for destroying the surfactant template. This procedure can cause opening of mesopores, increasing the pore size and improvement in the uniformity of the disordered structure, or even in transformation from the disordered to hexagonal structure. It has been reported that the template molecules in MCM-41 could also be removed by extraction with the ethanol solution of an acid or salt. A new method, for rapid removal of the surfactant template from the as-synthesized MCM-41 material and simultaneous implantation of transition metal complexes into mesopores, using supercritical CO_2 modified with CH_2Cl_2/MOH mixture as a solvent was proposed [45]. In this process the complex of di-N-[3-(trimethoxysilyl)-propylethylene]diamine copper(II) dichloride i.e. $(TPED)_2CuCl_2$ was used (Figure 1).

It was shown that cationic silane modified MCM-41 materials exhibited increased sorption capacity for aromatic organic compounds such as trichlorethylene and tetrachloroethylene [46,47] and nonionic organic contaminants (NOCs) such as benzene, toluene, and phenol [48] from contaminated water. The examples of typical materials belong to ordered mesoporous silica (OMS) of MCM-41 and then SBA-15 series are presented in Table 3. The great interest of this class of materials also results from the possibility of changing the nature of the bridging organic groups.

As mentioned earlier, periodic mesoporous organosilicas (PMOs) obtaining in this way possess unique properties compared to the first generation of mesoporous silica materials [60-66]. Consequently, PMOs combine the characteristics of the ordered mesoporous silicas with the organic functional groups.

These incorporated bridging groups are a requisite for preparation of very promising materials for the design of highly selective adsorbents, catalysts, and sensing devices as well as for immobilization and encapsulation of biomolecules. PMOs are characterized by homogeneously dispersed organic moieties inside the channel walls with a maximum loading which does not block the pores, additionally the organic precursors are easily varied to obtain a wide range of materials.

Table 3. Typical materials belong to ordered mesoporous silica (OMS) of MCM-41 and SBA-15 series

Type	Synthesis	Characteristic	Application	Reference
MCM-41	Silica source: TEOS Template: CTMABr Functionalisation: grafting RNH_2 Functionalisation: grafting RSH	SH-MCM-41 Surface area 774 m^2/g Pore diameter 2.82 nm NH_2-MCM-41 Surface area 990 m^2/g Pore diameter 3.02 nm	Sorption of Ag(I) Sorption of Cu(II)	[49]
MCM-41	Silica source: TEOS Template: CTMABr Functionalisation: grafting RNH_2 Functionalisation: hydrolysis and ion exchanged with sodium nitrate to give COONa-MCM-41	NH_2-MCM-41 Surface area 750 m^2/g Pore diameter 2.92 nmCOONa-MCM-41 Surface area 679 m^2/g Pore diameter 2.85 nm	Sorption of Cr(VI) below pH 3.5. Sorption of Cu(II) at pH 5	[50]
MCM-41	Silica source: TBOS Template: CTMABr Functionalisation: 3-mercaptopropyltrimethoxysilane	wt% MPTS Surface area 874.3 m^2/g 10.0 wt% MPTS Surface area 1375 m^2/g 17.0 wt% MPTS Surface area 883 m^2/g 20.0 wt%GMS surface area 949.3 m^2/g	Sorption of Hg(II) in the presence of Ag(I)	[51]
SBA-15	Silica source: TEOS Template: CTMABr Functionalisation: N-[3-(triethoxysilyl)propyl]-4,5-dihydroimidazol	IMI-SBA-15 Surface area 161 m^2/g Pore diameter 5.8 nm	Sorption of Pt(II) Sorption of Pd(II)	[52]
SBA-15	Silica source: TEOS Template: CTMABr Functionalisation: polytiol	SH-SBA-15 Surface area 161 m^2/g Pore diameter 5.8 nm	Sorption of Hg(II) in the presence of Cu(II), Pb(II) and Cd(II)	[53]
SBA-15	Silica source: TEOS Template: CTMABr Functionalisation: grafting (G) and co-condensation (C) of aminopropyl (N), [2-aminoethylamino]-propyl (NN), [(2-aminoethylamino)-ethylamino]-propyl (NNN)	C-SBA-15-N Surface area 572 m^2/g Pore diameter 8.2 nm C-SBA-15-NN Surface area 508 m^2/g Pore diameter 8.3 nm C-SBA-15-NNN Surface area 477 m^2/g Pore diameter 8.1 nm G-SBA-15-N Surface area 565 m^2/g Pore diameter 9.0 nm G-SBA-15-NN Surface area 246 m^2/g Pore diameter 7.3 nm G-SBA-15-NNN Surface area 48 m^2/g Pore diameter – nm	Sorption of Cu(II), Ni(II), Pb(II), Cd(II), and Zn(II)	[54]

SBA-15	Silica source: TEOS Template: poly (alkaline oxide) triblock copolymer Functionalisation: mercaptopyrimidine	MPY-SBA-15 Surface area 330 m^2/g Pore diameter 54.7 nm	Sorption of Cd(II) and Pb(II)	[55]
SBA-15	Silica source: TEOS Template: poly (alkaline oxide) triblock copolymer Functionalisation: mercaptopyridine	MPY-SBA-15 Surface area 342 m^2/g Pore diameter 0.37 nm	Sorption of Cr(VI)	[56]
SBA-15	Silica source: TEOS Template: puronic 123 Au source: – $HAuCl_4$	GMS-1048 wt% GMS Surface area 1067 m^2/g 1.5 wt% GMS Surface area 822 m^2/g wt%GMS Surface area 818 m^2/g 2.5 wt%GMS Surface area 759 m^2/g 10 wt % GMS Surface area 643 m^2/g	Removal of trichloroethylene (the efficient catalyst for reducing air pollution)	[57]
SBA-1	Silica source: TEOS Template: CTMABr Functionalisation: 3-mercaptopropyltrimethoxysilane	M-SB-1 Surface area 930 m^2/g Pore size 2.3 nm	Sorption of Hg(II)	[58]
SBA-15	Silica source: TEOS Template: poly (alkaline oxide) triblock copolymer Functionalisation: 3-mercaptopropyl groups and Functionalisation: primary 3-aminopropyl groups	SB-15-SH Surface area 461 m^2/g SB-15-NH_2 Surface area 279 m^2/g	Sorption of Hg(II) in the presence of Cu(II), Zn(II), Cr(III) and Ni(II)	[59]

Preparation of PMOs involves the catalyzed hydrolysis of bis(trialkoxysilyl) organic precursors of the general formula $[(R'O)_3Si–R–Si(OR')_3]$ in the presence of a surfactant template. The presence of the organic moiety greatly influences the nature of the hydrolysis reaction. To achieve the desired surface properties for catalytic, adsorption and other related applications the multifunctional PMOs are also prepared by co-condensation of organosilane mixtures or by post-synthesis modification of PMOs. In these processes the single bridging groups such as methane, ethane, ethene, benzene, 2,5-dimethylbenzene, thiophene, *p*-xylene; large bridging groups such as biphenyl, isocyanurate (ICS) rings or two and three small bridging groups in the presence of nonionic surfactants are incorporated. The materials containing bis-silylated ethylene, methylene and ethane precursors show great stability to strongly acidic or basic conditions.

There are also some attempts to introduce large organic groups such as ferrocene (for electrochemical properties), diazo units (for nonlinear optical properties) or 1,4,8,11-tetraazacyclotetradecane units (for chelating of heavy metal ions) [67-69]. POMs containing tris[3-(trimethoxysilyl)propyl]isocyanurate (ICS) have been examined for adsorption of heavy metal ions such as mercury(II) and platinium(IV) nanoparticles [70-73]. In the case of more complicated bridging organic precursors, however, aqueous hydrolysis often results in significant cleavage, as well as a decrease in the degree of order in the material.

2.2. Organic-inorganic Composites Based on Silica – Functionalised Materials

Sol-gel derived composite materials have found numerous applications in the areas of chemistry, biochemistry, engineering, and material science [74-79]. In particular, the functionalized amorphous and mesoporous silica materials have attracted more attention for the application in removal and recovery of metal ions (mainly for the adsorption of Hg(II), Cr(VI) and Cd(II) (Table 3) adsorption of organics and dyes, radionuclides, in immobilization of enzymes or protein that all relate to the surface functional groups which can be appropriately modified. Modification of silica surface leads to the change in chemical composition of the surface. Surface can be modified either by physical treatment (thermal or hydrothermal) that leads to the change in the ratio of silanol and siloxane concentration of the silica surface or by chemical treatment that leads to the change in chemical characteristics of silica surface.

The modification of amorphous and mesoporous silicas by chemical treatment with organic ligands can be carried out by two conventional methods. Heterogeneous methods involve the reaction of surface hydroxyl groups with a commercial silylating agent, usually 3-chloropropylsilane that acts as a precursor for further immobilization of the molecule containing the donor atom, whereas in the homogeneous method, the ligand is pre-synthetized by the reaction of a commercial silylating agent with the molecule containing the donor atom. The resulting ligand is then allowed to react with the silanol groups of activated silica. Nowadays, the heterogeneous method has been used preferably to the homogeneous one.

Silica polyamine composites (SPCs) have been developed as one among many remediation technologies including organic polymer chelator resins and surface silanized silica gels [80,81]. SPCs are composed of a silica gel support covalently bonded to the linear or branched water soluble chelating polyamines. They are synthesized by the following steps: acid washing of the silica gel surface (1), humidification (2), silanization with a mixture of methyltrichlorosilane (MTCS) and chloropropyltrichlorosilane (CPTCS) (3), and finally the addition of a polyamine and a modifying ligand (4). The example of this approach is presented in Figure 2.

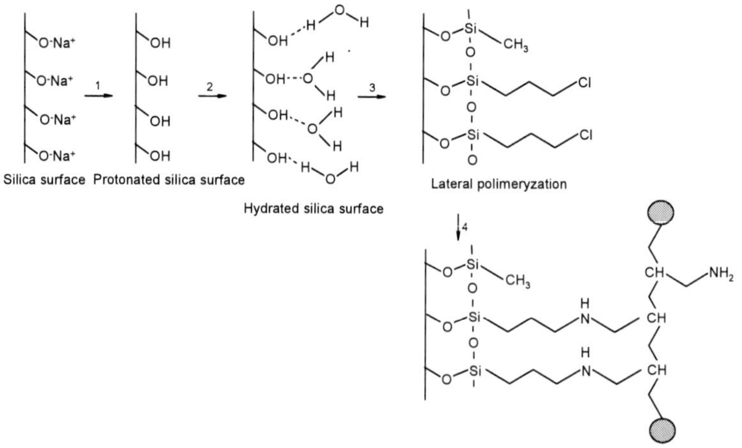

Figure 2. Synthesis of silica polyamine composite [80]. Reproduced with permission from Wong, Y. O. et al. *J. App. Polym. Sci.* 2010, *115*, 2855-2864. Copyright © 2010, Wiley Periodicals, Inc.

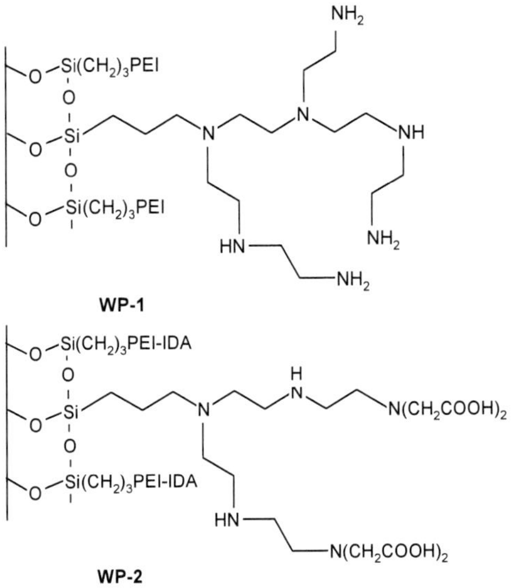

Figure 3. Structure of WP-1 and WP-2 [82]. Reproduced with permission from Beatty, S. T. et al. *Ind. Eng. Chem. Res.* 1999, *38*, 4402-4408. Copyright © 1999, American Chemical Society.

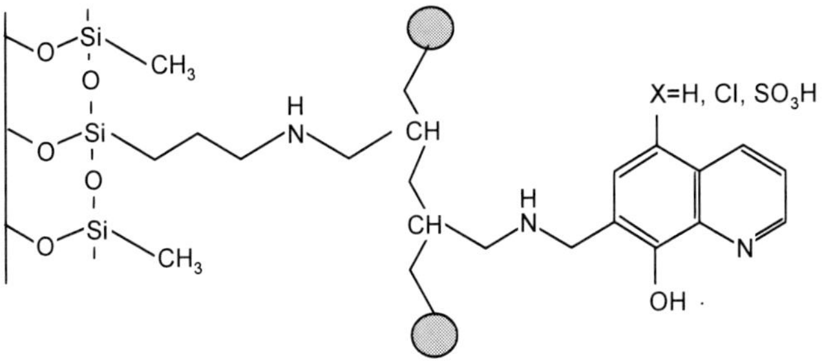

Figure 4. Covalent bonding of 5-X, 8-hydroxy quinoline ligand to BP-1 [80]. Reproduced with permission from Wong, Y. O. et al. *J. App. Polym. Sci.* 2010, *115*, 2855-2864. Copyright © 2010, Wiley Periodicals, Inc.

The group of Beatty and Allen [82-84] studied the synthesis of different silica polyamine composites, amog others, preparation, kinetics, and pH profile of a poly(ethyleneimine) (PEI)-silica composite material denoted as WP-1 (Figure 3). The sorption of Cu(II) ions on WP-1 proceeds according to the reaction:

$$Cu^{2+} + nR_xN^+H_y \rightleftarrows (R_xN^+H_{y-2})_n\, Cu^{2+} + 2H^+ \quad (1)$$

WP-1 exhibits better kinetics and capacity than the iminodiacetate chelating resin Amberlite IRC-718. Comparison of WP-1 and Amberlite IRC-718 in the presence of ethylenediaminetetraacetic acid (EDTA), reveals that WP-1 possesses a higher metal capacity than Amberlite IRC-718. The syntheses WP-2 (silica-PEI-CH_2-COOH) and WP-3 (silica-PEI-EtSH) were also reported and studied from the point of view of removal of low levels of Hg(II) and Pb(II) or separation of transition metal ions from the abandoned mine water.

In the paper by Wong [80] the synthesis of WP-4, SB-1 and CB-1 composites prepared by the Mannich reaction of the immobilized polymer amine groups in the poly(allylamine) composite BP-1 with 8-hydroxyquinoline, 5-sulfonic acid-8-hydroxy quinoline, and 5-chloro-8-hydroxy quinoline, respectively was presented (Figure 4).

These modified composites showed significant selectivity for Fe(III) ions over Cu(II) and Ni(II) (selectivity factors 2,8; 2.4 and 4.3 for WP-4, SB-1 and CB-1 in the case of Cu(II) and Fe(III) mixture as well as 45,36 and 45 for WP-4, SB-1 and CB-1 in the case of Ni(II) and Fe(III) mixture). They also showed selectivity for Ga(III) over Al(III). A larger ionic radius for Ga(III) (76 pm) than Al(III) (67.5 pm) can explain the selectivity of Ga(III) over Al(III) on WP-4. In the case of WP-4 and CB-1 each metal is bound to one ligand, whereas for SB-1 two oxine ligands are bound to each metal ion by electrostatic interaction with two neighboring anionic sulfonate groups (Figure 5).

The maximum sorption capacity of Fe(III) at pH 2 was equal to 24 mg/g and similar to that obtained for Ni(II) with the EDTA ligand on silica polyamine composites [85]. This type of composite was obtained by the modification of silica gel, MCM-41 and aluminum with EDTA and diethylenetriaminepentaacetic acid (DTPA) (Figure 6). It was found that the sorbents modified with DTPA used for the sorption of Cu(II) from aqueous solution possess lower chelating ability than those modified with EDTA.

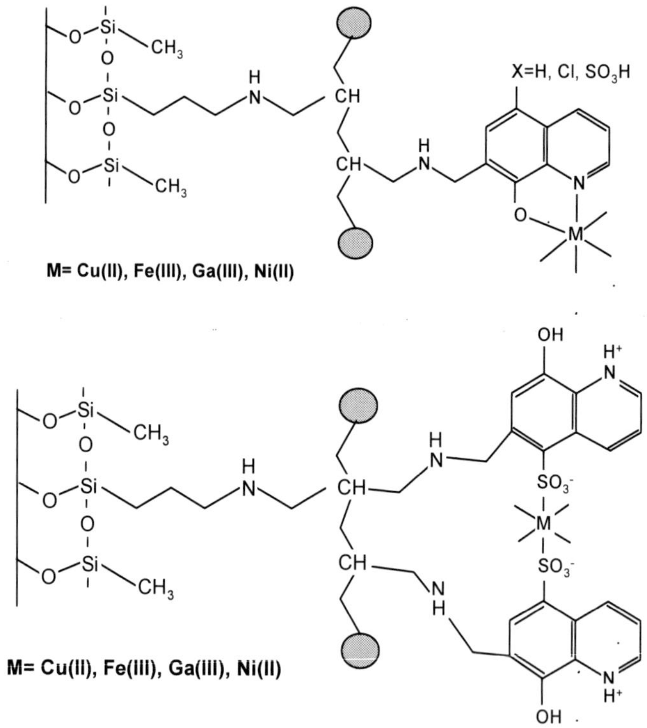

Figure 5. Metal coordination to WP-4 (a) and CB-1 as well as SB-1 (b) [80]. Reproduced with permission from Wong, Y. O. et al. *J. App. Polym. Sci.* 2010, *115*, 2855-2864. Copyright © 2010, Wiley Periodicals, Inc.

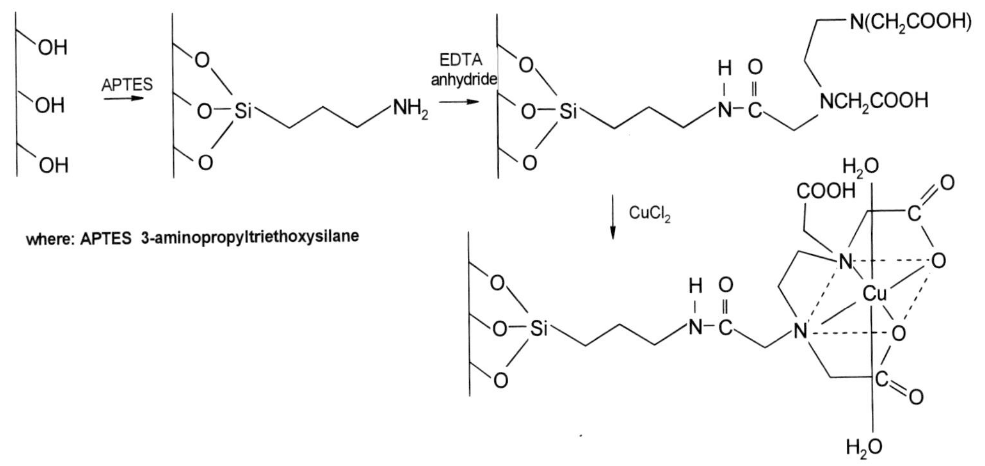

where: APTES 3-aminopropyltriethoxysilane

Figure 6. Synthesis of the adsorbent modified by EDTA [85]. Reproduced with permission from Shiraishi, Y. et al. *Ind. Eng. Chem. Res.* 2002, *41*, 5065-5070. Copyright © 2002, American Chemical Society.

Metal-ligand complexes are effectively formed on the adsorbents in the region of 1-2 pH but scarcely at pH >4. Moreover, the silica gel modified with EDTA (silicaEDTA) was shown to be the most effective adsorbent for the extraction of metals and to form specific metal-EDTA complexes, with the stability constants in the order Cu(II) > Ni(II) > VO^{2+} > Zn(II) > Co(II) > Mn(II). The metals coordinated on the silicaEDTA could be desorbed completely by contacting with 1 mol/l aqueous HCl solution.

The report concerning the sensing and ion exchange properties of a luminescent organic-inorganic hybrid supermicroporous material LHMM-1, containing a fluorophore diimine moiety covalently grafted inside the framework under strongly acidic pH conditions was done by Chandra et al. [86]. Based on the ion exchange efficiency and distribution coefficient (K_d) for Fe(III), Zn(II), Cd(II) and Hg(II), it was found that LHMM-1 showed very high cation removal efficiency (89-98%). Formation of a chelate-type complex due to the coordination of metal cations with phenolic OH and imine-N donors could be responsible for the high K_d values for all these metal cations. It should be mentioned that the synthetic strategy for hybrid frameworks through a non-templating pathway would be widely applicable for the synthesis of a large variety of related porous materials, which could have many other potential applications.

Molecular imprinting is an attractive method for the preparation of selective sorbents. Recently, molecularly imprinted sol gel materials (MISGMs) have been extensively studied. MISGMs are obtained in the process of conventional sol-gel synthesis and incorporation of the template molecules into inorganic or inorganic-organic networks. After removal of the template, molecular cavities with distinct pore size, shape, or chemical functionality remain in the cross-linked host.

The most important type of the imprinted sol-gel material synthesis is surface imprinting on mesoporous material supports. It was used to increase metal loading capacities and selectivity, as compared to those sorbents using conventional silica. A novel double imprinting method was proposed by Dai et al. [87] which combines molecular imprinting and micelle templating synthesis. This type of hybrid materials was obtained by self-hydrolysis, self-condensation, and co-condensation of the cross-linking agent such as tetraethoxysilicate and the functional precursor 3-(2-aminoethylamino)-propyltrimethoxysilane in an alkaline solution followed by gelation. The obtained material was used for sorption of Cd(II) and Zn(II) mixture. The largest relative selectivity coefficient between Cd(II) and Zn(II) was found to be over 200 [88].

Chemically modified silica materials with amine functionalities are an attractive candidate not only to adsorb heavy metal ions but also they have high CO_2 capacity due to the combination of their high surface areas and the chemical specificity of amines. Alkylamine groups can be grafted on to silica by the condensation of appropriate aminoalkoxysilanes. For example, silica gels and mesoporous silica grafted with 3-aminopropyl groups were reported to be able to adsorb CO_2 reversibly. Ethylenediamine (EDA)-modified silica gel and mesoporous silica were also reported to adsorb CO_2. EDA-modified SBA-15 adsorbs around 20 mg/g of CO_2 at 25 °C, which is comparable to the adsorption capacity of aminopropyl-modified mesoporous silica [89]. Such sorbents can offer the advantages of lower toxicity and corrosiveness and can potentially reduce equipment cost and energy consumption.

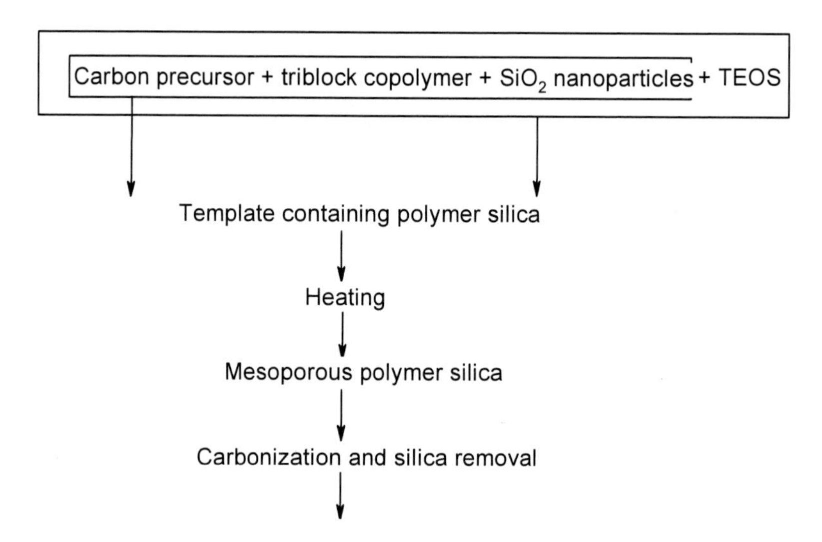

Figure 7. The scheme of the soft-templating synthesis of mesoporous carbons in the presence of SiO_2 and without and with TEOS [102]. Reproduced with permission from Jaroniec, M. et al. *Carbon* 2009, *47*, 3034-3040. Copyright © 2009, Elsevier B.V.

3. Carbon Composites

As mentioned earlier, among numerous achievements in the area of OMS there was the hard templating synthesis of ordered mesoporous carbons (OMCs) [39]. However, the soft-templating synthesis of OMCs [90,91] which involves polymerization of carbon precursors e.g. formaldehyde and phenol derivatives in the hydrophilic domains of block copolymer template turned out to be a very promising way for large-scale preparation of OMCs. To this end Liang et al. [90] used polystyrene-block-poly(4-vinylpyridine) (PS-P4VP) as a structure directing agent - soft template for the preparation of OMC films. Then the use of the thermally decomposable poly(ethylene oxide)-poly(propylene oxide)-poly(ethylene oxide) triblock copolymers (PEO-PPO-PEO) was reported [90-93]. Additionally, the use of solvents such as water and ethanol makes this synthesis strategy attractive from the green chemistry viewpoint. Since that time the soft-templating synthesis of mesoporous carbons has attracted a lot of attention. The obtained carbons possess large and uniform mesopores with the geometries analogous to the well known siliceous materials such as SBA-15, SBA-16, which distinguish them from such mesoporous carbon structures as CMK-3 and CMK-5. They are inverse replicas of the silica templates used. They consist of ordered carbon pipes instead of carbon rods formed due to incomplete filling of the ordered mesopores of SBA-15 or SBA-16 with the carbon precursor. Also CMK-1 carbons templated by cubic structural mesoporous molecular sieves, MCM-48 have been extensively studied [94].

In terms of microporosity, the soft-templated carbons derived from phenolic resins alone without additional treatment possess a relatively small amount of pores (below 2 nm). Therefore, the attempts to obtain new ordered mesoporous carbons include mainly searching for new materials used as soft templates (the use of copolymers with a different ratio of PEO-PPO blocks, nonionic difunctional block polymer F108, triblock copolymer Pluronic P123, PPO_{53}-PEO_{136}-PPO_{53}, diblock copolymer poly(ethylene oxide)-b-poly(methyl methacrylate) (PEO-b-PMMA) and others) and using new carbon precursors (resorcinol, phloroglucinol, triethyl orthoacetate (TEOA) together with resorcinol-formaldehyde or mixtures of these polymerized with formaldehyde) which can be easily carbonised. Also the other carbon precursors such as furfuryl alcohol, pitch, polyacrylonitrile (PAN) and polyaniline have been employed [95-100]. In the papers by Górka et al. for example the triblock copolymer Pluronic F127 – PEO_{106}-PPO_{70}-PEO_{106} was used [101,102].

Mesoporous carbons with incorporated inorganic species or carbon-inorganic nanocomposites are often required in such fields as adsorption, catalysis and separations. Inorganic species can be incorporated into mesoporous carbons during their synthesis or by post-synthesis infiltration. Generally, in comparison to hard templating the soft-templating synthesis is better suited for the incorporation of inorganic nanoparticles as well as for their formation from suitable inorganic precursors. The addition of for example iridium, titanium and nickel components during the self-assembly synthesis of carbons resulted in the formation of ordered mesostructures with inorganic nanoparticles. Moreover, the addition of inorganic nanoparticles such as silica and alumina colloids into the mixture of block copolymer template and carbon precursors does not disintegrate the self-assembly process and results in mesoporous carbons with incorporated nanoparticles [102]. The scheme of the soft-templating synthesis of mesoporous carbons in the presence of colloidal silica (to create spherical mesopores) without and with TEOS to increase both microporosity and surface area as well as to improve pore connection and accessibility to inorganic nanoparticles is presented in Figure 7.

It was found that the total pore volume and mesopore volume of the carbons obtained after silica dissolution depend strongly on the amount of the silica nanoparticles, whereas the observed microporosity is formed due to the removal of the TEOS-generated silica. In a further attempt to improve the conductive properties of these carbons, doping with a heteroatom such as nitrogen was used. Sulfur-doped carbons obtained by polymerization and carbonization of thiophene in SBA-15 have also been studied [103].

3.1. Functionalised Carbons

To enhance the capacity of activated carbons (ACs) to adsorb heavy metal ions from different wastewaters, many functional and surface modification methods have been introduced. In the literature, there are some examples of active carbon composites used for removal of heavy metal ions. Among them, metal, metal oxides, zeolite, anionic surfactants and chitosan modified ACs can be effective adsorbents for heavy metal ions.

In the paper by Yanagisawa et al. as well as Suzuki et al. [104,105], the result of obtaining AC-Mg composite by magnesium addition onto the activated carbon surface by two oxygen atoms bridge was presented. The obtained composite shows greater adsorption of Zn(II) than the outgassed active carbon (AC-OG). The difference between AC-OG and AC-Mg was compared for Cd(II) adsorption, even though the physical properties of AC-OG such as BET surface area, meso- and macropore surface area, number of basic sites were better than those for AC-Mg. It was found that the obtained AC-Mg composite could effectively adsorb Cd(II) ions from the low Cd(II) concentration and was saturated at 0.09 mmol/g. The temperature does not affect Cd(II) adsorption to AC-Mg. Cd(II) ions adsorption would be principally progressed not only by ion exchange with Mg(II) but also by electrostatic interaction with Cπ electrons on the graphite layer. The ion exchange on AC-Mg was estimated to be as follows:

$$\text{(Carbon at graphite edge-O}^-)_2\text{Mg}^{2+} + \text{Cd}^{2+} \rightleftarrows \text{(Carbon at graphite edge-O}^-)_2\text{Cd}^{2+} + \text{Mg}^{2+} \quad (2)$$

The iron-containing mesoporous carbon (IMC) was prepared from a silica template (MCM-48) for effective removal of arsenic from drinking water. It was obtained by in situ polymerization of resorcinol with formaldehyde (RF) in the porous structure of the silica template in a basic aqueous solution, followed by carbonization in an inert atmosphere and template removal. The adsorption of As(III) and As(V) from drinking water by IMC followed the Langmuir adsorption model, with the adsorption maximum reaching 5.96 mg As(III)/g and 5.15 mg As(V)/g [106]. Besides metal ions, metal (hydr)oxides have also been combined with such support materials as granulated activated carbon (GAC) or polymeric resin to fabricate adsorbent media capable of simultaneous removal of contaminants.

The effect of zinc oxide loading to the granular activated carbon (GAC) on Pb(II) adsorption from aqueous solution was also studied and compared to zinc oxide and oxidized activated carbon [107]. Cu(II), Cd(II) and nitrobenzene were used as the reference adsorbates to investigate the adsorption. The heavy metal adsorption was improved by both the zinc oxide loading and the oxidation of activated carbon. In contrast, the adsorption of nitrobenzene was considerably reduced by the oxidation, and slightly decreased by the zinc oxide loading. The GAC-ZnO composite was found to be effectively used for the Pb(II) adsorption whereas only a part of surface functional groups was used for the zinc oxide particles and the oxidized activated carbon. As follows from the experimental results, the surface functional groups responsible for the Pb(II) adsorption on the GAC-ZnO were considered to be hydroxyl groups.

Composite prepared by precipitating of FeO onto granular activated carbon (GAC-FeO) was also evaluated as an adsorbent for the removal and recovery of Cu(II) from water [108].

Cu(II) adsorption capacity increased compared to the adsorption onto uncoated GAC. It was found that increasing the amount of FeO from 1.4 mg/g GAC to 5 mg/g GAC, the sorption at low concentration solutions was significantly effective in the column process. 100 μg Cu(II)/l solution was reduced to 3 μg/l through 1000 BV (bed volumes) processed.

As(V) removal by an iron oxide-impregnated activated carbon was modelled by Vaughan and Reed [109] using the surface complexation model (SCM) approach given by Dzombak and Morel [110] and Reed and Matsumoto [111]. Daus et al. [112,113] studied As(III) and As(V) adsorption onto activated carbon (AC), zirconium-loaded activated carbon (Zr-AC), absorptionsmittel 3 (AM3), zero-valent iron (Fe0), and granulated iron hydroxide (GIH). The sorption of arsenate followed the sequence Zr-AC ~ GIH = AM3 > Fe0 > AC. A different order was found for arsenite (AC~Zr-AC = AM3 = GIH = Fe0). Activated carbon was pretreated with iron-salt solutions to improve arsenic adsorption [114]. The salt type and concentration, pH, and treatment time were examined to improve removal capacity. Peraniemi et al. [115] used zirconium-loaded activated carbon and successfully removed arsenic, selenium, and mercury.

Composite of activated carbon-zeolite formed by the zeolite growth on porous carbon supports can possess the bifunctional properties of both carbon and zeolite [116,117]. For example uptake experiments for Ni(II), Cu(II), Cd(II) and Pb(II) were performed on the composite of activated carbon and zeolite A-X [117]. The relative selectivity for various ions was Pb(II) > Cu(II) > Cd(II) > Ni(II) with the equilibrium uptake capacities of 2.65 mmol/g, 1.72 mmol/g, 1.44 mmol/g and 1.20 mmol/g, respectively.

Functionalized carbons can also be used to remove perchlorate. Chen et al. [118] conducted research on GAC impregnated with ammonia to enhance its capacity for perchlorate removal. According to the authors, perchlorate adsorption by impregnated GAC is directly related to the surface chemical properties of the tailored carbon. Na et al. [119] evaluated GAC beds preloaded with iron-oxalic acid for perchlorate adsorption.

3.2. Functionalized Carbon Nanotubes

Compared to other forms of carbon, carbon nanotubes (CNTs) have large surface to volume and surface to weight ratios, high thermal and electrical conductivity, and chemical inertness at high temperatures. Some attempts are also made to functionalize the CNTs with transition metals (TM). In this case CNTs can act as templates for growing metal nanowires and nanoparticles [120-123]. Transition metals-carbon nanotubes composites are formed by two main routes: nanoparticles are grown or deposited directly through van der Waals interactions onto the CNTs surface by using precursors during a reduction process (electroless deposition and electrodeposition of metal nanoparticles or nanowires on CNTs in the absence of any additional reducing agent have also been established) or the chemically-modified nanoparticles of metals are linked to CNT surfaces through organic fragments, or the nanoparticles are linked to chemically-modified CNT surfaces. An alternative route was proposed by An and Turner [124,125]. It involves substitutional doping of CNTs, for example by boron and nitrogen, obtaining B-CNT and N-CNT templates for the formation of TM-CNT. It was found that the TM nanostructures demonstrate particularly strong adsorption on B-CNTs, leading to the self-assembly of TM atoms with well-defined covalent bonds with the substrate and TM/B-CNT composites exhibit high stability and a variety of electronic and magnetic properties.

The obtained W/B-CNT was used as sorption material for H_2, O_2, CO, CO_2, NO, NO_2, H_2O, CH_3OH and NH_3. Among the mentioned gases CO, NO and NO_2 are common gaseous pollutants released from the combustion processes, CO_2 is a product of oxidation of CO as well as a major greenhouse gas.

It was found that the W/B-CNT composite is highly catalytically active exposed to O_2 and NO and moderately adsorbs H_2O, CH_3OH, and NH_3. Therefore, it can find application in catalysis, electrocatalysis, and gas sensing. CNT–Fe-Al_2O_3 composites have been prepared by hot pressing [126]. CNT–metal-oxide composites have been extruded at high temperatures [127]. Carbon nanotubes (CNTs) were also grown on the surface of Al_2O_3. They were used for adsorption of Pb(II), Cu(II) and Cd(II) from the water solutions and the results were compared with active carbon powders, commercial carbon nanotubes, and Al_2O_3 particles. It was found that the adsorption capacity of CNTs on Al_2O_3 is superior to the above-mentioned adsorbents and the preference order of adsorption on the composite can be as follows: Pb(II) > Cu(II) > Cd(II). The calculated saturation amounts adsorbed by 1 g of CNTs on Al_2O_3 are 67.11 mg/g, 26.59 mg/g, and 8.89 mg/g for Pb(II), Cu(II) and Cd(II), respectively [128].

4. Polymer Supported Nanoparticles (PSNs)

Ion exchangers have been widely used in wastewaters treatment and most of them were commercially available. They are characterized by excellent hydraulic properties, good sorption capacities, high selectivity towards certain metal ions (especially in the case of chelating ion exchangers), but on the other hand, they also possess some disadvantages such as poor thermal stability and mechanical strength, depression of their sorption potential at high temperatures and radiation doses. However, inorganic sorbents do not possess these drawbacks as they are charcterized by high surface area, high thermal and radiation stability, favourable sorption, good oxidation-reduction kinetics and magnetic properties but they are quite difficult to obtain in the form suitable for the column operation e.g. of granular beads.

The above-mentioned limitations are overcome by introducing the composite organic-inorganic resins. This new class of organic-inorganic materials denoted as the polymer supported nanoparticles (PSNs) combining advantages of porous polymeric materials and nanoscaled inorganic particles was introduced.

As a base of these materials phenolic, phenolsulfonic formaldehyde (PSF), sulfonated polystyrene-divinylbenzene (S-PS-DVB), hydrophilic polyacrylic hydrazid (PAH) and polyacrylonitrile (PAN) resins were used. Also, a wide variety of nanoscale inorganic particles (NIPs) were adopted in preparation of composite resins. In this group hydrated Fe(III) oxides particles (HFO), Mn(IV) oxides, Fe_3O_4 (magnetite) crystals and elemental Zn^0 or Fe^0 should be mentioned.

Polymer supported nanoparticles are scientifically analogous to SIRs (Solvent Impregnated Resins) introduced for the first time by Warshawsky [129]. Both of them integrate two separate phases for enhanced separation (in both porous polymeric substances are used to retain active ingredients i.e. extractants in the case of SIR and inorganic nanoparticles in the case of PSNs). According to Cumbal et al. [130] two classes of PSNs can be distinguished: Magnetically active polymeric particles (MAPPs) and hybrid ion exchangers (HIXs).

4.1. Magnetically Active Polymeric Particles (MAPPs)

Recently, there has been increased interest in the use of magnetically active polymeric particles (MAPPs). Magnetic beads typically consist of a polymer with embedded superparamagnetic nanoparticles of magnetic iron oxides. They are commercially available with diameters ranging from 50 nm to 10 microns and a large range of surface functionalizations. They were mainly used in the fields of biosorption and bioseparation in the heterogeneous system with its easy control and effectiveness (protein and biomolecule separations, enzyme immobilization, sell sorting, protein adsorption, nucleic acid detachment and drug delivery) [131-135]. Magnetic polymer beads with the uniform size distribution, in particular those with a diameter in the range from 1 to 10 μm, have been recognized as one of the most promising materials in the field of biomedicine. Magnetic polymer beads, which were used in both *in-vivo* and *in-vitro* biomedicines, should be biocompatible and nontoxic, and also should provide a large number of surface functional groups suitable for biomolecule sorption and separation. Magnetically active polymeric particles are also used for inorganic and organic ligands, chlorophenols and pesticides removal as well as for wastewater treatment, desalination, colour imaging and information storage [136-141].

Until now, various techniques have been available to prepare magnetically active polymer beads, including emulsion polymerisation, miniemulsion polymerization, dispersion polymerization, suspension polymerization and crosslinking, as well as two-step swelling method [142-146]. The most promising method is the spraying suspension polymerization (SSP) technique. Different hydrophobic magnetic polymer beads have been prepared by the SSP technique, including magnetic polystyrene beads, magnetic poly(methyl methacrylate) [poly(MMA)] beads and magnetic poly(methylmethacrylate-divinylbenzene-glycidylmethacrylate) [poly(MMA-DVB-GMA)] beads [147-150].

Magnetically active polymeric materials are prepared by dispersing of nanocrystals of magnetite, Fe_3O_4 within the polymeric beads, in order to obtain a magnetically active sorbent from the non-magnetic polymeric one. The magnetite part facilitates their separation from treated solutions. During synthesis the presence of extremely low concentration of dissolved oxygen to oxidize Fe(II) to Fe_3O_4 without forming non-magnetic $Fe(OH)_3$ according to the following reaction should be ensured.

$$4Fe_3O_4 + O_2 + 18H_2O \rightleftarrows 12Fe(OH)_3\downarrow \quad (3)$$

Also pH value of the process is very important. Additionally, particles must not be too small because at less than 1-2 nm, their magnetic properties vanish. However, if the particles are too large, magnetic interactions predominate, and the particles agglomerate.

The process of dispersing of magnetite can be applied to different types of polymers such as those containing sulfonic (Purolite C-145), diphosphonic/sulfonic/carboxylic (Diphonix Resin®), bis(2-pyridylmethyl) amine also known as the bis-picolylamine (Dowex M 4195, formerly known as DOW3N or XFS 4195), quaternary ammonium e.g. trimethylammonium

(Amberlite IRA 900) as well as iminodiacetic (Amberlite IRC-718) functional groups. For sulfonic Purolite C-145, the process of preparing MAPPs can be as follows:

Loading of Fe(II) at acidic pH

$$R\text{-}SO_3^-H^+ + Fe^{2+} \rightleftarrows R\text{-}(SO_3^-)_2Fe^{2+} + 2H^+ \quad (4)$$

Magnetite formation mechanism i.e. slow oxidation of Fe(II) into crystalline Fe_3O_4 at alkaline pH:

$$R\text{-}(SO_3^-)_2Fe^{2+} + 2Na^+ \rightleftarrows 2R\text{-}SO_3^-Na^+ + Fe^{2+} \quad (5)$$

$$3Fe^{2+} + 0.5O_2 + 3H_2O \rightleftarrows Fe_3O_4\downarrow + 6H^+ \quad (6)$$

$$Fe^{2+} + 2OH^- \rightleftarrows Fe(OH)_2\downarrow \quad (7)$$

$$Fe(OH)_2\downarrow + 0.5O_2 \rightleftarrows Fe_3O_4\downarrow + 3H_2O \quad (8)$$

where: R- resin matrix.

Finnally, washing with a solvent of a low dielectric constant and drying of residual material.

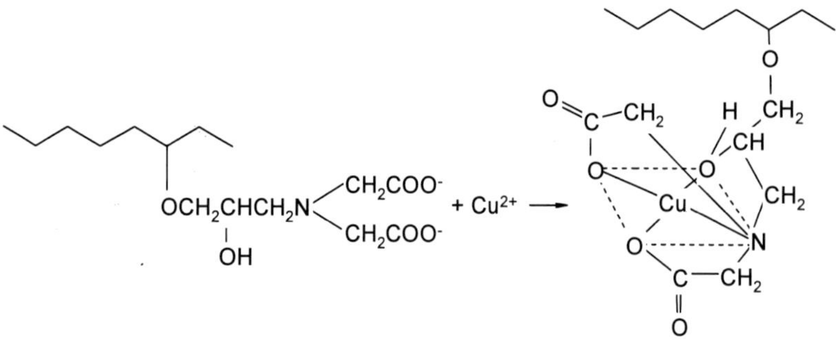

Fiure 8. Cu(II) sorption on the M-PVAC-IDA resin [160]. Reproduced with permission from Tseng, J. Y. et al. *J. Hazard. Mater.* 2009, *171*, 370-377. Copyright © 2009, Elsevier B.V.

The magnetic properties of these sorbents are greatly influenced by kind of the functional groups. They were found to change in the following order: sulfonic > diphosphonic/sulfonic/carboxylic > bis-picolylamine > iminodiacetic.

Moreover, different surface functional groups such as poly(ethylene glycol), polymethylmethacrylate, poly(vinyl alcohol), polyacrylamide (PAM) and alginate can be tailored to use specific applications [151-155]. They are also used for water treatment, heavy metal ions, radionuclides, anionic ligands, and metalloids removal.

According to Bajpai et al. high gradient magnetic separation (HGMS) technology has been successfully exploited for the separation of metal ions [156-158]. The major advantage of using magnetic particles as sorbent is that the sorbent particles can be easily removed from the adsorption system. They use the strongly cationic resin Seralite SRC-120 a potential sorbent for removal of Cu(II) ions. Cu(II) ions removal proceeds accordingly to the mechanism of ion-exchange process between Na(I) ions present within the sorbent particles and Cu(II) ions present in the sorbate solution. In addition, within the resin matrix, Cu(II) ions may also co-ordinate with the electron rich oxygen of magnetite nanoparticles.

In the research group of Tseng [159,160] the super-paramagnetic Fe_3O_4 was prepared by the chemical co-precipitation method and coated with polyvinyl acetate (PVAC) to form magnetic-PVAC (M-PVAC) by the suspension polymerization with vinyl acetate (VAC). The sequential treatments of alcoholysis, epoxide activation, and coupling of iminodiacetic acid (IDA) further yielded the magnetic-polyvinyl alcohol (M-PVOH), magnetic-polyvinyl propenepoxide (M-PVEP), and magnetic-polyvinyl acetate-IDA (M-PVAC-IDA), respectively. In the papers [159,160] the syntheses of magnetic-polyvinyl acetate-iminodiacetic acid (M-PVAC-IDA) and the chemical modification to enhance the capability of adsorbing Cu(II) from the printed circuit board were described. The results indicated that the adsorption capacities are higher at higher pH values and Cu(II) concentrations of solution. This may be attributed to the formation of surface complexes between Cu(II) ion and the chelating ligands of M-PVAC-IDA. As for the present system at pH 4.5 with C_0 of 5-1000 mg/l, about 0.012-1.334 mmol/g of Cu(II) ion were chelated on M-PVAC-IDA. It was found that due to the super-paramagnetic property of M-PVAC-IDA, it is convenient to recover the magnetic adsorbent from the treated solution. Magnetically separated adsorbents can be regenerated by using EDTA and reused. In this case EDTA reacts with Cu(II) ion forming the octahedral complex more stable than the square planar complex of M-PVAC-IDA with Cu(II) ion (Figure 8), suggesting that the EDTA is a good regenerator for removal of Cu(II) from a M-PVAC-IDA.

The analogous M-PVAC-IDA resin was obtained by Yang et al. [135] by replacing hydroxyl groups by amino ones and then with their carboxymethylation by sodium salt of chloroacetic acid. This reaction resulted in the transformation of amino groups to iminodiacetic acid (IDA) groups, which is a tridentate chelator and can coordinate with such metal ions as Cu(II).

Some synthesized polymer adsorbents such as methylmethacrylate-glycidyl methacrylate copolymer coupling with IDA (MMA-GMA-IDA) and magnetic poly(methylmethacrylate) incorporating with ethylenediamine (M-PMMA-EDA) [161] have also been applied for the adsorption of Cu(II) ions. For MMA-GMA-IDA, about 0.0127 mmol/g of Cu(II) was chelated on the IDA-bound particles. Cu(II) adsorbed on M-PMMA-EDA was about 0.09-0.201 mmol/g for C_0 in the range of 5-60 mg/l.

Adsorption of heavy metal ions on the unmodified MPMMA resin was equal to 0.0042 mmol/g for Pb(II), 0.0046 mmol/g for Cd(II) and 0.0029 mmol/g for Hg(II). EDA-incorporation significantly increased the heavy metal adsorption to 0.186 mmol/g for Pb(II), 0.162 mmol/g for Cd(II), and 0.150 mmol/g for Hg(II). It was also found that the competitive adsorption capacities were equal to 0.079.8 mmol/g for Cu(II), 0.0587 mmol/g for Pb(II), 0.0524 mmol/g for Cd(II), and 0.0453 mmol/g for Hg(II). Based on the obtained results the observed affinity order in adsorption was found to be: Cu(II) > Pb(II) > Cd(II) > Hg(II). The adsorption of heavy metal ions increases with the increasing pH and reached a plateau value at around pH 5.0. The optimal pH range for heavy metal ions removal was shown to be from 5.0 to 8.0. Desorption was achieved using 0.1 M HNO_3. The maximum elution value was as high as 98%.

Table 4. Comparison of the sorption capacity of M-GMA-DVB-IDA and GMA-DVB-IDA [162]

Metal ion	Sorption capacity [mmol/g]	
	M-GMA-DVB-IDA	GMA-DVB-IDA
Pb(II)	2.30	1.00
Cd(II)	2.00	1.30
Zn(II)	1.65	1.10
Ca(II)	1.60	1.00
Mg(II)	1.48	0.90

Different types of glycidyl methacrylate resins with different hydrophilic/hydrophobic cross-linkers, different functionalities, and different embedded metal oxides were also prepared and investigated towards the removal of various metal ions from aqueous solutions. One of them is magnetic resin of glycidyl methacrylate with iminodiacetate functionality (M-GMA-DVB-IDA) [162,163]. The preparation process requires successive treatment of magnetic glycidyl methacrylate-divinylbenzene resin with tetraethylenepentamine and potassium chloroacetate. The obtained resin shows sorption behaviour towards Pb(II), Cd(II), Zn(II), Ca(II) and Mg(II) comparable to the commercial chelating resin with the iminodiacetate functional groups such as Lewatit TP 207. For M-GMA-DVB-IDA composite the affinity order was as follows: Pb(II) > Cd(II) > Zn(II) > Ca(II) > Mg(II). However, the obtained resin was characterised by faster kinetics than Lewatit TP 207. For example, after 20 min, 90% of the maximum uptake of the metal ions was obtained whereas 25% was obtained for Lewatit TP 207. This may be attributed to the increase in the active site concentration of the magnetic resin through: immobilization with pentamine moiety and formation of stretched resin film over the magnetite particles which give higher exposed active sites for interaction with the metal ions. The comparison of the results obtained by the authors is presented in Table 4. As an effective regenerating agent 0.2 M EDTA was used. The efficiency of regeneration achieved 96% for five adsorption/desorption cycles.

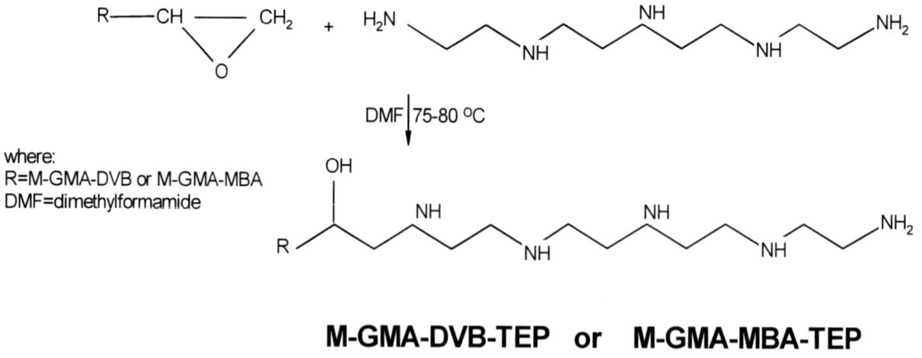

M-GMA-DVB-TEP or M-GMA-MBA-TEP

Figure 9. Synthesis of the M-GMA-DVB-TEP and M-GMA-MBA-TEP resins [164]. Reproduced with permission from Atia, A. A. et al. *J. Hazard. Mater.* 2008, *155*, 100-108. Copyright © 2008, Elsevier B.V.

The analogous glycidyl methacrylate (GMA) resins were synthesized through polymerization in the presence of divinylbenzene (DVB) or methylenebisacrylamide (MBA) as the hydrophobic or hydrophilic crosslinkers, respectively and in the presence of suspended magnetite particles. They were immobilized with tetraethylenepentamine (TEP) to give the amino resins, denoted as M-GMA-DVB-TEP and M-GMA-MBA-TEP (Figure 9) [164].

There was studied the uptake behaviour of these two resins towards molybdate anions and the following uptake capacities 4.24 mmol/g and 6.18 mmol/g Mo(VI) were obtained at pH 2. It was found that depending on the total metal concentration and pH, molybdenum ions in aqueous solutions occur as $[Mo_7O_{21}(OH)_3]^{3-}$, $[Mo_7O_{22}(OH)_2]^{4-}$, $[Mo_7O_{23}(OH)_5]^{5-}$ and $[Mo_7O_{24}]^{6-}$. The ion $[MoO_4]^{2-}$ occurs at pH >6. This anion can be protonated at low pH values forming H_2MoO_4. At pH <6 and concentrations above 1×10^{-4} M, there is found the $[Mo_7O_{24}]^{6-}$ ion singly, doubly or even triply protonated exists. The maximum uptake of molybdate at pH 2 proceeds through the ion exchange mechanism:

$$6RNH^+Cl^- + [Mo_7O_{24}]^{6-} \rightleftarrows [RNH]_6[Mo_7O_{24}]^{6-} + 6Cl^- \quad (9)$$

where: R- resin matrix.

At pH <2, the decrease in adsorption capacity may be attributed to the competitive adsorption between Cl^- anions (with higher concentration) and molybdate anion $[Mo_7O_{24}]^{6-}$. The observed decrease in the adsorption capacity above pH 2 may be attributed to the partial deprotonation of the amino groups.

Similar to poly(glycidyl methacrylate) (PGMA) microparticles, the novel magnetic poly(2-hydroxyethyl methacrylate) (PHEMA) offers a number of properties important for sorption and bioseparation such as biocompatibility, hydrophilicity, and protein-friendly surfaces with low nonspecific interaction. Its great advantage is their easy chemical modification [165]. For example, the magnetic poly(2-hydroxyethylmethacrylate) (MPHEMA) adsorbent with covalently attached Alizarin Red was used for the removal of Al(III) ions from drinking and dialysis water. The maximum Al(III) adsorption was 0.722 mmol/g at pH 5.0. Non-specific Al(III) adsorption was about 0.023 mmol/g polymer under the same conditions. High desorption ratios (98%) were achieved by using 0.1 M HNO_3 [166].

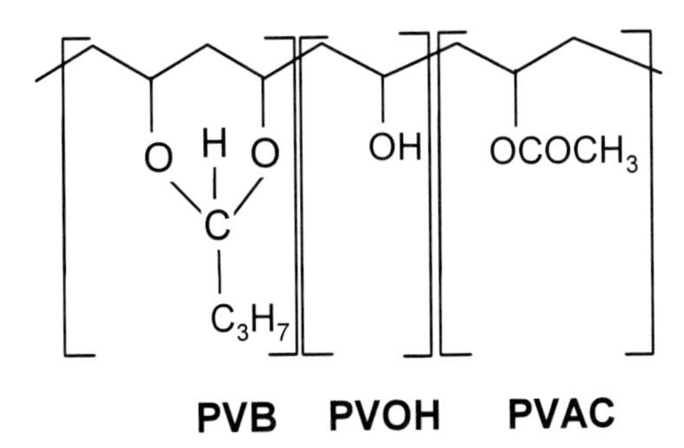

Figure 10. Structure of the resin containing vinylbutyral, vinyl alcohol and vinyl acetate [167]. Reproduced with permission from Denizli, A. et al. *J. Chromatogr. A* 1998, *793*, 47-56. Copyright © 1998, Elsevier B.V.

In the paper by Denizli et al. [167] the procedure of obtaining a new magnetic sorbent based on polyvinylbutyral (PVB) microbeads with Cibacron Blue F3GA dye covalently attached was presented. The structure of the resin containing polyvinylbutyral (PVB), polyvinyl alcohol (PVOH) and polyvinyl acetate (PVAC) is presented in Figure 10.

In the case of Cibacron Blue F3GA dye binding sites for heavy metal ions are $-SO_3H$, $=NH$, $-NH_2$ and triazine groups. In the case of sorption of Cu(II), Cd(II) and Pb(II) from aqueous media the obtained adsorption rates were high and the maximum adsorption capacities for the above mentioned ions onto the Cibacron Blue F3GA-attached to microbeads from their single solutions were 9.6 mg/g for Cu(II), 38.6 mg/g for Cd(II), 86.2 mg/g for Pb(II). When the heavy metal ions competed, the amounts of adsorption were 6.5 mg/g for Cu(II), 20.3 mg/g for Cd(II) and 41.6 mg/g for Pb(II). The affinity order of ions was Pb(II) > Cd(II) > Cu(II). Pb(II) was adsorbed much faster than Cd(II) and Cu(II) due to much higher affinity of the ligand. Desorption of heavy metal ions was achieved using 0.1 M HNO_3. Denizli et al. [168] also used copolymer of ethyleneglycoldimethacrylate (EGDMA) and hydroxyethylmethacrylate (HEMA) with attached Alkali Blue 6B in which the maximum adsorption capacities were 2.3 mg/g for Cu(II), 5.5 mg/g for Cd(II) and 125 mg/g for Pb(II).

As follows from the literature data, the resins containing amine and mercaptan as chelating groups show good selectivity for noble metal ions. However, only a few magnetic resins, consisting of amino and mercapto groups that selectively sorb noble metal ions have been reported. In the paper by Zhang and Pang [169] a series of magnetic chelating resins containing amino and mercapto groups was presented and their adsorption properties for Au(III), Pd(II), Ag(I), Pt(IV), Hg(II), Cu(II), Pb(II), and Zn(II) were investigated. The resins were prepared by the suspended polymerization of 2-chloroethoxymethyl thiirane (CEMT) and diamines H_2NRNH_2, where R = C_4H_8, C_6H_{12}, $C_{10}H_{20}$ (Figure 11). The obtained resins were denoted as TBD, THD and TDD.

The experimental results show that the magnetic resins have high affinity for Hg(II) and noble metal ions as well as Cu(II) and less for Pb(II) and Zn(II). The adsorptive ability of the magnetic resins increases in the order TBD > THD > TDD, corresponding to the N and S contents. It was found that due to the sorption Au(III) was reduced to Au(0). In the competitive adsorption, the resins predominantly adsorbed Hg(II) or Pd(II) in the presence of Cu(II), Zn(II), and Mg(II). Desorption of Pd(II) loaded on the resins was sufficient using the 2 M HCl solution containing 1% thiourea.

The glycidyl methacrylate-methylenebis-acrylamide resins (GMA-MBA) modified or not by iron oxide Fe_2O_3 with amino and thiol functional groups were obtained by Donia et al. [170]. Their structures are presented in Figure 12.

They were investigated towards Ag(I) sorption. It was found that the uptake of Ag(I) reached 0.91 mmol/g for GMA-MBA resin with amine functional groups (GMA-MBA-A), 1.46 mmol/g for magnetic-GMA-MBA resin with amine functional groups (M-GMA-MBA-A), 1.5 mmol/g for GMA-MBA resin with thiol functional groups (GMA-MBA-SH) and 2.37 mmol/g for M-GMA-MBA resin with thiol functional groups (M-GMA-MBA-SH). The higher and fast uptake of Ag(I) by M-GMA-MBA-A and M-GMA-MBA-SH resins relative to GMA-MBA-A and GMA-MBA-SH may be attributed to the resin films extending over the Fe_2O_3 particles. It was found that resins with thiol groups (GMA-MBA-SH and M-GMA-MBA-SH) give higher K_d values than amine containing ones (GMA-MBA-A and M-GMA-MBA-A).

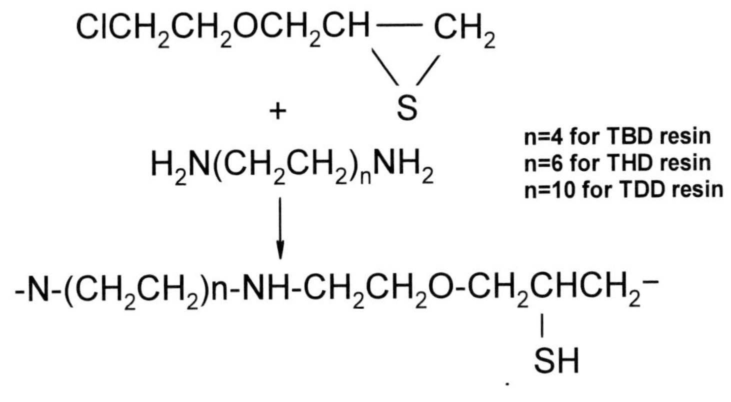

$ClCH_2CH_2OCH_2CH—CH_2$ (S)

\+

$H_2N(CH_2CH_2)_nNH_2$

n=4 for TBD resin
n=6 for THD resin
n=10 for TDD resin

$-N-(CH_2CH_2)n-NH-CH_2CH_2O-CH_2CHCH_2-$ (SH)

Figure 11. Scheme of the synthesis of magnetic chelating resins with thiol groups [169]. Reproduced with permission from Zhang, C. C. et al. *J. Appl. Polym. Sci.* 2001, *82*, 1587-1592. Copyright © 2001, Wiley Periodicals, Inc.

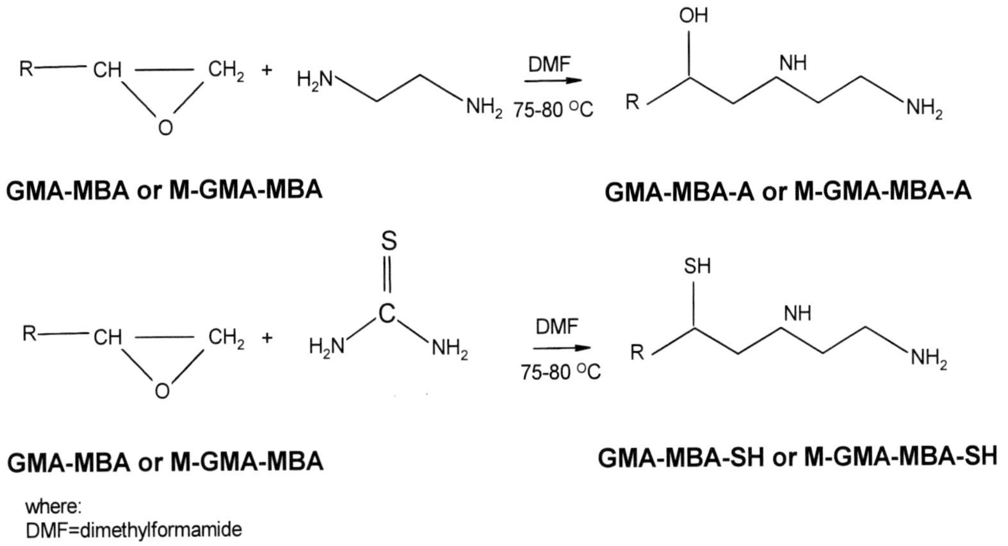

R—CH—CH_2 (O) + H_2N—NH_2 → (DMF, 75-80 °C) R—(OH)—NH—NH_2

GMA-MBA or M-GMA-MBA **GMA-MBA-A or M-GMA-MBA-A**

R—CH—CH_2 (O) + H_2N—C(=S)—NH_2 → (DMF, 75-80 °C) R—(SH)—NH—NH_2

GMA-MBA or M-GMA-MBA **GMA-MBA-SH or M-GMA-MBA-SH**

where:
DMF=dimethylformamide

Figure 12. Scheme of the synthesis of magnetic chelating resins with amine and thiol groups [170]. Reproduced with permission from Donia, A. M. et al. *Sep. Pur. Technol.* 2006, *48*, 281-287. Copyright © 2006, Elsevier B.V.

For all resins, the uptake increases with the increasing temperature, which could be explained by the increase in the active sites available for interaction with Ag(I) as well as higher dehydration of these ions and the active sites, consequently more favourable their interaction. As for the desorption process, almost 98% of Ag(I) was eluted from GMA-MBA-A, M-GMA-MBA-A, GMA-MBA-SH, M-GMA-MBA-SH using acidified thiourea with 1 M HNO_3.

Loading with $FeCl_3$ (solution pH < 2.0)

$$3R{-}SO_3^- H^+ + FeCl_3 \longrightarrow (R{-}SO_3^-)_3Fe^{3+} + 3HCl$$

Desorption and simulataneous hydroxide precipitation

$$(R{-}SO_3^-)_3Fe^{3+} + 3Na^+ \longrightarrow 3R{-}SO_3^- Na^+ + Fe^{3+}$$

$$Fe^{3+}_{(aq)} + 3OH^- \longrightarrow Fe(OH)_{3(s)}$$

Alcohol wash and mild thermal treatment

$$Fe(OH)_{3(s)} \xrightarrow[12\ h]{50\text{-}60\ ^{o}C} FeOOH + \text{amorphous HFO particles}$$

Figure 13. Scheme of the synthesis of HIXs [189]. Reproduced with permission from DeMarco, M. J. et al. *Water Res.* 2003, *37*, 164-176. Copyright © 2003, Elsevier B.V.

The same authors also obtained the Co_3O_4-containing resin with magnetic properties for the sorption of Hg(II), Cu(II) and Ni(II) [171]. It was prepared by the polymerization of GMA using DVB in the presence of finely divided Co_3O_4 powder. The investigated resin exhibits larger uptake values for Hg(II) ions equal to 2.1 mmol/g and Cu(II) 2.0 mmol/g relative to the resin without Co_3O_4 (1.75 mmol/g for Hg(II) and 1.24 mmol/g for Cu(II)). Therefore, the affinity series was found to be as follows: Hg(II) ~ Cu(II) >> Ni(II). The process is also pH dependent. At natural pH the metal salt-resin complex formation occurs, whereas the observed lower uptake in the acidic medium may be attributed to the partial protonation of the amino groups of the resin. At pH <2, no appreciable uptake was detected for these metal ions indicating complete protonation of the active sites.

4.2. Hybrid Ion Exchangers (HIXs)

It is well-recognized that oxides of polyvalent metals, namely, Al(III), Fe(III), Ti(IV), Mn(IV) and Zr(IV), as well as $M(HPO_4)_2$, where M = Zr, Ti and Sn exhibit ligand sorption properties through formation of inner-sphere complexes. For example iron(III) oxides have high sorption affinity towards both As(V) or As(III) ions. In this case the species are selectively bound to the oxide surface by the coordinate bonding. It has also been proved to exhibit specific removal of other anionic pollutants, such as selenite, or phosphate from contaminated waters. These hydrous metal oxides (HMO) are usually present as fine or ultrafine particles and can not be employed for direct use in fixed-bed or any flow-through systems due to the excessive pressure drop and poor mechanical strength.

To overcome these problems hybrid sorbents were designed by impregnating metal oxides onto the conventional porous materials – alginate, activated carbon, zeolite, diatomite, cellulose, sand and porous polymer. Chitoson also has been used as host materials to support HMO nanoparticles [172-187].

At the beginning of 2000 the simple chemical-thermal technique to produce a hybrid polymeric-inorganic sorbents selective towards both As(III) and As(V) compounds and at the same time compatible with the fixed-bed column operation was proposed by Cumbal and SenGupta [175,188]. The hybrid sorbents consisting of spherical macroporous polymeric beads with uniformly dispersed hydrated Fe(III) oxide proved especially effective for As(III) and As(V) sorption. Introducing the hydrated Fe(III) oxide (HFO) into the network made it possible to obtain excellent mechanical strength and hydraulic properties of polymer beads. HIXs were prepared according to the procedure presented in paper [189]. Their particles were found to vary between 20 and 100 nm. In the case of their synthesis, the first step of preparation includes loading of Fe(III) onto the functional groups of ion exchanger (for example sulfonic one), then desorption of Fe(III) and simultaneous precipitation of Fe(III) hydroxides within the gel and pore phase of the exchanger through passage of a solution containing both NaCl and NaOH, and finally rinsing and washing with a ethanol-water solution followed by a mild thermal treatment. High concentration of the functional groups allowed high and fairly uniform loading of HFO particles (approximately 9-12% of Fe by mass) within the polymeric beads (Figure 13).

In the case of column sorption experiments using the influent containing 0.050 mg/l of As(V), 129 mg/l of SO_4^{2-}, 70 mg/l of Cl^- and 100 mg/l of HCO_3^- at pH 7.1, it was found that anions broke through almost immediately after the start of the column run while 10 mg/l As(V) was observed after about 4000 bed volumes (BV). It is worth mentioning that no commercially available anion exchangers are capable of simultaneous removal of these ions. Both protonated and electrically neutral iron oxides selectively sorb As(V) and As(III) species through the Lewis acid-base interaction, i.e., formation of inner sphere complexes. For protonated and electrically neutral iron oxides the protonation constants in the resin phase are as follows [190]:

$$FeOH_2^+ \rightleftarrows H^+ + FeOH \quad pK_{a,1} = 5.3 \tag{10}$$

$$FeOH \rightleftarrows H^+ + FeO^- \quad pK_{a,2} = 8.8 \tag{11}$$

These oxides are capable of removing As(III,V) ($FeOH_2^+$ possesses high affinity for $H_2AsO_4^-$, whereas FeOH for $HAsO_2$) and heavy metal ions (FeO^- is especially selective for Zn^{2+}).

On the contrary, competing Cl^-, SO_4^{2-} and HCO_3^- ions form only outer sphere complexes with only weak affinities toward HFO and they are characterized by smaller selectivity. The regeneration process can be achieved by 10% NaOH solution in less than 10 BV. Subsequent rinsing sparged treated water with carbon dioxide for less than 10 BV is sufficient to make the column ready for the sorption cycle. It should be mentioned that the total iron content of HIX following the complete cycle of removal, regeneration and rinsing remained practically unchanged.

In the paper by Greenleaf et al., [191] synthesis of HFO nanoparticles incorporated into the structure of a cylindrical polymeric ion exchange fiber for removal of As(III,V) ions was described. By dispersing HFO nanoparticles into ion exchange fibers, excellent hydraulic properties and arsenic selectivity can be effectively intergrated. The obtained composite shows sorption capacities equal to 15 mg As(V)/g.

Iron nanomaterials commonly used for arsenic removal can be also impregnated into the pores of perchlorate selective resins. Cumbal and SenGupta [188,192] studied the applicability of hybrid ion exchange resins capable of simultaneous removal of arsenate and perchlorate. Crosby et al. [193] found that ferric hydroxide nanoparticles prepared in aqueous environments can have spherical or rod like shapes depending on whether they are derived from Fe^{2+} or Fe^{3+} ions. Therefore, six commercially available, perchlorate selective ion exchange resins Amberlite PWA2, Purolite A-530E, SIR-110, CalRes 2103, Purolite A-520E and SIR-100 were impregnated with ferric hydroxide nanoparticles using different chemical treatments. In this process the following stages can be distinquished:

Fe(II) oxidation by $KMnO_4$ to obtain ferric hydroxide nanoparticles

$$5Fe^{2+} + MnO_4^- + 6H_2O \rightleftarrows 5FeOOH_{(s)} + 7H^+ + Mn^{2+} \quad (12)$$

Fe(III) precipitation under alkaline conditions

$$Fe^{3+} + 3OH^- \rightleftarrows FeOOH_{(s)} + H_2O \quad (13)$$

For the studied HIX resin, As(V) removal was generally quite uniform except for Purolite-A520. However, commercial resins were characterized with the highest separation factors for perchlorate ions in relation to its HIX counterparts, which was probably due to blocking of the functional groups by the iron hydroxide nanoparticles. This could be a result of the blocking of these groups by iron hydroxide nanoparticles and oxidation of the polymer ion-exchange material by the permanganate.

Gu et al. [194,195] used the tetrachloroferrate ($FeCl_4^-$) ions to displace perchlorate from resins. They observed that perchlorate was desorbed rapidly, with close to 100% recovery and with a very small volume of water required for regeneration. This technology was applied in a pilot scale reactor in an Air Force base. Pan et al. [196,197] proposed a novel process to immobilize nanoparticulate hydrated ferric oxide (HFO) within a macroporous anion exchange resin D-201, and obtained a hybrid adsorbent HFO-201 for enhanced phosphate removal from aqueous systems. The process was based on the following steps:

Preparation of $FeCl_3$-HCl solution and loading of the column packed with the D-201 ion exchanger by $FeCl_4^-$ anions onto D-201 due to its much lower hydration energy than that of chloride:

$$R\text{-}N^+(CH_3)_3Cl^- + FeCl_4^- \rightleftarrows R\text{-}N^+(CH_3)_3\ FeCl_4^- + Cl^- \quad (14)$$

Decomposition and simultaneous in situ precipitatation of preloaded $FeCl_4^-$ ions on D-201 onto the inner-pore surface of D-201 by rinsing the D-201 column with NaCl-NaOH solution:

$$R\text{-}N^+(CH_3)_3\ FeCl_4^- + OH^- \rightleftarrows R\text{-}N^+(CH_3)_3\ OH^- + Fe(OH)_3 + Cl^- \quad (15)$$

Thermal treatment of the resin to obtain the hybrid adsorbent HFO-201:

$$Fe(OH)_3 \rightleftarrows \text{Nanosized HFO} \quad (16)$$

HFO-201 possesses two types of adsorption sites for phosphate removal, the ammonium groups bound to the D-201 matrix and the loaded HFO nanoparticles. The coexisting sulfate anions strongly compete for ammonium groups, which bind phosphates through electrostatic interaction. For the other competing ions the following affinity series was found: sulfate > chloride > hydrocarbonate. HFO-201 exhibits a little larger capacity for phosphate than the commercially available phosphate-specific adsorbent ArsenX^{np}, which possesses a similar structure of HFO-201. HFO-201 was amenable to efficient in situ regeneration with a binary NaOH-NaCl solution for the repeated use without any significant capacity loss.

Hydrated Fe(III) oxide nanoparticles were also dispersed within the ion exchanger fibres containing covalently attached quaternary ammonium functional groups. The obtained hybrid anion exchange fiber (HAIX-F) was used for perchlorate and arsenate removal in the presence of sulfate(VI) and chloride ions. It was found that hydrophobic perchlorate was selectively bound to the quaternary ammonium functional groups while arsenate removal proceeded by ligand exchange [198]. For selective desorption of these ions sodium hydroxide and sodium chloride solutions were effective. These unique properties of HAIX-F can be extended to pentachlorophenate, uranium(VI) oxoanion, benzenesulfonate removal with simultaneously present ligands such as phosphate, oxalate or vanadate. In the case of hydrous Mn(IV) oxides (HMOs), they have been impregnated onto polystyrene anion exchangers for anionic arsenic removal from waters [196,197]. In the literature there are not much data for efficient removal of heavy metal ions.

A new hybrid adsorbent of this group was obtained by impregnating nanosized hydrous manganese dioxide (HMO) within a porous polystyrene cation exchanger D-001 (HMO-001). Its preparation procedure [199] consists of the following steps:

Mn(II) loading onto a polymeric cation exchanger D-001:

$$2R\text{-}SO_3^-Na^+ + Mn^{2+} \rightleftarrows (R\text{-}SO_3^-)_2Mn^{2+} + 2Na^+ \quad (17)$$

Oxidation of the loaded Mn(II) as hydrous manganese dioxide (HMO). Sodium oxychloride (NaOCl) was used as the oxidant and the reaction can be as follows:

$$(R\text{-}SO_3^-)_2Mn^{2+} + NaOCl + NaOH \rightleftarrows R\text{-}SO_3Na + HMO + NaCl \quad (18)$$

Dispersion of the HMO particles within the inner surface of D-001 and finally the step of the thermal treatment. This type of hybrid material was used for selective sorption of Pb(II) ions. The process is pH dependent:

$$\text{-}MnOH + Pb^{2+} \rightleftarrows \text{-}Mn\text{-}O\text{-}Pb^+ + H^+ \quad (19)$$

$$\text{-}Mn(OH)_2 + Pb^{2+} \rightleftarrows \text{-}Mn\text{-}(O)_2\text{-}Pb + 2H^+ \quad (20)$$

Compared to a polystyrene cation exchanger D-001, HMO-001 exhibited more favourable Pb(II) adsorption in the presence of competing cations such as Na(I), Ca(II), and Mg(II). The fixed-bed column results showed that Pb(II) adsorption on HMO-001 could result in a conspicuous decrease of this toxic metal from 1.0 to below 0.01 mg/l. Moreover, the exhausted HMO-001 material is amenable to an efficient regeneration by mixture of sodium acetate and acetic acid solution for the repeated use without any significant capacity loss.

A strong base anion exchange resin Amberlite IRA-400 and its composites with $Mn(OH)_2$ and $Cu(OH)_2$ are used for the removal of Cr(VI) from the synthetic spent tannery bath [200]. The kinetic results of Amberlite IRA-400 are compared with its composite anion exchange resins IRA-400-$Mn(OH)_2$ and IRA-400-$Cu(OH)_2$. However, it was found that the composite ion exchangers have greater removal ability and fast kinetics as compared to Amberlite IRA 400.

4.3. Composite Adsorbers with the Organic Polymer Binder

The use of organic-inorganic composites is especially important in the case of radionuclide wastewater treatment. Many inorganic and organic ion exchangers have been used to this end. Especially inorganic ion exchangers such as zeolites have been studied extensively due to their selectivity for the specific ions, as well as thermal and radiation stabilities. However, the slow mass-transfer rate in the column operation has been the impeding factor for extensive applications. On the other hand, the application of organic ion exchangers is also limited due to possible release of the radionuclides by decomposition of organic matrix when disposed underground.

In composite ion exchangers, inorganic materials are active components to which all the radionuclides are bound and organic materials are simply inert binders. Among the required properties for the binding materials are: sufficient aggregation force, chemical and radiation stability, no influence on adsorbent, no sorption properties and high permeability in aqueous solution. Polyacrylonitrile (PAN) is reported as one of the most favourable organic binders taking into account its excellent pelletizing property, good solubility for organic solvents, strong adhesive forces with inorganic materials and chemical stability [201]. Besides radioactive sewages [202], such materials are used for removal of heavy metal ions, for example, in the paper [203] magnesium oxide-polyacrylonitrile composite (MgO–PAN) was described. The obtained results show that with the increase of pH, the amount of adsorption of such cations as Sr(II), Co(II), La(III), Ni(II) and Zr(IV) on the surface of magnesium oxide composite increases due to amphoteric nature of the ion exchanger. Magnesium oxide has an isoelectric point at pH 12. In the paper by Moon [201] preparing the poly(acrylonitrile) PAN-potassium titanate $K_2Ti_4O_9$ and PAN-nickelferrocyanate $Ni_2Fe(CN)_6$ composite ion exchangers and evaluation of their adsorption characteristics for the Ag(I) and Sr(II) ions in acidic waste solutions were presented. It was found that for the Ag(I) ions PAN-$Ni_2Fe(CN)_6$ ion exchanger was more selective than PAN-$K_2Ti_4O_9$ with the maximum adsorption capacity at 1.945 meq/g, while the PAN-$K_2Ti_4O_9$ was more selective for Sr(II) ions (maximum capacity at 2.103 meq/g).

Composite magnetic ion exchangers were applied by Szeglowski et al. [204] in sorption of radioisotope ^{137}Cs from the soil surface, sandy or volumetric samples of sediments. Their preparation process was for the first time patented by Narbutt et al. [205] and consists in introducting a phenylosulfone resin, sparsely soluble hexacyanateferrates of cobalt, nickel, copper or zinc as well as magnetite during the polycondensation synthesis. The obtained composite is characterized by great chemical and mechanical stability and small specific gravity. Large disintegration of the specific sorbent and magnetite in the matrix material with large cross-linking of phenosulfone resin affects favourably sorption and desorption kinetics.

Kubica et al. described [206] the application of the composite ion exchanger obtained by the direct mixing of nickel(II)-potassium hexacyanoferrate(II) and a resin matrix synthesized by polycondensation of sulfonated phenol and formaldehyde in the sorption process of ^{212}Pb. In the paper [207] the optimum conditions for isolations of Zr, Hf and Nb from nuclear reaction products were also described.

Recently, the preparation of ion exchange sorbents comprising anion exchange resin, ferromagnetic substance and a water permeable organic polymer was developed [208-211]. Contrary to the earlier discussed composites, in this class of sorbents water permeable organic polymer binder holds the anion exchange resin with the ferromagnetic material together in a granular form applicable in removal of different contaminants from wastewaters. As follows from the literature data these composites were successfully applied in removal of anions, remediation of water contaminated with heavy metal ions, and above all in separation of radionuclides from radioactive and mixed waste. They are used for analytical, preparative or other purposes.

In the paper by Sheha and El-Zahhar [208], the synthesis of the composite based on immobilization of magnetite (Fe_3O_4) within the ion exchange resin Amberlite IR120 with polymeric binder polyacrylonitrile (PAN) was presented. The sorption properties of the obtained material were tested towards Cr(VI) ions. The presence of magnetite particles in the sorbent increases their granular strength, thermal stability and facilitates their separation from aqueous solutions. The values of equilibrium sorption capacity of IR 120-PAN-Fe_3O_4 attain the range 893-951 mg/g for Cr(VI) at pH 5.1. Kinetically, both pore diffusion and film diffusion contribute to sorption of Cr(VI) ions. The isotherm parameters related to the heat of sorption are in the range 8-16 kJ/mol which is the range of bonding energy for ion exchange interactions and so suggest an ion exchange mechanism for removal of Cr(VI) by the composite sorbent.

5. Zwitterionic Hybrid Materials

One of the most interesting types of inorganic-organic charged hybrid materials are zwitterionic hybrid polymers which has drawn much attention in recent years. This type of hybrid polymer not only combines the advantages of organic and inorganic materials, but also exhibits some distinguished properties, such as structural flexibility, thermal and mechanical stability. Liu et al. prepared a series of zwitterionic hybrid polymers from the ring-opening polymerization of pyromellitic acid dianhydride (PMDA) and phenylaminomethyl trimethoxysilane (PAM-TMS), and a subsequent sol-gel process. Their application for Cu(II) removal from aqueous solution was examined. It is indicated that their adsorption for Cu(II) ions followed the Lagergren second-order kinetic model and the Langmuir isotherm model [212-216]. The similar results were obtained in the case of Pb(II) ions sorption. As for the desorption efficiency 2.0 M HNO_3 solution reaches up to 64.4%, indicating an effective regeneration cycle. The obtained adsorbent can find potential applications in the separation and recovery of divalent metal ions from wastewater of lead-acid rechargeable battery.

6. Hybrid Sorbents Based on Inorganic Support

6.1. Hybrid Materials Based on Barium Sulfate

New hybrid materials were developed by the template-free hybridization of organic substances, e.g. dyes on $BaSO_4$ [217,218]. Owing to the present complexity and difficulty of concentrated dye wastewater treatment, the synthesis of sorbent materials for the treatment of wastewater by forming the dye-conjugating complex hybrid is very important. These materials were prepared by immobilizing waster dye-Mordant blue 9 (MB) or weak acidic pink red B (APRB) with barium sulfate ($BaSO_4$). In the case of MB-$BaSO_4$ hybrid the adsorption of cationic dye-basic blue BO (BB) and Cu(II) ion was investigated. It was found that the adsorption of BB on the MB-$BaSO_4$ hybrid was probably attributed to ion-pair equilibrium and that of Cu(II) may result from the complexation [217]. The treatment of dye and heavy metal ion wastewaters indicated that the MB hybrid material removed 99.8% of BB and 97% of Cu(II). In the case of APRB, it was shown that it has good electrophilicity and hydrophobicity, due to containing two negative sulfonic acid groups and a long hydrophobic alkyl chain [218]. In contrast to the conventional sorbents, this hybrid material has a specific surface area of 0.89 m^2/g, but it contains lots of negative charges and lipophilic groups as the basis of specific adsorption. The efficient removal of cationic dyes and persistent organic pollutants (POPs) indicates that it has an improved adsorption capacity and selectivity with a short removal time. Instead of using the APRB reagent, an APRB-producing wastewaters were reused to prepare this cost-effective sorbent. It was used for removal among others of the weak acidic green GS (WAG), acidic brilliant blue 6B (ABB), ethyl violet (EV), methylene blue (MB) and basic pink red B (BPR). The obtained equilibrium adsorption capacities reached 222 mg/g and 160 mg/g for EV and BPR, respectively. The sorbents were used to treat wastewater samples with satisfactory results of over 97% decolonization and 88% COD (chemical oxygen demand) decreasing. In addition, the hybrid sorbent was regenerated from sludge over five cycles, and its adsorption capacity was not appreciably changed. Also polycyclic aromatic hydrocarbon (PAHCs), phenanthrene (Phe), biphenyl (Bip) and biphenol A (BPA) were used to examine the adsorption capacity of the hybrid material. POPs treated with the hybrid material were removed much more quickly than those treated with the $BaSO_4$ and APRB-$BaSO_4$ surface-modified materials. This indicates that the in situ chemical co-precipitation of APRB with $BaSO_4$ particles may produce a much more hydrophobic surface than surface modification with APRB.

6.2. Hybrid Materials Based on Aluminum, Zirconium, and Titanium Oxides

Inorganic solid adsorbents based on the oxohydrates of transition and main-group metals: iron, aluminium, zirconium, and titanium are well characterized by their high mechanical properties and strong resistance to thermal degradation compared to the traditional polymeric ion exchangers. Their additional advantages include existence in several structures, high surface area and amphoteric properties. For example, hydrous zirconium dioxide (denoted also as zirconia) was used for the removal of fission products (^{137}Cs and ^{90}Sr) with relatively long half-life (of approximately 30 years) from radioactive waste solutions. It was also chosen for Cr(VI) removal [219]. Impregnating zirconium phosphate (ZrP) was used to obtain a new hybrid adsorbent ZrP-001 with the polystyrene cation exchanger D-100 with the sulfonic functional groups [220]. Compared to D-001, ZrP-001 composite exhibited more favourable Pb(II) adsorption even in the presence of Ca(II) or Mg(II) at greater levels. Such unique performance of ZrP-001 was mainly attributed to the presence of the sulfonic acid groups bound to the D-001 matrix which enhances the preconcentration of Pb(II) ions from solution to inner surface of the polymeric phase. Moreover, Pb(II) ion can be selectively sequestrated by ZrP particles through possible inner-sphere complexation of Pb(II) and ZrP. Similarly, other inorganic particles exhibiting specific affinity toward heavy metal ions, namely, hydrated ferric oxides (HFOs), $Zr(HPO_3S)_2$, $Ti(HPO_4)_2$, hydrated manganese oxide (HMO) can be also used [221,222].

PAN and the block copolymer mesogen (F-127) together with a cationic surfactant (palmitic acid) or nonionic surfactant (Tergitol) were used for obtaining zirconium titanite composite for uranium removal by Sizgek et al. [223]. It was found that it displays fast uranium uptake, similar to that of the zirconium titanate xerogel powders. The accessibility to adsorption sites in the mesopores has been achieved with the open hierarchical pore structure. Uranyl adsorption isotherms for ZrTi obtained in the range of concentrations from 10 ppm to 1000 ppm show that the maximum sorption capacity from the Langmuir model is equal to 0.14 mmol/g which is somewhat higher than that reported for the ion exchangers such as Dowex 1x8, Purolite A520 E and Dowex 21 K.

Bang et al. [224] investigated the adsorption performance of nano and granular titanium dioxide towards As(III) and As(V) and they found that TiO_2 had a relatively high adsorption capacity for As(V). As for the sorption mechanism of arsenate and arsenite on the surface of TiO_2 it was also explained by formation of bidentate binuclear inner-sphere surface complexes. Kasuga et al. [225] recently reported the preparation of TiO_2-derived nanotubes (TNT) by hydrothermal treatment of TiO_2 powders. This method does not require any templates, and the prepared nanotubes have a small diameter of ca. 10 nm, and high crystallinity. They found that TNT is a good adsorbent for the removal of Cu(II) ions from aqueous solution with the adsorption capacity reaching 120 mg/g. Also TiO_2 materials with PVA, PAN and polyacrylic hydrazide (PAH) were used mainly for radionuclides adsorption [226,227].

6.3. Hybrid Materials Based on Molybdates

Organic-inorganic composite cation-exchanger polyaniline Ce(IV) molybdate was prepared and characterized for ion exchange properties towards sorption of metal(II)/(III) ions. It was obtained by mixing gel of polyaniline and Ce(IV) molybdate [228]. The chemical formula of the obtained ion exchanger was proposed to be: $[(-C_6H_4NH-)-Ce(OH)_2(HMoO_4)_2]\cdot nH_2O$.

Table 5. K_d values of Cd(II), Zn(II), Pb(II), Mn(II), Co(II), Cu(II) and Hg(II) obtained for poly-o-methoxyaniline Zr(IV) molybdate in different solvents [229]

Metal ion	DMW	10^{-2} M HNO_3	10^{-1}M HNO_3	10^{-2} M HCl	10^{-1} M HCl	Buffer pH 3.75	Buffer pH 5.75	Buffer pH 10.0
Cd(II)	1820	1350	1237	1152	1489	1352	948	824
Zn(II)	850	630	548	874	670	673	534	424
Pb(II)	784	884	745	721	649	748	611	613
Mn(II)	421	514	67	845	24	219	112	65
Co(II)	367	213	214	345	647	142	345	89
Cu(II)	670	670	577	458	890	360	268	459
Hg(II)	1733	1244	1221	1014	900	673	516	382

Reproduced with permission from Inamuddin and Ismail, Y. A. *Desalination* 2010, *250*, 523-529.

As follows from the research for the metal ions Mg(II), Ca(II), Sr(II), Ba(II), Pb(II), Mn(II), Co(II), Cd(II), Cu(II), Hg(II), Zn(II), Ni(II), Fe(III) and Al(III) the sorption effectiveness varies with the composition and nature of the contacting solvents (HCl, HNO_3, bufer at pH 3.75; 5.75 and 10). It was observed from the K_d value studies that the uptake of Cd(II) is exceptionally large, while the remaining metal ions are poorly sorbed. Thus, the composite cation-exchanger can be very well utilized for the determination and separation of cadmium ions from waste effluents.

Another example can be the cation exchange composite material prepared by the sol-gel mixing of organic polymer, poly-*o*-methoxyaniline with the inorganic ion-exchanger precipitate of Zr(IV) molybdate [229]. The composite cation exchanger possesses better ion-exchange capacity (2.56 meq/g) as compared to Zr(IV) molybdate (1.75 meq/g), which may be due to the presence of binding polymer i.e. poly-*o*-methoxyaniline. The ion-exchange capacity for alkali and alkaline earth metal ions increases with the decrease in their hydrated ionic radii. It also indicates large uptake of Cd(II) and Hg(II) ions (Table 5).

Ion exchange behaviour and analytical applications of an organic-inorganic composite cation exchanger - acrylamide stannic silicomolybdate was also described by Khan et al. [230]. This hybrid ion exchanger was obtained by mixing the solution of acrylamide, sodium molybdate and sodium metasilicate with aqueous solution of stannic chloride pentahydrate. It shows superiority in terms of ion exchange properties compared to stannic silicomolybdate [231]. The ion exchange capacities of alkali and alkaline earth metal ions decrease in the following order: K(I) > Na(I) > Li(I) while for the alkaline earth metal ions the analogous sequence can be as follows: Ba(II) > Sr(II) > Ca(II) > Mg(II). Based on the K_d values of metal ions in varied solvent systems such as acetic acid (AA), trichloroacetic acid (TCA), dimethylsulfoxide (DMSO), formic acid (FA) as well as their mixtures (e.g. AA-TCA, DMSO-AA, DMSO-FA), a number of binary separations, namely Cd(II)-Pb(II), Cd(II)-Cu(II), Al(III)-Pb(II), Al(III)-Cu(II), Zn(II)-Pb(II), Zn(II)-Cu(II) were achieved quantitatively. The most important property of the material was found to be high selectivity towards Pb(II) and Cu(II), also the separation of Pb(II) and Cu(II) was practically achieved using acrylamide stannic silicomolybdate from the synthetic mixtures containing Pb(II), Hg(II), Cd(II), Ba(II) and Mg(II) as well as Cu(II), Mg(II), Cd(II), Hg(II), Zn(II) and Ba(II), respectively.

6.4. Hybrid Materials Based on Tungstons and Tungstonphosphates

As mentioned earlier in recent years, synthetic inorganic exchangers based on tetravalent metals have been the objects of considerable study because of their selectivity towards the removal of toxic heavy metal ions and radionuclides [232,233]. Using intercalation and/or grafting of organic or polymeric molecules within a mineral lamellar network of clays, phosphates, phosphonates, oxides etc. allowed to obtain the composite materials of ion exchange properties better or comparable with the initial materials.

For example, the polyacrylonitrile Sn(IV) tungstate $[(SnO_2)(H_2WO_4)_4(CH_2CHCN)_3]\cdot H_2O$ cation exchanger has been explored for the quantitative and selective separations of Pb(II) from the mixtures of metal ions [234]. The separation capability of the material has been demonstrated by achieving some important binary separations such as Hg(II)-Pb(II), Zn(II)-Fe(II), Cu(II)-Pb(II), Cd(II)-Pb(II), Fe(III)-Pb(II), Ni(II)-Pb(II) and Mg(II)-Pb(II). High K_d values of Pb(II) enable also their selective separation from the mixture of metal ions such as Pb(II), Zn(II), Fe(III), Cu(II), Cd(II), Ni(II) and Al(III).

The powdered zirconium(IV) tungstophosphate (ZrWP) cannot be directly used in fixed bed columns or any other flow-through systems because it causes excessive pressure drops. To avoid this problem, polymeric hybrid composites were prepared by dispersing powdered ZrWP particles onto porous polymeric substrates. For example, styrene supported zirconium(IV) tungstophosphate composite was prepared and used in ion exchange of cations and sorption of pesticides [235]. The main objective of the research presented by Viswanathan and Meenakshi [236] was to synthesize chitosan (CS) supported ZrWP by dispersing ZrWP into the chitosan biopolymeric matrix for fluoride removal. ZrWPCS composite was synthesized by mixing the solutions of $ZrOCl_2\cdot 8H_2O$, $Na_2WO_4\cdot 2H_2O$, H_3PO_4 and chitosan. The effects of interfering ions Cl^-, SO_4^{2-}, NO_3^- and HCO_3^- on fluoride sorption by ZrWPCS was studied and no remarkable influence on the defluoridation capacity except of the HCO_3^- ion was found. The decrease in the defluoridation capacity of ZrWPCS in the presence of bicarbonate ion may be due to hydrolysis of $NaHCO_3$. However, the values of maximum sorption capacities increase with the increasing temperature. It was also established that the positive charge of the composite surfaces attracts the negatively charged fluoride ions by means of electrostatic attraction as well as complexation.

In the papers by Khan et al. [237-239] besides the presentation of polypyrrole Th(IV) phosphate and polyaniline Sn(IV) arsenophosphate composite materials, the polystyrene-Zr(IV) tungstophosphate used for the selective separation of Pb(II), Cd(II), and Hg(II) ions as well as adsorption of pesticides was described. Pandit et al. [240] have also synthesized such type of ion exchange materials, i.e. *o*-chlorophenol Zr(IV) tungstate and *p*-chlorophenol Zr(IV) tungstate. Polyaniline Zr(IV) tungstophosphate has been synthesized by Gupta et al. [241], which was used for the selective separation of La(III) and Th(IV).

6.5. Hybrid Materials Based on Phosphonates

Metal-organic compounds with open frameworks, such as metal phosphonates and chalcogenides could be considered as excellent ion sorbents and ion exchangers due to their great variety of chemical and structural compositions as well as high thermal and radiation resistance [242,243]. Metal phosphonate open-framework materials are located between the porous zeolite-like materials and the metal-organic framework (MOF) materials. However, compared with their phosphate analogs, the organic moiety of metal phosphonates can be designed to provide different networks and specific properties. Their structures depend on the metal type, the function of the organic ligand and the reaction conditions [244].

Clearfield and Wang [245] as well as Wu et al., [246] have reported zirconium phosphate-phosphonates, which have been successfully applied, among others in sorption and ion exchange processes. These composites are characterized by larger ion exchange capacities and thermal stabilities than the commonly used organic ion exchangers. More importantly, their selectivity for polyvalent cations is very high, which can be basis of analytical separations or concentrations. To this end, a large number of organo-monophosphonic acids with second or even third and forth functional groups have been used. The functional groups include –COOH, –OH, $–SO_3H_2$, pyridine and $–NH_2$. The use of the diphosphonic acids H_2O_3P-R-PO_3H_2 or functionalized phosphonic acids e.g. $H_2O_3PCH_2CH_2COOH$ in the preparation of pillared layered structure of zirconium, vanadium, zinc, bismuth and mixed manganese-zinc phases was also described in the literature [247-250]. By utilizing diphosphonic acids, in the place of monophosphonic acids, an interesting series of new compounds with three-dimensional structures was obtained. Metal bisphosphonate compounds may be described as pillared compounds that contain pores of definite shape and size in the interlayer space. A variety of phosphonic acids, both in terms of the length and nature of the organic group, could be utilized in preparing of these compounds. For example, in the paper by Johnson et al. [251] vanadium organophosphonates $VO(RPO_3)\cdot 2H_2O$ analogous to the zirconium compounds having an organic group replacing the hydroxyl of the hydrogen phosphate, which have layered inorganic-organic structures was described, whereas the structures of Cu(II) and Zn(II) alkylenebis(phosphonates) were presented in [252]. A mesoporous aluminum organophosphonate (denoted as AOP-2) was synthesized by using methylene diphosphonic acid ($(HO)_2OPCH_2PO(OH)_2$) in the presence of octadecyltrimethylammonium (C18TMA) surfactant. It should be mentioned that despite the potential and attractive applications of metal phosphonates as ion exchangers, catalysts, and nonlinear optics, diphosphonates have been used as only linkers through the substitution of phosphates in the layered compounds [253,254].

Metal-tetraphosphonates, based on tetraphosphonic acids $(H_2O_3PCH_2)_2N$-R-$N(CH_2PO_3H_2)_2$, where R = $(CH_2)_2$, $(CH_2)_4$ and C_6H_4 (as organic building blocks), as stable organic-inorganic hybrid materials display both microporous structures and metal phosphonates properties in conjunction with strong inorganic backbones were presented. Compared with the traditional ion exchange materials, besides the earlier mentioned high sorption capacity towards metal ions and thermal stability, they are characterized by the porosity with large surface areas and tunnels (which facilitates the movement of the adsorbed metal ions inside without steric effects), the negative charge concentrated on the proton-bearing oxygens (responsible for strong attraction to cations) as well as possibility of easy modification and tailoring the dimensions inside the tunnels.

In the paper by Wu et al. [246] the use of tetraphosphonic acid $(H_2O_3PCH_2)_2N$-$(CH_2)_2N(CH_2PO_3H_2)_2$ (H_8EDTP) to prepare two novel divalent metal tetraphosphonates, namely, $[Pb_7(HEDTP)_2(H_2O)]\cdot 7H_2O$ and $[Zn_2(H_4EDTP)]\cdot 2H_2O$ was reported. The prepared materials have high ion sorption capacities for Fe(III) (log K_d 4.75 and 4.18, respectively). It was found that in the case of the $[Pb_7(HEDTP)_2(H_2O)]\cdot 7H_2O$ the sorption takes place through the tunnel network. On the other hand, in the case of $[Zn_2(H_4EDTP)]\cdot 2H_2O$, where the layers are held together by hydrogen bonds, ions can diffuse into the interlamellar spacing.

7. Zeolite Composites

In the group of inorganic or hybrid networks obtained in the templated growth by using organic molecules and macromolecules as structure directing agents and templates (amines, alkyl ammonium ions, amphiphilic molecules or surfactants) materials of the zeolites families are among the most intensively investigated systems.

Although the fundamental understanding of mineral networks formation in the presence of organic templates is well described in the literature, the application of these new micro or mesoporous hybrids with tailored porosity in size, shape and function for selective separations, sensors, catalysis and low dielectric constant materials is of great importance.

In the paper by Anirudhan and Suchithra [255] the preparation of a new polyacrylamide-bentonite composite with amine functionality (PAM-B-A) by direct polymerization technique was described. The obtained hybrid was applied as an adsorbent for the removal of cationic dyes such as Malachite Green (MG), Methylene Blue (MB) and Crystal Violet (CV) from aqueous solutions. Polyacrylamide hydrogel bentonite composite is characterized by greater complexity, strength and reduced production cost. Moreover, by immobilization of humic acid (PAM-B-A-HA) its adsorption capacity was further enhanced. Introduction of HA with deprotonated carboxylic and phenolic groups in weakly acidic to basic media and possessing negative charge enabled the sorption of the above-mentioned dyes through electrostatic interactions.

It was proved that the removal capacity of PAM-B-A-HA increased with the increase in the solution pH and reached the maximum at pH 6.0. The maximum adsorption onto PAM-B-A-HA for 2.0 mmol/l initial dye concentration was 99.7 % for MG, 99.3 % for MB, and 98.8 % for CV which decreased to 84.0 % for MG, 78.1 % for MB, and 74.3 % for CV when initial dye concentration increased to 4.0 mmol/l. As follows from the adsorption studies the maximum sorption capacity is equal to 0.199 mmol/g for MG, 0.193 mmol/g for MB, and 0.187 mmol/g for CV. It was also proved that MG had posed greater interference on the adsorption of CV than CV on the MG adsorption. A novel and simple method to synthesize a new class of magnetic zeolite composites with surface-supported silver nanoparticles as sorbents for Hg(II) removal was developed [230].

8. Biopolymer Composites Based on Chitin, Chitosan and Alginic Acid

Cellulose, the most abundant biopolymer, gives strength and stiffness to plant fibres. It is a renewable polysaccharide with unique properties which has been widely exploited as a source of materials (cotton and paper) or after appropriate chemical modifications (cellulose acetate and carboxymethyl cellulose) [256]. Also preparation of organic-inorganic hybrid materials combining cellulose with silica (CSH) compounds has been reported fitting well into the concept of green chemistry, since cellulose is a biodegradable natural polymer [257-259].

A new adsorbent, bead cellulose loaded with iron oxyhydroxide (BCF) was prepared and applied for the adsorption and removal of arsenate and arsenite from aqueous systems. The adsorption capacity for As(III) and As(V) was 99.6 mg/g and 33.2 mg/g BCF at pH 7.0. As(V) removal was favoured at acidic pH, whereas the adsorption of As(III) by BCF was found to be effective in a wide pH range of 5-11. The addition of sulfates(VI) had no effect on arsenic adsorption, whereas phosphate(V) greatly influenced the elimination of both As(III) and As(V) [179]. BCF is compatible with fixed-bed column processes with excellent mechanical strength and attrition resistance properties.

Chemical polymerization of polypyrrole (Ppy) in the presence of $FeCl_3$ or phosphomolybdic acid on cellulose fibers gives new possibilities for the manufacturing of conductive paper. This novel composite, which was prepared in the form of a paper sheet, exhibited excellent mechanical stability and could be bent, twisted, or folded without disrupting its mechanical integrity. Due to its large surface area, the composite exhibited a high exchange capacity for chloride ions [260]. Figure 14 shows schematically the pyrrole polymerization processes with iron(III) chloride.

Of natural polymers, besides cellulose, lignin and proteins, chitin (CT) and its derivative chitosan (CS) are of significant importance (Figure 15). Chitosan being a deacetylation product of chitin at present one of the most promising natural polymers.

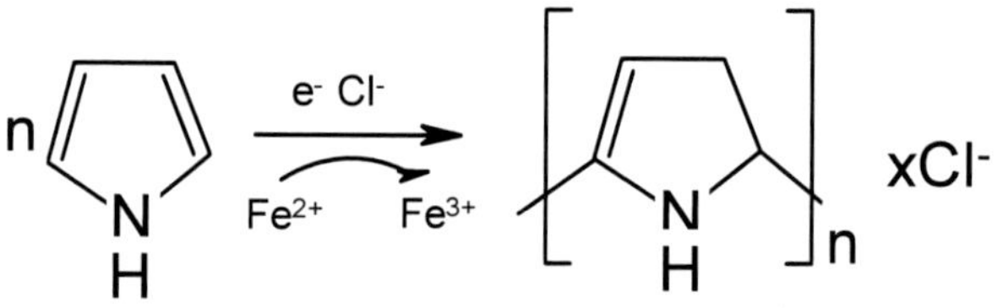

Figure 14. Schematic presentation of pyrrole polymerisation with $FeCl_3$ [260]. Reproduced with permission from Razaq, A. et al. *J. Phys. Chem. B*, 2009, *113*, 426-433. Copyright © 2009, American Chemical Society.

It possesses some functional qualities, among which, one can enumerate: high value of secondary swelling index, capability of polymer membrane formation (directly from the suspension), high adhesion ability, controlled bioactivity, biocompatibility and non-toxicity as well as good stability and miscibility with many substances (including polymers). Owing to these properties it can be readily formed into various shapes: globules, fibres, capsules, membranes and others. The most frequently found are chitosan globules characterized by remarkable porosity.

The research results reported in the literature also confirm complex creating action of chitosan towards metal ions. Therefore, chitosan is mainly applied as sorbent for removal of the heavy metal ions: Cu(II), Zn(II), Cd(II), Pb(II), Hg(II), Cr(III,VI), As(III,V), and recovery of the noble metals: Pt(IV), Pd(II), Ag(I) and Au(III) [261-266]. Guibal et al. [265,266] have extensively studied the effect of chitosan properties on the adsorption of metals, dyes and organic compounds such as phenols. The use of chitosan based material to remove anionic dyes has been recently reviewed by Crini and Badot [264]. Moreover, chitin and chitosan imprinted matrices have been prepared with different purposes by using different chitosan samples [267,268], among others chitin-phosphate composites used as bone substitutes and chitosan-bonded hydroxyapatite as bone-filling paste should be mentioned. As follows from the literature data, the slow rate of adsorption of commercially available activated alumina or magnesia as well as nanohydroxyapatite (HAP) limits their use in the removal of fluoride ion from waters. However, the biocompatible composite sorbent of chitin with HAP (HAP-CT) can be used for this purpose [269].

The maximum defluoridation capacity of HAP-CT as well as HAP was found to be 2840 and 1296 mg F^-/kg, respectively. Below pH 5.9, which is pH_{pzc} of the HAP-CT composite, where the surface acquires positive charge, the defluoridation capacity was found to be maximum. The dependence of defluoridation capacity of the sorbent in the presence of other common anions which are commonly present in water reveals that bicarbonate anion interferes with the fluoride sorption contrary to sulfate, nitrate and chloride anions. The main advantages of HAP-CT composite are: biocompatibility, low cost material, indigenous synthesis and effective utilization as a promising defluoridating agent. The proposed mechanism of F^- ions removal based on the electrostatic attraction can be as follows:

Various chitosan-based composites have been tested as adsorbents for dyes and metal ions [270-277]. The study on the removal of arsenic from real life groundwater using the iron-chitosan composites is presented by Gupta et al. [270]. They were obtained by coating iron hydroxide to chitosan flakes (ICSF) or doping iron(III) to chitosan and then casting it as granules (ICSG). Removal of As(III) and As(V) was studied at pH 7.

$$\text{HAP-CT-Ca}^{2+} + 2\text{F}^- \rightleftarrows \text{HAP-CT-Ca}^{2+}\text{--}2\text{F}^- \quad (21)$$

In the case of ICSF the maximum sorption capacity was found to be 22.47 mg/g for As(V) and 16.15 mg/g for As(III) that is higher than that obtained for ICSG. The interfering anions e.g. phosphates(V), sulfates(VI) and silicates at the levels present in the groundwater did not cause serious interference in the adsorption behaviour of ICTF and ICSB. In order to remove As(III) and As(V) from water, a composite chitosan biosorbent was prepared by coating chitosan on ceramic alumina by Veera et al. [273]. In this case the sorption capacity for As(III) and As(V) was found to be higher than for ICSF and equal to 56.5 and 96.5 mg/g, respectively. As(V) showed more adsorption capacity than As(III), which was explained by speciation of arsenic at pH 4.0.

A number of studies have focused on the use of natural organic polymers as support materials for catalytic compounds, especially in the preparation of homogeneous and heterogeneous catalysts used in one of the most important new advanced oxidation technologies in water purification. Torres et al. [274] reported the preparation of chitosan-niobium(V) oxide composite applied to the photodegradation of indigo carmine dye from contaminated waters. They found that the CS-Nb composite exhibiting high catalytic efficiencies for the degradation of indigo carmine dye (above 95%), was readily recovered from the reaction mixture and maintained high catalytic capacity after 15 cycles of reuse.

Chitosan-kaolin-nanosized γ-Fe_2O_3 composites were also used for the dyes adsorption [275]. It was proved that methyl orange (MO) could be removed within 180 min of adsorption from a 20 mg/l MO solution at pH = 6.0 by 1.0 g/l composites. In the case of interfering ions (Cl^-, NO_3^-, SO_4^{2-}, CO_3^{2-} and PO_4^{3-}) the order of decolorization was: no addition >> Cl^- > NO_3^- > SO_4^{2-} > CO_3^{2-} > PO_4^{3-}, indicating that the addition of competitive anions decreased dye adsorption. Procion Brown MX 5BR (Figure 16) immobilizing poly(hydroxyethyl-methacrylate/chitosan) composite membranes was used for removal of Cd(II), Pb(II) and Hg(II) from aquatic systems [276].

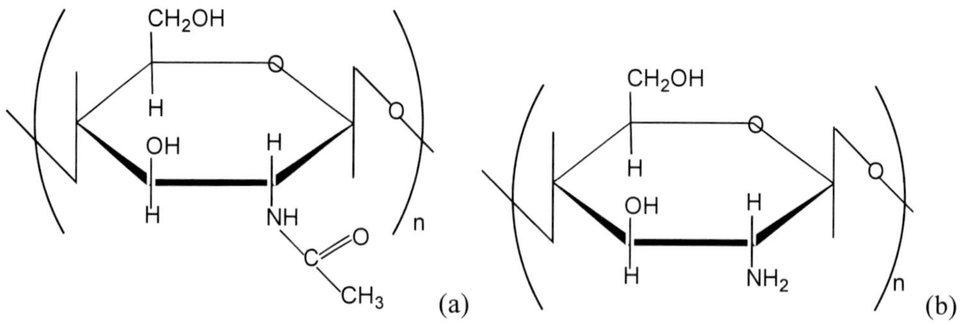

Figure 15. Chemical structures of chitin (a) and chitosan (b).

At pH 5.0 the maximum adsorption capacities of Procion Brown MX 5BR immobilized PHEMA-chitosan membranes (PHEMA-CS) were found to be as 18.5 mg/g for Cd(II), 22.7 mg/g for Pb(II) and 68.8 mg/g for Hg(II). The following metal ion affinity order, based on a weight uptake was established: Hg(II) > Pb(II) > Cd(II). Under competitive conditions, the maximum amounts of metal ions adsorbed were equal to: 1.8 mg Cd(II)/g, 2.2 mg Pb(II)/g and 52.6 mg Hg(II)/g. These results show the high selective affinity of the Procion Brown MX 5BR immobilized PHEMA-CS for Hg(II) ions. Novel cross-linked composite membranes were also synthesized to investigate their applicability in the anion exchange membrane fuel cells by Xiong et al. [277]. These membranes consist of quaternized poly(vinyl alcohol) (QA-PVOH) and quaternized chitosan (2-hydroxypropyltrimethyl ammonium chloride chitosan (HAC-CS) with glutaraldehyde as the cross-linking reagent. So far HAC-CS has been used as a flocculant in water treatment [278].

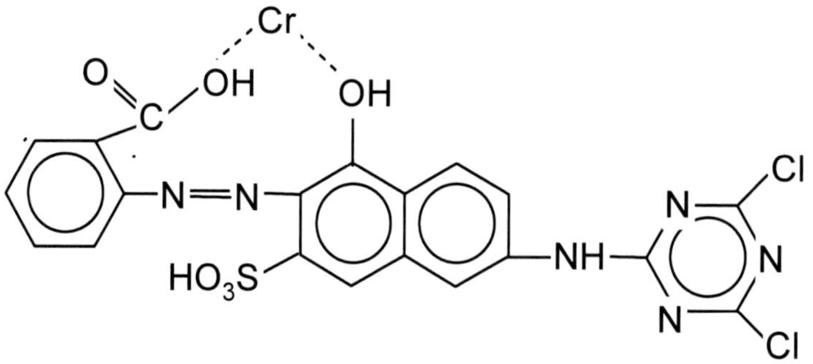

Figure 16. Structure of Procion Brown MX 5BR [276]. Reproduced with permission from Genc, Ö. et al. *Hydrometallurgy* 2002, *67*, 53-62. Copyright © 2002, Elsevier B.V.

Alginic acid also called algin or alginate is an anionic polysacharide distributed in the cell walls of brown alge. It is used not only in pharmaceutical applications but also as a sorbent for example of rare earth elements [279]. The same authors studied removal of Zn(II), Cd(II) and La(III) by gel beads of alginic acid [280].

Apart from the Ca-Fe-doped alginate beads, Ca-Fe-coated alginate beads were also examined for the removal of As(III) and As(V) [173]. The maximum amount of arsenic sorbed onto these materials was found to be 2.6 μg of As/g and 4.4 μg of As/mg, respectively. Arsenic removal was also examined using alginate beads prepared by a combination of doping and coating with iron oxides. In this case the removal of arsenic was found to be more efficient. The main mechanism responsible for arsenic removal is the chemical reaction between iron and arsenic. As(V) is known to form stable coordination compounds with iron oxides, by specific adsorption, through the following reaction:

$$\text{M-FeOH} + H_3AsO_4 \rightleftarrows \text{M-Fe-}H_2AsO_4 + H_2O \tag{22}$$

The product of this reaction is ferric arsenate, which has a very low solubility (10^{-20} mol/l).

Polyisocyanate, which is capable of interacting with alginate to give flexible polyurethane foams of high mechanical strength was used to obtain new material for selective removal of heavy metal ions e.g. alginate-polyurethane composite foams (ALG-PUCF) [281]. The obtained foams showed high capability for adsorbing lead(II) ions from the model water samples, demonstrating their applicability for their elimination from contaminated waters. The ALG-PUCFs were reusable by regenerating with ethylenediaminetetraacetic acid, disodium salt (EDTA-2Na). The ability of the ALG-PUCF to adsorb Pb(II) was found to depend strongly on the pH and at pH 3.96 it attains the maximum capacity. The authors also determined the sorption parameters towards Cd(II), Co(II), Mn(II), Mg(II) and Ca(II). They found that the K_d value for Pb(II) was 11.01, whereas for Cd(II)' Ca(II), Co(II), Mn(II) and Mg(II) they were 8.41, 7.63, 4.99, 3.09, and 1.15, respectively. Therefore, the affinity series was found to be as follows: Pb(II) > Cd(II) > Ca(II) > Co(II) > Mn(II) > Mg(II). Compared to the conventional ALG-Ca beads, ALG-PUCF shows a unique behavior for the adsorption of Pb(II) ions based on the electrostatic interaction model. It can be concluded that ALG-PUCF behaved as an ion-exchanger with the carboxyl functional groups.

CONCLUSION

Water occupies the firts position in the hierarchy of all human life needs. However, water contamination is caused by introducing excessive amonuts of inorganic, organic and radioactive substances which limit or enable exploitation of water for drinking and industrial purposes. The presence of oils, petrol, lubricants, petroleum, plant protection chemicals – pesticides, artificial fertilizers, aromatic hydrocarbons, phenols, acids, bases and also heavy metal salts is decisive for the choice of purification method. Among all the approches proposed, adsorption processess are by far the most popular methods which are currently considered as effective, efficient and economic in the environmental area for removal of the above mentioned different pollutants from water.

Nowadays, in the adsorption processess not only commerially available traditional ion exchangers and different types of sorbents are used but also new functional composite materials. It should be remembered that this type of materials displays synergy effects of their precursors resulting from specific morphological interactions between the two phases at the micro or nano-scale. Their preparation is the ultimate goal of research programs carried out by numerous scientific centres all over the world and the specific areas of interest include:

- Preparation of high capacity composite materials for the selective removal and recovery of heavy metal ions as well as removal of radionuclides from water,
- Design of hydrophobic composite materials with controlled pore dimensions for the selective removal of organic contaminants from water,
- Synthesis of nanoporous composite materials for the decomposition of specific pollutants in both water and air environments,

In this chapter the main concern was presenting the issues concerning preparation, advantages and applicability of the organic-inorganic composites used in the environmental protection processes. Among others, their specific chemical composition and structure make their self-destruction impossible, high surface area, high reactivity, easy dispersability, and rapid diffusion as well as their efficiency and effectivenes in sorption processess were specified. The chapter shows how these features can be tailored to address some of the environmental remediation and purification problems faced today. The examples of the removal of heavy metal ions, particularly including Pb(II), Cd(II), and Hg(II) as well as As(III,V) and Cr(III,VI) were presented. Some examples of perchlorate, pesticide and fluoride, chlorane removal were also included.

ACKNOWLEDGMENTS

The author expresses her gratitude to Prof. Zbigniew Hubicki for his valuable remarks and suggestions during preparation of this chapter.

REFERENCES

[1] Nordberg, G. F.; Fowler, B. A.; Nordberg, M.; Friberg, L. *Handbook of the toxicology of metals*, Academic Press: Amsterdam, 2007; pp 1-975.

[2] Hubicki, Z.; Jakowicz, A.; Łodyga, A. In *Adsorption and its Applications in Industry and Enironmental Protection. Studies in Surface Science and Catalysis*; Editor, A. Dąbrowski, Elsevier: Amsterdam, 1999; Vol. 120, pp 497-531.

[3] Gomółka, E.; Szaynok, A. Chemia Wody i Powietrza. Oficyna Wydawnicza Politechniki Wrocławskiej: Wrocław, 1997 (In Polish).

[4] U.S. Envionmental Protection Agency. Ground Water and Drinking Water. http://www.epa.gov/safewater/mcl.html#/mcls

[5] Hristovski, K.; Westerhoff, P.; Möller, T.; Sylvester, P.; Condit, W.; Mash, H. *J. Hazard. Mater.* 2008, *152*, 397-406.

[6] Mohan, D.; Pittman, Ch. U. *J. Hazard. Mater.* 2007, *142*, 1-53.

[7] Toxicological profile for arsenic. Agency for Toxic Substances and Disease Registry (ATSDR), U.S. Department of Health & Human Services, Atlanta, Georgia 2000.

[8] Selene, C. H; Chou, J.; De, R., Ch. T. *Int. J. Hygiene Environ. Health* 2003, 206, 381-386.

[9] Bissen, M.; Frimmel, F. H. *Acta Hydrochim. Hydrobiol.* 2003, *31*, 9-18.

[10] Volpe, S. L.; Huang, H. W.; Larpadisorn, K.; Lesser, I. I. *J. Am. Coll. Nutr.* 2001, *20*, 293-306.

[11] Schrauzer, G. N. *Crit. Rev. Biotechnol.* 2009, *29*, 10-17.

[12] Mahmoud, M. E.; Hafez, O. F.; Alrefaay, A.; Osman, M. M. *Desalination* 2010, *253*, 9-15.

[13] Febrianto, J.; Kosasih, A. N.; Sunarso, J.; Ju, Y. H.; Indraswati, N.; Ismadji, S. *J. Hazard. Mater.* 2009, *162*, 616-645.

[14] Sanchez, C.; Ribot, F.; Lebeau, B. *J. Mater. Chem.*, 1999, *9*, 35-44.

[15] Sanchez, C.; Lebeau, B.; Chaput, F.; Boilot, J. P. *Adv. Mater.* 2003, *15*, 1969-1994.

[16] Sanchez, C.; Julian, B.; Belleville, P.; Popall, M. *J. Mater. Chem.* 2005, *15*, 3559-3592.

[17] Srikulkit, K.; Chen, J. I.; Chareonsak, R.; Puengpipat, V. *J. Met. Mater. Miner.*, 1998, *8*, 1-10.

[18] Landry, C.; Cortrain, B.; Wesson, J.; Zumbulyasis, N.; Lippert, J. *Polymer* 1992, *33*, 1496-1506.

[19] Fitzgerald, J.; Landry, C.; Pochan, J. *Macromol.* 1992, *25*, 3715-3722.

[20] Brennan, A.; Wikes, G. *Polymer* 1991, *32*, 733-739.

[21] Coltrain, B.; Ferra, W.; Landry, C.; Molaire, T.; Zumbulyasis, N. *Chem. Mater.*1992, *4*, 358-364.

[22] Rahimpour, A.; Jahanshahi, M.; Mortazavian, N.; Madaeni, S. S.; Mansourpanah, Y. *Appl. Surf. Sci. 2010, 256,* 1657-1663.

[23] Bequet, S.; Abenoza, T.; Aptel, P.; Espenan, J. M.; Remigy, J. C.; Ricard, A. *Desalination* 2000, *131*, 299-305.

[24] Liu, Y.; Liang, P.; Guo, L. *Talanta* 2005, *68*, 25-30.

[25] Reynes, J.; Woignier, T.; Phalippou, J. *J. Non-Crys. Solids*, 2001, *285*, 323-327.

[26] Marinin, D. V.; Brown, G. N. *Waste Manage.* 2000, *20*, 545-553.

[27] Murphy, M. A.; Marken, F.; Mocak, J. *Electrochim. Acta* 2003, *48*, 3411-3417.

[28] Khan, A. A.; Inamuddin. *Sensor Actuat. B*, 2006, *120*, 10-18.

[29] Jungbauer, A.; Hahn, R. *J. Chromatogr. A*, 2008, *1184*, 62-79.

[30] Davis, S. A.; Dujardin, E.; Mann, S. *Curr. Opin. Solid State Mater. Sci.* 2003, *7*, 273-281.

[31] Beaucage, G.; Ulibarri, T. A.; Black, E. P.; Schaefer, D. W. In: *Hybrid organic-inorganic composites;*Mark, J. E.; Lee, C. Y. C.; Bianconi, P. A. ACS Symposium Series 585, American Chemical Society: Washington 1995; pp 97-111.

[32] Oubaha, M.; Etienne, P.; Calas, S.; Coudray, P.; Nedelec, J. M.; Moreau, Y. *J. Sol-Gel Sci. Technol.* 2005, *33*, 241-248.

[33] Takahashi, M.; Niida, H.; Tokuda, Y.; Yoko, T. *J. Non-Cryst. Solids* 2003, *326&327*, 524-528.

[34] Chou, T. P.; Chandrasekaran, C.; Cao, G. Z. *J. Sol-Gel Sci. Technol.* 2003, *26*, 321-327.

[35] Arkles, B. *Mater. Res. Soc. Bull.* 2001, *5*, 402-407.

[36] Schmidt, H. *J. Non-Cryst. Solids*, 1985, *73*, 681-691.

[37] Haas, K. H.; Rose, K. *Rev. Advan. Mater. Sci.* 2003, *5*, 47-52.

[38] Beck, J. S.; Vartuli, J. C.; Roth, W. J.; Loenowicz, M. E.; Kresge, C. T.; Schimitt, K. D.; Chu, C. T. W.; Olson, D. H.; Sheppard, E. W.; McCullen, S. B.; Higgins, J. B.; Schlenker, J. L. J. *J. Am. Chem. Soc.* 1992, *114*, 10834-10843.
[39] Katiyar, A.; Yadav, S.; Smirniotis, P. G.; Pinto, N. G. *J. Chromatogr. A*, 2006, *1122*, 13-20.
[40] Grandsire, A. F.; Laborde, C.; Lamaty, F.; Mehdi, A. *App. Organomet. Chem.* 2009, *24*, 179-183.
[41] Alauzun, J.; Mehdi, A.; Rey´e, C.; Corriu, R. J. P. *New J. Chem.* 2007, *31*, 911-915.
[42] Kao, H. M.; Tsai, Y. Y.; Chao, S. W. *Solid State Ionics*, 2005, *176*, 1261-1270.
[43] Yang, Q.; Ma, S.; Li, J.; Xiao, F.; Xiong, H. *Chem. Commun.* 2006, 2495-2497.
[44] Kruk, M.; Jaroniec, M.; Ko, Ch. H.; Ryoo, R. *Chem. Mater.* 2000, *12*, 1961-1968.
[45] Lu, X. B.; Zhang, W. H.; He, R. *Chinese Chem. Lett.* 2002, *13*, 480-483.
[46] Huang, L.; Huang, Q.; Xiao, H.; Eić, M. *Micropor. Mesopor. Mater.* 2007, *98*, 330-338.
[47] Huang, L.; Xiao, H.; Ni, Y. *Colloid. Surface. A*, 2004, *247*, 129-136.
[48] Ghiaci, M.; Abbaspur, A.; Kia, R.; Seyedeyn-Azad, F. *Sep. Pur. Technol.* 2004, *40*, 217-229.
[49] Lam, K. F.; Yeung, K. L.; Mckay, G. *Langmuir* 2006, *22*, 9632-9641.
[50] Lam, K. F.; Yeung, K. L.; Mckay, G. *Micropor. Mesopor Mater.* 2007, *100*, 191-201.
[51] Nooney, R. I.; Kalyanaraman, M.; Kennedy, G.; Maginn, E. J. *Langmuir* 2001, *17,* 528-533.
[52] Kang, T.; Park, Y.; Choi, K.; Lee, J. S.; Yi, J. *J. Mater. Chem.* 2004, *14*, 1043-1049.
[53] Brown, J.; Mercier, L.; Pinnavaia, T. J. *Chem. Commun.* 1999, 69-70.
[54] Aguado, J.; Arsuaga, J. M.; Arencibia, A.; Lindo, M.; Gascón, V. *J. Hazard. Mater.* 2009, *163*, 213-221.
[55] Pérez-Quintanilla, D.; del Hierro, I.; Fajardo, M.; Sierra, I. *J. Mater. Chem.* 2006, *16*, 1757-1764.
[56] Pérez-Quintanilla, D.; del Hierro, I.; Fajardo, M.; Sierra, I. *Mater. Res. Bull.* 2007, *42*, 1518-1530.
[57] Magureanu, M.; Mandache, N. B.; Hu, J.; Richards, R.; Florea, M.; Parvulescu V. I *App. Catal. B: Environ.* 2007, *76*, 275-281.
[58] Kao, H. M.; Shen, T. Y.; Wu, J. D.; Lee, L. P. *Micropor. Mesopor. Mater.* 2008, *110*, 461-471.
[59] Liu, A. M.; Hidajat, K.; Kawi, S.; Zhao, D. Y. *Chem. Commun.* 2000, 1145-1146.
[60] Burleigh, M. C.; Markowitz, M. A.; Wong, E. M.; Lin, J. S.; Gaber, B. P. *Chem. Mater.* 2001, *13*, 4411-4412.
[61] Corriu, R.; Mehdi, A.; Reyé, C. *J. Organomet. Chem.* 2004, *689*, 4437-4450.
[62] Yoshina-Ishii, Ch.; Asefa, T.; Coombs, N.; MacLachlan, M. J.; Ozin, G. A. *Chem. Commun.* 1999, 2539-2540.
[63] Lee, Ch. H.; Park, S. S.; Choe, S. J.; Park, D. H. *Micropor. Mesopor. Mater.* 2001, *46*, 257-264.
[64] Olkhovyk, O.; Jaroniec, M. *J. Am. Chem. Soc.* 2005, *127*, 60-61.
[65] Zhang, W. H.; Zhang, X.; Zhang, L.; Schroeder, F.; Harish, P.; Hermes, S.; Shi, J.; Fischer, R. A. *J. Mater. Chem.* 2007, *17*, 4320-4326.
[66] Wu, H. Y.; Liao, Ch. H.; Pan, Chi.Y.; Yeh, Ch. L.; Kao, H. M. *Micropor. Mesopor. Mater.* 2009, *119*, 109-116.

[67] Zhang, T.; Gao, Ch.; Yang, H.; Zhao, Y. *J. Porous Mater.* 2010, *17*, 643-649.
[68] Cerveau, G.; Corriu, R. J. P.; Costa, N. *Journal Non-Cryst. Solids* 1993, *163*, 226-235.
[69] Dubois, G.; Corriu, R. J. P.; Reyé, C.; Brandés, S.; Denat, F.; Guilard, R. *Chem. Commun.* 1999, 2283-2284.
[70] Zhang, W. H.; Zhang, X. N.; Hua, Z. L.; Harish, P.; Schroeder, F.; Hermes, S.; Cadenbach, T.; Shi, J. L.; Fischer, R. A. *Chem. Mater.* 2007, *19*, 2663-2670.
[71] Cho, E. B.; Kim, D.; Jaroniec, M. *Langmuir* 2009, *25*, 13258-13263.
[72] Grudzień, R. M.; Grabicka, B. E.; Pikus, S.; Jaroniec, M. *Chem. Mater.* 2006, *18*, 1722-1725.
[73] Grudzień, R. M.; Blitz, J. P.; Pikus, S.; Jaroniec, M. *J. Colloid Interf. Sci.* 2009, *333*, 354-362.
[74] Naushad, Mu. *Ion Exch. Lett.* 2009, *2*, 1-14.
[75] Collinson, M. M. *Crit. Rev. Anal. Chem.* 1999, 29, 289-311.
[76] Philipp, G.; Schmidt, H. *J. Non-Cryst. Solids* 1984, *63*, 283-292.
[77] Jal, P. K.; Patel, S.; Mishra, B. K. *Talanta* 2004, *62*, 1005-1028.
[78] Srisuda, S.; Virote, B. *J. Environ. Sci.* 2008, *6*, 379-384.
[79] Wu, Z.; Joo, H.; Ahna, Ik. S.; Haama, S.; Kima, J. H.; Lee, K. *Chem. Eng. J.* 2004, *102*, 277-282.
[80] Wong, Y. O.; Miranda, P.; Rosenberg, E. *J. App.Polym. Sci.* 2010, *115*, 2855-2864.
[81] Hughes, M. A.; Wood, J.; Rosenberg, E. *Ind. Eng. Chem. Res.* 2008, *47*, 6765-6774.
[82] Beatty, S. T.; Fischer, R. J.; Hagers, D. L.; Rosenberg, E. *Ind. Eng. Chem. Res.* 1999, *38*, 4402-4408.
[83] Beatty, S. T.; Fischer, R. J.; Rosenberg, E.; Pang, D. *Sep. Sci. Technol.* 1999, *34*, 2723-2739.
[84] Allen, J.; Rosenberg, E.; Chierotti, M. R.; Gobetto, R. *Inorg. Chim. Acta* 2010, *363*, 617-624.
[85] Shiraishi, Y.; Nishimura, G.; Takayuki, H.; Komasawa, I. *Ind. Eng. Chem. Res.* 2002, *41*, 5065-5070.
[86] Chandra, D.; Dutta, A.; Bhaumik, A. *Eur. J. Inorg. Chem.* 2009, 4062-4068.
[87] Dai, S.; Burleigh, M. C.; Ju, Y. H.; Gao, H. J.; Lin, J. S.; Pennycook, S. J.; Barnes, C. E.; Xue, Z. L. *J. Am. Chem. Soc.* 2000, *122*, 992-993.
[88] Lu, Y. K.; Yan, X. P. *Anal. Chem.* 2004, *76*, 453-457.
[89] Zheng, F.; Tran, D. N.; Busche, B.; Fryxell, G. E.; Addleman, R. S.; Zemanian, T. S.; Aardahl, Ch. L.; *Preprints Pap. Am. Chem. Soc., Div. Fuel Chem.* 2004, *49*, 261-262.
[90] Liang, C.; Dai, S. *J. Am. Chem. Soc.* 2006, *128*, 5316-5317.
[91] Meng, Y.; Gu, D.; Zhang, F.; Shi, Y.; Cheng, L.; Feng, D.; Wu, Z.; Chen, Z.; Wan, Y.; Stein, A.; Zhao, D. *Chem. Mater.* 2006, *18*, 4447-4464.
[92] Liang, C., Hong, K. L.; Guiochon, G. A.; Mays, J. W.; Dai, S. *Angew. Chem. Int. Edit.* 2004, *43*, 5785-5789.
[93] Liang, C.; Li, Z.; Dai, S. *Angew. Chem. Int. Edit.* 2008, *47*, 3696-3717.
[94] Fulvio, P. F.; Jaroniec, M.; Liang, Ch.; Dai, S. *J. Phys. Chem. C*, 2008, *112*, 3126-13133.
[95] Yao, J.; Wang, H.; Liu, J.; Chan, K. Y.; Zhang, L.; Xu, N. *Carbon* 2005, *43*, 1709-1715.
[96] Aggarwal, R. K.; Bhatia, G.; Bahl, O. P. *J. Mater. Sci.* 1990, *25*, 4604-4606.

[97] Inagaki, M.; Kato, M.; Morishita, T.; Morita, K.; Mizuuchi, K. *Carbon* 2007, *45*, 1121-1124.

[98] Chen, J. C.; Harrison, R. I. *Carbon* 2002, *40*, 25-45.

[99] Sreekumar, T. V.; Liu, T.; Min, B. G.; Guo, H.; Kumar, S.; Hauge, R. H.; Smalley, R. E. *Adv. Mater.* 2004, *16*, 58-61.

[100] Yang, Ch. Ch.; Gu, W. T.; Wu, K. H.; Pen, Y. H. Organic siloxane composite material containing polyaniline/carbon black and preparation method thereof. US Patent 524588.

[101] Górka, J.; Fenning, Ch.; Jaroniec, M. *Colloids Surf. A. Physicochem. Eng. Aspects* 2009, *352*, 113-117.

[102] Jaroniec, M.; Górka, J.; Choma, J.; Zawislak, A. *Carbon* 2009, *47*, 3034-3040.

[103] Shin, Y.; Fryxell, G. E.; Um, W.; Parker, K.; Mattigod, S. V.; Skaggs R. *Adv. Funct. Mater.* 2007, *17*, 2897-2901.

[104] Yanagisawa, H.; Matsumoto, Y.; Machida, M. *App. Surface Sci.* 2010, *256*, 1619-1623.

[105] Suzuki, N.; Machida, M.; Fujimura, Y.; Aikawa, M.; Tatsumoto, H. *TANSO* 2007, *229*, 242-248.

[106] Zhimang, G.; Baolin, D. *Environ. Eng. Sci.* 2007, *24*, 113-121.

[107] Kikuchi, Y.; Qian, Q.; Machida, M.; Tatsumoto, H. *Carbon* 2006, *44*, 195-202.

[108] Wang, T. C.; Chandra, K. P.; Anderson, P. R. *National Conference on Environmental Engineering*, 1994, 225-232.

[109] Vaughan, R. L.; Reed, B. E. *Water Res.* 2005, *39*, 1005-1014.

[110] Dzombak, D. A.; Morel, F. M. M. Surface complexation modeling hydrous ferric oxide. Wiley-Interscience: New York, NY 1990; pp 1-416.

[111] Reed, B. E.; Matsumoto, M. R. *J. Environ. Eng.* 1993, *119*, 332-348.

[112] Daus, B.; Wennrich, R.; Weiss, H. *Water Res.* 2004, *38*, 2948-2954.

[113] Daus, B.; Weiss, H. Testing of sorption materials for arsenic removal from waters, In *Environmental Science Research in Chemistry for the Protection of the Environment 4, Series: Environmental Science Research;* Mournighan, R.; Dudzinska, M. R.; Barich, J.; Gonzalez, M. A.; Black, R. K. Springer, New York, 2005; pp 23-28.

[114] Huang, C. P.; Vane, L. M. *J. Water Poll. Cont. Fed.* 1989, *61*, 1596-1603.

[115] Peraniemi, S.; Hannonen, S.; Mustalahti, H.; Ahlgren, M. *Anal. Bioanal. Chem.* 1994, *349*, 510-515.

[116] Zhang, X.; Zhu, W.; Liu, H.; Wang, T. *Mater. Lett.* 2004, *58*, 2223-2226.

117] Jha, V. K.; Matsuda, M.; Miyake, M. *J. Hazard. Mater.* 2008, *160*, 148-153.

[118] Chen, W.; Cannon, F. S.; Rangel-Mendez, J. R. *Carbon* 2005, *43*, 573-580.

[119] Na, C.; Cannon, F. S.; Hagerup, B. *J. Am. Water Works Assoc. 2002, 94,* 90-102.

[120] Khare, R.; Bose, S. *J. Miner. Mater. Character. Eng.* 2005, *4*, 31-46.

[121] Homma, Y.; Kobayashi, Y.; Ogino, Y.; Takagi, D.; Ito, R.; Jung, Y. J.; Ajayan, P. M. *J. Phys. Chem. C* 2003, *107*, 12161-12164.

[122] Kar, S.; Bindal, R. C.; Prabhakar, S.; Tewari, P. K.; Dasgupta, K.; Sathiyamoorthy, D. *Int. J. Nucl. Des.* 2008, *3*, 143-150.

[123] Shan, G.; Yan, S.; Tyagi, R. D.; Surampalli, R. Y.; Zhang, T. C. *Pract. Periodical Hazard. Toxic Radioact. Waste Manag.* 2009, *13*, 110-119.

[124] An, W.; Turner, C. H. *Chem. Phys. Lett.* 2009, *482*, 274-280.

[125] An, W.; Turner, C. H. *J. Phys. Chem. C* 2009, *113*, 7069-7078.

[126] Flahaut, E.; Peigney, A.; Laurent, Ch.; Marlie`Re, Ch.; Chastel, F.; Rousset, A. *Acta Mater.* 2000, *48*, 3803-3812.
[127] Peigney, A.; Flahaut, E.; Laurent, Ch.; Chastel, F.; Rousset, A. *Chem. Phys. Lett.* 2002, *352*, 20-25.
[128] Hsieh, S. H.; Horng, J. J. *Miner. Metall. Mater.* 2007, *14*, 77-84.
[129] Warshawsky, A. Extraction with Solvent-Impregnated Resins; In *Ion Exchange and Solvent Extraction,* Editors J. Marinsky, Y. Marcus, Vol. 8, Marcel Dekker: New York, NY, 1981; pp 1-438.
[130] Cumbal, L.; Greenleaf, J.; Leun D.; SenGupta, A. K. *React. Funct. Polym.* 2003, *54*, 167-180.
[131] Rudge, S. R.; Kurtz, T. L.; Vessely, C. R.; Catterall, L. G.; Williamson, D. L. *Biomaterials* 2000, *21*, 1411-1420.
[132] Peng, Z. G.; Hidajat, K.; Uddin, M. S. *J. Colloid Interf. Sci.* 2004, *271*, 277-283.
[133] Pourfarzaneh, M.; White, G. W.; Smith, D. S. *Clin. Chem.* 1995, *26*, 730-733.
[134] Sonti, S. B.; Bose, A. J. *Colloids Interf. Sci.* 1995, *170*, 575-585.
[135] Yang, Ch.; Guan, Y.; Xing, J.; Shan, G.; Liu, H. *J. Polym. Sci. A: Polym. Chem.* 2008, *46*, 203-210.
[136] Chang, C. F.; Chang, C. Y.; Hsu, T. L. *Colloids Surf. A: Physicochem. Eng. Aspects* 2008, *327*, 64-70.
[137] Chang, C. F.; Lin, P. H.; Höll, W. *Colloids Surf. A: Physicochem. Eng. Aspects* 2006, *280*, 194-202.
[138] Ziolo, R. F.; Giannelis, E. P.; Weinstein, B. A.; Òhoro, M. P.; Ganguly, B. N.; Mehrotra, V.; Russell, M. W., Huffman D. R. *Science* 1992, *257*, 210-223.
[139] SenGupta, A. K.; Zhu, Y.; Hauze, D. *Environ. Sci. Technol.* 1991, *25*, 481-488.
[140] Zhao, D.; SenGupta, A. K. *Water Res.* 1998, *32*, 1613-1625.
[141] Li, P.; SenGupta, A. K. *Environ. Sci. Technol.* 1998, *32*, 3756-3766.
[142] Yanase, N.; Noguchi, H.; Asakura, H.; Suzata, T. *J. App. Polym. Sci.* 1993, *50*, 765-776.
[143] Lu, S.; Forcada, J. *J. Polym. Sci. A: Polym. Chem.* 2006, *44*, 4187-4203.
[144] Horák, H.; Benedyk, N. *J. Polym. Sci. A: Polym. Chem.* 2004, *42*, 5827-5837.
[145] Lee, Y.; Rho, J.; Jung, B. *J. App. Polym. Sci.* 2003, *89*, 2058-2067.
[146] Müller-Schulte, D.; Brunner, H. *J. Chromatogr. A* 1995, *711*, 53-60.
[147] Yang, C.; Guan, Y.; Xing, J.; Liu, J.; An, Z.; Shan, G.; Liu, H. *AIChE J.* 2005, *51*, 2011-2015.
[148] Liu, X.; Guan, Y.; Liu, H.; Ma, Z.; Yang, Y.; Wu, X. *J. Magnetism Magnetic Mater.* 2005, *293*, 111-118.
[149] Yang, C.; Xing, J.; Guan, Y.; Liu, H. *App. Microbiol. Biotechnol.* 2006, *72*, 616-622.
[150] Yang, C.; Liu, H.; Guan, Y.; Xing, J.; Liu, J.; Shan, G. *J. Magnetism Magnetic Mater.* 2005, *293*, 187-192.
[151] Takahashi, K.; Tamura, Y.; Kodera, Y.; Mihama, T.; Saito, Y.; Inada, Y. *Biochem. Biophys. Res. Commun.* 1987, *142*, 291-296.
[152] Clark, D. S.; Bailey, J. E.; Yen, R.; Rembaum, A. *Enzyme Microbial Technol.* 1984, *6*, 317-320.
[153] Hu, T. T.; Wu, J. W. *Chem. Eng. Res. Design* 1987, *65*, 238-242.
[154] Burns, M. A.; Graves, D. J. *Biotechnology. Progress* 1985, *1*, 95-103.

[155] Moeser, G. D.; Roach, K. A.; Green, W. H.; Laibinis, P. E.; Hatton, T. A. *Ind. Eng. Chem. Res.* 2002, *41*, 4739-4749.
[156] Bajpai, S. K.; Armo, M. K.; Namdeo, M. *Acta Chim. Slovenica* 2009, *56*, 254-261.
[157] Namdeo, M.; Bajpai, S. K. *Colloids Surf. A: Physicochem. Eng. Aspects* 2008, *320*, 161-168.
[158] Namdeo, M.; Bajpai, S. K. *Electronic J. Environ. Agricultural Food Chem.* 2008, *7*, 3082-3094.
[159] Tseng, J. Y.; Chang, Ch. Y.; Chen, Y. H.; Chang, C. F.; Chiang, P. Ch. *Colloids Surf. A: Physicochem. Eng. Aspects* 2007, *295*, 209-216.
[160] Tseng, J. Y.; Chang, Ch. Y.; Chang, Ch. F.; Chen, Y. H.; Chang, Ch. Ch.; Ji, D. R.; Chiang, P. Ch. *J. Hazard. Mater.* 2009, *171*, 370-377.
[161] Denizli, A.; Özkan, G.; Arica, M. Y. *J. App. Polym. Sci.* 2000, *78*, 81-89.
[162] Atia, A. A.; Donia, A. M.; Yousif, A. M. *Sep. Pur. Technol.* 2008, *61*, 348-357.
[163] Atia, A. A.; Donia, A. M.; Elwakeel, K. Z. *Sep. Pur. Technol.* 2005, *43*, 43-48.
[164] Atia, A. A.; Donia, A. M.; Awed, H. A. *J. Hazard. Mater.* 2008, *155*, 100-108.
[165] Horák D.; Semenyuk N.; Lednický F. *J. Polym. Sci. A: Polym. Chem.* 2003, *41*, 1848-1863.
[166] Denizli, A.; Say, R. *J. Biomater. Sci. Polym. Ed.* 2001, *12*, 1059-1073.
[167] Denizli, A.; Tanyolaç, D.; Salih, B.; Özdural, A. *J. Chromatogr. A* 1998, *793*, 47-56.
[168] Denizli, A.; Salih, B.; Pişkin, E. *React. Funct. Polym.* 1996, *29*, 11-19.
[169] Zhang, C. C.; Li, X.; Pang, J. X. *J. App. Polym. Sci.* 2001, *82*, 1587-1592.
[170] Donia, A. M.; Atia, A. A.; El-Boraey, H. A.; Mabrouk, D. H. *Sep. Pur.Technol.* 2006, *48*, 281-287.
[171] Atia, A. A.; Donia, A. M.; Shahin, A. E. *Sep. Pur. Technol.* 2005, *46*, 208-213.
[172] Min, J. M.; Hering, J. *Water Res.* 1998, *32*, 1544-1552.
[173] Zouboulis, A. I.; Katsoyiannis, I. A. *Ind. Eng. Chem. Res.* 2002, *41,* 6149-6155.
[174] Chen, K. L.; Mylon, S. E.; Elimelech, M. *Langmuir* 2007, *23*, 5920-5928.
[175] SenGupta, A. K.; Cumbal, L. H. Method of manufacture and use of hybrid anion exchanger for selective removal of contaminating fluids, US Patent 20,050,156,136, 2005.
[176] Jang, M.; Chen, W. F.; Cannon, F. S. *Environ. Sci. Technol.* 2008, *42*, 3369-3374.
[177] Zhuang, J. M.; Hobenshield, E.; Walsh, T. *Environ. Technol.* 2008, *29*, 401-411.
[178] Hana, R. P.; Zou, W. H.; Li, H. K.; Li, Y. H.; Shi, J. *J. Hazard. Mater.* 2006, *137*, 934-942.
[179] Jang, M.; Min, S. H.; Kim, T. H.; Park, J. K. *Environ. Sci. Technol.* 2006, *40*, 1636-1643.
[180] Jang, M.; Min, S. H.; Park, J. K.; Tlachac, E. J. *Environ. Sci. Technol.* 2007, *41*, 3322-3328.
[181] Guo, X. J.; Chen, F. H. *Environ. Sci. Technol.* 2005, *39*, 6808-6818.
[182] Brandao, M.; Galembeck, F. *Colloids Surf. A: Physicochem. Eng. Aspects* 1990, *48*, 351-362.
[183] Hansen, B. O.; Kwan, P.; Benjamin, M. M.; Li, C. W.; Korshin, G. V. *Environ. Sci. Technol.* 2001, *35*, 4905-4909.
[184] Tsadilas, C. D.; Dimoyiannis, D.; Samaras, V. *Commun. Soil Sci. Plant Anal.* 1998, *29*, 2347-2353.

[185] Zachara, J. M.; Smith, S. C.; Kuzel, L. S. *Geochim. Cosmochim. Acta* 1995, *59*, 4825-4844.

[186] Lenoble, V.; Chabroullet, C.; Shukry, R.; Serpaud, B.; Deluchat, V.; Bollinger, J. C. *J. Colloid Interf. Sci.* 2004, *280*, 62-67.

[187] Lenoble, V.; Laclautre, C.; Serpaud, B.; Deluchat, V.; Bollinger, J. C. *Sci. Total Environ.* 2004, *326*, 197-207.

[188] Cumbal, L.; SenGupta, A. K. *Environ. Sci. Technol.* 2005, *39*, 6508-6515.

[189] DeMarco, M. J.; SenGupta, A. K.; Greenleaf, J. E. *Water Res.* 2003, *37*, 164-176.

[190] Chou, S.; Huang, C.; Huang, Y. *Environ. Sci. Technol.* 2001, *35*, 1247-1251.

[191] Greenleaf, J. E.; Lin, J. Ch.; SenGupta, A. K. *Environ. Prog.* 2006, *25*, 300-311.

[192] Greenleaf, J. E.; Cumbal, L.; Staina, I.; SenGupta, A. K. *Chem. Eng. Res. Design* 2003, *81*, 1-12.

[193] Crosby, S. A.; Glasson, D. R.; Cuttler, A. H.; Butler, I.; Turner, D. R.; Whitfield, M.; Millward, G. E. *Environ. Sci. Technol.* 1983, *17*, 709-713.

[194] Gu, B.; Brown, G. M.; Maya, L.; Lance, M. J.; Moyer, B. *Environ. Sci. Technol.* 2001, *35*, 3363-3368.

[195] Gu, B.; Ku, Y. K.; Brown, G. M. *Federal Facilities Environ. J.* 2003, *14*, 75-94.

[196] Pan, B.; Wu, J.; Pan, B.; Lv, L.; Zhang, W.; Xiao, L.; Wang, X.; Tao, X.; Zheng, S. *Water Res.* 2009, *43*, 4421-4429.

[197] Pan, B. C.; Chen, X. Q.; Zhang, W. M.; Pan, B. J.; Shen, W.; Zhang, Q. J.; Du, W.; Zhang, Q. X.; Chen, J. L. A process to prepare a polymer-based hybrid sorbent for arsenic removal. Chinese Patent: CN 200510095177.5 (2007).

[198] Lin, J. CH.; SenGupta, A. K. *Environ. Eng. Sci.* 2009, *26*, 1673-1683.

[199] Pan, B. C.; Su, Q.; Zhang, W. M.; Zhang, Q. X.; Ren, H. Q.; Zhang, Q. R. A process to prepare a hybrid sorbent by impregnating hydrous manganese dioxide (HMO) nanoparticles within polymer for enhanced removal of heavy metals, Chinese Patent No. ZL 200710134050.9 (2007).

[200] Mustafa, S.; Ahmad, T.; Naeem, A.; Shah, K. H.; Waseem, M. *Water Air Soil Poll.* 2010, *210*, 43-50.

[201] Moon, J. K.; Kim, K. W.; Jung, Ch. H.; Shul, Y. G.; Lee, E. H. *J. Radioanal. Nucl. Chem.* 2000, *246*, 299-307.

[202] John, J.; Šebesta, F.; Molt, A. *Radiochim. Acta* 1997, *78*, 131-135.

[203] Nilchi, A.; Hadjmohammadi, M. R.; Rasouli Garmarodi, S.; Saberi, R. *J. Hazard. Mater.* 2009, *167*, 531-535.

[204] Szeglowski, Z.; Kubica, B.; Tuteja-Krysa, M.; Fijalkowski, R. Kompozytowy Magnetyczny Wymieniacz Jonowy. PL 196361.

[205] Narbutt, J.; Bartos, B.; Bilwewicz, A.; Szeglowski, Z. Method of obtaining composite ion exchangers. US Patent, 4,755,322 (1988).

[206] Kubica, B.; Godunowa, H.; Tuteja-Krysa, M.; Stobiński, M.; Misiak, R. *J. Radioanal. Nucl. Chem.* 2004, *262*, 721-724.

[207] Kubica, B.; Tuteja-Krysa, M.; Szeglowski, Z. *J. Radioanal. Nucl. Chem.* 1999, *242*, 541-544.

[208] Sheha, R. R.; El-Zahhar, A. A. *J. Hazard. Mater.* 2008, *150*, 795-803.

[209] Šebesta F. *J. Radioanal. Nucl. Chem.* 1997, *220*, 77-88.

[210] Šebesta, F. Exchanger composed from an active component and binding organic matrix and the way of its production. Czech Patent A.O. 181 605 (1980).

[211] Yates, S. F.; Bedwell, W. Magnetically stabilized fluidized particles in liquid media. US Patent, 5,230,805 (1993).
[212] Liu, J.; Xu, T.; Fu, Y. *J. Non-Cryst. Solids* 2005, *351*, 3050-3059.
[213] Liu, J.; Xu, T.; Fu, Y. *J. Membrane Sci.* 2005, *252*, 165-173.
[214] Liu, J.; Wang, X.; Xu, T.; Shao, G. *Sep. Pur. Technol.* 2009, *66*, 135-142.
[215] Liu, J. S; Ma, Y.; Zhang, Y.; Shao, G. *J. Hazard. Mater.* 2010, *173*, 438-444.
[216] Liu, J.; Ma Y.; Xu, T.; Shao, G. *J. Hazard. Mater.* 2010, *178*, 1021-1029.
[217] Gao, H. W.; Lin, J.; Li, W. Y.; Zhang, J. H.; Zhang, Y. L. *Environ. Sci. Poll. Res.* 2010, *17*, 78-83.
[218] Hu, Z. J.; Xiao, Y.; Zhao, D. H.; Shen, Y. L.; Gao, H. W. *J. Hazard. Mater.* 2010, *175*, 179-186.
[219] Dzyazko, Y.; Vasilyuk, S.; Rozhdestvenskaya, L.; Belyakov, V.; Stefanyak, N.; Kabay, N.; Yüksel, M.; Arar, Ö.; Yüksel, Ü. *Chem. Eng. Commun.* 2009, *196*, 22-38.
[220] Pan, B. C.; Zhang, Q. R.; Zhang, W. M.; Pan, B. J.; Du, W.; Lv, L.; Zhang, Q. J.; Xu, Z. W.; Zhang, Q. X. *J. Colloid Interf. Sci.* 2007, *310*, 99-105.
[221] Zhang, Q. R.; Pan, B. C.; Pan, B. J.; Zhang, W. M.; Jia, K.; Zhang, Q. X. *Environ. Sci. Technol.* 2008, *42*, 4140-4145.
[222] Pan, B. C.; Su, Q.; Zhang, W. M.; Zhang, Q. X.; Ren, H. Q.; Zhang, Q. R. A process to prepare a hybrid sorbent by impregnating hydrous manganese dioxide (HMO) nanoparticles within polymer for enhanced removal of heavy metals. Chinese Patent No. ZL 200710134050.9 (2007).
[223] Sizgek, G. D.; Grifffh, Ch. S.; Sizgek, E.; Luca, V. *Langmuir* 2009, *25*, 11874-11882.
[224] Bang, S.; Patel, M.; Lippincott, L.; Meng, X. *Chemosphere* 2005, *60*, 389-397.
[225] Kasuga, T.; Hiramatsu, M.; Hoson, A.; Sekino, T.; Niihara, K. *Adv. Mater.* 1999, *11*, 1307-1311.
[226] Ooi, K.; Ashida, K.; Katoh, S.; Sugasaka, K. *J. Nucl. Sci. Technol.* 1987, *24*, 315-322.
[227] Nilchi, A.; Khanchi, A.; Atashi, H.; Bagheri, A.; Nematollahi, L. *J. Hazard. Mater.* 2006, *137*, 1271-1276.
[228] Alam, Z.; Inamuddin, Nabi, S. A. *Desalination* 2010, *250*, 515-522.
[229] Inamuddin, Ismail, Y. A. *Desalination* 2010, *250*, 523-529.
[230] Khan, A. M.; Ganai, S. A.; Nabi, S. A. *Colloids Surf. A: Physicochem. Eng. Aspects* 2009, *337*, 141-145.
[231] Nabi, S. A.; Khan, A. M. *React. Funct. Polym.* 2006, *66*, 495-508.
[232] Zhang, B.; Poojary, D. M.; Peng, A. C. G. *Chem. Mater.* 1996, *8*, 1333-1340.
[233] Ferragina, C.; Cafarelli, P.; De Stefanis, A.; Di Rocco, R.; Giannoccaro, P. *Mater. Res. Bull.* 2001, *36*, 1799-1812.
[234] Nabi, S. A.; Naushad, Mu.; Bushra, R. *Chem. Eng. J.* 2009, *152*, 80-87.
[235] Niwas, R.; Gupta, U.; Khan, A. A.; Varshney, K. G. *Colloids Surf. A: Physicochem. Eng. Aspects* 2000, *164*, 115-119.
[236] Viswanathan, N.; Meenakshi, S. *J. Hazard. Mater.* 2010, *176*, 459-465.
[237] Khan, A. A.; Inamuddin; Alam, M. M. *Mater. Res. Bull.* 2005, *40*, 289-305.
[238] Niwas, R.; Khan, A. A.; Vershney, K. G. *Colloids Surf. A: Physicochem. Eng. Aspects* 1999, *150*, 7-14.
[239] Khan, A. A.; Niwas, R.; Alam, M. M. *Indian J. Chem. Technol.* 2002, *9*, 256-260.
[240] Pandit, B.; Chudasma, U. *Bull. Mater. Sci.* 2001, *24*, 265-261.
[241] Gupta, A. P.; Agarwal, H.; Ikram, S. *J. Indian Chem. Soc.* 2003, *80*, 57-59.

[242] Maeda, K. *Micropor. Mesopor. Mater.* 2004, *73*, 47-55.
[243] Fang, Ch.; Chen, Z.; Liu, X.; Yang, Y.; Deng, M.; Weng, L.; Jia, Y.; Zhou, Y. *Inorg. Chim. Acta* 2008, *361*, 3785-3799.
[244] Cecconi, F.; Ghilardi, C. A.; Gili, P.; Midollini, S.; Luis, P. A. L.; Lozano-Gorrín, A. D.; Orlandini, A. *Inorg. Chim. Acta* 2001, *319*, 67-74.
[245] Clearfield, A.; Wang, Z. *J. Chem. Soc. Dalton Trans.*, 2002, 2937-2947.
[246] Wu, J.; Hou, H.; Han, H.; Fan, Y. *Inorg. Chem.* 2007, *46*, 7960-7970.
[247] Katz, H. E. *Chem. Mater.* 1994, *6*, 2227-2232.
[248] Soghomonian, V.; Chen, Q.; Haushalter, R. C.; Zubieta, J. *Angew. Chem. Int. Edit.* 1995, *34*, 223-226.
[249] Drumel, S.; Janvier, P.; Barboux, P.; Bujoli-Doeuff, M.; Bujoli, B. *Inorg. Chem.* 1995, *34*, 148-156.
[250] Drumel, S.; Bujoli-Doeuff, M.; Janvier, P.; Bujoli, B. *New J. Chem.* 1995, *19*, 239-242.
[251] Johnson, J. W.; Jacobson, A. J.; Lewandowski, J. T. *Inorg. Chem.* 1984, *23*, 3842-3844.
[252] Poojary, D. M.; Zhang, B.; Clearfield, A. *J. Am. Chem. Soc.* 1997, *119*, 12550-12559.
[253] Kimura, T. *Chem. Mater.* 2003, *15*, 3742-3744.
[254] Clearfield, A. *Current Opinion Solid State Mater. Sci.* 1996, *1*, 268-278.
[255] Anirudhan, T. S.; Suchithra, P. S. *J. Environ. Sci.* 2009, *21*, 884-891.
[256] Cunha, A. G.; Freire, C. S. R.; Silvestre, A. J. D.; Neto, C. P.; Gandini, A. *Carbohydrate Polym.* 2010, *80*, 1048-1056.
[257] Barud, H. S.; Assunção, R. M. N.; Martines, M. A. U.; Dexpert-Ghys, J.; Marques, R. F. C.; Messaddeq, Y.; Riberio, S. J. L. *J. Sol-Gel Sci. Technol.* 2006, *46*, 363-367.
[258] Portugal, I.; Dias, V. M.; Duarte, R. F., Evtuguin, D. V. *J. Phys. Chem. B,* 2010, *114*, 4047-4055
[259] Gill, R.; Marquez, M.; Larsen, G. *Micropor. Mesopor. Mater.* 2005, *85*, 129-135.
[260] Razaq, A.; Mihranyan, A.; Welch, K.; Nyholm, L.; Strømme, M. *J. Phys. Chem. B*, 2009, *113*, 426-433.
[261] Aranaz, I.; Mengíbar, M.; Harris, R.; Paños, I., Miralles, B.; Acosta, N.; Galed, G.; Heras, Á. *Curr. Chem. Biology* 2009, *3,* 203-230.
[262] Sicupira, D.; Campos, K.; Vincent, T.; Leao, V. A.; Guibal, E. *Adv. Mater. Res.* 2009, *71-73*, 733-736.
[263] Saitoh, T.; Sugiura, Y.; Asano, K.; Hiraide, M. *React. Funct. Polym.* 2009, *69*, 792-796.
[264] Crini, G.; Badot, P. M. *Progress Polym. Sci.* 2008, *33*, 399-447.
[265] Guibal, E. *Sep. Purif. Technol.* 2004, *38*, 43-74.
[266] Guibal, E.; Van Vooren, M.; Dempsey, B. A.; Roussy, J. *Sep. Pur.Technol.* 2006, *41*, 2487-2514.
[267] Roussy, J.; Van Vooren, M.; Dempsey, B. A.; Guibal, E. *Water Res.* 2005, *39*, 3247-3258.
[268] Muzzarelli, C.; Muzzarelli, R. A. A. *J. Inorg. Biochem.* 2002, *92*, 89-94.
[269] Sundaram, C. S.; Viswanathan, N.; Meenakshi, S. *J. Hazard. Mater.* 2009, *172*, 147-151.
[270] Gupta, A.; Chauhan, V. S.; Sankararamakrishnan, N. *Water Res.* 2009, *43*, 3862-3870.
[271] Miretzky, P.; Cirelli, A. F. *J. Hazard. Mater.* 2009, *167*, 10-23.

[272] Copello, G. J.; Varela, F.; Vivot, R. M.; Díaz, L. E. *Bioresource Technol.* 2008, *99*, 6538-6544.

[273] Veera, M.; Boddu, V. M.; Abburi, K.; Talbott, J. L.; Smith, E. D.; Haasch, R. *Water Res.* 2008, *42*, 633-642.

[274] Torres, J. D.; Faria, E. A.; SouzaDe, J. R.; Prado, A. S. G. *J. Photochem. Photobiology A: Chem.* 2006, *182*, 202-206.

[275] Zhu, H. Y.; Jiang, R.; Xiao, L. *App. Clay Sci.* 2010, *48*, 522-526.

[276] Genc, Ö.; Arpa, C.; Bayramoğlu, G.; Arıca, M. Y.; Bektaş, S. *Hydrometallurgy* 2002, *67*, 53-62.

[277] Xiong, Y.; Liu, Q. L.; Zhang, Q. G.; Zhu, A. M. *J. Power Sources* 2008, *183*, 447-453.

[278] Cai, Z. S.; Song, Z. Q.; Shang, S. B.; Yang, C. S. *Polym. Bull.* 2007, *59*, 655-665.

[279] Konishi, Y.; Shimaoka, J. I.; Asai, S. *React. Func. Polym.* 1998, *36*, 197-206.

[280] Konishi, Y.; Asai, S.; Midoh, Y.; Oku, M. *Sep. Sci. Technol.* 1996, *28*, 1691-1699.

[281] Sone, H.; Fugetsu, B.; Tanaka, S. *J. Hazard. Mater.* 2009, *162*, 423-429.

In: Advanced Organic-Inorganic Composites
Editor: Inamuddin

ISBN 978-1-61324-264-3

Chapter 4

ORGANIC-INORGANIC COMPOSITES FOR SOLAR CELLS

***Jing-Shun Huang*[a]*, Chen-Yu Chou* [a] *and Ching-Fuh Lin*[a,b,*]**
[a]Graduate Institute of Photonics and Optoelectronics and
[b]Department of Electrical Engineering, National Taiwan University,
Taipei 106, Taiwan

ABSTRACT

As the search for alternative sources of energy other than fossil fuels continues to expand, photovoltaic technology (direct conversion of solar energy into electrical energy) has been identified as one of the promising technologies. Solar cells based on blends of conjugated polymers and fullerene derivatives have recently attracted significant attention due to their great promise for the realization of printable, portable, flexible, and low-cost renewable energy sources. However, it is not easy to precisely control the nanoscale morphology of photoactive layer which seriously affects the carrier transport. In addition, control of the charge transport at organic-electrode or organic-inorganic interface is also challenging in organic-based solar cells. Quality of the electrode interface is also critical for the device stability. Development of the organic-inorganic composites for solar cells offers an alternative way to realize high performance and low cost solar cells. This chapter introduces the organic-ZnO nanorod composite solar cells. The environmentally friendly ZnO nanorod arrays can be grown in an aqueous solution with vertical alignment and small rod-to-rod spacing. This provides a solution-based route to the fabrication of low-cost organic-inorganic photovoltaic devices with highly oriented ZnO nanorod arrays. In order to achieve better exciton dissociation and charge transport, three types of interfacial modifications are demonstrated here. The addition of PCBM (6,6-phenyl-C_{61}-butyric acid methyl ester) clusters can enhance the phase separation and optical absorption. Inserting TiO_2 nanoparticles leads to a formation of double heterojunction, providing efficient exciton dissociation and charge transfer. The insertion of V_2O_5 nanopowder can suppress the leakage current and enhance the absorption. With the PCBM clusters, TiO_2 nanoparticles, or V_2O_5 nanopowder, their power conversion efficiencies can be significantly improved. Furthermore, these

[*] Corresponding author's e-mail: cflin@cc.ee.ntu.edu.tw

interfacial modifications are all fabricated by solution approaches. Compared to the vacuum-deposited techniques, these approaches are simple, expeditious, and effective. They are also advantageous for potential applications to mass production of various large-area printed electronics and photonics with a very low cost.

ABBREVIATIONS

η_{abs}	Photon absorption efficiency
η_{cc}	The fraction of carriers collected at the electrodes
η_{diff}	Exciton diffusion process
η_{tc}	The fraction of excitons that dissociate into free carriers at a donor-acceptor interface before recombining
η_{tr}	Carrier transport process
λ	Wavelength of the incident light
ν	Frequency of the incident light
σ_{RMS}	Root-mean-square roughness
1D	One-dimensional
AFM	Atomic force microscopy
Al	Aluminum
Ag	Silver
BHJ	Bulk-heterojunction
Btu	British thermal units
c	Speed of light in vacuum
CO_2	Carbon dioxide
CTC	Charge-transfer complexes
e	Elementary charge
EDX	Energy dispersive x-ray spectroscopy
EQE	External quantum efficiency
FESEM	Field-emission scanning electron microscopy
FF	Fill factor
h	Plank's constant
HOMO	Highest occupied molecular orbital
IPCE	Incident photon-to-current efficiency
IQE	Internal quantum efficiency
ITO	Indium tin oxide
J_d	Dark current density
J_M	Current density at maximum power point
J_{ph}	Photocurrent,
J_s	Saturation current density
J_{SC}	Short-circuit current density
J-V	Current density-voltage
k	Boltzmann constant
LUMO	Lowest unoccupied molecular orbital
MEH-PPV	Poly [2-methoxy-5-(2'-ethyl-hexyloxy)-1,4-pheny-lene vinylene]
MoO_3	Molybdenum oxide

NiO	Nickel oxide
NIR	Near-infrared
NREL	National Renewable Energy Laboratory
OSCs	Organic solar cells
P_{in}	Incident solar power density
P_{MAX}	Maximum power point of operation
P3HT	Poly(3-hexylthiophene)
PBDTTT	Poly[4,8-bis(2-ethylhexyloxy)-benzo[1,2-b:4,5-b']dithiophene-2,6-diyl-alt-(4-octanoyl-5-fluoro-thieno[3,4-b]thiophene-2-carboxylate)-2,6-diyl]
$PC_{70}BM$	6,6-phenyl-C_{71}-butyric acid methyl ester
PCBM	6,6-phenyl-C_{61}-butyric acid methyl ester
PCDTBT	Poly[N-9'-hepta-decanyl-2,7-carbazole-alt-5,5-(4',7'-di-thienyl-2'1',3'-b3nzothiadizaole)
PCE	Power conversion efficiency
PEDOT:PSS	Poly(3,4-ethylenedioxythiophene): polystyrene sulfonate
PFO-DBT	Poly[2,7-(9,9-dioctyl-fluorene)-alt-5,5-(4,7'-di-2-thienyl-2', 1', 3'-benzothiadiazole)
PL	Photoluminance
PV	Photovoltaic
R_{SH}	Shunt resistance
R_S	Series resistance
RR	Rectification ratio
T	Temperature
TiO_2	Titanium oxide
TiO_x	Titanium suboxide
UPS	Ultraviolet photoemission spectroscopy
UV-vis	Ultra violet visible
V	Voltage
V_M	Voltage at maximum power point
V_{OC}	Open-circuit voltage,
V_2O_5	Vanadium oxide
WO_3	Tungsten oxide
XRD	X-ray diffraction
XPS	X-ray photoemission spectroscopy
ZnO	Zinc oxide

1. Introduction

1.1. The Demand for Solar Cells

Global warming is causing changes in local and regional climates all over the world evidently. Carbon dioxide (CO_2), a by-product of burning fossil fuels, is perhaps the greatest contributor to the global warming. As a result, it is becoming clear that we have to find non-CO_2-releasing ways to create, transport, and store electricity. One promising way is photovoltaic (PV) technology. A PV device, or solar cell, directly converts sunlight into electrical energy. It has very little impact on the environment because it does not produce air pollution, hazardous waste, or noise. Furthermore, according to the projection reported by the U.S. Department of Energy [1], global energy consumption increases from 472 quadrillion British thermal units (Btu), 1 quadrillion Btu = 2.9×10^{11} kWh) in 2006 to 552 quadrillion Btu in 2015 and 678 quadrillion Btu in 2030 - a total increase of 44% over the projection period. The growing global energy needs and the finite supply of fossil fuel sources underscore the urgency of developing PV technology as a renewable energy resource.

In 1954, Chapin et al. developed the first solar cell at Bell Laboratories [2]. It was based on crystalline Si and exhibited an efficiency of 6%. Over the years the efficiency has reached 25% for crystalline Si solar cells in the laboratory [3]. Although the cost per watt of crystalline silicon solar cells has dropped significantly in recent years, they are still too expensive to compete with conventional grid electricity due to the need for high-purity materials, high temperature processing, and high vacuum equipments. Thus, PVs still account for less than 0.1% of the total world energy production. This is currently the main barrier that prevents PV technology from providing a large fraction of our electricity. For solar cells to gain widespread acceptance, the cost per watt of solar energy must be decreased. One potential alternative to crystalline silicon is conjugated (semiconducting) polymers. The conjugated polymers contain side chains that make them soluble in common organic solvents. This allows the polymers to be easily cast onto flexible substrates over a large area using solution-processing techniques, such as spin casting, dip coating, ink-jet printing [4], screen printing [5], and roll-to-roll processing [6]. These techniques represent an enormously attractive route for producing large-area PV cells at low cost. Being a part of third generation PV devices (Figure 1), polymer-based solar cells have great potential to provide higher efficiency and lower costs per watt of electricity generated.

1.2. Fundamentals of Organic Solar Cells (OSCs)

1.2.1. Brief History of OSCs

As early as 1959, the first investigation of OSCs was reported based on an anthracene single crystal. The solar cell exhibited a photovoltage of 200 mV with an extremely low efficiency [8]. However, such PV devices based on single (or homojunction) organic materials exhibit typical power conversion efficiencies below 0.1%, making them unsuitable for any possible application. Primarily, this is attributed to the fact that light absorption in organic materials almost always results in the strongly bound electron–hole pairs (excitons), rather than free electron–hole pairs as produced in inorganic semiconductors. In organic materials, the weak intermolecular forces localize the exciton on the molecules [9]. Because electron and hole wave functions are localized and the relative dielectric constants in polymers are low (typically 3) [10], the Coulomb attraction between the electron and hole is enhanced, leading to a large binding energy of excitons (0.1-1 eV [9]). This is in contrast to inorganic materials, where the exciton binding energy is only a few milli-electron volts and the relative dielectric constant is ~10. As a result, a much higher energy input than the thermal energy (kT ~0.0259 eV) is required to dissociate these excitons [11].

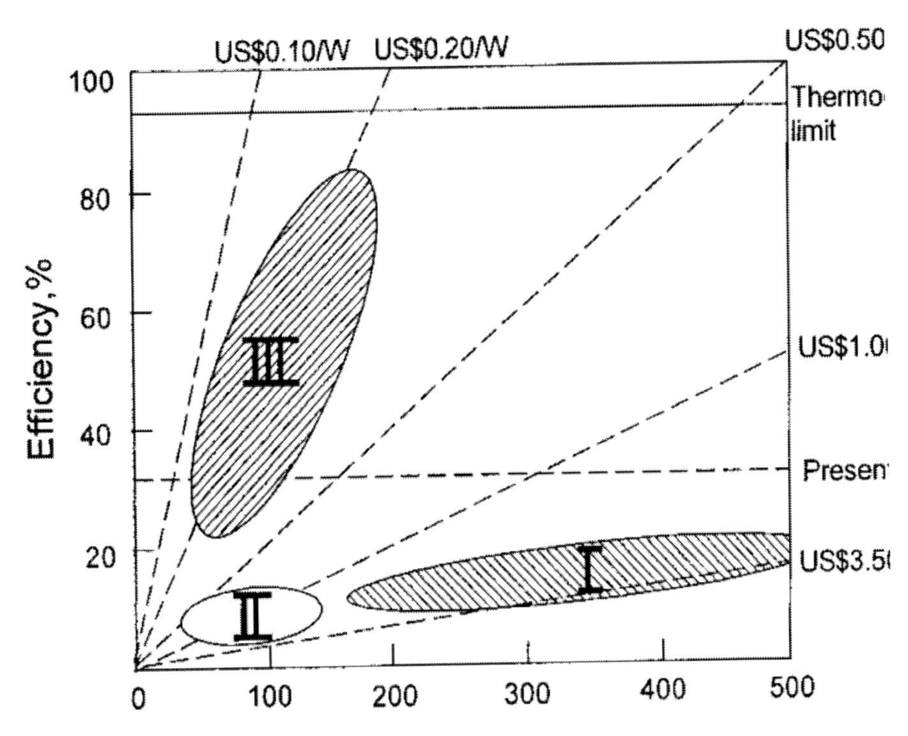

Figure 1. Efficiency and cost projections for first-, second-, and third-generation photovoltaic technology (wafers, thin-films, and advanced thin-films, respectively). Reproduced with permission from Green, M. A. *Prog. Photovoltaics: Research and Applications* 2001, *9,* 123-135. Copyright © 2001, Wiley & Sons, Inc.

In 1986, Tang discovered that much higher efficiencies (about 1%) can be achieved when an electron-donor material and an electron-acceptor material are brought together in one solar cell [12]. This heterojunction concept was a major breakthrough in cell performance at that time. The idea behind a heterojunction is to use two materials with different electron affinities (and/or ionization potentials). At the interface between the two materials, the resulting potentials are strong. Provided that the difference in potential energy is larger than the exciton binding energy, the electron will be accepted by the material with the larger electron affinity and the hole will be accepted by the material with the lower ionization potential. Thus, the excitons are dissociated into free charge carriers. In such PV cells, the excitons diffuse within the organic layers until they reach the interfaces. Since the exciton diffusion length is quite short (typically <10 nm) in organic materials [13,14], the excitons should be formed close to the interface (within the diffusion length). Otherwise, the excitons will decay and recombine instead of contributing to the photocurrent. As a result, the organic layer must be thin, leading to insufficient absorption most of the solar radiation flux.

In 1995, Yu et al. first introduced so-called bulk-heterojunction (BHJ) structure [15], where the donor and acceptor materials are blended together. This BHJ concept is so far the most successful device architecture for polymer solar cells because exciton harvesting is made near-perfect by creating a highly folded architecture such that all excitons are generated near the interfaces. In 2009, Park et al. reported BHJ polymer solar cells with internal quantum efficiency approaching 100% [16], indicating that nearly every absorbed photon leads to a separated pair of charge carriers and that almost all photogenerated carriers are collected at the electrodes. Chen et al. reported BHJ polymer solar cells based on a low band gap polymer (poly[4,8-bis(2-ethylhexyloxy)-benzo[1,2-b:4,5-b']dithiophene-2,6-diyl-alt-(4-octanoyl-5-fluoro-thieno[3,4-b]thiophene-2-carboxylate)-2,6-diyl]) (PBDTTT) with a power conversion efficiency (PCE) of 7.7% and an NREL (National Renewable Energy Laboratory, Boulder Colorado)-certified world-record PCE of 6.77% [17]. These exciting developments strongly push the polymer solar cells towards commercial viability.

1.2.2. Principles of Organic Solar Cells (OSCs)

Figure 2 (a) shows the device architecture of a typical BHJ OSC. Figure 2 (b) schematically shows the fundamental physical processes of the BHJ solar cell in an energetic diagram. The process of converting light into electric current in a BHJ polymer solar cell has four fundamental steps [11]:

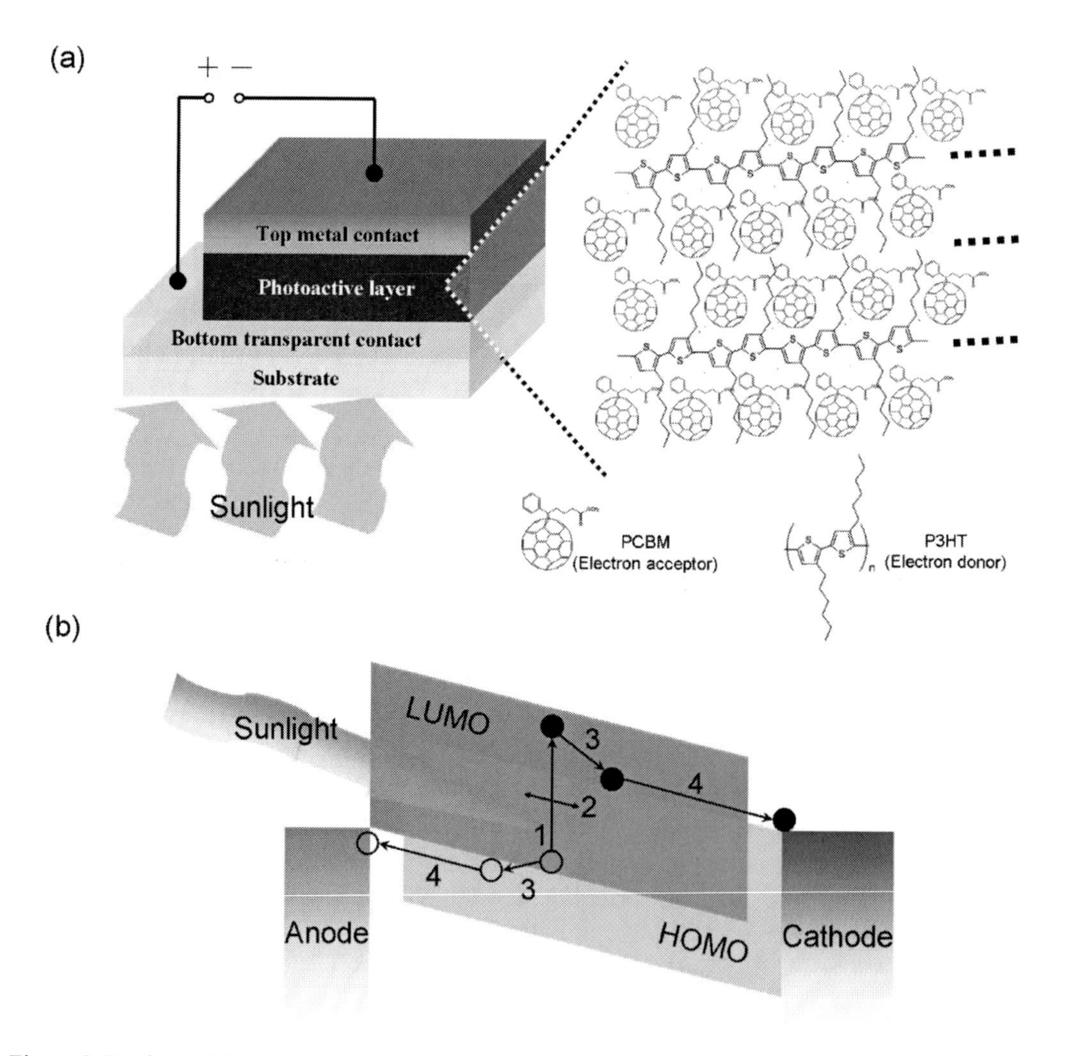

Figure 2. Device architecture (a) and schematic band diagram of a PCBM (6,6-phenyl-C_{61}-butyric acid methyl ester):P3HT (poly(3-hexylthiophene)) BHJ solar cell (b). The numbers refer to the operation processes explained in the text. HOMO: highest occupied molecular orbital; LUMO: lowest unoccupied molecular orbital.

(1) Absorption of photons leading to the generation of excitons,
(2) Diffusion of the excitons,
(3) Dissociation of the excitons with generation of free charge carriers, and
(4) Transport and collection of charge carriers to supply a direct current.

In the first step, the photon absorption depends on the value of the optical absorption coefficient and on the thickness of the donor material (polymer). Because of the high absorption coefficient ($\sim 10^5$ cm^{-1}), conjugated polymers absorb light very efficiently at the maximum of their absorption spectrum. On photon absorption, an electron is excited from the highest occupied molecular orbital (HOMO) to the lowest unoccupied molecular orbital (LUMO), forming a strongly bound electron-hole pair (exciton) by the Coulomb attraction in a singlet exciton state. Because coupling between neighboring molecules in molecular solids is low, the molecular excitations are localized and there is no band to band transition, unlike in inorganic semiconductors. Concomitantly, the low relative dielectric constant results in strongly bound Frenkel-like localized excitons. Due to the high exciton binding energy, the thermal energy at room temperature ($kT = 0.0259$ eV) is not sufficient to dissociate a strongly bound exciton into free charge carriers.

In the second step, the exciton diffuses inside the polymer as long as the recombination processes do not take place. In the third step, Förster (long range) or Dexter (between adjacent molecules) transfers can take place between an excited donor (material with low electron affinity) and an acceptor (material with high electron affinity) that receives the excitation. In other words, the bound exciton must migrate to a donor/acceptor interface where there is a sufficient chemical potential energy drop to drive dissociation into a free electron-hole pair. Although the exciton diffusion length in polymers is very short, the BHJ structure with an interpenetrating donor: acceptor network offers numerous donor/acceptor interfaces, resulting in almost every exciton formed within the diffusion distance from the interface. As a result, the dissociation efficiency of excitons is greatly improved in such BHJ architecture.

Finally, the free electrons and holes must be transported *via* percolated acceptor and donor pathways towards the electrodes while avoiding traps and recombination. These charges are transported primarily by drift caused by the built-in field. The current that reaches the contacts with no applied field is known as the short-circuit current density, J_{SC}. The maximum potential generated by the device is known as the open-circuit voltage, V_{OC}. Of course, for the current to do work, it must be generated with some potential. The ratio between the maximum power generated and the product of J_{SC} and V_{OC} is known as the fill factor (FF). The FF, describing the “squareness” of the current-voltage characteristics in a solar cell, is related to the quality of the device.

Figure 3 shows the illuminated current density-voltage (J-V) curve defining the primary quantities used to validate the performance of a solar cell. The power conversion efficiency (PCE) of a solar cell is the electrical power density (W/cm^2) divided by the incident solar power density (P_{in}), multiplied 100 to obtain a percent value. The electrical power density is determined by the maximum power point of operation (P_{MAX}) and occurs at some voltage (V_M) and current density (J_M) where a maximum rectangle meets the J–V curve. The PCE can be related to these key parameters, J_{SC}, V_{OC}, and FF, by $PCE = FF \cdot J_{SC} \cdot V_{OC} / P_{in}$.

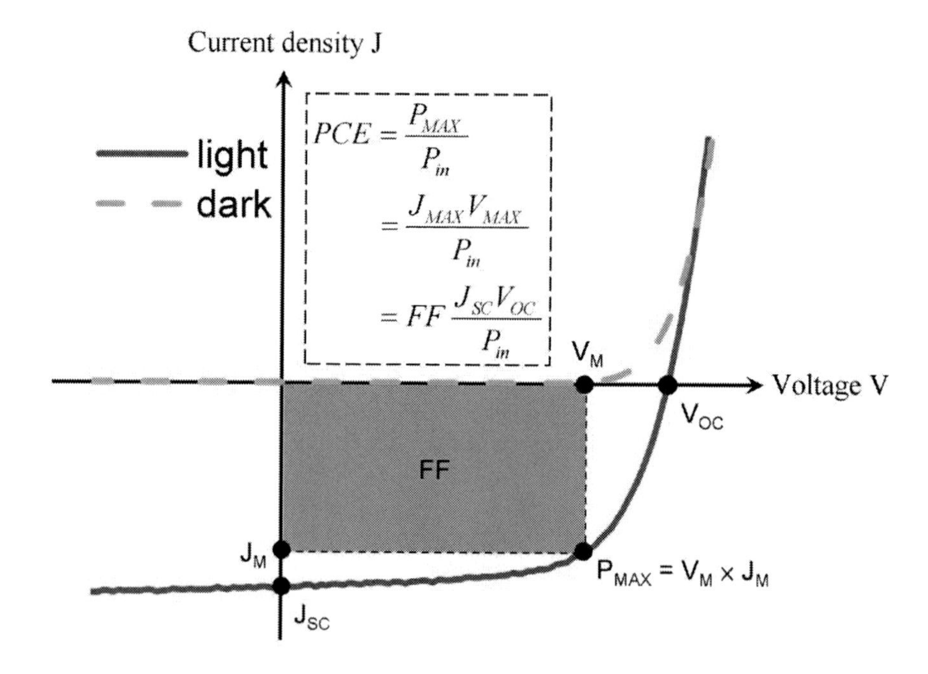

Figure 3. The typical current–voltage characteristics for dark and light current in a solar cell illustrate the important parameters for such devices: Jsc is the short-circuit current density, Voc is the open-circuit voltage, J_M and V_M are the current and voltage at the maximum power point, and FF is the fill factor. The PCE is defined, both simplistically as the ratio of maximum power out (P_{MAX}) to power in (P_{in}), as well as in terms of the relevant parameters derived from the current-voltage relationship.

1.2.3. Equivalent Circuits for Solar Cells

The J-V characteristics of an ideal solar cell can be seen as the sum of the currents from a diode in the dark and from a reverse current source (photocurrent)

$$J = J_d - J_{ph} = J_s\left[e^{eV/kT} - 1\right] - J_{ph}, \tag{1}$$

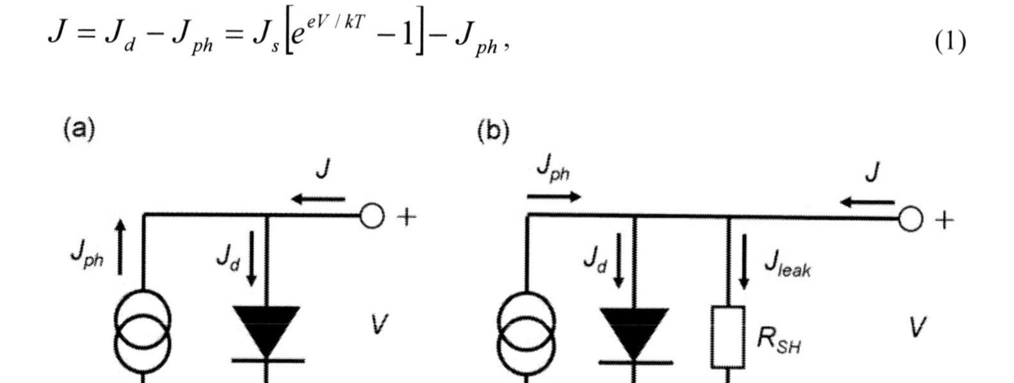

Figure 4. Equivalent circuits of an ideal solar cell (a) and a real solar cell with an additional shunt resistance R_{SH} and series resistance R_S (b).

where the J_d is the dark current density, J_{ph} is the photocurrent, J_s is the saturation current density, e is the elementary charge, V means the voltage, k is the Boltzmann constant, T is the temperature [18]. This leads to the equivalent circuit of an ideal solar cell sketched in Figure 4 (a).

However, in a real solar cell, power is dissipated through the contact resistances and through leakage current around the side of the device. These effects are equivalent electrically to two parasitic resistances in series resistance R_S and in shunt resistance R_{SH} with the device, as shown in Figure 4 (b). The series resistance arises from the contact resistances (electrode resistivity and metal-material interfaces) and the ohmic losses (due to the bulk resistivity of the cell materials). The shunt resistance arises from the leakage current across the cell, around the edges of the device and between contacts of different polarity. The shunt resistance also indicates the degree of lost charges due to recombination and trapping. When these factors are included, the total current density of a real solar cell is now composed of three contributions, the diode current, the leakage current (J_{leak}), and the photocurrent [18]:

$$J = J_d + J_{leak} - J_{ph} = J_s\left[e^{\frac{e}{kT}(V-JAR_S)} - 1\right] + \frac{V - JAR_S}{R_{SH}} - J_{ph}, \qquad (2)$$

where A is the device area. As the series resistance $R_S \rightarrow 0$ and shunt resistance $R_{SH} \rightarrow \infty$, the eq. 2 becomes the eq. 1.

In solar cells, the J_{SC} is proportional to the incident photon-to-current efficiency (IPCE), namely external quantum efficiency (EQE). The IPCE represents the ratio between the measured photocurrent and the intensity of the incident light. The former can be expressed as the number of collected electrons per unit time over a unit area. The latter can be expressed as the number of the incident photons of a certain wavelength (λ) per unit time and area. Thus,

$$IPCE(\lambda) = \frac{J_{SC}(\lambda)/e}{P_{in}(\lambda)/h\nu} = 1240\frac{J_{SC}(\lambda)}{P_{in}(\lambda)\cdot\lambda(nm)}, \qquad (3)$$

or

$$J_{SC} = \frac{e}{hc}\int_{\lambda_1}^{\lambda_2} P_{in}(\lambda)\cdot IPCE(\lambda)\cdot\lambda\cdot d\lambda, \qquad (4)$$

where h is the Plank's constant, c is the speed of light in vacuum, v is the frequency of the incident light, and λ is the wavelength of the incident light.

The IPCE of a polymer solar cell based on exciton dissociation at a donor-acceptor interface is

$$IPCE = \eta_{abs}\cdot\eta_{diff}\cdot\eta_{tc}\cdot\eta_{tr}\cdot\eta_{cc}, \qquad (5)$$

where η_{abs} is the photon absorption efficiency, η_{diff} is the exciton diffusion process, η_{tc} is the fraction of excitons that dissociate into free carriers at a donor-acceptor interface before recombining, η_{tr} is the carrier transport process, and η_{cc} is the fraction of carriers collected at the electrodes [19]. The product of the last four parameters ($\eta_{diff} \cdot \eta_{tc} \cdot \eta_{tr} \cdot \eta_{cc}$) is the so-called internal quantum efficiency (IQE). By improving the efficiency of any of these processes, the overall device efficiency can also be improved by an increase in J_{SC}.

Under open circuit conditions, the resulting current density $J = 0$, the open-circuit voltage derived from the eq. 4 becomes

$$V_{OC} = \frac{kT}{e} \ln\left(\frac{J_{ph} - \dfrac{V_{OC}}{R_{SH}}}{J_s} + 1 \right), \tag{6}$$

showing that the V_{OC} is increases logarithmically with light intensity. In this equation, a small R_{SH} caused by the leakage current leads to a reduced V_{OC}. In addition, it can be observed that for a low value of J_s, there is a large value of V_{OC}. The dark current originates from thermal excitations of ground-state charge-transfer complexes (CTC) between the donor and the acceptor [20]. The CTC is defined as having energy equal to the energy difference between the donor HOMO and the acceptor LUMO. The ground-state CTC which is generated by the interaction of the donor and the acceptor can absorb and emit light.

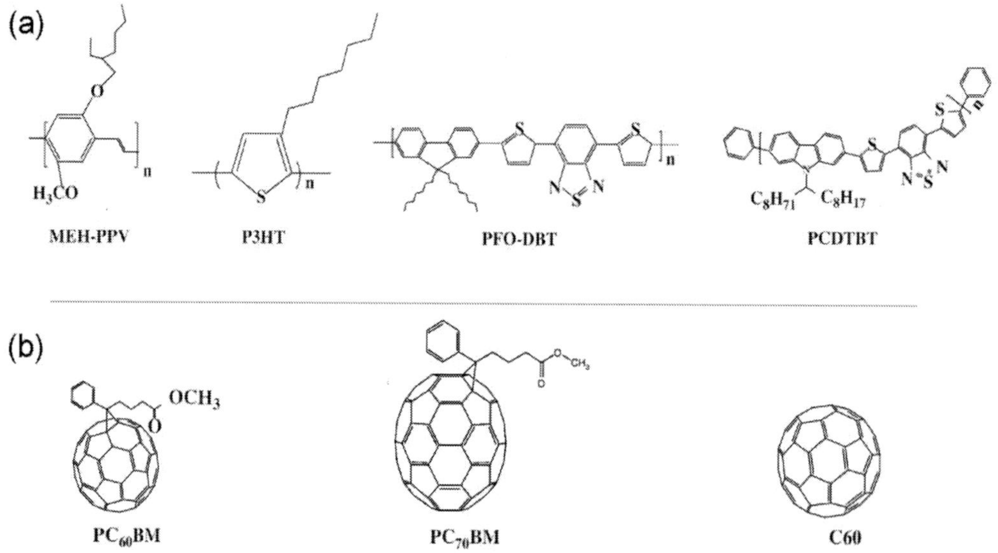

Figure 5. Conjugated polymers (a) and fullerenes commonly used in OSCs (b). The full names of above representative organic materials are:
MEH-PPV: poly [2-methoxy-5-(2'-ethyl-hexyloxy)-1,4-pheny-lene vinylene],
P3HT: poly(3-hexylthiophene),
PFO-DBT: poly[2,7-(9,9-dioctyl-fluorene)-alt-5,5-(4,7'-di-2-thienyl-2', 1', 3'-benzothiadiazole),
PCDTBT: poly[N-9'-hepta-decanyl-2,7-carbazole-alt-5,5-(4',7'-di-thienyl-2'1',3'-b3nzothiadizaole),
$PC_{60}BM$: 6,6-phenyl-C_{61}-butyric acid methyl ester, $PC_{70}BM$: 6,6-phenyl-C_{71}-butyric acid methyl ester.

Because the work function of the electrode is pinning to the corresponding material energy level, the V_{OC} of the BHJ cells is directly related to the energy difference between the HOMO level of the electron donors and the LUMO level of the electron acceptors [21]. The V_{OC} is empirically described by following formula [22]

$$V_{OC} = \frac{1}{e}\left(\left|E_{HOMO}^{Donor}\right| - \left|E_{LUMO}^{Acceptor}\right|\right) - 0.3 \text{ V}, \tag{7}$$

where the value of 0.3 V is an empirical factor which is caused by the dark current-voltage curve of the diode which is determined by and the reverse dark current J_d of the diode.

1.2.4. Device Design

1.2.4.1. Materials for Photoactive Layer

In traditional polymers such as polyethylene, the valence electrons are bound in sp^3 hybridized covalent bonds. Such "sigma-bonding electrons" have low mobility and do not contribute to the electrical conductivity of the material. Conjugated polymers are a class of semiconducting materials discovered in 1977 by Shirakawa, MacDiarmid, and Heeger. Conjugated polymers have backbones of contiguous sp^2 hybridized carbon centers. One valence electron on each center resides in a p_z orbital, which is orthogonal to the other three sigma-bonds. The electrons in these delocalized orbitals have high mobility when some of these delocalized electrons are removed. That is, the p-orbitals form a band where the electrons become mobile when this band is partially emptied. In a word, the conjugated polymers are electronically active because of their highly polarizable π-systems, which are hybridized orbitals based on the constituent p atomic orbitals. The π–π* optical transitions are strong (absorption coefficients greater than ~10^5 cm^{-1}), therefore the conjugated polymers can absorb all the light at their peak wavelength absorption in a few hundred nanometer. Generally, organic photovoltaic materials having delocalized π electron system can absorb sunlight, create photogenerated charge carriers, and transport these charge carriers [23]. These materials are classified to the electron donors and the electron acceptors.

Most of the conjugated polymers are hole-conductors. This kind of semiconducting polymers was named as the electron donor polymers. Figure 5 (a) shows some representative conjugated polymers for polymer solar cells. Four important representatives of electron donor polymers are MEH-PPV, P3HT, PFO-DBT [24], and PCDTBT [16,25]. In the past years, P3HT was a prominent donor polymer in the photoactive layer of BHJ polymer solar cells due to its advantage of easy self-organization [26]. There are number of low band gap polymers also under the investigation for polymer solar cells [27-33]. The electron acceptor molecules, C_{60} and soluble derivatives of C_{60} and C_{70}, namely $PC_{60}BM$ and $PC_{70}BM$, are shown in Figure 5 (b). Fullerene derivatives are wildly used acceptor molecules in the photoactive layer due to its good solubility [34].

Fullerenes are considered to be the best electron acceptors so far, because the following three reasons:

(i) Ultrafast (~50 fs) photo-induced charge transfer is happened between the donor polymers and fullerenes.

(ii) Fullerenes exhibit high mobility, for example, $PC_{60}BM$ shown electron mobility up to 1 $cm^2V^{-1}s^{-1}$ [35] measured by field effect transistors.
(iii) Fullerenes show a better phase segregation in the blend film.

1.2.4.2. Morphology of Photoactive Layer

Nanoscale phase separation of the donor and acceptor components leads to charge-separating heterojunctions through-out the bulk of the BHJ photoactive material. Separated carriers then move toward the electrodes on bicontinuous interpenetrating networks. As a result, the morphology of the nanoscale phase-separation in the photoactive layer plays a decisive role linking the optoelectronic properties and device performance to the fabrication processes. The control of nanoscale morphology of the BHJ photoactive layer is therefore critical. One of the earliest and most widely used methods of morphology control in BHJ solar cells is thermal annealing [36]. This technique has proven most effective in P3HT:PCBM system, where P3HT have demonstrable crystalline order due to its advantage of easy self-organization [26].

Thermal annealing can improve J_{SC} and FF significantly, thus leading to higher PCE [37,38] due to the improvement in the nanoscale lateral phase separation of both the crystalline P3HT aggregates and PCBM domains [36]. Thermal annealing can be applied either on the final device (post-annealing) or on the BHJ photoactive layer only (pre-annealing). The annealing temperature and time are the two most critical processing conditions in this approach, which influences the polymer self-organization and the corresponding optical and electrical properties. By this method, the crystallinity of P3HT can be increased, resulting in the formation of crystallites with the conjugated main chain parallel to and side chains perpendicular to the substrate [39]. Enhanced hole mobility of the P3HT phase in the BHJ blend from annealed devices is an important factor which result in an improved PCE. Other contributions for enhancement in the performance were strong red-shift of the near-infrared (NIR)-region absorption [40] and reduced charge recombination due to the improved percolation pathway [41].

It was also found that the performance of P3HT:PCBM BHJ solar cells were affected by molecular weights [42], regioregularity [43,44], mixed solvents without heat treatment [45,46], applying external electric field [37], solvent selection [47], solvent annealing [48], slow drying [49], processing additives [50]. All these results indicate that it was very complex to control the morphology of the BHJ film.

1.2.4.3. Stability

While the PCE of the OSC is being improved gradually, the request from commercialization of the OSC has forced the scientists and engineers to address the stability. Similar to other organic electronic devices, the OSCs without encapsulation have very short lifetime [51]. Common structure of BHJ device is ITO/ poly(3,4-ethylenedioxythiophene): polystyrene sulfonate (PEDOT:PSS)/P3HT:PCBM/Aluminium (Al). Jørgensen et al. had deeply investigated the mechanisms of the degradation of OSCs through difference pathways as shown in Figure 6 [52]. The organic materials and the electrodes (Al and indium tin oxide (ITO)) electrochemically react with oxygen and/or water (i.e. ambient air) which are diffused from both electrodes or lateral of the device are believed the major reason causing short lifetime of OSCs.

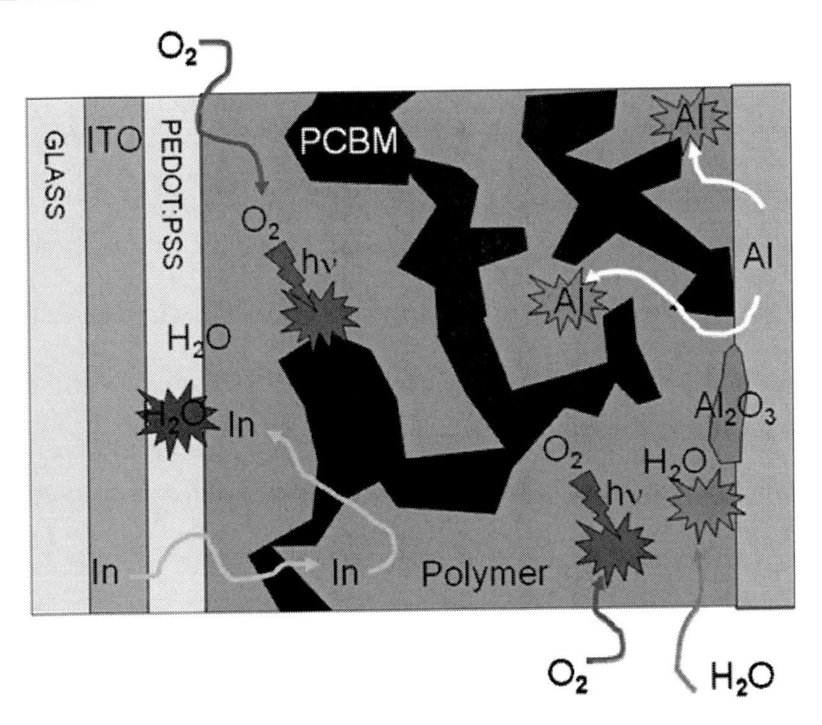

Figure 6. The degradation processes in OSCs. Reproduced with permission from Jørgensen, M. et al . *Sol. Energy Mater. Sol. Cells* 2008, *92*, 686-714. Copyright © 2008, Elsevier B.V.

The oxygen has been demonstrated to diffuse into the Al electrode through Al grains and microscopic holes on the Al film [53]. Inside the device, oxygen is progressively diffused in the lateral and vertical plane until reaching the counter electrode. The use of a low work function metal, Al, as the cathode is likely to from a dielectric layer (i.e. aluminum oxide layer) at the metal-polymer interface, resulting in significant degradation mechanisms and instability [52]. Another prominent degradation pathway is the diffusion of indium atoms from the ITO into the organic layer of the devices. The stability of the interface between ITO and PEDOT:PSS have been studied by de Jong et al. [54]. Kawano et al. also found that the resistance of ITO was increased due to the hygroscopic PEDOT:PSS layer absorbed the moisture from the ambient atmosphere [55]. Their investigations showed that the ITO-PEDOT:PSS interface is very sensitive to air and that the hygroscopic nature of PSS allows absorption of water that facilitate etching of the ITO layer due to its strong acidic nature.

1.2.4.4. Inverted Structure

It has been proven that acidic PEDOT:PSS layer was detrimental to the photoactive polymer layer, and the low work function metal cathode was easily to be oxidized in air even with a delicate encapsulation. In 2006, an alternative approach using inverted structures was proposed to overcome above shortages [56,57]. Figure 7 shows schematic device architectures of a conventional OSC and an inverted OSC. The inverted OSC utilizes an air-stable high work-function electrode as the back contact anode to collect holes and ITO as the cathode to collect electrons. By introducing an n-type interlayer, cesium carbonate, between the ITO and photoactive layer, Liao et al. have achieved 4.2% of PCE from P3HT:PCBM solar cells [58]. Other common materials used as the n-type interlayer are zinc oxide (ZnO)

[57], titanium oxide (TiO_2) [59], and TiO_x [60]. For example, Hau et al. had reported air-stable inverted BHJ OSCs using ZnO nanoparticles as electron selective layer. This inverted structure possessed much better stability under ambient conditions retaining over 80% of its original conversion efficiency after 40 days while the conventional one showed negligible photovoltaic activity after 4 days [61].

In addition to the n-type interlayer, the p-type interlayer between the organic photoactive layer and the anode has been demonstrated to be beneficial to device performance. Several materials have been reported for this application, such as PEDOT:PSS, vanadium oxide (V_2O_5) [62], tungsten oxide (WO_3) [63], molybdenum oxide (MoO_3) [64], and nickel oxide (NiO) [65]. However, only the deposition of PEDOT:PSS layer can be easily processed by solution-based coating techniques. Nevertheless, PEDOT:PSS solution is difficult to wet the surface of the photoactive layer because most organic photoactive layers are hydrophobic while PEDOT:PSS is hydrophilic. Most transition metal oxides as the anode interlayers are deposited by the vacuum evaporation, which could detract from the advantage of the ease of OSC fabrication. Furthermore, the PCEs from most of the inverted P3HT:PCBM OSCs were still lower than these from the conventional P3HT:PCBM OSCs (PCE>5%) [66].

1.2.4.5. Inverted Structure with Ordered Nanostructures

The nanoscale morphology of the BHJ photoactive layer is not easy to control precisely. The ultimate efficiency of the BHJ OSC will eventually be limited by the random network. The variations in the processing can lead to vastly different morphologies, and then severely affect exciton harvesting. Phase segregation kinetics during casting is a complicated process that may produce dead ends and isolated domains that trap charge carriers and prevent them from being collected.

In order to solve this problem and optimize the device performance, an ideal device structure based on the BHJ OSCs was proposed by introducing ordered one-dimensional (1D) nanostructured oxides. The ideal nanostructure for OSC is shown in Figure 8. In this structure, nanostructured n-type and p-type phases with spacing close to the exciton diffusion length (5-20 nm) would facilitate efficient exciton dissociation. The 1D nanostructured oxides can be vertically aligned rods, tubes, or pores which can ensure unhindered charge carrier transport to the electrodes.

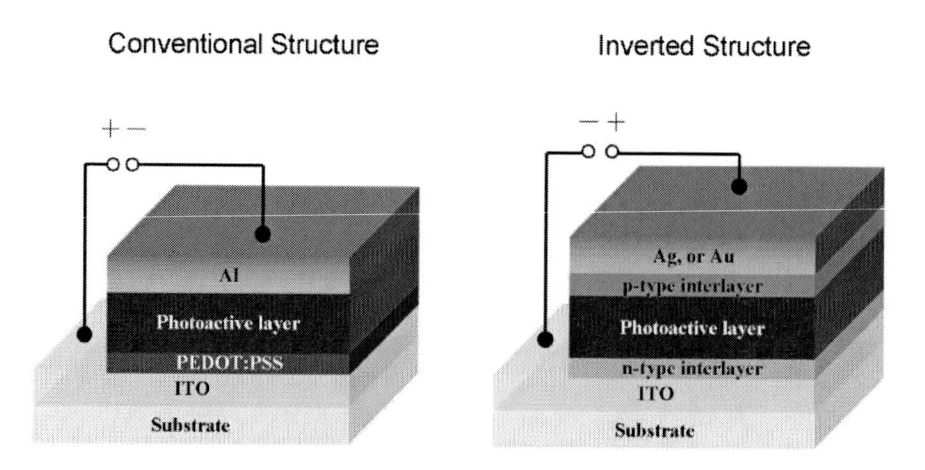

Figure 7. Schematic device architectures of a conventional OSC and an inverted OSC.

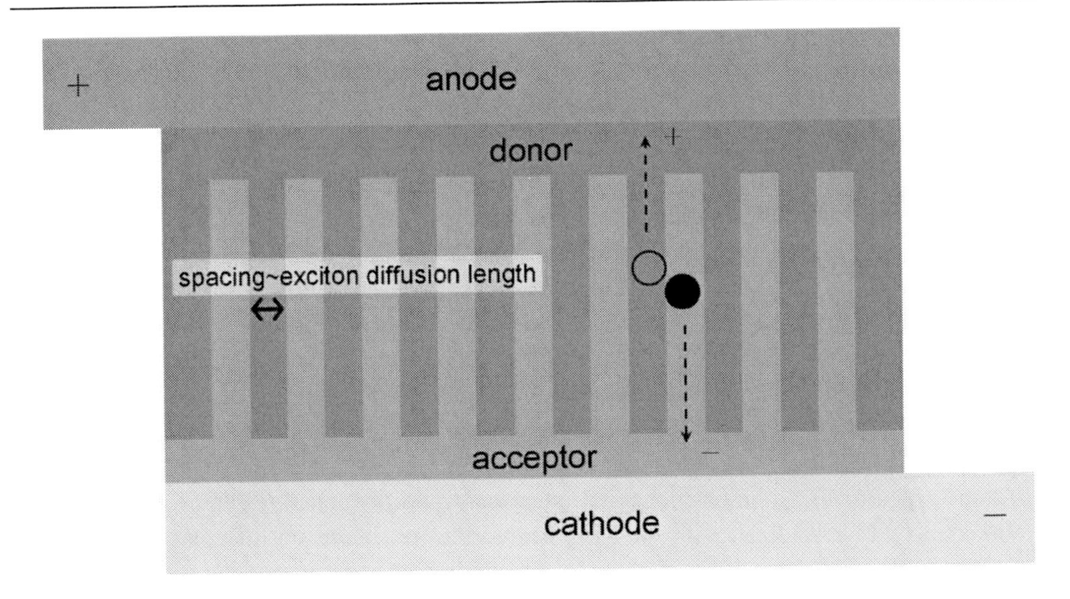

Figure 8. An ideal structure of a BHJ OSC.

Takanezawa et al. had reported that the ZnO nanorods have beneficial effects of collecting and transporting electrons in the inverted OSCs hybridized with the ZnO nanorods [62]. Their efforts indicated that the FF was remarkably improved from 50% to 65%, along with a small change in J_{SC} and V_{OC}. Chou et al. had also report that the presence of these ZnO nanorods allows the thickness of the photoactive layer thickening up to 400 nm, leading to efficient light harvesting without decreasing the possibility for charge transport [49]. In addition, Mor et al. had reported a double heterojunction formed by vertically oriented TiO_2 nanotube arrays and P3HT:PCBM BHJ, which provides two interfaces for exciton dissociation and charge transfer [67].

2. Organic-ZnO Nanorod Composite Solar Cells

Organic solar cells have attracted significant attention due to the great potential for large-area, light-weight, flexible, and low-cost devices. Recently, BHJ solar cells based on P3HT and PCBM with PCE exceeding 5% have been reported [66]. However, several issues are unsolved. For example, most of the fabrication and testing of these devices were performed in an inert environment due to the facile oxidation of the low work-function metal [52]. In addition, PEDOT:PSS layer has been demonstrated to have a side effect on the device performance due to its corrosion to ITO and electrical inhomogeneities [68]. Third, the transport of the charges after their separation simply relies on the percolation network of the components of the mixture films which is not easy to control precisely through dynamic coating or simple thermal annealing.

To solve the three problems, we investigate the way of changing the nature of charge collection by using a more air-stable high work function electrode as the back contact to collect holes [68,69] while using metal oxide nanostructures at the ITO to select and collect electrons [49,62,70,71]. It integrates the advantages of the two materials: solution processability, high hole mobility and photosensitivity of polymers, and high electron mobility of inorganic semiconductors [72,73]. The environmentally friendly and low-cost ZnO nanorod arrays are particularly well suited for this application as they can be grown at low temperature (<100 °C) using a hydrothermal method [74,75]. In addition, ZnO nanorod arrays have excellent electron mobility and small rod-to-rod spacing [76]. Olson et al. had reported the hybrid solar cells using the combination of the P3HT:PCBM and ZnO nanofibers [77]. Hashimoto et al. had investigated the performance dependence of P3HT:PCBM/ZnO hybrid solar cells on the length of the ZnO nanorods and reported a PCE of 2.7% [78]. Their investigations show that the ZnO nanorods have beneficial effects of collecting and transporting electrons. However, compared to the polymer solar cells, this type of hybrid solar cells tends to have lower PCE due to poor electrical coherence at the organic-inorganic and/or organic-metal interfaces. One approach to improve the performance and charge selectivity of hybrid solar cells is to modify the organic-inorganic and/or organic-metal interfaces with functional modifications.

Three types of the solution-processed interfacial modifications are reported to improve the inverted OSC hybridized with the ZnO nanorod arrays based on different mechanisms. They are PCBM, TiO_2, and V_2O_5. The additional PCBM clusters as cathode modification on the ZnO nanorod arrays can lead to enhanced phase separation and thus absorption. Depositing TiO_2 nanoparticles on the ZnO nanorod arrays can create a double heterojunction which provides two interfaces for exciton dissociation and charge transfer. V_2O_5 as anode modification on the photoactive layer can suppress the leakage current and enhance the absorption.

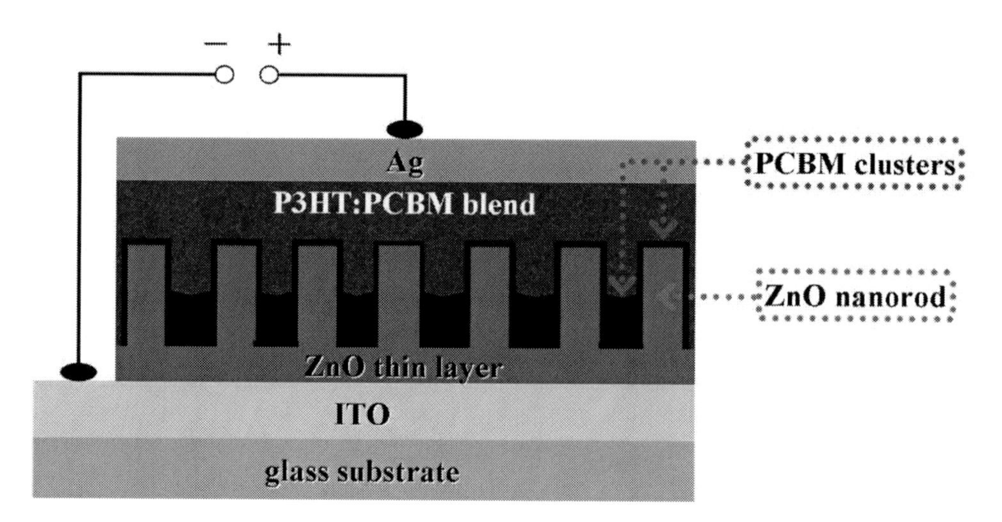

Figure 9. Hybrid device structure of ITO/ZnO nanorod arrays/PCBM clusters/ P3HT:PCBM/Silver (Ag).

2.1. Effect of Additional PCBM Clusters

This effect of additional PCBM clusters was reported by Huang et al. [79]. The PCBM clusters were spin-coated in air from a dichloromethane solution before depositing a blend of P3HT:PCBM. Figure 9 shows the device structure of ITO/ZnO nanorod arrays/PCBM clusters/ P3HT:PCBM/Ag.

2.1.1. PCBM Clusters on Zinc Oxide (ZnO) Nanorod Arrays

Figure 10 (a) shows a field-emission scanning electron microscopy (FESEM) image of the ZnO nanorod arrays, where the highly uniform, vertically aligned, and densely packed arrays of ZnO nanorods are successfully formed on ITO glass substrates. The rod diameter is about 40-50 nm and the rod-to-rod spacing is in the range of 10-40 nm. Figure 10 (b) shows a FESEM image of the PCBM clusters on the ZnO nanorod arrays. It clearly shows that the PCBM molecules fill the ZnO nanostructures. Figure 10 (c) gives three X-ray diffraction (XRD) spectra measured by using Cu $K\alpha$ radiation. The black curve represents pure PCBM powder, revealing its characteristic peaks. The red curve is the XRD spectrum of the ZnO nanorod arrays, where a peak at $2\theta = 30.3°$ from ITO and a strong peak at $2\theta = 34.4°$ from ZnO are observed [74]. After the PCBM coated on the ZnO nanorod arrays, a narrow peak at $2\theta = 7.03°$ from the PCBM molecules emerges abruptly in the XRD spectrum (blue curve). The corresponding lattice constant is 1.26 nm. Comparing with literature, it can be concluded that the PCBM crystallites present a layered structure due to the van der Waals interaction [80]. This clearly shows that the PCBM molecules aggregate orderly in the ZnO nanorod spacing so it maybe beneficial for electron transport. Figure 10 (d) shows the typical energy dispersive x-ray spectroscopy (EDX) obtained from Figure 10 (b). The peaks at 0.5, 0.9, 8.6, and 9.55 keV are from ZnO and the peak at 3.29 keV is from the ITO substrate. The peak at 0.26 keV is due to carbon which is from the PCBM. The XRD and EDX spectra both confirm the presence of the PCBM clusters on the ZnO nanorods.

2.2.1. Photovoltaic Performance

Solar cells were measured with a Keithley 2400 source measurement unit, under illumination at 100 mW/cm^2 from a 150 W Oriel solar simulator with AM 1.5G filters. The light intensity was calibrated with a calibrated standard solar cell with a KG5 filter which was traced to the NREL. The device area was defined as 10 mm^2 by a shadow mask. The J-V characteristics of the photovoltaic devices are plotted together in Figure 11 for comparison. Device A denotes as a hybrid solar cell with the PCBM clusters, while Device B represents a hybrid solar cell without the PCBM clusters. Device B exhibits a J_{SC} of 10.13 mA/cm^2, a V_{OC} of 0.51 V, a FF of 45.5%, and a PCE of 2.35%. Insertion of the fullerene clusters at the interface between P3HT:PCBM and ZnO nanorod arrays (Device A) provides significant improvement in the device performance. The J_{SC} increases to 11.67 mA/cm^2, V_{OC} to 0.55 V, and FF to 49.9%. This results in a PCE of 3.2%, a 36% improvement.

Table 1. Photovoltaic parameters of the hybrid solar cell with and without PCBM clusters

Hybrid solar cells	PCE (%)	V_{OC} (V)	J_{SC} (mA/cm^2)	FF (%)	R_S* (Ωcm^2)	R_{SH}** (Ωcm^2)
No PCBM clusters	2.35	0.51	10.13	45.5	5.83	251
With PCBM clusters	3.2	0.55	11.67	49.9	1.67	206

* R_S: series resistance.

** R_{SH}: shunt resistance.

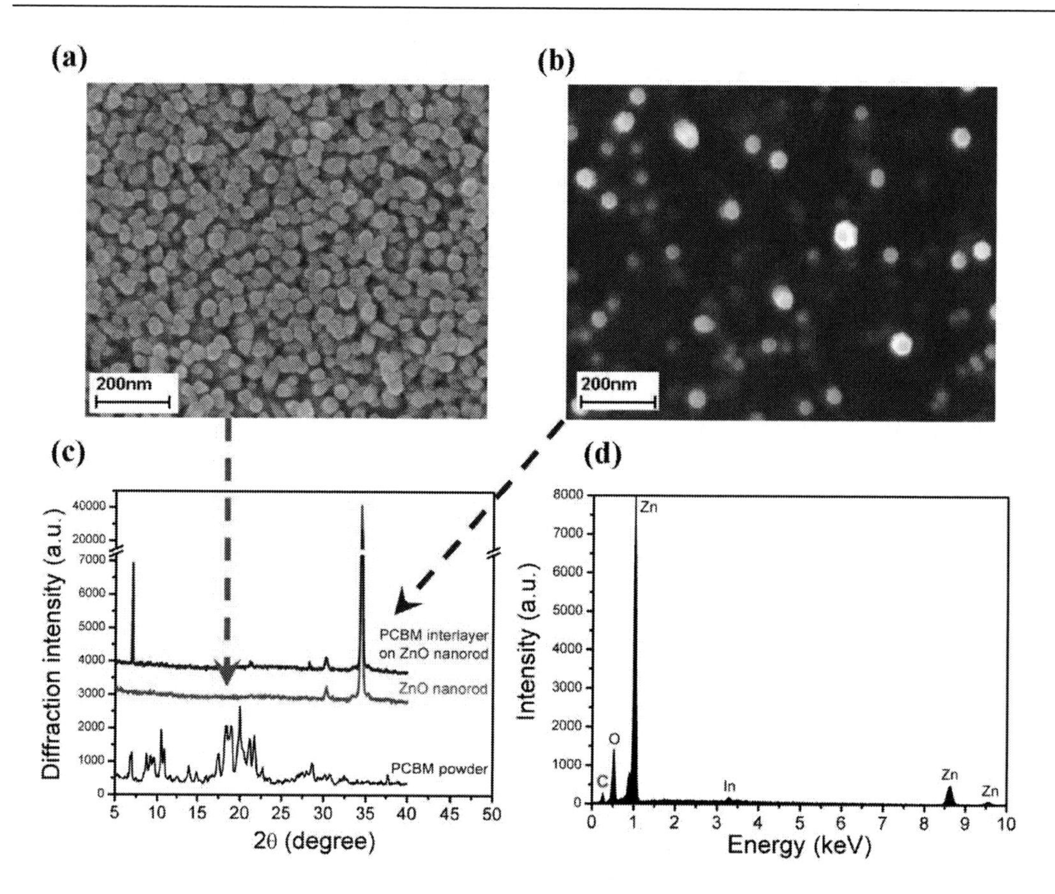

Figure 10. FESEM image of ZnO nanorod arrays on the ITO glass substrate (a), FESEM image of PCBM clusters on the ZnO nanorod arrays (b), XRD spectra of pure PCBM powder (black curve), ZnO nanorod arrays on ITO (red curve), and PCBM clusters on the ZnO nanorod arrays (blue curve) (c) and EDX spectrum of the PCBM clusters on the ZnO nanorod arrays (d).

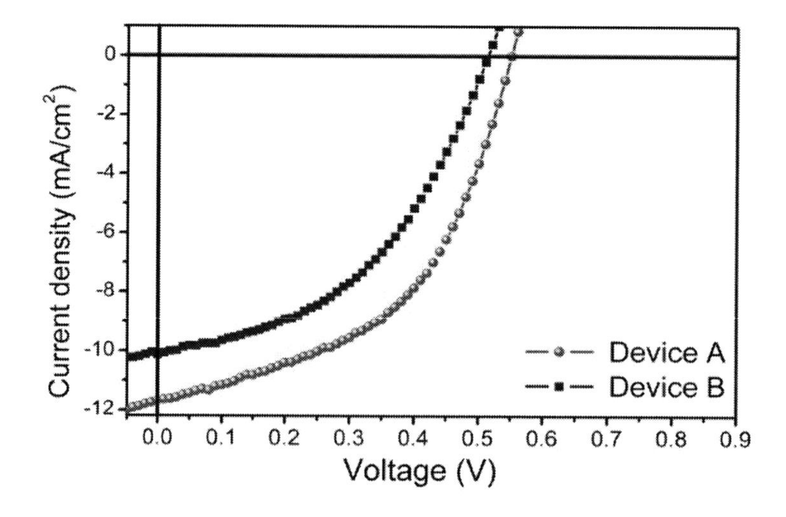

Figure 11. The J-V characteristics of the photovoltaic devices under 100 mW/cm^2 AM 1.5G irradiation. Device A represents the hybrid solar cells with the solution-processed PCBM clusters. Device B denotes the hybrid solar cells without the PCBM clusters.

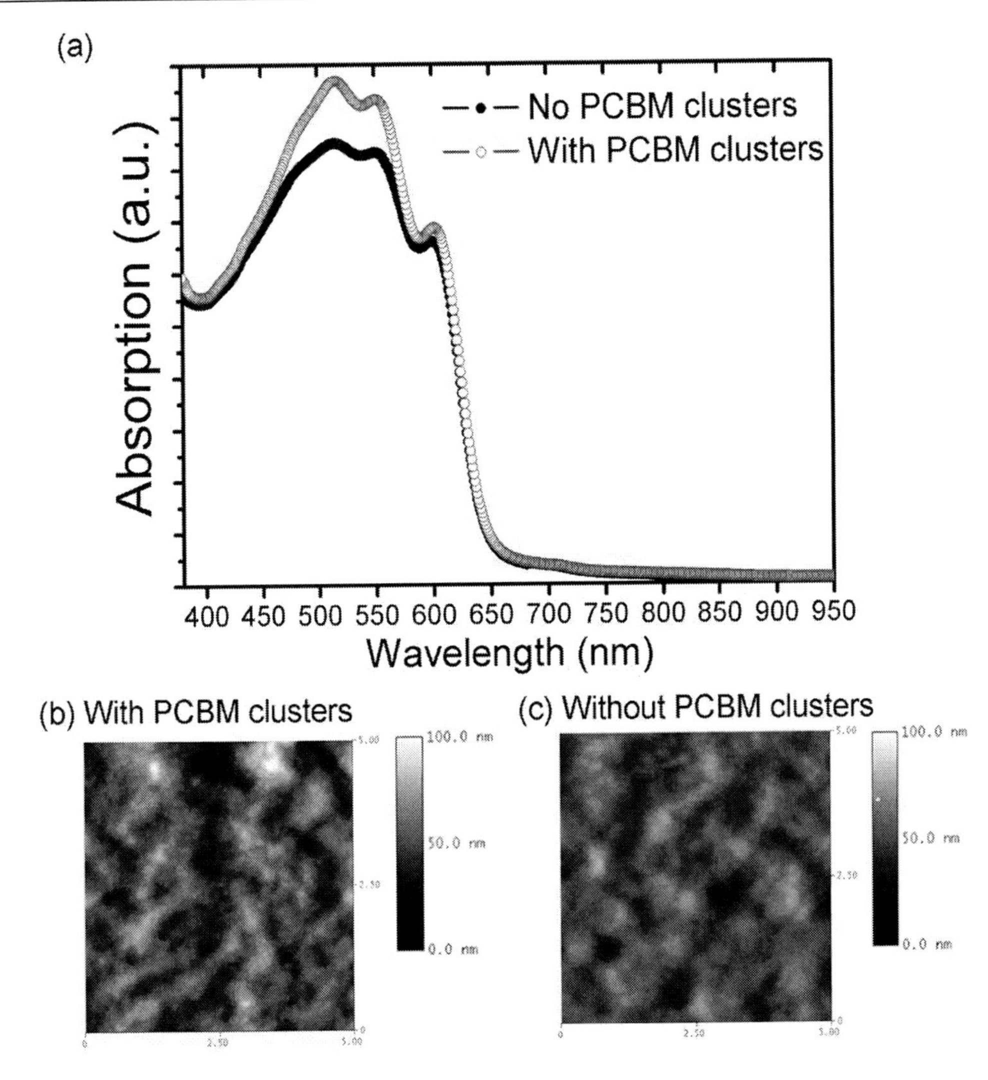

Figure 12. Ultra violet visible (UV-vis) spectra of hybrid solar cells with and without the PCBM clusters (a), AFM images of the photoactive layers (b) with and (c) without the PCBM clusters. AFM image scans are 5 × 5 μm. Reproduced with permission from Huang, J.-S et al. *Sol. Energy Mater. Sol. Cells* 2010, *94*, 182-186. Copyright © 2010, Elsevier B.V.

These results clearly show that PCBM clusters can serve as a functional modification to enhance the solar cell efficiency. Without the PCBM clusters, the device has similar performance compared to other reports on hybrid solar cells with ZnO nanorod arrays [77,78]. With the PCBM clusters, the J_{SC} increases from 10.13 to 11.67 mA/cm^2. Table 1 summaries the photovoltaic parameters of the both devices.

In order to further understand the reason behind the improvement of the J_{SC}, UV-vis absorption spectra and cross-sectional FESEM were examined. Figure 12 (a) shows the absorption spectra of the P3HT:PCBM BHJ films with and without the PCBM clusters prepared on glass substrates. Both spectra show the same absorption pattern with a peak at 515 nm and two shoulders at 550 and 600 nm. These vibronic features all arise from P3HT molecules [43]. After the introduction of the PCBM clusters, the three vibronic absorption peaks become more pronounced, indicating enhanced ordering of P3HT. It is suggested that the PCBM clusters assists the self-organization of P3HT chains during the solidification of the blend films. Figure 12 (b) and (c) show the surface topography, measured by atomic force microscopy (AFM), of the photoactive layers with and without the PCBM clusters, respectively. With the PCBM clusters, the surface is very rough with root-mean-square roughness (σ_{RMS}) of 15 nm, as shown in Figure 12 (b). For the photoactive film without the PCBM clusters, the smooth surface with σ_{RMS} ~7 nm is observed in Figure 12 (c). It has been reported that the rough surface is a signature of polymer self-organization [81]. In Figure 12 (b), the rough surface also suggests a higher degree of ordering of P3HT. This ordered structure reduces the internal series resistance of the device, thus increasing the photocurrent.

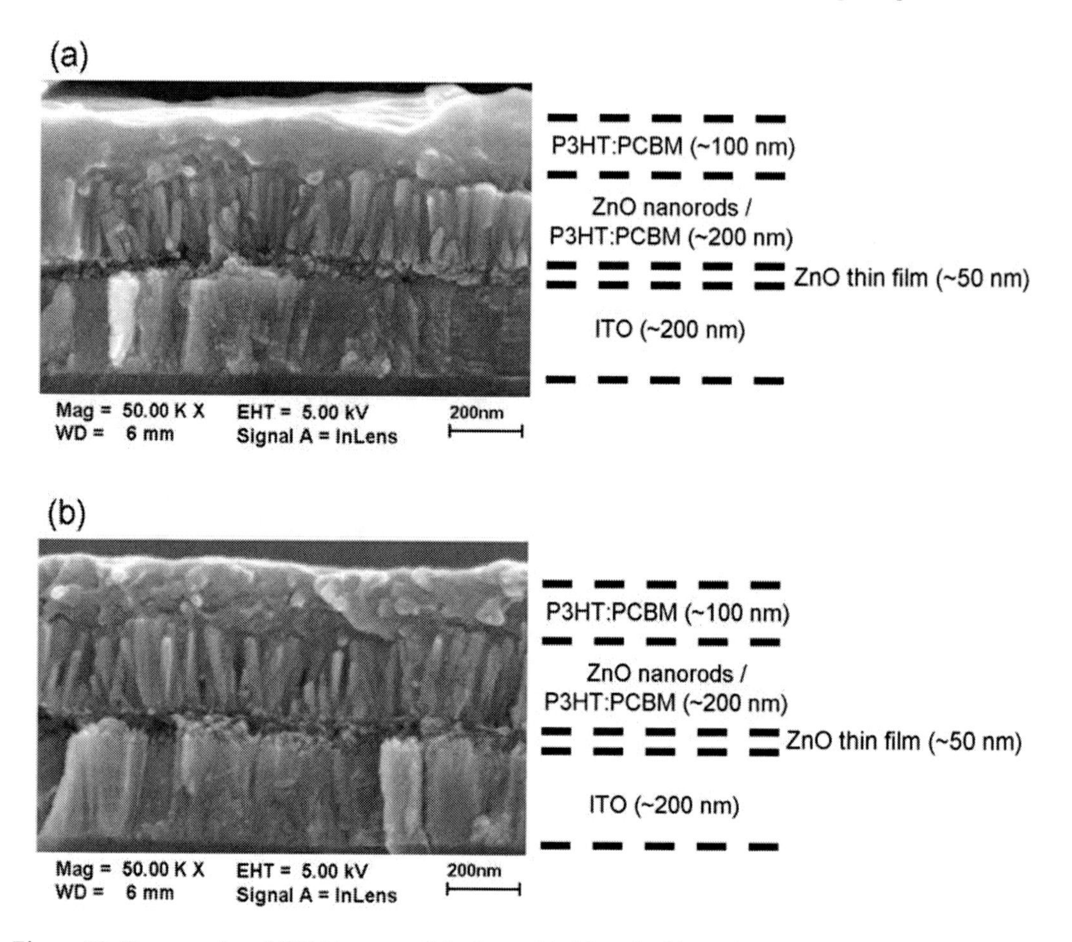

Figure 13. Cross-sectional SEM images of devices with (a) and without the PCBM clusters (b). Reproduced with permission from Huang, J.-S. et al. *Sol. Energy Mater. Sol. Cells* 2010, *94*, 182-186. Copyright © 2010, Elsevier B.V.

Figure 13 shows the cross-sectional FESEM images of the devices with and without the PCBM clusters. With the fullerene modification, the organic materials in the device infiltrates well, as shown in Figure 13 (a). In contrast, the infiltration of the organic materials into the ZnO nanorods becomes poor for the devices without the PCBM clusters, as shown in Figure 13 (b). Some void spaces are observed between the ZnO nanorods in the device without the fullerene modification. This leads to the reduction of the interfacial area between the ZnO nanorod arrays and the organic materials. Therefore, we may conclude that the PCBM clusters infiltrates into ZnO nanorod arrays, forms well ordering crystalline structures, creates short and continuous pathways for electron transport, and increases the contact area between the ZnO nanorod arrays and the organic materials, hence resulting in a high photocurrent. Note that the FF is also improved from 45.5% to 49.9% by introducing the fullerene clusters. The improvement in FF could be attributed to the reduction of the series resistance (R_S, defined from the J-V curves near 1.5 V under light illumination). The R_S of the device without the fullerene clusters is 5.83 Ωcm^2, while the R_S of the device with the fullerene clusters significantly decreases to 1.67 Ωcm^2. This result agrees with the enhanced absorption spectra (Figure 12 (a)) and the rough surface as shown in the AFM image (Figure 12 (b)).

Eight samples for each individual structure of the devices are tested to study the consistency of the device photovoltages. Without the PCBM clusters, the mean V_{OC} is 0.51 V with a standard deviation of 0.0014 V. With the PCBM clusters, the mean V_{OC} increases to 0.55 V with a standard deviation of 0.0015 V. Because both standard deviations are very small, it is reliable that the V_{OC} is improved from 0.51 V to 0.55 V by introducing the PCBM clusters. The reason for the increased V_{OC} could be attributed to the reduced contact area between the P3HT and the ZnO nanorod arrays due to the insertion of the PCBM clusters. Refer to the energy band diagram shown in Figure 14. Theoretically, it has been reported that the V_{OC} of the P3HT/PCBM BHJ solar cells is directly related to the energy difference between the HOMO level of the P3HT ($HOMO_{P3HT}$) and the LUMO level of PCBM ($LUMO_{PCBM}$) [21]. The energy difference between the CB of ZnO and the $HOMO_{P3HT}$ is 0.4 eV, which is smaller than the energy difference ($\Delta E_{PCBM,P3HT}$ = 1.1 eV) between the $LUMO_{PCBM}$ and the $HOMO_{P3HT}$. Thus, the theoretical V_{OC} of the P3HT/ZnO device is smaller than that of P3HT:PCBM device. Therefore, the experimentally measured V_{OC} of the P3HT/ZnO nanorod arrays solar cells is 0.4-0.45 V [82-84], typically lower than that of the P3HT:PCBM BHJ solar cells (0.5-0.6 V). From the theoretical perspective and the experimental results, the introduction of the PCBM clusters has beneficial effects on hybrid solar cells with ZnO nanorod arrays.

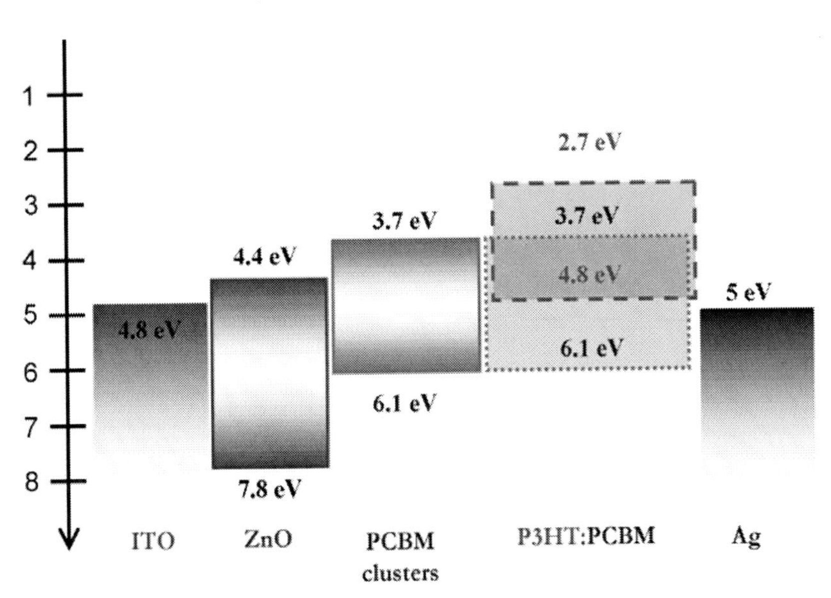

Figure 14. Energy band diagram for the hybrid solar cell with a structure of ITO/ZnO nanorod arrays/PCBM clusters/P3HT:PCBM/Ag.

In the absence of ZnO nanorod arrays, we observe that devices with and without the fullerene clusters have similar photovoltaic performances. The optical absorption spectra also show that the no difference between the devices with and without the fullerene clusters. In the absence of ZnO nanorod arrays, fullerene clusters coated on the flat ZnO film is very thin and the amount of PCBM molecules is very small. It is possible that the pre-coated PCBM molecules are almost re-blended into wet P3HT:PCBM film during the coating of photoactive layer. Thus, the fullerene clusters nearly vanishes. As a result, the fullerene clusters can not function well in this type of devices as in the nanorod-type devices.

2.2. Formation of Double Heterojunction by Titanium Oxide (TiO_2) Nanoparticles

The TiO_2 nanoparticles (<100 nm, Sigma-Aldrich, 99.5%) were homogeneously dispersed and suspended in isopropanol at different concentrations by using ultrasonic agitation. The structure of TiO_2 nanoparticles used here is rutile. After the ultrasonic agitation, the color of the TiO_2 colloidal solution is uniformly semitransparent white. Then the TiO_2 colloidal solution was spin-coated in air on top of the ZnO nanorod arrays. Subsequently, a blend of P3HT:PCBM (1:1 by weight) was deposited on the TiO_2 nanoparticles in air. The thickness of the photoactive layer is about 300 nm. Finally, silver film (~150 nm) was deposited on top of the photoactive layer. Figure 15 schematically shows the device structure of ITO/ZnO nanorod arrays/TiO_2 nanoparticles/P3HT:PCBM/Ag.

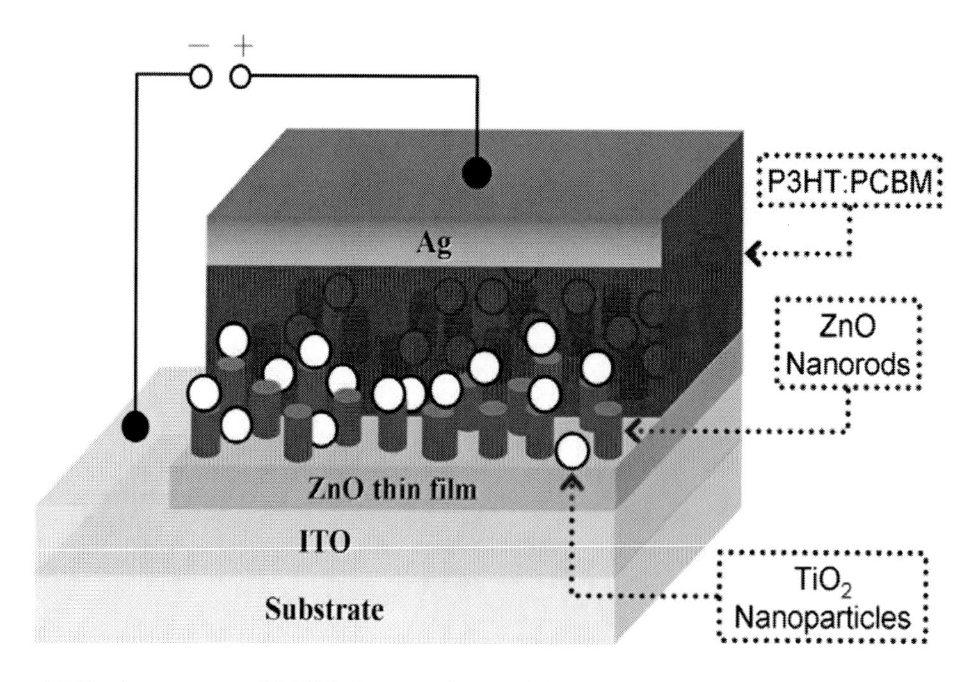

Figure 15. Device structure of ITO/ZnO nanorod arrays/TiO_2 nanoparticles/P3HT:PCBM/Ag.

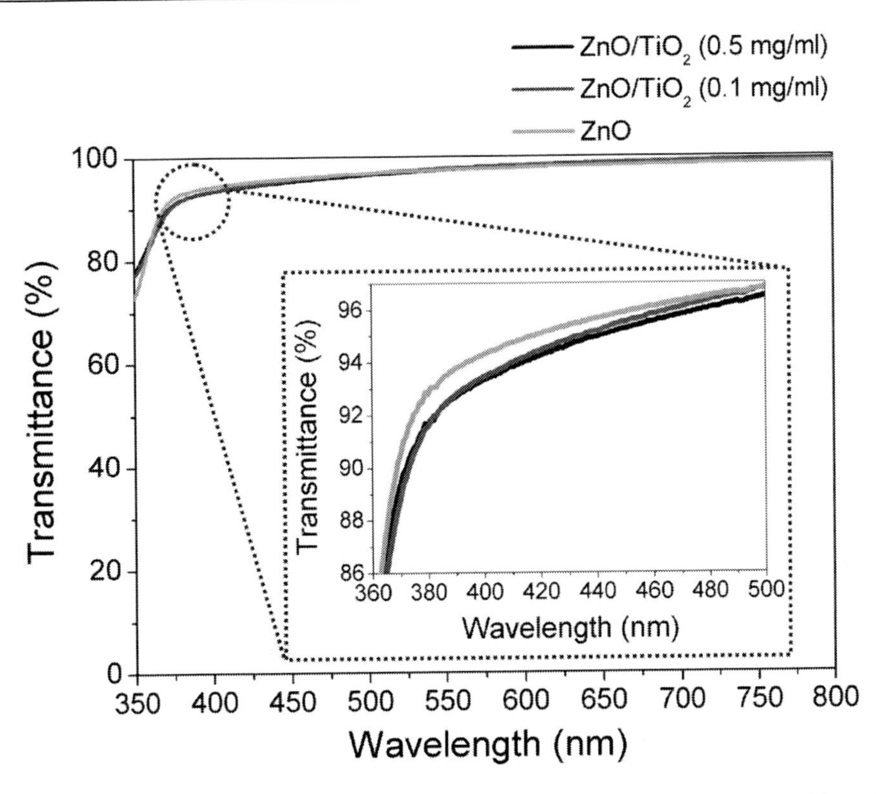

Figure 16. Transmittance spectra of the ZnO nanorods covered by the TiO_2 nanoparticles with various concentrations (0, 0.1, and 0.5 mg/ml). Substrates: glass.

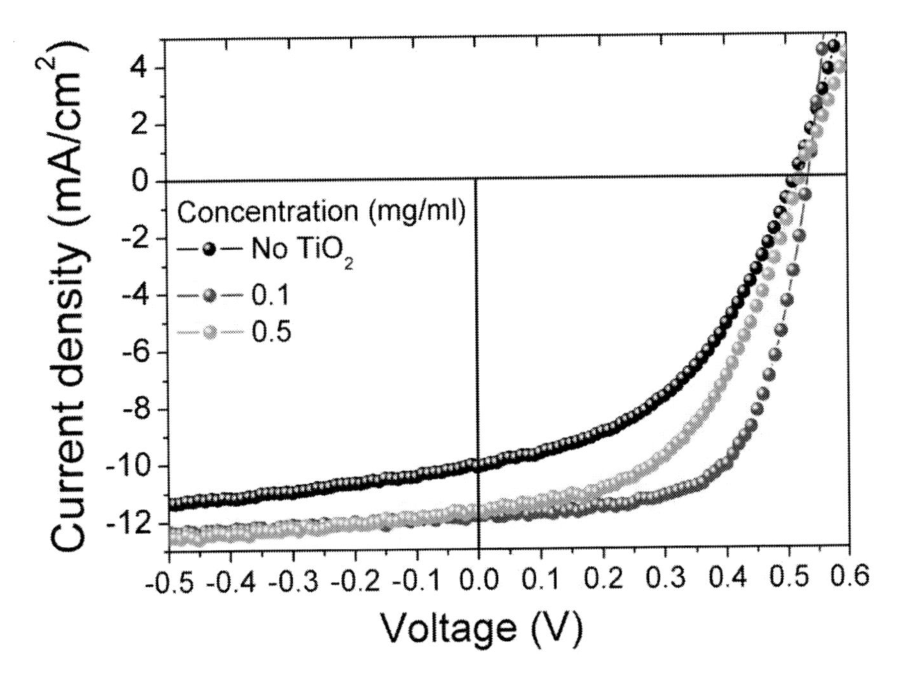

Figure 17. The J-V characteristics of the photovoltaic devices with various concentrations of TiO_2 under 100 mW/cm^2 AM 1.5G irradiation.

Figure 16 shows the transmittance spectra of the ZnO nanorods covered by the TiO_2 nanoparticles with various concentrations (0, 0.1, and 0.5 mg/ml). The three samples were prepared on glass substrates. The transmittance spectra of these films were measured using a Perkin-Elmer Lambda 35 UV-vis spectrophotometer. The three films are highly transparent (transmittance >90%) in the whole visible region. The TiO_2 nanoparticles on ZnO nanorod arrays are found to be highly transparent at the concentration of either 0.1 or 0.5 mg/ml, although their transmittance is slightly small than that of ZnO nanorods in 360-500 nm (inset of Figure 16). This result shows the TiO_2 nanoparticles could allow the maximum photon flux to reach the photoactive layer for photocurrent generation.

2.1.3. Photovoltaic Performance

Solar cells were measured with a Keithley 2400 source measurement unit, under illumination at 100 mW/cm^2 from a 150 W Oriel solar simulator with AM 1.5 G filters. The light intensity was calibrated with a calibrated standard solar cell with a KG5 filter which was traced to the NREL. The device area was defined as 10 mm^2 by a shadow mask. Figure 17 shows the J-V characteristics of the devices with various concentrations of TiO_2. The device without the TiO_2 nanoparticles exhibits a J_{SC} of 10.13 mA/cm^2, a V_{OC} of 0.51 V, and a FF of 45.5%, resulting in a PCE of 2.35%. Introducing the TiO_2 nanoparticles at a concentration of 0.1 mg/ml, the device has a significant improvement with J_{SC} of 11.83 mA/cm^2, V_{OC} of 0.535 V, and FF of 63.8%. This results in a considerable improvement of PCE to 4.04%, a 72% improvement. When the concentration of the TiO_2 nanoparticles further increases to 0.5 mg/ml, the PCE is conversely decreased to 3.04% with J_{SC} of 11.68 mA/cm^2, V_{OC} of 0.522 V, and FF of 49.9%. However, it is still better than that without the TiO_2 nanoparticles. These results show that the TiO_2 nanoparticles can act as a functional modification to enhance the photovoltaic performance. Photovoltaic parameters of these devices are summarized in Table 2.

Table 2. Photovoltaic parameters of the devices with TiO_2 nanoparticles of different concentrations

Concentration (mg/ml)	PCE (%)	Voc (V)	Jsc (mA/cm^2)	FF (%)	Rs (Ωcm^2)	Rsh (Ωcm^2)
0	2.35	0.51	10.13	45.5	5.83	251
0.1	**4.04**	**0.535**	**11.83**	**63.8**	**1.26**	**917**
0.5	3.04	0.522	11.68	49.9	2.51	370

As shown in Figure 17, the introduction of the TiO_2 nanoparticles (with the concentrations of 0.1 or 0.5 mg/ml) notably enhances the photocurrents. The J_{SC} is increased from 10.13 mA/cm^2 to 11.83 mA/cm^2 for 0.1 mg/ml and to 11.68 mA/cm^2 for 0.5 mg/ml, respectively. In order to further understand the reason behind the improvement in the J_{SC}, photoluminance (PL) spectra were investigated. The room temperature PL spectra were measured by using a violet laser diode (120 mW, Nichia) at 405 nm as the excitation source. Figure 18 shows the PL spectra of the various composited films of the P3HT:PCBM layer covering the TiO_2 nanoparticles with different concentrations (0, 0.1, and 0.5 mg/ml). These three samples all exhibit PL spectra peaked at 650 nm with similar emission feature to that of P3HT. With the excitation of 405 nm, only P3HT is excited. Hence, the two TiO_2-composited films exhibit the PL emission features similar to the P3HT:PCBM only, indicating that the luminescence predominantly results from the radiative recombination of the photogenerated excitons in P3HT.

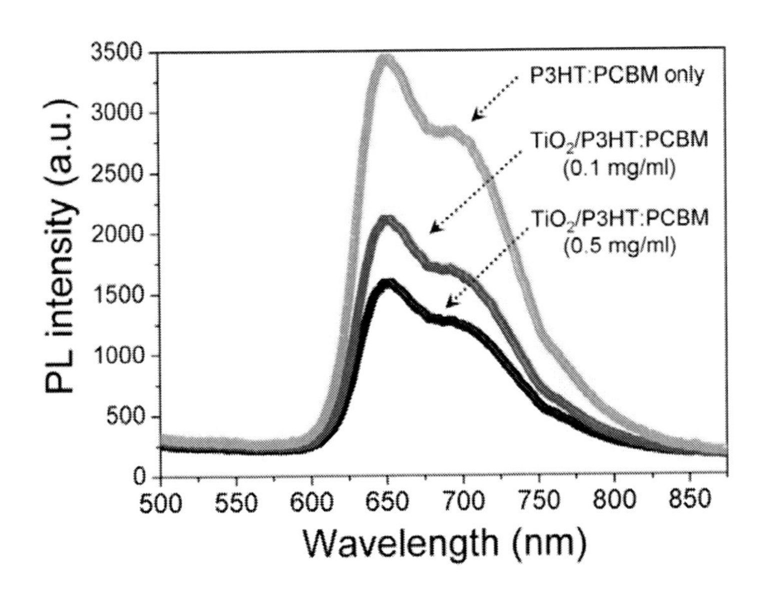

Figure 18. PL spectra of various composited films of the P3HT:PCBM layer covering the TiO_2 nanoparticles with different concentrations (0, 0.1, and 0.5 mg/ml).

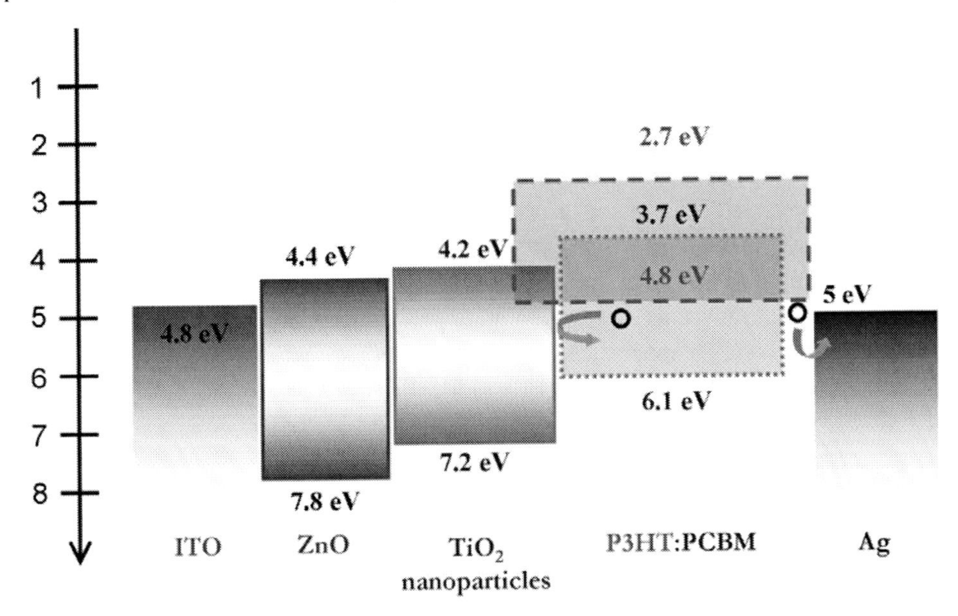

Figure 19. Energy band diagram for the OSC with a structure of ITO/ZnO nanorod arrays/TiO_2 nanoparticles/P3HT:PCBM/Ag.

It is obvious that the PL intensity is substantially decreased with the increasing concentration of the TiO_2 nanoparticles, suggesting the occurrence of significant PL quenching in the TiO_2-composited film [85]. The quenching of the PL emission of P3HT can be attributed to the TiO_2 nanoparticles. Like PCBM, the TiO_2 nanoparticles can act as electron acceptors, resulting in the enlargement of donor-acceptor interfacial areas. This is beneficial for exciton dissociation to takes place, therefore the radiative recombination process can be suppressed. As a result, this produces a charge-separated state with an electron on the TiO_2 nanoparticle and a hole on the P3HT, which is responsible for the PL emission quenching of P3HT.

Due to the efficient charge separation in the TiO_2-composited film, more charge carriers can be transported towards the electrodes for collecting, leading to the enhancement in the photocurrent. Accordingly, the J_{SC} is increased from 10.13 mA/cm^2 to 11.83 mA/cm^2 for the TiO_2-composited film with the TiO_2 nanoparticles at the concentration of 0.1 mg/ml. However, as the concentration increases to 0.5 mg/ml, the J_{SC} is not further increased but slightly decreased to 11.68 mA/cm^2. This value of 11.68 mA/cm^2 is still higher than that without the TiO_2 nanoparticles (10.13 mA/cm^2), showing that the introduction of the TiO_2 nanoparticles is also probitable for the photocurrent at a high concentration. However, at such a high concentration, it is suggested that the TiO_2 nanoparticles may form large clusters, resulting in the considerable resistance and thus in the reduced photocurrent. The R_S of the device with 0.5 mg/ml TiO_2 nanoparticles is 2.51 Ωcm^2, which is 2 times larger that that of the device with 0.1 mg/ml TiO_2 nanoparticles (1.26 Ωcm^2). This agrees with the suggestion of the formation of the large TiO_2 clusters. It is worthy to note that the R_S of the device with 0.1 mg/ml TiO_2 nanoparticles is 4-times smaller that that of the device without any TiO_2 nanoparticles (5.83 Ωcm^2). This evidence shows that the TiO_2 nanoparticles are profitable for carriers separation and transportation.

It is notable that the FF is improved from 45.5% to 63.8% as the 0.1 mg/ml TiO_2 nanoparticles is introduced. Figure 19 shows the energy band structure of the OSC with a structure of ITO/ZnO nanorod arrays/TiO_2 nanoparticles/P3HT:PCBM/Ag. Since the conduction band minimum of TiO_2 (4.2 eV) [86] is similar to the LUMO level of PCBM (3.7 eV), electrons can easily transport to ZnO then to ITO electrode through TiO_2. Moreover, because the valance band maximums of TiO_2 (7.2 eV, the optical band gap of rutile TiO_2 is 3.0 eV) and ZnO (7.8 eV) are lower than the HOMO level of P3HT (4.8 eV), TiO_2 and ZnO both can block the reverse hole transfer from P3HT to ITO, thus forcing hole to transport toward Ag. Thereby, the leakage current is efficiently suppressed and the shunt resistance (R_{SH}, defined from the J-V curves near 0 V under light illumination) is improved accordingly. With the 0.1 mg/ml TiO_2 nanoparticles, the R_{SH} is improved from 251 Ωcm^2 to 917 Ωcm^2. The high R_{SH} and low R_S together contribute to the improved FF and V_{OC} [19].

2.3. Anode Modification Using Vanadium Oxide (V_2O_5)

V_2O_5 powder (Riedel-de Haën, 99%) was homogeneously dispersed and suspended in isopropanol at different concentrations by using ultrasonic agitation. During the process of the ultrasonic agitation, it was observed that the V_2O_5 powder was pulverized to smaller particles. After the ultrasonic agitation, the color of the V_2O_5 colloidal solution is uniformly orange. Then the V_2O_5 colloidal solution was spin-casted in air on top of the organic photoactive layer. Finally, silver film (~200 nm) was deposited on top in a vacuum of 2×10^{-6} torr. Figure 20 shows the OSC with a device structure of ITO/ZnO nanorod arrays/P3HT:PCBM/V_2O_5/ Ag.

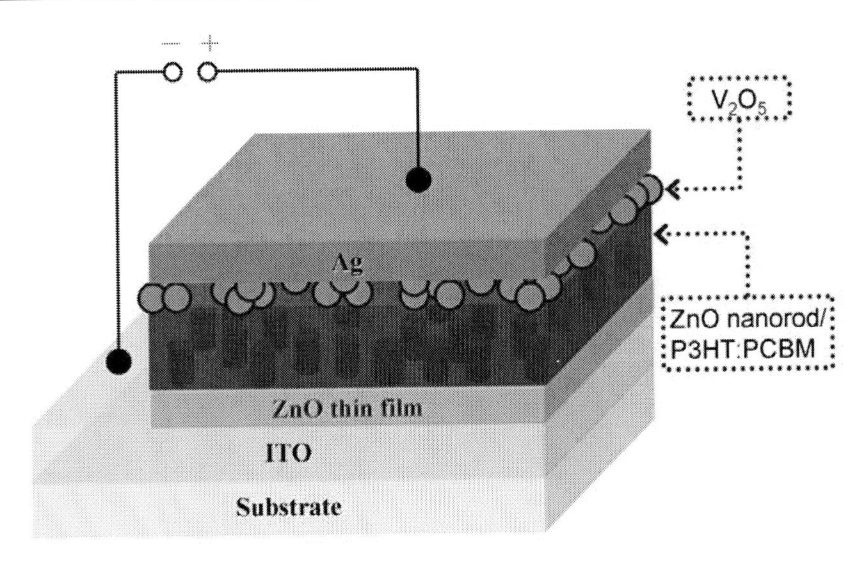

Figure 20. Schematic device structure of ITO/ ZnO nanorod arrays/P3HT:PCBM/V_2O_5/Ag.

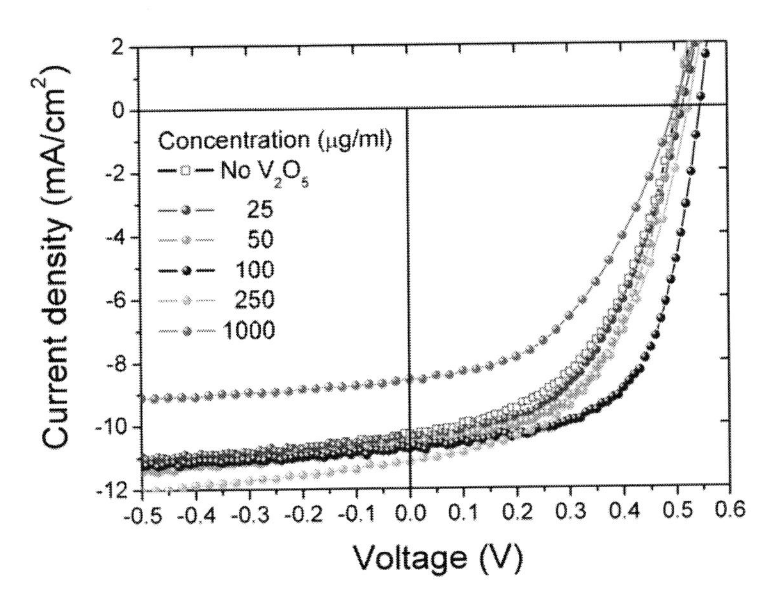

Figure 21. The J-V curves of the photovoltaic devices with the V_2O_5 modification from various concentrations under 100 mW/cm^2 AM 1.5G irradiation. Reproduced with permission from Huang, J.-S. et al. *Org. Electron.* 2009, 10, 1060-1065. Copyright © 2009, Elsevier B.V.

2.3.1. Photovoltaic Performance

Devices were unencapsulated, stored in air, and illuminated at 100 mW/cm^2 from a ThermoOriel 150 W solar simulator with AM 1.5G filters. The solar simulator was calibrated using a reference Si solar cell. All electrical measurements were carried out in air at room temperature. The device area was defined as 10 mm^2 by a shadow mask, so no extra current outside of the defined area was collected. J–V curves were measured with a Keithley 2400 source measurement unit. Figure 21 shows the J–V characteristics of the devices with various concentrations of V_2O_5. The device without the V_2O_5 modification exhibits a J_{SC} of 10.21 mA/cm^2, a V_{OC} of 0.5 V, and a FF of 49.36%, resulting in a PCE of 2.52%. The performance is similar to other reports on the same structure without the V_2O_5 modification [77,78]. Introducing a 25 μg/ml V_2O_5 modification, PCE is slightly improved to 2.67% with J_{SC} of 10.49 mA/cm^2, V_{OC} of 0.51 V, and FF of 49.91%. When the concentration of V_2O_5 further increases to 100 μg/ml, the device has a significant improvement with J_{SC} of 10.75 mA/cm^2, V_{OC} of 0.55 V, and FF of 60.21%. This results in a considerable improvement of PCE up to 3.56%, a 41% improvement. These results show that the V_2O_5 can act as a functional modification to enhance the photovoltaic performance.

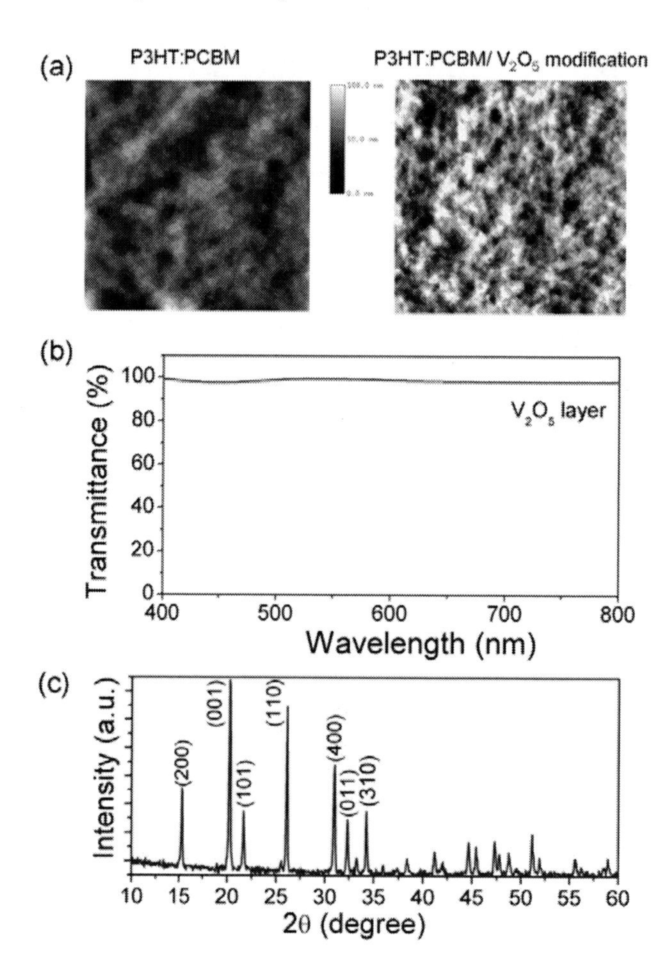

Figure 22. AFM images of the photoactive layers covered with and without the optimum V_2O_5 modification (a). AFM image scans are 5 × 5 μm. Transmission spectrum of the V_2O_5 layer (from the 100 μg/ml V_2O_5 colloidal solution) on a glass substrate (b). XRD spectrum of V_2O_5 (c). Reproduced with permission from Huang, J.-S. et al. *Org. Electron.* 2009, 10, 1060-1065. Copyright © 2009, Elsevier B.V.

A series of V_2O_5 concentrations (25, 50, 100, 250, and 1000 μg/ml) is further investigated and summarized in Table 3. The J_{SC}, V_{OC}, and FF increase with the V_2O_5 concentration from 25-100 μg/ml. The highest PCE of 3.56% is achieved at the concentration of 100 μg/ml, showing that the optimum V_2O_5 modification is obtained. As a low-concentration V_2O_5 modification is introduced (50 μg/ml or less), the improvement of device performance is not obvious. It is suspected that the concentration is too low to cover the photoactive layer completely. As a result, the leakage current will not be efficiently reduced. However, as the concentration of V_2O_5 increases to 1000 μg/ml, most V_2O_5 particles cluster together (~2 μm in average), resulting in the increased contact resistance and thus leading to a low photocurrent (J_{SC} ~8.55 mA/cm^2).

Figure 22 (a) shows the atomic force microscopy (AFM) images of the surface morphologies of photoactive layer covered with and without the optimum V_2O_5 modification. The V_2O_5 particles can be clearly observed. The root-mean-square roughness (σ_{RMS}) of the photoactive layer with the optimum V_2O_5 modification is ~18.4 nm, which is about the 2 times of that without V_2O_5 (σ_{RMS} ~10.6 nm).

Table 3. Photovoltaic parameters and efficiencies of inverted OSCs with the V_2O_5 modification from various concentrations. Reproduced with permission from Huang, J.-S. et al. *Org. Electron.* 2009, 10, 1060-1065. Copyright © 2009, Elsevier B.V

Concentration (μg/ml)	J_{SC} (mA/cm^2)	V_{OC} (V)	FF (%)	PCE (%)	R_S (Ωcm^2)	R_{SH} (Ωcm^2)
No V_2O_5	10.21	0.50	49.36	2.52	2.24	394
25	10.49	0.51	49.91	2.67	2.12	570
50	10.61	0.51	53.96	2.92	2.10	579
100	*10.75*	*0.55*	*60.21*	*3.56*	*1.35*	*620*
250	11.16	0.52	51.35	2.98	3.4	431
1000	8.55	0.50	46.78	2	6.3	250

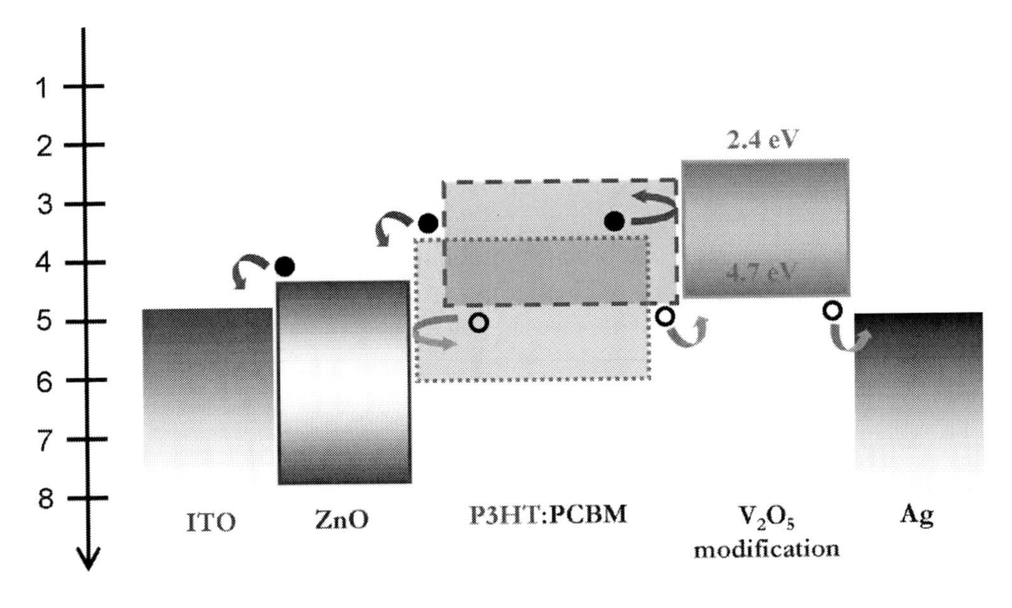

Figure 23. Energy band structure for the OSC with a structure of ITO/ZnO nanorod arrays/P3HT:PCBM/V_2O_5 modification/Ag. Other energy level values can be referred to Figures 15.

The AFM images clearly indicate that at the concentration of 100 μg/ml the photoactive layer is almost fully covered with V_2O_5. Figure 22 (b) shows the transmission spectrum of the V_2O_5 layer deposited from the 100 μg/ml V_2O_5 colloidal solution on a glass substrate. The transmission spectrum of the V_2O_5 layer was measured using a Perkin-Elmer Lambda 35 UV-vis spectrophotometer. It shows that the V_2O_5 layer is almost transparent in the visible region (transmittance >97%). The crystallinity of V_2O_5 was analyzed at room temperature by x-ray diffraction (XRD) using Cu Kα radiation. Figure 22 (c) shows the XRD spectrum of V_2O_5. The diffraction peaks appears in the 2θ range from 10-35° characterizing the orthorhombic crystalline structure of V_2O_5.

Note that the performance of the device without the V_2O_5 modification in our experiments has similar performance to other reports [77,78]. By introducing the optimum V_2O_5 modification, the device has significant improvements in FF (from 0.49 to 0.6) and V_{OC} (from 0.5 to 0.55 V). This indicates that V_2O_5 efficiently suppresses the leakage currents at the organic/Ag interface. Considering the device without V_2O_5, both P3HT and PCBM are in direct contact with Ag. It is possible for electrons to transfer from PCBM to Ag, thereby increasing the leakage currents. However, incorporating a V_2O_5 modification introduces two additional interfaces, organic/V_2O_5 and V_2O_5/Ag. As shown in Figure 23, the conduction band of V_2O_5 (2.4 eV) [87] is higher than the lowest unoccupied molecular orbital level of PCBM (3.7 eV), showing that V_2O_5 can block the reverse electron flow from PCBM to Ag. Thereby, V_2O_5 can effectively prevent the leakage current at the organic/Ag interface. In addition, the valence band of V_2O_5 (4.7 eV) [87] is close to the highest occupied molecular orbital level of P3HT (4.8 eV), revealing that V_2O_5 will help collecting holes.

Moreover, with the optimum V_2O_5 modification, the photovoltaic device has significant improvements in the R_S (from 3.09 Ωcm^2 to 1.35 Ωcm^2) and R_{SH} (from 376 Ωcm^2 to 610 Ωcm^2). It is known that the high R_{SH} indicates less leakage current across the cell and contributes to the improved FF and V_{OC} [19].

Another evidence for less leakage current is the rectification ratio (RR), defined as the current ratio at $\pm$1.5 V from the J-V curves measured in the dark. The RR of the device without V_2O_5 is 4.37×10^2, while that of the device with the optimum V_2O_5 modification increases to 1.77×10^4. The high RR and elevated R_{SH} both are strong evidences showing that the V_2O_5 modification can serve as an electron-blocking layer to effectively prevent the leakage currents, resulting in the dramatic improvement in PCE.

Figure 24 (a) compares the incident photon-to-current conversion efficiency (IPCE) spectrum of the devices with and without the optimum V_2O_5 modification. The IPCE is defined as the number of photogenerated charge carriers contributing to the current per incident photon. The IPCE spectra are measured in air under illumination by a 150 W Xenon lamp with a monochromator. The device with V_2O_5 modification shows the typical spectral response of P3HT:PCBM blend with a maximum IPCE of ~69% at 515 nm, while for the device without V_2O_5 modification, the peak reaches ~65% only. The IPCE spectra are consistent with the measured J_{SC} in the devices. The insertion of V_2O_5 demonstrates a substantial enhancement of ~6% at 515 nm in the IPCE. This enhancement agrees with the increase in J_{SC} (~5% increase in the device with V_2O_5). It indicates that the V_2O_5 modification also contributes to the increase in photocurrent.

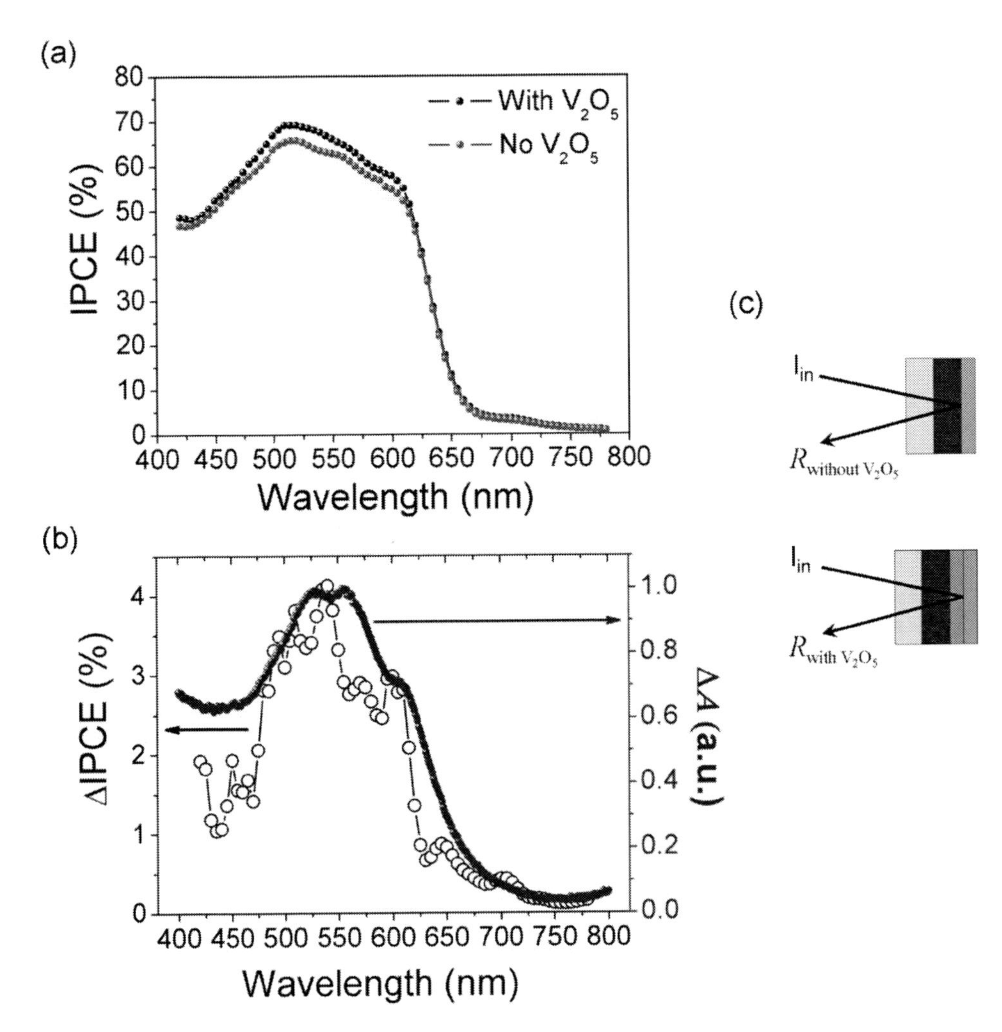

Figure 24. IPCE spectra for the devices with and without the optimum V_2O_5 modification (a). The change in absorption spectrum [$\Delta A(\lambda)$] and the difference in IPCE spectrum [$\Delta IPCE(\lambda)$] resulting from the insertion of the optimum V_2O_5 modification (b). Schematic of the optical beam path in the both samples (c). The variables are defined in the text. Reproduced with permission from Huang, J.-S. et al. *Org. Electron.* 2009, 10, 1060-1065. Copyright © 2009, Elsevier B.V.

To further clarify the role of the V_2O_5 modification, we measured the reflectance (R) spectra of the devices with and without the optimum V_2O_5 modification. The reflectance spectra of the devices were obtained using a Perkin-Elmer Lambda 35 UV-vis spectrophotometer. Since the two devices are identical except the addition of the V_2O_5 modification, comparison of the reflectance yields information on the additional absorption, $\Delta A(\lambda)$, in the photoactive layer as a result of the spatial redistribution of the light intensity by the V_2O_5 modification [Figure 24 (b)]. The $\Delta A(\lambda)$ is given by [88]

$$\Delta A(\lambda) \approx -\frac{1}{2\sqrt{2}d}\ln\left(\frac{R_{\text{with }V_2O_5}(\lambda)}{R_{\text{without }V_2O_5}(\lambda)}\right),- \quad (8)$$

where $R_{\text{with } V_2O_5}(\lambda)$ is the reflection from a device with the V_2O_5 modification, $R_{\text{without } V_2O_5}(\lambda)$ is the reflection from an identical device without the V_2O_5 modification, and d is the thickness of the photoactive layer (d is ~300 nm in both). Figure 24 (c) schematically shows the optical beam path in the devices. The result shows a clear increase in absorption over the spectral region of the interband transitions. Since the spectral features of the P3HT absorption are evident in the ΔA spectrum, the increased absorption arises from a better match of the spatial distribution of the light intensity to the position of the photoactive layer.

Figure 24 (b) also shows the difference in IPCE spectrum, $\Delta IPCE(\lambda)$, between the devices with and without the optimum V_2O_5 modification. This spectrum reveals three peaks at 510, 540, and 600 nm, respectively. It implies that the contribution of the V_2O_5 modification in photocurrent is mainly at the three peaks which are vibronic features from the P3HT molecules [43]. Moreover, the feature of the $\Delta IPCE$ spectrum is analogous with ΔA spectrum, showing that the increased optical absorption is nearly transferred to the photocurrent. Evidently, the V_2O_5 modification functions as an optical spacer to increase the optical absorption by spatially redistributed the light intensity and thereby increase the photocurrent. Alternatively, V_2O_5 could give rise to scattering of reflected light into photoactive layer and thereby increase the fraction of absorbed light. As a result, the amount of absorbed light is increased by the V_2O_5 modification, as shown in Figure 24 (b).

Although PEDOT:PSS layer can be solution processed, its hygroscopic nature is likely to form insulating patches due to the water adsorption, thus degrading the devices [55]. In contrast, V_2O_5 is relatively insensitive to water and stable in air. The solution-processed V_2O_5 modification can serve as a barrier preventing oxygen or water from entering and degrading the photoactive layer.

In addition, this approach does not need annealing treatment like PEDOT:PSS nor vacuum equipments, so it is simple, expeditious, and effective. This is very important for commercial realization of low-cost and large-area printed solar cells.

Conclusions and Future Work

This chapter focuses on the development of highly efficient, air-stable, and low-cost ZnO nanostructure/organic composite solar cells. The background of OSCs is reviewed in the beginning. In section 2, three types of the solution-processed interfacial modifications in the inverted OSCs hybridized with the ZnO nanorod arrays are investigated. These solution-processed modifications are PCBM clusters, TiO_2 nanoparticles, and V_2O_5 nanopowder, which can notably enhance the photovoltaic device performance based on different mechanisms. The PCBM clusters can lead to enhanced phase separation and thus absorption. The formation of a double heterojunction by TiO_2 nanoparticles provides efficient exciton dissociation and charge transfer. V_2O_5 nanopowder can suppress the leakage current and enhance the absorption. Compared to vacuum-deposited techniques, these solution approaches are simple, expeditious, and effective. They are also advantageous for potential applications to mass production of various large-area printed electronics and photonics with a very low cost. As for further research, the following aspects are suggested.

3.1. Formation of Double Heterojunction

Formation of a double heterojunction by TiO_2 nanoparticles (<100 nm) has been demonstrated to provide efficient exciton dissociation and charge transfer. However, the donor-acceptor interfaces can be even more enlarged by incorporating with TiO_2 nanoparticles smaller than 50 nm. Another formation of the double heterojunction is by using one-dimensional TiO_2 nanorods or nanotubes which can directly transfer electrons toward cathode. In addition, PCBM is not always suitable as a PV material because of its very weak light absorption in the visible and infrared regions. The high structural symmetry in C_{60} fullerene molecule forbids low-energy transitions and renders the absorption spectrum limited in the visible range [89]. An analogous soluble C_{70} derivative, [6,6]-phenyl-C_{71}-butyric acid methyl ester ($PC_{70}BM$), has wider and stronger absorption in the visible light region than PCBM [90]. Using the $PC_{70}BM$ clusters to take place of the additional PCBM clusters is another strategy, because $PC_{70}BM$ has larger absorption coefficient than $PC_{60}BM$. The energy level diagram for this architecture is shown in Figure 25.

3.2. Morphology of Zinc Oxide (ZnO) Nanostructures for Polymer Solar Cells

The ZnO nanorod arrays reported in the previous chapters have a rod-to-rod spacing ranging from 10 to 50 nm. However, the spacing is still larger than exciton diffusion length. Future research should focus on fabrication of ZnO nanorod arrays with a small rod-to-rod spacing no more than 20 nm. Other ZnO nanostructures, such as nanotubes, are suitable for this application. In addition, organic surfactant ligands [91], such as pyridine and anthracene-9-carboxylic acid, can be used to cap on the ZnO nanorod surface and control the characteristics of the interfaces between ZnO and polymers, thus possibly improving the carrier transportation there.

3.3. Metal Oxides at the Anode

On the other hand, the interactions between metal oxides and Ag should be studied. As silver is deposited, the characteristics of the metal oxides, such as work functions, electron affinities, and ionization potentials, may be slightly modified and tuned, thus affecting the carrier transport at the interfaces. The formation of alloy or compound may also accompany these changes. Employing ultraviolet photoemission spectroscopy (UPS) and X-ray photoemission spectroscopy (XPS) is useful for looking into these changes and their mechanisms behind.

3.4. Low Band Gap Polymer

Recently, highly efficient OSCs based on BHJ composites of low band gap polymers as donors and $PC_{70}BM$ as acceptors have been reported with PCE approaching 6% [16,32,17].

To achieve high efficiency, the band gap of conjugated polymers should be reduced to ~1 eV to enable absorption in the infrared and near-infrared regions of the solar spectrum [92]. In addition, a donor polymer with a lower HOMO level or an acceptor with a higher LUMO level should be used to achieve higher V_{OC}. Future research should focus on the system based on low-band-gap polymers and $PC_{70}BM$. Combined with the sandwiched structure developed here, the efficiency of low-cost polymer solar cell above 10% will be realized in the future.

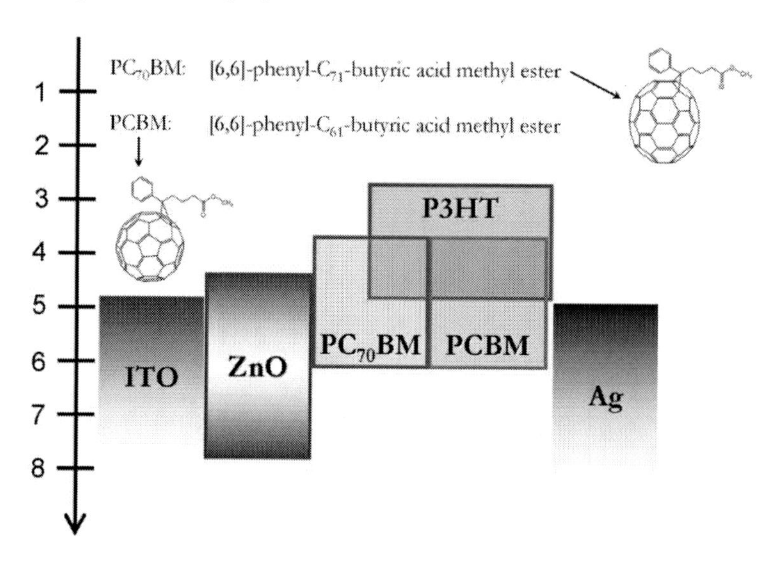

Figure 25. Energy band diagram for double-heterojunction with $PC_{70}BM$ clusters. The HOMO and LUMO levels of $PC_{70}BM$ are the same with PCBM [16].

3.5. Others

The stability of the photovoltaic devices should be further investigated by continuous testing in ambient. Replacing the ZnO film by TiO_x film may be another investigation direction in the inverted structures.

Acknowledgment

The authors thank Mr. Ing-Jye Wang for his assistance in article arrangement. This work was supported by the National Science Council, Taiwan under Grant NSC96-2221-E-002-277-MY3, NSC97-2221-E-002-039-MY3, and NSC98-2218-E-002-002.

References

[1] Krebs, F. C.; Gevorgyan, S. A.; Gholamkhass, B.; Holdcroft, S.; Schlenker, C.; Thompson, M. E.; Thompson, B. C.; Olson, D.; Ginley, D. S.; Shaheen, S. E.; Alshareef, H. N.; Murphy, J. W.; Youngblood, W. J.; Heston, N. C.; Reynolds, J. R.; Jia, S.; Laird, D.; Tuladhar, S. M.; Dane, J. G. A.; Atienzar, P.; Nelson, J.; Kroon, J.

M.; Wienk, M. M.; Janssen, R. A. J.; Tvingstedt, K.; Zhang, F.; Andersson, M.; Ilnganäs, O.; Lira-Cantu, M.; de Bettignies, R.; Guillerez, S.; Aernouts, T.; Cheyns, D.; Lutsen, L.; Zimmermann, B.; Würfel, U.; Niggemann, M.; Schleiermacher, H.-F.; Liska, P.; Grätzel, M.; Lianos, P.; Katz, E. A.; Lohwasser, W.; Jannon, B. *Sol. Energy Mater. Sol. Cells* 2009, *93*, 1968-1977.

[2] Chapin, D. M.; Fuller, C. S.; Pearson, G. L. *J. Appl. Phys.* 1954, *25*, 676-677.

[3] Martin, A. G.; Keith, E.; Yoshihiro, H.; Wilhelm, W. *Prog. Photovoltaics* 2009, *17*, 320-326.

[4] Krebs, F. C. *Sol. Energy Mater. Sol. Cells* 2009, *93*, 394-412.

[5] Krebs, F. C.; Jørgensen, M.; Norrman, K.; Hagemann, O.; Alstrup, J.; Nielsen, T. D.; Fyenbo, J.; Larsen, K.; Kristensen, J. *Sol. Energy Mater. Sol. Cells* 2009, *93*, 422-441.

[6] Krebs, F. C.; Gevorgyan, S. A.; Alstrup, J. *J. Mater. Chem.* 2009, *19*, 5442-5451.

[7] Green, M. A. *Prog. Photovoltaics: Research and Applications* 2001, *9*, 123-135.

[8] Kallmann, H.; Pope, M. *J. Chem. Phys.* 1959, *30*, 585-586.

[9] Alvarado, S. F.; Seidler, P. F.; Lidzey, D. G.; Bradley, D. D. C. *Phys. Rev. Lett.* 1998. *81*, 1082-1085.

[10] Heeger, A. J. *Rev. Mod. Phys.* 2001, *73*, 681-700.

[11] Blom, P. W. M.; Mihailetchi, V. D.; Koster, L. J. A.; Markov, D. E. *Adv. Mater.* 2007, *19*, 1551-1566.

[12] Tang, C. W. *Appl. Phys. Lett.* 1986, *48*, 183.

[13] Kroeze, J. E.; Savenije, T. J.; Vermeulen, M. J. W.; Warman, J. M. *J. Phys. Chem. B* 2003, *107*, 7696-7705.

[14] Shaw, P. E.; Ruseckas, A.; Samuel, I. D. W. *Adv. Mater.* 2008, *20*, 3516-3520.

[15] Yu, G.; Gao, J.; Hummelen, J. C.; Wudl, F.; Heeger, A. J. *Science* 1995, *270*, 1789.

[16] Park, S. H.; Roy, A.; Beaupre, S.; Cho, S.; Coates, N.; Moon, J. S.; Moses, D.; Leclerc, M.; Lee, K.; Heeger, A. J. *Nature Photon.* 2009, *3*, 297-302.

[17] Chen, H.-Y.; Hou, J.; Zhang, S.; Liang, Y.; Yang, G.; Yang, Y.; Yu, L.; Wu, Y.; Li, G. *Nature Photon.* 2009, *3*, 649-653.

[18] Nelson, J. *The physics of solar cells*; World Scientific Pub Co Inc: 2003.

[19] Moliton, A.; Nunzi, J. M. *Polym. Int.* 2006, *55*, 583-600.

[20] Vandewal, K.; Tvingstedt, K.; Gadisa, A.; Inganas, O.; Manca, J. V. *Nature Mater.* 2009, *8*, 904-909.

[21] Brabec, C. J.; Cravino, A.; Meissner, D.; Sariciftci, N. S.; Fromherz, T.; Rispens, M. T.; Sanchez, L.; Hummelen, J. C. *Adv. Funct. Mater.* 2001, *11*, 374-380.

[22] Scharber, M. C.; Mühlbacher, D.; Koppe, M.; Denk, P.; Waldauf, C.; Heeger, A. J.; Brabec, C. J. *Adv. Mater.* 2006, *18*, 789-794.

[23] Günes, S.; Neugebauer, H.; Sariciftci, N. S. *Chem. Rev.* 2007, *107*, 1324-1338.

[24] Zhou, Q.; Hou, Q.; Zheng, L.; Deng, X.; Yu, G.; Cao, Y. *Appl. Phys. Lett.* 2004, *84*, 1653.

[25] Blouin, N.; Michaud, A.; Gendron, D.; Wakim, S.; Blair, E.; Neagu-Plesu, R.; Belletete, M.; Durocher, G.; Tao, Y.; Leclerc, M. *J. Am. Chem. Soc.* 2007, *130*, 732-742.

[26] Campoy-Quiles, M.; Ferenczi, T.; Agostinelli, T.; Etchegoin, P. G.; Kim, Y.; Anthopoulos, T. D.; Stavrinou, P. N.; Bradley, D. D. C.; Nelson, *Nature Mater.* 2008, *7*, 158-164.

[27] Wang, X. J.; Perzon, E.; Oswald, F.; Langa, F.; Admassie, S.; Andersson, M. R.; Inganas, O. *Adv. Funct. Mater.* 2005, *15*, 1665-1670.

[28] Blouin, N.; Michaud, A.; Leclerc, M. *Adv. Mater.* 2007, *19*, 2295-2300.

[29] Muhlbacher, D.; Scharber, M.; Morana, M.; Zhu, Z. G.; Waller, D.; Gaudiana, R.; Brabec, C. *Adv. Mater.* 2006, 18, 2884-2889.

[30] Liang, Y.; Wu, Y.; Feng, D.; Tsai, S.-T.; Son, H.-J.; Li, G.; Yu, L. *J. Am. Chem. Soc.* 2009, *131*, 56-57 (2009).

[31] Peet, J.; Kim, J. Y.; Coates, N. E.; Ma, W. L.; Moses, D.; Heeger, A. J.; Bazan, G. C. *Nature Mater.* 2007, *6*, 497-500.

[32] Kim, J. Y.; Lee, K.; Coates, N. E.; Moses, D.; Nguyen, T. Q.; Dante, M.; Heeger, A. J. *Science* 2007, *317*, 222-225.

[33] Bundgaard, E.; Krebs, F. C. *Sol. Energy Mater. Sol. Cells* 2007, *91*, 954-985.

[34] Wudl, F. *Accounts Chem. Res.* 1992, *25*, 157-161.

[35] Singh, T. B.; Marjanovi, N.; Matt, G. J.; Gunes, S.; Sariciftci, N. S.; Montaigne Ramil, A.; Andreev, A.; Sitter, H.; Schwodiauer, R.; Bauer, S. *Org. Electron.* 2005, *6*, 105-110.

[36] Ma, W. L.; Yang, C. Y.; Gong, X.; Lee, K.; Heeger, A. J. *Adv. Funct. Mater.* 2005, *15*, 1617-1622.

[37] Padinger, F.; Rittberger, R. S.; Sariciftci, N. S. *Adv. Funct. Mater.* 2003, *13*, 85-88 (2003).

[38] Kim, Y.; Choulis, S. A.; Nelson, J.; Bradley, D. D. C.; Cook, S.; Durrant, J. R. *Appl. Phys. Lett.* 2005, *86*, 063502.

[39] Erb, T.; Zhokhavets, U.; Gobsch, G.; Raleva, S.; Stuhn, B.; Schilinsky, P.; Waldauf, C.; Brabec, C. J. *Adv. Funct. Mater.* 2005, *15*, 1193-1196.

[40] Mihailetchi, V. D.; Xie, H. X.; de Boer, B.; Koster, L. J. A.; Blom, P. W. M. *Adv. Funct. Mater.* 2006, *16*, 699-708.

[41] Savenije, T. J.; Kroeze, J. E.; Yang, X. N.; Loos, J. *Adv. Funct. Mater.* 2005, *15*, 1260-1266.

[42] Schilinsky, P.; Asawapirom, U.; Scherf, U.; Biele, M.; Brabec, C. J. *Chem. Mater.* 2005, *17*, 2175-2180.

[43] Kim, Y.; Cook, S.; Tuladhar, S. M.; Choulis, S. A.; Nelson, J.; Durrant, J. R.; Bradley, D. D. C.; Giles, M.; McCulloch, I.; Ha, C. S.; Ree, M. *Nature Mater.* 2006, *5*, 197-203.

[44] Urien, M.; Bailly, L.; Vignau, L.; Cloutet, E.; de Cuendias, A.; Wantz, G.; Cramail, H.; Hirsch, L.; Parneix, J. P. *Polym. Int.* 2008, *57*, 764-769.

[45] Moule, A. J.; Meerholz, K. *Adv. Mater.* 2008, *20*, 240-245.

[46] Zhang, F. L.; Jespersen, K. G.; Bjorstrom, C.; Svensson, M.; Andersson, M. R.; Sundstrom, V.; Magnusson, K.; Moons, E.; Yartsev, A.; Inganas, O. *Adv. Funct. Mater.* 2006, *16*, 667-674.

[47] Al-Ibrahim, M.; Ambacher, O.; Sensfuss, S.; Gobsch, G., *Appl. Phys. Lett.* 2005, *86*, 201120.

[48] Li, G.; Yao, Y.; Yang, H.; Shrotriya, V.; Yang, G.; Yang, Y., *Adv. Funct. Mater.* 2007, *17*, 1636-1644.

[49] Chou, C.-Y.; Huang, J.-S.; Wu, C.-H.; Lee, C.-Y.; Lin, C.-F., *Sol. Energy Mater. Sol. Cells* 2009, *93*, 1608-1612.

[50] Hwang, I. W.; Cho, S.; Kim, J. Y.; Lee, K.; Coates, N. E.; Moses, D.; Heeger, A. J., *J. Appl. Phys.* 2008, *104*, 033706.

[51] Krebs, F. C.; Spanggaard, H. *Chem. Mater.* 2005, 17, 5235-5237.
[52] Jørgensen, M.; Norrman, K.; Krebs, F. C. *Sol. Energy Mater. Sol. Cells* 2008, *92*, 686-714.
[53] Krebs, F. C.; Norrman, K. *Prog. Photovoltaics* 2007, *15*, 697-712.
[54] de Jong, M. P.; van Ijzendoorn, L. J.; de Voigt, M. J. A. *App. Phys. Lett.* 2000, 77, 2255-2257.
[55] Kawano, K.; Pacios, R.; Poplavskyy, D.; Nelson, J.; Bradley, D. D. C.; Durrant, J. R. *Solar Energy Materials and Solar Cells* 2006, *90*, 3520-3530.
[56] Li, G.; Chu, C. W.; Shrotriya, V.; Huang, J.; Yang, Y. *Appl. Phys. Lett.* 2006, *88*, 253503.
[57] White, M. S.; Olson, D. C.; Shaheen, S. E.; Kopidakis, N.; Ginley, D. S. *Appl. Phys. Lett.* 2006, *89*, 143517.
[58] Liao, H.-H.; Chen, L.-M.; Xu, Z.; Li, G.; Yang, Y. *Appl. Phys. Lett.* 2008, *92*, 173303.
[59] Qiao, Q. Q.; Xie, Y.; McLeskey, J. T. *J. Phys. Chem. C* 2008, *112*, 9912-9916.
[60] Waldauf, C.; Morana, M.; Denk, P.; Schilinsky, P.; Coakley, K.; Choulis, S. A.; Brabec, C. J., *Appl. Phys. Lett.* 2006, *89*, 233517.
[61] Hau, S. K.; Yip, H. L.; Baek, N. S.; Zou, J.; O'Malley, K.; Alex, K. Y. J. *Appl. Phys. Lett.* 2008, *92*, 253301.
[62] Takanezawa, K.; Tajima, K.; Hashimoto, K. *Appl. Phys. Lett.* 2008, *93*, 063308.
[63] Tao, C.; Ruan, S. P.; Xie, G. H.; Kong, X. Z.; Shen, L.; Meng, F. X.; Liu, C. X.; Zhang, X. D.; Dong, W.; Chen, W. Y. *Appl. Phys. Lett.* 2009, *94*, 043311.
[64] Kim, D. Y.; Subbiah, J.; Sarasqueta, G.; So, F.; Ding, H.; Irfan; Gao, Y. *Appl. Phys. Lett.* 2009, 95, 093304.
[65] Irwin, M. D.; Buchholz, B.; Hains, A. W.; Chang, R. P. H.; Marks, T. J. *Proc. Natl. Acad. Sci. U.S.A.* 2008, *105*, 2783-2787.
[66] Liang, C.-W.; Su, W.-F.; Wang, L. *Appl. Phys. Lett.* 2009, 95, 133303.
[67] Mor, G. K.; Shankar, K.; Paulose, M.; Varghese, O. K.; Grimes, C. A. *Appl. Phys. Lett.* 2007, *91*, 152111.
[68] Krebs, F. C. *Sol. Energy Mater. Sol. Cells* 2008, *92*, 715-726.
[69] Krebs, F. C.; Thomann, Y.; Thomann, R.; Andreasen, J. W. *Nanotechnology* 2008, *19*, 424013.
[70] Yu, B. Y.; Tsai, A.; Tsai, S. P.; Wong, K. T.; Yang, Y.; Chu, C. W.; Shyue, J. J. *Nanotechnology* 2008, *19*, 255202.
[71] Huang, J.-S.; Chou, C.-Y.; Liu, M.-Y.; Tsai, K.-H.; Lin, W.-H.; Lin, C.-F. *Org. Electron.* 2009, 10, 1060-1065.
[72] Huynh, W. U.; Dittmer, J. J.; Alivisatos, A. P. *Science* 2002, *295*, 2425-2427.
[73] Günes, S.; Sariciftci, N. S. *Inorg. Chim. Acta* 2008, 361, 581-588.
[74] Vayssieres, L. *Adv. Mater.* 2003, *15*, 464-466.
[75] Huang, J.-S.; Lin, C.-F. *J. Appl. Phys.* 2008, *103*, 014304 (2008).
[76] Law, M.; Greene, L. E.; Johnson, J. C.; Saykally, R.; Yang, P. D. *Nature Mater.* 2005, *4*, 455-459.
[77] Olson, D. C.; Piris, J.; Collins, R. T.; Shaheen, S. E.; Ginley, D. S., *Thin Solid Films* 2006, *496*, 26-29.
[78] Takanezawa, K.; Hirota, K.; Wei, Q. S.; Tajima, K.; Hashimoto, K. *J. Phys. Chem. C* 2007, *111*, 7218-7223.
[79] Huang, J.-S.; Chou, C.-Y.; Lin, C.-F. *Sol. Energy Mater. Sol. Cells* 2010, *94*, 182-186.

[80] Nápoles-Duarte, J. M.; Reyes-Reyes, M.; Ricardo-Chavez, J. L.; Garibay-Alonso, R.; López-Sandoval, R. *Phys. Rev. B* 2008, *78*, 035425.

[81] Li, G.; Shrotriya, V.; Huang, J. S.; Yao, Y.; Moriarty, T.; Emery, K.; Yang, Y. *Nature Mater.* 2005, *4*, 864-868.

[82] Olson, D. C.; Lee, Y. J.; White, M. S.; Kopidakis, N.; Shaheen, S. E.; Ginley, D. S.; Voigt, J. A.; Hsu, J. W. P. *J. Phys. Chem. C* 2007, *111*, 16640-16645.

[83] Olson, D. C.; Shaheen, S. E.; Collins, R. T.; Ginley, D. S. *J. Phys. Chem. C* 2007, *111*, 16670-16678.

[84] Olson, D. C.; Lee, Y. J.; White, M. S.; Kopidakis, N.; Shaheen, S. E.; Ginley, D. S.; Voigt, J. A.; Hsu, J. W. P. *J. Phys. Chem. C* 2008, *112*, 9544-9547.

[85] Lin, Y.-Y.; Chen, C.-W.; Chang, J.; Lin, T. Y.; Liu, I. S.; Su, W.-F. *Nanotechnology* 2006, *17*, 1260-1263.

[86] Kim, J. Y.; Kim, S. H.; Lee, H. H.; Lee, K.; Ma, W. L.; Gong, X.; Heeger, A. J. *Adv. Mater.* 2006, *18*, 572-576.

[87] Shrotriya, V.; Li, G.; Yao, Y.; Chu, C. W.; Yang, Y. *Appl. Phys. Lett.* 2006, *88*, 073508.

[88] Lee, J. K.; Coates, N. E.; Cho, S.; Cho, N. S.; Moses, D.; Bazan, G. C.; Lee, K.; Heeger, A. J. *Appl. Phys. Lett.* 2008, *92*, 243308.

[89] Yao, Y.; Shi, C.; Li, G.; Shrotriya, V.; Pei, Q.; Yang, Y. *Appl. Phys. Lett.* 2006, *89*, 153507.

[90] Wienk, M. M.; Kroon, J. M.; Verhees, W. J. H.; Knol, J.; Hummelen, J. C.; Hal, P. A. v.; Janssen, R. A. J. *Angew. Chem.-Int. Edit.* 2003, *115*, 3493-3497.

[91] Lin, Y.-Y.; Chu, T.-H.; Chen, C.-W.; Su, W.-F., *Appl. Phys. Lett.* 2008, *92*, 053312.

[92] Shrotriya, V. *Nature Photon.* 2009, *3*, 447-449.

In: Advanced Organic-Inorganic Composites
Editor: Inamuddin

ISBN 978-1-61324-264-3

Chapter 5

ORGANIC-INORGANIC COMPOSITE COATINGS

***Eram Sharmin*[a], *Fahmina Zafar*[a,*] *Manawwar Alam*[b] *and Sharif Ahmad*[a]**

[a]Department of Chemistry, Faculty of Natural Science, Jamia Millia Islamia (A Central University), New Delhi 110025, India

[b]Research Centre-College of Science, King Saud University, Riyadh, Kingdom of Saudi Arabia

ABSTRACT

Organic-inorganic composites have become a promising area of research, presently in the world of coatings. Such composites have gained considerable attention since they bear synergistic characteristics of both the organic and inorganic components. They exhibit good scratch resistance, impact resistance, adhesion, ultra violet (UV) radiation resistance, transparency, anti-microbial and self-cleaning activity, corrosion/chemical resistance, and gas barrier properties. Consequently, they find greater applications as antifog, antifouling, optical glass, biocompatible coatings and others. The chapter provides a brief overview of organic-inorganic hybrid coatings, metal, conducting polymer, UV curable and waterborne as well as some vegetable oil based composite coatings. A few characterization methods of the composite coatings have also been exemplified.

ABBREVIATIONS

AB	Aluminium-sec-butoxide
ACP	Amorphous calcium phosphate
AEAPS	*N*-(2-Aminoethyl)-3-aminopropyltri-methoxysilane
AgNP	Silver nanoparticle
APTEA	3-Aminopropyltriethoxy silane
AuNP	Gold-nanoparticle

[*] Corresponding author's e-mail: fahmzafar@gmail.com

CNTs	Carbon nanotubes
CP	Conducting polymer
DMA	Dynamic mechanical analysis
DSC	Differential scanning calorimetry
DTA	Differential thermal analysis
DTG	Differential thermal gravimetry
2EHA	2-Ethylhexylacrylate
EIS	Electrochemical impedance spectroscopy
FRCC	Fibre reinforced composite coatings
FTIR	Fourier transform infrared
GPTMS	Glycidoxypropyltrimethoxysilane
HDDA	Hexanedioldiacrylate
ICPTES	Isocyanatopropyltriethoxysilane
LSV	Linear sweep voltammetry
MA	Methacrylic acid
MMC	Metal matrix composite
MMA	Methylmethacrylate
MOPTMOS	Methacryloxypropyl trimethoxysilane
MTMS	Methyltriethoxysilane
MWCNT	Multiwalled carbon nanotubes
OIHM	Organic-inorganic hybrid materials
ORMOSIL	Organically modified silicates
ORMOCERS	Organically modified ceramics
PANI	Polyaniline
PEML	Polyelectrolyte multilayers
PhTES	Phenyltriethoxysilane
PMMA	Polymethylmethacrylate
PU	Polyurethane
SEM	Scanning electron micrography
TEM	Transmission electron microscopy
TEOS	Tetraethoxysilane
TGA	Thermo gravimetric analysis
TMOS	Tetramethoxysilane
TPOZ	Zirconium-n-propoxide
UV	Ultra violet
UV–vis	Ultra violet-vissible
VO	Vegetable oils
VTES	Vinyltriethoxysilane
VTMA	Vinyltrimethoxysilane
WB	Waterborne
WC	Tungsten carbide
XRD	X-ray diffraction

1. INTRODUCTION

Gone are the days when man's sole object of prime concern was food, clothing and shelter. Today, the advancements in knowledge, development of sophisticated analytical instruments and techniques and the advent of nanotechnology have brought manifold changes in the life of man in every sphere. The world of coatings has also witnessed drastic changes globally in the past decades out of three major driving forces

(i) the ever growing consumer expectations of good quality and performance coupled with lower cost,
(ii) concerns related to energy consumption, and
(iii) environmental contamination

Due to regular predictions, strict regulations and quest for innovations, the researchers in industries and academics are actively engaged to explore and formulate new strategies to meet the demands of the coatings' world. As a consequence, several technological advancements in the field have come up, e.g., environment friendly (high solids, powder, waterborne, UV curable) coatings, high performance (organic-inorganic hybrids and composites) coatings and others. Efforts are also in progress to develop high performance and cost effective organic-inorganic composite coatings through "environment friendly" route, that is, high solids, waterborne or UV-curable organic-inorganic composite coatings.

The chapter briefly discusses the role of organic-inorganic composites as coatings, their characterization methods and also their applications in different fields such as for medicine, anti-fog, sensor, packaging materials and others. However, prior to this, a short introduction to coatings, in general, has been provided in the proceeding section.

1.1. Coatings

Metals or alloys corrode due to chemical or electrochemical reactions with their environment, which often results in drastic deterioration in their properties. Corrosion occurs on a matal surface, due to the development of anodic and cathodic areas, through oxidation and reduction reactions, leading to the formation of oxides of metals/alloys [1,2]. Atmospheric corrosion is caused by the reaction of metals/alloys with oxygen and water or moisture. In the absence of either one of these, e.g., in polar regions and deserts (humidity < 30%), corrosion does not take place. Corrosion control is based upon the prevention of such chemical reactions. As a simple and effective method of corrosion control or protection, coatings form a barrier on the metal substrate. They primarily confer aesthetic appearance as well as protection to the substrate by the prevention of diffusion of oxygen and water.

Coatings consist of binders or resins, pigments, solvents, extenders and additives. They provide protection either by the barrier action of the coating itself or by corrosion inhibition by the pigments in coatings [2-5]. The most common resins are vinyls, acrylics, epoxies, polyurethanes, polyesters and others.

The type of coatings depends upon the type of substrate and environment. In industrial and marine environments, high corrosivity is caused by soot, sand, sulphate salts, chloride ions and acidic environments, respectively. Structures situated near the waterline of the sea (e.g., offshore plants, wind turbines) referred as splash zone, experience oxygen-rich atmosphere, electrolytes from the sea, coupled with UV radiations and mechanical stresses due to alternating weather or climate conditions. The substrates buried under water or soils are influenced by temperature, salinity, pH, dissolved gases, humidity, bacteria, sand, gravels and stones. Thus, the choice of coatings for a particular substrate actually depends upon the type and impact of various corrodents to which it is exposed [1,6]. This governs the type of binder to be used in the formulation of coatings. Binders have a preponderant impact on coatings performance. These polymers are the most important constituent of coatings influencing the adhesion of coatings to metals, cohesion within the coatings, high mechanical strength, low permeability, aesthetic appearance, anticorrosive performance, drying characteristics and others. The structural attributes that affect the coating properties of polymers are polarity, unsaturation, steric hindrance, crosslinking, crystallinity, glass transition temperature and others. Polymers used as binders are polyepoxies, polyacrylics (polymethylmethacrylate)(PMMA), polyvinyls (polystyrene), polyurethanes (PU), polyesters (PE), and others. With view to develop high performance coatings, such polymers are further reinforced with micro/nano materials, which provide enhanced corrosion protection as organic-inorganic composite coatings [3-14].

1.2. Organic-Inorganic Composite Coatings

Nature has remarkable ability to combine biobased organic and inorganic components at nanoscale. Natural organic-inorganic composites include crustacean carapaces or mollusc shells, bone or teeth tissues in vertebrates, silicic skeleton of a diatom, in radiolaria and others. The prospects to combine properties of organic and inorganic components for materials design and processing is an old challenge faced by man since ages, e.g. Egyptian inks, green bodies of china ceramics, prehistoric frescos and others [15]. Maya blue is another important example of man made hybrid material. It combines the color of the organic pigment, natural blue indigo, encapsulated within inorganic clay mineral. Thus, the concept of organic-inorganic hybrids is not new. Infact, paints also belong to the class of hybrids. The area however, gained huge significance in eighties [15].

The term "composite coatings" basically refers to coating materials formed by the combination of two or more constituents that results in better properties of the final material relative to those of the individual components (Figure 1). Here, each material retains its separate chemical, physical and mechanical properties. Organic-inorganic composite coatings consist of organic polymer "matrix" with inorganic "reinforcements" that serve to strengthen the composites and improve the overall mechanical properties, adhesion to substrate and chemical resistance of coatings. The properties of composite coatings are strongly dependent on the properties of their constituent materials, their distribution and interactions among them and the geometry of the reinforcement (shape, size, uniformity and size distribution). The concentration, distribution and orientation of the reinforcement also affect the final performance characteristics [9-15].

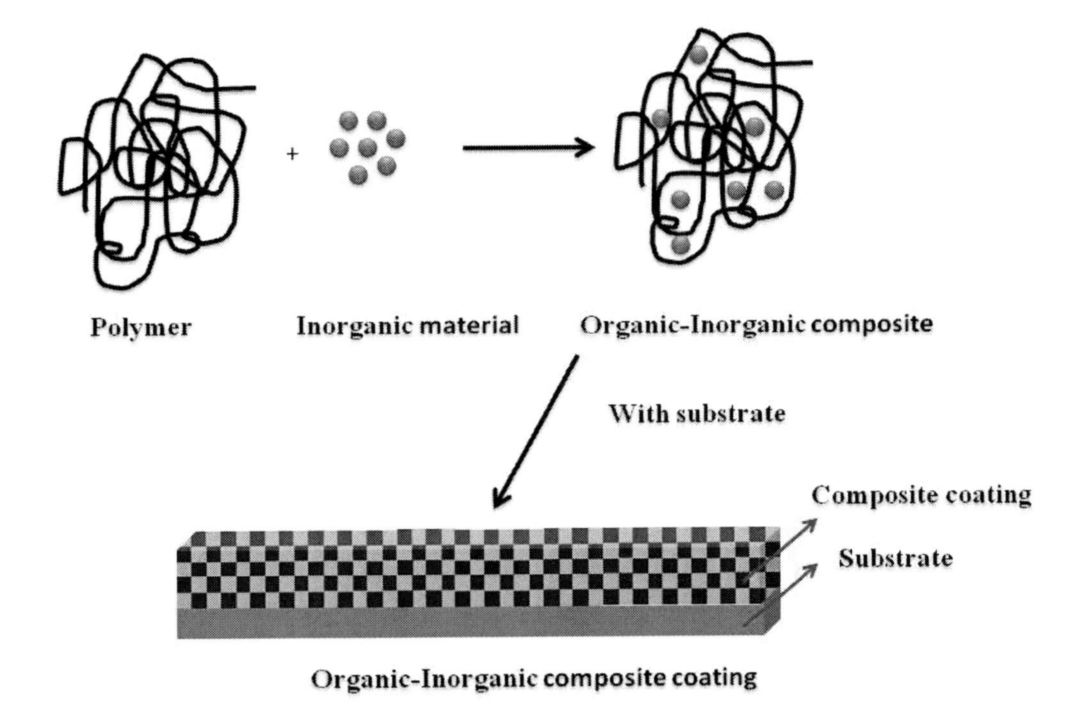

Figure 1. Organic-inorganic composite coatings.

Composite coatings can be classified on the basis of the type of matrix used. These are

(a) *Metal matrix composite coatings*
(b) *Ceramic matrix composite coatings*
(c) *Polymer Matrix composite coatings*

Another classification is based on the type of reinforcement used. These are

(a) *Particulate reinforced composite coatings (*reinforcements may be spherical, cubic, tetragonal, platelet, or particles of other shape)
(b) *Fibrous reinforced composite coatings* (long fibers or short fibres)

1.2.1. Organic-inorganic Hybrid Materials (OIHM) Coatings

It is literally "something that is obtained by mixing different types of materials", and becomes a new material that can be called a "mongrel". Therefore, an organic-inorganic hybrid is a combination of organic and inorganic materials.

Organic-inorganic hybrids consist of (1) *the inorganic matrix,* where organic materials are embedded in an inorganic polymer. Example, zinc silicates (inorganic) coatings modified with organic products, (2) *the organic matrix,* where inorganic materials are embedded in an organic polymer. Example in paint formulations where inorganic pigments are used in organic binders, (3) *the interpenetrating network*, where inorganic and organic polymeric networks are independently formed without mutual chemical bonds. Example, the incorporation of inorganic macro-molecular sol-gel networks into organic polymer structures,

(4) *true hybrids,* where inorganic and organic polymeric systems with mutual chemical bonds are formed.

The main objective behind the development of organic-inorganic composite coatings is the combination of both organic and inorganic constituents, to achieve synergistic combination of both the components. The most important method to achieve this goal is the sol-gel method. The sol-gel technique of alkoxysilanes is one of the useful methods available for preparing organic-inorganic hybrid materials. The advantage of the sol-gel technique is that the reaction proceeds at ambient temperature to form ceramic materials, as compared to the traditional methods performed at high temperature [7,9,17]. These yield organic-inorganic hybrid coatings which are generally scratch resistant, impact resistant, solvent resistant, thermally stable and tough coatings [9,11,13,15,17]. These can be prepared by the insitu formation of inorganic species in polymer matrix, by co-condensation of functionalised oligomers and polymers with metal alkoxides (Si, Ti, Zr) by simultaneous formation of organic and inorganic networks (interpenetrating networks), using organic functional group containing alkoxysilanes, which can later form hybrid network on curing thermally or photochemically [9,11,13,15,17].

The sol-gel reaction is generally divided into four stages: (i) hydrolysis of metal alkoxides to produce hydroxyl groups (in acidic or basic media) followed by (ii) condensation of the resulting hydroxyl and residual alkoxy groups to form a three-dimensional network, (iii) growth of particles, and (iv) agglomeration of polymer by network formation (gel) (Figure 2) [9-17].

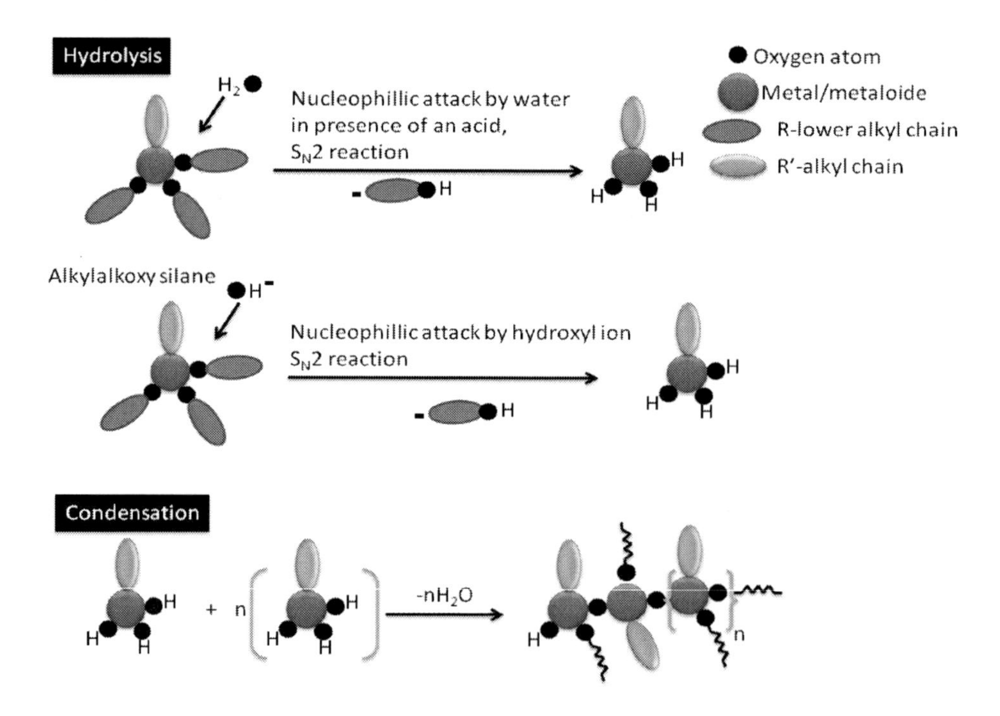

Figure 2. Sol-gel process.

These reactions are governed by several factors such as the nature of alkyl groups, solvent, reaction temperature, water to alkoxide molar ratio and others. These factors not only affect the structure and morphology of the hybrid materials but also drastically influence their coating properties such as physico-mechanical, thermal, chemical behavior, gas barrier properties, optical properties and others. Metal alkoxides generally used for the purpose are those of Al, Zr, Ti, Sn, Si and others [17].

OIHM coatings generally show good chemical stability, oxidation control, corrosion resistance, thermal stability, transparency, super-hydrophobicity, antimicrobial behavior and gas barrier property. These may undergo heat or radiation curing. These show high thermal stability compared to the pristine resins. The glass transition temperatures of OIHM also shift to higher values with incorporated inorganic component. At higher content of metal alkoxides, generally due to agglomeration or phase separation, these coatings show deterioration in performance [10-14,17,18]. Different silane precursors are used to prepare OIHM as given in Figure 3.

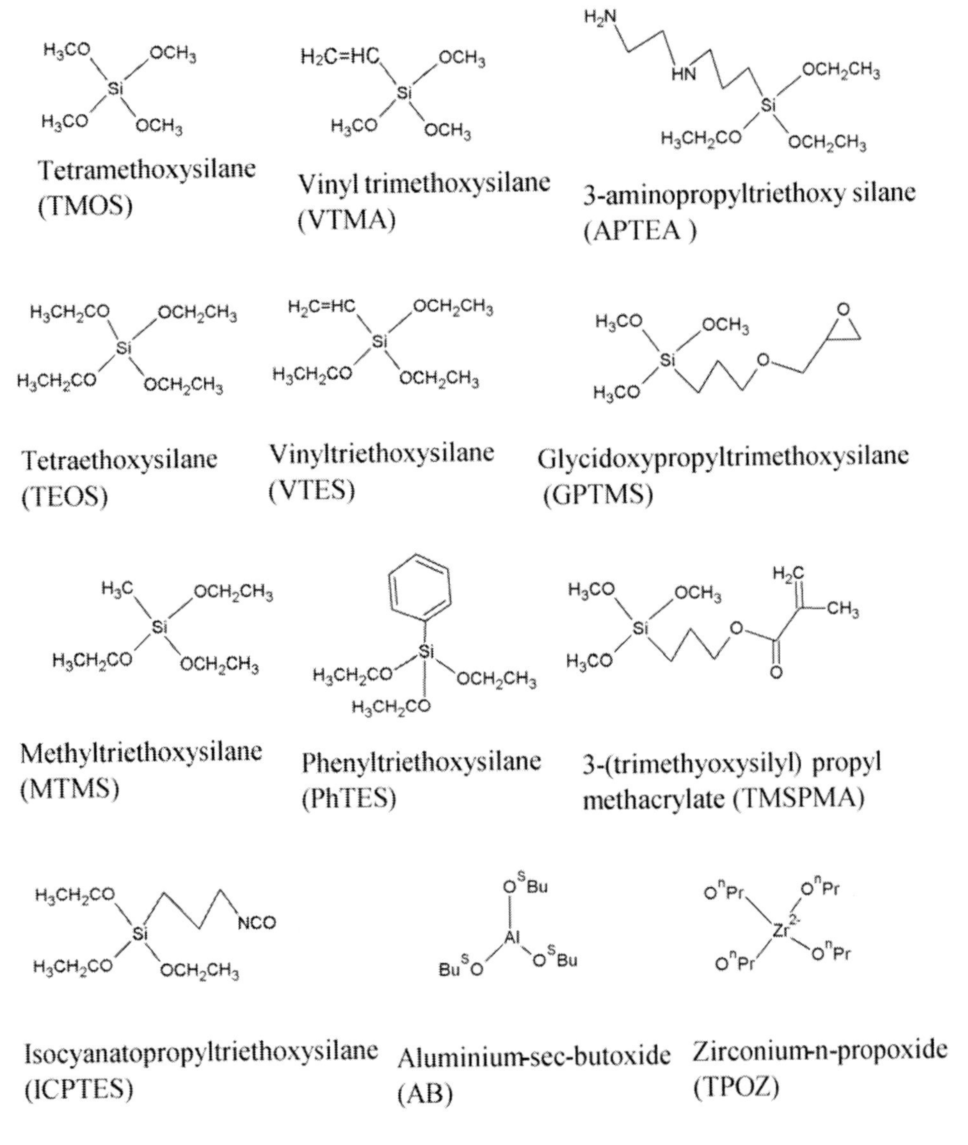

Figure 3. Some examples of inorganic precursors for OIHM.

A lot of work has been carried out on OIHM coatings. Medda and De [19] prepared OIHM coatings with tetraethoxysilane (TEOS), 3-GPTMS at pH = 1.3. Hydrolysis-condensation generated silica nanoparticles, -Si-O-Si- network, are protected by the organic functionality of GPTMS. The coatings were very hard attributed mainly to the in situ generated silica nanoparticles chemically bonded with the highly cross-linked silica-polyethylene oxide network [19]. Raju and Jena prepared hyperbranched polyurethane hybrid coatings using TEOS as crosslinker. They studied the effect of SiO_2 loading in different wt% on the coating properties [20,21]. They have also developed waterborne hyperbranched polyurethane-urea OIHM coatings, which showed good thermal and mechanical properties [22].

1.2.2. Organic-inorganic Organoclay Nanocomposite Coatings

Organic-inorganic clay composite coating material comprises of polymer clay combinations. These are prepared by solvent intercalation, in situ intercalation and melt intercalation process [23-25]. Layered silicates or nanoclays consist of stacks of sheet like platelets with thickness of ~1 nm and extremely large surface areas and aspect ratios. Clays are generally treated or functionalized to be compatible with the host polymer. The functionalization as well as processing techniques governs the resulting differences in clay morphologies, which finally determine the efficiency of the resulting polymer-clay composites. The platelets in the clay stacks are separated by the penetration of polymer into these platelets; consequently, intercalated and exfoliated clay morphologies are obtained as provided in Figure 4.

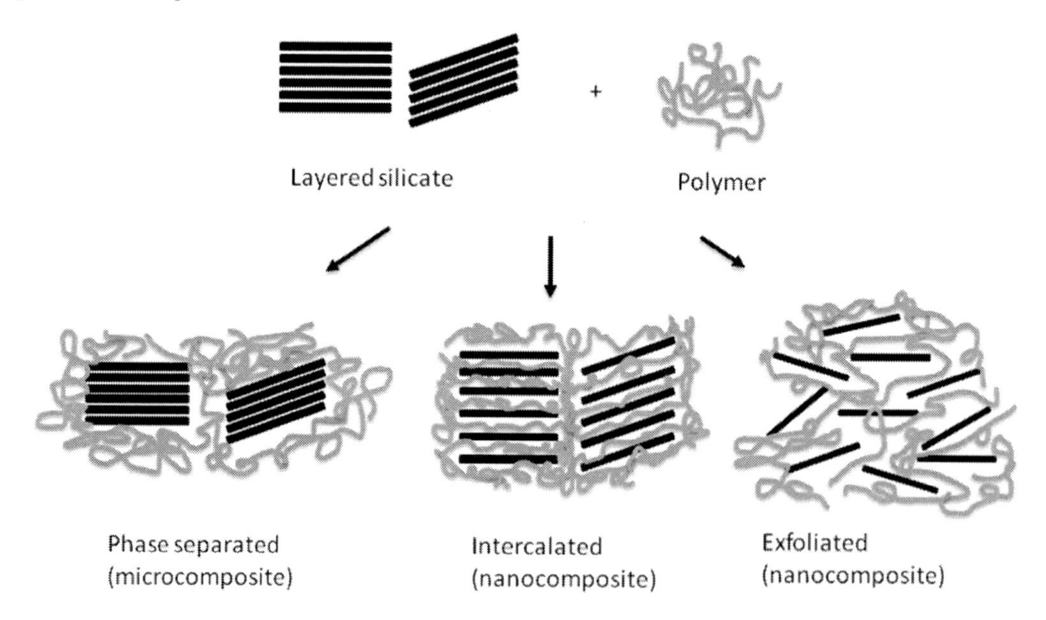

Figure 4. Polymer clay composites.

In exfoliated clay morphology, the individual platelets in the stack are completely separated. By intercalation, the penetration of the polymer in between clay platelets occurs to the extent that the stacks/galleries of clay platelets expand without separation and an increase in d-spacing (clay platelet interlayer spacing) is observed. Sometimes, the clay layers remain stacked with no polymer intercalation; these are microcomposites with distinct phase separation. They may also show tactoid formation, where the layered silicates are not individually well dispersed rather they consist of several stacked silicate monolayers [23-25].

Okamoto and co-workers [26-28] proposed an interpretation of the nanocomposites structure related to the aspect ratio and the organomodifier chain lengths. According to them, the smaller the size of silicate layers (smectite, montmorillonite), the lower the physical jamming, restricting the conformation of organomodifier alkyl chains, and thus, the lower the coherency of the organoclay. In organomodified mica, due to its stacked structure, the polymer chains can hardly penetrate up to the core of the silicate layers, contrary to their smaller size as in smectite and montmorillonite.

The clay morphologies play very important role in governing the physico-mechanical and corrosion resistance performance of their coatings. At very low dosage levels of organoclay, generally 1 wt% to 5 wt%, drastic improvement in properties is reported such as high scratch resistance, impact resistance, good flexibility retention, good adhesion, transparency and toughness [26-32]. These are a consequence of hydrogen bonding between polar groups residing on polymer and clay, leading to very strong interactions between matrix and silicate layers. However, at higher loading, it is observed that the coating performance generally deteriorates due to agglomeration or aggregation [23-28].

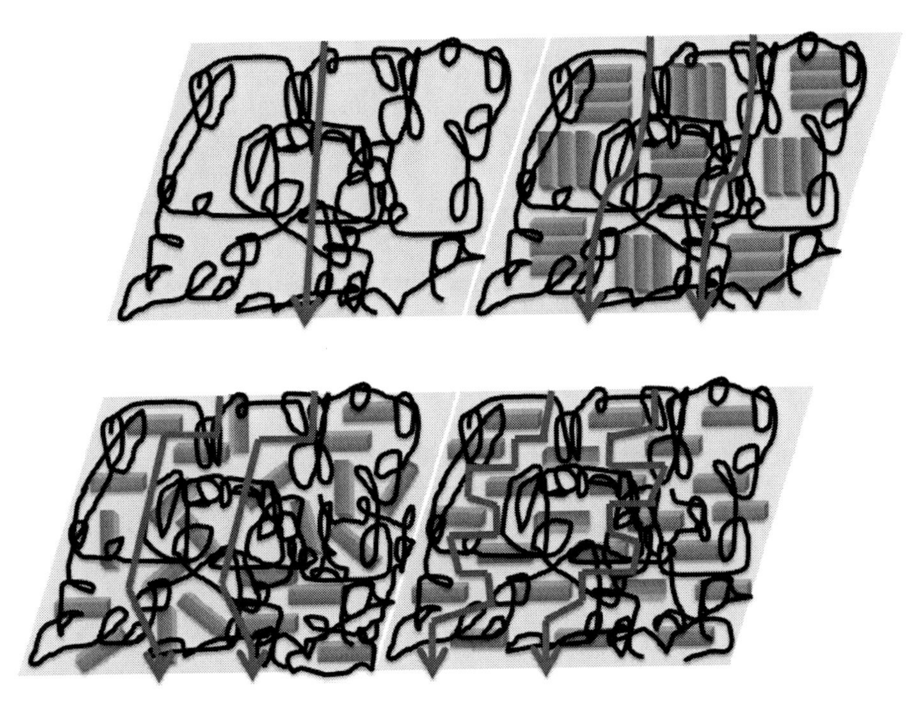

Figure 5. Diffusion path way for corrodents in polymer clay composite coatings.

Literature also reports the formation of biodegradable and antimicrobial clay composite coatings at an optimum loading of organoclay [24]. However, at higher levels of clay concentration, the biodegradability decreased, since the presence of dispersed clay layers with large aspect ratio makes the diffusion path more tortuous for micro-organisms to reach the bulk matrix. Similarly, polymer organoclay composite coatings have great potential to combat corrosion. The nanoclay layers fill the micro voids and crevices in the pristine polymer coatings. They increase the length of diffusion paths of corroding agent through the coating thus, making it more difficult for the corroding agent to seep through the coating [32]. Thus, in other words, the nanoclay layers dispersed in polymer matrix increase the tortuisity of diffusion pathways of corrosive agents such as oxygen, hydrogen and hydroxide ions as shown in Figure 5 [33,34].

Yeh et al. [33] have reported the preparation, characterization and corrosion studies of environment friendly PU/organoclay nanocomposite coatings where upto 3wt% clay loading, greatest corrosion protection performance was achieved relative to neat PU. Polyaniline (PANI)/organoclay conducting nanocomposite coatings prepared by Olad and Rashidzadeh [34] showed good anticorrosion efficiency and conductivity relative to pure polyaniline.

1.2.3. Metal Organic-inorganic Composite Coatings

A composite material consisting of metal as matrix is called metal matrix composite [MMC]. It has higher specific strength and modulus, lower coefficient of thermal expansion, better properties at elevated temperatures and is considered for wide range of applications. Example combustion chamber nozzle (in rocket, space shuttle), tubings, cables, heat exchangers, structural members, housings and others. MMC offer unique and superior characteristics including hardness, wear resistance and oxidation protection at high temperatures with respect to a metal or an alloy. A metal matrix composite (MMC) coating can improve the surface properties of a material beyond the capability of the substrate, the composite material or the metal alone [35]. The methods of application of MMC on metal substrates include E-beam, sputtering, thermal spray and chemical vapor deposition, which require high temperatures, high vacuum or high melting-point for the substrate. Electroplating is a convenient and effective method for applying a MMC coating onto a base metal.

Electrodeposition of composite coatings, based on hard particles dispersed in a metallic matrix, is gaining importance for engineering applications. The second phase can be hard oxide (Al_2O_3, TiO_2, SiO_2), carbides particles (SiC, WC), diamond, solid lubricate (poly terafluoro ethylene, graphite, MoS_2) and others [36-39].

Micrometer-sized tungsten carbide (WC) particles were incorporated into an electroplated copper matrix. WC particles had a hardening effect on the metal matrix. The wear resistance as well as durability of these metal matrix composites improved. The inclusion of tungsten carbide particles promoted the formation of a smooth crystalline structure of copper matrix [37].

The incorporation of gold- nanoparticles in metal-matrix-systems is useful for electronic applications [40]. The application of core shell systems with a core of gold nanoparticles and polymer shells around is a completely new concept for the deposition of nanocomposite films [41]. The corrosion studies of copper and copper-matrix composite coatings, with zirconium oxide particles embedded during electrodeposition from an acid copper sulphate plating bath have been carried out. As compared to electrodeposited pure copper the zirconium oxide particles embedded in the electroplated copper increase the polarisation resistance and decrease the corrosion rates with three times in sulphuric acid [42].

Ni–Al_2O_3 composite coatings possess better oxidation resistance than unreinforced nickel in an oxygen rich atmosphere [38]. Electrochemical behavior of electrodeposited Ni–WC composite coatings has been carried out by R. Balasubramaniam and co-workers [43]. Electrodeposited Ni–SiC composites show improvement in the wear resistance [44].

Successful codeposition of hard particles (like Al_2O_3, TiO_2, SiC, WC, Cr_3C_2, TiC and diamond) in a range of metal matrices such as Ni, Cr, Co, Re has been carried out [45,46].

Another approach involves the dispersion of metal nanoparticles in the matrix. These nanoparticles are generally incorporated into the coating materials to improve their scratch resistance, UV resistance, conductivity and other properties. A lot depends on processing methods of surface treatment of metal nanoparticles to render compatibility between the nanoparticle and polymer matrix, to control their average particle size and particle size distribution. These coatings find applications in wood floors, safety glasses, electronic displays, automotive finishes and corrosion protection [47,48].

Generally, metal nanocomposite coatings undergo loss in transparency, flexibility, impact resistance, abnormal increase in material's viscosity, and other defects. It is imperative that a filler material should impart improved scratch resistance without causing the aforementioned deterioration. This can be achieved by monitoring the film thickness, filler concentration, filler particle size, and the difference in refractive index between the matrix and the filler particles. The refractive indices of nanoparticles as well as the resin should closely match each other to minimize light scattering so as to achieve transparent coatings.

Zinc oxide nanoparticle-containing organic-inorganic composite coatings as reported by Hong et al. [48] have shown good mechanical and antibacterial properties. Biocompatible two-layer tantalum/titania polymer hybrid coatings have been reported by Scandola and coworkers [49]. Silver nanocluster–silica matrix composite coatings have shown good adhesion as well as antibacterial properties [50].

1.2.4. Conducting Polymer (CP) Composite Coatings

When an electrically conducting phase is dispersed in sufficient quantity in a polymeric resin, a conducting composite is formed. These possess superior properties than alternative materials. CP polymer blends and composites are prepared by dispersion and chemical in situ polymerization of CP in polymer matrix, electrochemical polymerization, polymer grafting to a CP surface, solution blending, dry blending and others. The development methods to process CP are based on the difficulties encountered in its processibility and aromatic structure, interchain hydrogen bonding and effective charge delocalization in its structure [51].

CP such as polyphenylene, PANI, polypyrrole, have been very efficient in coating applications since these can store and transport charge which features their ability to anodically protect metals against rapid corrosion [9,52]. The corrosion protection mechanism of CP is very complex, consequently, several mechanisms have been proposed for the same such as protection by barrier mechanism, by the formation of a passive oxide film on metal surface through an oxidation-reduction mechanism or by anodic mechanism, where PANI film withdraws charge from the metal and passivates its surface against corrosion [53-55]. PANI has been very effective in coating applications. It combats corrosion by accepting electrons from the metal and in turn donates them to oxygen, creating a two-step reaction that forms a layer of pure iron oxide [56]. Under controlled conditions, PANI prevented rust 10 times more than Zn and in the field, PANI was found to be 3-10 times more effective. The substituted PANI like polyanisidine, polymethoxyaniline have also proved to be efficient corrosion inhibitors [57].

The incorporation of nanoclay in CP provides enhanced corrosion protection by decreasing the coating porosity and permeability of gases and liquids as well as by stabilizing their electronic structure [9]. The corrosion protection by PANI/clay nanocomposites was first presented by Chang and coworkers [58]. The preparation of PANI/clay nanocomposite coatings has been carried out by Olad et al [34]. Ahmad and coworkers have prepared waterborne poly(1-naphthylamine)/poly(vinylalcohol)–resorcinol formaldehyde-cured corrosion resistant composite coatings with superior protection performance in various corrosive media [57].

1.2.5. Carbon Nanotubes (CNTs) Organic-inorganic Composite Coatings

Carbon nanotubes (CNTs) were discovered by Iijima in 1991. Since then, these have been used as reinforcement for polymer, ceramic and metal matrix composites [59]. Polymer based composites including CNTs have attracted considerable attention in the academic research and industry worldwide. These show good electrical conductivity, high stiffness and strength at relatively low CNTs content [60]. Multiwalled carbon nanotube (MWCNT) reinforced aluminum nanocomposite coatings were prepared using cold gas kinetic spraying by Agarwal et al [59]. Recently, Chen et al. codeposited Ni–CNTs composites from conventional plating nickel bath [61]. Ni–P–CNTs composite coating and CNTs/copper matrix composites were prepared by electroless plating and powder metallurgy techniques. These revealed a lower wear rate and friction coefficient compared with pure copper, and their wear rates and friction coefficients showed a decreasing trend with increasing volume fraction of CNTs within the range upto 12 vol% due to the effects of the reinforcement and reduced friction of CNTs [62].

1.2.6. Fibre Reinforced Composite Coatings (FRCC)

Research is underway to develop fibre reinforced composite materials with varied combinations of fibres, fillers and matrices with tailormade properties. The final properties are governed by the length, cross-sectional dimensions, orientation and content of fibres, shape, size, volume fraction, specific surface area of filler, and the nature of matrix. A number of investigations have been conducted on several types of fibres such as glass, carbon, kenaf, hemp, flax, bamboo, and jute to study the effect of these fibres on the mechanical properties of composite materials [35]. FRCC exhibit higher strength, improved flexibility, crack resistance and good service life. These find applications in architectural paints, antifouling coatings, epoxy ballast tank coatings and others. Carbon-fiber-reinforced silicon carbide composites are used as barrier coatings [63]. Nickel- and copper-coated carbon fibre reinforced tin-lead alloy composites have been reported by Ho. The composites containing coated treated carbon fibres had higher tensile and shear strength relative to those containing coated pristine carbon fibres [64]. Carbon fibers reinforced hydroxyapatite show high elastic modulus (close to that of human bone), good biocompatibility and biostability (such as no release of dissoluble products), these cannot be corroded under stress or degraded with time [65].

1.2.7. Ultra Violet (UV) Curable Organic-inorganic Composite Coatings

UV curable OIHM coatings have also been developed as an alternate to environment friendly coating technology. The technique is used to enhance condensation and polymerisation reactions in OIHM. There are many chemical reactions that take place in OIHM during UV curing, the main being the silanol condensation [66-69].

OIHM coatings obtained are generally amorphous and optically transparent. The degree of interaction between the organic and inorganic phases, the physico-mechanical and chemical resistance behavior of coatings and their thermal stability were all found to depend strongly on the exposure or UV curing time. For the particular proportions of inorganic and organic components used in OIHM coatings, an optimum UV curing time is required to obtain coatings of desirable performance, e.g., optical transparency, adhesion to substrate, scratch resistance, impact resistance, abrasion resistance, flexibility, and others [66,70].

Earlier unsaturated polyesters were used extensively in wood finishing as fillers, sealers, and topcoats where styrene was incorporated into the coating formulations as a reactive diluent to reduce viscosity, increase the crosslink density, and decrease cost. Presently, UVcurable coatings are mostly based on acrylated oligomers due to their relatively higher reactivity and lower volatility. Acrylated resins are preferred over methacrylated ones due to higher cure rates at room temperature and lower oxygen inhibition in the former. Some examples of acrylate oligomers are: epoxy acrylates, polyether acrylates, polyurethane acrylates, polyester acrylates and silicone acrylates [71].

In the preparation of hybrid coatings by UV curing, the two-step curing process is commonly performed. In the first step, the photocurable resin is mixed with the metal oxide (MOx) precursor in presence of HCl (catalyst). Next, the photoinitiator is added and the formulation coated onto the substrate. As the system is exposed to irradiation, the organic network is formed. Then, it is heated in an oven at 80 ^{o}C in humid atmosphere to induce the formation of the inorganic network by hydrolysis and condensation of the organic precursor [66,70].

UV curable organic–inorganic hybrid coatings were prepared by the sol–gel method. The covalent links occurred between the inorganic and the organic networks. UV curable coatings obtained were hard, and transparent. These hybrid coatings were synthesised using a commercially available, acrylate end-capped polyurethane oligomeric resin, hexanedioldiacrylate (HDDA) as a reactive solvent, 3-(trimethoxysilyl)propoxymethacrylate (MPTMS) as a coupling agent between the organic and inorganic phase, and a metal alkoxide, TEOS [72].

UV curing hybrid coatings on polycarbonate using HDDA as the reactive diluent, and MPTMS as the coupling agent were prepared by Gilberts and coworkers [73]. Here, the introduction of prehydrolyzed TEOS significantly improved the abrasion resistance of coatings. Hybrid films based on vinyltriethoxysilane, TEOS and polyfunctional acrylates are reported by Gigant et al. [74]. These showed structure-property correlations between microhardness and normalized Raman intensities. Hybrid sol-gel materials have also been prepared by Soppera et al [75]. Colloidal silica acrylates and methacrylates were also used by various researchers to form abrasion resistant inorganic/organic hybrid coatings [71,76].

1.2.8. Waterborne (WB) Organic-inorganic Composite Coatings

Solvent borne coatings are replaced by WB coatings for environmental, health and safety reasons. They are generally prepared by sol-gel method, in which nanometer-size fillers are dispersed in a polymer matrix. In polymer/silica hybrid films, the silica component acts as a hard segment in a soft polymer matrix to achieve excellent performance of coatings. Silane coupling agents are commonly used to achieve miscibility of the polymer and silica [77]. Polymer/silica hybrid nanoparticles have been prepared by emulsion and miniemulsion polymerizations. The sol–gel process seems to be applicable to hybrid emulsion coatings because the condensation of alkoxysilanes occurs under room temperature to produce crosslinking between the polymer particles and silica. Watanabe and coworkers prepared acrylic polymer/silica hybrid emulsions by adding hydrolyzed alkoxysilane to polymer emulsions both in the absence and presence of silane coupling agents methacryloxypropyl trimethoxysilane (MOPTMOS) [77]. The films with coupling agents showed high solvent resistance relative to their counterparts without coupling agents.

WB coatings show high weather durability, water resistance, and mechanical properties. However, satisfactory stain resistance could not be achieved in WB coatings. WB stain preventive coating materials were prepared by Wada et al. [78] with organic-inorganic composites in which colloidal silica was employed as an inorganic component. The organic component of the composites included conventional acrylic resin emulsions prepared by the emulsion polymerization of methylmethacrylate (MMA), 2-ethylhexylacrylate (2EHA), methacrylic acid (MA), and an organic silane hybridized acrylic resin emulsion prepared by the core/shell emulsion polymerization of MMA, 2EHA, MA, and 3MOPTMOS.

WB organo clay based nanocomposite coatings also provide good corrosion protection for use in coating applications. WB PU/organoclay nanocomposite coatings as reported by Yeh and co-workers, showed good corrosion protection effect relative to plain WB PU [33]. The use of water-based resins in combination with the UV curing technique is also finding attraction since the process combines important features of both the techniques. Such coatings also exhibit dual cure process as in case of UV curable coatings. However, when a water-based formulation is employed, the order of steps is reverse relative to that in UV curing due to the necessity to remove water, in basic medium. Here, heat treatment is carried out first to evaporate water and also to carry out the formation of inorganic network, followed by UV irradiation.

1.2.9. Renewable Resource Based Organic-inorganic Composite Coatings

Recent advances in the field have been focused on the development of cost effective, environment friendly and high performance coatings from biobased resources that are renewable, cost effective, eco-friendly, and may in some respects, compete with their petro-based counterparts, e.g., starch, cellulose, chitosan, vegetable oils, and others. Such biobased organic-inorganic composite coatings may find applications in packaging materials, as edible coatings, in biomedicine and others.

Vegetable oils (VO) have found immense practical implications in the field of organic-inorganic hybrid coatings. Organic-inorganic hybrid materials from linseed, soyabean, castor and others have been prepared [30,79-83]. VO polymers in their virgin form lack the properties of rigidity and strength. VO organic-inorganic hybrids show good performance as scratch resistant, impact resistant, transparent and thermally stable hybrid coatings relative to pristine oil based coatings. These have been prepared by sol-gel process where the metal alkoxide (Si, Ti, Zr) forms the inorganic precursor and the epoxidized or hydroxylated oil serves as the organic component. The hybrids occur as "tethered" structures, where "tethering" usually occurs between epoxide or hydroxyl groups of polymers. The main disadvantage associated with these coating materials is their complex, multi-step, high time consuming synthesis and curing process, which occurs at elevated temperature [30,79-83].

PU and PE-bisphenol A epoxy/clay composite coatings have been prepared from Mesua ferrea L. oil by Karak et al. Such clay composites are obtained as transparent coatings that showed good adhesion, flexibility retention, impact resistance, scratch resistance and antimicrobial behavior [29,30,84].

Nanostructured PANI dispersed smart anticorrosive composite coatings were prepared from methyl orange doped PANI/castor oil PU. These coatings were found to act as ''corrosion sensors'' by exhibiting different colors when placed in acid as well as alkaline media. The protection behavior of coatings was attributed to the formation of a passive iron oxide/dopant layer at the metal-coating interface that obstructs the penetration of the corrosive ions [85]. VO based silver nanoparticle embedded antimicrobial paints have been reported by Kumar and co-workers [86].

Advantage is taken of the free radicals generated during the natural drying process of drying oils/alkyd paints for the preparation of silver- and gold-nanoparticle (AgNP and AuNP) embedded paints (*in situ*). Highly branched PE/clay silver nanocomposites based on VO with different loadings of silver were prepared *via* reduction of silver salt by employing dimethylformamide as solvent as well as reducing agent at room temperature. Fourier transform infrared (FTIR), ultra violet-vissible (UV–vis), X-ray diffraction (XRD), scanning electron microscopy (SEM) and transmission electron microscopy (TEM) studies substantiate the formation of well-dispersed silver nanoparticles within the clay gallery with an average size of 15 nm. High antibacterial activity was observed against gram negative bacteria (*Escherichia coli* and *Psuedomonas aeruginosa*).

Excellent chemical resistance in various chemical media except in alkali has also been noticed. Polymer silver nanocomposites have been prepared by using sunflower oil based macromonomer. The prepared silver nanocomposites showed good film properties and antimicrobial efficiency at low concentration (1 wt%) of silver nitrate [87]. Several examples are available in literature where improved coating characteristics are obtained in VO polymer organic inorganic composite coatings.

2. Characterization Methods for Organic-Inorganic Composite Coatings

As discussed earlier, the performance of organic inorganic composite coatings depends upon the processing methods as well as the structural characteristics of the material, the morphology, the type or particle size of the modifier and other parameters. These can be assessed with the help of some important characterization methods. To determine the electrostatic interactions between constituents such as hydrogen bonding, FTIR spectroscopy is the principal tool. Second one is the morphology of the prepared composites, especially in polymer clay composites, where with the help of XRD and TEM one can ascertain the degree of intercalation or exfoliation.

The techniques given in Figure 6 are considered important even in case of in situ created or mechanically added metal oxides in polymer matrix as well as in the dispersed polymer/ metal nanoparticle composites. The performance evaluation of coatings may be carried out by other methods also provided in Figure 6.

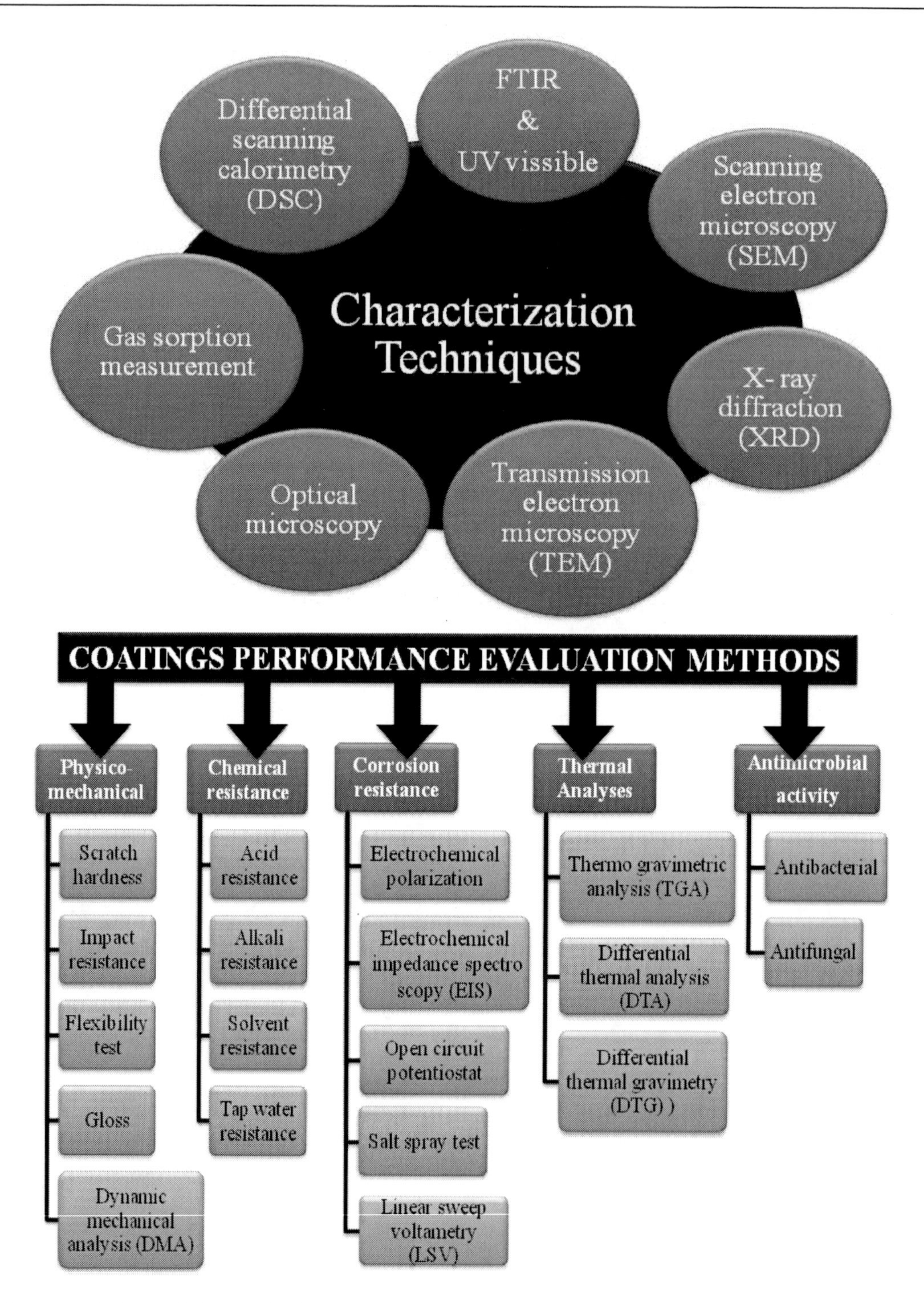

Figure 6. Methods of characterization.

3. Applications of Organic-inorganic Composite Coatings

3.1. Decorative Coatings

Organically modified ceramics (ORMOCERS) have been used as decorative coatings on crystal glasses [88]. The main advantages with these coatings are easy reproducibility of the colours and the finishing by conventional wet painting procedures (low curing temperatures < 200 °C) [15].

3.2. Antifouling Coatings

Kuroda et al. [89], developed organic-inorganic hybrid composites from linseed oil by mixing with octylsilyl titanium dioxide particles in volatile silicone for both highly hydrophobic and highly hydrophilic coatings. These fluorine free composites may be applied as self-cleaning and antifouling coatings. New silicone/phosphorus/sulphur containing nano-hybrid epoxy coatings have demonstrated marked antifouling behavior [90].

3.3. Optical Glass Coatings

SiO_2/TiO_2/ORMOSIL (Organically modified silicates) coatings prepared from TEOS, γ-glycidoxypropyltrimethoxysilane and tetrapropylorthotitanate are obtained as hard optical coating, high hardness of the coatings being attributed to the carbon and titanium content in the film [88, 92].

3.4. Scratch and Abrasion Resistant Coatings

Nano/micro hybrid polyacrylate composite material (18 wt.% silica + 15 wt.% corundum) on parquet substrate may be employed as scratch and abrasion resistant coatings. The coatings containing both silica nanoparticles and corundum microparticles exhibit highly improved performance such as heat, scratch and abrasion resistance relative to the neat acrylate coatings [93]. Hybrid coatings of silica-PMMA were prepared on glass substrate by a modified sol-gel process using different molar relationships for TEOS:MMA and constant quantities (0.5 M in relation to TEOS) of 3MOPTMOS as bonding agent. The hybrid coatings showed good hardness over plain polyacrylate, due to the formation of SiO_2 rich layer at the surface of the coatings during the drying process [94].

3.5. Biocompatible Coatings

TEOS and polyethylene glycol, heparin, dextran sulfate, nafion or polystyrene sulfonate based and sol-gel derived biocompatible coatings were prepared with practical applications as implantable glucose sensors. The dextran sulfate containing coating was found to be the most promising for future in vivo glucose measurements [95]. Biomimetic organic–inorganic nanocomposite coatings for titanium implants have been prepared by a biomimetic three-step procedure: (1) embedding amorphous calcium phosphate (ACP) particles between organic polyelectrolyte multilayers (PEML), (2) in situ transformation of ACP to octacalcium phosphate and/or poorly crystalline apatite nanocrystals by immersion of the material into a metastable calcifying solution and (3) deposition of a final PEML. The coating procedure is simple, energy saving, environmentally friendly, and production can be easily and cost effectively scaled up for factory use. Coatings with final PEML possess well-adapted sites for cell adhesion and proliferation. It is possible to apply the coatings to a wide variety of bioinert implant materials of different shapes and sizes because the deposition of PEML is largely independent of the type and topology of the substrate [96].

3.6. Biocidal Coatings

Li et al. [48], developed polyurethane-based coatings reinforced by ZnO nanoparticles (about 27 nm) *via* the solution blending. The performance of coatings improved significantly by the addition of (2.0 wt%) ZnO nanoparticles. The coatings showed good abrasion resistance, UV light fastness and the climate resistance. The antibacterial experiments show that the ZnO doped PU films exhibit excellent antibacterial activity, especially for *E. coli* [48].

Silver bromide nanoparticle/polymer composites bear potent antibacterial activity toward both gram-positive and gram-negative bacteria. The materials are capable of forming good coatings on surfaces and destroy both airborne and waterborne bacteria. These composites may find potential applications as antimicrobial coatings in a wide variety of biomedical and general use applications [97].

4. Future Research Directions

Organic-inorganic composites are generally synthesized by complex multi-step processes. These processes are often very cumbersome and require elevated temperatures and longer reaction time, in the presence of volatile organic solvents. Besides these, what adds to the misery is their prolonged curing times, multi-step cure schedules, at very high temperatures. For use as environment friendly coatings, these features should be taken care of. In this context, another question of prime concern is the degree of "environment friendliness" of these materials, the fate of their coatings after service-life and speculations about their effect on the atmosphere and human life. This requires the use of clean and green synthesis processes such as low/no solvent methods, microwave irradiation techniques and others. Attempts may also be taken to prepare biodegradable coatings, so that their after-use degradation gets expedited. The enhanced use of renewable resources such as starch, cellulose, polyhydroxyalkanoates and others in the field is also an alternative.

CONCLUSION

Organic inorganic composites have become increasingly important in coatings due to their drastically improved performance characteristics at very low dosages of modifiers. Consequently, these find versatile applications in the field. With several sophisticated methods and equipments for coating characterization, studies may be carried out to investigate their structure and morphology for improvement in their coating performance. The area has wider scope for exploration and research.

ACKNOWLEDGMENTS

Dr. Eram Sharmin is thankful to CSIR, New Delhi, India for Research Associateship against Grant No. 9/466(0102) 2K8-EMR-I. Dr Fahmina Zafar (Pool Officer) acknowledges CSIR, New Delhi, India for Senior Research Associateship against Grant No. 13(8385-A)/Pool/2010. They are also thankful to the Head, Dept of Chemistry, Jamia Millia Islamia, for providing facilities to carry out their research work.

REFERENCES

[1] Sorensen, P. A.; Kiil, S.; Johansen, K. D.; Weinell, C. E. *J. Coat. Technol. Res.* 2009, *6,* 135–176.
[2] Sangraj, N. S.; Malshe, V. C. *Prog. Org. Coat.* 2004, *50,* 28–39.
[3] Parka, H.; Na, M.; Ahn, M.; Kang, D.; Bae, D. *J. Non. Cryst. Solids.* 2001, *290,* 153–162.
[4] VanOoij, W. J.; Zhou, D.; Stacy, M.; Seth, A.; Mugada, T.; Gandhi, J.; Puomi, P. *Tinghua. Sci. Technol.* 2005, *10,* 639–664.
[5] Zandi-zand, R.; Ershad-langroudi, A.; Rahimi, A. *Prog. Org. Coat.* 2005, *53,* 286–291.
[6] Bounor-legare, V.; Angelloz, C.; Blanc, P.; Cassagnau, P.; Michel, A. *Polymer* 2004, *45,* 1485–1493.
[7] Soucek, M. D.; Zong, Z.; Johnson, A. J. *J. Coat. Tech. Res.* 2006, *3,* 133–140.
[8] Wang, D.; Bierwagen, G. P. *Prog. Org. Coat.* 2009, *64,* 327–338.
[9] Zaarei, D.; Sarabi, A. A.; Sharif, F.; Kassiriha, S. M. *J. Coat. Technol. Res.* 2008, *5,* 241–249.
[10] Schmidt, H.; Krug, H. *Inorganic and Organometallic Polymers II;* ACS Symposium Series, ACS, Chapter 15, 1994; Vol. 572, pp 183–194.
[11] Fernando, R. H. *Nanotechnology Applications in Coatings*; ACS Symposium Series, ACS, Chapter 1, 2009; Vol. 1008, pp 2–21.
[12] Mark, J. E. *Hybrid Organic-Inorganic Composites*; ACS Symposium Series, ACS, Chapter 1, 1995; Vol. 585, pp 1–4.
[13] Zhou, S.; Wu, L.; You, B.; Gu. Guangxin. *Smart Coatings II*; *ACS Symposium Series*, ACS, Chapter 10, 2009; Vol. 1002, pp 193–219.
[14] Nik, A. V.; Jerman, I.; Vuk, A. U.; Elj, M. K.; Orel, B.; Tom, B.; Simon, B.; Kova, J. *Langmuir*, 2009, *25*, 5869–5880.

[15] Sanchez, C.; Julian, B.; Belleville, P.; Popall, M. *J. Mater. Chem.* 2005, *15,* 3559–3592.

[16] Chujo, Y. KONA 2007, *25,* 255–260. Originally published in Japanese in The Micrometrics, 2006/2007, *50*, 11-15.

[17] Chattopadhyay, D. K.; Raju, K. V. S. N. *Prog. Polym. Sci.* 2007, *32,* 352-418.

[18] Xu, Q. F.; Wang, J. N.; Sanderson, K. D. *Am. Cnem. Soc. Nano*. 2010, *4,* 2201–2209.

[19] Medda, S. K.; De, G. *Ind. Eng. Chem. Res.* 2009, *48*, 4326–4333.

[20] Jena, K. K.; Raju, K. V. S. N. *Ind. Eng. Chem. Res.* 2008, *47*, 9214–9224.

[21] Jena, K. K.; Raju, K. V. S. N. *Ind. Eng. Chem. Res.* 2007, *46*, 6408–6416.

[22] Florian, P.; Jena, K. K.; Allauddin, S.; Narayan, R.; Raju, K. V. S. N. *Ind. Eng. Chem. Res.* 2010, *49*, 4517–4527.

[23] Kornmann, X.; Lindberg, H.; Berglund, L. A. *Polymer* 2001, 42, 4493-4499.

[24] Bordes, P.; Pollet, E.; Avérous, L. *Prog. Polym. Sci.* 2009, *34,* 125–155.

[25] Haq, M.; Burgueño, R.; Mohanty, A. K.; Misra, M. *Compos. Part A.* 2009, *40,* 540–547.

[26] Maiti, P.; Yamada, K.; Okamoto, M.; Ueda, K.; Okamoto, K. *Chem. Mater.* 2002, *14,* 4654–4661.

[27] Sinha, R. S.; Yamada, K.; Okamoto, M.; Ueda, K. *Nano Lett.* 2002, *2,* 1093–1096.

[28] Sinha R. S.; Yamada, K.; Ogami, A.; Okamoto, M.; Ueda, K. *Macromol. Rapid Commun.* 2002, *23,* 943–947.

[29] Deka, H.; Karak, N. *Nanoscale Res. Lett.* 2009, *4,* 758–765.

[30] Deka, H.; Karak, N. *Polym. Adv. Technol.* 2009, DOI:10.1002/pat. 1603.

[31] Tsujimoto, T.; Uyama, H.; Kobayashi, S. *Polym. Degrad. Stab.* 2010, *95,* 1-7.

[32] Allie, L.; Thorn, J.; Aglan, H. *Corr. Sci.* 2008, *50,* 2189-2196.

[33] Yeh, J. M.; Yao, C. T.; Hsieh C. F.; Lin, L. H.; Chen, P. L.; Wu, J. C.; Yang, H. C.; Wu, C. P. *Eur. Polym. J.* 2008, *44,* 3046-3056.

[34] Olad, A.; Rashidzadeh, A. *Prog. Org. Coat.* 2008, *62,* 293-298.

[35] "Study on mechanical behaviour of polymer based composites with and without wood dust filler", A thesis submitted in partial fulfilment of the requirements for the degree of Bachelor of Technology in Mechanical Engineering by S. K. Behera in the Dept of Mechanical Eng., National Institute of Technology, Rourkela, India, May 2010.

[36] Surender, M., Balasubramaniam, R.; Basu, B. *Surf. Coat. Technol.* 2004, *187,* 93–97.

[37] Medelienė, V.; Kosenko, Aleksandr. *Mater. Sci. (Medžiagotyra)* 2008, *14,* 29–33.

[38] Gheorghies, C.; Carac, G.; Stasi, I. V. J. *Optoelectron. Adv. Mater.* 2006, *8,* 1234 – 1237.

[39] Low, C. T. J.; Wills, R. G. A.; Walsh, F. C. *Surf. Coat. Technol.* 2006, *201,* 371–383.

[40] Freudenberger, R.; Zielonka, A.; Funk, M.; Servin, P.; Haag, R.; Valkova, T.; Landau, U. *Gold Bull.* 2010, *43,* 169–180.

[41] Keilitz, J.; Radowski, M.; Marty, J. D.; Haag, R.; Gauffre, F.; Mingotaud, C. *Chem. Mater.* 2008, *20,* 2423–2425.

[42] Benea, L.; Mitoseriu, O.; Galland, J.; Wenger, F.; Ponthiaux, P. *Mater. Corros.* 2000, *51,* 491–495.

[43] Surender, M.; Basu, B.; Balasubramaniam, R. *Tribol. Int. 2004, 37,* 743–749.

[44] Stott, F. A.; Asby, D. J.; *Corros. Sci.* 1978, *18,* 183–198.

[45] Kedward, E. C. *Metal. Finish. Rev.* 1972, 79–89.

[46] Malone, M. A.; Report No. SCL. DR, 720090, Bell Aerospace Div.

[47] Cayton, R. H. *NSTI-Nanotech*, 2004, 3, 312–315.(www.nsti.org, ISBN 0-9728422-9-2).
[48] Li, J. H.; Hong, R. Y.; Li, M. Y.; Li, H. Z.; Zheng, Y.; Ding, J. *Prog. Org. Coat.* 2009, *64,* 504–509.
[49] Cortecchia, E.; Pacilli, A.; Pasquinelli, G.; Scandola, M. *Biomacromolecules* 2010, *11,* 2446–2453.
[50] Ferraris, M.; Perero, S.; Miola, M.; Ferraris, S. Gautier, G.; Maina, G.; Fucale, G.; Verne, E. *Adv. Eng. Mater.* 2010, *12,* B276–B282.
[51] Macinnes Jr, D.; Funt, B. L. *Synth. Met.* 1988, *25,* 235–242.
[52] Ramakrishnan, S. *Resonance* 1997, *2,* 48–58.
[53] Twite, R. L.; Bierwagen, G. P. *Prog. Org. Coat.* 1998, *33,* 91–100.
[54] Kinlen, P. J.; Silverman, D. C.; Jeffreys, C. R. *Synth. Met.* 1997, *85,* 1327–1332.
[55] Epstein, A. J.; Jasty, S. G. United States Patent, 5972518, October 1999.
[56] Schauer, T.; Joos, A.; Duo, L.; Eisenbach, C. D. *Prog. Org. Coat.* 1998, *3,* 20–27.
[57] Ahmad, S.; Ashraf, S. M.; Riaz, U.; Zafar, S. *Prog. Org. Coat.* 2008, *62,* 32–39.
[58] Chang, K. C.; Lai, M. C.; Peng, C. W.; Chen, Y. T.; Yeh, J. M.; Lin, C. L.; Yang, J. C. *Electrochim. Acta* 2006, *51,* 5645–5653.
[59] Bakshi, S. R.; Singh, V.; Balani, K.; McCartney, D. G.; Seal, S.; Agarwal, A. *Surf. Coat. Technol.* 2008, *202,* 5162–5169.
[60] Chen, Li.; Pang, Xiujiang.; Yu, G; Zhang, J. *Adv. Mat. Lett.* 2010, *1,* 75–78.
[61] Surender, M.; Balasubramaniam, R.; Basu, B. *Surf. Coat. Technol.* 2004, *187,* 93–97.
[62] Chen, W. X.; Tu, J. P.; Wang, L. Y.; Gan, H. Y.; Xu, Z. D.; Zhang, X. B. *Carbon* 2003, *41,* 215–222.
[63] Latzel, S.; Vaßen, R.; Stöver, D. *J. Ther. Spr. Technol.* 2005, 14, 268–272.
[64] Ho, C. T. *J. Mater. Sci.* 1996, *31,* 5781–5786.
[65] Su-ping, H.; Ke-chao, Z. Zhi-you, L. *Trans. Nonferous. Mat. Soc. China*, 2008, 18, 162–166 31 (19
[66] Han, Y.; Taylor, A.; Mantle, M. D.; Knowles, K. M. *J. Sol-Gel Sci. Technol.* 2007, *43,* 111–123.
[67] Medda, S. K.; Kundu, D.; De, G. *J. Non-Cryst. Solids* 2003, *318,* 149–156.
[68] Innocenzi, P.; Brusatin, G. *J. Non-Cryst. Solids* 2004, *333,* 137–142.
[69] Van de Leest, R. E. *Appl. Surf. Sci.* 1995, *86,* 278–285.
[70] Gianni, A. Di.; Bongiovanni, R.; Turri, S.; Deflorian, F.; Malucelli, G.; Rizza, G. *J. Coat. Technol. Res.* 2009, *6,* 177–185.
[71] Nebioglu, A.; Soucek, M. D. *J. Coat. Technol. Res.* 2007, *4,* 425–433.
[72] Wouters, M. E. L.; Wolfs, D. P.; Van der Linde, M. C.; Hovens, J. H. P.; Tinnemans, A. H. A. *Prog. Org. Coat.* 2004, *51,* 312–319.
[73] Gilberts, J.; Tinnemans, A. H. A.; Hogerheide, M. P.; Koster, T. P. M. *J. Sol-Gel Sci. Technol.* 1998, *11,* 153–159.
[74] Gigant, K.; Posset, U.; Schottner, G. *J. Sol-Gel Sci. Technol.* 2003, 26, 369–373.
[75] Soppera, O.; Moreira, P. J.; Leite, A. P.; Marques, P. V. S. *J. Sol-Gel Sci. Technol.* 2005, *35,* 27–39.
[76] Nebioglu, A.; Soucek, M. D. *Euro. Polym. J.* 2007, *43,* 3325–3336.
[77] Watanabe, M.; Tamai, T. *J. Polym. Sci. Part A Polym. Chem.* 2006, *44,* 4736–4742.
[78] Wada, T.; Uragami, T. *J. Coat. Technol. Res.* 2006, *3,* 267–274.
[79] Brasil, M. C.; Gerbase, A. E.; De Luca, M. A.; Gregorio, J. R. *J. Am. Oil Chem. Soc.* 2007, *84,* 289–295.

[80] De Luca, M. A.; Martinelli, M.; Jacobi, M. M.; Becker, P. L. *J. Am. Oil Chem. Soc.* 2006, *83,* 147–1511.
[81] Sailor, R. A.; Wegner, J. R.; Hurtt, G. J.; Janson, J. E.; Soucek, M. D. *Prog. Org. Coat.* 1998, *33,* 117–125.
[82] Sailor, R. A.; Soucek, M. D. *Prog. Org. Coat.* 1998, *33,* 36–43.
[83] Tsujimoto, T.; Uyama, H.; Kobayashi, S. *Macromol. Rapid Commun.* 2003, *24,* 711–714.
[84] Konwar, U.; Karak, N.; Mandal, M. *Prog. Org. Coat.* 2010, *68*, 265-273.
[85] Alam, J.; Riaz, U.; Ahmad, S. *Curr. Appl. Phys.* 2009, *9,* 80-86.
[86] Kumar, A.; Vemula, P. K.; Ajayan, P. M.; John, G. *Nat. Mate*. 2008, *7,* 236–241.
[87] Eksik, O.; Tuncer Erciyes, A.; Yagci, Y. *J. Macromol. Sci. Part A*, 2008, *45,* 698–704.
[88] Haas, K. H. Rose, K. *Rev. Adv. Mater. Sci.* 2003, *5,* 47–52.
[89] Kuroda, A.; Joly, P.; Shibata, N.; Takeshige, H.; Asakura, K. *J. Am. Oil Chem. Soc.* 2008, *85,* 549–553.
[90] Kumar, A.; Sasikumar, A. *Prog. Org. Coat.* 2010, *68,* 189–200.
[91] Que, W. Y. Sun, Z.; Zhou, Y.; Lam, L.; Cheng, S. D.; Chan, Y. C.; Kam, C. H. *Mater. Lett.* 2000, *42,* 326–330.
[92] Que, W.; Zhang, Q. Y.; Chan, Y. C.; Kam, C. H. *Compos. Sci. Technol.* 2003, *63,* 347–351.
[93] Bauer, F.; Flyunt, R.; Czihal, K.; Buchmeiser, M. R.; Langguth, H.; Mehnert, R. *Macromol. Mater. Eng*. 2006, *291,* 493–498.
[94] Almaral-Sanchez, J. L.; Lopez-Gomez, M.; Ramirez-Bon, R.; Munoz-Saldana, J. *J. Mater.* 2006, *2,* 1–10.
[95] Kros, A.; Gerritsen, M.; Sprakel, Vera S. I.; Sommerdijk, Nico A. J. M.; Jansen, J. A.; Nolte, R. J. M. *Sens. Actuat. B: Chem*. 2001, *81,* 68–75.
[96] Sikiric, M. D.; Gergely, C.; Elkaim, R.; Wachtel, E.; Cuisinier, F. J. G.
[97] Furedi-Milhofer, H. *J. Biomed. Mater. Res.* 2009, *89A,* 759–771.
[98] Sambhy, V.; MacBride, M. M.; Peterson, B. R., Sen, A. *J. Am. Chem. Soc.*, 2006, *128,* 9798–9808.

In: Advanced Organic-Inorganic Composites
Editor: Inamuddin

ISBN 978-1-61324-264-3

Chapter 6

POLYMER ELECTROLYTE MEMBRANES AND ELECTRODES FABRICATED USING PLASMA METHOD AND THEIR APPLICATIONS IN PROTON EXCHANGE MEMBRANE FUEL CELLS

Zhongqing Jiang[a,b,*]***, Xinyao Yu***[a]***, Zhong-Jie Jiang***[c,*]***,***
and Yuedong Meng[a]

[a]Institute of Plasma Physics, Chinese Academy of Sciences, Hefei 230031, Anhui, China
[b]Department of Chemical Engineering, Ningbo University of Technology, Ningbo 315016, Zhejiang, China
[c]Department of Nature and Sciences, University of California, Merced 95348, U. S.

ABSTRACT

Proton exchange membrane fuel cells (PEMFCs), which are identified as an important power source in applications ranging from portable electronic devices to fuel cell vehicles, have been the subject of intensive research. Currently, research in this field is largely focused on the preparation of proton-exchange membranes (PEMs) with high performance of ion exchange and catalysts with high performance of fuel electro-oxidation. Plasma technique, as a novel approach combines physics, chemistry, biology and engineering, has been widely applied to the preparation and modification of materials and shows great potentials in improving the performance of materials. On the one hand, the plasma modified PEMs exhibit superior properties, such as improved fuel antipermeability, ion selectivity, and reduced fouling. The PEMs obtained by the plasma polymerization usually hold a highly crosslinked structure, which improves their thermal and chemical stability compared with conventional polymerized membranes and thus increases the service lifetime in their applications to fuel cell systems. Morever, the membranes obtained from plasma polymerization usually have controllable thicknesses, which allows them an easily miniaturization and a strong adhesion to electrode surfaces. On the other hand, plasma modification of the cathode and anode surface can enhance the cell performance of PEMFCs, and plasma techniques are also shown to be able to reduce

* Corresponding author's e-mail: zhongqingjiang@hotmail.com

the catalyst loading and increase the catalyst utilization efficiency of electrodes for PEMFCs. The obtained electrodes exhibit higher CO and/or methanol tolerance, easy proton or oxygen adsorption/desorption and fast kinetics of oxygen reduction reaction. Therefore, plasma technique has great potentials for use in PEMFCs.

ABBREVIATIONS

AAO	Anodized aluminum oxide
AFCs	Alkaline fuel cells
Ar	Argon
CC	Carbon cloth
CCP	Capacitively coupled plasma
CF_4	Carbon tetrafluoride
CF_3CH_2Cl	2-chloro-1,1,1-trifluoroethane
CF_3SO_3H	Trifluoromethane sulfonic acid
C_6H_{14}	Hexane
$ClSO_3H$	Chlorosulfonic acid
CNFs	Carbon nanofibers
CNTs	Carbon nanotubes
CrN	Chromium nitride
CV	Cyclic voltammograms
CVD	Chemical vapor deposition
DC	Direct current
DMFC	Direct methanol fuel cell
E_a	Activation energy
ECSA	Electrochemically active surface area
FTIR	Fourier-transform infrared
GDLs	Gas diffusion layers
GLAD	Glancing-angle deposition
H_2	Hydrogen
HRTEM	High-resolution transmission electron microscopy
IC	Integrated circuit
ITO	Indium-tin oxide
MEAs	Membrane electrode assemblies
MeOH	Methanol
MIS	Metal–insulator–semiconductor
MOR	Methanol oxidation reaction
MSP	Magnetron sputtering
MWCNTs	Multi-wall carbon nanotubes
NCI	Nafion-carbon ink
NSTF	Nanostructured thin film
NWAs	Nanowire arrays
OCV	Open circuit voltage
ORR	Oxygen reduction reaction
PABS	Plasma assisted bias sputtering

PEMFCs	Proton exchange membrane fuel cells
PEMs	Proton-exchange membranes
pp-HMDSO	Plasma-polymerized hexamethyldisiloxane
PSD	Plasma sputter deposition
PtO_2	Palatinum oxide
PTFE	Poly(tetrafluoroethylene)
RF	Radio frequency
RRDE	Rotating ring-disk electrode
SAED	Selected area electron diffraction
SAFCs	Solid alkaline fuel cells
SEM	Scanning electron microscope
TaO_xN_y	Tantalum (oxy) nitrides
TCP	Transformer coupled plasma
TEM	Transmission electron microscopy
TFCE	Trifluorochloroethylene
t_{H^+}	The proton transport numbers
TiNTs	Titania nanotubes
TiO_2	Titanium oxide
TM−C−N (TM =Fe, V, Cr, Mn, Co, and Ni)	Transition metal-carbon-nitrogen
TONT	Titanium oxide nanotube
XPS	X-ray photoelectron spectroscopy
ZrO_xN_y	Zirconium oxynitride

1. INTRODUCTION

The ever-rising global energy demand and the environmental impact of energy use from traditional sources pose serious challenges to human health, energy security, environmental protection and the sustainability of natural resources. The continued growth in global population, steadily increasing per-capita energy consumption in developing countries, and dependence on fossil fuel resources make the fossil fuel-based economy unsustainable. This scenario leads to intensive research to find alternate sources of energy. PEMFCs as alternate renewable energy sources with a solid polymer membrane as electrolyte, pose no serious issues to environment caused by use of fossil fuels. It is demonstrated that PEMFC systems can provide an order of magnitude higher power density than any other fuel cell systems. The use of a solid polymer electrolyte eliminates the corrosion and safety concerns associated with liquid electrolyte fuel cells. Due to their high conversion efficiency and power density, PEMFCs are identified as important enegy sources with great potentials for applications in portable electronic devices, such as cellular phones, laptops, digital cameras or other conventionally battery driven devices and long-term stationary monitoring electronics. However, some problems associated with PEMs, such as high cost, low operation temperature and high fuel permeability, and electrocatalysts, such as the limited supply and high cost, low activity and poor durability, in PEMFCs have been the main reasons inhibiting the commercialization of PEMFCs in practical applications. Research in the exploitation of PEMs with high performance of ion exchange and high fuel antipermeability and the development of electrocatalysts with low loadings and higher utilization efficiency or other nonplatium electrocatalysts are therefore of great importance. Up to now, various methods and techniques ranging from the modification of existing PEMs [1-9] and catalysts [10-14] to the development of new types of PEMs and catalysts [15-19] have been reported. Among them, plasma technique, as a novel approach which combines physics, chemistry, biology and engineering, shows great potentials for fabrication of the key materials of PEMFCs, e.g. catalysts and electrolyte membranes, with improved properties. Application of the membranes [20,21] and catalysts [22,23] obtained by the plasma process usully give rise to an increase in the performance of PEMFCs.

As the requisite component of PEMFCs, PEM which usually serves dual functions of conducting protons and separating fuels and oxidant in some extent determines the performances of PEMFCs. The development of PEMs with improved performance in PEMFCs has been the focus of research for decades. Recent advances in the PEM technology have demonstrated the great promising of the plasma technique in this area. These PEMs can be either surface modified by a plasma technique or directly synthesized by a plasma polymerization. The modification of PEMs involves exposure of the membrane to inert gas plasma [24] and/or reactive gas plasma [25]. In this case, the surface chemical and physical properties of membranes are altered by plasma surface treatments or deposition of additional polymer layers. The plasma modified PEMs usually exhibit superior properties, such as improved fuel antipermeability, ion selectivity, and reduced fouling. While in a plasma polymerization, the formation of PEMs is initialized by a plasma discharge. In the process of plasma discharge, the monomers are activated, dissociated and ionized by exposure of them to electronic bombardment. By recombination, the gas molecules fragments (radicals) form the polymer film on the desired substrates. The PEMs obtained by the plasma polymerization usually hold a highly crosslinked structure, which improves their thermal and chemical stability compared with conventional polymerized membranes and thus increases the service lifetime in their applications to fuel cell systems [26-28]. Morever, the membranes obtained from plasma polymerization usually have controllable thicknesses, which allows them an easily miniaturization and a strong adhesion to electrode surfaces.

Catalyst is the heart of a PEMFC, which initiates the redox reaction of fuels at the surface of electrode. The activation of a catalyst directly determines the proformance of PEMFCs, such as the energy conversion efficiency. In this sense, the development of catalysts with improved performance and reduced cost in PEMFCs are very important. Plasma sputter deposition (PSD) method has been approved to an effective way to address these problems, because it has the following advantages: (a) catalyst content and thickness, as well as microstructure morphology (such as Pt particle size and particle distribution) can be precisely controlled; (b) catalysts can be deposited onto different components of the membrane electrode assemblies (MEAs) such as PEM and gas diffusion layers (GDLs) and so on; (c) the obtained MEAs have higher CO and/or methanol tolerance, easy proton or oxygen adsorption/desorption and fast kinetics of oxygen reduction reaction (ORR) and exhibits higher performance in PEMFCs. The deposition of catalyst onto various types of substrates, such as catalyzed [10-14] and un-catalyzed [29-43] GDLs, multi-wall carbon nanotubes (MWCNTs) [44-49], carbon nanofibers (CNFs) [50-52], carbon nanorods [18,53,54], carbon nanotubes (CNTs) and carbon black composites [55], Ti films [56-59] and nanorods [60], TiO_2 nanotubes [19,22,61,62], Pd nanorods [63], vertically oriented CrN nanoparticles [64], 3 M's nanostructured thin film (NSTF) [65], etc. by a PSD method has been reported. Pt co-sputtered with carbon [66-71], other metals [72-90], metal oxides [91-101] or metal phosphates [102] have been reported. The obtained composites are used as electrodes for PEMFCs and exhibit improved performance due to strong interaction between catalysts and supports, the uniform dispersion of catalysts, the porous structure of catalyst layers and possible synergistic effects of multi-elemental catalysts, etc.

The integration of plasma-polymerized PEMs and/or catalysts prepared by a PSD method allows fabrication of ultrathin MEAs for the application of PEMFCs. It is demonstrated ultrathin MEAs can decrease the internal resistances of PEMFCs, making the process more efficient, and provide an improved water management due to the enhanced back diffusion of water produced in the chemical process from the cathode to the anode side. Water management is of great importance in fuel cell operation. Water needs to be present in sufficient quantities to hydrate the membrane for sufficient proton conduction [103], but also need to be removed at a sufficient rate at the cathode so as not to choke the fuel cell. Fuel and oxidant are often humidified to provide water for the reaction while PEMs and GDLs contain hydrophobic polymers to expel excess water. Poor water management in the cell can result in drop out of current. In general, thinner MEAs are appreciated because of the lower resistance, making the process more efficient. Thus, the integration of plasma-polymerized PEMs and electrode prepared by PSD method in PEMFCs as MEAs seems promising.

2. Application of Plasma Techinique in the Synthesis of Polymer Electrolyte Membrenes (PEMs)

2.1. Plasma Surface Modification of PEMs

Generated by continuous electrical discharge in either an inert or a reactive gas, plasma contains a certain portion of ions, electrons, neutral particles, free radicals, and/or photons. Due to high energy and mobililty, it has been widely used to modify the materials with complicated geometry. Plasma surface modification of PEMs can be conducted as follows:

- Crosslinking the top layer and reducing the pore size of PEMs. PEMs used for PEMFCs usually consist of networked organic polymer with a porous structure. Plasma treatment with inert gases, such as argon or helium, can lead to ablation of the PEMs by the energetic plasma gases and the subsequent redeposition of the ablated material will result in the formation of a highly crosslinked layer on the surface and the reduction of pore size in the PEM.
- Introducing functional groups to the surface. It has been demonstrated that plasma treatment with air, oxygen and water vapor could introduce oxygen-containing functional groups onto the surface of PEMs, while with nitrogen, ammonia and alkyl amine could introduce nitrogen-containing functional groups.
- Grafting and depositing a thin selective layer on a porous substrate. During the plasma discharge, the introduced monomers are activated, dissociated and ionized. By recombination, the gas molecules fragments (radicals) form the polymer film on the desired substrates. Hydrophilization can be achieved by plasma-induced graft polymerization of PEMs with the incorporation of hydrophilic monomers. Due to

free of solvents or any hazardous liquid that might be destructive to PEMs, plasma surface modification, as a dry and an environmental benign process, has become an attractive method to modify the surface properties of PEMs. The modified membranes usually exhibit enhanced wettability, printability, and adhesion, and improved fuel antipermeability and selectivity among co-ions of PEMs.

2.1.1. Plasma Reactors for Plasma Surface Modification

Plasma used for surface treatment are required to have a sufficiently high electron number density to provide useful fluxes of active species, but not to cause the damage of the materials being treated. This constraint rules out dark or coronas discharges for most surface treatment applications, because of their low production rates of active species, and arc or torch plasmas, which have power densities and active-species flux intensities high enough to damage most exposed materials. Glow discharge plasmas which produce the appropriate electron number density and active-species flux are therefore commonly adopted for plasma surface treatment applications.

Figure 1 shows a typical low pressure radio frequency (RF) argon plasma reactor expanding from a helicon source in a horizontal system called 'Chi Kung' [104]. It consists of a helicon source attached contiguously to an earthed aluminium diffusion chamber in which the holder is placed, with the sample electrically floating. The antenna is fed from a RF matching network/generator system. The argon feed gas is introduced at the closed end of the source to give a constant pressure and the turbo-molecular/rotary pumping system is connected to the side wall of the chamber. The base pressure is kept constant, the pressure being measured with an ion gauge and a baratron gauge, both attached to the diffusion chamber. Two solenoids situated around the source are used to create an expanding magnetic field in the source centre decreasing to a few tens of Gauss in the diffusion chamber.

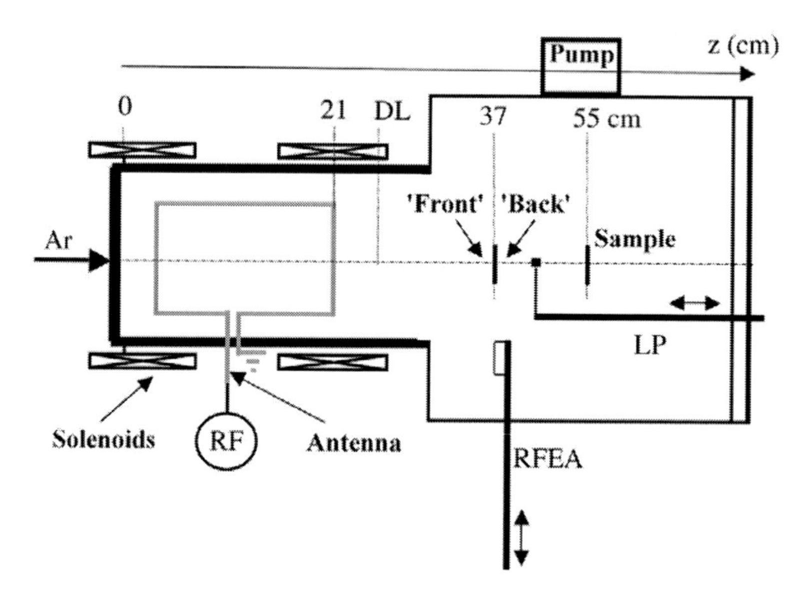

Figure 1. Schematic of 'Chi Kung', the helicon plasma reactor used for the plasma treatment of the Nafion samples, showing major components (DL: Double-layer, RFEA: Retarding field energy analyzer, LP: Langmuir probe) [104]. Reproduced with permission from Charles, C. et al. Plasma Phys. Controlled Fusion, 2007, 49, A73-A79. Copyright © 2007, Institute of Physics Publishing Ltd.

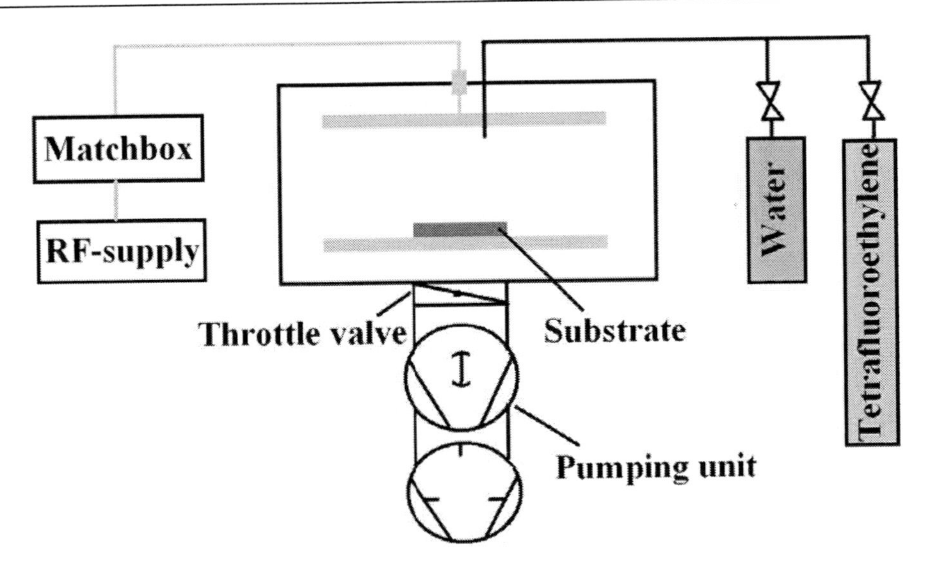

Figure 2. Schematic diagram of plasma reactor [105]. Reproduced with permission from Gruber, D. et al. J. Fuel Cell Sci. Technol. 2005, 2, 186-189. Copyright © 2005, American Society of Mechanical Engineers.

Depositing additional polymer layer on the surface of Nafion® membranes can be performed in a glow discharge between two capacitively coupled inner disk electrodes which is shown in Figure 2 [105]. The apparatus consists of a glass reactor equipped with capacitively coupled inner parallel disk electrodes to which an alternating voltage is applied, and a vacuum system. Pretreated-PEM sheets can be placed on a RF electrode. The reactor is evacuated to a pressure of about <10 Pa before starting the polymerization. By activation, dissociation and ionization processes in the plasma, monomers and gas molecules form fragments, groups and clusters which deposit on the substrate surface and form the polymer film. The process is controlled by the parameters, such as pressure, monomer flow, power density, and temperature, which determine the degree of fragmentation of the monomers and density of the fragments and gases in the plasma.

A schematic view of the plasma source called Duo-Plasmaline® is shown in Figure 3 [5,106]. In this reactor, the plasma is excited by microwaves at a pressure between 10 and 100 Pa. The device in principle works like a coaxial waveguide. A quartz tube (G) is mounted through the vacuum chamber (A), so that the inside of the quartz tube is at atmospheric pressure whereas the outside is at low pressure. The microwaves generated by two magnetrons (I) are fed from both sides into a coaxial waveguide (J) that is centred in the quartz tube. The waveguide is formed by two cylindrical copper rods. The inner conductor (H) leads through the whole device whereas the outer conductor ends within the vacuum chamber. When the microwaves reach this point where the outer copper conductor ends, they enter the low pressure region through the quartz tube. In the low pressure region the field strength of the microwaves exceeds the breakdown field strength and plasma are formed by ionizing collisions of accelerated electrons with neutral species. The so-formed plasma in the low pressure regime outside the quartz tube then acts as the outer conductor of the coaxial waveguide. The device generates linearly extended plasma with cylindrical symmetry around the quartz tube. The length of the plasma is controlled by the microwave power. The membranes are placed on a sample stage (F) movable perpendicular to the plasma source. The working gases are supplied to the plasma through mass flow controllers (D) and a following gas inlet formed like a shower (E) with equidistant holes.

2.1.2. Depositing a Polymer Layer on the Surface of Polymer Electrolyte Membranes (PEMs) by Plasma Surface Modification

The essential property of PEMs is to selectively permeate protons through the membrane. An increase of proton selectivity of PEMs is of great importance to improve the performance of PEMFCs. Research has demonstrated that plasma surface modification can effectively enhance the proton selectivity of PEMs by deposition of a polymer layer on the surface of PEMs. Work by Ogumi et al. [6,7,107] showed that the Nafion® membrane modified by a glow discharge plasma polymerization exhibited a high selectivity of proton permeation. The shortcoming of this work was that the obtained PEMs had a high resistance. In order to reduce the resistance of membranes, Yasuda et al. [9] improved the preparation method, where they prepared novel monovalent cation perm-selective membranes. In this case, Nafion was firstly sputtered with oxygen plasma to produce radical sites on its surface, and an ultra-thin plasma polymerized films was then deposited on the Nafion surface by reactions between the radical sites and 4-vinylpyridine vapor. The resulting membranes showed high monovalent cation perm-selectivity and the reduced resistance. Zeng et al. [108] used other monomers, e.g. ethylene and ammonia as reaction gas, and deposited an ultra-thin anionic exchange layer on the surface of a Nafion® membrane by a glow–discharge plasma polymerization technique. The ion selectivity coefficient for H^+ ions of the obtained PEMs exhibited a linear Nernst response and the selectivity of proton was enhanced. The resistance of modified Nafion® (about 1 Ωcm^2) was only slightly higher than that of the Nafion® membrane.

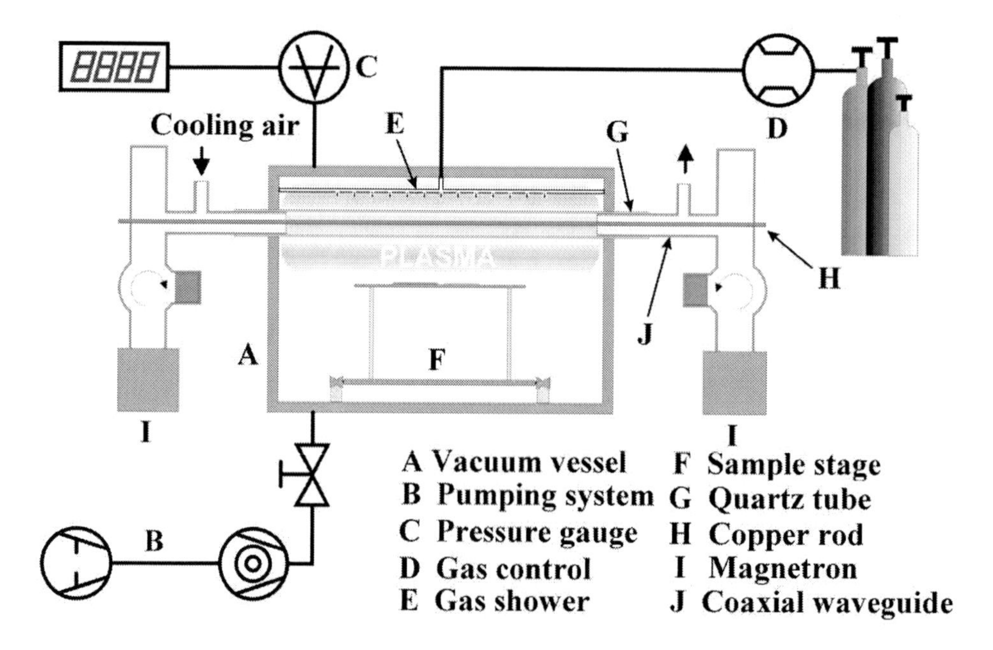

Figure 3. Schematic diagram of the plasma device [5]. Reproduced with permission from Walker, M. et al. J. Appl. Polym. Sci. 1999, 74, 67–73. Copyright © 1999, John Wiley and Sons.

The crossover of methanol through the membrane which causes the reduction of fuel utilization efficiencies and cathode performance is particularly limiting disadvantage in PEMFCs. As forementioned, plasma surface modification of PEMs can lead to the reduction in the sizes of pores in membranes, which can be implemented to improve the fuel antipermeability of PEMs. For example, Lue et al. [25] investigated the application of Nafion modified by a plasma technique in a direct methanol fuel cell (DMFC). The results indicated that the modification of Nafion by either argon (Ar) or carbon tetrafluoride (CF_4) plasma did not cause the alteration of the bulk properties of PEMs, but the suppression of methanol (MeOH) permeability. By adjusting the plasma operating conditions, the proton conductivity could be maintained at a satisfactory level. It showed that the Nafion surface exposed to CF_4 plasma exhibited a more hydrophobic property and an even lower MeOH permeability than did the Ar-treated membrane due to the hydrophobic property of CF_4. Feichtinger and Walker et al. [4,109] deposited thin hydrocarbon films on Nafion™ membranes in a low-pressure plasma excited by microwaves, in which gas mixtures of hexane (C_6H_{14}) with hydrogen (H_2) were used as monomers. The modified PEMs could reduce the methanol permeability by a factor of about 15. Actually, further reduction of the fuel permeability in PEMs can be realized by deposition of inorganic materials onto these plasma modified membranes [2]. The deposited inorganic film can act as a barrier to methanol crossover. However, the deposition inorganic film may induce a reduction in the proton conductivity of PEMs, due to the block of proton pathway through the PEMs [110,111]. For example, Ha et al. [110] fabricated composite PEMs by depositing Pd films on the surface of Nafion™ membranes by sputtering. The Pd film acted as a barrier to methanol crossover, but at the same time it also reduced proton conductivity. Both the methanol permeability and the protonic conductivity through the modified PEMs decreased with increasing Pd thickness.

2.1.3. Modification of the Nafion® Membrane by Exposure to Inert Gas Plasma

Although the deposition of a polymer layer on the surface of Nafion® membranes by the plasma technique can effectively improve the ions selectivity and the fuel antipermeability of PEMs. In some cases, however, the deposited hydrophobic layer can not well adhere to the surface of Nafion due to the differences of stress and the water uptake between the Nafion and the deposited layers. In this sense, a bunch of work has been done to directly modify the Nafion® membranes by exposing them to inert gas plasma, without introducing of exotic polymer layer on the surface of Nafion. In these cases, the improvement of PEMs in performance of PEMFCs can be accomplished by increasing the effective contact area between an electrolyte membrane and an electrode catalyst layer. As well known, the interfacial structure between an electrolyte membrane and an electrode catalyst layer plays an important role in determining the cell performance since the electrochemical reactions take place on the interfaces that are in contact with hydrogen or oxygen, so-called three phase boundaries. With enlarging effective area of the interfaces, the performance of PEMFCs can be improved and material dosage reduced, leading to cost reduction of PEMFC stacks. When the Nafion® membrane is exposed to inert gas plasma, energetic particles will bombard the surface of the membrane. Due to the bombardment of PEMs by the energetic plasma particles, the roughness of membranes can be increased, leading to an increased area for the contact of electrode catalyst layer. Cho et al. [112] demonstrated that plasma bombardment of the PEM surface could increase the maximum power density of a single cell when the obtained membrane was used in a PEMFC. They attributed increased power density of a single cell to the increased contact area between the membrane, catalyst and reactants, a view that has been futher confirmed by Prasanna et al. [113] and Hobson et al. [1], where higher performance of PEMFCs with lower catalyst loadings was obtained when roughened membranes were used.

2.2. Plasma Polymerized Polymer Electrolyte Membranes (PEMs)

2.2.1. Sulfonic Acid Functionalized Plasma Polymerized PEMs

Since the membranes obtained from plasma polymerization usually have dense structures and controllable thicknesses, which provide them an improved water management and fuel antipermeaility and allow them an easily miniaturization and a strong adhesion to electrode surfaces, research in preparation of PEMs by plasma technique has stimulated great interests. Among various types of PEMs polymerized by the plasma techinique, sulfonic acid functionalized PEMs are received particular attention. These sulfonated PEMs usually possess higher proton conductivity. They can be prepared by plasma copolymerization of trifluoromethane sulfonic acid (CF_3SO_3H), chlorosulfonic acid ($ClSO_3H$), SO_2, CO_2, vinylphosphonic acid, etc., with different fluorocarbons, and vinylbenzenes. Approaches that are used in plasma polymerization mainly include glow discharge plasma polymerization, ion beam assisted plasma polymerization, and low-frequency after-glow capacitively coupled plasma (CCP) discharge technique.

2.2.1.1. Plasma Polymerization in the Glow Discharge

In a glow discharge plasma polymerization reactor, substrates on which plasma-polymerized PEMs are deposited are placed in the discharge region. Schematic diagram of the reactor for plasma polymerization in the glow discharge is shown in Figure 4 [114]. The reactor is usually made of stainless steel. Pressure gauges and a vacuum system allow the partial pressures of the monomers and the total pressure in the reactor to be controlled. All monomers are stored in Pyrex glass vessels, whose temperatues are manipulated using oil baths.

A liquid nitrogen trap is placed between the plasma chamber and the vacuum pump to avoid contamination of the pump oil from unreacted monomers. Glow discharge is usually generated using Ar as a carrier gas. The plasma polymerized membranes are deposited in a glow discharge between two capacitively coupled inner disk electrodes.

The system has gas lines to introduce the reactants around the edges of the electrodes to ensure a uniform distribution. Using this synthetic protocol, the monomer molecules are prone to be dissociated during the glow discharge plasma polymerization reaction. Thus, the intrinsic ionic conductivities of plasma polymerized membranes are lower than that of Nafion® membranes, but their conduction ability is observed to be competitive due to their lower thickness [115,116]. Owing to highly cross-linked structure, the plasma-polymerized PEMs exhibit methanol permeability much lower than Nafion® membranes [117-119], and thus, with a high proton permselectivity [120,121].

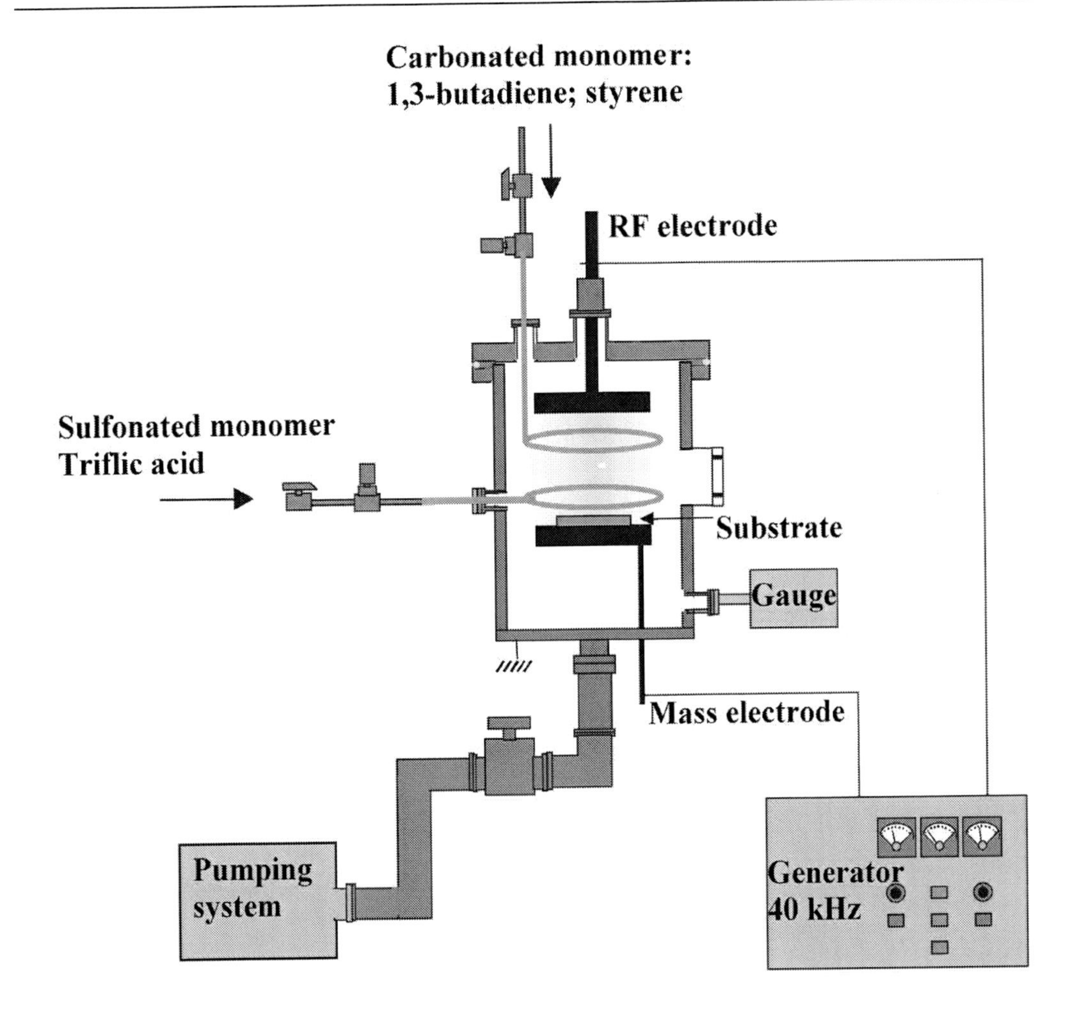

Figure 4. Schematic diagram of a reactor for plasma polymerization in the glow discharge [114]. Reproduced with permission from Mahdjoub, H. et al. Fuel Cells 2005, 5, 277-286. Copyright © 2005, Wiley-VCH Verlag GmbH & Co. KGaA.

Pioneering work of plasma polymerized PEMs prepared by the glow discharge was done by Inagaki and co-workers [122-124], where they used SO_2 as the source of sulfonic acid group to plasma polymerize with pentafluorobenzene, tetrafluorobenzene or perfluorobenzene. Followed by this work, Ogumi and co-workers [120,121] prepared fluorinated plasma PEMs using trifluoromethane sulfonic acid (CF_3SO_3H) and 2-chloro-1,1,1-trifluoroethane (CF_3CH_2Cl) as starting materials. Uchimoto et al. [125,126] synthesized the plasma PEMs by plasma polymerization of 1,3-butadiene and methylbenzene sulfonate.

The thin cation-exchange films with fixed sulfonic acid groups by plasma polymerization were also synthesized by Uchimoto et al. [116] and Yasuda et al. [118,119]. It was demonstrated that these PEMs synthesized by a glow discharge plasma polymerization have highly cross-linked structures, low methanol permeabilities and high proton permselectivity. However, the intrinsic ionic conductivities of plasma polymerized membranes were lower than that of Nafion® membranes.

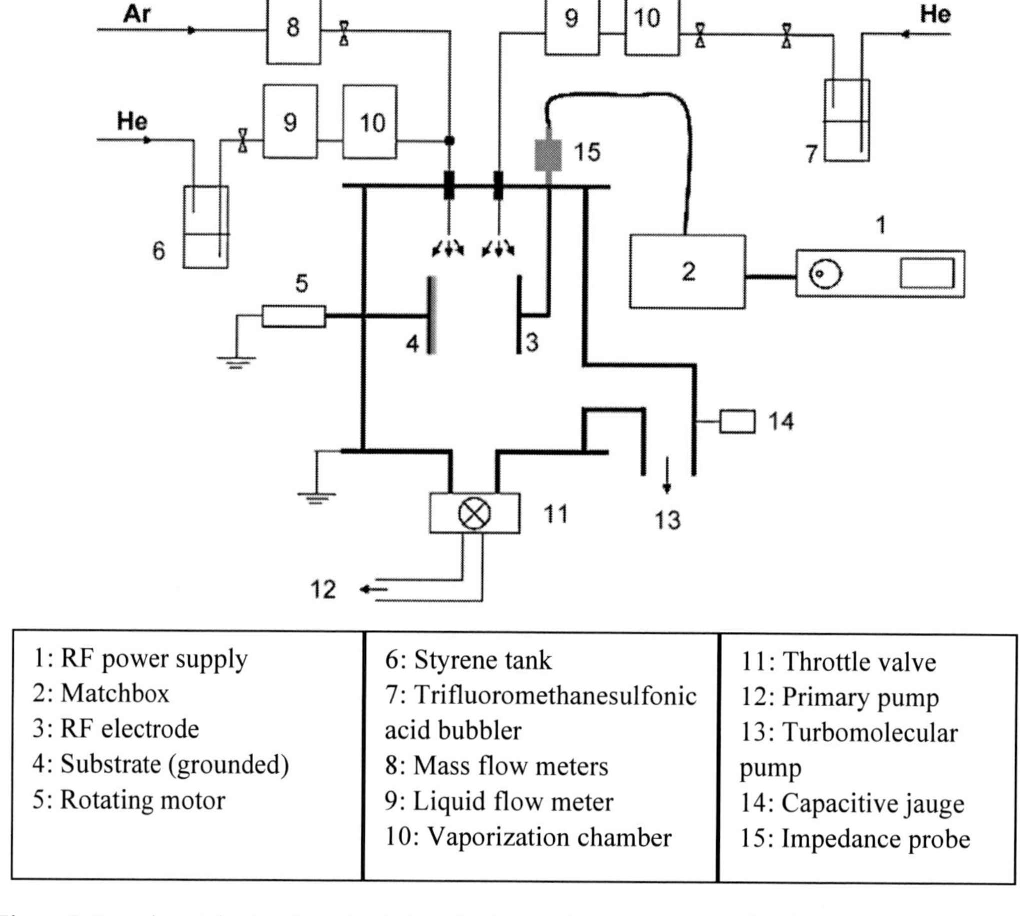

1: RF power supply	6: Styrene tank	11: Throttle valve
2: Matchbox	7: Trifluoromethanesulfonic acid bubbler	12: Primary pump
3: RF electrode	8: Mass flow meters	13: Turbomolecular pump
4: Substrate (grounded)	9: Liquid flow meter	14: Capacitive jauge
5: Rotating motor	10: Vaporization chamber	15: Impedance probe

Figure 5. Experimental setup for pulsed glow discharge plasma polymerization [128]. Reproduced with permission from Ennajdaoui, A. et al. J. Power Sources 2010, 195, 232-238. Copyright © 2010, Elsevier Science Ltd.

In order to improve the intrinsic ionic conductivities of plasma polymerized membranes, Brumlik et al. [117] prepared sulfonated fluorochlorocarbon ionomer films by a grow discharge plasma polymerization of trifluorochloroethylene (TFCE) and CF_3SO_3H. It showed the films synthesized using TFCE as the carrier gas are an order of magnitude more conductive than that synthesized using Ar as the carrier gas. The simplest explanation for the higher conductivities of the films using TFCE as the carrier gas was that the sulfonation levels were significantly higher in these films than in the films prepared using Ar as the carrier gas. However, plasma polymerized PEMs prepared under continuous RF mode show surface buckling, due to the film relaxation. This kind of structure was not compatible with the other fuel cell parts. The plasma conditions were too hard and led to an important cross-linking of the structure and low content of sulfonic acid groups.

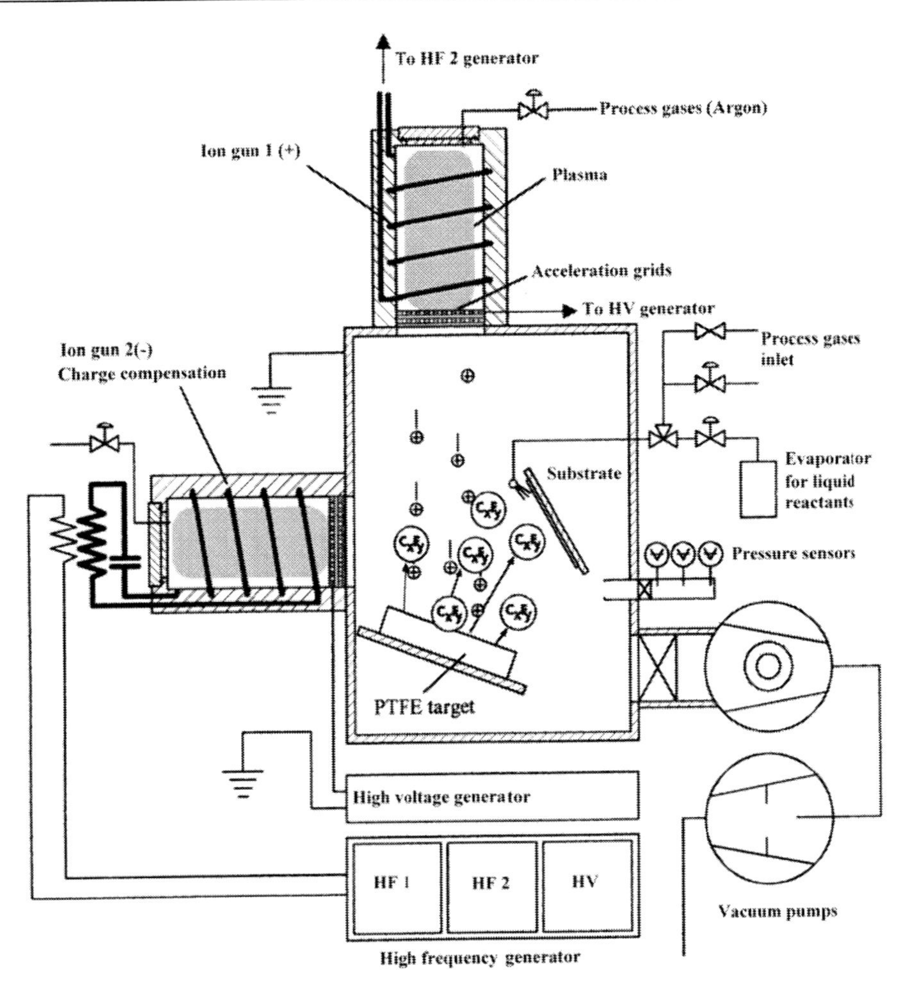

Figure 6. Schematic diagram of an ion beam assisted plasma polymerization reactor [129]. Reproduced with permission from Finsterwalder, F. et al. J. Membr. Sci. 2001, 185, 105-124. Copyright © 2001, Elsevier Science Ltd.

Recently, Roualdes et al. [127] presented a big asset of pulsed plasma glow discharge for the synthesis by plasma polymerization of PEMs for micro fuel cells applications. The assets of such process involved pulsing the discharge, the originality of the precursor injection system and the deposition on a large scale area. Experimental set-up is shown in Figure 5 [128] using a parallel plate electrodes configuration. The plasma reactor made of stainless steel is a cylindrical chamber. The RF power is fed to the RF electrode through a matching box using a power supply, allowing the pulsing or not. They prepared sulfonated polystyrene-like membranes by plasma polymerization of styrene, CF_3SO_3H and Ar mixtures in a low pressure glow discharge. Membranes synthesized under pulsed plasma glow discharge [127] were thin, uniform and adherent to the substrates such as silicon wafer and carbon conductive cloth (E-TEK). The surface and the cross section of such thin films were free of any defects. The SEM picture of a PEM deposited on a typical electrode used for micro fuel cells (Platinum sputtered onto conductive GDLs) [127] showed it had a good compatibility at the interface between the pulsed deposited PEM and the catalyst layer. The plasma membrane diffuses into the first nanometers of the electrode. Like this, a good quality of the triple point between fuel, catalyst and membrane can be obtained. They reported that the incorporation of sulfonic acid groups was tuneable as a function of plasma parameters and notably under pulsed conditions. By increasing the duration of after glow (t_{off}) and the peak power P value, the content of sulfonic acid groups could be improved. It was demonstrated that in a certain range of plasma conditions, a pulsed plasma discharge was better than a continuous plasma discharge enabling to deposit plasma polymers with the best monomer structure retention and a higher deposition rate. Nevertheless, a high limitation of the pulsed plasma discharge was the difficult priming and stability of the electric discharge out of a quite reduced range of plasma conditions [128].

2.2.1.2. Plasma Polymerization by an Ion Beam Assisted Method

A type of reactor used for an ion beam assisted plasma polymerization is outlined in Figure 6 [129]. In this reactor, two identical cylindrical ion sources, e.g. ion guns, are mounted onto the reactor chamber, one pointing to a poly(tetrafluoroethylene) (PTFE) target, the other pointing to a substrate for PEM deposition. The ion gun pointing to the PTFE target extracts positive ions whilst the second gun is negatively polarized and acts as electron shower without being supplied by any process gases, which ensures charge compensation, reduces the sparks as a consequence of discharging, and therefore, makes the process more stable. The gas pressure in the reaction chamber is recorded by a Pirani sensor, a Baratron capacity sensor and a penning tube. During the plasma polymerization, Ar gas is firstly introduced into the ion gun pointing to a PTFE and ionized in a high frequency plasma chamber. Liquid reactants are heated and processed as vapors.

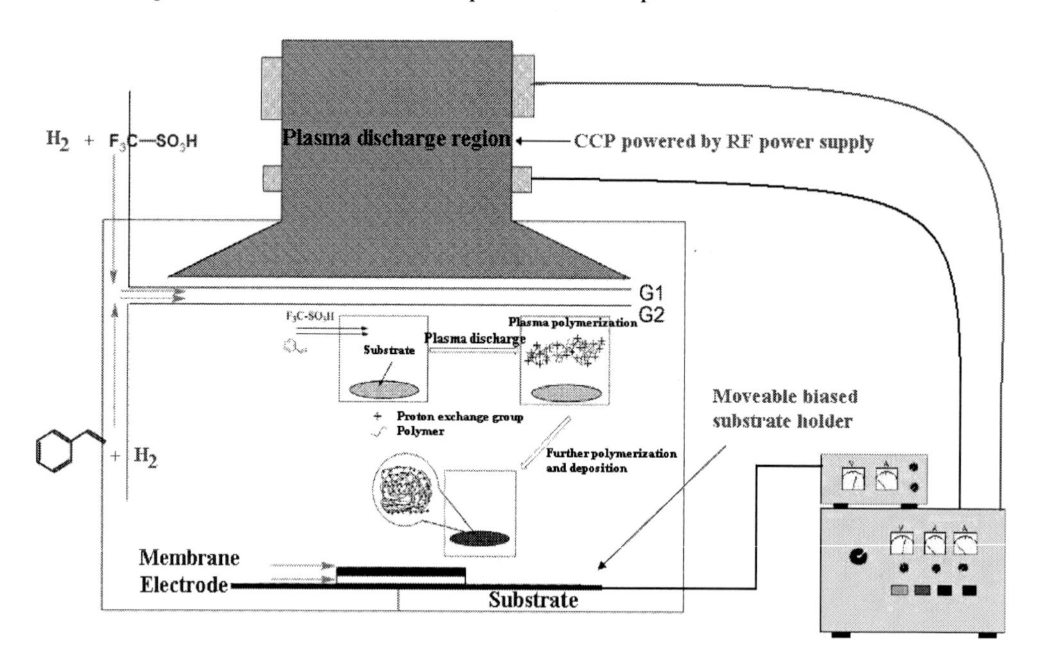

Figure 7. Schematic diagram of the after glow plasma polymerization reactor and simplified strategy for preparation of the proton-exchange composite membranes [20]. Reproduced with permission from Jiang, Z. Q. et al. Plasma Processes and Polym. 2010, 7, 382-389. Copyright © 2010, Wiley-VCH Verlag GmbH & Co. KGaA.

The films are usually polymerized onto cleansed glass substrates 15 cm × 15 cm in size. The positive charged Ar^+ ions are then accelerated by a high voltage applied on a metal grid and are directed down to the PTFE target. At sufficiently high ion energies the PTFE target is eroded by bombardments with the positive charged Ar^+ ions and releases carbon fluorine fragments. These fragments can recombine and eventually polymerize to form a solid film on the substrate. In some cases, the introduction of the reactive gases, such as oxygen or sulfur dioxide, during the sputtering process, is used to alter the composition of the deposited layer. These reactive gases can either be supplied straightly into the ion gun or into the reactor chamber or in close distance over the substrate (see Figure 6 [129]). The preparation of the PEM containing sulfonic acid groups by an ion beam assisted plasma polymerization process was first done by Finsterswalder et al. [129], where various sulfur components (SO_2, CF_3SO_3H or $ClSO_3H$) were added to achieve proton conductivity by the formation of sulfonic acid groups. The C_xF_y fragments combined with the sulfur components to form a coherent thin film on a substrate. The best membrane conductivities (>10^{-4} S/cm) and highest ion exchange capacities (0.15 mmol/g) were achieved with $ClSO_3H$ involved in the plasma polymerization process.

2.2.1.3. Plasma Polymerization in the after Glow Discharge

Due to the simultaneous occurrence of the polymer formation and degradation of monomers in the process of the glow discharge and ion beam assisted plasma polymerization [130-132], which makes the introduction of functional groups into the polymer membranes difficulty, the obtained plasma polymerization membranes usually hold a low ionic conductivity. Several studies [115,119] have shown that under drastic plasma conditions, plasma polymers are likely to contain very little and even no sulfonic acid groups as the sulfonic acid groups are completely decomposed in the plasma. Although the formation mechanism of plasma PEMs is not yet well understood due to the complexity of plasma reactions, it is generally accepted that active species, such as excited molecules, free radicals and ions, are formed by collisions of energetic electrons with monomer molecules during plasma discharge. Such collisions result in the excitation, fragmentation, and ionization of the discharge gas, and produce active species. These active species then recombine to form polymer membranes. In conventional plasma polymerization reactors, e.g. glow discharge and ion beam assisted plasma polymerization reactors [26,133], the monomer molecules are prone to dissociate due to their random collisions with the energetic electrons and the nonuniformity of energy distribution of these energetic electrons, and thus the preservation of monomer groups is rather difficult [134]. In this regard, the development of new plasma polymerization reactors that can be utilized to fabricate polymer membranes with high contents of proton exchange groups is of great interests. It is found that a low-frequency CCP discharge reactor can be used to synthesize highly functionalized polymer membranes.

In this type of plasma reactor, the monomer structure can be effectively preserved during the plasma discharge and the obtained membranes possess higher contents of proton exchange groups. Figure 7 schematically illustrates the reactor for the after glow discharge plasma polymerization [20]. The reactor is a stainless steel vessel connected with a vacuum system consisting of a mechanical booster pump, a rotary pump, and a cold trap. The plasma discharge is sustained by the a.c. power supplier between two external electrodes in a Pyrex glass tube, using Ar (Air Liquid) as working gas. The reactor is comprised of two regions: the upper part is plasma discharge region for the production of RF glow discharge, and the lower part is the plasma polymerization region for the deposition of PEMs, between them are two screen grids (G1 and G2), through which electrons or ions of the plasma can be effectively extracted from the discharge volume to the down stream region. The bias voltage applied to the screen grids can be easily controlled by adjusting transformers connecting to them. In addition, a controllable bias voltage to substrate is also applied. During the preparation, polarization of substrate holder and/or screen grid G2 is maintained at certain values. The polarization of substrate can facilitate the deposition of PEMs onto the substrate by the electrostatic interaction.

As an example, Roualdes et al. [114] compared the PEMs prepared by the glow discharge and after glow discharge plasma polymerization. It showed that PEM formed in the after glow discharge were predominant with sulfonic acid groups whereas it was sulfone groups in plasma polymers formed in the glow discharge, due to a too high fragmentation of the CF_3SO_3H monomer in this latter configuration, although both of them were flat, uniform and free from defects. Proton conductivity was a very important parameter for evaluation of PEMs. It showed that the PEMs formed in the glow discharge show the lower proton conductivities than those prepared in the after glow discharge. Taking into account the conclusions drawn from structural analysis of materials, they concluded that the proton conduction capacity of a plasma material was increased when its density was low (high water and proton mobility) and its sulfonic acid content was high (enhancement of proton exchange). Based on this work, Roualdes et al. [135] further synthesized a series of sulfonated polystyrene-type membranes by plasma polymerization of a mixture of styrene and trifluoromethane sulfonic acid monomers in a low-frequency after-glow discharge plasma reactor. It was confirmed that such a deposition process enabled the preservation of the monomers structure and the synthesized plasma-polymerized PEMs were dense and uniform with controllable thicknesses. The structure determination by fourier-transform infrared (FTIR) spectroscopy and X-ray photoelectron spectroscopy (XPS) showed that the synthesized plasma-polymerized PEMs were very rich in sulfonic acid groups. Although not very conductive compared to Nafion® 117, these membranes exhibited specific resistances similar (2 Ωcm^2 for 2 μm thick films) and even lower (1 Ωcm^2 for 1 μm thick films) than Nafion® (1.9 Ωcm^2) because of their adjustable low thickness. The activation energy (E_a) for proton conduction of the plasma-polymerized PEM, calculated from the linear least-square fits to the experimental temperature dependence of conductivity was equal to 4.54 kJ/mol, a value that is lower than that for Nafion® (10–40 kJ/mol). The fact is that the energy barrier for proton conduction was lower for plasma polymers and explained by an easier structural reorganization under the effect of temperature, due to the existence of more electrostatic driving forces evidenced by FTIR analysis. Becaused of their highly cross-linked structure, these membranes exhibited a reduction of the methanol crossover in a factor of 10 in comparison to Nafion®.

With aim to improve proton conduction of PEMs, our group [20] fabricated sulfonated ultra-thin PEMs by plasma polymerization of a styrene/trifluoromethane sulfonic acid monomer mixture in a low-frequency after-glow CCP discharge reactor based on the work by Roualdes et al. [135]. As with PEMs prepared by Roualdes et al. [135], the synthesized polymer membranes were dense and uniform with controllable thicknesses. FTIR and XPS analysis demonstrated that the obtained membranes possess a chemical structure similar to Nafion 117, while the proportion of proton exchange groups (4.84%) was much higher than that of Nafion 117 (1.23%).

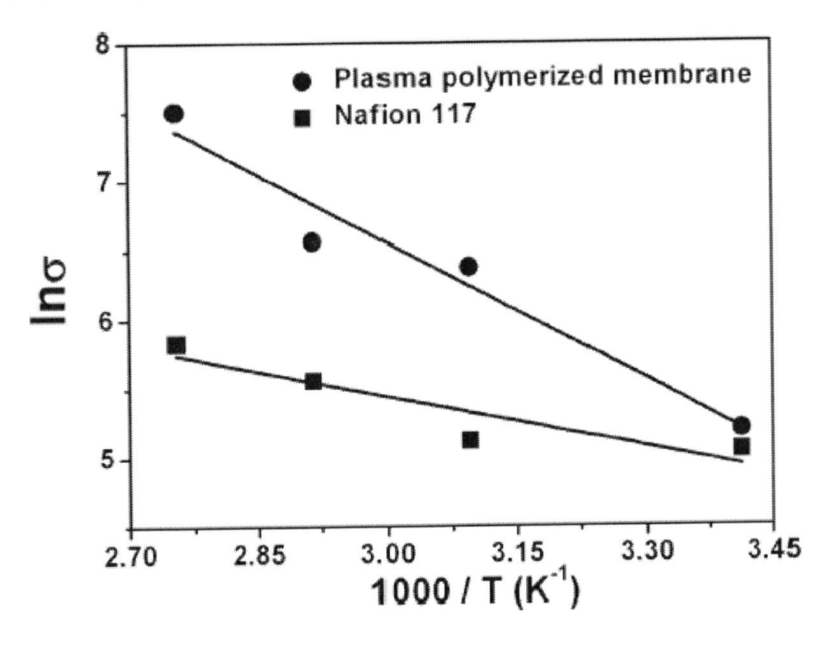

Figure 8. Arrhenius plot showing temperature dependence of membranes proton conductivity. The dots are the experimental data [20]. Reproduced with permission from Jiang, Z. Q. et al. Plasma Processes and Polym. 2010, 7, 382-389. Copyright © 2010, Wiley-VCH Verlag GmbH & Co. KGaA.

Due to their high proportion of proton exchange groups and highly crosslinked structure, the proton conduction capacity of plasma polymerized membrane (181.0 mS/cm) was higher than Nafion 117 (155.3 mS/cm) in the same experimental cell, the permeabilities of methanol is 6.64×10^{-12} m^2/s for the plasma polymerized membrane which was lower than that of Nafion 117 (2.06×10^{-10} m^2/s). Figure 8 presented the temperature dependence of the conductivity for the plasma polymerized membrane and Nafion 117 (temperature range between 20 and 90 °C) [20]. The activation energy (E_a) for proton conduction of the plasma-polymerized PEMs, calculated from the linear least-square fits to the experimental temperature dependence of conductivity (Figure 8 [20]), was equal to 27.29 kJ/mol. However, the activation energy for Nafion 117 is only 10.17 kJ/mol, which was comparable to the values reported previously [136,137]. The higher energy barrier to proton conduction for plasma polymers than for Nafion was due to their highly crosslinked structure, which made the structural reorganization under the effect of temperature much difficulty. Table 1 displayed a simple comparison of the performance of membranes prepared by our group [20] and Roualdes et al. [135] and Nafion 117 for their application in DMFCs. As shown in Table 1, both PEMs prepared by an after glow plasma polymerization methods exhibited improved features over Nafion 117. Worth noting was that the membranes synthesized in our group [20] exhibited higher conductivity, activation energy, and resistance, compared to that reported by Roualdes et al. [135], although they had the comparable percentage contents of proton exchange groups. Higher activation energy and resistance of membrane could be easily explained by its relative higher thickness [20]. Undoubtedly, an increase in the thickness of membranes would increase the length of protons through them, and correspondingly, increase their activation energy and resistance. It seemed that the membranes prepared by Roualdes et al. [135] had a much denser structure due to its lower methanol permeability even though they had a lower thickness. Such dense structure would render the migration of proton and some other molecules through the membrane difficult, and decrease its ability of water uptakes as shown in Table 1 [20]. As described above, plasma polymerization method was a complicated process during which polymer formation and degradation of monomers occur simultaneously. Any variation in experimental conditions, such as electron density, input power, monomer flow rates and ratio of the partial pressure of CF_3SO_3H to styrene, etc, would greatly influence the properties of the resulting membranes. Therefore, an optimization process was usually required to establish an optimal condition for the preparation of proton exchange membranes to improve their performance in the application of DMFCs.

Table 1. Performance of plasma polymerized proton exchange membrane and Nafion 117 membrane as PEM electrolyte [20]

	Thickness μm	SO_3H %	Water uptake %	Conductivity (σ) mS/cm	Activation energy kJ/mol	Specific resistance Ω cm^2	Methanol permeability m^2/s
Plasma polymerized membrane	8.82	4.84	64.4	181	27.29	6.6	7.45×10^{-12}
Nafion 117	183	1.23	34.6	155.3	10.17	7.7	2.08×10^{-10}
Membranes reported by Roualdes et al. [a)]	0.22	5	Lower water uptake	9.8×10^{-2}	4.54	2[b)]	2.30×10^{-13}

[a)] Values for plasma polymerized membrane reported by Roualdes et al. obatined from reference [135]; [b)] Around 2 Ωcm^2 for 2 μm thick membranes.

Reproduced with permission from Jiang, Z. Q. et al. Plasma Processes and Polym. 2010, 7, 382-389. Copyright © 2010, Wiley-VCH Verlag GmbH & Co. KGaA.

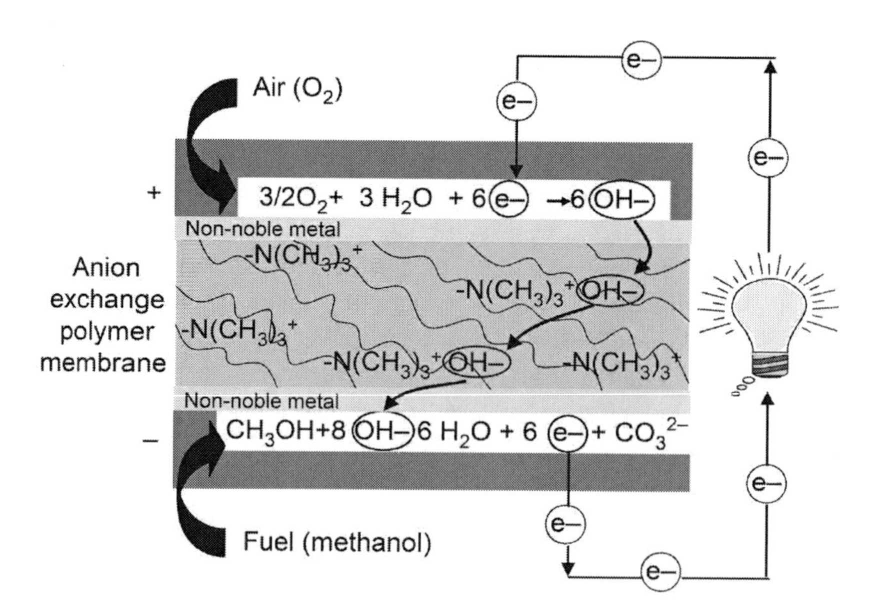

Figure 9. Operating principle of SAFC [148]. Reproduced with permission from Schieda, M. et al. Desalination 2006, 199, 286-288. Copyright © 2006, Elsevier Science Ltd.

Pyridine groups tethered to the polymer backbone can act as a medium through the basic nitrogen for transfer of protons between the sulfonic acid groups of PEM even at low relative humidity conditions and higher temperatures. The sulfonic acid group of plasma polymerized PEM can protonate the nitrogen site of pyridine and thus facilitates the hopping of the proton bound to another basic site of the pyridine unit or to the oxygen of another sulfonate anion group. The swaying vibration of pendant pyridine promotes long-range proton motion in the polymer system. Thus, the membranes are expected to be used in DMFC at low relative humidity conditions and higher temperatures. Our group [138] prepared sulfonated ultra-thin PEM carrying pyridine groups from a plasma polymerization of styrene, 2-vinylpyridine, and trifluoromethanesulfonic acid by after-glow capacitively coupled discharge technique.

2.2.2. Other Functionalities Based Plasma Polymerized Polymer Electrolyte Membranes (PEMs)

Up to date, research in PEMs are mainly focus on the synthesis of membranes carrying terminal sulfonic acid groups due to their high conductivities. However, high level of swelling effect, poor chemical and mechanical stabilities, and high cost of sulfonic acid grafted membrane have prompted researcher to exploit PEMs with other functionalities, such as phosphonic acid, carboxyl acid and hydroxyl, etc. [139-141]. Actually, the proton exchange membranes with other functionalities might also exhibit a good ionic conductivity. For example, Mex and Muller [142-144] have developed phosphonic acid functionalized plasma polymerized PEMs with tetrafluoroethylene for the polymeric backbone and vinylphosphonic acid for acid groups responsible for the proton conductivity. Depending on the process parameters these films exhibited ion conductivities ranging from 100 mS/cm to 200 mS/cm (at 80 °C), determined by ac-impedance measurements. For use in DMFCs, these films could be optimized to achieve a high ion conductivity and high thermal resistance.

Fluorocarboxylic acid functionalized PEMs were developed by Le Van-Jodin et al. [145]. It showed that the obtained fluorocarboxylic acid functionalized PEMs exhibited a good ionic conductivity (the highest conductivity leads to 160 mS/cm) near the ionic conductivity of Nafion®. In parallel works, we also prepared carboxylic acid based PEMs [146,147]. The ultra-thin PEMs were fabricated by an after-glow CCP discharge technique in a plasma polymerization reactor, where styrene and acrylic acid are used as starting materials. During the preparation, the energy of the ionized particles extracted from the radio frequency glow discharge region to the plasma polymerization region could be easily controlled by adjusting the bias voltage applied to the screen grids and substrate. Therefore, the degradation of monomers could be effectively avoided, and the contents of the proton exchange groups on the obtained membranes could reach to a higher extent. The synthesized membranes were dense with uniform structure and are demonstrated as good proton conductors. The proton transport numbers (t_{H^+}) of the obtained membranes showed the value high as 0.961, which was comparable to that of Nafion 117 membrane (t_{H^+} = 0.950, China ShenMa Group). This result indicated that the obtained membranes were of better proton conductivity.

New hydroxyl exchange membranes are the core of solid alkaline fuel cells (SAFCs), whose working principle is shown in Figure 9 [148]. SAFC is a hybrid concept between those of both DMFCs and alkaline fuel cells (AFCs). The expected advantages of SAFC, when compared to classical AFC, are the possibility to use easy handling and stocking liquid fuel and the easy control of carbonatation due to the solid nature of the electrolyte (membrane).

When compared to DMFC, the advantages of SAFC include the decrease of the fuel cross-over and the possibility to use less expensive nonprecious metal catalysts such as Ni or Ag. For such fuel cells, one of the main challenges is the development of new hydroxyl exchange membranes (typically based on quaternary amine functions) showing high OH^- conductivity, high chemical stability and low fuel permeability. Very few studies have been devoted to the development of such membranes yet. Roualdès et al. [148] demonstrated the synthesis of dense materials containing hydroxyl exchange amine functions by plasma polymerization. Starting from a monomer containing tertiary amine groups (triethylamine or triallylamine), the membrane synthesis procedure consisted of two steps. In the first step, the monomer vapour was plasma-polymerized in a glow discharge to form a thin film partially composed of amine functions (Figure 10 [148]). In the second step, the methylation of the amine functions into quaternary amine functions was realized by immersion of the plasma film into a methyliodide solution. The optimization of the synthesis parameters had enabled to manufacture micrometric membranes showing satisfying anionic conductivity. It showed that these membranes had surprisingly higher conductivities than that of Nafion at 30 °C or even higher temperatures when used in fuel cells.

A 4-vinylpyridine-based anion-exchange membrane (poly-4-VP) by plasma polymerization was recently developed by Ogumi et al. [149]. Scanning electron microscope (SEM) and FTIR showed that the membrane had uniformly thin and highly cross-linked form. The resistances of plasma-polymerized poly-4-VP and commercial membrane (NEOSEPTA AHA, Tokuyama Corp., Japan) were measured. The resistance and ionic conductivity of plasma-polymerized membranes were about 1.9 Ωcm^2 and 5.4×10^{-4} S/cm, respectively. Although plasma-polymerized membrane showed lower ionic conductivity than the commercial membrane (4.3×10^{-3} S/cm), the effective resistance of polymerized membrane was less than that of commercial membrane (3.6 Ωcm^2 of OH form) due to the difference in the thickness of the membranes.

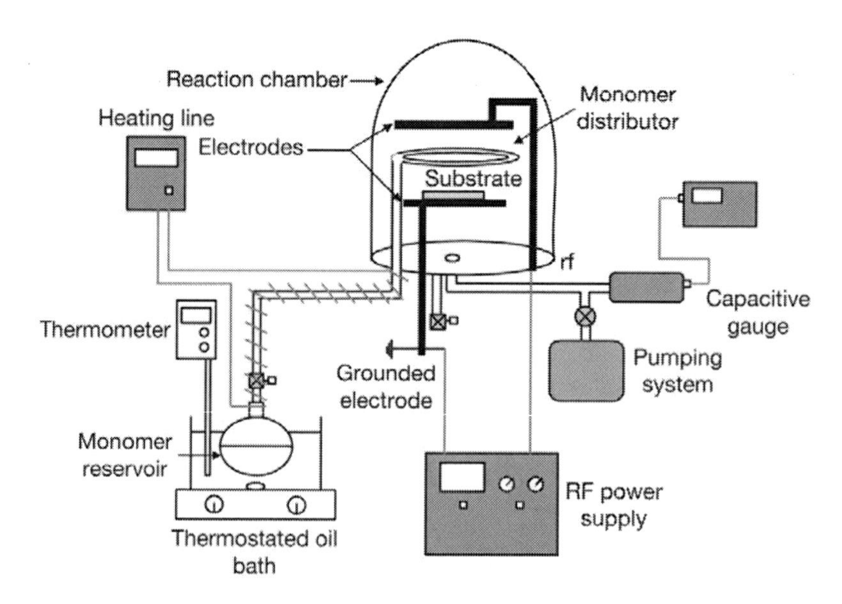

Figure 10. Scheme of a plasma polymerization set up [148]. Reproduced with permission from Schieda, M. et al. Desalination 2006, 199, 286-288. Copyright © 2006, Elsevier Science Ltd.

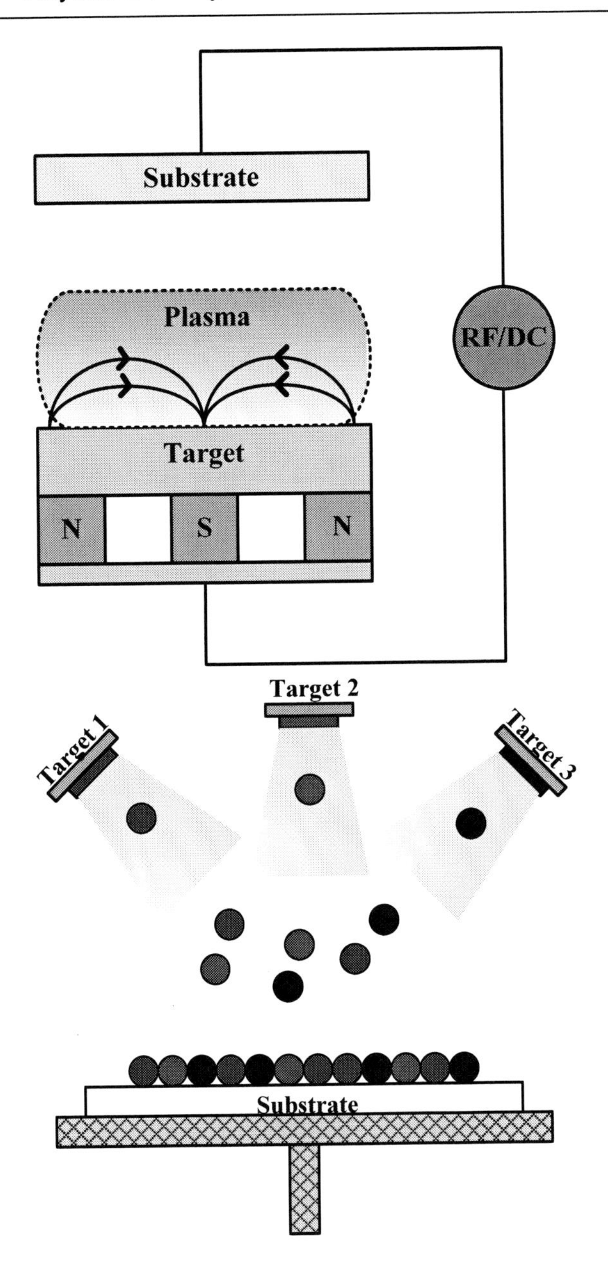

Figure 11. Schematic plasma confinement of a conventional MSP system (left) and multi-target MSP system (right).

The low ionic conductivity might be associated with the lower flexibility of the ion-exchange group due to high cross-linking. In addition, modification of carbon-supported platinum with the plasma-membrane enhanced hydrogen oxidation at the gas diffusion electrode due to an increase in the triple-phase boundary. Therefore, it is believed that plasma treatment is very useful for improving the performance of alkaline fuel cell using anion-exchange membrane.

3. Electrodes Fabricated Using Plasma Methods and their Applications in Polymer Electrolyte Membranes Fuel Cells (PEMFCs)

For higher performance of PEMFCs, catalysts which are usually worked as electrodes in PEMFCs are required to meet the following requirements

- Higher CO and/or methanol tolerance,
- Easy proton or oxygen adsorption/desorption and
- Fast kinetics of the ORR.

There are two methods for preparing the high performance electrocatalysts: (1) Modification the surface of catalysts supported GDLs with metal or metal catalysts by PSD method which localize the electrocatalysts in the immediate neighborhood between the electrode and electrolyte; (2) *In situ* electrocatalysts preparation by PSD method which involves forming catalyst directly on the GDLs, various electronically conducting supports, and PEM, etc. PSD is the most used *in situ* electrocatalysts preparation methods which is widely used for integrated circuit (IC) manufacturing and has been used in PEMFC electrocatalysts preparation for decades.

3.1. Sputtering Theory and Apparatus

In a typical sputter process, a target plate is bombarded by energetic ions generated in glow discharge plasma. The bombardment causes the removal of the target atoms which condense on various substrates as a thin film. In the process of the ion bombardment, secondary electrons are also emitted from the target surface and these electrons play an important role in maintaining the plasma [150,151]. For the fabrication of electrodes for PEMFCs, two techniques, i.e. magnetron sputtering (MSP) and plasma assisted bias sputtering (PABS), are commonly used.

MSP has been extensively applied over the last two decades for the deposition of a wide range of industrially important coatings including hard, wear-resistant coatings, low friction coatings, corrosion-resistant coatings, decorative coatings and coatings with specific optical or electrical properties. It allows for the versatile and precise control of the microstructure and related properties of the film such as density, adhesion, surface roughness and crystallinity [150,151]. In the process of MSP system, a magnetic field configured parallel to the target surface is usually used to constrain secondary electron motion to the vicinity of the target surface and trapping the electrons. In this way, the probability of an ionizing electron-atom collision occurring is increased, which results in a dense plasma in the target region, and in turn, leads to increased ion bombardment of the target and higher sputtering and deposition rates [150]. Schematic plasma confinement of a conventional MSP system is shown in Figure 11 (left). In practical applications, a multi-target MSP system is generally adopted as in Figure 11 (right). In this case, it is feasible to introduce different types of materials. By varying the input power of different targets, the atom ratio of materials deposited onto the substrate can be well controlled. Although conventional MSP systems can be a powerful tool to modify some substrates with some specified surfaces, it is impossible to modify the surface of materials uniformly. In order to modify the entire surface of powdery materials, a polygonal barrel-MSP system has been developed as shown in Figure 12. The rotation or swing motion of the substrate vessel (so called as barrel) leads to continuous stirring of the substrates and crushing of the aggregated substrates during sputter deposition [152].

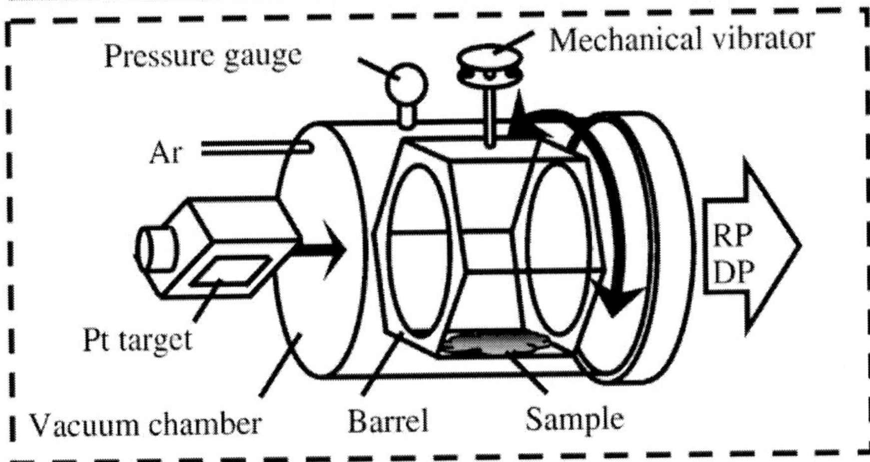

Figure 12. A photograph and a diagram of the polygonal barrel-magnetron sputtering system [152]. Reproduced with permission from Yamamoto, H. et al. *Mater. Lett.* 2008, *62,* 2118-2121. Copyright © 2008, Elsevier Science Ltd.

A schematic representation of a typical PABS device is shown in Figure 13 [37,153]. In this reactor, a magnet is used to confine the plasma so that high density plasma can be achieved. Low pressure (5 mTorr) RF helicon plasma is usually employed as it can produce high density plasmas with great efficiency in rather large volumes [18,36-40,50-53]. During the working process, low pressure RF plasma is firstly initiated in argon gas. The obtained Ar^+ ions are then attracted to the negatively biased metal target and gain sufficient energy to induce sputtering. When there are two or more targets, the atom ratio of different target materials deposited onto the substrate can be easily manipulated through varying the negative bias of different targets.

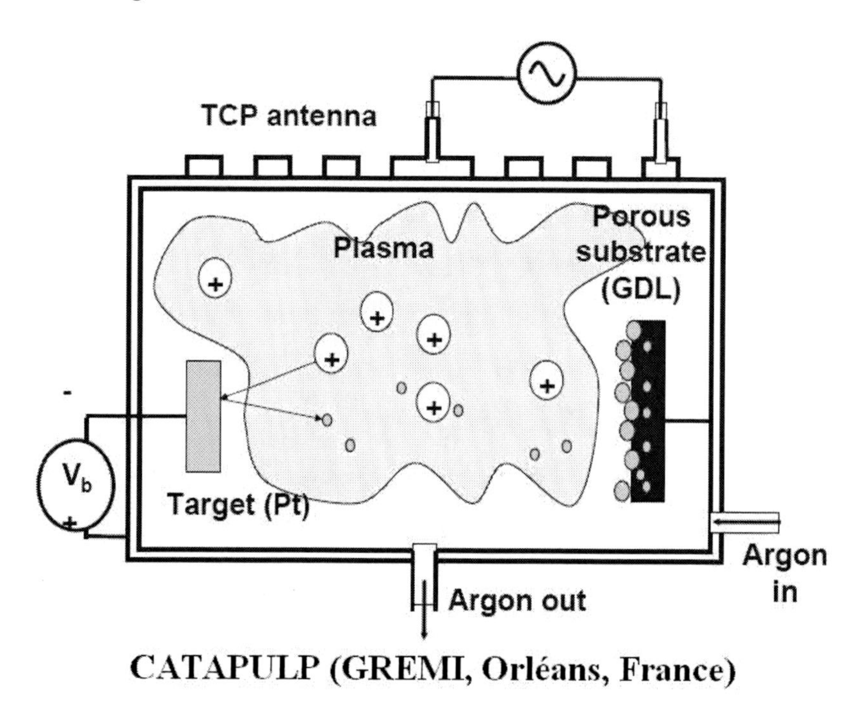

Figure 13. A schematic representation of the CATAPULP system [37]. Reproduced with permission from Caillard, A. et al. *Surf. Coat. Technol.* 2005, *200,* 391-394. Copyright © 2005, Elsevier Science Ltd.

3.2. Modification of Catalyst Supported Gas Diffusion Layers (GDLs) Using Plasma Methods

Modification the surface of catalysts supported GDLs by PSD method has been widely used to improve the performance of electrodes, which can be conducted as follows:

- Plasma modification of catalysts with metal oxide. The modified catalysts usually possess the higher reduction rate of oxygen and improved methanol tolerant characteristics. For example, Lee et al. [13] investigated the influence of modification of a platinum cathode with copper oxide (CuO_m) on the performance of DMFCs. Comparing with an unmodified Pt cathode, the authors stated that $PtCuO_m$ prepared by a PSD technique exhibited a higher ORR rate and a suppressed methanol oxidation reaction (MOR). Due to the higher reduction rate of oxygen and methanol tolerant characteristics, the power performance of $PtCuO_m$ in a DMFC was increased up to 195 mW/cm^2 which was about 25% larger than that of an unmodified Pt cathode.
- Plasma deposition of a thin metal film onto the surface of the electrodes which was used for localizing the catalysts near the front surface of them. The sputtered layers can increase the activity of catalysts for cathode reaction (improved ORR kinetics)

and increase the three-phase zone (the area of the triple point between fuel, catalyst and membrane), thus, improving the electrochemically active surface area (ECSA) for the electrode. Ticianelli et al. [10] demonstrated that the deposition of a thin layer of Pt onto the top of the catalyst layer of the electrodes using a PSD method effectively improved the performance of PEMFCs, and an optimal configuration for PEMFCs, that was an electrode with 20 wt% Pt/C electrocatalyst and a 50 nm sputtered Pt film. Based on Ticianellis' work, Mukerjee and co-workers [11] examined the effects of the sputtered Pt film on the electrodes with 20 wt% Pt/C electrocatalyst on the electrode kinetics of ORR in PEMFCs. The cyclic voltammograms (CV) for both the sputtered and un-sputtered electrodes revealed that the ECSA for the electrode with a sputtered film was about 80% higher than that for the un-sputtered one. The improvement of ORR kinetics by the sputter-deposition of a thin Pt film was manifested by a 4 fold improvement of current density and of the exchange current density in a PEMFC. The effects of the thickness of the deposition layer on the ORR were investigated by Tang et al. [14], where they fabricated a series of MEAs coated with different thicknesses of Co layers prepared by a PSD method. The polarization curve indicated that the MEA with 5 nm Co layer exhibited the best cell performance with a high current density of 750 mA/cm^2 at 0.6 V and a maximum output power of 0.49 W/cm^2. The high cell performance might originate from the formed Co-Pt network with many nano-sized clusters and the porosity of the obtained electrode. The SEM images showed that when a 3 nm Co layer was sputtered on the electrode, only small Co clusters was formed on the surface of the electrode and this thickness might not be adequate to promote the catalytic effect for cathode reaction and increase the three-phase zone. As the deposition layer was increased to 5 nm, the Co clusters became bigger and some coalesce to form Co-Pt network with many nano-sized clusters and pores, and this contributed to enhance the cell performance. Further increase of the deposition layer to 10 nm induced the cover of the electrode surface by a dense Co film where fewer metal nano-particles were present and the porosity of the electrode decreased, which ultimately resulted in the deterioration of the cell performance as shown in Figure 14 [14]. The work mentioned above describes the effects of the deposition of metal film on the cathodes. Actually, the deposition of a thin metal film onto the surface of the anodes can also improve the performance of PEMFCs. For example, Xiu et al. [12] investigated the effects of sputtered Pt layer onto anodes on the cell performance of DMFCs. It showed the sputtered Pt layer with thickness ranging from 6-30 nm significantly improved the cell performance, e.g., the power output increased 25-34% at 353 K. The top power density of 170 mW/cm^2 was achieved by a 6 nm sputtered Pt layer on the anode at 363 K. The thicker the sputtered Pt layer the more the cell performance in the low current density range improved, but was substantially limited by a concentration over-voltage in the high current density range.

3.3. Preparation of Catalyst by Plasma Sputtering

3.3.1. Plasma Sputtering of Pt onto Gas Diffusion Layers (GDLs)

For commercialization of PEMFCs, it is compulsively required to lower amounts of catalyst loading and improve mass activities of them, since these catalysts, which are usually noble metals, such as Pt, Ru and Au, are extremely expensive. Much effort has been devoted to improving the utilization efficiency of the catalysts. PSD techniques which can be used to localize noble catalyst within the (membrane carbon interface) regions responsible for electrochemical reactions in the MEAs, have shown great promise in this area. In comparison to the other methods, PSD can facilitate the uniform dispersion of catalysts in the whole electrodes and increase the catalyst utilization efficiency by diminishing inactive catalyst sites. Thus, the advantages of such electrodes are: (a) the localization of the catalyst on the uppermost surface, which makes it active in the immediate neighborhood to both the PEM and the electrode; (b) thin active layer, which are not only active closest to the gas supply at high cell current densities; (c) ultra-low level of Pt loading, which can decrease the cost of PEMFCs.

Pioneering work of plasma sputtering of Pt onto GDLs was done by Hirano et al. [29] Although the PEMFC with this cathode exhibited a linear variation of potential with current density in the intermediate current density region, and minimal mass transport overpotential in the high current density region, its cell performance was not as good as that with the E-TEK electrodes. The authors of this chapter attributed the reason to the high anodic overpotentials in the high current density region which was probably due to hindered water transport in the sputtered Pt layer. In order to understand the effect of Pt loadings on the performance of PEMFCs, Gruber et al. [30] deposited thin Pt layers onto two types of GDLs, i.e. SIGRACET GDL-HM (SGL Technologies, SGL Carbon Group) and E-TEK, and compared the performance of the PEMFCs with these two different electrodes. It was found that the performance of the PEMFCs was not proportional to the Pt loading for both SGL and E-TEK. The best performances for the SGL and E-TEK samples were obtained at Pt loadings of 0.054 mg/cm^2 (191 mW/cm^2, 285 mA/cm^2 at 0.6 V) and 0.107 mg/cm^2 (203 mW/cm^2, 301 mA/cm^2 at 0.6 V), respectively.

The improvement of fuel cells performance can be further achieved by insertion of other metal catalyst layers into the Pt sputtered layers. Gruber et al. [30] found that the PEMFCs performances were improved by additional insertion of Cr or Pd thin layers into the Pt sputtered layers. In their following work, the effects of adding additional sublayers on the performance of PEMFCs were systematically investigated [31]. According to their results, it was said that the insertion of the hydrogen-permeable materials Pd or plasma-polymerized hexamethyldisiloxane (pp-HMDSO) sublayers underneath the anode catalyst thin film could improve the performance of the PEMFCs up to 18%, and the addition of a thin layer of Cr (~1 nm) underneath the catalyst layer could improve the maximum power density of the PEMFCs to 259 mW/cm^2. They attribute the effectiveness of the Pt catalyst to its more evenly distributed growth on the Cr seeds and more catalytically active sites available.

For a conventional sputtering, the obtained catalyst layers are usually nonporous, and display a small ECSA, due to their compact and dense structures. The increase of Pt loading always results in the decrease of mass activity. Optimization of the electrode efficiency requires distributing the catalyst over a sufficient depth in the GDL near the PEM or preparing catalysts with porous structures. Brault et al. [36] pioneered to sputter Pt onto GDLs using PABS technology.

Figure 14. SEM images of the sputter-deposited Co layer on 0.2 mg/cm^2 Pt catalyzed electrodes: Top surface of electrode before Co sputtering (a), 3 nm Co sputtered (b), 5 nm Co sputtered (c) and 10 nm Co sputtered (d) [14]. Reproduced with permission from Tang, Z. et al. *J. Appl. Electrochem.* 2009, *39,* 1821-1826. Copyright © 2009, Springer.

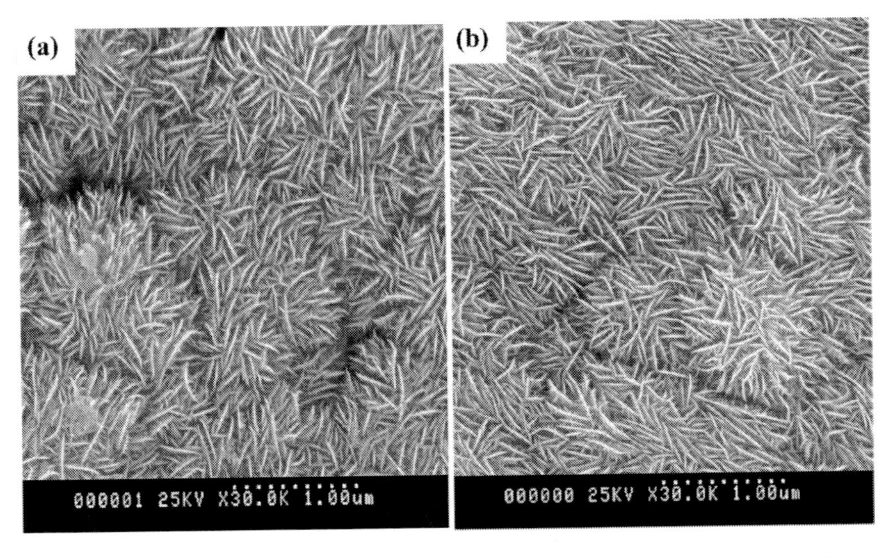

Figure 15. SEM photographs of the dendritic Pt film: before the reduction treatment (a), after the reduction treatment (b) [154]. Reproduced with permission from Yamada, K. et al. *J. Power Sources,* 2008, *180,* 181-184. Copyright © 2008, Elsevier Science Ltd.

The resulting platinum concentration profile extended up to 2 μm into the GDLs and was well fitted by a generalized stretched Gaussian function. The Pt layer was found to consist of nano-clusters with a maximum size of 4 nm. Although the power density of the PEMFC with plasma sputtering prepared electrodes (platinum loading of 0.08 mg/cm2) was still lower than that of the PEMFC with reference commercial E-TEK (Pt loading of 0.35 mg/cm2), the Pt utilization of plasma prepared electrode was much higher than that of the commercial E-TEK electrode. The porous structure of catalyst layers can improve the ability of fuel gas diffusion and the corresponding mass activity. For the prapartion of porous structure, Yoo et al. [41] used a high-pressure sputtering technique with a mixture of gaseous Ar and He to sputter Pt onto GDLs. A high-pressure (>100 mTorr) sputtering technique can produce nanoscale particles, which form in the vapor as a result of the high pressure. In comparison with conventional sputtering technique, the high pressure sputtering technique produced a more porous Pt nano-catalyst layer and the ECSA was enhanced 250%. This was due to the smaller particle size of Pt catalyst fabricated in the high pressure with a mixture of Ar and He. The Pt nano-catalyst layer sputtered at 200 mTorr pressure consisted of Pt nanoparticles with an average size of 8.9 nm. The PEMFCs performance of Pt nanocatalysts fabricated using high-pressure sputtering deposition showed a maximum power density of 420 mW/cm2 and efficiency per total weight of 10.5 W/mg.

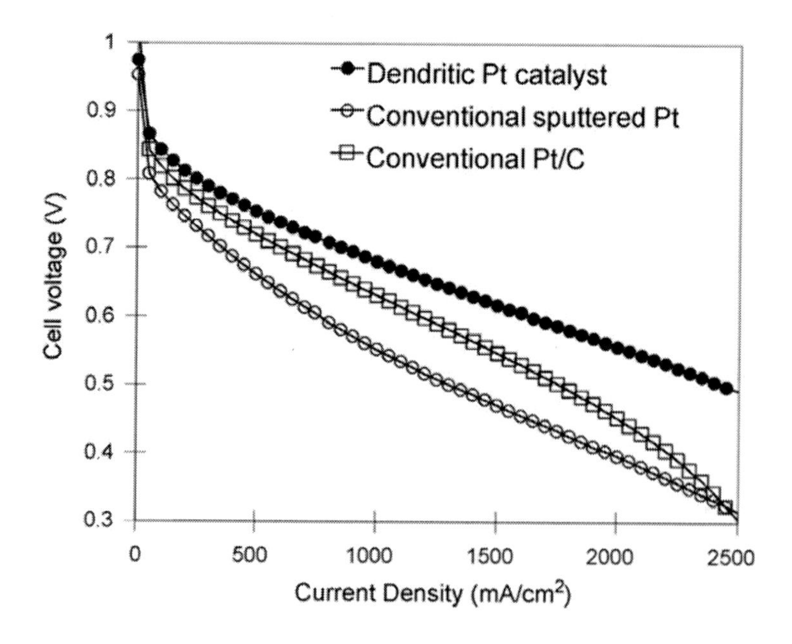

Figure 16. Single cell performances for H_2/O_2 operation at 80 °C of the dendritic Pt, the conventional Pt catalyst, and the conventional Pt/C [154]. Reproduced with permission from Yamada, K. et al. *J. Power Sources,* 2008, *180,* 181-184. Copyright © 2008, Elsevier Science Ltd.

Recently, Yamada et al. [154] reported a Pt catalyst film with dendritic microcrystalline structure prepared by reducing palatinum oxide (PtO_2) deposited by reactive sputtering method for PEMFC cathode. They prepared PtO_2 film under 100% O_2 partial pressure at room temperature. Then the PtO_2 film was subjected to a hydrogen reduction treatment using 2% H_2/He (0.1 MPa) at room temperature to obtain a Pt catalyst film. The SEM micrographs of the as-deposited Pt dioxide and the reduced Pt film are shown in Figure 15 a and b, respectively. The density of the dendritic Pt film was 3.3 g/cm^3, while the density of the conventional sputtered Pt film was about 13.2 g/cm^3. The ECSA of the dendritic Pt catalyst, the conventional sputtered Pt catalyst and the Pt/C were about 14, 3 and 53 m^2/g, respectively. Due to improved diffusion characteristics and the higher ECSA of the dendritic Pt catalyst, the cathode of dendritic Pt catalyst showed better performance than that of the cathode of conventional sputtered Pt catalyst, as shown in Figure 16.

3.3.2. Sputtering of Pt onto Various Electronically Conducting Supports

It has been demonstrated that the supporting materials which are uitilized for the stability of the electrocatalyst have a great impact on the electrochemical performance of PEMFCs. These supporting materials are usually electronically conductive such as CNFs, MWCNTs, carbon nanocolumns, carbon nanorods, columnar titanium, TiO_2 nanotubes, titania nanotubes (TiNTs), etc. and have a strong interaction with supported catalysts. Well understanding of interactions between the supporting materials and the supported catalysts is the intergral parts of design process for the preparation of higher performance of PEMFCs.

CNFs have already been proposed as a replacement for traditional carbon particle powders in PEMFC electrode owing to their high electrical conductivity, unique surface structure, large surface area, and chemical inertia. These nanostructures are widely used in powders spread on the carbon backing, but in some cases the results did not show many advantages over conventional carbon particle powders due to the formation of the agglomerates. Caillard et al. [50] first used the PABS system to prepare aligned CNFs on carbon paper and GDLs. These aligned CNFs were then worked as supports for the deposition of Pt catalyst. The results showed that the MEA based on Pt/CNFs nanostructures grown on hydrophobic GDLs led to the best maximum power density 230 mW/cm^2, which was 40% higher than the case in the absence of CNFs on GDLs and 65% higher than the case without hydrophobic GDLs as catalyst support. In their later work, [51] a Pt/CNFs/GDLs electrode with a Pt loading of 0.1 mg/cm^2 Pt was compared with a higher Pt loaded standard chemically prepared electrode (0.5 mg/cm^2 Pt) on the cathode side of a MEA. Although the activity of the sputtered catalyst on CNFs was lower than that obtained with a standard cathode in terms of maximum achieved power density (300 mW/cm^2 versus 370 mW/cm^2), the platinum utilization efficiency in the oriented Pt/CNFs nanostructures based cathode was at least four times higher than that of the conventional one.

Mployedually lowtion efficiency ition of metal catalysts MWCNTs are tubular nanomaterials, which have attracted great interests in both theoretical and experimental studies. Due to their unique geometrical, electronic and mechanical properties, it is reported to have potential applications in field emission, superconductivity, hydrogen storages. Deposition onto the surface of MWCNTs can effectively improve catalytic activities of metal nanoparticles, which make them suitable for applications in heterogeneous catalysis and nanoeletrodes in the next generation of high performance fuel cells and batteries, etc. For example, in the work of Tang et al. [48], an integrated Pt/CNTs layer was prepared for PEMFCs electrodes by *in situ* growth of a dense CNT layer on carbon paper using a thermal chemical vapor deposition (CVD) technique followed by direct sputter-deposition of Pt nanodots onto the CNTs layer. Most of the sputter-deposited Pt nanodots were less than 4 nm in diameter and had a size distribution of 2–3 nm. A high maximum power density of 595 mW/cm^2 was observed for a low Pt loading of 0.04 mg/cm^2 at the cathode. However, due to the chemical inertness of CNTs, the adhesivity of the metal NPs to the surface of CNTs is weak [155], so the loading of catalysts on CNTs are usually low. Surface modification and functionalization are therefore employed to enhance the adhesivity of the metal NPs, such as the addition of radicals, nitrenes, or carbenes, supramolecular complexation with detergents, proteins, or polymers [156-159]. These modifications of carbon supports could effectively increase the surface binding sites of carbon nanotubes, avoid the aggregation of metal nanoparticles, improve the dispersion of metal nanoparticles, and reduce the average size of metal nanoparticles deposited. Nevertheless, it is inevitably accompanied with some other problems, such as uneven distribution of the surface functional groups, structural damage, and blockage of the direct touch between metal nanoparticles and MWCNTs, and thus partial loss in electrical conductivity of the carbon supports and hindrance of electron migration from metal to nanoparticles [160,161]. Furthermore, in those methods, large amounts of chemicals were used, which made it easier to cause environmental pollution. In order to avoid the use of toxic solvents and/or extreme conditions and minimize the above disadvantages during the preparation, it is highly desirable to develop a mild surface functionalization process to introduce high density and homogeneous surface functional groups but cause less structural damage to the carbon nanomaterials (and thus retain good electrical conductivity). Various dry processes including both nonreactive and reactive plasmas [162-168] and low-energy ion beam bombardment in a vacuum [169] have been found as good candidates. Compared to wet approaches, dry vacuum processing may be easier to control, with relatively less contamination. Work by our group modified the surface of CNT with an after glow plasma method where N_2 was used [170]. The results show that plasma surface modification treatment does not cause structural damage of carbon nanotubes, but provides effective functional groups on the surface of MWCNTs for the deposition of Pt nanoparticles. The Pt nanoparticles deposited on these plasma processed MWCNTs had a much higher catalytic activity even with low Pt nanoparticles loadings. The surface structures of carbon nanotubes and the catalyst/CNTs interactions may play important roles in determining the performance of the catalyst. For comparison, a strong acid oxidation method readily causes structural damage of MWCNTs. The Pt nanoparticles deposited on the stong acid treated MWCNTs showed a much lower performance of Pt-MWCNTs in methanol oxidation reaction, due to the decrease in the conductivity of the MWCNTs caused by structural damage. In addition, it was found that an insertion of impurities between Pt nanoparticles and MWCNTs would cause a decrease in electron migration from the metal to MWCNTs and gave rise to the decrease of electrochemical performance of Pt-MWCNTs in methanol oxidation reaction [170]. Sun et al. [45] deposited Pt nanoparticles on nitrogen-containing MWCNTs (N-MWCNTs) for micro direct methanol fuel cells (μDMFC) application. The N-MWCNTs were grown on substrate through a microwave plasma enhanced chemical vapor deposition using CH_4, N_2 and H_2 gases and subsequent Pt deposition was done by a direct current (DC) sputtering technique. Some well-separated Pt nanoparticles with an average diameter of 2 nm were formed on the sidewalls of N-MWCNTs (a transmission electron microscopy (TEM) image of Pt/N-MWCNT was shown in Figure 17). In this work, the nitrogen incorporation in the MWCNTs may play a critical role in the self-limited growth of the Pt nanoparticles. CV results showed that Pt/N-MWCNTs catalyst had electrochemical activity towards methanol oxidation and thus promising for a future μDMFC device.

Carbon nanocolumns and carbon nanorods are a new kind of carbon nanostructures with a high slenderness ratio. They have potential use as dielectrics in metal–insulator–semiconductor (MIS) devices and field emission planes. Carbon layer with carbon nanocolumns or carbon nanorods is uniform and has a columnar, pillar shape. Due to their large layer porosity, the reactants and products of the fuel cells can easily penetrate through the nanocoloums which may decrease the mass transfer resistance. When they are used as catalysts support in sputtering process, the catalyst sputtered can penetrate into the pores of the nanocolumns with a sufficient depth. Rabat et al. [18,53] used a transformer coupled plasma (TCP) type PABS devices to prepare carbon nanocolumns and then sputtered Pt on them. TEM analysis showed that platinum could diffuse into the nanocolumns layer and formed nano-sized particles (mean diameter ca. 3 nm) along and around the carbon nanocolumns and down to the film/support interface. The fabricated MEAs using this carbon nanocolumns supported Pt electrodes exhibited the increased open circuit voltage (OCV) with 12% higher than using commercial MEAs, although it might not be reasonable to say that the new catalyst was more efficiency than that of the commercial one. Further electrochemical measurements and fuel cell performance tests should be taken to support the authors' conclusion. The use of carbon nanorods as catalyst support was done by Gasda et al. [54], where they built PEMFCs cathodes by growing carbon nanorods by glancing-angle deposition (GLAD) technique onto silicon wafer substrates, coating the nanorods with 0.10 mg/cm^2 Pt by a MSP method, and transferring them from the Si wafer to PEMs. The rods were etched within fully assembled cells by applying a potential above the reversible H_2/O_2 voltage, which led to polarization curves that showed a 4–7 times higher current at 0.40 V. The current increase was attributed to the opening of pores within the electrode, which facilitated easy oxygen transport and led to a reduction in mass transport resistance by a factor of 360. Etching sequences with increasing voltage V_E indicated that $V_E \leq 1.6$ V led to water electrolysis and Pt oxidation that facilitated Pt agglomeration and migration of Pt ions into the electrolyte, while $V_E = 1.7$ V resulted in removal of C and the formation of pores within rods that facilitated oxygen transport to reaction sites, yielding a 400–700% increase in fuel cell output current at low potential. This study suggests that the controlled etching of temporary scaffolds to create pores in an operating fuel cell might be an effective approach to reduce mass transport limitations.

Columnar titanium structures have the common characteristics of Ti, with high electronic conductivity and acidic stability. Also, due to their high porosity, the Pt sputtered onto it may have greater ECSA than a flat Pt surface. They are stable during a wide range of processing, annealing and operating conditions which are preferable for a catalyst support. Bonakdarpour et al. [60] fabricated columnar titanium structures on smooth glassy carbon disks using the GLAD technique and then platinum films ranging from 10 to 90 nm in thickness were deposited onto these posts by MSP technique. The ORR activity of these electrocatalysts was measured by the rotating ring-disk electrode (RRDE) method in O_2-saturated 0.1 M $HClO_4$ at room temperature. The ECSA of the catalysts was about 10–15 times higher than smooth Pt films.

Titanium oxide (TiO_2) nanotube has attracted significant interest because of its advantages such as large specific surface, unique chemical and physical properties, stability in PEMFC operation atmosphere, low cost, and commercial availability. Its specific geometry can improve the dispersion of catalysts, catalyst loading used, and easy permeation of reactant gas into the highly porous structure, leading to the enhancement of the catalytic activity. Ordered arrays of TiO_2 nanotubes can be served as support for oxygen reduction catalysts with controlled sites to enhance the rate of the electrochemical reaction.

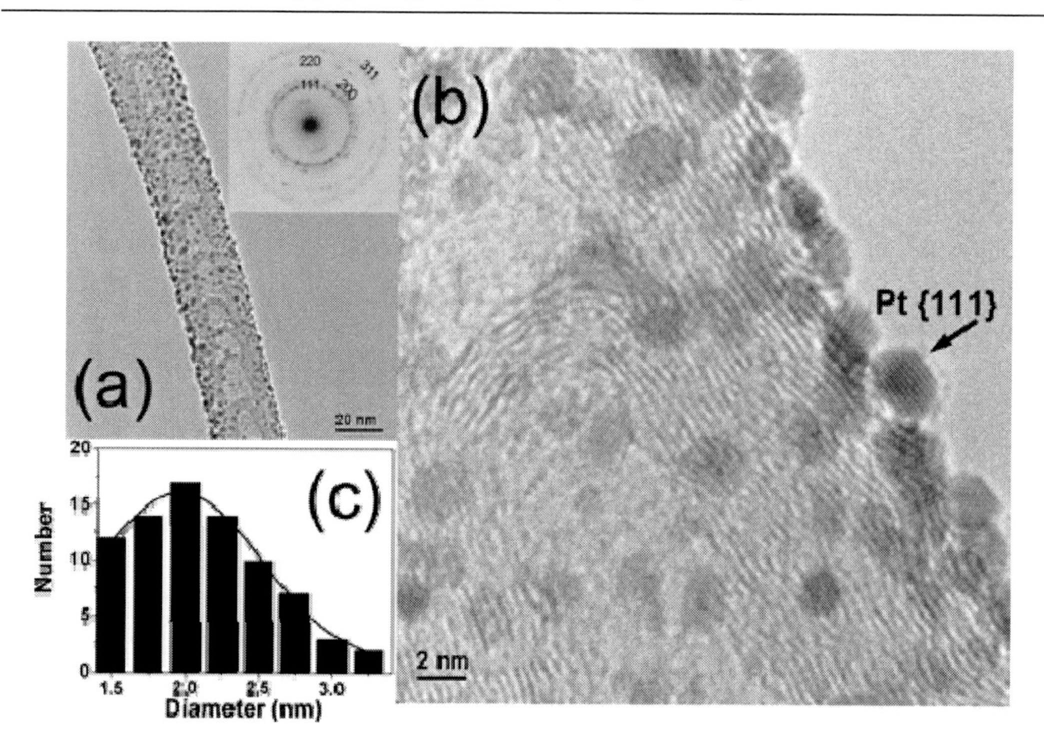

Figure 17. Transmission electron microscopy (TEM) image of Pt nanoparticles on the side wall of one single N-MWCNT with the corresponding selected area electron diffraction (SAED) pattern in the inset (a), high-resolution transmission electron microscopy (HRTEM) image of Pt nanoparticles on the side wall of one single N-MWCNT (b) and a size histogram of Pt nanoparticles estimated from (b) (c) [45]. Reproduced with permission from Sun, C. L. et al. *Chem. Mater.* 2005, *17,* 3749-3753. Copyright © 2005, American Chemical Society.

With control of the site disposition on the arrays, one may expect to lower the amount of catalyst needed for efficient performance and to facilitate design of MEAs that have effective catalyst locations. This would have a major impact on fuel cell applications in renewable energy. Robust titanium oxide nanotube (TONT) arrays were formed by electrochemical anodization on various sample shapes (rod, screw, and foil) [22]. A Pt layer was deposited by DC sputtering and evaporation. They found that the sputtered Pt was more conformal to the TONT surface morphology than the evaporated one, and this preserved the surface area increased by TONT formation. The CV obtained from the sputtered Pt/TONT showed a somewhat higher potential value for initiating ORR than for the evaporated Pt/TONT and the authors attributed this to better electrochemical catalytic activity enhanced by the high surface area of the Pt layer. Pt sputtered on TONT arrays also showed high performance and durability in the performed single cell tests.

TiNT is a wide bandgap semiconducting material, hence it has many important applications to photocatalysts, gas sensors, and photovoltaic cells. The combination of its intrinsic bulk properties with the nano-tubular structures is expected to lead to achieving higher performance in PEMFCs. The nanotubes array allows for the precise design and control of the geometrical features, allowing one to achieve a material with specific light absorption and propagation characterization with the aligned porosity, crystallinity, and oriented nature of the nanotubes arrays making them attractive electron percolation pathways for vectorial charge transfer between interfaces. Hassan et al. [19] prepared TiNTs by electrochemical anodization method. Then Pt particles were sputtered and deposited on the as-anodized samples by using PSD method. The prepared TiNT/Pt/C electrode was used as anodes for DMFCs. The catalytic activities of TiNT/Pt/C catalyst for MOR were enhanced due to the large surface area of a low Pt loading on the TiNTs support. The CO stripping led to the increase in the current peak of methanol oxidation and this was due to the activation of the catalyst surface by point defect formation. The ratio of forward anodic peak current to the reverse anodic peak current was higher and this indicated more effective removal of poisonous species.

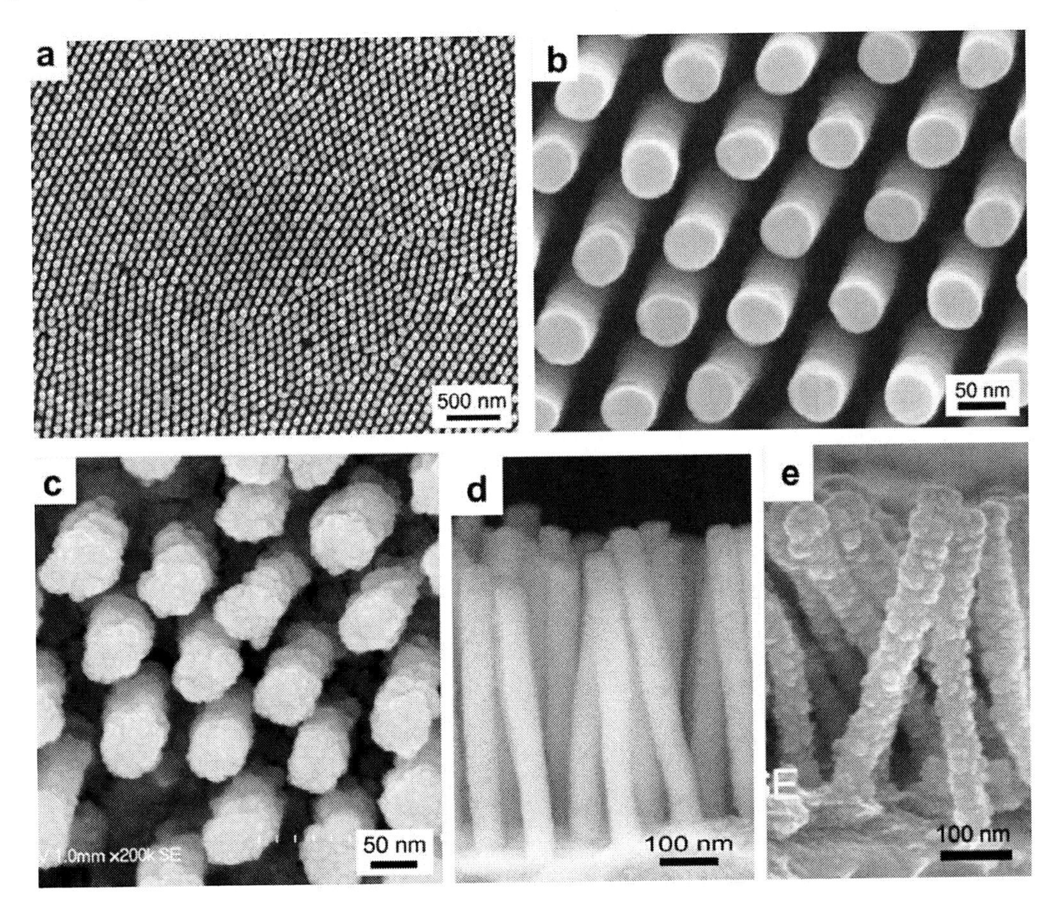

Figure 18. SEM micrographs of surface of Pd NWAs (a) and (b), surface of Pd/Pt core–shell NWAs (c), cross-section of Pd NWAs (d) and cross-section of Pd/Pt core–shell NWAs (e) [63]. Reproduced with permission from Wang, H. et al. *Electrochem. Commun.* 2008, *10,* 1575-1578. Copyright © 2008, Elsevier Science Ltd.

In order to achieve the high catalytic activity and utilization efficiency of Pt electrocatalysts, core-shell nanostructures with high surface-to-volume ratio by producing Pt nanosphere shell have been widely investigated. MSP techniques combined with other technology can also be used to prepare core-shell nanostructure electrocatalysts. Wang et al. [63] prepared a new highly ordered Pd/Pt core–shell nanowire arrays (NWAs) (Pd/Pt NWAs) by an anodized aluminum oxide (AAO) template-electrodeposition method in combination of MSP techniques.

Figure 19. Plan-view SEM micrographs showing GDL substrates that were successively coated with 0.29 mg/cm^2 chromium nitride (CrN) and 0.05 mg/cm^2 Pt (a) and 0.25 mg/cm^2 Pt (b) [64]. Reproduced with permission from Gasda, M. D. et al. *J. Electrochem. Soc.* 2010, *157,* B71-B76. Copyright © 2010, The Electrochemical Society.

In this work, Pd core nanowire was firstly fabricated using AAO template-electrodeposition method, and Pt with loading of 0.03 mg/cm^2 was then sputtered onto the Pd core nanowire. The SEM images of Pd NWAs and Pd/Pt core–shell NWAs are shown in Figure 18. The mass specific anodic peak current density was 756.7 mA mg^{-1} Pt for the methanol oxidation on the Pd/Pt NWA electrode, substantially higher than the conventional E-TEK PtRu/C electrocatalysts. The results showed that the significant promotion effect of ordered Pd NWA core to the electrooxidation of methanol was due to the oxophilic nature of Pd as compared to Pt and the high surface-to-volume ratio and excellent current collector of the Pd NWA support.

Many compounds have been demonstrated to have electrocatalytic activities for PEMFC electrodes. These compounds can also be used as catalyst supports for Pt electrocatalysts. CrN is an attractive candidate to supplement or replace Pt in lower cost PEMFC electrodes due to its well-known wear and corrosion resistance. Gasda et al. [64] prepared vertically oriented CrN nanoparticles onto GDLs using GLAD and then sputtered Pt onto this CrN nanostructures, as shown in Figure 19. The reason that they chose CrN was they expected the hydrophobic CrN-supported two-phase electrode could change the proton conductivity of the catalyst layer and the current density generated for a given electrode potential. However, water could not fill the entire two-phase catalyst layer for the Pt/CrN materials system. And they suggested that the design of thin film electrodes that contain hydrophobic materials should consider the proton conduction throughout the electrode by controlled water management.

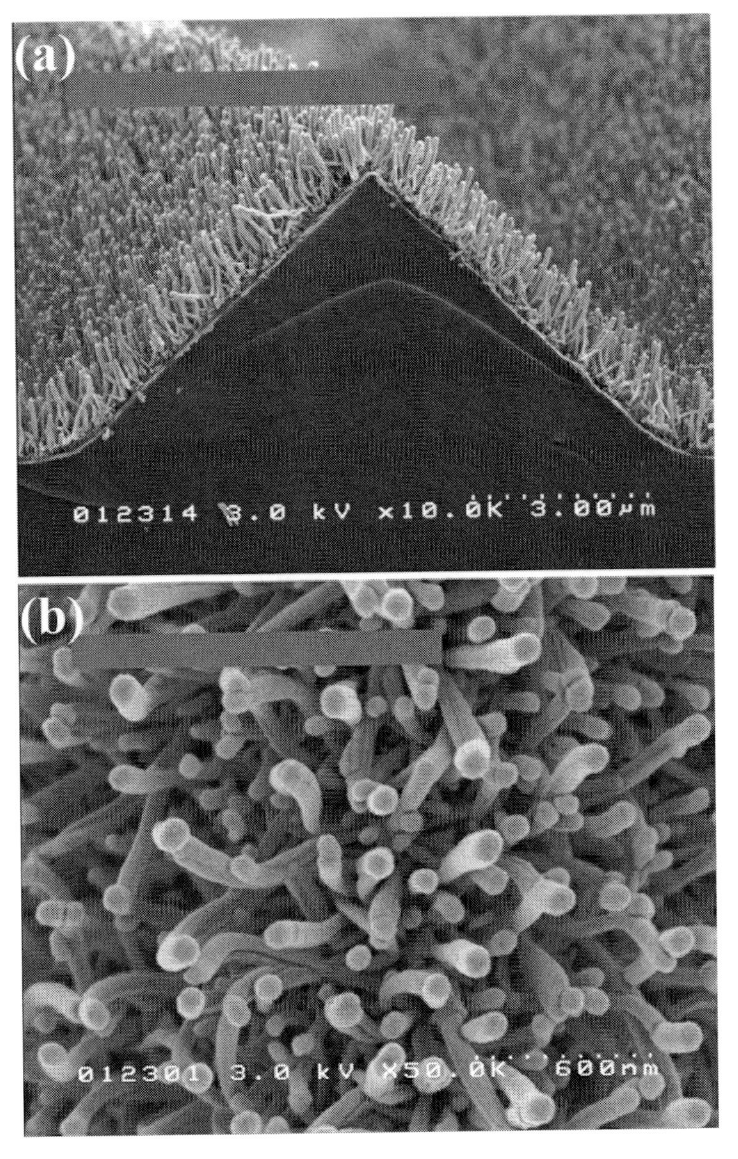

Figure 20. SEM images of typical NSTF catalysts as fabricated on a microstructured catalyst transfer substrate in cross-section with original magnification of ×10,000 (a) and in plan view with original magnification of ×50,000 (b). The dotted scale-bar is shown in each micrograph [65]. Reproduced with permission from Debe, M. K. et al. *J. Power Sources* 2006, *161,* 1002-1011. Copyright © 2006, Elsevier Science Ltd.

A new class of support material, nanostructured thin film (NSTF) for PEMFCs applications has been developed at 3M Inc. (3M Fuel Cell Components Program, 201-2N-19, 3M Center, St. Paul, MN 55144, USA) [65]. The NSTF support layer consists of a monolayer of oriented, crystalline, organic, whiskers structure that are obtained by thermal sublimation and the subsequent annealing of a red organic pigment. The catalysts are coated on top of the whiskers by PSD method. The SEM images of 3M NSTF catalyst are shown in Figure 20. The organic whiskers are highly inert thermally, chemically, and electrochemically. The thin film catalyst coating encapsulates the crystallized pigment whisker support particles, eliminating issues with oxidatively unstable supports. The thin film catalyst coatings consist of relatively large crystallite domains of nanoscopic particles, which give to the NSTF catalysts enhanced specific activity and resistance to loss of surface area by Pt dissolution. Most notable is the five-fold or greater gain in specific activity of the NSTF catalysts over high surface area dispersed Pt/carbon. However, one potential drawback of these supports is that the organic whiskers decompose above about 350 °C [60].

3.3.3. Sputtering of Pt onto Nafion Membrane

The sputtering deposition of thin film catalyst layers onto the surface of PEMs has been reported as a way to improve the mass-transportation and to obtain a more active catalyst layer. O'Hayre et al. [171] sputtered Pt directly onto Nafion 117 membrane and investigated the effect of Pt film thickness and morphology on the fuel cell performance. It showed that the fuel cell performance was extremely sensitive to sputtered platinum thickness and membrane texture. For the smooth Nafion 117 membranes, peak performance was achieved when a sputtered platinum film thickness of 5 nm was used. The increase or decrease in the thickness of platinum films would lead to the significant reduction of the fuel cell performances due to the changes in the film morphology. In addition, they also studied the effects of membrane surface roughening on the performance of fuel cell. However, the roughened membrane surfaces required increased platinum loading levels in order to achieve good performance and the improved performance did not exceed that of smooth Nafion membrane. In addition, the performance of a sputter-deposited membrane with a Pt loading of 0.04 mg/cm^2 based on MEA was compared to that of a commercial MEA with a Pt loading of 0.4 mg/cm^2. The maximum power output of the sputtered MEA was three-fifths that of the commercial MEA, but used one-tenth the platinum.

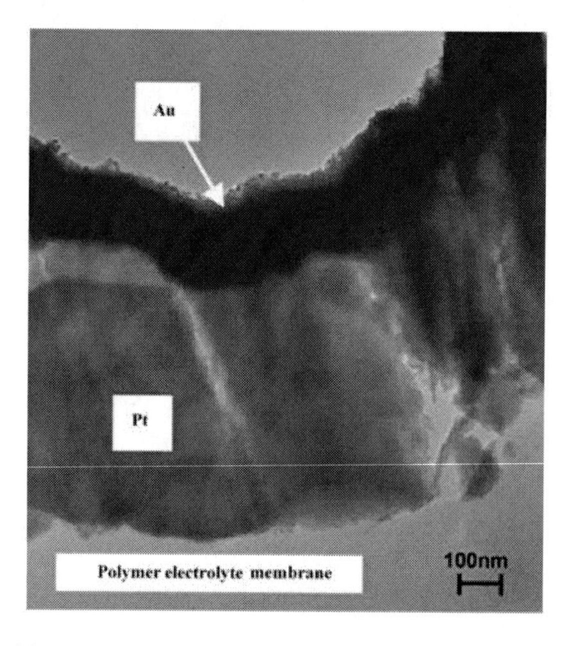

Figure 21. TEM image of the cross-section of a MEA with sputter-deposited and transferred catalyst layer. 600 nm thick Pt layer was deposited on the electrolyte membrane and 200 nm thick Au layer was deposited on it [172]. Reproduced with permission from Nakakubo, T. et al. *J. Electrochem. Soc.* 2005, *152,* A2316-A2322. Copyright © 2005, The Electrochemical Society.

Although the direct sputtering of Pt onto the membrane interface can improve the utilization efficiency of Pt catalyst, the sputtered catalyst layer cracks up when the layer gets thicker as the electrolyte membrane absorbs water in the air and swell after the vacuum process. In order to solve this problem, Nakakubo et al. [172] sputtered Pt and Au onto a PTFE blank sheet in low-pressure air with ionomer brushed on the sputtered layer and then transferred to a PEM by hot press. By brushing the electrolyte solution on the surface of the catalyst layer before the transferring process, the ionic connection between the PEM and catalyst layer was improved. By sputtering a gold layer before platinum deposition the electrical connection between the catalyst layer and the electrodes was improved. Sputtering gold was also effective to support the catalyst structure and to collect current. Figure 21 shows the TEM of Au, Pt and PEM structures. PEMFC prepared using this method exhibited higher performance than that of PEMFCs fabricated with other methods.

3.3.4. Co-sputtering of Pt and C, Other Metal, Metal Oxides and Metal Phosphate

Carbon is the most used catalyst support for PEMFC electrodes. Co-sputtering of Pt and C is a convenient way to fabricate Pt/C catalysts. Using this method, uniform platinum-cluster dispersed Pt/C composite catalyst film can be prepared and the agglomeration of the Pt nanoparticles can be avoided. The Pt nanoparticles embedded in the carbon matrix have a large surface-to-volume ratio and consequently a high utilization efficiency of Pt catalyst. Hajbolouri et al. [66,67] first investigated the application of co-sputtered Pt/C layers in PEMFCs electrodes. In that work, graphite with attached Pt wires or graphite with pasted Pt pellets on its top were used for the plasma co-sputtering targets. CV and X-ray absorption measurements showed that the obtained Pt/C layers were electrochemically active and the specific active surface area was comparable to the commercial Pt electrodes. Nechitaĭlov et al. [68] incorporated the co-sputtered Pt/C nanocomposites using a similar method onto polyaniline film for PEMFCs electrodes and found that these electrodes exhibited higher catalytic activity with a specific activity of 430 mW/mg Pt and 290 mW/mg Pt on the cathode and anode, respectively. In their following work, the effect of parameters such as porosity (density) and thickness of the Pt/C layer on the activity of catalyst and the specific power provided by this layer in PEMFCs were studied [70]. It showed that the introduction of platinum into an amorphous carbon matrix led to a substantial increase in the porosity of the structure and the porosity decreased as Pt/C grew. These Pt/C layers exhibited high catalytic activities in PEMFCs, which was sensitive to porosity and steeply grew due to the decrease in the diffusion resistance with increase of the layer porosity. According to these results, it was suggested the most efficient method for raising the specific power was to produce thicker layers with a porosity exceeding 60%. With aim to improve utilization of catalyst, Cavarroc and coworkers [71] made a slight modification to the preparation protocol. In their work, the deposition chamber equipped with two targets (one of pure carbon and one of $Pt_{0.01}Carbon_{0.99}$) and the two targets tilted 45° with respect to the substrate were used. They found that both the efficiency and the utilization of platinum were increased by this preparation method of catalyst. The higher performance of fuel cell was benefited from the small particle size co-deposition of Pt and C and localization of platinum close to the membrane surface and the low penetration depth of platinum in the electrode did not require the needs for brushing the electrodes with Nafion solution as usual.

The alloy of Pt with non-precious metals has been reported to enhance performance of catalysts. It has been reported that the ORR or the MOR rate are significantly enhanced when

Pt is alloyed with Ru, Sn, Mo, etc. Changes in short-range atomic order, particle size, Pt *d*-band vacancy, Pt skin effects, and Pt–OH inhibition are considered as some of the reasons that are attributed to the enhanced performance by these alloys [173]. The fabrication of Pt alloyed catalysts can be easily realized using PSD methods, and the composition of these Pt alloys can be controlled by varying the ratio of the sputtering powers for Pt and the second metal or more metals. The improved PEMFC performance of alloys was demonstrated by Caillard et al. [75] where they compared the performance of anodes prepared by Pt and Ru multilayer deposition with that prepared by simultaneous deposition of Pt and Ru (alloyed). The best efficiency of the DMFCs was reached when both metals Pt and Ru were simultaneously deposited (alloyed) with a Ru atomic ratio of 30% or 40%. The ORR activity of Pt alloys with various types of metals such as Ni, Co, and Fe was investigated by Toda et al. [72,73]. They found that the catalytic activity of alloys was dependent upon the relative contents of alloyed metals. The maximum activity was observed at ca. 30, 40, and 50% content of Ni, Co, and Fe, respectively. The XPS analysis of the surfaces after the reaction revealed that the active surfaces were covered by a few monolayers of Pt. They attributed the enhanced performance of ORR to an increased d-electron vacancy of the thin Pt surface layer caused by underlying alloy. These thin Pt skin layer covered alloy catalyst also exhibit a higher tolerance to poisonous CO gas, as demonstrated by Watanabe and coworkers [74]. They believed that the less leakage of noble metal such as Fe out of the alloys in acidic solutions resulted in the formatiom of a thin Pt skin layer, which gave rised to the catalyst with a modified electronic structure and exhibited high CO-tolerance towards H_2 oxidation.

Recently, a great of synthetic work have been devoted to the preparation of catalysts with multi-elements with expectation for possible synergistic effects [88]. For example, Wang et al. [49] used microwave plasma-enhanced CVD system to synthesize CNTs on carbon cloth (CC) using $CH_4/H_2/N_2$ as precursors. The anode electrocatalysts, Pt and Ru, were coated on CNTs by sputtering to form Pt-Ru/CNTs-CC. Pt-Ru electrocatalysts were uniformly dispersed on the CNTs, as indicated by HRSEM and TEM which are shown in Figure 22, because the nitrogen doped in the CNTs acted as active sites for capturing electrocatalysts. The Pt-Ru/CNTs-carbon electrode exhibited excellent electrochemical activity of methanol oxidation even at a low metal loading according to CV measurement. The high performance of the Pt-Ru/CNTs-CC anode was attributable to the smallness of the Pt-Ru particles, the good dispersion of electrocatalysts, the thin electrocatalyst layer, the highly conductive supports, and the low internal and interfacial resistances. Kang et al. [61,62] sputtered PtCo and PtNi catalysts on TiO_2 nanotube arrays for the ORR using the co-sputtering technique. ORR tests showed that the Pt alloy catalyst incorporating 30 atom% Co metal showed the maximum activity and there was also a remarkably reduced onset potential and dominantly enhanced activity in the kinetic control region, compared to that of Pt/TONT which was attributed to Pt electronic modification by an alloying effect and the shortening of the Pt–Pt interatomic distance. As to PtNi catalysts, the as-deposited PtNi on TONT had poor ORR activity with a significant over-potential, while the ORR activity of annealed PtNi/TONT was improved and the maximum ORR activity, with a significant increase (120 mV) of the half-wave potential, was achieved for 400 °C annealed PtNi/TONT. The authors attributed this to two factors: the formation of a PtNi alloy catalyst and a strong metal-support interaction effect. Umeda et al. [86] investigated the sputtered ternary Pt-Ru-W alloys for methanol oxidation. Compared to the binary alloys such as Pt-Ru and Pt-W, higher current densities and a noticeable cathodic shift in the onset potential for methanol oxidation were observed at

Pt–Ru–W electrode. The author of that work believed that Pt-Ru-W electrodes facilitated the generation of M-OH_{ad} (M=Ru, W) more than Pt–Ru or Pt–W and Au substrate electronically affected the Pt–Ru–W surface in terms of M–OH_{ad} creation. Based on Umeda's work, the characteristics of ethanol electrooxidation on Pt-Ru-W alloys catalyst was investigated by Tanaka et al. [87]. The performance of Pt–Ru–W was desirable in comparison to that of binary alloys, such as Pt–W, Pt–Sn and Pt–Ru.

Pt/oxides nanocomposites for use in methanol electro-oxidation have been intensively studied due to the interesting role of metal oxides and their possible use as an electrode in thin-film fuel cells. For example, the use of WO_3, TiO_2, Ta_2O_5, CuO, CeO_2, RuO_2 and so on was shown to be effective in improving the catalytic activities for methanol electro-oxidation. Co-sputtering of Pt and metal oxides is a useful way to fabricate Pt-metal oxides nanocomposite structures. Using this method, the agglomeration of the Pt nanoparticles can also be avoided. The Pt nanoparticles embedded in the metal oxides matrix have a large surface-to-volume ratio and thus exhibit a high utilization efficiency of Pt catalyst. Park et al. [91] first fabricated this kind of nanocomposite (Pt-WO_x) using a co-sputtering method. The coexistence of a polycrystalline Pt nanosized phase and an amorphous tungsten oxide phase in the electrode layer was confirmed. Synergy effects such as the spill-over effect caused by porous tungsten oxides made this co-sputtered two-phase electrode excellent performance in comparison to the Pt one-phase electrode, as shown in Figure 23. Encouraged by this work, the fabrication of PtRu-WO_x electrode was also done by the same group [92-94]. It showed that PtRu–WO_x nanostructured alloy electrode had the best performance compared with a PtRu thin-film or a Pt one-phase electrode. The enhanced catalytic activity in the nanocomposite electrode was closely related to proton transfer in tungsten oxide. Park et al. [95] investigated the use of Pt-RuO_2, Pt-TiO_2, Pt-Ta_2O_5 Pt-CuO [98] and Pt-CeO_2 nanocomposite as electrode. The electrodes made of these composite catalysts exhibited an enhanced performance for methanol oxidation than the Pt only thin film electrode.

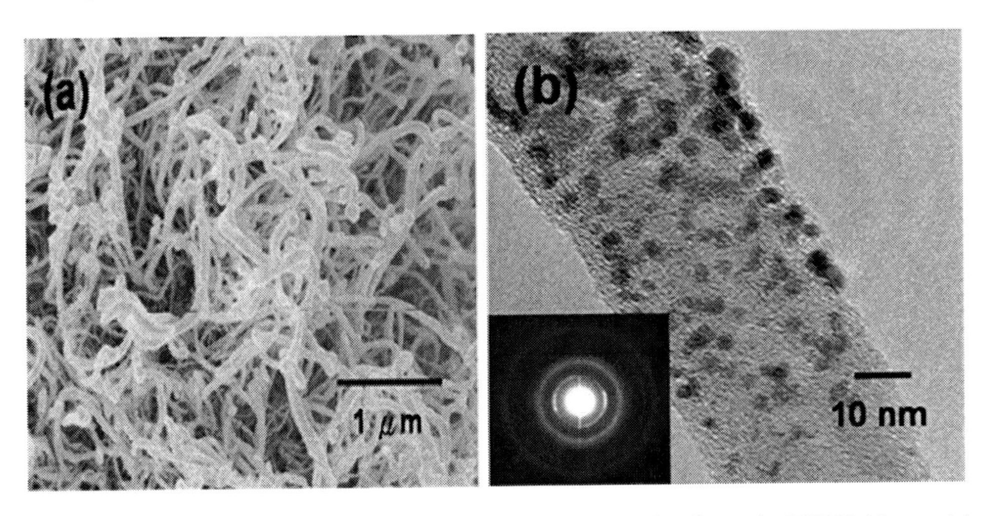

Figure 22. Pt-Ru electrocatalysts on the CNTs-carbon cloth composite electrode: HRSEM image (a) and TEM image with a selected area electron diffraction (SAED) pattern in the inset (b) [49]. Reproduced with permission from Wang, C. H. et al. *J. Power Sources* 2007, *171,* 55-62. Copyright © 2007, Elsevier Science Ltd.

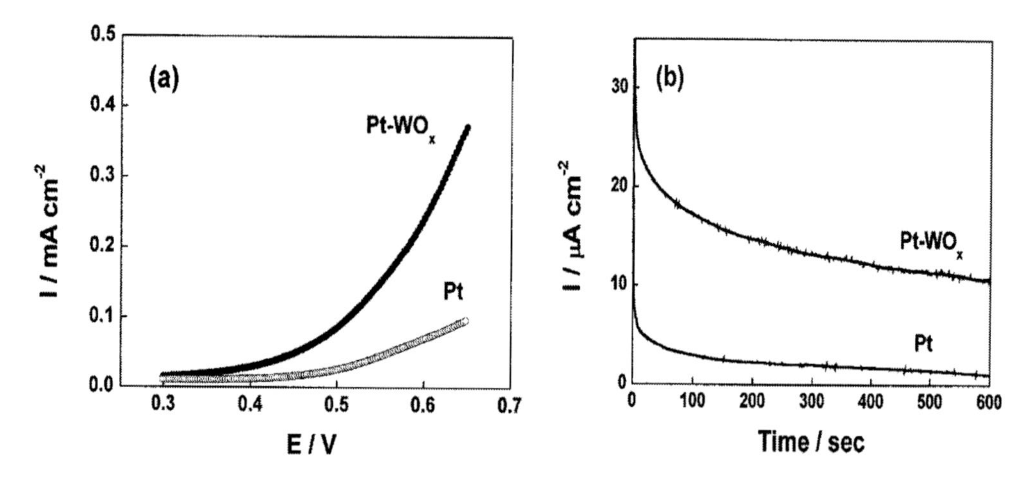

Figure 23. Current density accelerating potentials (a) and current density vs time at the oxidation potential for the Pt–WO$_x$ two-phase and the Pt one-phase electrode (b) [91]. Reproduced with permission from Park, K. W. et al. *Appl. Phys. Lett.* 2000, *81,* 907-909. Copyright © 2007, American Institute of Physics.

Metal phosphates are of great interest for catalytic applications due to their high proton conductivity, acid stability, open framework structures and catalytic activity and so on. Lee et al. [102] fabricated nanostructured Pt-$FePO_4$ thin film electrode. The obtained Pt-$FePO_4$ thin film electrode showed higher performance than the single Pt thin film electrode and they attributed this to an increase in the ECSA, the possible ability of $FePO_4$ for the effective transfer of protons and the CO oxidation during methanol oxidation. All these above Pt-metal oxide/metal phosphate thin film electrodes showed enhanced MOR activity than the Pt only thin film electrode. The catalytic activity improvements of Pt-metal oxide/metal phosphate thin film electrodes can be attributed to physical effects (the increase in surface area) and/or catalytic effects (such as spill-over effects) of the oxide matrix/phosphate. However, all of these nanocomposites were just sputtered onto indium-tin oxide (ITO) glasses for testing catalytic activity and were not used in the real thin film fuel cells.

3.3.5. Multi-layered Catalysts

As most of sputtered films deposited onto GDLs have poor surface area when the Pt loading increases, many investigations focus on the development of the multi-layer to improve the performance of fuel cells. In these studies, a Nafion-carbon ink layers were often used as the sputtered Pt catalyst supports. For example, Cha and Lee [174] first used PSD techniques to fabricate multi-layer catalysts. They obtained the highest utilization of Pt catalyst by alternating layers of sputtered Pt (5 nm thick) and a painted mixed electron and proton-conducting layer of carbon black (XC-72) particles in an ink of the Nafion monomer. It was found that repeating successive applications of these layers up to five times continued to improve the V-I characteristics of the PEMFCs. PEMFCs made in this way achieved ~90% of the power density available in a conventionally manufactured H_2-air PEMFC with ~10% of the Pt catalyst. Huag et al. [175] suggested that the cell performance could be further improved by reducing the amount of Pt on each layer and the thickness of each Nafion-carbon ink (NCI) layer to monolayer thickness. They investigated the sputter-depositing alternating layers of Pt and NCI onto the membranes. MEAs built from multilayered sputter-deposited Pt and spray-deposited NCI layers exhibited improved performance over single-layer, sputter-deposited MEAs of equivalent or greater Pt loadings. Based on their experimental results, the optimal performance achieved from the MEA containing six-layer Pt + dilute NCI anode and cathode was 0.17 A/cm^2 at 0.6 V, and using the optimal MEA structures, catalyst activities of greater than 2650 A/g at 0.6 V and 5500 A/g at 0.4 V were achieved.

Actually, the cell performance is correlated with the sputtering process parameters, especially the bias voltage, operating pressure and sputtering time. For example, Wan et al. [176] investigated the sputtering parameters on the performance of MEAs with multi-layered catalysts. The results indicated that three layers of Pt sputter-deposited on the GDLs (the structure of which is shown in Figure 24) provided better performance (324.4 mA/cm^2 at 0.6 V) than sputtering single Pt layer in the same loading. The reason is that the obtained performance exhibited better than that of similar catalyst layer structure done by Huag et al. [175] was that an appropriate amount of the Nafion solution using the brushing method to each sputtered Pt layer was used, which produced the completely continuous three-phase interface.

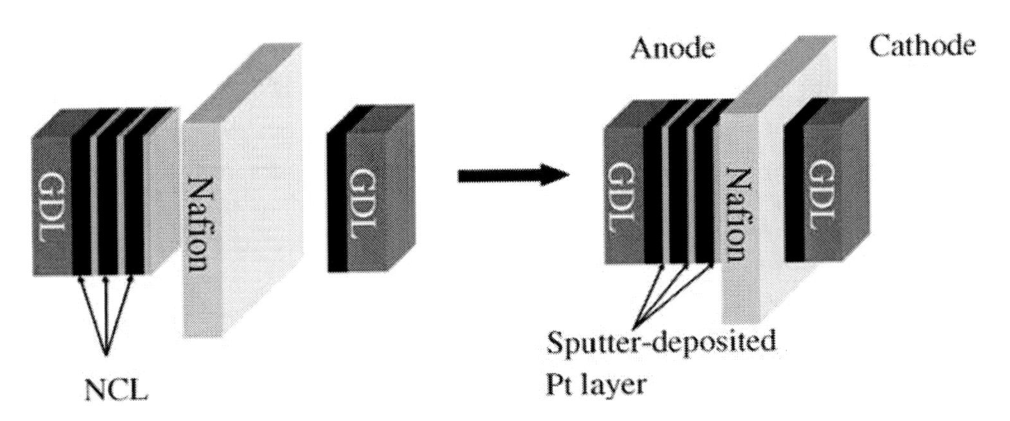

Figure 24. Diagrams for the proposed catalyst layer structure (anode) and MEA [176]. Reproduced with permission from Wan, C. H. et al. *Surf. Coat. Technol.* 2006, *201,* 214-222. Copyright © 2006, Elsevier Science Ltd.

The placement of a layer of Ru catalyst before the Pt electrode can be used as a filter layer to increase the CO tolerance ability of anodes. As demonstrated by Haug et al. [177], MEAs containing Pt + Ru filter anodes showed increased CO tolerance compared to a Pt:Ru alloy containing similar amounts of Pt and Ru, as shown in Figure 25. Among the Ru filtered anodes, the MEA with a single sputter-deposited Ru filter exhibited greater CO tolerance than the two and the three-layered sputter-deposited Ru filter. The author of that work suggested that the anode configuration consisting of a sputter-deposited Ru filter placed in front of and adjacent to a Pt:Ru alloy would provide optimal tolerance to CO. Based on Haug's work, Wan et al. [178] investigated some other novel composite anode with CO filter layers for PEMFCs. In that work, the anode electrode structures was designed to make the poisonous CO react in a separate layer with a CO active electrocatalyst (Pt-Ru alloy) in advance, and then make the main hydrogen to react at another layer using a platinum electrocatalyst. Therefore, all of these anode layers offered superior CO-tolerance capability to that of conventional and Huag's structures.

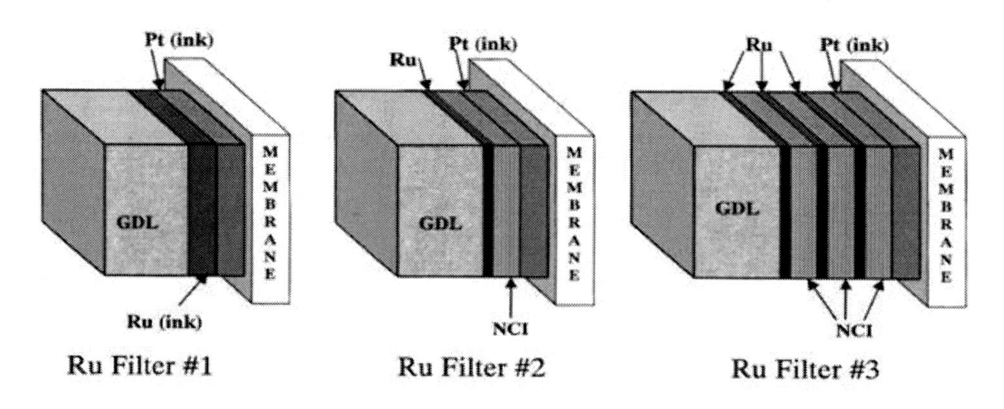

Figure 25. Diagram of the Ru filtered anode. The filter is a layer of Ru catalyst placed between the Pt catalyst and the gas-diffusion layer for oxidizing CO present within the anode. Filter 1 is the standard Ru filter prepared from a catalyst ink. Filter 2 is a single sputter-deposited layer of Ru separated from the Pt electrode by a layer of NCI. Filter 3 is a three-layer sputter-deposited Ru filter separated from the Pt electrode by a layer of NCI. All filters have a loading of 0.08 mg/cm^2 Ru and were placed a top a Pt electrode with a loading of 0.15 mg/cm^2 Pt [177]. Reproduced with permission from Haug, A. T. et al. *J. Electrochem. Soc.* 2002, *149,* A868-A872. Copyright © 2002, The Electrochemical Society.

3.3.6. Non Pt Catalysts

As well known, the high cost, limited world's supply of Pt and the slow ORR and MOR kinetics on Pt catalysts are the major problems inhibiting the widespread commercialization of PEMFCs and the improvement of their energy conversion efficiency, which have stimulated enormous research interests in the search for inexpensive, more efficient electrocatalysts as alternative materials for Pt. Many kinds of non-platinum cathode and anode electrocatalysts have been investigated for PEMFCs application using the MSP technology. Savadogo's group [179] first fabricated Pd and Pd alloy catalysts for the ORR using a MSP techinique. The preliminary results showed that Pd–Co 28% alloy may exhibit better performances for the ORR than Pt. Palladium and palladium-alloy cathode catalysts also showed a high degree of tolerance to ethanol during the ORR in an acid medium [180]. They also studied the ORR on Pd–Cu catalyst in acidic media. The ORR on this catalyst proceeded through four transferred electrons mechanism and the $Pd_{50}Cu_{50}$ exhibited the highest activity for the ORR [181]. Lee et al. [182] studied the Pd-based alloys such as Pd-Co, Ni and Cr to develop a novel methanol-tolerant oxygen reduction electrocatalyst for DMFCs. It showed that the Pd alloys had no electrocatalytic activity for methanol oxidation in the presence of methanol although they showed a higher ORR electrocatalytic activity than Pd. They attributed this to the filling of the Pd d-band by alloying which decreased the density of states at the Fermi level and inhibited the formation of Pd oxide on the surface of the electrocatalyst. Ye et al. [183] sputtered Pd nanoparticles onto MWCNTs and used this catalyst for oxygen reduction. For the Pd–MWCNTs deposited carbon electrode, oxygen reduction shifted to close to 0.5 V, and the peak current varied linearly with the square root of the scan rate, which indicated that the ORR at Pd–MWCNTs carbon electrode was a diffusion-controlled process.

Transition metal oxides (oxynitride) catalysts have also attracted some attention due to their superior electrochemical properties and high stabilities. The use of transition metal oxides such as ZrO_{2-x} films, Co_3O_{4-x}, TiO_{2-x}, tantalum (oxy) nitrides (TaO_xN_y), zirconium oxynitride (ZrO_xN_y), etc., have been investigated as new cathodes for PEMFCs without platinum by Ota et al. [184-188] However, their electrocatalytic activities for the ORR were not very high. The factor which affected the catalytic activity has not yet been clarified. For the ORR applications, these catalysts still need to be improved.

Transition metal-carbon-nitrogen (TM−C−N, TM = Fe, V, Cr, Mn, Co, and Ni) compounds are another kind of potential non-noble metal catalyst for the ORR. The electrocatalytic activity of sputtered TM−C−N films which were prepared by combinatorial magnetron sputtering techniques for the ORR in acid and alkaline electrolytes were studied by Dahn's group [189-194] and Lin et al. [195]. The choice of TM and the heat-treatment temperature played important roles in determining the activity of the sputtered $TM_xC_{1-x-y}N_y$ films for ORR in acid and alkaline electrolytes. Since hexagonal boron–nitride is isostructural and isoelectronic with graphite, Garsuch et al. [190] modified the Co−C−N and Fe−C−N thin film catalysts by controlled doping with boron in an attempt to increase the nitrogen content and hence the catalytic activity. Corresponding novel thin film catalysts Co−C−N−B and Fe−C−N−B were synthesized by combinatorial magnetron sputter deposition in an Ar/N_2 gas mixture followed by subsequent heat-treatment between 700 and 1000 °C in an argon atmosphere. The nitrogen content of the as-prepared thin film catalysts could be increased by the addition of boron. Furthermore, the amount of remaining nitrogen in heat-treated catalyst samples was significantly higher in case of boron containing samples. The catalytic activity was found to decrease with the boron content in the thin film catalysts even though the N-content increased. Kim et al. [196,197] prepared Ta-N-C and Cr−N−C electrocatalysts using reactive sputtering with heat treatment method. The catalytic activity for the ORR of the sputtered Ta−C−N and Cr−N−C increased with the increasing heat treatment temperature. In particular, the current of the ORR on the Ta−C−N with the heat treatment temperature of 800 °C was observed at about 0.73 V vs. RHE. It was found that the crystallinity of the Ta−C−N would affect the catalytic activity for the ORR. Lin et al. [195] fabricated Ti–Cu binary films for the oxygen reduction using co-sputtering method. The activity for the ORR increased with increasing the Cu-content in the Ti-Cu films and it reached to a maximum with the copper composition up to 90 atom%. In the cyclic voltammograms, the strong broad peak should have arisen from the oxidation of Cu to Cu^+ and Cu^{2+} ions. This oxidation indicated that the Ti−Cu films were unstable and the Cu-component was susceptible to dissolution in 0.5 M H_2SO_4. This dissolution caused a loss of catalytic activity in the films. Preparing the Ti−Cu films enriched in Ti will stabilize these films to prevent the Cu-dissolution.

CONCLUSION

As shown above, PEMs modified or synthesized by the plasma technique usually possess superior properties in comparison to those obtained by other techniques and commercially available Nafion®. Suppressed MeOH permeability can be achieved while maintaining the proton conductivity at a satisfactory level by adjusting the plasma operating conditions and depositing thin barrier films on Nafion membranes in low-pressure plasma excited by microwaves and/or radio frequency. Roughening the surface with plasma surface modification can increase the maximum power density of a single cell. This is attributed to the larger contact area between the membrane, catalyst and reactants. These plasma polymerized membranes are dense and highly cross-linked with uniform structure and rich with proton exchange groups, which increase their lifetime and improve the power density, energy conversion efficiency, and fuel utilization when used in fuel cells. Due to their high proportion of proton exchange groups and highly cross-linked structure, the obtained plasma membranes have higher thermal and chemical stabilities, exhibiting higher proton conductivity and a lower methanol permeability than conventional polymerized membranes. However, plasma polymerization is a complicated process during which polymer formation and degradation of monomers occur simultaneously. Any variation in experimental conditions, such as electron density, input power, monomer flow rates and ratio of the partial pressure of precursors will greatly influence the properties of the resulting membranes. Therefore, an optimization process is usually required to establish a condition for the preparation of optimal PEMs to improve their performance in the application of DMFCs.

Modifying the surface of cathodes and anodes with metal and metal oxide films both enhance the cell performance of PEMFCs. PSD methods are shown to be able to reduce the catalyst loading and increase the catalyst utilization efficiency of electrodes for PEMFCs. The electrode fabricated by the PSD method usually has higher tolerance to poisnous gases (such as CO and methanol), easy proton or oxygen adsorption/desorption and fast kinetics of ORR and therefore exhibit improved performance in PEMFC applications due to strong interaction between catalysts and supports, the uniform dispersion of catalysts, the porous structure of catalyst layers and possible synergistic effects of multi-elemental catalysts, etc.

In conclusion, plasma technique is a promising approach with great potential in improving the performance of key materials for PEMFCs. It is expected that plasma technique can address more problems associated with PEMFCs and promote the commercialization and large-scale industrialized production of PEMFCs.

REFERENCES

[1] Hobson, L. J.; Ozu, H.; Yamaguchi, M.; Hayase, S. *J. Electrochem. Soc.* 2001, *148*, A1185-A1190.

[2] Choi, W. C.; Kim, J. D.; Woo, S. I. *J. Power Sources* 2001, *96*, 411-414.

[3] Yoon, S. R.; Hwang, G. H.; Cho, W. I.; Oh, I. H.; Hong, S. A.; Ha, H. Y. *J. Power Sources* 2002, *106*, 215-223.

[4] Walker, M.; Baumgartner, K. M.; Feichtinger, J.; Kaiser, M.; Rauchle, E.; Kerres, J. *Surf. Coat. Technol.* 1999, *116*, 996-1000.

[5] Walker, M.; Baumgartner, K. M.; Kaiser, M.; Kerres, J.; Ullrich, A.; Rauchle, E. *J. Appl. Polym. Sci.* 1999, *74*, 67-73.

[6] Ogumi, Z.; Uchimoto, Y.; Tsujikawa, M.; Takehara, Z.; Foulkes, F. R. *J. Electrochem. Soc.* 1990, *137*, 1430-1435.

[7] Ogumi, Z.; Uchimoto, Y.; Tsujikawa, M.; Yasuda, K.; Takehara, Z. *Bull. Chem. Soc. Jpn.* 1990, *63*, 2150-2153.

[8] Ogumi, Z.; Uchimoto, Y.; Tsujikawa, M.; Yasuda, K.; Takehara, Z. I. *J. Membr. Sci.* 1990, *54*, 163-174.

[9] Yasuda, K.; Yoshida, T.; Uchimoto, Y.; Ogumi, Z.; Takehara, Z. *Chem. Lett.* 1992, 2013-2016.

[10] Ticianelli, E. A.; Beery, J. G.; Srinivasan, S. *J. Appl. Electrochem.* 1991, *21*, 597-605.

[11] Mukerjee, S.; Srinivasan, S.; Appleby, A. J. *Electrochim Acta* 1993, *38*, 1661-1669.

[12] Xiu, Y. K.; Nakagawa, N. *J. Electrochem. Soc.* 2004, *151*, A1483-A1488.

[13] Lee, J. K.; Choi, J.; Kang, S. J.; Lee, J. M.; Tak, Y.; Lee, J. *Electrochim Acta* 2007, *52*, 2272-2276.

[14] Tang, Z.; Poh, C. K.; Chin, K. C.; Chua, D.; Lin, J. Y.; Wee, A. *J. Appl. Electrochem.* 2009, *39*, 1821-1826.

[15] Mehta, V.; Cooper, J. S. *J. Power Sources* 2003, *114*, 32-53.

[16] Elabd, Y. A.; Napadensky, E.; Sloan, J. M.; Crawford, D. M.; Walker, C. W. *J. Membr. Sci.* 2003, *217*, 227-242.

[17] Carretta, N.; Tricoli, V.; Picchioni, F. *J. Membr. Sci.* 2000, *166*, 189-197.

[18] Rabat, H.; Andreazza, C.; Brault, P.; Caillard, A.; Beguin, F.; Charles, C.; Boswell, R. *Carbon* 2009, *47*, 209-214.

[19] Hassan, F. M. B.; Nanjo, H.; Venkatachalam, S.; Kanakubo, M.; Ebina, T. *J. Power Sources* 2010, *195*, 5889-5895.

[20] Jiang, Z. Q.; Jiang, Z.; Yu, X. Y.; Meng, Y. D. *Plasma Processes Polym.* 2010, *7*, 382-389.

[21] Ennajdaoui, A.; Roualdes, S.; Brault, P.; Durand, J. *J. Power Sources* 2010, *195*, 232-238.

[22] Lee, W. J.; Alhosan, M.; Yohe, S. L.; Macy, N. L.; Smyrlz, W. H. *J. Electrochem. Soc.* 2008, *155*, B915-B920.

[23] Stevens, D. A.; Rouleau, J. M.; Mar, R. E.; Atanasoski, R. T.; Schmoeckel, A. K.; Debe, M. K.; Dahn, J. R. *J. Electrochem. Soc.* 2007, *154*, B1211-B1219.

[24] Millet, P.; Alleau, T.; Durand, R. *J. Appl. Electrochem.* 1993, *23*, 322-331.

[25] Lue, S. J.; Shih, T. S.; Wei, T. C. *Korean J. Chem. Eng.* 2006, *23*, 441-446.

[26] Yasuda, H. *Plasma Polymerization*; Academic Press: New York, 1985.

[27] Yasuda, H. *J. Membr. Sci.* 1984, *18*, 273-284.

[28] Brault, P.; Roualdes, S.; Caillard, A.; Thomann, A. L.; Mathias, J.; Durand, J.; Coutanceau, C.; Leger, J. M.; Charles, C.; Boswell, R. *Eur. Phys. J. Appl. Phys.* 2006, *34*, 151-156.

[29] Hirano, S.; Kim, J.; Srinivasan, S. *Electrochim Acta* 1997, *42*, 1587-1593.

[30] Gruber, D.; Ponath, N.; Muller, J.; Lindstaedt, F. *J. Power Sources* 2005, *150*, 67-72.

[31] Gruber, D.; Muller, J. *J. Power Sources* 2007, *171*, 294-301.

[32] Alvisi, M.; Galtieri, G.; Giorgi, L.; Giorgi, R.; Serra, E.; Signore, M. A. *Surf. Coat. Technol.* 2005, *200*, 1325-1329.

[33] Makino, K.; Furukawa, K.; Okajima, K.; Sudoh, M. *Electrochim Acta* 2005, *51*, 961-965.

[34] Makino, K.; Furukawa, K.; Kajima, K.; Sudoh, M. *J. Power Sources* 2007, *166*, 30-34.

[35] Huang, K. L.; Lai, Y. C.; Tsai, C. H. *J. Power Sources* 2006, *156*, 224-231.

[36] Brault, P.; Caillard, A.; Thomann, A. L.; Mathias, J.; Charles, C.; Boswell, R. W.; Escribano, S.; Durand, J.; Sauvage, T. *J. Phys. D Appl. Phys.* 2004, *37*, 3419-3423.

[37] Caillard, A.; Brault, P.; Mathias, J.; Charles, C.; Boswell, R. W.; Sauvage, T. *Surf. Coat. Technol.* 2005, *200*, 391-394.

[38] Caillard, A.; Charles, C.; Boswell, R.; Meige, A.; Brault, P. *Plasma Sources Sci. Technol.* 2008, *17*, 035028: 1-7.

[39] Brault, P.; Josserand, C.; Bauchire, J. M.; Caillard, A.; Charles, C.; Boswell, R. W. *Phys. Rev. Lett.* 2009, *102*, 45901: 1-4.

[40] Caillard, A.; Charles, C.; Ramdutt, D.; Boswell, R.; Brault, P. *J. Phys. D Appl. Phys.* 2009, *42*, 045207: 1-9.
[41] Yoo, S. J.; Cho, Y. H.; Park, H. S.; Lee, J. K.; Sung, Y. E. *J. Power Sources* 2008, *178*, 547-553.
[42] Akyalcin, L.; Kaytakoglu, S. *J. Power Sources* 2008, *180*, 767-772.
[43] Gasda, M. D.; Teki, R.; Lu, T. M.; Koratkar, N.; Eisman, G. A.; Gall, D. *J. Electrochem. Soc.* 2009, *156*, B614-B619.
[44] Chen, C. C.; Chen, C. F.; Hsu, C. H.; Li, I. H. *Diamond Relat. Mater.* 2005, *14*, 770-773.
[45] Sun, C. L.; Chen, L. C.; Su, M. C.; Hong, L. S.; Chyan, O.; Hsu, C. Y.; Chen, K. H.; Chang, T. F.; Chang, L. *Chem. Mater.* 2005, *17*, 3749-3753.
[46] Prehn, K.; Adelung, R.; Heinen, M.; Nunes, S. P.; Schulte, K. *J. Membr. Sci.* 2008, *321*, 123-130.
[47] Soin, N.; Roy, S. S.; Karlsson, L.; Mclaughlin, J. A. *Diamond Relat. Mater.* 2010, 19, 595-598.
[48] Tang, Z.; Poh, C. K.; Lee, K. K.; Tian, Z. Q.; Chua, D.; Lin, J. Y. *J. Power Sources* 2010, *195*, 155-159.
[49] Wang, C. H.; Dub, H. Y.; Tsai, Y. T.; Chen, C. P.; Huang, C. J.; Chen, L. C.; Chen, K. H.; Shih, H. C. *J. Power Sources* 2007, *171*, 55-62.
[50] Caillard, A.; Charles, C.; Boswell, R.; Brault, P.; Coutanceau, C. *Appl. Phys. Lett.* 2007, *90*, 223119: 1-3.
[51] Caillard, A.; Charles, C.; Boswell, R.; Brault, P. *Nanotechnology* 2007, *18*, 305603: 1-9.
[52] Caillard, A.; Charles, C.; Boswell, R.; Brault, P. *J. Phys. D: Appl. Phys.* 2008, *41*, 185307: 1-10.
[53] Rabat, H.; Brault, P. *Fuel Cells* 2008, *8*, 81-86.
[54] Gasda, M. D.; Eisman, G. A.; Gall, D. *J. Electrochem. Soc.* 2010, *157*, B113-B117.
[55] Kim, H. T.; Lee, J. K.; Kim, J. *J. Power Sources* 2008, *180*, 191-194.
[56] Radev, I.; Slavcheva, E.; Budevski, E. *Int. J. Hydrogen Energy* 2007, *32*, 872-877.
[57] Wang, H.; Zhang, M.; Cheng, F. L.; Xu, C. W. *Int. J. Electrochem. Sci.* 2008, *3*, 946-952.
[58] Slavcheva, E.; Ganske, G.; Topalov, G.; Mokwa, W.; Schnakenberg, U. *Appl. Surf. Sci.* 2009, *255*, 6479-6486.
[59] Ozturk, O.; Ozdemir, O. K.; Ulusoy, I.; Ahsen, A. S.; Slavcheva, E. *Int. J. Hydrogen Energy* 2010, *35*, 4466-4473.
[60] Bonakdarpour, A.; Fleischauer, M. D.; Brett, M. J.; Dahn, J. R. *Appl. Catal. A* 2008, *349*, 110-115.
[61] Kang, S. H.; Jeon, T. Y.; Kim, H. S.; Sung, Y. E.; Smyrl, W. H. *J. Electrochem. Soc.* 2008, *155*, B1058-B1065.
[62] Kang, S. H.; Sung, Y. E.; Smyrl, W. H. *J. Electrochem. Soc.* 2008, *155*, B1128-B1135.
[63] Wang, H.; Xu, C. W.; Cheng, F. L.; Zhang, M.; Wang, S. Y.; Jiang, S. P. *Electrochem. Commun.* 2008, *10*, 1575-1578.
[64] Gasda, M. D.; Eisman, G. A.; Gall, D. *J. Electrochem. Soc.* 2010, *157*, B71-B76.
[65] Debe, M. K.; Schmoeckel, A. K.; Vernstrorn, G. D.; Atanasoski, R. *J. Power Sources* 2006, *161*, 1002-1011.

[66] Vad, T.; Hajbolouri, F.; Haubold, H. G.; Scherer, G. G.; Wokaun, A. *J. Phys. Chem. B* 2004, *108*, 12442-12449.
[67] Hajbolouri, F.; Scherer, G. G.; Vad, T.; Horisberger, M.; Schnyder, B.; Wokaun, A. *PSI Scientific Report*, 2002.
[68] Nechitailov, A. A.; Astrova, E. V.; Goryachev, D. N.; Zvonareva, T. K.; Ivanov-Omskii, V. I.; Remenyuk, A. D.; Sapurina, I. Y.; Sreseli, O. M.; Tolmachev, V. A. *Tech. Phys. Lett.* 2007, *33*, 545-547.
[69] Zvonareva, T. K.; Sitnikova, A. A.; Frolova, G. S.; Ivanov-Omskii, V. I. *Semiconductors* 2008, *42*, 325-328.
[70] Nechitailov, A. A.; Zvonareva, T. K.; Remenyuk, A. D.; Tolmachev, V. A.; Goryachev, D. N.; El'Tsina, O. S.; Belyakov, L. V.; Sreseli, O. M. *Semiconductors* 2008, *42*, 1249-1254.
[71] Cavarroc, M.; Ennadjaoui, A.; Mougenot, M.; Brault, P.; Escalier, R.; Tessier, Y.; Durand, J.; Roualdes, S.; Sauvage, T.; Coutanceau, C. *Electrochem. Commun.* 2009, *11*, 859-861.
[72] Toda, T.; Igarashi, H.; Uchida, H.; Watanabe, M. *J. Electrochem. Soc.* 1999, *146*, 3750-3756.
[73] Toda, T.; Igarashi, H.; Watanabe, M. *J. Electrochem. Soc.* 1998, *145*, 4185-4188.
[74] Watanabe, M.; Igarashi, H.; Fujino, T. *Electrochemistry* 1999, *67*, 1194-1196.
[75] Caillard, A.; Coutanceau, C.; Brault, P.; Mathias, J.; Leger, J. M. *J. Power Sources* 2006, *162*, 66-73.
[76] Kim, T. W.; Park, S. J.; Jones, L. E.; Toney, M. F.; Park, K. W.; Sung, Y. E. *J. Phys. Chem. B* 2005, *109*, 12845-12849.
[77] Shibamine, M.; Yamada, A.; Umeda, M.; Tanaka, S. *Electr. Eng. Jpn.* 2008, *163*, 14-21.
[78] Umeda, M.; Sugii, H.; Uchida, I. *J. Power Sources* 2008, *179*, 489-496.
[79] Inoue, M.; Shingen, H.; Kitami, T.; Akamaru, S.; Taguchi, A.; Kawamoto, Y.; Tada, A.; Ohtawa, K.; Ohba, K.; Matsuyama, M.; Watanabe, K.; Tsubone, I.; Abe, T. *J. Phys. Chem. C* 2008, *112*, 1479.
[80] Inoue, M.; Nishimura, T.; Akamaru, S.; Taguchi, A.; Umeda, M.; Abe, T. *Electrochim Acta* 2009, *54*, 4764-4771.
[81] Witham, C. K.; Chun, W.; Valdez, T. I.; Narayanan, S. R. *Electrochem. Solid-State Lett.* 2000, *3*, 497-500.
[82] Bommersbach, P.; Mohamedi, M.; Guay, D. *J. Electrochem. Soc.* 2007, *154*, B876-B882.
[83] Yoo, S. J.; Park, H. Y.; Jeon, T. Y.; Park, I. S.; Cho, Y. H.; Sung, Y. E. *Angew. Chem. Int. Ed.* 2008, *47*, 9307-9310.
[84] Brown, B.; Wolter, S. D.; Stoner, B. R.; Glass, J. T. *J. Electrochem. Soc.* 2008, *155*, B852.-B859.
[85] Slavcheva, E.; Ganske, G.; Topalov, G.; Mokwa, W.; Schnakenberg, U. *Appl. Surf. Sci.* 2009, *255*, 6479-6486.
[86] Umeda, M.; Ojima, H.; Mohamedi, M.; Uchida, I. *J. Power Sources* 2004, *136*, 10-15.
[87] Tanaka, S.; Umeda, M.; Ojima, H.; Usui, Y.; Kimura, O.; Uchida, I. *J. Power Sources* 2005, *152*, 34-39.
[88] Sun, C. L.; Hsu, Y. K.; Lin, Y. G.; Chen, K. H.; Bock, C.; Macdougall, B.; Wu, X. H.; Chen, L. C. *J. Electrochem. Soc.* 2009, *156*, B1249-B1252.

[89] Tsai, C. F.; Wu, P. W.; Lin, P.; Chao, C. G.; Yeh, K. Y. *Jpn. J. Appl. Phys.* 2008, *47*, 5755-5761.
[90] Tsai, C. F.; Yeh, K. Y.; Wu, P. W.; Hsieh, Y. F.; Lin, P. *J. Alloys Compd.* 2009, *478*, 868-871.
[91] Park, K. W.; Ahn, K. S.; Choi, J. H.; Nah, Y. C.; Kim, Y. M.; Sung, Y. E. *Appl. Phys. Lett.* 2002, *81*, 907-909.
[92] Park, K. W.; Ahn, K. S.; Choi, J. H.; Nah, Y. C.; Sung, Y. E. *Appl. Phys. Lett.* 2003, *82*, 1090-1092.
[93] Park, K. W.; Choi, J. H.; Ahn, K. S.; Sung, Y. E. *J. Phys. Chem. B* 2004, *108*, 5989-5994.
[94] Park, K. W.; Sung, Y. E.; Toney, M. F. *Electrochem. Commun.* 2006, *8*, 359-363.
[95] Park, K. W.; Sung, Y. E. *J. Appl. Phys.* 2003, *94*, 7276-7280.
[96] Park, K. W.; Sung, Y. E. *J. Vac. Sci. Technol. B* 2004, *22*, 2628-2631.
[97] Park, K. W.; Han, S. B.; Lee, J. M. *Electrochem. Commun.* 2007, *9*, 1578-1581.
[98] Ahn, H. J.; Shim, H. S.; Kim, W. B.; Sung, Y. E.; Seong, T. Y. *J. Alloys Compd.* 2009, *471*, L39-L42.
[99] Ahn, H. J.; Jang, J. S.; Sung, Y. E.; Seong, T. Y. *J Alloys Compd.* 2009, *473*, L28-L32.
[100] Vaclavu, M.; Matolinova, I.; Myslivecek, J.; Fiala, R.; Matolin, V. *J. Electrochem. Soc.* 2009, *156*, B938-B942.
[101] Matolin, V.; Cabala, M.; Matolinova, I.; Skoda, M.; Vaclavu, M.; Prince, K. C.; Skala, T.; Mori, T.; Yoshikawa, H.; Yamashita, Y.; Ueda, S.; Kobayashi, K. *Fuel Cells* 2010, *10*, 139-144.
[102] Lee, B.; Kim, C.; Park, Y.; Kim, T. G.; Park, B. *Electrochem. Solid-State Lett.* 2006, *9*, E27-E30.
[103] Springer, T. E.; Zawodzinski, T. A.; Gottesfeld, S. *J. Electrochem. Soc.* 1991, *138*, 2334-2342.
[104] Charles, C.; Ramdutt, D.; Brault, P.; Caillard, A.; Bulla, D.; Boswell, R.; Rabat, H.; Dicks, A. *Plasma Phys. Controlled Fusion* 2007, *49*, A73-A79.
[105] Gruber, D.; Ponath, N.; Muller, J. *J. Fuel Cell Sci. Technol.* 2005, *2*, 186-189.
[106] Feichtinger, J.; Kerres, J.; Schulz, A.; Walker, M.; Schumacher, U. *J. New Mater. Electrochem. Syst.* 2002, *5*, 155-162.
[107] Ogumi, Z.; Uchimoto, Y.; Tsujikawa, M.; Takehara, Z. *J. Electrochem. Soc.* 1989, *136*, 1247-1248.
[108] Zeng, R.; Pang, Z. C.; Zhu, H. S. *J. Electroanal. Chem.* 2000, *490*, 102-106.
[109] Feichtinger, J.; Galm, R.; Walker, M.; Baumgartner, K. M.; Schulz, A.; Rauchle, E.; Schumacher, U. *Surf. Coat. Technol.* 2001, *142*, 181-186.
[110] Yoon, S. R.; Hwang, G. H.; Cho, W. I.; Oh, I. H.; Hong, S. A.; Ha, H. Y. *J. Power Sources* 2002, *106*, 215-223.
[111] Kim, D. J.; Scibioh, M. A.; Kwak, S.; Oh, I. H.; Ha, H. Y. *Electrochem. Commun.* 2004, *6*, 1069-1074.
[112] Cho, S. A.; Cho, E. A.; Oh, I. H.; Kim, H. J.; Ha, H. Y.; Hong, S. A.; Ju, J. B. *J. Power Sources* 2006, *155*, 286-290.
[113] Prasanna, M.; Cho, E. A.; Kim, H. J.; Lim, T. H.; Oh, I. H.; Hong, S. A. *J. Power Sources* 2006, *160*, 90-96.
[114] Mahdjoub, H.; Roualdes, S.; Sistat, P.; Pradeilles, N.; Durand, J.; Pourcelly, G. *Fuel Cells* 2005, *5*, 277-286.

[115] Uchimoto, Y.; Yasuda, K.; Ogumi, Z.; Takehara, Z.; Tasaka, A.; Imahigashi, T. *Ber. Bunsen Ges. Phys. Chem.* 1993, *97*, 625-630.

[116] Uchimoto, Y.; Endo, E.; Yasuda, K.; Yamasaki, Y.; Takehara, Z.; Ogumi, Z.; Kitao, O. *J. Electrochem. Soc.* 2000, *147*, 111-118.

[117] Brumlik, C. J.; Parthasarathy, A.; Chen, W. J.; Martin, C. R. *J. Electrochem. Soc.* 1994, *141*, 2273-2279.

[118] Yasuda, K.; Uchimoto, Y.; Ogumi, Z.; Takehara, Z. *Ber. Bunsen. Ges. Phys. Chem.* 1994, *98*, 631-635.

[119] Yasuda, K.; Uchimoto, Y.; Ogumi, Z.; Takehara, Z. *J. Electrochem. Soc.* 1994, *141*, 2350-2355.

[120] Ogumi, Z.; Uchimoto, Y.; Takehara, Z. *J. Electrochem. Soc.* 1990, *137*, 3319-3321.

[121] Ogumi, Z.; Uchimoto, Y.; Yasuda, K.; Takehara, Z. I. *Chem. Lett.* 1990, 953-954.

[122] Inagaki, N.; Tasaka, S.; Kurita, T. *Polym. Bull.* 1989, *22*, 15-20.

[123] Inagaki, N.; Tasaka, S.; Horikawa, Y. *J. Polym. Sci. Part A Polym. Chem.* 1989, *27*, 3495-3501.

[124] Inagaki, N.; Tasaka, S.; Chengfei, Z. *Polym. Bull.* 1991, *26*, 187-191.

[125] Uchimoto, Y.; Yasuda, K.; Ogumi, Z.; Takehara, Z.; Tasaka, A.; Imahigashi, T. *Ber. Bunsen Ges. Phys. Chem.* 1993, *97*, 625-630.

[126] Yasuda, K.; Uchimoto, Y.; Ogumi, Z.; Takehara, Z. *Denki Kagaku* 1993, *61*, 1438-1441.

[127] Ennajdaoui, A.; Larrieu, J.; Roualdes, S.; Durand, J. *Eur. Phys. J. Appl. Phys.* 2008, *42*, 9-15.

[128] Ennajdaoui, A.; Roualdes, S.; Brault, P.; Durand, J. *J. Power Sources* 2010, *195*, 232-238.

[129] Finsterwalder, F.; Hambitzer, G. *J. Membr. Sci.* 2001, *185*, 105-124.

[130] Yasuda, H. *J. Macromol. Sci. Chem.* 1976, *A 10*, 383-420.

[131] Poll, H. U.; Arzt, M.; Wickleder, K. H. *Eur. Polym. J.* 1976, *12*, 505-512.

[132] Yasuda, H. *Macromol. Rev. Part D J. Polym. Sci.* 1981, *16*, 199-293.

[133] Francombe, M. H.; Vossen, J. L. *Physics of Thin Films*; Academic Press: Orlando, 1994.

[134] Chen, M.; Yang, T. C.; Ma, Z. G. *J. Polym. Sci. Part A Polym. Chem.* 1998, *36*, 1265-1270.

[135] Roualdes, S.; Topala, I.; Mahdjoub, H.; Rouessac, V.; Sistat, P.; Durand, J. *J. Power Sources* 2006, *158*, 1270-1281.

[136] He, R. H.; Li, Q. F.; Xiao, G.; Bjerrum, N. J. *J. Membr. Sci.* 2003, *226*, 169-184.

[137] Lobato, J.; Canizares, P.; Rodrigo, M. A.; Linares, J. J.; Manjavacas, G. *J. Membr. Sci.* 2006, *280*, 351-362.

[138] Jiang, Z. Q.; Meng, Y. D.; Jiang, Z. J.; Shi, Y. C. *Surf. Rev. Lett.* 2009, *16*, 297-302.

[139] Savadogo, O. *J. Power Sources* 2004, *127*, 135-161.

[140] Genova-Dimitrova, P.; Baradie, B.; Foscallo, D.; Poinsignon, C.; Sanchez, J. Y. *J. Membr. Sci.* 2001, *185*, 59-71.

[141] Zhang, X. Ph.D. Thesis, Universitat Rovirai Virgili (URV), Spain, 2005.

[142] Mex, L.; Sussiek, M.; Muller, J. *Chem. Eng. Commun.* 2003, *190*, 1085-1095.

[143] Mex, L.; Müller, J. *Membr. Technol.* 1999, *1999*, 5-9.

[144] Mex, L.; Ponath, N.; Müller, J. *Fuel Cells Bull.* 2001, *4*, 9-12.

[145] Le, V. J. L.; Martin, S.; Gaillard, F. *Ionics* 2008, *14*, 403-406.

[146] Jiang, Z. Q.; Meng, Y. D.; Jiang, Z. J.; Shi, Y. C. *Surf. Rev. Lett.* 2007, *14*, 1165-1168.
[147] Jiang, Z. Q.; Meng, Y. D.; Shi, Y. C. *Jpn. J. Appl. Phys.* 2008, *47*, 6891-6895.
[148] Schieda, M.; Roualdes, S.; Durand, J.; Martinent, A.; Marsacq, D. *Desalination* 2006, *199*, 286-288.
[149] Matsuoka, K.; Chiba, S.; Iriyama, Y.; Abe, T.; Matsuoka, M.; Kikuchi, K.; Ogumi, Z. *Thin Solid Films* 2008, *516*, 3309-3313.
[150] Kelly, P. J.; Arnell, R. D. *Vacuum* 2000, *56*, 159-172.
[151] Han, J. G. *J. Phys. D Appl. Phys.* 2009, *42*, 043001: 1-16.
[152] Yamamoto, H.; Hirakawa, K.; Abe, T. *Mater. Lett.* 2008, *62*, 2118-2121.
[153] Charles, C.; Brault, P.; Caillard, A.; Ramdutt, D.; Boswell, R.; Rabat, H.; Corr, C.; Li, W.; Ladewig, B.; Dicks, A. China Australia Energy Symposium, Sydney, 2006.
[154] Yamada, K.; Miyazaki, K.; Koji, S.; Okumura, Y.; Shibata, M. *J. Power Sources* 2008, *180*, 181-184.
[155] Balasubramanian, K.; Burghard, M. *Small* 2005, *1*, 180-192.
[156] Holzinger, M.; Vostrowsky, O.; Hirsch, A.; Hennrich, F.; Kappes, M.; Weiss, R.; Jellen, F. *Angew. Chem. Int. Ed.* 2001, *40*, 4002-4005.
[157] Chen, R. J.; Zhang, Y. G.; Wang, D. W.; Dai, H. J. *J. Am. Chem. Soc.* 2001, *123*, 3838-3839.
[158] Chen, R. J.; Bangsaruntip, S.; Drouvalakis, K. A.; Kam, N.; Shim, M.; Li, Y. M.; Kim, W.; Utz, P. J.; Dai, H. J. *Proc. Natl. Acad. Sci.* 2003, *100*, 4984-4989.
[159] Star, A.; Stoddart, J. F.; Steuerman, D.; Diehl, M.; Boukai, A.; Wong, E. W.; Yang, X.; Chung, S. W.; Choi, H.; Heath, J. R. *Angew. Chem. Int. Ed.* 2001, *40*, 1721-1725.
[160] Hsin, Y. L.; Hwang, K. C.; Yeh, C. T. *J. Am. Chem. Soc.* 2007, *129*, 9999-10010.
[161] Anderson, M. L.; Stroud, R. M.; Rolison, D. R. *Nano Lett.* 2002, *2*, 235-240.
[162] Plank, N.; Jiang, L. D.; Cheung, R. *Appl. Phys. Lett.* 2003, *83*, 2426-2428.
[163] Khare, B. N.; Wilhite, P.; Quinn, R. C.; Chen, B.; Schingler, R. H.; Tran, B.; Imanaka, H.; So, C. R.; Bauschlicher, C. W.; Meyyappan, M. *J. Phys. Chem. B* 2004, *108*, 8166-8172.
[164] Plank, N.; Cheung, R.; Andrews, R. J. *Appl. Phys. Lett.* 2004, *85*, 3229-3231.
[165] Yan, Y. H.; Chan-Park, M. B.; Zhou, Q.; Li, C. M.; Yue, C. Y. *Appl. Phys. Lett.* 2005, *87*, 213101-213103.
[166] Brunetti, F. G.; Herrero, M. A.; Munoz, J. D.; Diaz-Ortiz, A.; Alfonsi, J.; Meneghetti, M.; Prato, M.; Vazquez, E. *J. Am. Chem. Soc.* 2008, *130*, 8094-8100.
[167] Chen, Q. D.; Dai, L. M.; Gao, M.; Huang, S. M.; Mau, A. *J. Phys. Chem. B* 2001, *105*, 618-622.
[168] Chen, Q. D.; Dai, L. M. *Appl. Phys. Lett.* 2000, *76*, 2719-2721.
[169] Yang, D. Q.; Rochette, J. F.; Sacher, E. *Langmuir* 2005, *21*, 8539-8545.
[170] Jiang, Z. Q.; Yu, X. Y.; Jiang, Z. J.; Meng, Y. D.; Shi, Y. C. *J. Mater. Chem.* 2009, *19*, 6720-6726.
[171] O'Hayre, R.; Lee, S. J.; Cha, S. W.; Prinz, F. B. *J. Power Sources* 2002, *109*, 483-493.
[172] Nakakubo, T.; Shibata, M.; Yasuda, K. *J. Electrochem. Soc.* 2005, *152*, A2316-A2322.
[173] Wakisaka, M.; Suzuki, H.; Mitsui, S.; Uchida, H.; Watanabe, M. *J. Phys. Chem. C* 2008, *112*, 2750-2755.
[174] Cha, S. Y.; Lee, W. M. *J. Electrochem. Soc.* 1999, *146*, 4055-4060.
[175] Haug, A. T.; White, R. E.; Weidner, J. W.; Huang, W.; Shi, S.; Stoner, T.; Rana, N. *J. Electrochem. Soc.* 2002, *149*, A280-A287.

[176] Wan, C. H.; Lin, M. T.; Zhuang, Q. H.; Lin, C. H. *Surf. Coat. Technol.* 2006, *201*, 214-222.
[177] Haug, A. T.; White, R. E.; Weidner, J. W.; Huang, W.; Shi, S.; Rana, N.; Grunow, S.; Stoner, T. C.; Kaloyeros, A. E. *J. Electrochem. Soc.* 2002, *149*, A868-A872.
[178] Wan, C. H.; Zhuang, Q. H.; Lin, C. H.; Lin, M. T.; Shih, C. *J. Power Sources* 2006, *162*, 41-50.
[179] Savadogo, O.; Lee, K.; Oishi, K.; Mitsushima, S.; Kamiya, N.; Ota, K. I. *Electrochem. Commun.* 2004, *6*, 105-109.
[180] Savadogo, O.; Varela, F. *J. New Mater. Electrochem. Syst.* 2008, *11*, 69-74.
[181] Fouda-Onana, F.; Bah, S.; Savadogo, O. *J. Electroanal. Chem.* 2009, *636*, 1-9.
[182] Lee, K.; Savadogo, O.; Ishihara, A.; Mitsushima, S.; Kamiya, N.; Ota, K. *J. Electrochem. Soc.* 2006, *153*, A20-A24.
[183] Ye, J. S.; Bai, Y. C.; Zhang, W. D. *Microchim Acta* 2009, *165*, 361-366.
[184] Liu, Y.; Ishihara, A.; Mitsushima, S.; Ota, K. *Electrochim Acta* 2010, *55*, 1239-1244.
[185] Ishihara, A.; Doi, S.; Mitsushima, S.; Ota, K. *Electrochim Acta* 2008, *53*, 5442-5450.
[186] Maekawa, Y.; Ishihara, A.; Kim, J. H.; Mitsushima, S.; Ota, K. I. *Electrochem. Solid-State Lett.* 2008, *11*, B109-B112.
[187] Liu, Y.; Ishihara, A.; Mitsushima, S.; Kamiya, N.; Ota, K. *J. Electrochem. Soc.* 2007, *154*, B664-B669.
[188] Liu, Y.; Ishihara, A.; Mitsushima, S.; Kamiya, N.; Ota, K. *Electrochem. Solid-State Lett.* 2005, *8*, A400-A402.
[189] Yang, R. Z.; Stevens, K.; Dahn, J. R. *J. Electrochem. Soc.* 2008, *155*, B79-B91.
[190] Garsuch, A.; Yang, R.; Bonakdarpour, A.; Dahn, J. R. *Electrochim Acta* 2008, *53*, 2423-2429.
[191] Liu, G.; Dahn, J. R. *Appl. Catal. A* 2008, *347*, 43-49.
[192] Yang, R. Z.; Bonakdarpour, A.; Easton, E. B.; Stoffyn-Egli, P.; Dahn, J. R. *J. Electrochem. Soc.* 2007, *154*, A275-A282.
[193] Easton, E. B.; Yang, R. Z.; Bonakdarpour, A.; Dahn, J. R. *Electrochem. Solid-State Lett.* 2007, *10*, B6-B10.
[194] Easton, E. B.; Bonakdarpour, A.; Dahn, J. R. *Electrochem. Solid-State Lett.* 2006, *9*, A463-A467.
[195] Lin, J. C.; Chuang, C. L.; Lai, C. M.; Chu, H. C.; Chen, Y. S. *Thin Solid Films* 2009, *517*, 4728-4730.
[196] Kim, J. H.; Ishihara, A.; Mitsushima, S.; Kamiya, N.; Ota, K. I. *Chem. Lett.* 2007, *36*, 514-515.
[197] Kim, J. H.; Ishihara, A.; Mitsushima, S.; Kamiya, N.; Ota, K. *Electrochemistry* 2007, *75*, 166-168.

In: Advanced Organic-Inorganic Composites
Editor: Inamuddin
ISBN 978-1-61324-264-3

Chapter 7

POLYMER/CERAMIC/METAL COMPOSITES FOR AUTOMOBILES

Mukesh Kumar[*]
Department of Chemistry, University of Delhi,
North Campus, Delhi 110007, India

ABSTRACT

Composites play a major role in the automobile industry and the present chapter overviews some of the basic understanding of ceramic matrix composites (CMCs), metal matrix composites (MMCs) and polymer matrix composites. CMCs are mainly used to make automobile engine products whereas MMCs and polymer matrix composites are used to make light weight automobile parts in order to reduce the weight of the automobile. By reducing the weight of the automobile, fuel efficiency can easily be increased. In addition to the basic understanding of the different types of matrix composites, preparation methods of matrix composites are also discussed in this chapter. Applications of matrix composites in automobile industries are also discussed in brief.

ABBREVIATIONS

ACA	Automotive Composites Alliance
ACC	Automotive composites consortium
AMG	Advanced material gas generator
BIW	Body in-White
CEC	Commission of the European Community
CFCC	Continuous fiber ceramic composite
CMCs	Ceramic matrix composite
CNT	Carbon Nanotube
COF	Coefficient of friction

[*] Corresponding author's e-mail: mknanocomposites@gmail.com

Cu-MMCs	Copper based metal matrix composites
CVD	Chemical vapor deposition
DMA	Dynamic mechanical analyzer
ELV	End of life vehicles
FRP	Fiber reinforced polymer
GMT	Glass mat thermoplastic
HSCT	High Speed Civil Transport
HSR	High Speed Research
IHPTET	Integrated High Performance Turbine Engine Technology
LSI	Liquid silicon infiltration method
LFR	Long Fiber Reinforced
MMCs	Metal matrix composites
Mg-MMCs	Magnesium based metal matrix composites
PBCM	Process-based cost models
PET	Polyethylene terephthalate
PLA	Poly(lactic acid)
PMC	Polymer matrix composites
PNGV	Partnership for a New Generation of Vehicles
PP	Polypropylene
PRP	Particle reinforced polymer
PVA	Polyvinyl-alcohol
PVD	Physical vapour deposition
SBS	Styrene-butadiene-styrene
SEM	Scanning electron microscope
SFR	Short Fiber Reinforced
SMC	Sheet molding compound
Ti-MMCs	Titanium based metal matrix composites
UF	Urea–formaldehyde
VOCs	Volatile organic compounds
XRD	X-ray diffraction

You cannot step twice into the same river
- Heraclitus[1]

1. Introduction

To show change is the order of the universe, Heraclitus (c. 500 BC) said the above observation and this is true also for the automobile industry. Ever since the discovery of the first automobile in 1769 by Nicolas Joseph Cugnot [1], the automobile industry traveled a long journey and underwent several changes. These changes, in larger context, have taken place to increase fuel efficiency, safety and also to reduce the weight and price of the automobile. During these modifications, many value-added products including different types

[1] Theodore de Laguna, The physiological Review, vol. 30. No. 3 (1921), p. 238-254.

of electronic parts and advanced materials such as composites have been introduced in the automobile industry. A composite material is made by combining two or more materials to give a unique combination of properties. The final properties of the composite materials are better than the properties of the constituent materials [2]. Composites are used to replace the existing automobile parts and they are also used to replace different multiple parts into single integrated one.

By replacing the existing automobile parts with the composites, we can achieve less weight, high fuel efficiency, less pollution, better performance by adding value-added product and can improve the safety of the travelers [3]. In comparison of monolithic components, automobile parts made from composite materials may be more durable, strong and lightweight. According to automotive Composites Alliance (ACA), automotive parts made up of composite materials may reduce automobile component weight by 35% and ACA believes these materials can also reduce the cost of making automotive parts into half.

According to Belyea and Deckman, composites exhibit high tensile strength and modulus, as a result, these composites can withstand significant structural loads and/or harsh operating environments. Continuous fiber reinforced and discontinuous or chopped fiber reinforced composites are two different types of composites and the former one shows superior properties than the later one. Carbon, S-2 glass, aramid fibers, boron, ceramics, metal, and extended-chain polyethylene are generally used to prepare fiber reinforced composites [4].

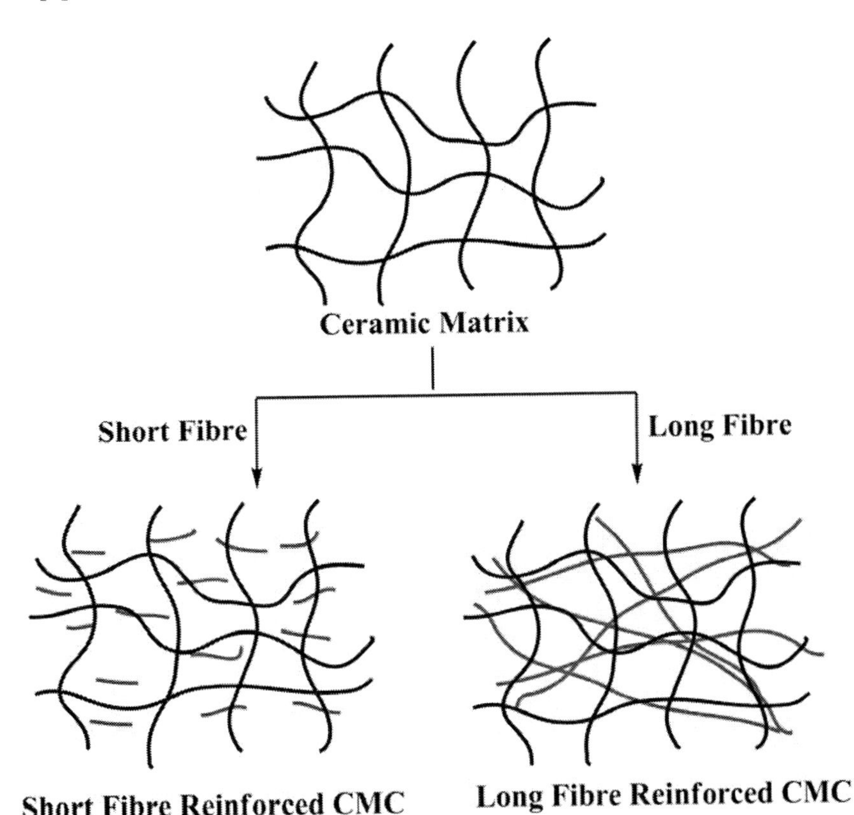

Scheme 1. Synthesis of CMCs.

As composites play a major role in the automobile industry, it is important to understand their preparation, characterization and applications. The role of polymer/ceramic/metal composites in automobile industry is discussed in this chapter.

2. Ceramic Matrix Composites for Automobiles

It is well known that gas turbine and diesel engines are the two different engines which are being used widely in automobiles. In both the types of engines, fuel efficiency is one of the important factors and to increase the fuel efficiency, reduction of weight of an automotive body is necessary. As steel cannot withstand engine's very high temperature (>1350 °C), gas turbine engine was initially fabricated using super alloys but they were more expensive and hence, they were not suitable for the automobile industries. To reduce the manufacturing cost and total weight of an automobile in 1970s, ceramic was proposed as a material to produce engine components [5]. But, properties such as thermal shock resistance, particle impact resistance and creep resistance should be good for the materials which can be used at high temperature. In addition to these properties, a good turbine rotor should perform without any failure for at least 10 years of operation over 1,00,000 km including 10,000 starts and stops [6]. But, ceramics could not satisfy these requirements as these materials are brittle, and develop cracks under stress, as a result, the idea of using these materials to make the automobile engines was not successful [5]. To utilize the high temperature withstanding capacity of ceramics and to overcome the cracking nature of these materials, ceramic matrix composite (CMC) was proposed.

2.1. Materials Selection for Ceramic Matrix Composite (CMC)

A CMC is a ceramic material that consists of reinforced ceramic phase into the ceramic matrix. Oxide and nonoxide are the two types of ceramic matrices used during the preparation of CMC. Alumina, silica, mullite, barium alumino silicate and calcium alumino silicate are some of the oxide ceramic matrices used for the preparation of CMC. As alumina and mullite are compatible with common reinforcing agents and as they have good thermal and chemical stability, these ceramic matrices are widely used. SiC, Si_3N_4, boron carbide and AlN are some of the non-oxide ceramic matrices and SiC is widely used among them. AlN is used whenever CMC with high thermal conductivity is required whereas Si_3N_4 is used when high strength CMC is required.

Ceramic reinforcements can be done using short or long fibers as shown in Scheme 1. In short fiber reinforcements, platelets, particulates and whiskers from SiC, Si_3N_4, titanium diboride, boron carbide can be used. Most of these materials are non-oxide ceramic reinforcing materials and oxide ceramic materials in short fiber reinforcement is rare as they are incompatible with the ceramic matrices. Among different types of non-oxide ceramic materials, SiC has been used widely as a reinforcing agent for the preparation of CMC. When the ceramic matrices are combined with special fibers, fiber reinforced CMCs are obtained. To prepare these composites, fiber materials which can withstand high temperatures can only be used. As organic polymer fiber materials degrade below 500 °C, these fibers are not suitable for making fiber reinforced CMCs. Conventional glass fibers which have melting or softening temperature below 700 °C cannot be used as the fiber material for making fiber reinforced CMCs. Polycrystalline or amorphous inorganic fibers are good fibers for making fiber reinforced CMCs. Al_2O_3, mullite, ZrO_2 are some of the oxide ceramic fibers and SiC, Si-C-O, S-C-N-O, S-N-C-N are some of the non-oxide ceramic fibers which can be used to make fiber reinforced CMCs. Though carbon fiber degrades above 450 °C at oxidizing atmosphere, these fibers can also be used in non-oxidizing conditions. At non-oxidizing conditions, it was found that carbon fibers are stable up to 2800 °C [7].

Though, short and long fiber reinforced CMCs can be used to make automobile engine components, it is important to understand which CMC show superior properties [8]. Figure 1 shows the comparison of impact strength of short and long fiber reinforced CMCs along with impact strength of monolithic SiC. Results from Figure 1 show that the impact strength of CMC is better than monolithic SiC. It is clear from the Figure 1 that the reinforcement of particulate and fibers increased toughness when compared to monolithic ceramic. Among different types of reinforcements present in CMC, fiber reinforced CMC is tougher than the particulate reinforced CMC [6]. In a relatively new study, carbon nanotube (CNT) based ceramic composites can also be prepared with good fracture toughness. Recently, $C_{fibre}/SiC_{filler}/Si\text{-}B\text{-}C\text{-}N_{matrix}$ composites have also been prepared and these CMCs were found to show good thermal stability and toughness [9].

As CMCs show high modulus at high temperatures, they were initially used to prepare turbine rotors, a back plate, orifice liner and an extension liner. Later on, it was extended to prepare different types of gas turbine engine materials. Table 1 shows some of the components of gas turbine engine which can be prepared using ceramic matrix composites [6,10]. Some of the automobile parts prepared from ceramic were recently accounted by Akira Okada [10]. CMCs are increasingly being used to make gas turbine engine in USA [11,12], Europe [13,14] and Japan [15-17].

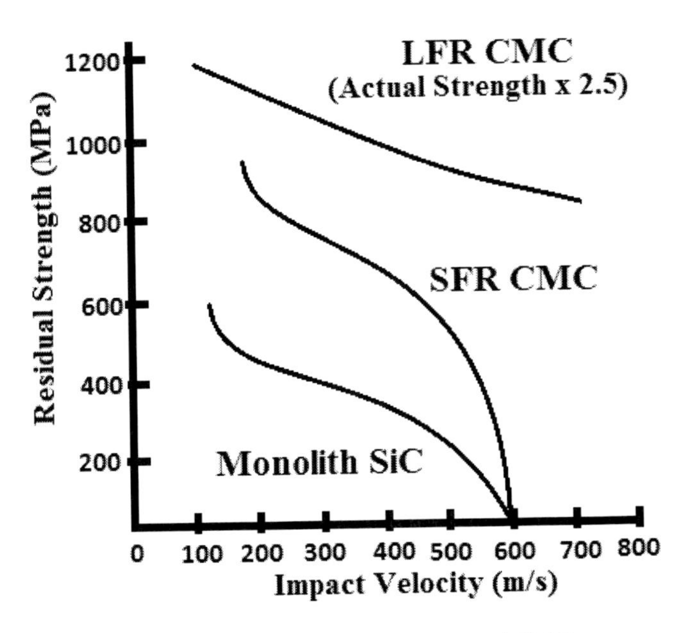

Figure 1. Comparison of strengths of long and short fiber reinforced CMCs.

Table 1. Some of the gas turbine components prepared from CMC [19]

S. No.	Automobiles parts	CMC used
1	Inner shroud	Carbon fiber reinforced SiC
2	Inner scroll support	Carbon fiber reinforced SiC Silicon nitride reinforced SiNC
3	Inner scroll	Carbon fiber reinforced SiC
4	Orifice liner	Milled carbon fiber and TiB_2 powder reinforced SiC
5	Extension liner	Chopped SiC fiber reinforced SiC
6	Combustion liner	Carbon fiber reinforced SiC
7	Outer shroud	SiC whisker reinfrorced SiAlON In-situ Si_3N_4 reinforced Si_3N_4
8	Outer shroud	SiC fiber reinforced SiNC SiN fiber reinforced SiNC
9	Turbine rotor	SiC whisker reinfrorced SiAlON In-situ Si_3N_4 reinforced Si_3N_4
10	Turbine rotor	Carbon fiber reinforced SiC
11	Back plate	SiC whisker reinfrorced SiAlON In-situ Si_3N_4 reinforced Si_3N_4

The typical projects related to gas turbines are Integrated High Performance Turbine Engine Technology (IHPTET) [11], High Speed Civil Transport (HSCT), propulsion system in High Speed Research (HSR) [18], Continuous Fiber Ceramic Composite (CFCC) Program [12] in the USA and the Novel Oxide Ceramic Composites project [13,14] funded by the Commission of the European Community (CEC) in Europe and Research Institute of Advanced Material Gas Generator (AMG) [16] in Japan.

2.2. Methods for the Preparation of Ceramic Matrix Composites (CMCs)

CMC can be prepared by combining reinforcing ceramic materials with ceramic particles, or whiskers or fibers. As Roso discussed, high processing temperature and crake formation are the two challenges face by the industries during the preparation of CMCs [20]. Continuous unidirectional reinforcements, discontinuous reinforcements or multi-layer, multi-directional reinforcements are some of the reinforcements used during the preparation of CMC. Processing methods commonly used include slip casting or injection molding followed by sintering to full density in a high-temperature-capable furnace. For making particulate or whiskers-based discontinuous CMCs, first, particles or whiskers of a ceramic are mixed with the matrix ceramic powder and then densifying the mixture using techniques such as sintering or hot pressing. In the case of fiber reinforced CMCs, first, constructing a pre-form of long and high-strength ceramic fibers are carefully carried out and then ceramic matrix are added by the methods such as chemical vapor deposition, sol-gel or polymer infiltration, or melt infiltration [10,21]. The shaping and sintering processes can also be combined using unidirectional hot pressing or hot isotatic pressing. Net or near-net shape processing can be achieved with final machining often limited to satisfying high-tolerance dimensions or surface finishes [22]. In the preparation of multilayer and multidirectional reinforcement composites, lamination is effectively used for the production of CMCs with high degree of strength, elastic and thermoplastic tailoring [23]. Figure 2 shows force displacement curve for a monolithic ceramic and CMCs illustrating the greater energy of fracture of the CMCs [24].

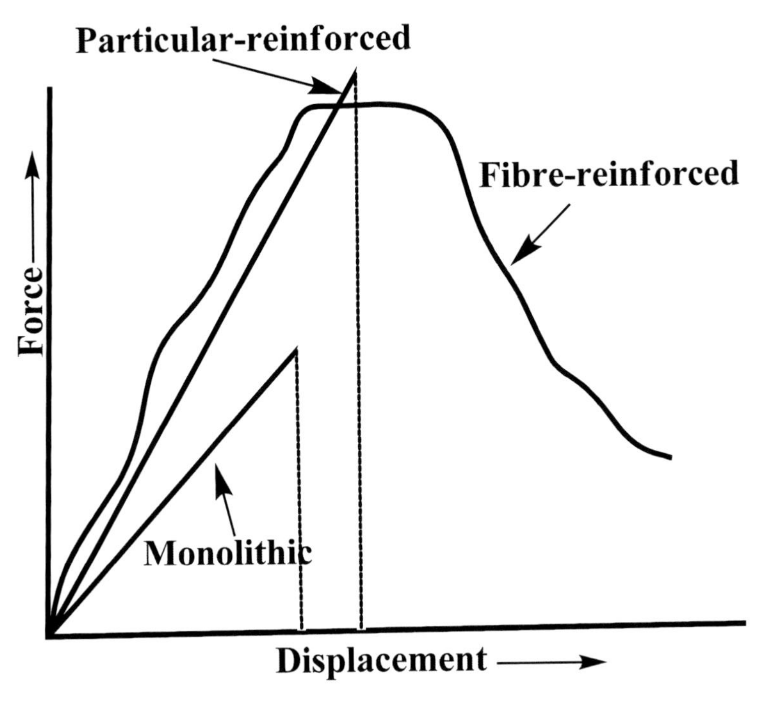

Figure 2. Force-displacement curves for a monolithic ceramic and CMCs illustrating the greater energy of fracture of the CMCs.

2.3. Ceramic Matrix Composites (CMCs) as Brake Materials

In the field of making brake material for automobile application until 1970s, asbestos based material was used. As asbestos is one of the cancer suspect agents, various composites materials were considered to make brake pads for automobiles [25]. A good brake material should have superior wear resistance, low specific gravity, good corrosive resistance and excellent external thermal stability under severe conditions. To meet these requirements, in 1920s, metal matrix composites based on aluminum were considered. As aluminum metal matrix alloys have low melting temperature, carbon-carbon composites have been used in aircraft brake systems and racing cars due to their low density and excellent thermal stability. But carbon-carbon composites show poor brake performance at low temperature and high humidity. Hence, recently ceramic matrix composites were proposed as an alternative material to make brake discs. Jang et al. proposed C/C-SiC composites as the material to make brake disc [26]. Krnel et al. prepared C/C-SiC dual-matrix composites for making automobile brake disc [27]. They proposed that the coefficient of friction (COF) of the composites material are same in the range of traditional steel disc, but COF is increasing with increasing operational temperature which is not the case with steel disc.

Jang et al. have studied the friction and the wearness of CMCs designed for the automotive break discs [26]. The ceramic composites were produced by reinforcing a SiC matrix with carbon fibers using a liquid silicon infiltration method (LSI). Fiber-reinforced CMCs have much potential to overcome the drawback of conventional high-temperature structural ceramics i.e., brittleness and low reliability [28]. A method to improve thermal stability of the carbon fiber /silicon carbide fibre/Si-B-C-N_{matrix} composites up to 1700 °C was reported by Lee et al. [29]. These composites retained its strength and displayed non-brittle fracture behavior up to 1500 °C. CMC also retained 98% of the original strength and fiber pull-out behavior after heating at 1700 °C for 10 h. The flexural creep strain of the CMC was only 0.25% after testing at 1400 °C for 60 h under a pressure of 100 MPa. These results indicate that carbon fiber/silicon carbide fibre/Si-B-C-N_{matrix} composites are the novel and potential brake materials for application at very high temperature [29].

3. Metal Matrix Composites for Automobiles

Metal matrix composites (MMCs) are composed of at least two parts, among them one part is a matrix and another part is a reinforced material. Among the two parts, one part may be a metal and another part may be some other material, such as ceramic or another metal. MMCs are generally prepared to replace heavy metals or alloys with improved properties. Based on the size, shape and amount of the second part, properties of the composites vary. Reinforcement of aluminum alloys with solid lubricants such as hard ceramic particles, short fibers and whiskers results in MMCs with precise balances of mechanical, physical and tribological characteristics. Generally, increasing the weight fraction of the reinforcement phase in the matrix leads to an increased stiffness, yield strength and ultimate tensile strength. When monolithic matrix is replaced with MMCs, improved strength [30], improved tensile strength [31], improved fatigue strength [32], improved thermal shock resistance [33], improved Young's modulus, improved corrosion resistance, reduced vehicle weight, improved wear resistance properties [30] etc. can be achieved. The biggest disadvantage of MMCs is its high cost of fabrication which reduces their actual applications [34]. The incorporation of ceramic particles with aspect ratios (length to diameter ratio) of near unity in aluminum alloy matrices typically decreases many of the good properties of alloys, and this can be offset by careful use of short ceramic fibers or whiskers with aspect ratios over 200.

3.1. Materials Selection for the Preparation of Metal Matrix Composites (MMCs)

To increase the fuel efficiency, automobile companies wanted to replace iron, steel and other heavy metals with light weight metals without compromising the performance of the materials. Aluminum is the most suitable popular material to be used for the preparation of MMCs in the field of automobiles due to its light weight, high strength to weight ratio, high stiffness or wear resistance properties (tribological properties), but pure aluminum cannot be used due to its low melting point and strength and also its extremely poor resistance to seizure and galling [35]. But, aluminum alloys possess good mechanical and physical properties that make them a good alternative to aluminum. For example, 16 % silicon containing aluminum alloys are quite attractive due to their suitable properties such as low density, good corrosion resistance, high thermal and electrical conductivity [36,37], and high damping capacity [5,38]. Aluminum-silicon eutectics are quite common in the fabrication of automobile engine components such as blocks and cylinder heads. The properties of alloys can further be improved by the reinforcing ceramic fibers in to light weight alloys which result the formation of MMCs [39]. The properties of the MMCs are the combination of reinforcing material with the metal matrix. Continuously reinforced MMCs are superior to the short-fiber/whisker-reinforced MMCs which in turn are superior to matrix metal/alloys. Currently light weight metal alloys, especially on magnesium alloys, aluminum alloys and titanium alloy matrices and high temperature withstanding nickel based super alloys are some of the alloys used to prepare MMCs [40]. In addition to this, metal carbides (SiC, TiC, ZrC, TaC, WC, B_4C), metal nitrides (AlN, BN, TaN, ZrN, Si_3N_4, TiN), metal borides (TaB_2, ZrB_2, TiB_2, WB) and metal oxides (ZrO_2, MgO, SiO_2, TiO_2, Al_2O_3, ThO_2) are also used. In the automobile industry, SiC, TiC and Al_2O_3 are widely being used as primary reinforcing materials in 5-30 % range of volume fraction. Cementite is known to be hard, brittle, stable at room temperature and it is one of the very good materials for the preparation of MMCs [41].

3.2. Methods for the Preparation of Metal Matrix Composites (MMCs)

MMCs can be prepared by various casting methods [42-47] such as melt stirring [48], pressure-less liquid metal infiltration [31,49,50], pressure filtration, semi-solid slurry processing [51], exothermic dispersion, co-spray deposition process of dispersion particles in atomized metal alloys [52-54], CVD, physical vapour deposition (PVD) and powder metallurgy methods [49,55]. During the preparation of MMCs through casting methods particles [56,57], platelets [42], non-continuous (short) fibers and continuous (long) fibers can be used as the reinforced materials. Generally non-continuous SiC particles and whiskers are being used with light weight Al for the production of low cost, isotropically composites structures. In powder metallurgy methods, composite materials can be prepared using the reinforced materials such as particles, platelets, non-continuous fibers and continuous fibers [40]. In the production of MMCs, melting metallurgy technique is more useful than powder metallurgy technique as the former technique can be carried out through casting methods which are more economical. Melting metallurgy involves three processing techniques namely melt stirring, gas pressure infiltration and squeeze casting or pressure casting., cold forging and thixoforging are the other two other methods used for the production of MMCs [58].

3.3. Applications of Metal Matrix Composites (MMCs) in Automobiles

MMCs have been used commercially in the automotive market for nearly 25 years. MMCs based on light metal alloys with very good mechanical properties find applications in the production of aircraft and cars. MMCs find applications to make automobile parts such as applications in engine part as brake rotors, pistons, connecting rods and integrally cast engine blocks, brake backing pads, drive shaft, oil pump rotor, suspension components, cylinder liners etc. Tribological application of the aluminum MMCs was discussed in detail by Prasad et al. [59]. The automobile industries have successfully used Al-based particulate composites such as SiC/Al and Al/Al_2O_3, for making pistons, engine blocks, disc rotor brakes, drums, calipers, connecting rods, drive shafts, snow tire studs and other automotive parts. In early 1990s, Honda developed a thin-walled cylindrical ceramic preform fibers for the preparation MMCs using squeeze infiltrated molten aluminum alloys which provided higher performance [60]. Engine blocks were prepared out of these MMCs. Diesel engine pistons containing Saffil (Al_2O_3) short fibers have been used by Toyota since 1985. Graphite reinforced in aluminum, silicon carbide reinforced in aluminum, aluminum oxide reinforced in aluminum are used to make automobile pistons, cylinder liners, piston rings, connecting rods which can reduce the wear, anti seizing, cold start, lighter, conserves fuel, improved efficiency. Silicon carbide reinforced aluminum MMCs can be used in turbocharger impellers at high temperature conditions.

Due to light weight and high strength, titanium composites are more attractive for aerospace and automotive industries. Among α, β, and metastable-β titanium alloys β-form alloy is most useful for the automotive applications. Titanium MMC is one of the promising materials which could be employed to enhance stiffness and wear resistance. These titanium alloys have higher tensile strength-to-weight ratios as well as better strength retentions at 400–500 °C than those of aluminum alloys. Titanium MMCs are used in applications where performance is demanded without regard to cost effectiveness.

4. Polymer Matrix Composites for Automobiles

Polymer matrix composite (PMC) consists of a polymer matrix and a reinforcing material such as fiber or particle or whisker. PMCs are considered to be a more prominent class of composites when compared to CMCs and PMCs. In the field of automobiles, most of polymer composite technologies came out in the early to mid 1990s through the Partnership for a New Generation of Vehicles (PNGV) project. The automotive plastics industry is now ready to develop innovative products with new standards in design, safety, and environmental performance [61]. As PMC is lighter than CMC and MMC, in general, the former class of materials is good alternative materials to reduce the weight of the automotive body. In addition, the processing of PMCs does not involve high pressure and high temperature, as a result, the equipments required for manufacturing polymer matrix composites are not complex. For this reason, PMCs can be developed easily.

4.1. Selection of Materials for the Preparation of Polymer Matrix Composites (PMCs)

The selection of polymer for the fabrication of automotive component depends on the physical properties of polymeric materials which cover an extremely broad range. In particular their strength and stiffness are low compared to metals and ceramics. These difficulties are overcome by reinforcing other materials with polymers. To strengthen the material, two types of PMCs can be prepared. Fiber reinforced polymer (FRP) and particle reinforced polymer (PRP) are being used for the automotive applications. In particular, thermoplastics-based FRPs have important role in preparing PMCs where high strength, high stiffness, corrosion resistance, parts integration, and energy absorption are required. Though the properties of the PMCs vary with the type of polymer matrix and reinforcing material used, in general, they exhibit high tensile strength and modulus. Depends upon the temperature at which the PMCs need to be used, the selection of polymer matrix also varies. The temperature range over which polymeric materials can be used depends on the glass transition temperature, the crystallization melt temperature, and the degradation temperature. Rubber modified plastics are available which remain ductile down to -40 °C while high-temperature plastics can remain rigid up to 300 °C. Unsaturated polyester matrices based PMCs can withstand 70 °C for longterm use whereas it can be used up to 100 °C for longer use. Similarly, epoxide matrices can withstand up to 125 °C for long time use whereas it can be used up to 200 °C if it is short time use. Phenolic resin matrices can withstand up to 600 °C it is exposed for very short time but for longer time use it is stable up to 250 °C only. Polyimides are another class of matrices which can be used between 315 °C and 400 °C.

Carbon fiber reinforced composites are an alternative to glass fiber reinforced composites because the former type of materials are stiffer than the latter type of materials, and therefore carbon fiber reinforced composites show better properties. As carbon fiber reinforced PMCs are lighter than the glass fiber reinforced PMCs, the former type of materials significantly reduces weight of the automobile. As a result, carbon fiber reinforced composites have been used by automobile industries for the past two decades. The length, shape, orientation and composition of the fiber and the integrity of the bond between the fiber and the matrix decide the performance of PMCs. Nylon and phenolic resins can be used where good mechanical properties are important. For instance, speedometer gears, window regulator rollers, water pump bearings, switch bodies, carburetor spacers, and distributor covers can be prepared using these types of PMCs.

In the past decade, natural fiber based PMCs using thermoplastic and thermoset matrices have been used by European car manufacturers to produce door panels, seat backs, headliners, package trays, dashboards, and interior parts. The widely used resin matrix for the preparation of natural fiber-based PMCs is thermoset epoxy resin. Other thermoset resins such as bismaleimides and polyimides are also increasingly used in automobiles for high temperature environments [62,63]. The processing, morphological, and thermo mechanical properties of wood fiber reinforced with polypropylene, and poly(lactic acid) were studied. Tensile and flexural properties were found to be higher when compared with the virgin samples. Dynamic mechanical analysis revealed that the incorporation of wood fiber lead to higher storage modulus and lower tan delta value. Morphological results also reveal that less fracture surface can be result [64].

The commonly used thermoplastic in the automotive applications are polypropylene, polyethylene, polystyrene and polyamide but most commonly used thermoplastic is polypropylene, particularly for nonstructural components. Polypropylene is widely used due to its low density, excellent processability, good mechanical properties, excellent electrical properties, and good dimensional stability and good impact strength [65]. Large numbers of polymeric materials, both thermosetting and thermoplastic, are used as matrix materials for the composites. Some of the major advantages of resin matrices are low density, good corrosion resistance, low thermal conductivity and low electrical conductivity. Polymeric matrices are translucent and they can be prepared in different colors with aesthetic look. But, low transverse strength and low operational temperature limit are some of the major disadvantages of them. Polymer matrices are generally selected on the basis of adhesive strength, fatigue resistance, heat resistance, chemical and moisture resistance. The resin matrix must be able to wet and penetrate into the fibers to get good PMCs. Shear, chemical and electrical properties of a PMC depend primarily on the resin and hence, it is important to select the suitable resin according to the end user needs. As already mentioned, the high temperature resistant properties of the composites are directly related more to the matrix, rather than the reinforced fibers. The higher range of operational temperature of PMCs can be achieved when engineering polymers such as polyimide, polyaminoxaline and polybenzthiazole matrices are used.

CONCLUSION

The preparation and applications of polymer/ceramic/metal composites have been discussed in the present chapter. Ceramic matrix composites are a good alternative to the monolithic ceramics in order to prepare automobile engine parts. Though CMCs can be prepared using either discontinuous or continuous fibres, CMCs prepared using continuous fibre reinforcements show superior properties. MMCs, in particular those based on Al-matrixes, are suitable materials for manufacturing gears, clutch parts, brakes callipers, turbochargers, pumphousing, valves, pulleys, etc. Both fibers reinforced and particle reinforced polymer composites can be used for the automotive applications. Polymer composite can be used to make secondary load-bearing aerospace structures, automotive parts, radio controlled vehicles, brake and clutch linings. CMCs can be used to make automobile engine parts whereas MMCs and polymer composites can be used to reduce the weight of the automobile, as a result fuel efficiency of the vehicle can be increased.

REFERENCES

[1] Clothier, R. A.; Fulton, N. L.; Walker, R. A. *J. Risk Res.* 2008, *11*, 999-1023.
[2] Mazumdar, S. K.; Composites Manufacturing Materials, Product and Process Engineering, CRC Press, Boca Raton London, New York, Washington, DC, 2002.
[3] Jacob, G. C.; Fellers, J. F.; Starbuck, J. M.; Simunovic, S. *J. App. Poly. Sci.* 2004, *92*, 3218-3225.
[4] Belyea, M. O.; Deckman, B. W. *Materials & Design* 1988, *9*, 78-84.
[5] Narula, C. K.; Allison, J. E.; Bauer, D. R.; Gandhi, H. S. *Chem. Mater.* 1996, *8*, 984-1003.
[6] Kaya, H. *Comp. Sci. Tech.* 1999, *59*, 861-872.
[7] Claub, B. In *Ceramic Matrix Composites*; Krenkel, W., Ed.; WILEY-VCH Verlag GmbH & Co. KGaA: Weinheim, 2008; pp 1-20.
[8] Krenkel, P. D. I. W., Ed.; Wiley-VCH Verlag GmbH & Co. KGaA, 2008.
[9] Lee, S.-H.; Weinmann, M. *Acta Materialia* 2009, *57*, 4374-4381.
[10] Okada, A. *Mat. Sci. Eng. B* 2009, *161*, 182-187.
[11] *IHPTET brochure, Wright-Patterson AFB, USA.*

[12] *Continuous Fiber Ceramic Composite (CFCC) Program, Office of Industrial Technologies, US Department of Energy, January 1997.*
[13] *Holmquist M, Lundberg R, Razzell T, Sude O, Molliex L, Adlerborn J. Development of ultra high temperature ceramic composites for gas turbine combustors. ASME, 97-GT-413, 1997.*
[14] MH, V. d. V.; MR, N. *Ceram. Eng. Scien. Proceed.* 1996, *3*, 3-21.
[15] Ohnabe H. *Potential application of composites and other advanced materials to SST/HST propulsion system in Japan. 11th Coordinating Committee of HYPR, May 1996.*
[16] Hiromastsu M, Seki S. *Status of Advanced Gas-Generator and Development Project, 95-YOKOHAMA-IGTC-134, 1995:I-203-I-210.*
[17] Hiromatsu M. Status *of Advanced Material Gas Generator, Proceedings of the 24th Meeting of the Gas Turbine Society of Japan. Tokyo: The Gas Turbine Society of Japan, 1996:1-8 (in Japanese).*
[18] Stephens R, Hecht RJ, Johnson AM. *Material requirements for the high speed civil transport. ISABE 1993;1:701-710.*
[19] Kaya, H. *Comp. Scien. Technol.* 1999, *59*, 861-872.
[20] Rosso, M. *J. Mat. Process. Technol.* 2006 *175*, 364-375.
[21] http://www.advancedceramics.org/clientuploads/pdf/ceramics_roadmap.pdf.
[22] Freitag, D. W.; Richerson, D. W. In *Opportunities for Advanced Ceramics to Meet the Needs of the Industries of the Future*: Prepared by U.S. Advanced Ceramics Association and Oak Ridge National Laboratory for the Office of Industrial Technologies Energy Efficiency and Renewable Energy U.S. Department of Energ.
[23] Peter, S. T.; Springer-Verlag, 1998.
[24] *F.L. Matthews, R.D. Rawlings, Composite Materials: Engineering and Science, Chapman & Hall, 1994.*
[25] Li, Z.; Xiuping, D.; Heng, Z. *J. Wuhan University of Technol. Mater. Sci.* 2009, *Ed. Feb.*, 91-94.
[26] Jang, G. H.; Cho, K. H.; Park, S. B.; Lee, W. G.; Hong, U. S.; Jang, H. *Met. Mater. Intl.* 2010, *16*, 61-66.
[27] Krnel, K.; Stadler, Z.; Kosma, T. *Mat. Manuf. Process.* 2008, *23*, 587-590.
[28] Krenkel, W.; Naslain, R.; Schneider, H.; Weinheim: Wiley-VCH, 2006.
[29] Lee, S. H.; Weinmann, M. *Acta Materialia* 2009, *57*, 4374-4381.
[30] Kumar, G. B. V.; Rao, C. S. P.; Selvaraj, N.; Bhagyashekar, M. S. *Journal of Minerals & Materials Characterization & Engineering,* 2010, *9*, 43-55.
[31] Wang, W.-X.; Takao, Y.; Matsubara, T. *Composites: Part A* 2008, *39*, 231-242.
[32] Tan, M.; Li, X. In *materials science forum*; Lai, M., Lu, L., Eds., 2003.
[33] Delannay, E.; Colin, C.; Marchal, Y.; Tao, L.; Boland, F.; Cobzaru, P.; Lips, B.; Dellis, M. A. *J. Physique IV Colloque C7, supplBment au J. de Physique III* 1993, *3*, 1675-1684.
[34] *T.W.Clyne, (2001), Metal Matrix Composites: Matrices and Processing, Encyclopedia of Materials : Science and Technology ,p- 8.*
[35] Prasad, S. V.; Asthana, R. *Tribology Letters* 2004, *17*, 445-453.
[36] Coleman, S. L.; Scott, V. D.; McEnaney, B. *J. Mat. Sci.* 1994, *29*, 2826-2834.
[37] Durai, T. G.; Das, K.; Das, S. *J. Alloys and Comp.* 2008, *462*, 410-415.
[38] Froes, F. H. *Materials Science and Engineering* 1994, *A 184*, 119-133.

[39] DELANNAY, E.; COLIN, C.; MARCHAL, Y.; TAO, L.; BOLAND, F.; COBZARU, P.; LIPS, B.; DELLIS, M.-A. *JOURNAL DE PHYSIQUE IV Colloque C7, supplBment au Journal de Physique III* 1993, *3*, 1675-1684.
[40] Kaczmara, J. W.; Pietrzak, K.; Wlosinski, W. *Journal of Materials Processing Technology* 2000, *106*, 58-67.
[41] Chaira, D.; Sangal, S.; Mishra, B. K.; Chatterjee, U. K.; Dhindaw, B. K. *International Conference on Advances in Materials and Materials Processing (ICAMMP-2006)* 3-5 February 2006, 79-85.
[42] RAY, S. *Bull. Mater. Sci.* 1995, *18*, 693-709.
[43] Pai, B. C.; Rohatgi, P. K.; Venaktesh, S. *Wearbol* 1974, *30*, 117.
[44] Rohatgi, P. K.; Pai, B. C. *Wearbol* 1980, *59*, 323.
[45] Krishnan, B. P.; Raman, N.; Narayanswamy, K.; Rohatgi, P. K. *Wearbol* 1980, *80*, 205.
[46] Biswas, S.; Rohatgi, P. K. *Tribol. Int.* 1983, *16*, 89.
[47] Froes, F. H. *Mat. Sci. Eng. A* 1994, *184*, 119-133.
[48] Skibo, M. D.; Schuster, D. M. 1988.
[49] Ramesh, K. C.; Sagar, R. *Int J Adv Manuf Technol* 1999, *15*, 114-118.
[50] Aghajanian, M. K.; Rocazella, M. A.; Burke, J. T.; Keck, S. D. *J. Mat. Sci.* 1991, *26*, 447-454.
[51] Mehrabian, R.; Rick, R. G.; Flemings, M. C. *Met. Trans.* 1974, *5*, 1899.
[52] Corbin, S. F.; Wilkonson, D. S. *Canad. Metall. Quart.* 1996, *35*, 189-198.
[53] Doel, T. J. A.; Bowen, P. *Composites A* 1996, *27*, 655-665.
[54] White, J.; Hughes, I. R.; Willis, T. C.; Jordan, R. M. *J. Phys. Colloq.* 1987, *C3*, 347.
[55] Hamiuddin, M. *Powder Metall. Inst.* 1987, *19*, 2.
[56] Corbin, S. F.; Wilkonson, D. S. *Canad. Metall. Quart.* 1996, *35*, 189-198.
[57] Gupta, M.; Lai, M. O.; Soo, C. Y. *Mater. Sci. Eng. A* 1996, *210*, 114-122.
[58] *http://www.uni-kassel.de/fb15/ipl/metform/hp_112004/page_de/publikationen/online/Thixoforming-3.pdf.*
[59] Prasad, S. V.; Asthana, R. *Trib. Lett.* 2004, *17*, 445-453.
[60] *M. Ebisawa, T. Hara, T. Hayashi and H. Ushio, 'Production Process for Metal Matrix Composite (MMC) Engine Block', SAE Special Paper Series, 910835.*
[61] *http://www.popsci.com/cars/article/2004-09/plastics-automotive-markets-vision-and-technology-roadmap.*
[62] Wambua, P.; Ivens, J.; Verpoest, I. *Composites Science and Technology* 2003, *63*, 1259-1264.
[63] Tserki, V.; Zafeiropoulos, N. E.; Simon, F.; Panayiotou, C. *Composites Part A: Applied Science and Manufacturing* 2005, *36*, 1110-1118.
[64] Huda, M. S.; Drzal, L. T.; Mohanty, A. K.; Mishra, M. *Paper presented in 5th annual SPE automotive composites conference, Sept. 12-14, 2005, Troy, Michigan*
[65] George, J.; Sreekala, M. S.; Thomas, S. *Poly. Eng. and Science* 2001, *41*, 1471-1485.

In: Advanced Organic-Inorganic Composites
Editor: Inamuddin

ISBN 978-1-61324-264-3

Chapter 8

LIGHT EMISSION FROM ORGANIC-INORGANIC COMPOSITE MATERIALS

Muhammd. Jamil[a,b,c], Farzana Ahmad[c,*], Jin Woo Lee[c], and Young Jae Jeon[c*]

[a]Division of International Affairs, University College, Konkuk University, Seoul 143-701, Korea
[b]High Energy Physics Lab: IAP, Department of Physics, Konkuk University, Seoul 143-701, Korea
[c] Liquid Crystal Research Center, Konkuk University, Seoul 143-701, Korea

ABSTRACT

In the beginning, thin film electroluminescence devices based on organic materials were very attractive among researchers because of having the advantageous features like color tunability, low operating voltages, and ease of fabrication (Kumar, N. D. et al. *App. Phys. Lett.* 1997, *71,* 1388-1390). However such devices fabricated with organic materials have noticeable drawbacks i.e they exhibit inferior charge transport properties and low chemical stability. Furthermore, at higher applied voltages, using device made of organic semiconductors can cause joule heating that damage the organic thin film/metal interface which seriously affects the stability and the operational lifetime of such devices. In recent years, organic-inorganic composite materials and multilayer structures are gaining interest because their better stability, electroluminescence and photoluminescence spectra. This chapter reviews and addresses the light emission characteristics of both the organic and inorganic composite materials as well as discusses about the organic and inorganic-hybrid materials. These comparative studies based on the organic and inorganic composite materials e.g., addressing their working principles, the basic characteristics and fabrication methods in detail are considered very important for their further development and future applications. Moreover, some of the applications of organic, inorganic and organic-inorganic hybrid composite materials are also presented.

[*] Corresponding author's e-mail: farzana@konkuk.ac.kr, yjjeon@konkuk.ac.kr

ABBREVIATIONS

Alq_3	Tris(8-hydroxyquinolinato) aluminum
B34	Bentone 34
BCP	Bathocuproine,[2,9-Dimethyl-4,7-diphenyl-1,10-phenanthroline], a derivative of phenanthroline
BGO	$Bi_4Ge_3O_{12}$
CIE	International Commission on Illumination
CRTs	Cathode ray tubes
EL	Electroluminescent
ELD	Electroluminescent display
EQE	External quantum efficiency
ETL	Electron transport layer
FWHM	Full wave half maximum
HOMO	Highest occupied molecular orbital
HTL	Hole transport layer
HTMs	Hole-transport materials
HyLEDs	Hybrid light-emitting devices
ILEDs	Inorganic light emitting diodes
IR	Infrared
ITO	Indium tin oxide
LCD	Liquid crystal display
LEDs	Light emitting diodes
LuAG	$Lu_3Al_5O_{12}$
LUMO	Lowest unoccupied molecular orbital
MEH-PPV	Poly[2-methoxy-5-(2'-ethylhexyloxy)-p-phenylene vinylene]
NBP	N,N′-diphenyl-1,1′-biphenyl-4,4′-diamine
Nc-PbS	Nano-crystal Lead sulfide
NIR	Near-infrared
NPB	N,N′-bis(lnaphthyl)- N,N′-diphenyl-1,1′-biphenyl-4,4□-diamine
N-RR	Non radiative relaxation
OLEDs	Organic light emitting diodes
OVPD	Organic vapor phase deposition
PBD	2-(4-biphenyl)-5-(4-tert-butylphenyl)-1,3,4-oxadiazole
PCEs	Power conversion efficiencies
PEDOT:PSS	Poly(3,4-ethylenedioxythiophene) poly(styrenesulfonate)
PL	Photoluminescence
PMMA	Polymethyl methacrylate
PPV	Poly(p-phenylene vinylene)
PPV-B34	Poly(p-phenylene vinylene)- Bentone 34
PVK	Poly(9-vinylcarbazole) polymer matrix
PZT	Lead–zirconate–titanate
QD-LEDs	Quantum dot light emitting diodes
RGB	Red green blue
SNOM	Near-field optical microscope

SOLED	Stacked OLED
TD	Thick dielectric
TDEL	Thick dielectric electroluminescent displays
TFEL	Thin-film electroluminescent displays
TPD	N,N′-diphenyl-N,N′-bis(3-methylphenyl)(1,1′ biphenyl)-4,4′ diamine
UV	Ultra violet
VGA	Video Graphics Array
Vis	Visible
VUV	Vacuum-ultraviolet
YAG	$Y_3Al_5O_{12}$

1. INTRODUCTION

Electroluminescent devices based on organic materials are of considerable interest due to their attractive characteristics and potential applications to flat panel displays [1]. Such types of electroluminescent devices are attaining more attention among researchers for color tenability [2-5], low operating voltages [6,7] and ease of fabrication. Since the first observation of electroluminescence from organic crystals for anthracene in 1963 [1,8], a number of research work has been reported from organic polymers [9-11] and fluorescent dye-doped polymeric systems [12,13]. Similarly poly(*p*-phenylene vinylene) and its derivatives showed much promise towards the realization of these devices [9,14,15].

The design of the organic electroluminescent (EL) devices is quite simple which consists of an active emitter layer sandwiched between two electrode materials of dissimilar work functions [5]. When a suitable bias voltage is applied to the electrodes, charge carriers are injected into the active layer. As a result such charge carriers then recombine within the active layer to form excitons. The excitons then undergo radiative decay to give visible radiation or excite other emitter molecules present in the layer [5,10]. In such devices, the active region may be made up of either molecularly doped polymers [5,13], pure polymeric systems, or polymeric composites [5,16-20]. Furthermore, multilayer structures with electron and hole transporting layers in addition to the emitter layer have also been reported to provide balanced transport of charges [4,5,7].

The most accessible advantage of such organic based materials is their high luminescence efficiency, due to the strong binding between the electron and the hole produced by either optical or electrical excitation [21,22-24]. However, the properties of these organic materials have been limited, owing to quenching of the excitation, charge trapping and the catalytic degradation reactions of components in the manufacturing process of devices by high vacuum vapor deposition [23]. Since the efficiencies and lifetimes of resulting devices were considerably lower than those obtained for inorganic systems at the same time, research activities were focused on the inorganic materials [21,23].

In recent years, developments in luminescence display technology have made this research area of great interest. Thus organic-inorganic composite materials and multilayer structures are gaining much attention due to their better stability, electroluminescence and photoluminescence spectra. Studies of organic, inorganic and hybrid organic-inorganic light emitting devices about their working principles, the basic characteristics and fabrication methods in details are important for their future applications. In this review work, we discuss some of the novel approaches that have been implemented to improve the organic, inorganic, and hybrid organic-inorganic light emitting devices. Additionally some of the efficient techniques are reviewed and their future prospects are presented.

2. Light Emission from Organic Composite Materials

With the spread of information technology equipment headed by computers, information display devices have attracted much attention as a man-machine interface, and the realization of solid state flat display devices is essential today [25]. In this area of research cathode ray tubes (CRTs) dominate the field, especially for television applications.

Nevertheless conventional CRTs have disadvantages involved i.e they are heavy and require large volume for large picture size [25]. In order to solve such problems, many other types of flat panel display have been developed which include liquid crystal display (LCD), plasma display, electroluminescent display (EL) and light emitting diodes (LEDs) [25]. Since LEDs are the main category in display technology, so a special emphasis is given to this type in this review.

Over the last decade, organic light emitting diodes (OLEDs) have been rapidly developed due to their potential applications in flat panel displays, domestic solid state lighting etc. [5].

In the 1987, Tang and VanSlyke [26] as well as Saito and Tsutsui et al. [27,28] revived this research on electroluminescence of organic compounds, developing a new generation of light-emitting diodes with organic fluorescent dyes [23]. Based on charge injection-type organic EL diodes research, many studies have been performed on EL devices based on organic thin films [25,29-31]

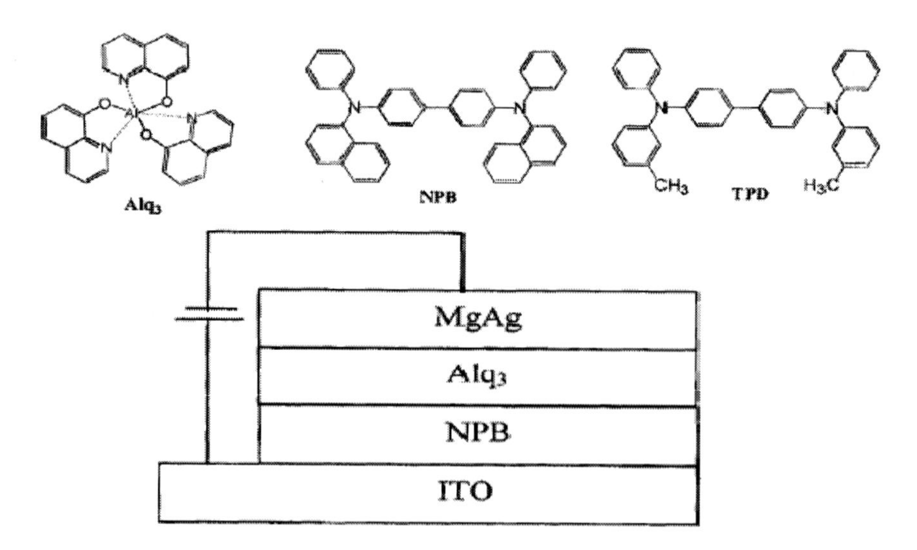

Figure 1. Common electron and hole-transport materials and a two-layer OLED. Reproduced with permission from Hung, L. S.; Chen C. H. *Mater. Sci. Engg.* 2002, *R39*, 143-222. Copyright © 2002, Elsevier Science Ltd.

The organic light emitting diode (OLED) has many advantages for flat panel display, e.g., it is very thin device with high luminance. It has also high luminous efficiency. Furthermore optional color is obtainable by using fluorescent molecules throughout the entire visible region to obtain full color display [25].

2.1. Organic Light Emitting Diode (OLED): Device Configuration and its Operation

The principle of light generation mechanism in organic light-emitting devices (OLEDs) is the radiative decay of molecular excited states, which are known as excitons [5,25,32-35].

The basic device configuration of a conventional OLED consists of glass substrate coated with transparent conducting oxide, an organic and polymeric light emitting layer sandwiched between hole transport layer (HTL) and electron transport layer (ETL) and on the top of the device a metal cathode is also deposited.

According to L.S. Hung et al. [1], when the voltage is applied between the anode and the cathode such that the anode is at a more positive electrical potential with respect to the cathode, injection of holes occurs from the anode into the hole-transport layer, whereas electrons are injected from the cathode into the electron-transport layer. The injected holes and electrons each migrate toward the oppositely charged electrode, and the recombination of electrons and holes occurs near the junction in the luminescent ETL. Upon recombination, energy is released as light, which is emitted from the light-transmissive anode and substrate [1].

In order to enhance the probability of exciton formation and recombination near the interface region, the hetero-junction must be designed to facilitate hole-injection from the HTL into the ETL and to block electron injection in the opposite direction. As can be seen in the Figure 1, the highest occupied molecular orbital (HOMO) of the HTL is slightly above that of the ETL, thus holes can readily enter into the ETL, whereas the lowest unoccupied molecular orbital (LUMO) of the ETL is significantly below that of the HTL, so that electrons are confined in the ETL. The lower hole mobility in the ETL causes an increase in hole density, and thus enhance the collision capture process. Moreover, by spacing this interface at a sufficient distance from the contact, the probability of quenching near the metallic surface can be greatly reduced [1].

Hung and Chen [1], suggests that the simple structure of OLED can be modified into a three-layer structure, in which an additional luminescent layer is introduced between the HTL and ETL to function primarily as the site for hole-electron recombination and thus electroluminescence. In this case, the functions of the individual organic layers are distinct and therefore can be optimized independently. Hence, the luminescent or recombination layer can be chosen to have a desirable EL color as well as a high luminance efficiency. Similarly, the ETL and HTL can be optimized primarily for the carrier-transport property.

The advantage of extreme thin organic EL medium is it offers reduced resistance, that permits higher current densities for a given level of electrical bias voltage. In view of the fact that light emission is directly related to current density through the organic EL medium, the thin layers coupled with increased charge injection and transport efficiencies have allowed acceptable light emission to be achieved at low voltages [1].

2.2. Organic Light Emitting Diode (OLED): Materials

Some of the notable advantages of organic materials over inorganic materials are their excellent color gamut and high fluorescence efficiency [1]. Light is produced in organic materials by the fast decay of excited molecular states. The color of light depends on the energy difference between those excited states and the molecular ground level. Many materials show intense photoluminescence with near unity quantum yield. However, the EL efficiency is limited by the probability of creating non-radiative triplet exited states in the electron–hole recombination [1].

Most of the organic materials support preferentially the transport of either electrons or holes with their mobilities in the range of 10^{-8} to 10^{-2} $cm^2/(Vs)$ [1]. Electron mobility in organic materials is usually orders of magnitude lower than hole mobility. The most important electron-transport material is tris(8-hydroxyquinolinato) aluminum (Alq_3) with its molecular structure (can be seen in Figure 1). The electron mobility in Alq3 strongly depends on electric field with a value of approximately 10^{-6} $cm^2/(V\ s)$ at 4×10^5 V/cm. Alq_3 can also be used as an emissive material, which emits in the green with a broad emission peaking at 530 nm. Other EL colors can be obtained by doping a small amount of specific guest molecules in Alq_3 or by choosing different organic fluorescent materials as emitters. Furthermore in some cases, doping can also enhance luminance efficiency by reducing non-radiative decay [1].

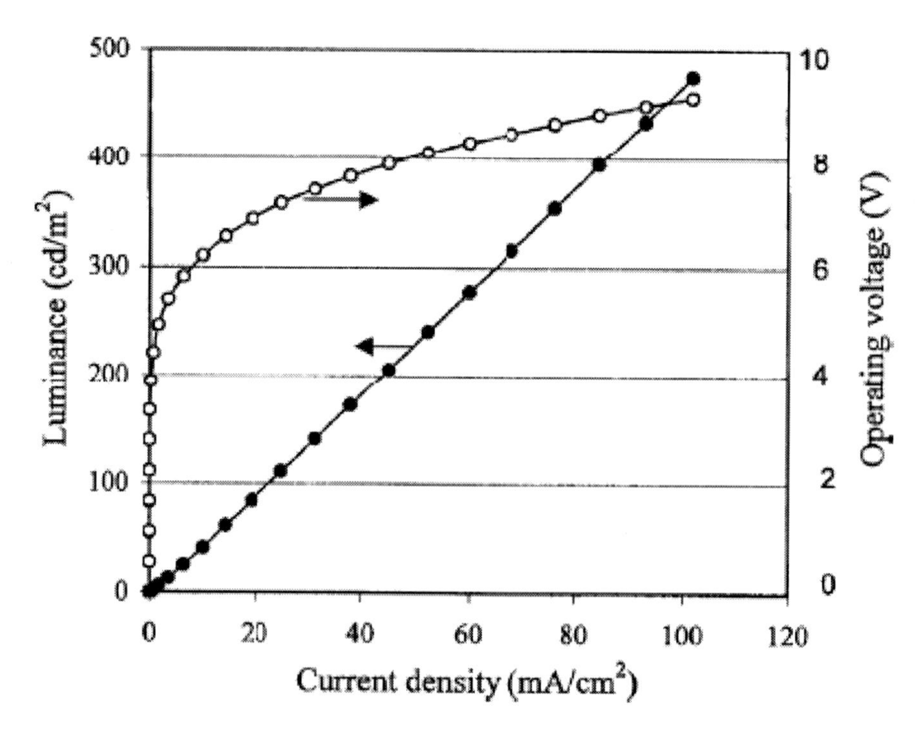

Figure 2. A schematic view showing luminance and operating voltage vs. current density measured on an Alq_3-based OLED. Reproduced with permission from Hung, L. S.; Chen, C. H. *Mater. Sci. Engg.* 2002, *R39*, 143-222. Copyright © 2002, Elsevier Science Ltd.

Till to date, several materials that have been proposed as hole-transport materials (HTMs). Among them, N,N′-diphenyl-N,N′-bis(3-methylphenyl)(1,1′ biphenyl)-4,4′ diamine (TPD) and N,N′-bis(lnaphthyl)- N,N′-diphenyl-1,1′-biphenyl-4,4′-diamine (NPB) have been employed extensively [1]. The holetransport materials in Figure 2 have a glass transition temperature below 100 °C and a hole mobility in the range of 10^{-3} to 10^{-4} $cm^2/(Vs)$. In OLEDs, the hole current dominates the total current, owing to efficient hole-injection and sufficiently high hole mobilities [1].

Generally in OLEDs, a barrier for electron injection is commonly present at the metal-organic contact when the work function of the metal is higher than the LUMO of the organic materials, and thus the use of a low work function metal is highly desirable to facilitate the injection of electrons [1]. Mg alloyed with a small amount of silver is a commonly used as cathode. Mg is a relatively stable metal with a work function of 3.66 eV. The sufficiently lower value of the work function makes it to be useful as an electron-injecting electrode. Further, the small amount of Ag assists the Mg deposition by presumably providing nucleating sites on the alloyed film during co-sublimation [1].

The hole-injecting contact requires a metal that has high work function to match with the HOMO of the organic material. Mostly all OLEDs rely on the transparent and conductive indium tin oxide (ITO) as the anodic material to facilitate hole-injection while permitting light to exit the device in an effective manner. The work function of ITO ranges from 4.5 to 5.0 eV, which strongly depends on the methods of surface treatment. The treatments of ITO glass substrates using ultra violet (UV) ozone or oxygen plasma substantially increase its work function and thus enhance hole-injection from the ITO anode into the HTL [1].

2.3. Organic Light Emitting Diode (OLED): Device Preparation

Vacuum evaporation by resistive heating is considered as the most appropriate for depositing molecular materials. Organic vapor phase deposition (OVPD) has also been demonstrated to deposit organic materials on large substrates [1,36].

Further it is also common to deposit cathode materials using the same vacuum evaporation from filaments. For the deposition of high-temperature metals one may employ e-beam evaporation or sputtering. Among these techniques, the latter is particularly useful for large substrates and high throughput production. However, as reported by Hung and Chen [1], OLEDs are extremely sensitive to radiation, and special care needs to be taken. In e-beam deposition, a magnetic field is applied across the substrate to repel electrons and ions. While in sputter deposition, a buffer layer is required to minimize the radiation damage inflicted on the OLED organic layer stack [1].

According to work reported by Hung and Chen [1], typical device fabrication occurs by the following sequence: devices are grown on glass slides pre-coated with transparent ITO with a sheet resistance of 15–100 Ω/cm. Substrates are ultrasonically cleaned in detergent solution, followed by thorough rinsing in deionized water. In the next step they are then cleaned in organic solvents and dried in pure nitrogen gas. After cleaning, the ITO glass is subject to an oxygen treatment either using UV ozone or oxygen plasma to enhance hole-injection. As can be seen from Figure 2, single heterostructure devices are formed by sequential high vacuum (10^{-5} to 10^{-6} Torr) vapor deposition of a hole-transport layer such as NPB, followed by an electron-transport layer of Alq_3, previously purified by temperature gradient sublimation. The deposition is carried out by thermal evaporation from a baffled Ta crucible at a nominal deposition rate of 0.2–0.4 nm/s. An electron injecting electrode of approximately 10:1 Mg:Ag volume ratio is subsequently deposited by co-evaporation from separate Ta boats at a vacuum of 10^{-5} Torr. In the last step the device preparation is completed with encapsulation in a dry argon box [1].

Usually OLEDs are constructed using glassy and amorphous organic films and consequently provide significant advantages in device fabrication and cost reduction. Such type of OLEDs are pronouncedly different in structures from inorganic LEDs consisting of epitaxial semiconductor thin films [1].

2.4. Organic Light Emitting Diode (OLED): Electrical and Optical Properties

Absorption, the excitation and luminescence spectra are determined with a spectrophotometer. Current–voltage and luminance-current characteristics are measured by using a radiometer and a digital voltmeter. Typical current-voltage and luminance–current characteristics of a device with 75 nm of NPB, and 70 nm of Alq_3 are presented in Figure 2. It is clear from the figure that the voltage required for light emission is strongly dependent on the thickness of the Alq_3 [1].

In OLEDs one can distinguish the quantum efficiency (Z) and the luminous efficiency (Zp). The quantum efficiency Z is defined as the ratio of the number of emitted quanta to the number of charge carriers [1]. Quantum efficiency is a well known quantity, that reflects the comprehensive result of the EL process, but the luminous efficiency has a more technical significance, which is the ratio of the luminous flux emitted by the device and the consumed electric power [1].

2.5. External Quantum Efficiency of Organic Light Emitting Diode (OLED)

Mostly in the conventional small molecule organic materials only the radiative decay of singlet excitons is responsible for producing light and the energy of triplet excitons is wasted which is due to the slow radiative decay and generates delayed fluorescence [1,5]. Consequently, the internal quantum efficiency η_{int} (defined as the ratio of the total number of photons generated within the organic emitter to the number of injected electrons) of fluorescent OLEDs is only 25% [1,5,25,37,41]. The internal quantum efficiency of OLEDs can be achieved near 100% by means of harvesting both singlet and triplet molecular excitation states using electro-phosphorescent materials, which is nearly fourfold increase in efficiency as compared to singlet – harvesting fluorescent organic materials [5,38-41]. Here the external quantum efficiency η_{ext} (known as the ratio of the total number of photons emitted by the OLED into the viewing direction to the number of electrons injected into organic emitter) of an OLED device is related to the internal quantum efficiency η_{int} and the external coupling efficiency $\eta_{coupling}$ (known as the ratio of the total number of photons coupled out in the forward direction to the number of injected electrons) by the following relation [42,43].

$$\eta_{ext} = \eta_{int}\,\eta_{coupling} = \gamma\,\eta_{exc}\,\Phi_p\,\eta_{coupling} \quad (1)$$

In the above expression γ is the electron–hole charge balance factor, η_{exc} is the fraction of total number of excitons formed which result in radiative decay (η_{exc} ~1/4, and ~1 for fluorescence, and electrophosphorescence based OLED materials, respectively). Similarly Φ_p is the intrinsic quantum efficiency for radiative decay (including both fluorescence and phosphorescence) [41,42-43]. It is important to note that the internal quantum efficiency of OLEDs is mainly affected by the non-radiative electron-hole recombination loss and the singlet–triplet branching ratio [37,41-43]. Most of the electro-phosphorescent OLED materials have very small non-radiative loss. On the other hand, despite achieving near 100% internal quantum efficiency, the external coupling efficiency ($\eta_{coupling}$) of the conventional OLED device remains very low. Supposing isotropic emission in the organic layer and a perfectly reflecting cathode, the fraction of generated light escaping from the substrate is [41,42-45]

$$\eta_{exc\text{ , coupling}} = \frac{1}{\xi n^2} \quad (2)$$

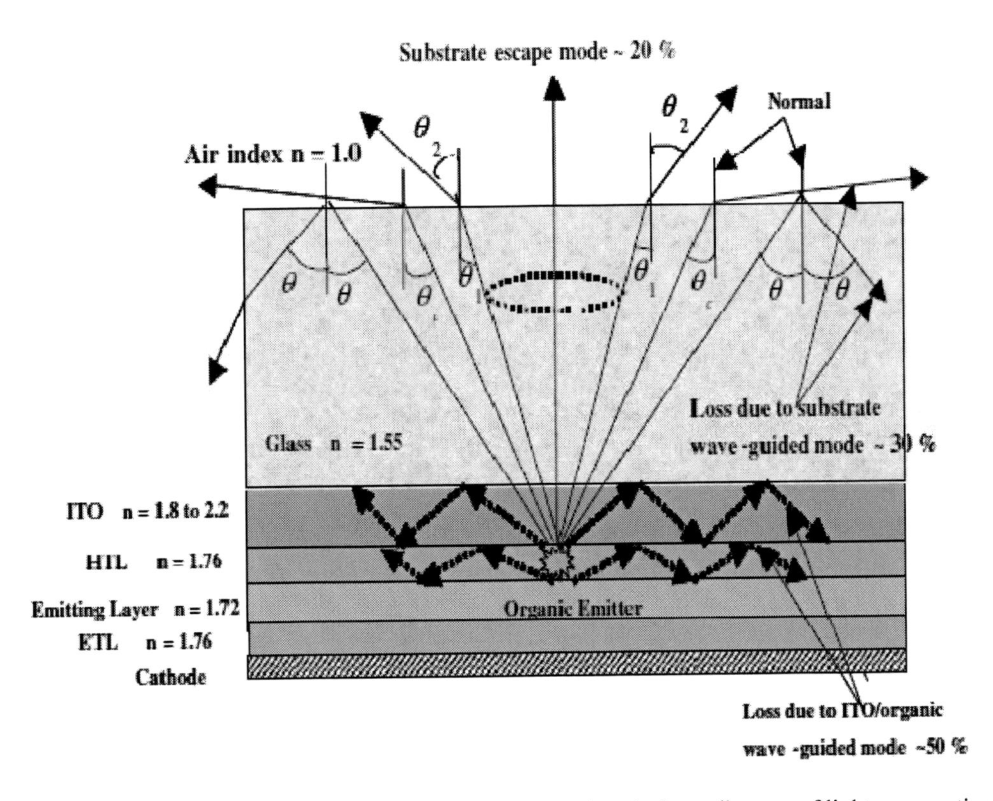

Figure 3. Schematic view of multi-layer OLED structure and optical ray diagram of light propagation *via* various modes, i.e., substrate escape, substrate wave-guided mode and ITO/organic wave-guided modes. Reproduced with permission from Saxena, K. et al. *Optical Mater.* 2009, *32*, 221-233. Copyright © 2002, Elsevier Science Ltd.

Table 1. A comparative study of various light out-coupling techniques implemented in OLEDs

S. No.	Light extraction techniques employed	Achievements in η_{ext}	Remarks	Ref. No.
1.	Conventional OLED structure	~20%	significant loss due to substrate and ITO/organic wave guided modes	[12-16]
2.	Controlling the thickness of ITO/organic	~52 %	Effective in reducing ITO/organic modes but substrate wave-guided modes remains unaffected layers	12
3.	Substrate modification by use of rough surface	~50%	One of the simplest and cost effective method but ITO/organic wave-guided mode remains unaffected	[12,13]
4.	Shaped mesas with truncated right 5.circular cone	By a factor of 2	Very effective method but for fabricating mesa structure at large scale is costly	[12,15]
5.	Spherically shaped patterns on the backside of the glass substrates	By a factor of 9.6	ITO/organic wave-guided modes remains unaffected; large scale production is costly and challenging	[18,19]
6.	Use of high-refractive index substrates with the backside substrate modification	53%	Requires high-refractive index resin substrate with embedded scattering particles; costly for large scale production	[22,23]
7.	Truncated square pyramid luminaire	Twofold enhancement	A commercially viable technique for large scale lighting applications but ITO/organic modes remains unaffected	[24]
8.	Insertion of low-refractive index silica	Twofold enhancement	Very effective and low cost method for extracting ITO/organic modes aerogel between glass substrate and ITO	[26]
9.	Addition of a diffusive layer at the backside of substrate	~50%	Requires an additional, optimized diffusive and optimized light-emitting layer and ITO/organic modes remain unaffected	[27]
10.	Structuring the substrates with small and lenses and pyramids	80%	Lambertian output distribution; extracted light output pattern is homogenous isotropic	[12,28]
11.	Texturing meshed structures on substrate surface and stamped Bragg grating	46%	Effective and low cost method and insensitive to optical wavelength and preserves color spectrum of OLED	[30,31]
12.	Incorporation of periodic dielectric structures of silica micro-spheres	Significant enhancement	Spectral changes due to scattering particles at different observation angles	[32-36]
13.	Phosphor particles coated on the glass Substrates	Strong enhancement	Efficient down-conversion/isotropic re-emission. Low cost method	[37,38]

S. No.	Light extraction techniques employed	Achievements in η_{ext}	Remarks	Ref. No.
14.	Ordered micro-lens array on the backside of the glass substrate	50%	Micro-lens diameter <200 lm has no significant influence on the extraction efficiency. Fabrication cost is high	[41-43]
15.	Micro-lens arrays on high-refractive index Glass	65%	High-cost of fabrication	[42]
16.	Weak micro-cavity OLED	22–55% Depending on the location of recombination zone.	Careful design and lost cost method	[48,49]
17.	Strong micro-cavity OLEDs	1.5–2-fold enhancement	Requires careful design of micro-cavity and location of emission zone. But may lead to spectral changes, such as, narrowing, shifting and modulations	[61-65]
18.	Micro-cavity two-unit tandem OLED	Fivefold enhancement	A fivefold enhancement in OLED luminance was achieved by Cho et al. using microcavity two-unit tandem OLED	[66]
19.	Photonic-crystals and nano-cavity	50% and 140%	Enhancement of 50% and 140% in the luminance efficiency and the peak intensity, respectively, were achieved from OLEDs with 2D-PC. High cost of fabrication	[79]
20.	Surface Plasmon-enhanced OLEDs	40%	One of the potential technique to enhance the out-coupling efficiency of OLEDs	[97]
21.	Nano-structured substrates	50% Optimized photonic crystal nano-pattern, 80% is expected theoretically and	over 50% was achieved experimentally	[107,109]
22.	Nano-wires and nano-particles from glass substrate	46% from glass; 80% from organic layer	Metallic nano-wires on the glass surface and indium tin oxide ITO anode. High cost of production and may lead to spectral changes	[113]
23.	A periodic dielectric mirrors	80%	Effective method but requires multilayer structure	[119]
24.	Low-refractive index embedded grids	By a factor of 2.3	Effective and commercially viable technique	[125]

Here in the above equation n is the refractive index of OLED material and ξ is a constant that depends on the dipole alignment and the geometry of the OLED device. For most of the organic materials n is about 1.7 and taking the value of 2 for ξ, the internal coupling efficiency is only about 20% [41]. Furthermore according to classical ray optics theory about 80% of generated light is lost in wave-guided modes due to glass substrate and ITO/organic material (Figure 3) which means that the majority of the light is either trapped inside the glass substrate and device, or emitted out from the edges of an OLED device [41,44-47]. The external quantum efficiency η_{ext} of OLED can be shown by [42]:

$$\eta_{ext} = \frac{q \int \lambda I_{det}(\lambda) d\lambda}{hcfI_{OLED} \int R(\lambda) d\lambda} \quad (3)$$

where q is electronic charge, k is wavelength, $I_{det}(k)$ is the detected photocurrent, h stand for Planck's constant, c is the speed of light in vacuum, f is OLED-to detector coupling factor, I_{OLED} is current and R(k) is the detector responsivity [41].

Additionally, a comparative study of the various light extraction techniques that have been implemented in OLEDs (Table 1). This table predicts that various techniques have their own merits and demerits. Modifications in the device structure can be done according to the applications of the OLED as well [41].

2.6. Molecular Organic Electroluminescent Materials

This phenomenon of organic electroluminescence was first discovered by Pope in 1963 [48]. However, the development of organic light-emitting device (or diode) known as today's OLED technology actually began in the Chemistry Division of Kodak Research Laboratories in the late 1970s by Tang and coworkers [49-51]. This research led eventually to the discovery of the first efficient multi-layered organic electroluminescent device based on the concept of the heterojunction architecture [1,49-50] which was followed by the disclosure of the doped emitter using the highly fluorescent organic dyes for color tuning and efficiency enhancement [50,51]. Since then, tremendous progress has been made in the field of organic electroluminescence. One of the key enablers in the history of OLED advancement can be attributed to the continuing discovery of new and improved electroluminescent materials which were made possible because of the dedication and ingenuity of many organic chemists who provide the design and skilled synthesis. Indeed, from small molecules, oligomers to conjugated polymers, intense research in both academia and industry has yielded OLEDs with remarkable color fidelity, device efficiencies and operational stability [41].

For further readings, the readers can refer to two other reviews published by Shirota [52], on organic materials for electronic and optoelectronic devices and by Mitschke and Bauerle [53], on the electroluminescence of organic materials, respectively. Similarly the readers who are more interested in polymeric electroluminescent materials can have a look at the excellent reviews [54-57].

The device fabrication consists of following categories:

2.6.1. Hole-injection Materials

Oxidation of the ITO surface by O_2 plasma, CF_4/O_2 plasma [1,58] or UV ozone treatment can reduce the carrier injection energy barrier, remove residual organic contaminants and get its work function up to near 5 eV which is still about 0.5 eV lower than most of the HOMO of the hole-transport materials. Further details can be seen in reference [1].

2.6.2. Hole-Transport Materials

The discovery of using tri-arylamines with a ''bi-phenyl'' center core as the holetransport layer have greatly improved both EL efficiency and operational stability of OLED [59]. Most of the new HTM developed seemed to have all evolved around such theme. The creativity of synthetic chemists and material scientists throughout the world continue to provide the OLED device community with their ever improved products having superb properties and elegant design [1].

Till now, one of the most widely used HTM in OLED is still NPB. One of the reasons for its popularity is because sublimed NPB can be manufactured readily and is thus abundantly available even though its Tg at 98 °C is a trifle low which may affect its morphological stability at high operating temperature. Thus, studies on the design and synthesis of new HTMs have been continually focused on finding materials with high thermal and thin film morphological stabilities and on finding ways to control and optimize carrier injection and transport. The details of such approaches can be seen in reference [1].

2.6.3. Electron-Transport and Host Emitting Materials

Till to date, the most widely used electron-transport and host emitting material in OLEDs is still Alq_3 [1]. This became possible as Alq_3 is thermally and morphologically stable to be evaporated into thin films, easily synthesized and purified, molecularly shaped to avoid exciplex formation (e.g. with NPB at the interface), and green fluorescent to be a good host emitter. Possibly it is still one of the most robust electron-transport backing layers in OLED, particularly with the help of the hole blocker to trap the hole carriers from injecting into Alq_3 [60] or doping with lithium or other alkali metals [61] to assist electron injection to lower the drive voltage [62]. But, this has also numerous shortcomings such as quantum efficiency, mobility, band gap and the ashing problem during sublimation. More about this can be seen in reference [1].

2.6.4. Fluorescent Dopants

A key developments in the advancement of OLED display technology can also be attributed to the discovery of the guest–host doped emitter system [1,63]. This is due to a single host material with optimized transport and luminescent properties may be used together with a variety of highly fluorescent guest dopants leading to EL of desirable hues with very high efficiencies. Additional advantage of the doped emitter system in OLED is the enhancement of its operational stability by transferring the electro-generated exciton to the highly emissive and stable dopant site thus minimizing its possibility for non-radiative decay [1,64]. This doping principle has recently been successfully extended to the exploitation of highly phosphorescent materials leading to nearly 100% internal EL efficiency [1,38]. Further detailed highlights of fluorescent red green blue (RGB) dopant developments and the review phosphorescent triplet emitter ref. [1] can be read.

2.6.5. Triplet Emitting Materials

The discovery of electrophosphorescence which lifts the upper limit of the internal quantum efficiency of the usual fluorescent dopant-based devices from 25% to nearly 100% is considered as one of the key developments in the advance of modern OLED sciences and technology. Phosphorescence is inherently a slower and less efficient process, but triplet states constitute the majority of electrogenerated excited states (~75%), thus the successful utilization of the triplet manifold to produce light should undoubtedly increase the overall luminance [1]. The design and synthesis of triplet emitting materials containing heavy-metal complexes, where strong spin-orbit coupling leads to singlet–triplet state mixing that removes the spin-forbidden nature of the radiative relaxation of the triplet state, are therefore particularly important in achieving high-efficiency electrophosphorescence in OLEDs [1]. For further studies reference [1] can be seen.

2.6.6. Commonly used Electroluminescent (EL) Materials for Organic Light Emitting Diodes (OLED)

The color purity (or International Commission on Illumination (CIE) coordinates) in OLED displays has great importance, which is controlled by the selection of organic materials and appropriate device structures. Some of other important requirements are efficiency and luminance stability. The performance of red, green, and blue OLEDs are summarized in Table 3 of ref. [65]. In such scheme the OLED is composed in order of an ITO anode, a NPB HTL, an emitter, an Alq_3 ETL, and a MgAg or LiF/Al cathode. Further the performance is characterized at 20 mA/cm^2. Some of the recently reported materials have CIE coordinates more close to NTSC standards [1].

3. Light Emission from Inorganic-Composite Materials

In recent years, light emitting devices have made great advances. Among them LEDs are considered very popular and attractive devices. LEDs are electronic devices constituted by a light-emitting semiconductor material sandwiched between two electrodes that glow when a voltage is applied. LEDs categorized into two based on types of semiconductor used: Organic LEDs (OLEDs) and inorganic LEDs (ILEDs). The light emitting diodes (LED) relay as the basic principle on the injection and radiative recombination of electrons and holes across a semiconductor p-n junction [66].

Many efforts on organic light emitting diodes (OLED) and inorganic light emitting diodes (ILED) particularly in active research and development have been made. OLEDs are found in many devices in everyday life, e.g. cell phones and portable mp3 players, and also in flat panel display industry [67]; as they offer faster refresh rates contrast ratios, power efficiencies and brighter colors [68,69] than conventional LCD screens. In OLEDs the light is emitted by an organic material, deposited on a substrate by a printing process. OLED is a very well established technology that makes manufacturing very cheap by printing of a computer screen at once.

In ILEDs an inorganic semiconductor used to emit light. This makes them brighter, more efficient, and more durable than OLEDs. There are most widespread solid-state lamps possessing long-life and good compatibility with integrated circuits, which broadens their area of application as well in display [70,71]. Examples include not only ILED displays for desktop monitors, home theater systems, and instrumentation gauging, but also, when implemented in flexible or stretchable forms, wearable health monitors or diagnostics and biomedical imaging devices. In micro scale sizes, such ILEDs can also yield semitransparent displays, with the potential for bidirectional emission characteristics, for vehicle navigation, heads-up displays, and related uses [72].

However unfortunately, the technology currently used to fabricate and assembling the semiconductors cannot be scaled effectively down to very small ILEDs to a few hundred microns – or scaled up to high-pixel count screens – over a few thousand pixels. That is the reason ILED computer screens do not yet exist in the market. Rogers et al. [73], approach overcomes such limitations by manufacturing all ILEDs needed for a certain screen at once. This allows for much smaller ILEDs, because these do not need to be manipulated one by one and hence for screens with many more pixels. According to Rogers, compared to conventional ILED technologies, the new ILEDs are much smaller, thinner and more easily integrated into high resolution arrays, in flexible or stretchable formats.

Recently, inorganic LEDs are assembled and connected on a flexible substrate [74] that can successfully developed for use in other flexible inorganic device technologies in electronics [75] and photovoltaics [76]. There outcome exceeds even the best levels of bend ability in literature reports of OLEDs where bending radii as small as several millimeters have been achieved [77-80]. It expected that in future it will finally allow the miniaturization of the technology, which beats OLEDs in brightness, energy efficiency [81-82], durability, and moisture resistance [74]. A schamtic view of transparent and wrapped ILED schemes can be seen in Figure 3 of reference [72].

LEDs based on inorganic materials emerge the most efficient source of lighting available today as signs, edge lighting or as element lit [83]. New materials and technologies, long-life and good compatibility with integrated circuits broadens the use of LEDs in areas other than indicator lighting. Element lit, signs are usually used for alpha numeric displays and the edge-lighting technology is often used for large area displays also used for advertising [84]. LEDs are now available in various colors and specifications including white. This has opened new vistas of applications for LEDs from indication to illumination [85].

3.1. Luminescence and Scintillation of Inorganic Phosphor Materials

3.1.1. Luminescence

It is defined as the capability of a material to emit light when exposed to electromagnetic radiation. Such luminescent materials are generally characterized by the emission of light with energy beyond thermal equilibrium. Luminescence can occur as a result of many different kinds of excitation, some of them are as photo-, electro-, chemi-, thermo-, sono-, or tribo- luminescence. Practically, most often the excitation is via X-rays, cathode rays, or UV emission of a gas discharge. Here the role of luminescent materials or phosphor is to convert the incoming radiation into visible light. Depending on the type of excitation luminescent materials can be classified into two terms. Both of them can be related to the decay time (τ): fluorescence (τ <10 ms) and phosphorescence (τ >0.1 sec) [86].

In most of the cases the emitted radiation appears in the visible (Vis) part of electromagnetic spectrum that may also occur in the infrared (IR) or UV and even vacuum-ultraviolet (VUV) regions. Mostly phosphors consist of a host material with added dopant ions or activators, which plays two distinct roles: as a passive matrix to define the spatial locations of the activator ions; and as an active participant in the luminescence process, exerting its own specific influence on the spectroscopic behavior of the activator. In the latter case, it helps to shape the structure of the energy levels of the activator and also introduces vibrations of the various energies, known as phonons, which influence the kinetics of the luminescence phenomena. Luminescent materials property depends on the chemical composition of the host lattice, its crystallographic structure, and the manner in which it accommodates defects [87].

Luminescence deals first with excitation of luminescence material by absorbing an amount of energy and then emission of light as photon from it. Basically the excitation supplies energy to the material raises an electron from its ground state to higher lying excited level, that is energetically unstable, once this occur the excited electron seeking a way to shed the excess energy by relaxation process [87].

There are two main ways to get rid of excess energy by the matter: 1) radiative and 2) non-radiative relaxation. In radiative relaxation process the phosphor material first jump to an excited state on higher energy level after getting the energy and then relaxed to ground state by emission of light. The energy difference between these two levels provides the energy of emitted photon of light. While in non radiative relaxation (N-RR) the most of the excitation energy moves to a different, lower-lying, state, with the remainder being dissipated with the emission of light [87].

Both of the mentioned processes can occur simultaneously, between the same two levels. In such a case, some of the excited electrons get rid of their excess energy gained during the excitation by generating optical photons, while other dissipates the extra energy non-radiatively. This is always accompanied by the liberation of heat, as the energy originally held by the excited electron becomes converted into lattice vibrations or phonons. The competition between the two processes is a rather common situation in luminescent materials, with the respective probabilities defining the overall emission efficiency: the more probable is the non-radiative process; here the less efficient becomes the radiate emission [87].

There are many different luminescent species and scintillation mechanisms possible in inorganic materials [88]. The luminescence may be intrinsic to the material and involve electron-hole recombination; free, self-trapped, and defect-trapped exciton luminescence; constituent transition group or post-transition group ion fluorescence; core-valence band transitions; or charge transfer transitions within a molecular complex. Or it may be extrinsic, such as luminescence associated with impurities or defects and additive dopant ions. In the role of an activator, the dopant ion may be the luminescence species or may promote luminescence as in the case of defect bound exciton emission.

3.1.1.1. The Characteristics of Ideal Luminescence

The selection of best photo detector to use, and the energy resolution depends upon the emission wavelength (~400 nm), the light yield and proportionality respectively. One of the highest light yields example is CsI (Tl), this has light yields (65,000 photons/MeV). The number of electron-holes pair production depends also on the proportionality of average energy for creating an electron-hole pair and to recoil the electron. If the average energy is proportional to the recoil electron energy, then mono-energetic gamma rays will produce the same number of electron-hole pairs independent of Compton interactions occur before the final photoelectric absorption. Variation in this proportionality result into a degraded energy resolution. $LaBr_3$ (Ce) is the one example, it has high light output (61,000 photons/MeV) and the best energy resolution for 662 keV gamma rays (2.9% full wave half maximum(FWHM)) reported so far [89]. Similarly the fast signal rise and decay times are important for good timing resolution and high counting rates or for time-of-flight modes of operation. As well the radiation damage is extremely important for detectors in high-energy physics experiments.

3.1.2. Scintillation

The excitation through irridation of luminescent material accomplished by number of different means such as gamma rays, X-rays, or β-radiation is known as scintillation. The materials capable of converting such energy into light photons are termed scintillators or X-rays phosphors, both of which find very important applications in industry, medicine, or science. The term scintillation is the ability of a material to produce visible or near visible radiation when radiated with any type of ionizing radiation, including γ–rays, X-rays, cathode rays, neutrons or others [87].

3.1.2.1. Types of Scintillators Materials

All the known scintillator materials can be divided into three groups [87]:

i. *Activated scintillators:* These are the materials consisting of a host lattice into which an impurity ion has been intentionally introduced as an activator. In such materials the incident radiation dissipates its energy by interaction with the lattice electrons it encounters in the host. The energy thereby deposited diffuses to and excites the activator ion, and finally radiate relaxation of the activator liberates a light photon. Thus, ideally, the emission results from a radiative recommendation of an excited electron of an impurity, exclusively. Some examples are as NaI:Tl, CsI:Tl, CaF_2:Eu, $YAlO_3$:Ce, $LaBr_3$:Ce, Lu_2S_3:Ce, $LuAlO_3$:Ce, Lu_2SiO_5:Ce.
ii. *Stoichiometric or self-activated (intrinsic) scintillators:* In such type of scintillators the penetration of the excitation particle into the lattice creates excited electron, that may return to the ground state in different ways. In one way they form excitons with holes left in the valance bond and then relax radiatively. In the second way, the excited electrons may, excite a particular constituent of the lattice and then relax radiatively. Some examples of this group are BaF_2, $CaWO_4$, CeF_3, $Bi_4Ge_3O_{12}$ (BGO), un-doped $Y_3Al_5O_{12}$ or $Lu_3Al_5O_{12}$ (YAG and LuAG, respectively), $PbWO_4$, un-doped CsI. Some of the listed stoichiometric scintillators become activated scintillators when doped with a proper activator.
iii. *Core-valance scintillators:* Here the mechanism of scintillation is quite different in these materials. The ionizing radiation hits an electron out of a core level of an atom and the created hole is filled with an electron of a valance bond of another atom. Such process is a radiative and extremely fast that produces scintillation flash. The major example of this class is BaF_2, which produce core-valance scintillation impulse that decays with time constant of about 0.8 msec [90]. The high efficiency flash, fats transition, and extremely short decay time makes them useful in practical applications. There are additional examples of these materials which can produce scintillation light *via* this mechanism: CsF, $RbCaF_3$, KF, RbF, RbCl, KLu_2F_7, $KLuF_4$, $LiBaF_3$, and others [91].

3.1.2.2. Scintillation Process

The process whereby a scintillator converts the energy deposited in it by a γ-ray into a pulse of visible or ultraviolet photons includes three main steps. It starts when an incident ionizing radiation particle hits the material and ends when all the created light escapes the scintillator. According to Lempicki et al. [92], the overall efficiency (η) of the conversion process is product of three factors: $\eta = \beta SQ$, where β is the conversion efficiency of the γ-ray energy to electron hole pairs, S is the transfer efficiency of the energy held by the electron-hole pairs to the activator ions or other luminescence centers, and Q is the quantum efficiency of the luminescence centers themselves.

3.1.2.3. Mechanisms in Inorganic Scintillators

Step 1: The ionization event starts by absorption of high energy photon (2-7 times the band gap energy of the crystal) which creates an inner shell hole and an energetic primary electron (e-h) pairs, followed by radiative decay (secondary x-rays), non-radiative decay (Auger processes. secondary electrons), and inelastic electron-electron scattering in the time period of $\sim 10^{-15}$- 10^{-13} sec [93-95].

Step 2: In the second step the energy transfers from the excited host to luminescence centers by relaxation process. When the electron energy becomes less than the ionization threshold, hot electrons and holes thermalize by intra band transitions and electron-phonon relaxation. In the case of semiconductors, the charge carriers can remain as diffuse band states and can trapped on defects and impurities, or self-trapped by the crystal lattice, or remain as free and impurity-bound excitons, within a short time scale of $\sim 10^{-12}$-10^{-11} sec. During this stage luminescent centers may excite by hot electrons to a chain process electron-hole capture or sequential hole-electron capture, and sensitizer-activator energy transfer processes over a time scale ranging from $<10^{-12}$ to $>10^{-8}$ sec. Depending on carrier mobility, this time is responsible for the intrinsic rise time of the scintillation light.

Step 3: In this step the excited luminescent species return to the ground state by non-radiative quenching processes or by emitting a photon. The radiative process can be as short as 10^{-9} sec for electron-hole recombination, free and bound exciton emission, and core-valence recombination, while it can take many minutes for the case of highly forbidden processes.

3.1.2.4. Characteristics of an Ideal Scintillator

For an ideal scintillator process a combination of physical and scintillation properties must be consider. High detection efficiency for the gamma rays; the high density, high atomic number and high stopping power are required for the crystal. High efficiency for neutron detection may achieve by the essential ion which constituent with a high neutron absorption section e.g. ^{6}Li, ^{10}B, or ^{157}Gd. For the purpose high crystal growth with low cost, the crystals should select that do not decompose or undergo a phase transformation between room temperature and their melting point.

The chemical stability and mechanical strength are the two factors that must be controlled. Stability includes environmental or chemical durability, ruggedness and mechanical shock resistant, and variation of light output with temperature and time. Insensitivity to air, moisture, and light and the absence of weak cleavage planes are highly desirable characteristics [93,96].

4. Light Emission from Organic–Inorganic Hybrid Composite Materials

From the beginning man has been exploring the properties of both inorganic and organic substances for millennia, whether it be in the pottery of ancient civilizations or the pharmaceutical properties of plant extracts. The ancient Greeks regarded these as very different, the pottery being of mineral origin and the plant being vegetable. This distinction has survived largely unscathed until relatively recently, with the two types of materials generally displaying radically different properties and finding quite different applications. While the attractions of combining, such as an inorganic glass with an organic polymer on a molecular level are manifold, it is equally self-evident that the conditions required for processing a glass are entirely incompatible with the much lower processing temperatures needed for polymers. This all changed with the realization in the 1980s that organic components could be combined with inorganic glasses using the sol-gel process. Around the same time, various research groups were exploring routes to combining polymers with clay minerals to produce different types of organic-inorganic hybrids. This area has flourished in the past 20 years or so and this article will highlight some of the advances made by us and others in recent years [97].

The reason behind all this activity is the unique combination of properties accessible via a hybrid approach. Organic-inorganic hybrids have attracted much attention because of the potential of combining distinct properties of organic and inorganic components within a single molecular composite. Organic materials offer structural flexibility, convenient processing, tunable electronic properties, photoconductivity, efficient luminescence, and the potential for semiconducting and even metallic behavior [98,99]. Inorganic compounds provide the potential for high carrier mobilities, band gap tunability, a range of magnetic and dielectric properties, and thermal and mechanical stability.

In addition to combining distinct characteristics, new or enhanced phenomena can also arise as a result of the interface between the organic and inorganic components. By synthesizing organic-inorganic composites, the best of both worlds can potentially be obtained within a single material. An example of an organic-inorganic LED (OILED), based on CdSe nanoparticles dispersed in different polymers has recently been fabricated, although such devices currently operate only at low temperatures or at room temperature with very low efficiencies (0.001-0.01%)[100,101].

In case of electroluminescence comparison, it was noticed that the electric field strength of inorganic EL and organic EL is similar [102-105] and subsequent possibility to fabricate a hybrid electroluminescence device from inorganic and organic semiconductors. Generally, most polymer materials transport holes preferentially which cause imbalanced carrier injection in organic electroluminescence device, thus holes may pass through the emission layer without forming excitons with the oppositely charged carriers and lead to ohmic losses [105]. Furthermore, the holes mobility is larger than that of electrons in most organic material, so the recombination zone of holes and electrons is close to cathode where excitons are easily quenched.

Conversely, inorganic semiconductors contain large numbers of carriers and the most important is that most inorganic materials have higher electron mobility. Hence some attempts have been made to fabricate organic–inorganic heterostructure taking advantage of both the organic and inorganic materials [105,106-107]. In fact, high external quantum efficiency (0.4%) and intense luminescence (2000 cd/m^2) have been demonstrated (shown in Figure 4) by Coe et al. [108] in an organic-inorganic hybrid device structures combining II–IV semiconductor nanocrystals and conducting polymers [105]. In the following, some of the notable organic–inorganic hybrid composite materials are reported as:

4.1. Nanomaterials

Many of these hybrids are examples of nanomaterials [109], where one or more component phases exists on a nanometre scale (<ca. 100 nm). There has been much excitement about nanomaterials, creating an impression that they are a panacea for all technical ills; the reality may sometimes be quite different, leading to a degree of creeping cynicism. In the context of organic-inorganic hybrids, a key question is whether it really matters whether they are true nanocomposites or not. For some applications (e.g. sensors), this can result in a transparent hybrid, which may be highly desirable. For further study reference [97] can be studied.

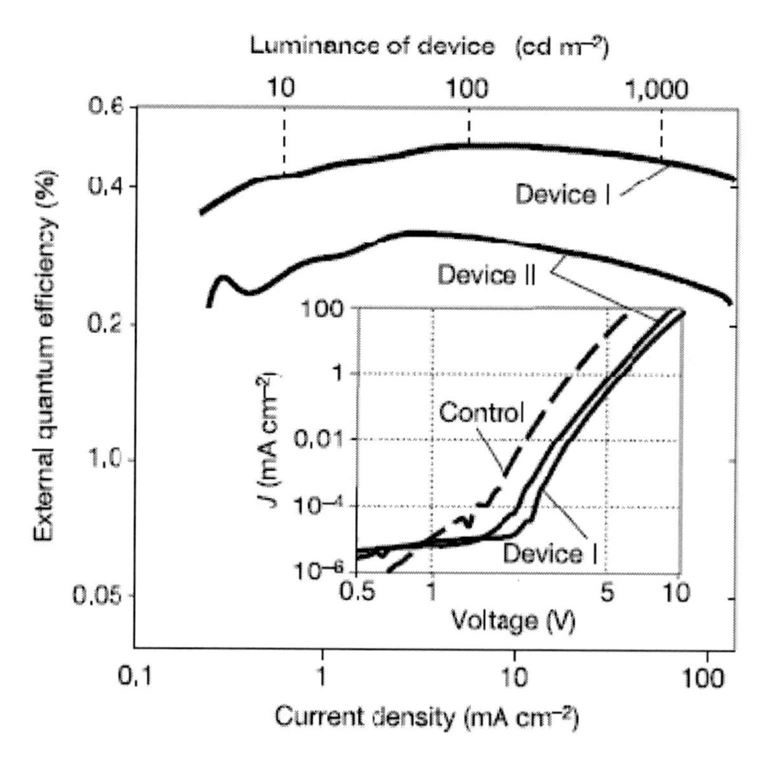

Figure 4. External quantum efficiency versus current density for the two quantum dot light emtting diodes (QD-LEDs), devices. The peak efficiency of device I is η= 0.52% at 10 mA cm^{-2} and 6.6 V, which corresponds to a luminance of 190 cdm^{-2}. Inset, current density versus voltage data for the same devices, as well as for a control device that consists of spin-cast TPD and vacuum-deposited Alq_3 of the same thicknesses as used in device I. In all cases $J \propto V^{9}$ for V > 3 V, whereas the devices that contain QDs have operating voltages that are 15–30% greater than the control device. Reproduced with permission from Coe, S. et al. Nature 2002, 420, 800-803. Copyright © 2002, Nature Publishing Group.

4.2. Clay Nanocomposites

There are two types of clay nanocomposites, which are referred to as either intercalated or exfoliated, depending on the degree of separation of the individual clay platelets. The main techniques of preparing these nanocomposites initially require the platelets to be made 'organophilic', by exchanging the interstitial metal ions with a charged organic species such as a cationic surfactant [97].

The optical properties of nanocomposites have attracted much interest because of their transparency and 'active' optical properties. Transparent products can be obtained from layered systems under certain conditions.

Poly(p-phenylene vinylene)s have been studied extensively for their electroluminescence and potential applications in light emitting diodes (LEDs). The synthesis of intercalated hybrids of smectic clay and a substituted poly(p-phenylene vinylene) has been reported. These hybrids exhibit electroluminescence and the luminescence appears to be colour tunable depending on the degree of intercalation. Some other additional types of such materials are sol-gel hybrids and biomimetics as well [97].

4.3. Previous and Recent Directions

Since these hybrid composites constitute a new class of materials, which could show properties characteristic of both constituent components with potential synergetic effects [110-113]. Much work in the form of their development has been done and is in progress.

In this framework, certain electroluminescent organic-inorganic hybrid materials have been developed by sol-gel chemistry [114-117] while various polymers including poly(ethylene oxides) [118,119], poly(olefines) [120], polyimide [121], polypyrrole [122], and polyaniline [123] have been incorporated into clay particles through either a solution or a melt intercalation process [108,124]. The organic-inorganic hybrid composites thus prepared have shown improved environmental stability, mechanical strength, and lower permeability for gases (e.g., O_2) with respect to corresponding pure polymers [114–124]. Here we review some of the previous and current ongoing work in this filed:

Winkler et al. in 1990 reported [110] the first solution intercalation of EO3-Poly(p-phenylene vinylene) into clay nanoparticles for light-emitting applications. Their obtained results from a single layer LED device with the structure ITO/EO3-PPV-B34 (0.95:0.05)/Al showed a similar EL spectrum (λ_{max} = D590 nm) to the PL emission. The LED showed a typical diode characteristic with a turn-on voltage of ca. 8V at the forward bias (Inset of Figure 3 of ref. [110]). According to their results the organic-inorganic hybrid composites based on EO-PPVs intercalated clay nanoparticles may be a class of very promising new materials of good color-tunability and environmental stability for both the PL and EL applications.

In 2003, Ermakov et al. exhibited the hybrid organic-inorganic LEDs based on GaN and InGaN blue emitters with some organic luminophores providing blue, green, red and white emission colors with high luminous efficiency [125]. Their results predicted that the color of the emitted light from the GaN-based LED can be changed by varying the current through the device due to change in the relation between the UV and visible components of GaN electroluminescence.

Table 2. Summary of the figures of merit for the different HyLEDs

	TPD contents [wt%]	Maximum Luminance [cd m^{-2}] [at voltage [V]]	Maximum Efficacy [cd A^{-1}] [at luminance[cd m^{-2}]	Maximum PCE(lmW^{-1}) [at luminance[cd m^{-2}]
Device A	5	4978(12)	9.0(65)	4.5(23)
Device B	10	10012(11)	11.1(458)	6.0(172)
Device C	20	5086(9)	15.1(608)	9.1(177)

Additional remarkable works in this category are performed [128-134], interested reader can get benefit from these works.

Reproduced with permission from Henk, J. B. et al. Adv. Mater. 2010, 22, 2198. Copyright © 2010, WILEY-VCH Verlag GmbH & Co.

Lin et al. [126], reported in 2007 the visible light emission of organic-inorganic light emitting devices by low-cost spin-coating. They employed different nanoparticles, CdSe, ZnO and Eu_2O_3, to form organic-inorganic hybrid composites and generate the R-G-B and white light emission, while TPD and polymethyl methacrylate (PMMA) were adopted as the organic matrices. Their evaluated SNOM analysis demonstrated that the phase segregation of hybrid ZnO composite enhances the blue EL, while the homogeneity of hybrid Eu_2O_3 improves the white light emission.

In the latest work performed in 2010 by Bolink et al. [127], presented a high efficiency phosphorescent HyLED based on a poly(9-vinylcarbazole) polymer matrix (PVK) doped with the green-light emitter iridium (III) tris(2-(4-totyl)pyridinato-N,C2), (Ir(mppy)3). PVK was additionally doped with TPD as the hole-transporting material and 2-(4-biphenyl)-5-(4-tert-butylphenyl)-1,3,4-oxadiazole (PBD) as the electron-transporting material. They evaluated three devices with decreasing PBD and increasing TPD contents e.g : A) PBD 30 wt%, TPD 5 wt%; B) PBD 25 wt%, TPD 10 wt%; C) PBD 15 wt%, TPD 20 wt%. Through the use of a novel ZnO:Cs electron injecting contact and by tuning the charge transport properties of the PVK matrix, efficacies up to 15 cd A^{-1} and PCE exceeding 9 lmW^{-1} have been achieved (Table 2). Their results demonstrated that HyLEDs can compete with polymer LEDs (PLEDs) and that the design rules of OLEDs can be successfully applied to this novel class of devices.

Table 3. Characteristics of the typical RGB OLED cells

EL color	Blue	Green	Red
Peak wavelength(nm)	460	525	615
External Quantum Efficiency (%)[a]	4.5	4.4	2.6
Luminanace Efficiency (cd/A)[a]	3.9	16.3	3.8
Luminanace Efficiency (lm/W)[a]	2.2	11.4	1.4

[a] At 300 cd/m^2.

Reproduced with permission from Kubota, H.; et al., J. Luminescence, 2000, 87-89, 56. Copyright © 2000, Elsevier Science Ltd.

5. Various Applications of Light Emission from Organic-Inorganic Composite Materials

5.1. Organic-electroluminescent Composite Materials and Devices

In this section we review some of the main work performed on organic composite materials and their results obtained from these techniques are discussed.

A notable work in the category of organic composite materials is performed by Shen et al. [135]. In this work an independently controlled, three-color, organic light-emitting device was constructed with a vertically stacked pixel architecture that allows for independent tuning of color, gray scale, and intensity. This 12-layer device structure consists of sequentially stacked layers of metal oxide, amorphous organic, crystalline organic, and metal thin films deposited by a combination of thermal evaporation and radio-frequency sputtering. Each of the three addressable colors was found sufficiently bright for flat panel video display applications. A novel inverted structure was utilized for the middle device in the stack to lower the maximum drive voltage of the compound pixel. Output spectra from each individual element of the SOLED can be seen in Figure 3 [135].

The first physical "doping", to use the traditional term, using small concentrations of multi-walled nanotubes in a conjugated luminescent polymer, poly(m-phenylenevinylene-co-2,5-dioctoxyp-phenylenevinylene), in a polymer/nanotube composite was reported by Curran et al. [136]. Such doping increased electrical conductivity of the polymer by up to eight orders of magnitude. It was noticed that the use of nanotubes for their physical attribute of strength and nanometric heat sinks is considered very important, as many other organic materials require such strengthening without the trade-off in electronic properties that is normally observed. The electroluminescence achieved from an organic light-emitting diode (LED) using the composite as the emissive layer in the device was also reported.

Kubota et al. [137] developed a full color 5.2 in 1/4 (Video graphics array) VGA passive-matrix organic LED display. The display was consisted of 320(×3) × 240 pixels with an equivalent pixel size of 0.33 × 0.33 mm^2, white peak luminance of over 150 cd/m^2, and power consumption of 6 W. A cathode patterning method was employed with cathode separators and an RGB emitters selective deposition method by a high accuracy mask moving system. Table 3 shows the typical characteristics of t RGB OLED cells. They attributed that this display can display high-resolution video rate pictures with virtually the same quality as a CRT could offer. Furthermore passivation film for OLED was also developed. They noticed that the EL characteristics remained almost identical and the dark spot growth were negligible for practical use after the storage tests.

Tang et al. [138] reported the photoemission study of the hole-injection enhancement in organic electroluminescent devices using Au/CFx anode. Here the ultraviolet and x-ray photoemission spectroscopic results indicated that chemically tailoring the Au surface with CFx can reduce the hole-injection barrier to ~1 eV with respect to bare Au. Their results suggested that CFx can function as a hole-injection enhancement layer for organic optoelectronic/electronic devices which use a metallic anode.

Sahu and Pal [139] proposed a multifunctionality in heterostructure devices based on poly(3-hexylthiophene) and copper phthalocyanine layers. Under solar light illumination, the devices yielded photovoltage in an open-circuit condition. These devices emitted light in the forward bias and could also detect a trace of light in the reverse-bias direction. According to this work, the highest occupied molecular orbitals and lowest unoccupied molecular orbitals of the two materials, along with the metal work functions, were found suitable for the three functionalities at different bias modes. The electroluminescence spectrum of the device was compared with the photoluminescence spectra of the components. They also observed that the spectral response of the photocurrent matched well with the electronic absorption spectrum of the poly(3-exylthiophene)/copper phthalocyanine hetero-junction.

In a latest study, Dinh et al. [146] to improve the photonic efficiency of an organic light emitting diode (OLED) and its display duration, both the hole transport layer (HTL) and the emitting layer (EML) prepared as nano-structured thin films. Here, the HTL nanocomposite films were prepared by spin coating solutions of poly(vinylcarbazole) (PVK) and PEDOT-PSS containing TiO_2 nanoparticles onto low resistivity ITO substrates; for the EML, TiO_2-embedded MEH-PPV (MEH-PPV + nc-TiO_2) conjugate polymers were spin-coated onto the HTL. OLEDs fabricated from above mentioned films have the structure of multilayers such as Al/MEH-PPV + nc-TiO2/PVK+nc-TiO2/ITO and Al/MEH-PPV + nc-TiO_2/PEDOT-PSS+nc-TiO_2/ITO. From the characterization of the nanocomposite films, it was evaluated that both the photoluminescence and I-V characteristics of the nanocomposite materials were significantly enhanced in comparison with the standard polymers and one can expect a large electroluminescent efficiency of the OLEDs made from these layers.

5.2. Inorganic Electroluminescent Materials and Devices

About thirty years of research on these intriguing TFEL devices, monochrome (amber) TFEL displays are currently commercially available, and full-color TDEL displays are being developed [141-146]. The development of electroluminescent displays EL device technology requires: high luminance, simple fabrication methods such as screen-printing [141,147] which reduces the manufacturing costs with higher contrast ratios [141,148] which enables the display to perform equally well under strong lighting conditions. In this section, a review of inorganic electroluminescent devices and materials is described.

Munasinghe et al. [141] presented a an improved thick dielectric (TD) layer for inorganic electroluminescent (EL) display devices through a composite high- κ dielectric sol-gel/powder route. They performed the device characterization for both full color (GaN:RE) and monochrome (ZnS:Mn) phosphor systems. The use of a composite TD film, composed primarily of lead–zirconate–titanate (PZT), resulted in a significantly higher charge (>3 μC cm^2) coupling to the phosphor layer. Accoording to this report, the reduction in porosity of the TD improved the homogeneity of electric field applied to the phosphor layer, resulting in a steeper luminance–voltage slope. Further from these ZnS:Mn TDEL and GaN:Eu devices, a high luminance levels of up to 3500 cd/ m^2 and 450/ cd m^2 have been achieved, respectively. Such results demonstrated that three critical requirements for practicality of the TDEL approach can be obtained by careful selection and design of the device materials, fabrication process and device structure.

On realizing that hybrid LEDs with n-type ZnO as an electron injector in many different systems, Chang et al. [149] in 2006 investigated electroluminescence from ZnO nanowire/polymer composite p-n junction. This work showed that with a simple process involving integration of existing n-type ZnO nanowires with hole conducting polymers, light emission can be achieved in a ZnO nanowire/semiconductor polymer matrix. Here the spin coating of polystyrene was used to electrically isolate neighboring nanorods and a top layer of transparent conducting indium tin oxide was used to contact the PEDOT/PSS. Multiple peaks were observed (can be seen in Figure 5) in the electroluminescence spectrum from the structure under forward bias, including ZnO band edge emission at ~383 nm as well as peaks at 430, 640, and 748 nm. Furthermore, the threshold bias for UV light emission was <3 V, corresponding to a current density of 6.08 A cm^{-2} through the PEDOT/PSS at 3 V. Such results predicted that a low-cost, low temperature process holds a strong potential for ZnO-based UV light emission and reduces the requirement for achieving robust p doping of ZnO films or substrates.

Addiotional applications of inorganic electroluminescent materials include their usage in the flexible inorganic device technologies especially in electronics and photovoltaics, bent or wrapped display technology. Interested reader can read the ref. [74-82].

5.3. Organic–inorganic Hybrid Electroluminescent Materials and Devices

Many studies have proven that organic-inorganic hybrid materials provide both the superior carrier mobility of inorganic semiconductors and the processability of organic materials [150]. Here we review some of the work performed on applications of organic–inorganic hybrid materials.

In 1990, Kagan, C. R. et al. [150] demonstrated a thin-film field-effect transistor having an organic-inorganic hybrid material as the semiconducting channel. In their scheme, hybrids based on the perovskite structure crystallized from solution to form oriented molecular-scale composites of alternating organic and inorganic sheets. The spin-coated thin films of the semiconducting perovskite $(C_6H_5C_2H_4NH_3)_2SnI_4$ formed the conducting channel, with fileld-effect mobilities of 0.6 square centimeters per volt-second and current modulation greater than 10^4. The molecular engineering of the organic and inorganic components of the hybrids improved device performance for low-cost thin-film transistors. This work [150] further suggest that the flexibility in the chemistry of organic-inorganic hybrid materials may provide a path to preparation of both n-type and p-type transporting materials, which are necessary for complementary logic and normally "on" or "off " organic-inorganic TFTs. In the following one class of organic-inorganic hybrid is based on the three-dimensional (3D) perovskite structure ABX3 (Figure 1) [150].

In the same trend organic-inorganic hybrid electroluminescence device was fabricated using conjugated polymer and ZnS:Mn by Yang, X. et al. [151]. The structure of the hybrid device consisted of two layers sandwiched between two injecting electrodes, one of which was a conjugated polymer layer and the other was an inorganic material layer. A schematic view of the EL spectra of the PPV/ZnS:Mn device is presented in Figure 6. Different mechanism of luminescence were observed for the PPV and ZnS:Mn layers. Further the emission from the PPV layer came from the recombination of holes and electrons, while in the ZnS:Mn layer emission was resulted from the impact excitation of Mn luminescent centers. This device showed an improved quantum efficiency of light emission from the PPV layer.

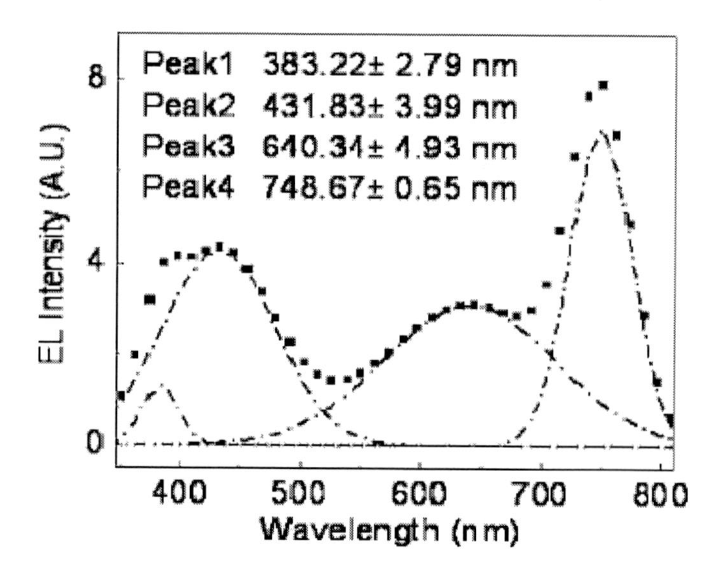

Figure 5. PL of vertical ZnO nanowires and EL of n-ZnO NW/p-polymer (inset). Reproduced with permission from Chang, C. Y. et al. Appl. Phys. Lett. 2006, 88, 173503-3. Copyright © 2006, American Institute of Physics.

Later on Wenge Yu et al. [152] obtained blue EL of ZnSe thin film using an organic–inorganic heterostructure with ZnSe as emission layer and PVK as holes transport layer. The EL mechanism of the hybrid device included the injection and recombination of the carriers, which is considered as basically identical to that of OLED. Here, the EL emission at 466 nm came from the ZnSe layer. These results proposed the possibility of using II–IV compounds in low voltage injection EL and provide a new way of obtaining blue emission using an organic–inorganic heterostructure.

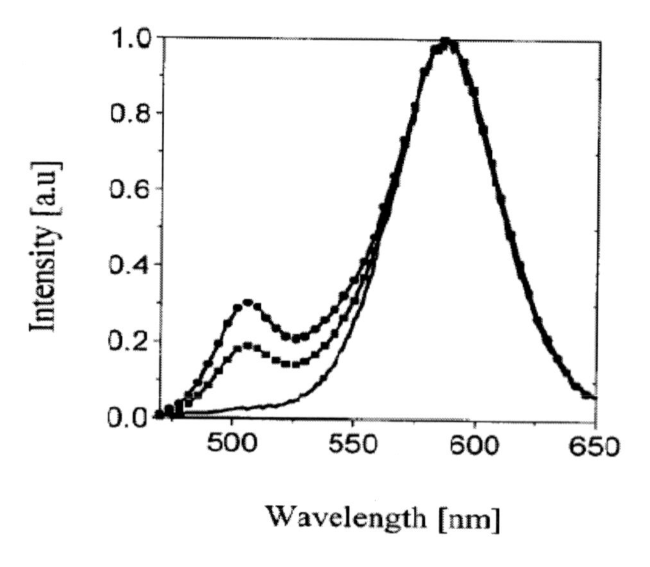

Figure 6. The EL spectra of the PPV/ZnS:Mn device. EL spectra under the positive bias: square: (14 V), dots (~17 V), and EL spectra under the negative bias. Reproduced with permission from Xiaohui Yang et al. Appl. Phys Lett. 2000, 77, 797-799. Copyright © 2000, American Institute of Physics.

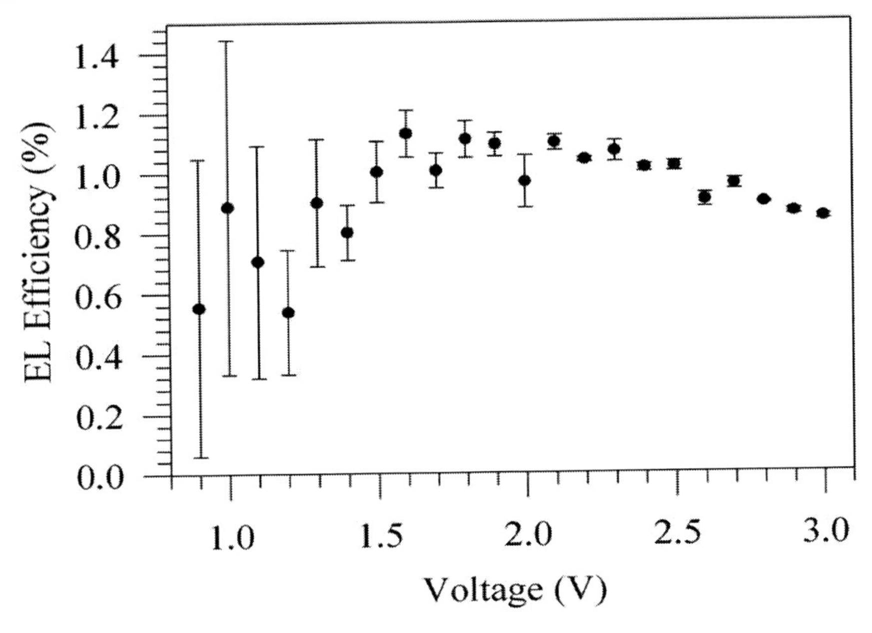

Figure 7. EQE of the ITO/pentacene/nc-PbS/BCP/devices as a function of voltage. Reproduced with permission from Bourdakos, K. N. et al. Appl. Phys. Lett. 2008, 92, 153311. Copyright © 2008, American Institute of Physics.

Bourdakos, K. N. et al. [153] reported the use of PbS nanocrystals within a hybrid device that emits 1.2 μm electroluminescence with an external quantum efficiency of 1.15% corresponding to an internal quantum efficiency of ~5%–12% thus demonstrating a viable, low-cost, highly efficient near infrared organic electroluminescent device. In the following Figure 7 shows direct measurements of the device EQE indicating a peak value of 1.15% at 1.6 V representing a significant improvement over previously reported devices. According to this work, direct generation of the excited state on the nanocrystal resulted in eliminating competing processes that have previously led to the low reported efficiencies in near-infrared light emitting devices. These electroluminescence devices represent a significant improvement in the state of the art and offer the potential for the development of low-cost NIR devices.

Summary and Conclusion

This review chapter aims to provide some insight into the light emission from organic and inorganic composite materials. Studies on organic-inorganic composite materials and multilayer structures are gaining much attention due to their better stability, electroluminescence and photoluminescence spectra. Among the electroluminescent materials organic light emitting diodes (OLEDs) are very popular and attractive because of their potential applications in flat panel displays, domestic solid state lighting. OLEDs are very thin devices with high luminance. One can attain high luminous efficiency using OLEDs. Furthermore by using OLEDs, optional color is obtainable by using fluorescent molecules throughout the entire visible region to obtain full color display. In OLEDs the light is emitted by an organic material, deposited on a substrate by a printing process. OLED is a very well established technology that makes manufacturing very cheap by printing of a computer screen at once.

The second category named as inorganic light emitting diodes (ILED), such technology is in the phase of active research and development. One can easily find OLEDs being utilized in many display devices in our daily life, e.g. cell phones and portable mp3 players, and also in flat panel display industry. ILEDs usually offer faster refresh rates contrast ratios, power efficiencies and brighter colors than the conventional LCD screens. In ILEDs an inorganic semiconductor is used to emit light. This makes them brighter, more efficient, and more durable than OLEDs. Some of the known examples of the ILEDs include their utilization in the desktop monitors, home theater systems, and instrumentation gauging. They can be also be implemented in flexible or stretchable forms, wearable health monitors or diagnostics and biomedical imaging devices. In the micro scale sizes, such ILEDs can also yield semitransparent displays, with the potential for bidirectional emission characteristics, for vehicle navigation, heads-up displays, and related uses. Additionally inorganic electroluminescent materials can be used in the flexible inorganic device technologies especially in electronics and photovoltaics and bent or wrapped display technology as well.

While the third technology known as organic-inorganic hybrids composite materials have attracted much attention because of the potential of combining distinct properties of organic and inorganic components within a single molecular composite. These materials offer structural flexibility, convenient processing, tunable electronic properties, photoconductivity, efficient luminescence, and the potential for semiconducting and even metallic behavior. The hybrid materials combine the distinct characteristics of both the organic materials (such as structural flexibility, convenient processing, tunable electronic properties, photoconductivity, efficient luminescence, the potential for semiconducting and even metallic behavior) and inorganic components (i.e., high carrier mobilities, band gap tunability, a range of magnetic and dielectric properties, and thermal and mechanical stability). By synthesizing organic-inorganic composites, the best of both worlds can potentially be obtained within a single material. Moreover, the organic-inorganic hybrid composites show improved environmental stability, mechanical strength, and lower permeability for gases (e.g., O2) with respect to corresponding pure polymers. In addition to such studies, the current chapter also highlights some of the previous and new applications of organic, inorganic and organic-inorganic hybrids composite materials and devices. This work is of both academic and commercial interest. The authors are hopeful that continuing research in the electroluminescent materials containing organic, inorganic and organic-inorganic hybrids composite materials will continue to further advance technologies and provide the road towards new discoveries in this filed.

ACKNOWLEDGMENTS

Authors are thankful to the publishers/authors for granting the permission for the reproduction of the following figures and the table taken from Elsevier Science Ltd., The Netherlands, WILEY-VCH Verlag GmbH & Co, American Institute of Physics and American Assoc. for the Adv. of Science USA.

REFERENCES

[1] Hung, L. S.; Chen, C. H. *Mat. Sci. and Eng.* R 2002, 39, 143.
[2] Parker, I. D.; Pei, Q.; Marrocco, M. *Appl. Phys. Lett.* 1994, 65, 1272.
[3] Hu, B.; Yang, Z.; Karasz, F. E. *J. Appl. Phys.* 1994, 76, 2419.
[4] Hosokawa, C.; Higashi, H.; Kusumoto, T. *Appl. Phys. Lett.* 1993, 62, 3238.
[5] Kumar, N. D.; Joshi, M. P.; Friend, C. S.; Prasad, P. N. *Appl. Phys. Lett.* 1997, 71, 1388.
[6] Berggren, M.; Gustafsson, G.; Inganas, O. *Appl. Phys. Lett.* 1994, 65, 1489.
[7] Burrows, P. E.; Forrest, S. R.; *Appl. Phys. Lett.* 1994, 64, 2285.
[8] (a) Pope, M.; Magnante, H. P. K. P. *J. Chem. Phys.* 1963, 38, 2042. (b) Helfrich, W.; Schneider, W. G. Phys. Rev. Lett., 1965, 14, 229.
[9] Burroughes, J. H.; Bradley, D. D. C.; Brown, A. R.; Marks, R. N.; Mackay, K.; Friend, R. H.; Burns, P. L.; Holmes, A. B. *Nature,* 1990, 374, 539.
[10] Braun, D.; Gustafsson, G.; McBranch, D.; Heeger, A. J. J. Appl. Phys. 1992, 72, 564.
[11] Braun, D.; Heeger, A. *J. Appl. Phys. Lett.* 1991, 58, 1982.
[12] Kido, J.; Kohda, M.; Okuyama, K.; Nagai, K. *Appl. Phys. Lett.* 1992, 61, 761.
[13] Vestweber, H.; Sander, R.; Greiner, A.; Heitz, W.; Mahrt, R. F.; Bassler, H.; *Synth. Met.* 1994, 64, 141.
[14] Gustafsson, G.; Cao, Y.; Treacy, G. M.; Klavetter, F.; Colaneri, N.; Heeger, A. J. *Nature* 1992, 357, 477.
[15] Greenham, N. C.; Moratti, S. C.; Bradely, D. D. C.; Friend, R. H.; Holmes, A. B. *Nature* 1993, 365, 628.
[16] Vincett, P. S.; Barlow, W. A.; Hann, R. A.; Roberts, G. G. *Thin Solid Films* 1982, 94, 171.
[17] Grem, G.; Leditzky, G.; Ullrich, B.; Leising, G. *Adv. Mater.* 1992, 4, 36.
[18] Karg, S.; Dyakonov, V.; Meier, M.; Rieb, W.; Paasch, G. *Synth. Met.* 1994, 67, 165.
[19] Braun, D.; Staring, E. G. J.; Demandt, R. C. J. E.; Rikken, G. L. J.; Kessener, Y. A. R.R.; Venhuizen, A. H. *J. Synth. Met.* 1994, 66, 75.
[20] Sokolik, I.; Yang, Z. ; Karasz, F. E.; Morton D. C. *J. Appl. Phys.* 1993, 74, 3584.
[21] Heo, J. S.; Park, N.H.; Ryu, J. H.; Choi, G.H.; Suh, K.D. *Macromol. Chem. Phys.* 2003, 204, 2002.
[22] Cleave, V.; Yahioglu, G.; Barny, P. L.; Friend, R. H.; Tessler, N. *Adv. Mater.* 1999, 11, 285.
[23] Mitschke, U.; Ba¨uerle, P. *J. Mater. Chem.* 2000, 10, 1471.
[24] Segura, J. L., *Acta Polym.* 1998, 49, 319.
[25] Nalwa, H. S; Rohwer, L. S., Handbook of luminescence, Display Materials and devices; ISBN 1-58883-010-1.; *American Scientific Publishers,* L. A, 2003, Vol. 1,Organic light emtting diodes, page 132.
[26] Tang, C. W.; VanSlyke, S. A. *Appl. Phys. Lett.,* 1987, 51, 913.
[27] Adachi, C.; Tsutsui, T.; Saito, S. *Appl. Phys. Lett*., 1990, 56,799;
[28] Adachi, C.; Tokito, S.; Tsutsui, T.; Saito, S. *Jpn. J. Appl. Phys*., 1988, 28, L269.
[29] Holonyak, N. Jr.; Bevacqua, S.F *App. Phys. Lett.* 1962, 1, 82.
[30] Craford M.G.; Steranaka, F.M, Light –emitting diodes, in *"Encyclopedia of Applied Physics"* (G. L. Grigg, Ed.) VCH, Weinheim, 1994, Vol. 8, p. 485.

[31] Craford, M.G. *Mater. Res. Soc. Bull.* 2000, 25, 27.

[32] Burroughes, J. H.; Bradey, D. D. C.; Brown, A. R.; Marks, R. N.; Mackay, K.; Friend, R. H.;Burns, P. L.; Holmes, A.B. *Nature* 1990, 347, 539.

[33] Sheats, J. R.; Antoniadis, H.; Hueschen, M.; Leonard, W.; Miller, J.; Moon, R.; Roitman, D.; Stocking, A. *Science* 1996, 273, 884.

[34] Baldo, M.A.; O'Brien, D.F.; You, Y.; Shoustikov, A.; Sibley, S.; Thompson, M.E.; Forrest, S.R. *Nature* 1998, 395, 151.

[35] Baldo, M.A.; Lamansky, S.; Borrows, P.E. *Appl. Phys. Lett.* 1999, 75, 4.

[36] Baldo, M.A.; Deutsch, M.; Burrows, P.E.; Gossenberger, H.; Gerstenberg, M.; Ban, V.; Forrest, S.R. *Adv. Mater.* 1998, 10, 1505.

[37] Cleave, V.; Yahioglu, G.; Barney, P. Le; Friend, R.; Tessler, N. *Adv. Mater.* 1999, 20, 285.

[38] Baldo, M.A. *Nature* 2000, 403, 750.

[39] Adachi, C.; Baldo, M.A.; Thompson, M.E.; Forrest, S.R. *J. Appl. Phys*. 2001, 90, 5048.

[40] Sun, Y.; Giebink, N.C.; Kanno, H., Ma, B.; Thompson, M.E.; Forrest, S.R. *Nature* 2006, 440, 908.

[41] Saxena, K.; Jain, V.K.; Mehta, D. S. *Opt. Mat.* 2009, 32, 221–233.

[42] Forrest, S.R.; Bradley, D.D.C.; Thomson, M.E. *Adv. Mater.* 2003,15, 1043.

[43] Patel, N. K.; Cina, S.; Burroughes, J.H. *IEEE J. Select. Top. Quant. Electron.* 2002, 8, 346.

[44] Mehta, D.S.; K. Saxena, in: *Proceedings of the Ninth Asian Symposium on Information Display* (ASID-06), 2006, 198.

[45] Forrest, S.R. *Org. Electron.* 2006, 4, 45.

[46] Gu, G.; Garbuzov, D.Z.; Burrows, P.E.; Venkatesh, S.; Forrest, S.R.; Thompson, M. E. *Opt. Lett.* 1997, 22, 396.

[47] Kim, J.S.; Ho, P. K. H.; Greenham, N.C.; Friend, R.H. *J. Appl. Phys*. 2000, 88, 1073.

[48] Pope, M.; Kallmann, H. P.; Magnante, P. *J. Chem. Phys.* 1963, 38, 2042.

[49] VanSlyke, S.A.; Tang, C.W. *US Patent* 1985, 4, 539,507.

[50] Tang, C.W.; VanSlyke, S.A. *Appl. Phys. Lett*. 1987, 51, 913.

[51] Tang, C.W.; Chen, C.H.; Goswami, R. *US Patent* 1998, 4,769,292.

[52] Shirota, Y. *J. Mater. Chem*. 2000, 10, 1.

[53] Mitschke, U.; Bauerle, P.; *J. Mater. Chem.* 2000, 10, 1471.

[54] Kraft, A.; Gimsdale, A.C.; Holmes, A.B. Angew. *Chem. Int. Ed.* 1998, 37, 402.

[55] Grell, M.; Bradley, D. D. C. *Adv. Mater*. 1999, 11, 895.

[56] Friend, R.H.; Gymer, R.W.; Holmes, A.B.; Burroughes, J.H.; Marks, R.N.; Taliani, C.; Santos, D. D. C. D; Bradley, D.A.; Bredas, J. L.; Logdlund, M.; Salaneck, W.R. *Nature* 1999, 397, 121.

[57] Thelakkat, M.; Schmidt, H.W. *Polym. Adv. Technol.* 1998, 9, 429.

[58] Chan, I.M.; Cheng, W.C.; Hong, F.C. *Asia Display/IDW'01*, 2001, p. 1483.

[59] VanSlyke, S.A.; Tang, C.W. *US Patent* 1991, 5,061,569.

[60] Hamada, Y.; Matsusue, N.; Kanno, H.; Fujii, H. *Jpn. J. Appl. Phys.* 2001, 40, L753.

[61] Kido, J.; Mizukami, T. *US Patent* 2000, 6,013,384.

[62] Liu, Z.; Pinto, J.; Soares, J.; Pereira, E. *Synth. Met.* 2001, 122, 177.

[63] Tang, C.W.; VanSlyke, S.A.; Chen, C.H. *J. Appl. Phys*. 1989, 65, 3610.

[64] Shi, J.; Tang, C.W. *Appl. Phys. Lett.* 1997, 70, 1665.

[65] VanSlyke, S.A.; Hettel, M.; Boroson, M.; Arnold, D.; Armstrong, N.; Andre, J.; Saito, Y.; Matsuki, H.; Kanno, H.; Takahas, H. *IDRC,* 2000, C4.
[66] Michio, A (Tokyo, JP), *Inorganic light emitting diode,* United States Patent, 6111274.
[67] Gu, G.; Forrest, S.R *IEEE J. of Sel. Topics in Q. Electr.,*1998,4 (1), 83.
[68] Lo, S.C.; Burn, P. L. *Chem. Rev*. 2007, 107, 1097.
[69] So, F.; Kido, J.; Burrows, P. *MRS Bull.* 2008, 33, 663.
[70] Gaul, D. A.; Rees Jr., W. S.; *Adv. Mater*. 2000, 12, 935.
[71] Nakamura, S.; Fasol, G.*; The Blue Laser Diode: GaN Based Light Emitters and Lasers,* Springer, New York, 1997.
[72] Park, Sang-Il; Xiong, Y.; Kim, R. H, *Science* 2009, 325 21, 977.
[73] Park, Sang-ll; Le, A. P.; Wu, J.; Huang, Y.; Li, X.; Rogers, J. A. *Adv. Mater.* 2010, XX, 1–5.
[74] IEEE Spectrum: Flexible Inorganic LED Displays, *Semiconductors Devices*; Willie Jones, 2009.
[75] Kim, D.H.; Ahn, J. H.; Choi, W. M.; Kim, H. S.; Kim, T. H.; Song, J.; Huang, Y. Y.; Liu, Z.; Lu, C.; Rogers, J. A. *Science* 2008, 320,507.
[76] Yoon, J.; Baca, A. J.; Park, S. I.; Elvikis, P.; GeddesIII, J. B.; Li, L.; Kim, R. H.; Xiao, J.; Wang, S.; Kim, T.H.; Motala, M. J.; Ahn, B. Y.; Duoss, E. B.; Lewis, J. A.; Nuzzo, R. G.; Ferreira, P. M.; Huang, Y.; Rockett, A.; Rogers, J. A. *Nat. Mater.* 2008, 7, 907.
[77] Cheng, Y. H.; Chen, C.M.; Cheng, C.H.; Lee, M. C. M. *Jpn. J. Appl. Phys*. 2009, 48, 021502.
[78] Cho, H.; Yun, C.; Park, J.W.; Yoo, S.; *Org. Electron.* 2009, 10, 1163.
[79] Kang, J.W. ; Jeong, W.I.; Kim, J.J.; Kim, H.K.; Kim, D.G.; Lee, G. H.; J. *Electrochem. Soc.* 2007, 10, J75.
[80] Kim, S.; Kim, K.; Hong, K.; Lee, J. L. *J. Electrochem. Soc.* 2009, 156, J 253.
[81] Gaul, D. A.; Rees, W. S.*; J. Adv. Mater*. 2000, 12, 935.
[82] Nakamura, S.; Fasol, G.*; The Blue Laser Diode: GaN Based Light Emitters and Lasers,* Springer, New York, 1997, 256.
[83] Steigerwald, D.A.; Bhat, J. C.; Collins, D.; Fletcher, R. M.; Holcomb, M. O.; M. J.; Ludowise, Martin, P. S.; Rudaz, S. L.*; IEEE J. Selected Topics in Quantum Electronics,* 2002, 8(2), 310.
[84] Sharma, P.; Kwok, H. in *International Optical Design, Technical Digest (CD) (Optical Society of America,* 2006, paper ME16.
[85] Bullough, J.D.; *Lighting answers: LED lighting systems.* Troy, N.Y.: Rensselaer Polytechnic Institute.2003, 1-47.
[86] Feldmann, C.; Jüstel, T.; Ronda, C.R.; Schmidt, P.J.; *Inorganic Luminescent Materials: 100 Years of Research and Application* 2003, 13(7), 511.
[87] Nalwa, H. S; Rohwer, L. S., " *Handbook of luminescence, Display Materials and devices"* ISBN: 1-58883-010-1, American Scientific Publisher, L.A, 2003, Volume 2, Inorganic Display Materials, Chapter 5, page 251.
[88] Blasse, G. ; Grabmaier, B. C.; *Luminescent Materials*, Springer-Verlag, New York, 1994.
[89] Loef, E. V.; Dorenbos, P.; Eijk, C. V.; *Appl. Phys. Lett.,* 2001,79,1573.
[90] Laval, M.; Moszynski, M.; Aallemand, R.; Cormoreche, E.; Guiner, P.; Odru, R.; Vacher, J. *Nucl. Instrum. Meth.* 1983, 206, 169.
[91] Van Eijk, C.E. E. J. Lumin. 1994, 60&61, 936.

[92] Lempicki, A,; Wojtowicz, A.J.; Berman, E.; *Nucl. Instr. Meth.* 1993, A333, 304.
[93] Derenzo, S.E.; Weber, M.J.; Bourret-Courchesne, E.; Klintenberg, M.K. *Nucle.Instr. Meth. Phys. Rese.* A. 2003,505, 111.
[94] Rodnyi, P. A.; Dorenbos, P.; Eijk, C. W. E. V. *Phys. Stat. Solidi* (b), 1995,187,15.
[95] Bartram, R. H.; Lempicki, A. J. *Lumin.,* 1996, 68, 225.
[96] Melcher, C. L., *J. Nucl. Med,* 2000,41,1051.
[97] Hay, J. N.; Shaw, S. *J. Euro. Phys. News,* 2003, 34 89.
[98] Mitzi, D. B. *Chem. Mater.* 2001, 13, 3283.
[99] Chondroudis, K.; Mitzi, D. B. *Chem. Mater.* 1999, 11, 3028.
[100] Colvin, V. L.; Schlamp, M. C.; Alivisatos, A. P. *Nature* 1994, 370, 354.
[101] Dabbousi, B. O.; Bawendi, M. G.; Onitsuka, O.; Rubner, M. F. *Appl. Phys. Lett.* 1995, 66, 1316.
[102] Burroughes, J. H.; Bradley, D. D. C.; Brown, A. R.; Marks, R.N.; Mackay, K.; Friend, R.H.; Burns, P.L.; Holmes, A.B. *Nature* 1990, 374, 539.
[103] Kalinowski, J. *J. Phys. D.* 1999, 32, 179.
[104] Rack, P.D.; Holloway, P.H. *Mater. Sci. Eng. R.* 1998, 21, 171
[105] Wenge, Yu; Zheng, Xu; Feng, Te.; Shengyi, Yang; Yanbing, Hou; Lei, Qian; Chong, Qu; Sanyu, Quan; Xurong Xu. *Physics Letters A.* 2005, 338, 402.
[106] Colvin, V.L.; Schlamp, M.C.; Allvisatos, A.P. *Nature.* 1994, 370, 354.
[107] Yang, X.; Xu, X. *Appl. Phys. Lett.* 2000,77, 797.
[108] Coe, S.; Woo, W. K.; Bawendi, M.; Bulovic, V. *Nature* 2002, 420,800.
[109] Stoneham, A. M. *Mater. Sci. Eng. C.* 2003, 23, 235.
[110] Dai, L.; Winkler, B.; Dong, L.; Tong, L.; Albert, W. H. M. *J. Materials Sci. Letts.* 1999, 18, 1539.
[111] Ozin, G. A. *Adv. Mater.* 1992, 4(10), 612.
[112] Fendler, J. H.; Meldrum, F. C. *Adv. Mater.* 1995,7(7), 607.
[113] Mark, J. E.; Lee, C. Y. -C.; Bianconi, P. A.; (Eds.), *"Hybrid Organic- Inorganic Composites"* ACS Symp. Ser. 585, ACS Washington, DC, 1995.
[114] Chang, W. P.; Whang, W. -T.; Polymer 1996, 37, 4229.
[115] De Morais, T. D.; Chaput, F.; Lahlil, K.; Boilot, J. -P. *Adv. Mater.* 1999, 11, 107.
[116] Carlos, L. D.; Bermudez, V. de Zea; Ferreira, R. A. Sá; Marques, L.; Assunção, M. *Chem. Mater.* 1999, 11, 581.
[117] Brinker, C. J.; Scherrer, G.; (Eds.), *"Sol-Gel Science: Physics And Chemistry Of Sol-Gel Processing",* Academic Press, SD, CA, 1989.
[118] Aranda, P.; Ruiz-Hitzky, E. *Chem. Mater.* 1992, 4, 1395.
[119] Ruiz-Hitzky, E.; Arando, P. *Adv. Mater.* 1990, 2, 545.
[120] Johnson, S. A.; Brigham, E. S.; Ollivier, P. J.; Mallouk, T. E.; *Chem. Mater.* 1997, 9, 2448.
[121] Yano, K.; Usuki, A.; Okada, A.; *J. Polym. Sci. A: Polym. Chem.* 1997, 35, 2289.
[122] Ramachandran, K.; Lerner, M. M.; *J. Electrochem. Soc.* 1997, 144, 3739.
[123] Carrado, K. A.; Xu, L.; *Chem. Mater.* 1998,10, 1440.
[124] Ruiz-Hitzky, E. *Adv. Mater.* 1993, 5(5), 334.
[125] Ermakov, O.N.; Kaplunov, M.G.; Efimov, O.N.; Yakushchenko, I.K.; Belov, M.Yu.; Budyka, M.F. *Microelectronic Engineering* 2003, 69, 208.
[126] Lin, C.-F.; Lee, C.-Y.; Lu, W.-B.; Su, W.-F.; Hui, Y. Y. *Proceedings of the 7th IEEE International Conference on Nanotechnology,* 2007, 858.

[127] Henk, J. B.; Hicham, B.; Eugenio, C.; Michele, S. *Adv. Mater.* 2010, 22, 2198.

[128] Cheng, C.-C.; Chien, C.-H.; Yen, Y.-C.; Ye, Y.-S. Ko, F.-H.; Lin, C.-H.; Chang, F.-C.; *Acta Materialia,* 2009, 57, 1938.

[129] Guha, S.; Haight, R. A.; Bojarczuk, N. A.; Kisker, D. W. *J. Appl. Phys.* 1997, 82 (8), 4126.

[130] Brown, A. R.; Pichler, K.; Greenham, N. C.; Bradley, D. C.; Friend, R. H.; Holmes, A. B. *Chem. Phys. Lett.* 1993, 61,210.

[131] Baldo, M. A.; O'Brien, D. F.; You, Y.; Shoustikov, A.; Sibley, S.; Thompson, M. E.; Forrest, S. R. *Nature* 1998, 395, 151.

[132] Bolink, H. J.; Coronado, E.; Sessolo, M. *Chem. Mater.* 2009, 21, 439.

[133] Yang, X. H.; Neher, D.; Hertel, D.; Da¨ubler, T. K. *Adv. Mater.* 2004, 16, 161.

[134] Yang, X. H.; Neher, D. *Appl. Phys. Lett.* 2004, 84, 2476.

[135] Shen, Z.; Burrows, P. E.; Bulovic, V.; Forrest, S. R.; Thompson, M. E.; *Science* 1997, 276, 2009.

[136] Curran, S. A.; Ajayan, P. M.; Blau, W. J.; Carroll, D. L.; Coleman, J. N.; Dalton, A. B.; Davey, A. P.; Drury, A.; McCarthy, B.; Maier, S.; Strevens, A. Adv. *Mater.* 1998, 10(14), 1091.

[137] Kubota, H.; Miyaguchi, S.; Ishizuka, S.; Wakimoto, T.; Funaki, J.; Fukuda, Y.; Watanabe, T.; Ochi, H.; Sakamoto, T.; Miyake, T.; Tsuchida, M.; Ohshita, I.; Tohma, T. J. *Luminescence,* 2000, 87-89, 56.

[138] Tang, J.X.; Li, Y.Q.; Hung, L.S.; Lee, C.S. *Appl. Phys. Lett.* 2004, 84(1),73.

[139] Sahu, S.; Pal, A. J. J. *Phys. Chem. C.* 2008, 112, 8446.

[140] Dinh, N. N.; Chil, L. H.; Long, N. T.; Thuy, T. T. C.; Trung, T. Q.; Kim, H.-K. *J. Phys.: Conference Series* 187, 2009, 012029.

[141] Munasinghe, C; Heikenfeld, J.; Dorey, R.; Whatmore, R.; Bender, J. P.; Wager, J. F.; Steckl, A. J. *IEEE Trans. Electron Devices,* 2005, 52(2) 194.

[142] Heikenfeld, J. C.; Steckl, A. J. I*nfo. Displ.,* 2003, 20.

[143] Ono, Y.A.; Electroluminescent Displays. In: H.L. Ong, Editor, *Series on Information Display*, World Scientific, NY, Vol. 1, 1995, 61.

[144] C. N. King, J. Vac. Sci. Technol. A, *Vac.Surf. Films,* 1996,14(3), 1729.

[145] Waldrip, K. E.; Lewis, J. S.; Zhai, Q.; Puga–Lambers, M.; Davidson, M. R.; Holloway, P. H. ; Sun, S. S. *J. Appl. Phys*., 2001, 89(3), 1664.

[146] Wu, X.; Carkner, D.; Hamada, H.; Yoshida, I.; Kutsukake, M.; Dantani, K.; *SID Tech. Dig.,* Seattle,WA, 2004, 1146.

[147] Heikenfeld, J.; Steckl, A. *J. IEEE Trans. Electron Devices,* 2002, 49(6), 557.

[148] Heikenfeld, J.; Jones, R. A.; Steckl, A. J. SID Tech. Dig., Baltimore, MD, 2003, 1098.

[149] Chang, C.-Y.; Tsao, F.-C.; Pan, C.-J.; Chi, G.-C.; Wang, H. –T.; Chen, J. –J.; Ren, F.; Norton, D. P.; Pearton, S. J.; Chen, K.-H.; Chen, L.-C. *Appl. Phys. Lett.,* 2006, 88, 173503.

[150] Kagan, C. R.; Mitzi, D. B.; Dimitrakopoulos C. D. *Science* 1999, 286, 945.

[151] Xiaohui, Yang.; Xurong, Xu. *Appl. Phys Lett.,* 2000, 77(6), 797.

[152] Chong, Q.; Quan, S.; Wenge, Y.; Zheng, Xu.; Shengyi, Y.; Feng, T.; Xurong, X.; Yanbing H.; Lei, Q. *Physics Letters A.* 2005, 338,402.

[153] Bourdakos, K. N.; Dissanayake, D. M. N. M.; Silva, S. R. P.; Curry, R. *J. Appl. Phys Lett.,* 2008, 92, 153311.

In: Advanced Organic-Inorganic Composites
Editor: Inamuddin
ISBN 978-1-61324-264-3

Chapter 9

BIMETAL COMPOSITES FOR APPLICATION TO A CLEANER ENVIRONMENT

Ahmed Hamood Ahmed Dabwan[a], Satoshi Kaneco[b,*], Hideyuki Katsumata[b], Tohru Suzuki[c] and Kiyohisa Ohta[b,c]

[a]Faculty of Chemical Engineering Technology, Tati University College, Jalan Panchor, Teluk Kalong, 24000 Kemaman, Terengganu, Malaysia

[b]Department of Chemistry for Materials, Graduate School of Engineering, Mie University, Mie 514-8507, Japan

[c]Environmental Preservation Center, Mie University, Mie 514-8507, Japan

ABSTRACT

The single metal and bimetallic metal composite were applied into the removal of trihalomethanes (THMs) viz. $CHCl_3$, $CHBrCl_2$, $CHBr_2Cl$, and $CHBr_3$, from the aqueous solution in the flow system. The degradation reaction was performed under mild temperature. All of THMs could not be detected in the treated sample solution by the analytical system used in the present work. Therefore, the degradation efficiencies of THMs were almost 100%. The present technology may allow for the development of an economic and environmentally benign method of THMs remediation.

ABBREVIATIONS

AC	Activated carbon
ECD	Electron capture detector
GAC	Granular activated carbon
GC	Gas chromatograph
HPLC	High performance liquid chromatograph
MCL	Maximum contaminant level

* Corresponding author's e-mail: kaneco@chem.mie-u.ac.jp

PRBs	Permeable reactive barriers
PTFE	Polytetrafluoroethylene
THMs	Trihalomethanes
USEPA	United States Environmental Protection Agency
XRD	X-ray diffraction

1. INTRODUCTION

Halogenated organic solvents have a wide range of uses including metal processing, electronics, dry cleaning and paint, paper and textile manufacturing [1]. They thus have the potential to contaminate almost every aspect of the environment, particularly water and soil. Since halogenated solvents are generally denser than water, they tend to sink and accumulate in groundwater sources.

Many of these halogenated aliphatics are listed as priority pollutants by the United States Environmental Protection Agency (USEPA) and known or suspected human carcinogens, mutagens, or toxins [2]. The USEPA drinking water regulations maximum contaminant level (MCL) is 80 μg/l for the total trihalomethanes (THMs) as a yearly average [3]. The THMs are primarily produced by the reaction of free chlorine with natural organic materials, such as natural organic polyelectrolytes containing humic and fulvic acids. The nonbiodegradable THMs were selected for the present work because they are found as groundwater pollutants, and a degradation method for their total removal from drinking water supplies is necessary, given that, in most physicochemical treatments used, the THMs are removed from the aqueous medium; however, they could not be destroyed [4,5].

The presently used methods for removing of these compounds from drinking water and groundwater are air stripping, adsorption onto activated carbon (AC), filtration, ozonation, and reverse osmosis [5]. The air stripping proved to be the most efficient treatment for the removal of trihalomethanes. However, the method removes halogenated compounds from the liquid and into the gas phase, therefore requiring a secondary treatment [5]. Granular AC (GAC) adsorption removes trihalomethanes only to a limited degree, and desorption is expected at very low trihalomethanes concentration. Furthermore, the GAC has to be renewed frequently when exhausted. If powdered AC is used, sludge disposal may be a setback. The studies concerning the decomposition of the volatile halogenated hydrocarbons have been mainly carried out by the photocatalytic method [6,7] and using gene-manipulated bacteria [8,9], although these methods have a low potential for the application to large-scale plants due to their complex nature.

Recent studies on the reduction of chlorinated aliphatics and aromatics by zero-valent iron (Fe^0) have shown promising potential for applying the technology to the in situ remediation of contaminated groundwater [10]. Numerous feasibility studies, pilot tests, and field-scale demonstration projects have been initiated using granular Fe^0 as the reactive medium in permeable reactive barriers (PRBs) [11]. PRB technology represents a valid alternative to traditional pump-and-treat systems, as its main advantages are no energy consumption and minimal operation and maintenance costs. Further interest in the use of zero-valent metals can be attributed to Reynolds and coworkers [12], who reported the aliphatic halide transformation in the presence of several metals, including iron, galvanized metals, aluminium, and copper.

Boronina et al. [13] studied the destruction of carbon tetrachloride in water using metallic particles of magnesium, tin, and zinc. Furthermore, Xie et al. [14] showed that Zn can be an effective reductant for the dechlorination/destruction of organic chlorocarbons that contaminate ground water.

In the present review article, we describe the removal of THMs in aqueous solution with metal and bimetal composites.

2. CHLOROFORM [15]

Zinc metal powders (Nacalai Tesque, Inc., Kyoto, Japan: diameter 7 μm; surface area 0.29 m^2/g) were packed into a SUS316 pipe reactor (4.34 mm internal diameter (i.d.)). The amounts of the zinc powders varied in the range of 1 to 4 g. The length of the reactor increased with the amount of zinc powders, i.e., the packing density was constant (4.51 g/cm^3). The powders were used as received without further purification. A water bath was used for the controlling temperature of the reactor. The temperature was examined from 25 to 50 °C. The column pressure was in the range of 5 to 20 kg/cm^2. The residence time (reaction time) increased with increasing zinc amounts, that is, the time per unit column length was the same (2.2 sec/cm). A chloroform sample solution (150 ppm) was fed into the reactor at a flow rate of 1.5 ml/min using high performance liquid chromatograph (HPLC) pump (Hitachi Ltd., pump L-7110, Tokyo, Japan). $CHCl_3$ was obtained from Nacalai Tesque, Inc. The continuous flow time was monitored for 9 h. After dechlorination, the samples were collected using a 25 ml glass syringe. An aliquot of the samples (10 ml) was poured into a 20 ml glass vial sealed tightly with polytetrafluoroethylene (PTFE) and then left for almost 2 h in a water bath at 25°C to equilibrate.

The analysis of the gaseous products collected in the head space of the collector was performed using a gas chromatograph (Shimazu Corp., Kyoto, Japan; GC-14A) equipped with an electron capture detector (ECD ^{63}Ni). A glass column (3 mm i.d., 3 m long) packed with silicone DC-550 (25% phenyl methyl silicone) was used at the following chromatographic conditions: 150 °C injector temperature, 70 °C column temperature, 150 °C detector temperature, and nitrogen carrier gas. The analysis of the liquid products in the collector was carried out using ion chromatography consisting of a Hitachi HPLC L-6000 pump, a column #2710-SA-IC, and an electrical conductivity detector. The separation temperature was 40 °C.

First, the effect of the amount of zinc metal powder on the continuous destruction of chloroform in the flow system was investigated at 25 °C. These results are shown in Figures 1 and 2. At zero time, the data were obtained from the initial sample solutions. At a temperature of 25 °C, the degradation efficiency tended to decrease with the treatment time for all of zinc amounts tested. The maximum destruction efficiency and chloride ion concentration were observed using 4 g of zinc powders. If the zinc amount was greater than 4 g, the degradation efficiency may increase. However, since the column pressure is higher, the system become very complex and costly. Therefore, 4 g of zinc metal powder was used for the study of the temperature influence.

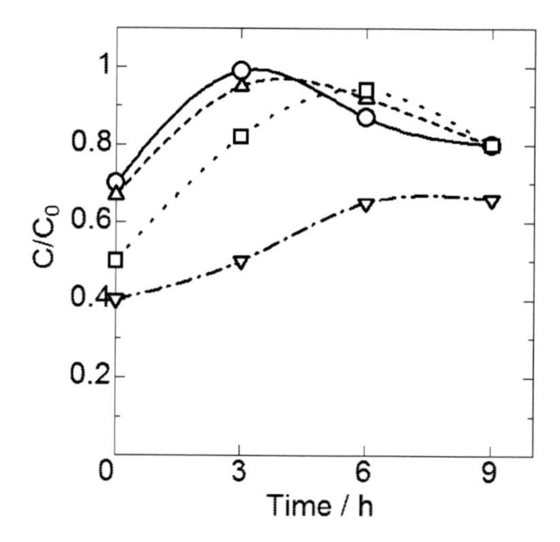

Figure 1. Effect of Zn powder amount on the continuous degradation of chloroform in the flow system. Temperature 25 °C, Zn amounts; 1 g (○), 2 g (Δ), 3 g (□), and 4 g (∇). Reproduced with permission from Dabwan, A. H. A. et al. *ITE Lett. Batt. New Technol. Med.* 2002, 3, 585-588. Copyright © 2002, ITE-IBA Inc.

The effect of temperature on the continuous degradation of chloroform in the flow system was studied using 4 g of zinc powder. The results are shown in Figures 3 and 4. At temperatures of 25 and 30 °C, the degradation efficiencies decreased with the treatment time and were less than 35% after 6 h. On the other hand, at 40 and 50 °C, the efficiencies increased with the treatment time. The reason for these phenomena is not clear. It may be attributed to passivation layers of zinc oxides and hydroxides being formed on the surface at a relatively low temperautre. At 50 °C, especially, the degradation efficiency was more than 93% after 3 h and the yield of chloride ion was almost 100%. Therefore, the continuous decomposition treatment using Zn powder was very effective for chloroform under mild conditions, taking the analytical error into consideration.

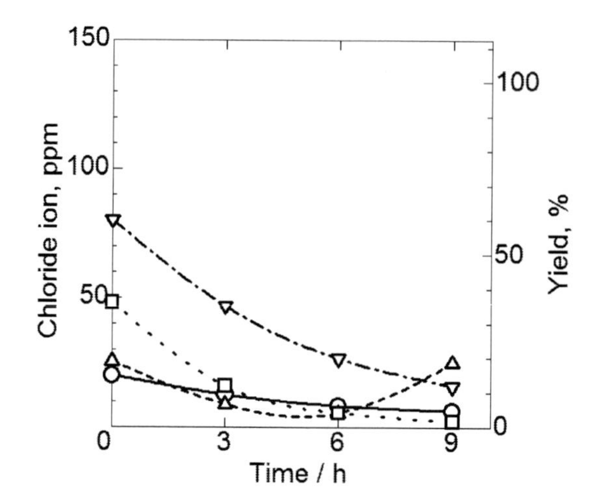

Figure 2. Effect of Zn powder amount on the concentration of chloride ion. Temperature 25 °C, Zn amounts; 1 g (○), 2 g (Δ), 3 g (□), and 4 g (∇). Reproduced with permission from Dabwan, A. H. A. et al. *ITE Lett. Batt. New Technol. Med.* 2002, 3, 585-588. Copyright © 2002, ITE-IBA Inc.

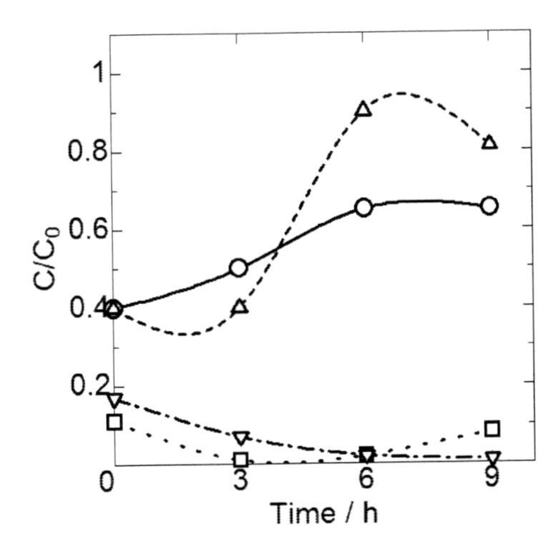

Figure 3. Effect of temperature on the continuous degradation of chloroform in the flow system. Zn amount 4 g, temperature; 25 °C (○), 30 °C (Δ), 40 °C (□), and 50 °C (∇). Reproduced with permission from Dabwan, A. H. A. et al. *ITE Lett. Batt. New Technol. Med.* 2002, 3, 585-588. Copyright © 2002, ITE-IBA Inc.

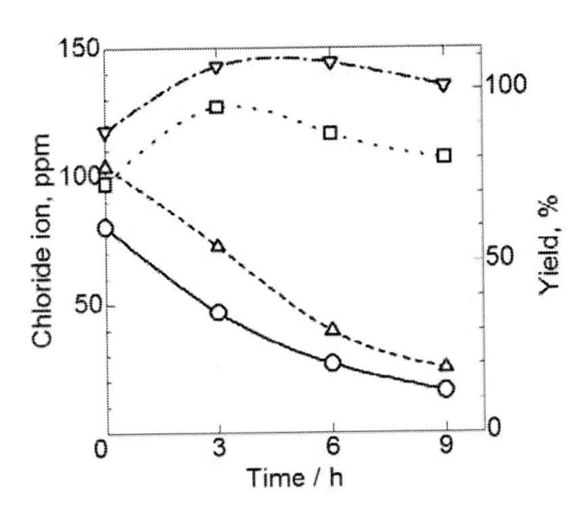

Figure 4. Effect of temperature on the concentration of chloride ion. Zn amount 4 g, temperature; 25°C (○), 30 °C (Δ), 40 °C (□), and 50 °C (∇). Reproduced with permission from Dabwan, A. H. A. et al. *ITE Lett. Batt. New Technol. Med.* 2002, 3, 585-588. Copyright © 2002, ITE-IBA Inc.

3. Bromoform [16]

The instrumentation and experimental conditions were the same as that in the case of the degradation of chloroform. A bromoform sample solution (125 ppm) was fed into the reactor using HPLC pump. The flow rate was set in the range of 0.25 to 2 ml/min. An oxygen gas in the sample vessel was removed by flowing hydrogen gas. $CHBr_3$ was obtained from Nacalai Tesque, Inc. After debromination, the samples were collected using a 25 ml glass syringe. An aliquot of the samples (10 ml) was poured into a 20 ml glass vial sealed tightly with PTFE and then left for almost 2 h in a water bath at 25 °C to equilibrate. The analysis methods for the gaseous and liquid products were the same as that for $CHCl_3$. After the continuous treatment, the surface of zinc metal was checked by X-ray diffraction (XRD) analysis. XRD measurements were carried out using a Rigaku RAD-RC (12kW) system with monochromated CuK_α radiation.

First, the effect of temperature on the continuous destruction of bromoform in the flow system was investigated using 3 g of zinc powder. The results are shown in Figures 5 and 6. At zero time, the data were obtained from the initially-treated sample solutions. As shown in Figure 5, the degradation efficiencies at 25 and 35 °C were relatively low. However, the efficiencies were very high (>90%) dramatically as temperatures above 40 °C. Therefore, it was found that bromoform could be almost completely degraded in the continuous flow system using the zinc powder above 40 °C. Since it could be confirmed that zinc oxides and hydroxides were formed on the surface of zinc metal after treatment by XRD analysis, the inactivation at relatively low temperature may be attributed to passivation layers of them. Here, we define the yield (%) as (bromide ion concentration after degradation treatment) × 100 / (the maximum bromide ion concentration 118.6 [ppm]). The yield for bromide ion increased with increasing temperature, as shown in Figure 6. Whereas the degradation efficiency at 40 °C was very high (>99%) after 3 hour treatment, the maximum yield for bromide ion was not so high (82%). These phenomena may be attributed to the formation of intermediate products (e.g. dibromoethane, dibromoethylene and tetrabromoethane). Consequently, the continuous decomposition treatment using Zn powder was very effective for bromoform under mild conditions (40 °C).

The effect of treatment rate (flow rate) of bromoform sample solutions on the continuous degradation in the flow system using 3 g of zinc powder was studied at 40 °C. The results were shown in Figures 7 and 8. The degradation efficiency for bromoform and the yield of bromide ion increased with decreasing the flow rate of sample solutions. The flow rates of 0.25, 0.50, 1.0 and 2.0 ml/min corresponded to the reaction time (residence time) of 39, 20, 9.8 and 4.9 sec, respectively. Therefore, when the reaction time was more than 20 sec, the degradation efficiency was relatively high (>90%) and the yield of bromide ion was above 93%.

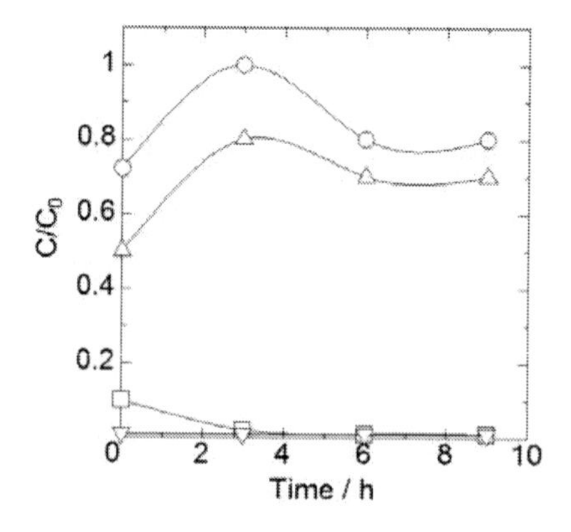

Figure 5. Effect of temperature on the continuous degradation of bromoform in the flow system using zinc powder. Zn amount 3 g, sample flow rate 1 ml/min, temperature; 25 °C (○), 35 °C (Δ), 40 °C (□), and 45 °C (∇). Reproduced with permission from Dabwan, A. H. A. et al. *Photo/Electrochem. Photobiol. Environ. Energy Fuel* 2005, *4*, 409-418. Copyright © 2005, Research Signpost.

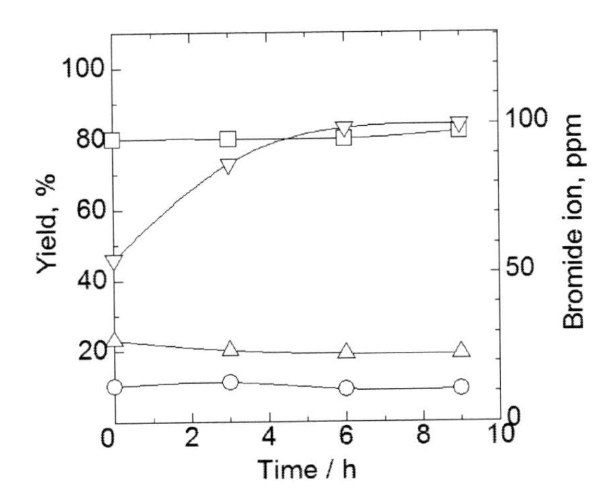

Figure 6. Effect of temperature of formation of bromide ions in the continuous degradation of bromoform in the flow system using zinc powder. Zn amount 3 g, sample flow rate 1 ml/min, temperature; 25 °C (○), 35 °C (Δ), 40 °C (□), and 45 °C (∇). Reproduced with permission from Dabwan, A. H. A. et al. *Photo/Electrochem. Photobiol. Environ. Energy Fuel* 2005, *4*, 409-418. Copyright © 2005, Research Signpost.

The effect of the amount of zinc metal powder on the continuous destruction of bromoform in the flow system was investigated at 40 °C. The treatment rate for the sample solution was 1.0 ml/min. These results were shown in Figures 9 and 10. The residence time (reaction time) increased with increasing zinc amounts because the packing density was constant. The reaction times were 3.3, 9.8, 13 and 19 sec in the case of zinc powder amounts of 1, 3, 4.5 and 6 g, respectively. With increase in the metal amounts, the degradation efficiency of bromoform and the bromide ion yield increased gradually. Especially, the high yield of bromide ion (>97%) was observed with more than 4.5 g of zinc metal amounts.

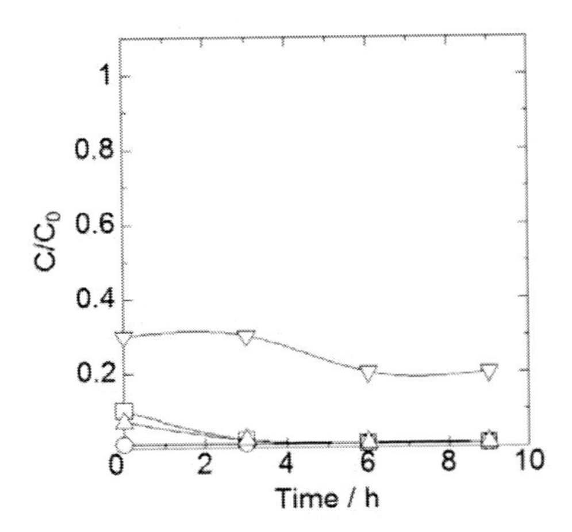

Figure 7. Effect of treatment rate on the continuous degradation of bromoform in the flow system using zinc powder. Zn amount 3 g, temperature 40 °C, treatment rate; 0.25 ml/min (○), 0.50 ml/min (Δ), 1.0 ml/min (□), and 2.0 ml/min (∇). Reproduced with permission from Dabwan, A. H. A. et al. *Photo/Electrochem. Photobiol. Environ. Energy Fuel* 2005, *4*, 409-418. Copyright © 2005, Research Signpost.

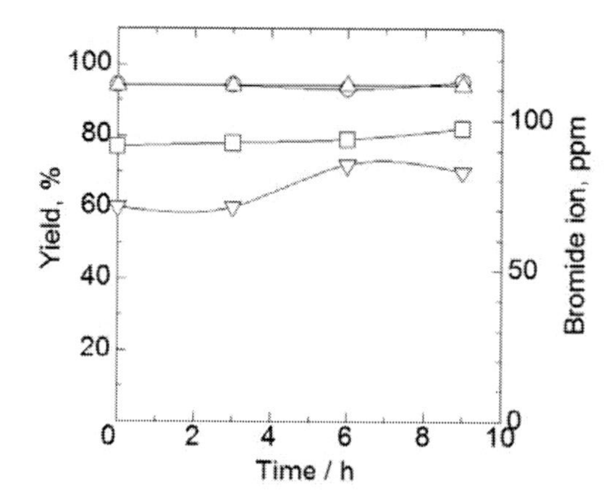

Figure 8. Effect of treatment rate on formation of bromide ions in the continuous degradation of bromoform in the flow system using zinc powder. Zn amount 3 g, temperature 40 °C, treatment rate; 0.25 ml/min (○), 0.50 ml/min (Δ), 1.0 ml/min (□), and 2.0 ml/min (∇). Reproduced with permission from Dabwan, A. H. A. et al. *Photo/Electrochem. Photobiol. Environ. Energy Fuel* 2005, *4*, 409-418. Copyright © 2005, Research Signpost.

The effect of treatment time on the continuous degradation of bromoform in the flow system using 6 g of zinc powder was studied at 45 °C. The continuous flow time was monitored for 250 hours. The results were shown in Figure 11. Bromoform seems to be completely degraded in the continuous flow system in the treatment time 250 hours since it in the sample solution after treatment could not be detected by gas chromatograph–electron capture detector (GC–ECD). The high yield of bromide ion was found to be maintained until 250 hours, although the results had a relatively large distribution. In consequence, in the continuous flow system using 6 g of zinc powder at 45 °C and 1 ml/min of treatment rate, bromoform could be almost effectively decomposed for the long treatment time (250 hours). The treatment at 1 ml/min flow rate of sample solution for 250 hours corresponded to 15000 ml of sample volume.

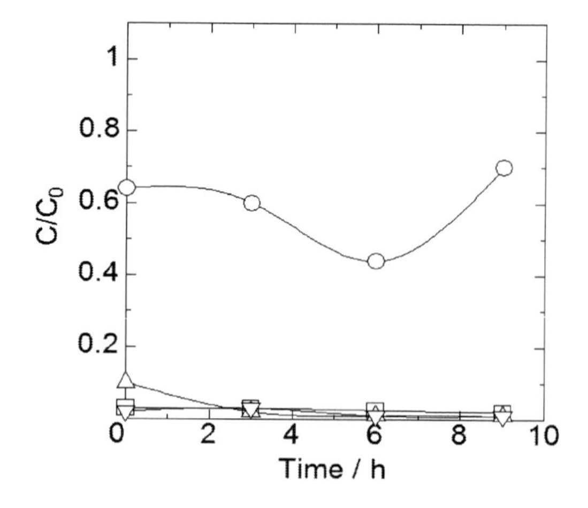

Figure 9. Effect of zinc powder amount on the continuous degradation of bromoform in the flow system using zinc powder. Temperature 40 °C, treatment rate 1 ml/min, Zn amounts; 1 g (○), 3 g (Δ), 4.5 g (□), and 6 g (∇). Reproduced with permission from Dabwan, A. H. A. et al. *Photo/Electrochem. Photobiol. Environ. Energy Fuel* 2005, *4*, 409-418. Copyright © 2005, Research Signpost.

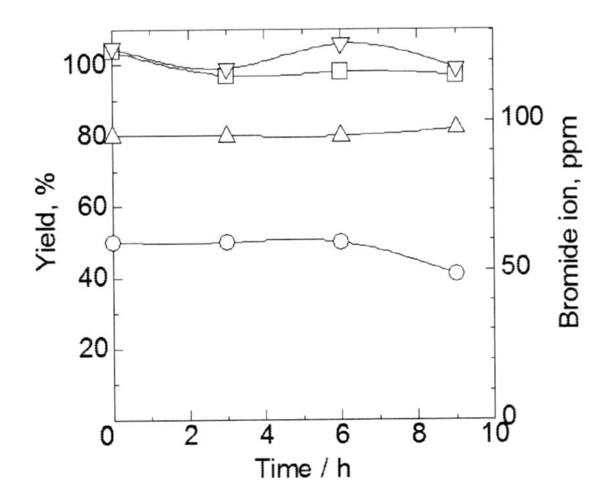

Figure 10. Effect of zinc powder amount on formation of bromide ions in the continuous degradation of bromoform in the flow system using zinc powder. Temperature 40 °C, treatment rate 1 ml/min, Zn amounts; 1 g (○), 3 g (Δ), 4.5 g (□), and 6 g (∇). Reproduced with permission from Dabwan, A. H. A. et al. *Photo/Electrochem. Photobiol. Environ. Energy Fuel* 2005, *4*, 409-418. Copyright © 2005, Research Signpost.

4. BROMODICHLOROMETHANE [17]

Nickel (diameter average 2.9-3.6 μm; surface area 5.8 m^2/g; 99% purity) and zinc (diameter 7 μm; surface area 0.29 m^2/g) powders were obtained from Nakalai Tesque Inc., Kyoto, Japan. The powders were used as received without further purification. Bimetallic Ag/Zn was prepared by the simple method. The zinc powder was put into a silver nitrate solution. The metallic silver was deposited onto the Zn surface, since silver ions underwent a fast reduction to elemental silver by the oxidation of Zn metal. The metal suspension was shaken at room temperature for 24 h. Assuming that all the Ag metal was reductively precipitated onto the Zn surface, the content of Ag in the Zn was calculated as 3% by weight.

The nickel, zinc and bimetallic silver/zinc powders were packed into a SUS316 pipe reactor (i.d. 4.34 mm) with the amount of 4.0, 3.0, and 3.0 g, respectively. A water bath was used for controlling the temperature of the reactor. The temperature was evaluated from 25 to 80 °C. The column pressure was in the range of 5 to 20 kg/cm^2. A bromodichloromethane sample solution (175 ppm) was fed into the reactor at the flow rate range of 0.5 to 1.0 ml/min using an HPLC pump (Hitachi, Ltd., pump L-7110, Tokyo, Japan). The continuous flow time was monitored for 12 h. After degradation, the samples were collected using a 25 ml glass syringe. An aliquot of the samples (10 ml) was poured into a 20 ml glass vial sealed tightly with PTFE and then left for almost 2 h in a water bath at 25 °C to equilibrate. The analysis of the gaseous and liquid products was the same as that in the case of the degradation of chloroform.

First, the continuous destruction of $CHBrCl_2$ in the flow system using nickel metal powder at a 0.5 ml/min flow rate was investigated in the range of 25 to 80 °C. The typical results are shown in Table 1. At zero time, the data were obtained from the initial sample solutions. At 25 °C, bromodichloromethane could not be totally destroyed in the continuous flow system. Even at the high temperature of 80 °C, the yields of bromide and chloride ions were relatively low (less than 71.3%). Therefore, although the surface area of the nickel powder used was relatively large, it was not effective for the continuous degradation of $CHBrCl_2$ in the flow system. The poor performance of Ni alone may be a consequence of the formation of a protective oxide layer on the Ni surface that inhibits the reducibility of Ni.

Next, zinc metal powder was studied for the continuous treatment of bromodichloromethane in the flow system at the flow rate of 1.0 ml/min (Table 1). The temperature was examined from 25 to 40°C. The degradation efficiencies at 25°C were relatively low. However, the efficiencies were very high (>95%) at temperatures above 30°C. Therefore, it was found that $CHBrCl_2$ could be almost completely degraded in the continuous flow system using the zinc powder above 30 °C. Because the yield of bromide ion was greater than that of chloride ion, the debrominaiton reaction preferentially proceeded vs. the dechlorination one in the continuous flow system using the zinc powder. Therefore, zinc powder alone was not effective for the dechlorination of $CHBrCl_2$ in the continuous flow system, though it was very useful for the degradation of chloroform, bromoform, and chlorodibromomethane.

Since the yield of chloride ion was not very high in the continuous flow system with zinc metal powder, the bimetallic Ag/Zn powder was then tested (Table 1). The flow rate was 1.0 ml/min. The yield of chloride ion was greater than that obtained with zinc alone. Hence, the combination of Zn with Ag significantly enhanced the chloride ion yield compared to Zn alone. Metallic silver is incapable of serving as an effective reductant because of its low reduction potential. The high activity of Ag/Zn may be attributed, at least in part, to the enhanced-bimetallic corrosion. In the bimetallic Ag/Zn system, galvanic cells are created by coupling active metal (i.e., Zn) with inert metal (i.e., Ag). The galvanic cells of Ag/Zn couples lead to a significantly high potential gradient (approximately 1.6 V). Elemental Zn serves as the anode and becomes preferably oxidized in the galvanic cells while Ag serves as the cathode. In other words, the bimetallic structure of Ag/Zn enhances the reducibility of Zn for the reductive dehalogenation by facilitating the Zn corrosion.

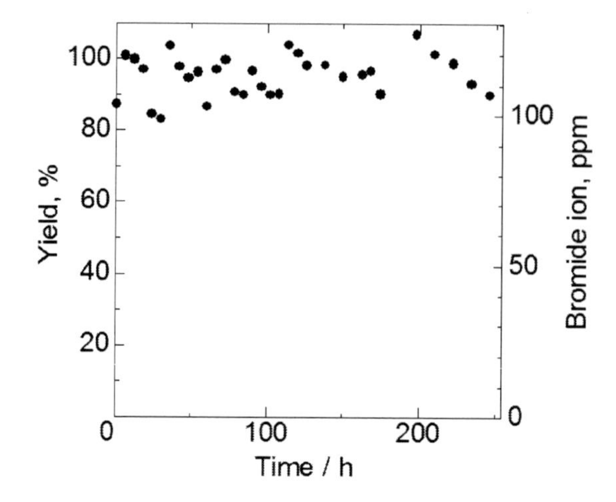

Figure 11. Effect of treatment time on formation of bromide ions in the continuous degradation of bromoform in the flow system using zinc powder. Zn amount 6 g, treatment rate 1 ml/min, temperature 40 °C. Reproduced with permission from Dabwan, A. H. A. et al. *Photo/Electrochem. Photobiol. Environ. Energy Fuel* 2005, *4*, 409-418. Copyright © 2005, Research Signpost.

Table 1. Continuous degradation of $CHBrCl_2$ in a flow system using Ni, Zn and Ag/Zn powders

Metal	Temperature (°C)	Time (h)	C/C_0	Chloride ion		Bromide ion	
				Concentration (ppm)	Yield (%)	Concentration (ppm)	Yield (%)
Ni	80	0	0.03	25.4	33.3	36.3	42.3
		3	0.01	30.2	39.6	60.3	70.3
		6	0.01	30.2	39.6	58.2	67.7
		9	0	29.7	39.0	51.8	60.4
		12	0	26.1	34.3	54.9	64.0
Zn	25	0	0.33	18.2	23.9	26.2	30.5
		3	0.88	2.4	3.2	13.8	16.1
		6	0.85	1.5	2.0	9.1	10.6
		9	0.85	1.5	2.0	9.2	10.7
		12	0.85	1.5	2.0	9.1	10.6
	30	0	0.02	14.1	18.5	45.3	52.8
		3	0.01	4.8	6.3	59.8	69.7
		6	0	4.7	6.2	60.5	70.5
		9	0	4.9	6.4	60.1	70.0
		12	0	5.2	6.8	51.4	59.9
	35	0	0.05	14.6	19.2	77.2	89.9
		3	0.02	5.4	7.7	84.6	98.6
		6	0.01	6.1	8.0	80.3	93.6
		9	0	6.1	8.0	77.8	90.6
		12	0	6.1	8.0	74.6	86.9
	40	0	0.02	14.2	18.6	75.6	88.1
		3	0	5.1	6.7	67.1	78.2
		6	0	4.3	5.6	87.3	102
		9	0	6.1	8.0	87.5	102
		12	0	5.9	7.7	81.2	94.6
Ag/Zn	40	0	0	72.3	94.9	85.6	99.7
		3	0	57.4	75.4	82.4	96.0
		6	0	55.2	72.5	85.2	99.3
		9	0	53.3	69.7	85.2	99.3
		12	0	53.4	70.1	82.3	95.9

Reproduced with permission from Dabwan, A. H. A. et al. *ITE Lett. Batt. New Technol. Med.* 2003, *4*, 461-464. Copyright © 2003, ITE-IBA Inc.

5. CHLORODIBROMOMETHANE [18]

The instrumentation and experimental conditions were the same as that in the case of the degradation of chloroform. A chlorodibromomethane sample solution (140 ppm) was fed into the reactor at a flow rate of 1.0 ml/min using HPLC pump. A dissolved oxygen gas in the sample solution was removed by flowing hydrogen gas. $CHClBr_2$ was obtained from Nacalai

Tesque, Inc. The continuous flow time was monitored for 12 h. After degradation, the samples were collected using a 25 ml glass syringe. An aliquot of the samples (10 ml) was poured into a 20 ml glass vial sealed tightly with PTFE and then left for almost 2 h in a water bath at 25 °C to equilibrate. The analysis methods for the gaseous and liquid products were the same as that for $CHCl_3$.

The effect of temperature on the continuous destruction of chlorodibromomethane in the flow system was investigated. The results are shown in Figure 12. At zero time, the data were obtained from the initially-treated sample solutions. At a temperature of 25 °C, although the degradation efficiency was very high (98%) in the initial treatment, the efficiency became worse (less than 40%) instantly. At 30 °C, the high degradation efficiency (>97%) was maintained until 3 h, and then the efficiency decreased with the treatment time. The destruction efficiency remained high (>94%) in the treatment time up to 12 h at the temperature of 35 °C. The results obtained at 40 and 50 °C were very similar to those at 35 °C. Therefore, it was found that chlorodibromomethane could be almost completely degraded in the continuous flow system using the zinc powder above 35 °C. The inactivation at relatively low temperature may be attributed to passivation layers of zinc oxides and hydroxides being formed on the metal surface. Consequently, the continuous decomposition treatment using Zn powder was very effective for chlorodibromomethane under mild conditions (35 °C).

The effect of temperature on the formation of ions (chloride ion and bromide ion) in the flow system was studied using 3 g of zinc powder. The results are shown in Figures 13 to 17. At temperature of 25 °C, the yield of chloride ion was approximately 100% in the initial treatment and the yield of bromide ion was 72%, as shown in Figure 13. However, both yields lowered instantly. These trends agreed with the results for degradation efficiency. As illustrated in Figure 14, the yield for chloride ion at 30 °C was relatively high until 3 h, and the value lessened with the treatment time. The tendency of bromide ion was the same as that of chloride one. At 35 °C, the yield curve for bromide ion had a convex pattern, as shown in Figure 15. The chloride ion yield decreased gradually with increasing the time. As shown in Figures 16 and 17, the tendencies observed at 40 and 50 °C were very similar to those obtained at 35 °C. Therefore, both dechlorination and debromination processes could occur at temperatures of more than 35 °C, though it was not clear which process proceeded in the continuous decomposition of chlorodibromomethane in a flow system using zinc powder.

6. TRIHALOMETHANE MIXTURES [19]

Zinc (diameter 7 μm; surface area 0.29 m^2/g) powders were obtained from Nakalai Tesque Inc., Kyoto, Japan. The powders were used as received without further purification. Bimetallic Ag/Zn was prepared by the simple method. The zinc powder was put into a silver nitrate solution. The metallic silver was deposited onto the Zn surface, since silver ions underwent a fast reduction to elemental silver by the oxidation of Zn metal. The metal suspension was shaken at room temperature for 24 h. Assuming that all the Ag metal was reductively precipitated onto the Zn surface, the content of Ag in the Zn was calculated as 3% by weight. The experimental system and conditions were the same as that in the case of the degradation of chloroform. Bimetallic silver/zinc powders were packed into a SUS316 pipe reactor (i.d. φ4.34 mm) with the amount of 3.0 g. The length of the pipe reactor was 45 mm, and the packing density was 4.51 g/cm^3. A water bath was used for controlling the temperature of the reactor. The reaction temperature was kept constant at 25 °C. The column pressure was approximately 10 kg/cm^2. Trihalomethane chemicals were obtained from Nacalai Tesque, Inc. A sample solution containing the mixtures of chloroform, bromodichlorometane, chlorodibromomethane and bromoform was fed into the reactor using HPLC pump (Hitachi Ltd., pump L-7110, Tokyo, Japan). Each of trihalomethane concentrations was 90 μg/ml. The flow rate was 1 ml/min. The residence time (reaction time) was 9.8 sec. (per the 45 mm reactor). An oxygen gas in the sample vessel was removed by flowing hydrogen gas. After the degradation, the samples were collected using a 25 ml glass syringe. An aliquot of the samples (10 ml) was poured into a 20 ml glass vial sealed tightly with PTFE and then left for almost 2 h in a water bath at 25 °C to equilibrate. The analysis methods for the gaseous and liquid products were the same as that for chloroform.

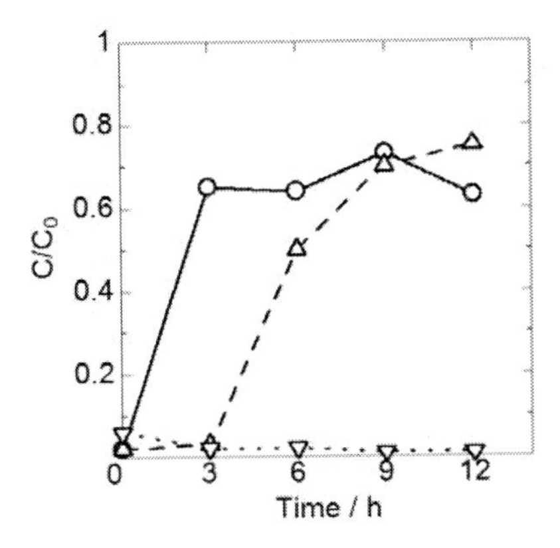

Figure 12. Effect of temperature on the continuous degradation of chlorodibromomethane in the flow system using zinc powder. Zinc amount 3 g, temperature; 25 °C (○), 30 °C (Δ), and 35 °C (∇). Reproduced with permission from Dabwan, A. H. A. et al. *Photo/Electrochem. Photobiol. Environ. Energy Fuel* 2003, *2*, 163-171. Copyright © 2003, Research Signpost.

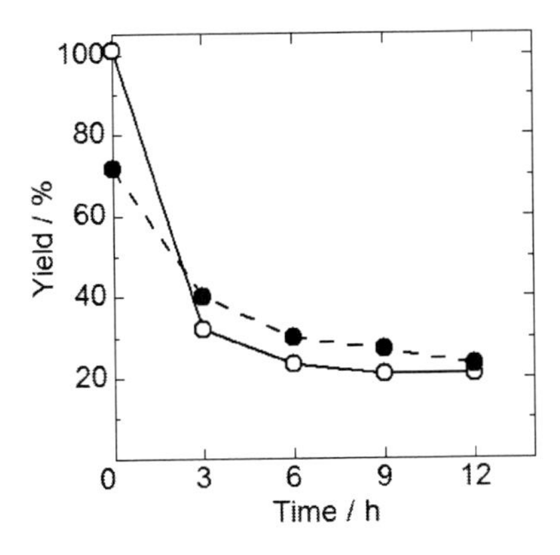

Figure 13. Formation of chloride and bromide ions in the the continuous degradation of chlorodibromomethane in the flow system using zinc powder at 25 °C. Zn amount 3 g, ○; chloride ion, ●; bromide ion. Reproduced with permission from Dabwan, A. H. A. et al. *Photo/Electrochem. Photobiol. Environ. Energy Fuel* 2003, *2*, 163-171. Copyright © 2003, Research Signpost.

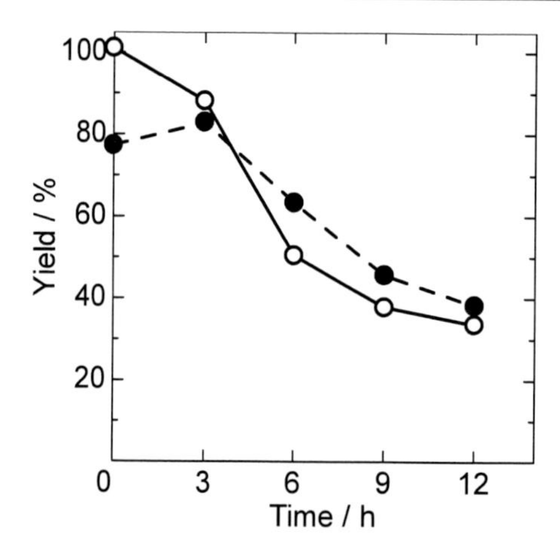

Figure 14. Formation of chloride and bromide ions in the continuous degradation of chlorodibromomethane in the flow system using zinc powder at 30 °C. Zn amount 3 g, ○; chloride ion, ●; bromide ion. Reproduced with permission from Dabwan, A. H. A. et al. *Photo/Electrochem. Photobiol. Environ. Energy Fuel* 2003, *2*, 163-171. Copyright © 2003, Research Signpost.

Metallic silver is incapable of serving as an effective reductant because of its low reduction potential. The high activity of Ag/Zn may be attributed, at least in part, to the enhanced-bimetallic corrosion. In the bimetallic Ag/Zn system, galvanic cells are created by coupling active metal (i.e., zinc) with inert metal (i.e., silver). The galvanic cells of Ag/Zn couples lead to a significantly large potential gradient (approximately 1.6 V) as compared to other bimetallic systems (e.g. 1.4 V of Pd/Fe bimetal). Elemental zinc serves as the anode and becomes preferably oxidized in the galvanic cells while Ag serves as the cathode. In other words, the bimetallic structure of Ag/Zn enhances the reducibility of zinc for the reductive dehalogenation by facilitating the zinc corrosion. It has found in the previous initial experiment that bimetallic Ag/Zn significantly proved the effective dechlorination for $CHBrCl_2$ compared with those obtained with nickel and zinc metal powders [17].

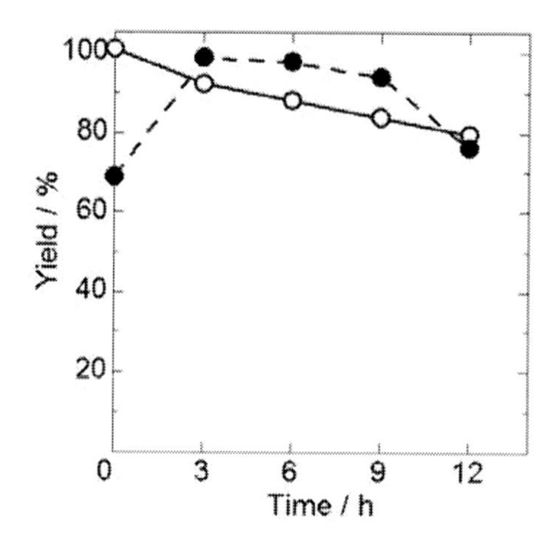

Figure 15. Formation of chloride and bromide ions in the the continuous degradation of chlorodibromomethane in the flow system using zinc powder at 35°C. Zn amount 3 g, ○; chloride ion, ●; bromide ion. Reproduced with permission from Dabwan, A. H. A. et al. *Photo/Electrochem. Photobiol. Environ. Energy Fuel* 2003, *2*, 163-171. Copyright © 2003, Research Signpost.

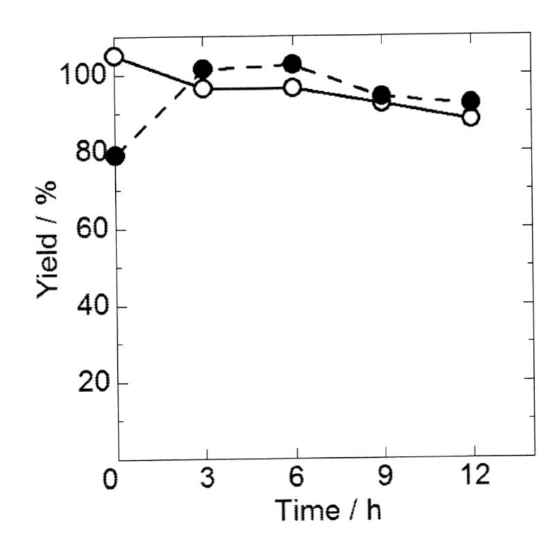

Figure 16. Formation of chloride and bromide ions in the the continuous degradation of chlorodibromomethane in the flow system using zinc powder at 40 °C. Zn amount 3 g, ○; chloride ion, ●; bromide ion. Reproduced with permission from Dabwan, A. H. A. et al. *Photo/Electrochem. Photobiol. Environ. Energy Fuel* 2003, *2*, 163-171. Copyright © 2003, Research Signpost.

The degradation efficiencies of trihalomethanes and the yields for the formation of bromide and chloride ions enabled the comparison of the removal potential of each compound from the aqueous solutions with bimetallic silver/zinc. The yield, θ(Br), is defined as (bromide ion concentration after the degradation treatment) × 100 / (the maximum bromide ion concentration 198.3 μg/ml). Similarly, the yield, θ(Cl), (chloride ion concentration after the degradation treatment) × 100 / (the maximum chloride ion concentration 134.4 μg/ml).

First, the degradation efficiencies of trihalomethanes with bimetallic silver/zinc were investigated for the treatment time of less than 12 hours. Consequently, all of trihalomethanes could not be detected in the treated sample solution. Since the detection limits for the trihalomethanes were c.a. 0.1 μg/ml, all of degradation efficiencies of trihalomethanes were almost 100% in the treatment time range.

Next, the influence of pH on the formation of bromide and chloride ions were studied in the pH range of 2.8 to 10, as depicted in Figures 18 and 19. Little effect of pH on the yields of bromide and chloride ions, θ(Br) and θ(Cl) could be observed in the present treatment system. Both of the yields for bromide and chloride ions increased slightly with the treatment time until 3 hours, and then decreased gradually with the time.

The formation of methane was observed in the early stage for the treatment time (for instance, until 3 hours), while ethane and ethylene formations in the addition of methane was obtained after 3hours. By comparing between Figures 18 and 19, bromine atoms seem to be released from trihalomethanes, compared with chlorine atoms. A more stable molecule (i.e. the stronger the bonds) is less likely to undergo a chemical reaction. The bond dissociation energy between the carbon and chlorine atoms, (397 ± 29) kJ/mol, is stronger than that of between carbon and bromine atoms ((280 ± 21) kJ/mol) [20]. Therefore, the results were reasonable by considering the bond dissociation energy. Since the dichloromethane appears to be less toxic than chloroform, the toxicity of the sample solution could be lowered by the present treatment.

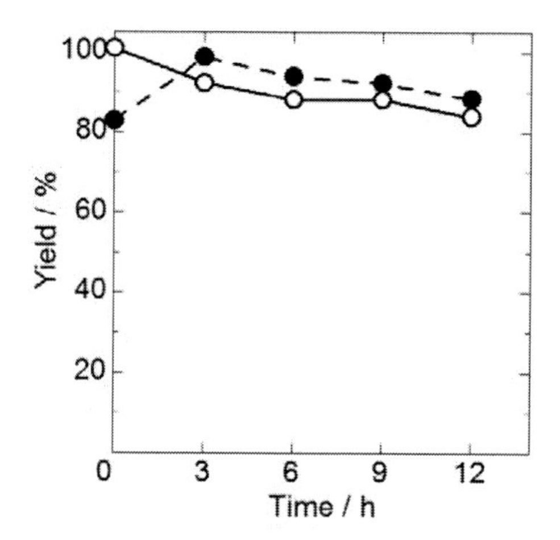

Figure 17. Formation of chloride and bromide ions in the the continuous degradation of chlorodibromomethane in the flow system using zinc powder at 50°C. Zn amount 3 g, ○; chloride ion, ●; bromide ion. Reproduced with permission from Dabwan, A. H. A. et al. *Photo/Electrochem. Photobiol. Environ. Energy Fuel* 2003, *2*, 163-171. Copyright © 2003, Research Signpost.

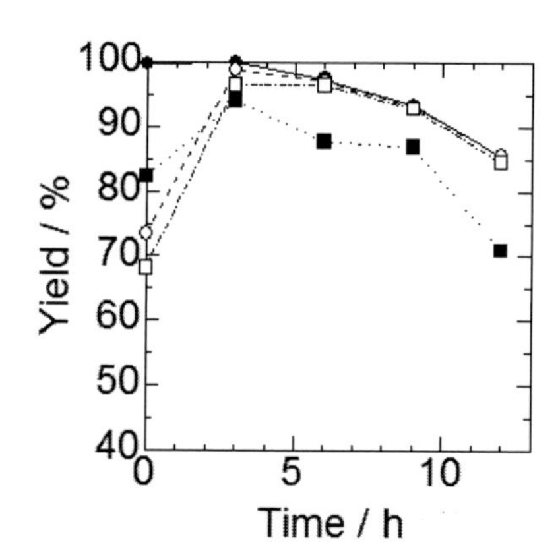

Figure 18. Effect of pH on the formation of bromide ion in the continuous degradation of trihalomethane mixtures in the flow system using bimetallic silver/zinc. Each trihalomethan concentration 90 μg/ml, pH; 2.8 (●), 3.5 (○), 7.6 (■) and 10 (□). Reproduced with permission from Dabwan, A. H. A. et al. *Photo/Electrochem. Photobiol. Environ. Energy Fuel* 2007, *6*, 313-319. Copyright © 2003, Research Signpost.

7. Reaction Mechanism for Bimetallic Silver/Zinc Composite

The mechanism for the dechlorination/debromination of trihalomethanes in a batch system using bimetallic silver/zinc was investigated. From our results and the literature [1,2,4,5,7,10–13,14–19,21], the degradation pathway of trihalomethanes using bimetallic silver/zinc can be estimated, as illustrated in Figure 20. It has been reported in the previous paper [4] that the sonodegradation efficiencies and rates follow the decreasing order $CHCl_3$ >

$CHBrCl_2$ > $CHBr_2Cl$ > $CHBr_3$. The estimated mechanism is that the catalytic silver enhances the formation of atomic hydrogen or hydride on the surface and alters the electronic properties on the zinc. Next, C–X (X = Cl, Br) bond of adsorbed halorinated compounds onto the surface of the Ag/Zn particles is broken, and the halogen atom is replaced with hydrogen. Because the production of hydrocarbons was observed in the present study, it seems reasonable to speculate a multistep reduction sequence. Here, CHX_3 (X = Cl, Br) oxidatively adds to Ag/Zn forming $X_2HC(Ag/Zn)X$, a reactive intermediate that would be susceptible to rapid protonation by Ag–H. In a similar way, CH_2X_2 and CH_3X could be attacked and protonated. Since trihalomethanes could be almost completely degraded by bimetallic silver/zinc, the final main products were hydrocarbons, chloride ion, and bromide ion. However, because the yield for chloride ion was not 100%, intermediate products (e.g., dichloromethane) may be formed.

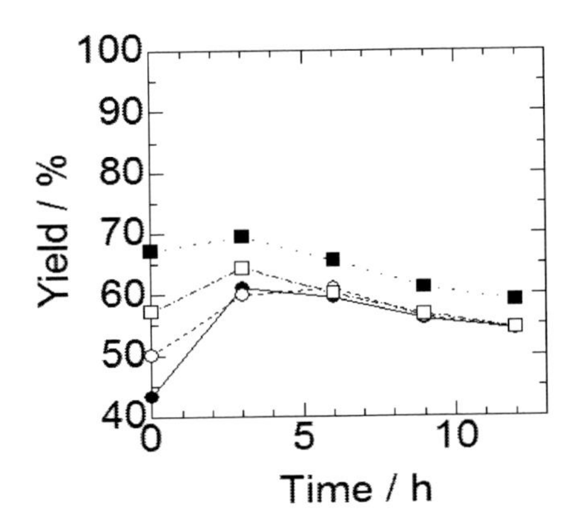

Figure 19. Effect of pH on the formation of chloride ion in the continuous degradation of trihalomethane mixtures in the flow system using bimetallic silver/zinc. Each trihalomethan concentration 90 μg/ml, pH; 2.8 (●), 3.5 (○), 7.6 (■) and 10 (□). Reproduced with permission from Dabwan, A. H. A. et al. *Photo/Electrochem. Photobiol. Environ. Energy Fuel* 2007, *6*, 313-319. Copyright © 2003, Research Signpost.

$$Zn^0 \longrightarrow Zn^{2+} + 2e^-$$

$$2H_2O + 2e^- \xrightarrow[Ag]{} 2H^* + 2OH^-$$

$$Ag + H^* \longrightarrow Ag\text{–}H$$

$$X_3C\text{–}H \xrightarrow[Ag/Zn]{} X_2HC(Ag/Zn)X \xrightarrow[Ag\text{–}H,\ +e^-]{} X_2H_2C\ (+\ X^-)$$

$$\xrightarrow[(X = Cl,\ Br)]{} \longrightarrow \longrightarrow CH_4,\ C_2H_4,\ C_2H_6$$

Figure 20. Proposed mechanism for the dechlorination and debromination of trihalomethanes using bimetallic silver/zinc.

Conclusion

The removal of THM compounds from the aqueous solution with metal and bimetallic composite was investigated in the flow system. Because the chloride ion yield was not 100%, some intermediate products (e.g., dichloromethane) might be formed. Although the Zn^{2+} ions were released in the remediation reaction system using Zn metal, the integrated technique combining the zinc and the microbial community may be beneficial to remove the possible secondary contamination of Zn^{2+} [21]. Consequently, the developed treatment system may be viewed as an environmentally desirable remediation.

Acknowledgments

This work was partially supported by the Ministry of Education, Culture, Sports, Science, and Technology of Japan. All experiments were conducted at both Mie University and Tati University College. Any opinions, findings, conclusions, or recommendations expressed in this paper are those of the authors and do not necessarily reflect the view of the supporting organizations.

References

[1] Ronald, J. K.; Gregory, A. S.; Thomas, R. C., Eric, A. B.; Robert, G. A. *Environ. Sci. Technol.* 1993, *27,* 2104–2111.

[2] Symons, J. M.; Stevens, A. A.; Clark, R. M.; Geldreich, E. E.; Love, O. T.; DeMarco, J. *Water Eng. Man.* 1981, *128*, 50, 52–53, 56, 61–64.

[3] United States Environmental Protection Agency. *National Primary Drinking Water Regulations: Disinfectants and Disinfection Byproducts. Federal Register*; 63(241), 63 FR 69390-69476, December 16, 1998; http://water.epa.gov/lawsregs/rulesregs/sdwa/stage1/ (accessed July 2010).

[4] Shemer, H.; Narkis, N. *Ultrason. Sonochem.* 2005, *12*, 495-499.

[5] Tang, W. Z.; Tassos, S. *Water Res.* 1997, *31*, 1117-1125.

[6] Pruden, A. L.; Ollis, S. F. *Environ. Sci. Technol.* 1983, *17*, 628-631.

[7] Glaze, W. H.; Kenneke, J. F.; Ferry, J. L. *Environ. Sci. Technol.* 1993, *27*, 177-184.

[8] Wackett, L. P.; Gibson, D. T. *Appl. Environ. Microbiol.* 1988, *54*, 1703-1708.

[9] Zylstra, G. J.; Wackett, L. P.; Gibson, D. T. *Appl. Environ. Microbiol.* 1989, *55*, 3162–3166.

[10] Gillham, R. W.; O'Hannesin, S. F. *Ground Water* 1984, *32*, 958-967.

[11] O'Hannesin, S. F.; Gillham, R. W. *Ground Water* 1998, *36*, 164-170.

[12] Reynolds, G. W.; Hoff, J. T.; Gillham, R. W. *Environ. Sci. Technol.* 1990, *24*, 135-142.

[13] Boronina, T. N.; Lagadic, I.; Sergeev, G. B.; Klabunde, K. J. *Environ. Sci. Technol.* 1998, *32*, 2614-2622.

[14] Xie, N.; Qiu, G.; Chen, S. *Huagong Huanbao (in Chinese)* 2007, *27,* 227-229.

[15] Dabwan, A. H. A.; Suzuki, T.; Kaneco, S.; Katsumata, H.; Ohta, K. *ITE Lett. Batt. New Technol. Med.* 2002, 3, 585-588.

[16] Dabwan, A. H. A.; Suzuki, T.; Kaneco, S.; Katsumata, H.; Ohta, K. *Photo/Electrochem. Photobiol. Environ. Energy Fuel* 2005, *4*, 409-418.

[17] Dabwan, A. H. A.; Suzuki, T.; Kaneco, S.; Katsumata, H.; Ohta, K. *ITE Lett. Batt. New Technol. Med.* 2003, *4*, 461-464.

[18] Dabwan, A. H. A.; Suzuki, T.; Kaneco, S.; Katsumata, H.; Ohta, K. *Photo/Electrochem. Photobiol. Environ. Energy Fuel* 2003, *2*, 163-171.

[19] Dabwan, A. H. A.; Suzuki, T.; Kaneco, S.; Katsumata, H.; Ohta, K. *Photo/Electrochem. Photobiol. Environ. Energy Fuel* 2007, *6*, 313-319.

[20] Lide, D. R. CRC Handbook of Chemistry and Physics. 85th ed. CRC Press, Boca Raton, FL, USA, 2005; pp 9-53.

[21] Ma, C.; Wu, Y. *Environ. Geol.* 2008, 55, 47-54.

In: Advanced Organic-Inorganic Composites
Editor: Inamuddin

ISBN 978-1-61324-264-3

Chapter 10

CeO_2 BASED ELECTRO-CERAMIC COMPOSITE MATERIALS FOR SOLID OXIDE FUEL CELL APPLICATIONS

Suddhasatwa Basu [*] ***and Rajalekshmi Chockalingam***
Department of Chemical Engineering, Indian Institute of Technology,
New Delhi 110 016, India

ABSTRACT

Recently, many researchers have shown interest in developing composite materials for the application of solid oxide fuel cell (SOFC) electrodes and electrolytes. The advantage of using composite material is that they provide the properties of both components compared to either component alone. An attempt has been made in this chapter to present a range of materials which are used for various applications in the area of SOFCs. Although doped ZrO_2 electrolyte is known for its better electrical and mechanical properties, it requires high operating temperature, greater than 800 °C and costly refractory for thermal confinement. This led to an interest in researchers to look for alternative electrolyte materials capable of operating at a lower temperature. Doped CeO_2 is considered one such material which is capable of operating at intermediate temperature range of 500 to 700 °C. The disadvantage of CeO_2 is its high electronic conductivity at lower partial pressures of oxygen and poor mechanical strength. Addition of an insulating second phase to CeO_2 matrix is suggested as one method to reduce the electronic conductivity. The other advantage of adding suitable second phase such as Al_2O_3 and ZrO_2 to ceria is the improvement in mechanical strength. An alternative method to reduce to electronic conductivity of CeO_2 is to reduce the operating temperature below 500 °C. This is possible by mixing a suitable carbonate phase to CeO_2 matrix.

The successful operation of SOFC also requires mixed electronic and ionic conducting electrodes with matching thermal expansion coefficient with that of electrolyte in order to reduce the thermal stresses during thermal cycling of the cells. Long term continuous

[*] Corresponding author's e-mail: sbasu@chemical.iitd.ac.in

operation of fuel cells also requires that the electrolyte, anode and cathode should be chemically and structurally stable under reduced conditions. CeO_2-ZrO_2 composite has not shown any improvement in ionic conductivity, whereas the mechanical properties improved significantly. Addition of PrO_2 into CeO_2 has shown a positive trend in terms of ionic conductivity. The ionic conductivity of CeO_2-TbO_2 composite increased with Tb addition up to $x = 0.25$. The composite exhibited 1-2 orders of magnitude higher oxygen ion conductivity than that of the most commonly used solid electrolyte, stabilized zirconia, at 650 °C. In the case of CeO_2-UO_2 solid solution, UO_2 rich end, behaves like a p-type semiconductor, whereas at the CeO_2 rich end behave like an n-type semiconductor such a way that overall the material will behave like a highly non-stoichiometric soild solution ideal to study defect chemistry of both an oxygen deficient composition and oxygen rich composition in a single crystal structure. CeO_2-Al_2O_3 composite has shown significant reduction in electronic conductivity due to the trapping of electrons at the interface between trace amounts of Co and Mn doped Al_2O_3 and Gd-CeO_2, which suppresses the electronic contribution to the total conduction. Based on the collective information one can argue that addition of second phase to the matrix is one of the methods to improve the electrical as well as mechanical properties of single phase SOFC electrodes and electrolytes.

ABBREVIATIONS

Bi_2O_3	Bismuth oxide
CeO_2	Cerium oxide
Coat B	Nanocomposite of 0.68% Mn, 0.68% Co, 39.05% Ce and 9.76% Gd coated on 49.84 at% alumina
Coat C	Nanocomposite of 0.33% Mn, 0.33% Co, 80% Ce and 20% Gd coated on 49.84 at% alumina
Ea	Activation energy
GDC	Gadolinium doped cerium oxide
HT-XRD	High temperature X-ray diffraction
LSGM	Sr and Mg doped $LaGaO_3$
MIEC	Mixed electronic and ionic conductor
PCO	Praseodymium-cerium oxide
P_{O2}	Oxygen partial pressures
Pr	Praseodymium
SCZ	Scandium doped zirconium oxide
SDC	Samarium doped cerium oxide
SEM	Scanning electron microscopy
SOFCs	Solid oxide fuel cells
UO_2	Uranium oxide
XANES	X-ray absorption near edge spectroscopy
XRD	X-ray diffraction
YSZ	Yttria stabilized zirconia
ZrO_2	Zirconium oxide
ZrO_2	Zirconium oxide

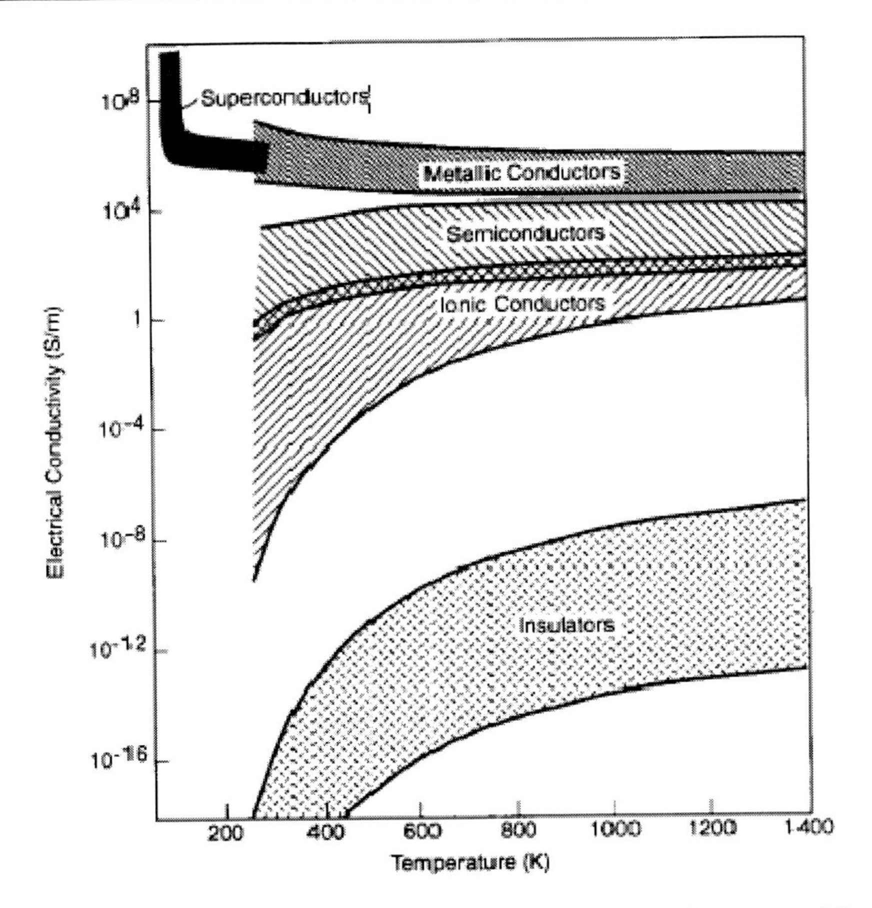

Figure 1. The electrical conductivity of various conductors as a function of temperature [6]. Reproduced with permission from Weber, W. J. et al. *Mater. Sci. and Eng. B* 1993, *18*, 52-71. Copyright © 1993, Elsevier Science Ltd.

1. INTRODUCTION

A Fuel cell is an energy conversion device that produces electricity directly from gaseous fuel by electrochemical reaction [1]. The first commercially successful ceramic solid oxide fuel cell, yttria stabilized zirconia (YSZ) demonstrated ionic conductivity of 0.14 Scm^{-1} at 1000 °C [2]. This fuel cell typically contains 8 mol% Y_2O_3 stabilized ZrO_2 as electrolyte, a ceramic-metal composite, Ni + YSZ as anode and $La_{1-x}Sr_xMnO_{3-\delta}$, as the cathode. Alkali doped $LaCrO_3$ with specific dopants such as Sr, Ca or Mg is used as interconnect material and the concentration is optimized to eliminate thermal expansion mismatch with the different fuel cell components [3]. Other prominent electrolyte materials in this regard are: Sc-doped ZrO_2 (SCZ), Gd or Sm doped CeO_2 (GDC or SDC) and Sr and Mg doped $LaGaO_3$ (LSGM) [4]. Glass and glass-ceramic sealant materials are also used in planar solid oxide fuel cells (SOFCs) to isolate anode and cathode chambers in a stacked configuration. Among the other oxygen ion conductors, stabilized bismuth oxide (Bi_2O_3) shows the highest ionic conductivity of 0.1 $\Omega^{-1}cm^{-1}$ at 700° C and 0.01 $\Omega^{-1}cm^{-1}$ at 500 °C, respectively [5]. The drawback of this material is that it easily reduces under low oxygen partial pressures and decomposes into Bi metal. Oxide-ion conducting perovskites such as $La_{0.9}Sr_{0.1}Ga_{0.8}Mg_{0.2}O_{3-\delta}$ (LSGM) and $BaZrO_3$ are other alternative fuel cell electrolytes [5]. The oxygen transport properties of LSGM are comparable to those of Sc doped ZrO_2 but less than Gd doped CeO_2. However, the problem associated with LSGM is the reactivity of lanthanum with nickel at the anode.

Recently, electro-ceramic materials attracted the attention of many researchers due to their mixed ionic and electronic conduction. Materials like cubic-ZrO_2, CeO_2, Bi_2O_3, Na-β-Al_2O_3, and α-AgI exhibit predominately ionic conduction with electrons as minority carriers, whereas, materials like TiO_2, WO_3, and TiS_2 exhibit predominantly electronic conductivity and small fraction of ionic conductivity. Not much information is available in the literature about mixed ionic-electronic conductors [4]. An attempt has been made in this article to briefly review the composite mixed ionic-electronic conductors used as electrolytes and electrodes. The review is organized as seven sections. Sections I, II and III are devoted to review the conductivity mechanism in general with special emphasis on conventional solid oxide fuel cells such as doped ZrO_2 and CeO_2. The rest of the sections from IV to VIII discuss in detail about the use of CeO_2 based mixed ionic-electronic conductors as electrolytes and electrodes. The cases considered for review are CeO_2–ZrO_2, CeO_2–PrO_2, CeO_2–TbO_2, CeO_2–UO_2, $GdCeO_2$–Al_2O_3, and $GdCeO_2$-$(LiNa)CO_3$.

2. The Electrical Conductivity of Materials

The electrical conductivity of electro-ceramic materials as a function of temperature is shown in Figure 1. The electrical conductivity of ionic conductors varies from 10^{-8} Sm^{-1} to 10^2 Sm^{-1} when the temperature increased from 0 to 1127 °C.

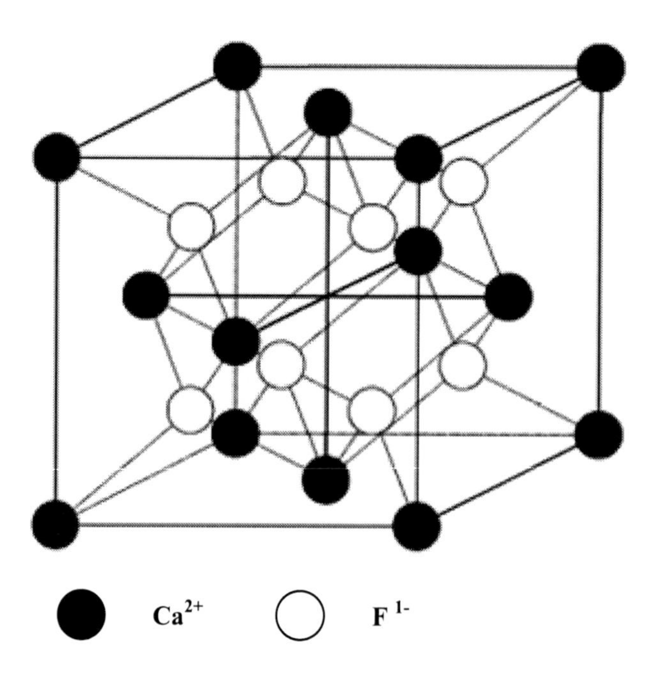

Figure 2. The fluorite crystal structure, solid circles represents cations and open circles represent anions.

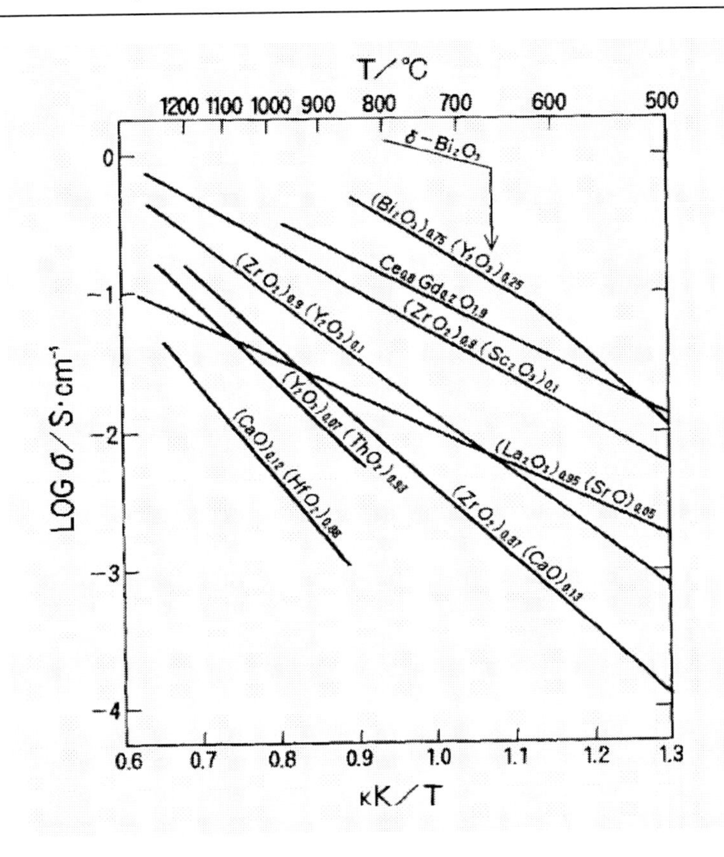

Figure 3. Ionic conductivities (σ) of several fluorite based oxide ion conductors [9]. Reproduced with permission from Inaba, H. and Tagawa, H. *Solid State and Ionics,* 1996, 1-16. Copyright © 1996, Elsevier Science Ltd.

At the same time ionic conductivity of the insulator vary from 10^{-18} to 10^{-9} in the same temperature range. The conductivity of metallic conductors decreases with increasing temperature. The conductivity of materials originates from their crystal structures. Oxides with high ionic conductivity have open structure, such as fluorite and pyrochlore. Most conventional solid state ionic conductors belong to fluorite structure which is relatively open and shows large tolerance for high level atomic disorder. This can be introduced by doping, reduction or oxidation. Some examples are: CeO_2, PrO_2, ThO_2, TbO_2 and UO_2. In addition, HfO_2 and cubic ZrO_2 can be stabilized to fluorite phase by doping with divalent or trivalent oxide with substantial addition of acceptor dopants [7].

Figure 2 represents the unit cell of a fluorite lattice, which consists of a face centered cubic array of cations with all tetrahedral interstices fixed with oxygen ions [8]. It consists of four cations and eight anions per unit cell. The cations are tetrahedrally coordinated and anions are 8-fold coordinated. The basic fluorite structure is very stable with respect to removal of oxygen ions from the structure. It is possible to replace 20% of the cations in the system with triple charged acceptors and the charge compensation is achieved through generation of oxygen vacancies, which are responsible for the movement of ions. The ionic conductivities of several fluorite based oxide ion conductors are shown in Figure 3. Where Bi_2O_3 and yttria-stabilized Bi_2O_3 show the highest conductivity followed by Gd doped CeO_2 and Sc doped ZrO_2, respectively at 800 °C. Homogeneous doping with appropriate elements is one approach to obtain high ionic conductivity. Electrical neutrality of the solid solution is maintained through supplementary defect formation. By using Kroger Vink nomenclature, creation of additional metal vacancies by doping with cations of higher valence can be expressed as follows:

$$N\,X_2\,(MX) \rightarrow N_M^{\bullet} + V_M' + 2X_X \ldots \quad (1)$$

Moreover, doping with +2 and +3 cations into Zr^{4+} introduces oxygen vacancies, which in turn enhances the conductivity. Table 1 shows the Kroger Vink notation for creation of oxygen vacancies in ZrO_2, CeO_2 and $LaGaO_3$ based electrolyte systems when doped with divalent and trivalent dopants, respectively. Where O_O^X and V_O^X represent regular lattice oxygen and vacancy of oxygen, respectively [10]. Oxides with fluorite type structure (CeO_2 and ThO_2) can form extended solid solutions with oxides of lower valence cations like CaO, Y_2O_3 etc., resulting in oxygen deficiencies as high as 10-15%.

3. Doped Zirconium Oxide (ZrO_2) Electrolytes

Owing to its high ionic conductivity, mechanical and chemical stability ZrO_2 based oxides are extensively used as electrolytes for SOFC operating above 800 °C. Badwal [11] extensively studied the oxygen ion conducting properties of cubic, partially stabilized and tetragonal zirconia and reported that the ionic conductivity depends on dopant type (size, valence), dopant concentration and phase assemblage. Dopants, such as Y_2O_3 are added to stabilize the high temperature cubic phase. Addition of 8 mol% Y_2O_3 stabilizes the cubic phase to room temperature. Addition of Y_2O_3 to ZrO_2 also creates oxygen vacancies. This mechanism can be expressed in defect chemical notation as:

$$Y_2O_3 \xrightarrow{ZrO_2} 2Y_{Zr}' + V_O^{\bullet\bullet} + 3O_O \quad (2)$$

Even though, the highest conducting ZrO_2 solid solutions were obtained with Sc_2O_3 doping (~8 mol% Sc_2O_3, $\sigma = 0.32\ Scm^{-1}$ at 1000 °C) [12], Y_2O_3 is most commonly preferred dopant due to its lower cost. It is well known that ZrO_2 exhibits three different crystal structures depending up on temperature. At room temperature, it exhibits monoclinic structure, which is a stable form. During heating at 1170 °C it transforms to tetragonal structure (t-ZrO_2). Further increasing the temperature to 2370 °C, the tetragonal phase transforms to cubic phase. The cubic phase is stable up to the melting point of ZrO_2 (2680 °C). These phase transformations are reversible in nature and as temperature reduces to 950-1000 °C, the tetragonal ZrO_2 transforms into monoclinic accompanied with a large volume change, which disintegrate the material. This can be avoided by stabilizing ZrO_2 with lower valence alkaline rare earth oxides. Additions of Y_2O_3 or Sc_2O_3 up to 8 mol% increase the conductivity of ZrO_2. However, increasing the amount of additives beyond 8 mol% reduces the conductivity due to association of point defects followed by reduction in defect mobility as shown in Figure 4.

4. Doped Ceria (CeO_2) Electrolytes

Doped ceria has been proposed as a potential alternative candidate for SOFC electrolyte for medium temperature applications (500-800 °C). Ceria has many advantages over zirconia: (i) Ceria exhibit a higher conductivity than zirconia for the same dopant concentration. (ii) Unlike ZrO_2 pure ceria has the fluorite crystal structure and follows simple mass action principles, making it less complicated to study the ceria electrolytes in the doped form. (iii) Electrical conductivity of gadolinia- stabilized ceria is about one order of magnitude larger than that of yttria- stabilized zirconia and it is about 0.1 Scm^{-1} at 800 °C. This is due to the larger ionic radius of Ce^{4+} (0.87 Å) than Zr^{4+} (0.72 Å), which produces a more open structure through which oxygen ions can easily migrate. Despite its favorable ion transport properties, ceria had not, until recently, been considered as a candidate for intermediate temperature fuel cell applications because of its high electronic conductivity at operating temperature. Numerous studies focus on methods by which the electronic conductivity in the system at low partial pressure of oxygen can be suppressed. Doping of CeO_2 with divalent or trivalent cations, create anion vacancies and enhances the ionic conductivity. Figure 5 shows the variation in conductivity as a function of samarium concentration in CeO_2 at various temperatures. The conductivity increases with increasing samaria content, reaches a maximum and then decreases. Similar behavior is observed for all acceptor dopants in ceria where the maximum conductivity varies with dopants. While doping with acceptors, the ionic conductivity increases at high pO_2 values, at lower pO_2 values and at high temperatures.

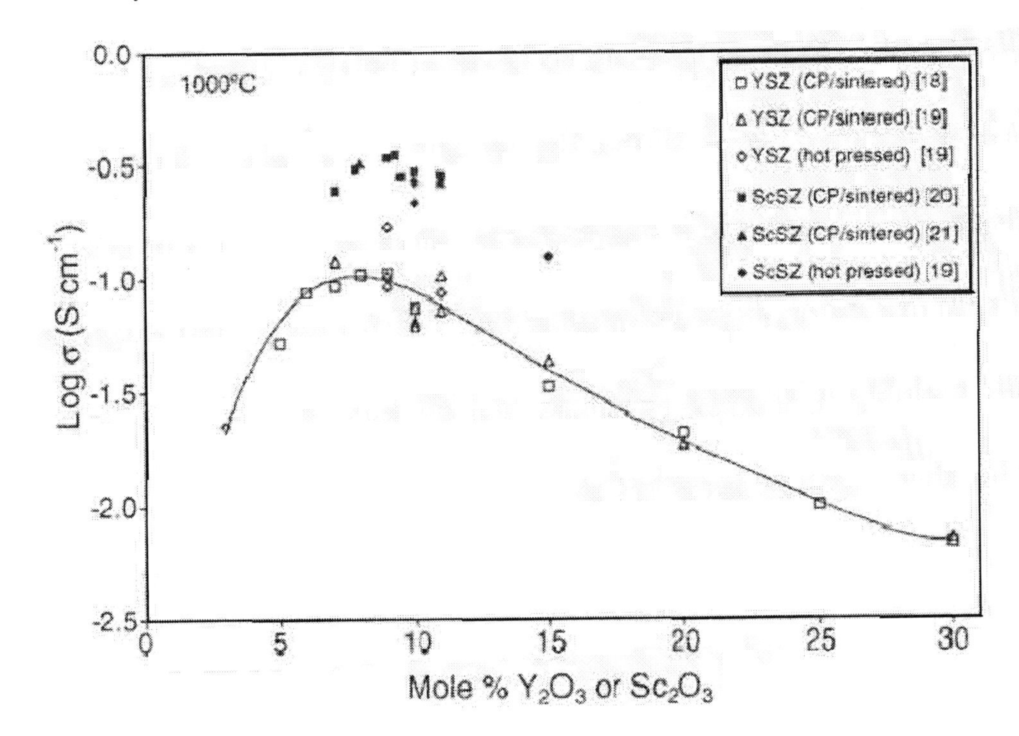

Figure 4. Conductivity of Y_2O_3 and Sc_2O_3 stabilized ZrO_2 in air at 1000 °C [13]. Reproduced with permission from Fergus, J. W. *J. Power Sources* 2006, *162,* 30-40. Copyright © 2006, Elsevier Science Ltd.

Table 1. Defect reactions illustrating creation of oxygen vacancies of ZrO_2, CeO_2 and $LaGaO_3$ based solid electrolyte system

Dopants	Divalent	Trivalent
ZrO_2-based	$MO \xrightarrow{ZrO_2} M''_{Zr} + O_O^x + V_O^{\bullet\bullet}$	$M_2O_3 \xrightarrow{ZrO_2} 2M'_{Zr} + 3O_O^x + V_O^{\bullet\bullet}$
CeO_2-based	$MO \xrightarrow{CeO_2} M''_{Ce} + O_O^x + V_O^{\bullet\bullet}$	$M_2O_3 \xrightarrow{CeO_2} 2M'_{Ce} + 3O_O^x + V_O^{\bullet\bullet}$
$LaGaO_3$-based	$SrO \xrightarrow{LaGaO_3} 2Sr'_{La} + O_O^x + V_O^{\bullet}$	$MgO \xrightarrow{LaGaO_2} 2Mg'_{Ga} + O_O^x + V_O^{\bullet\bullet}$

As observed in the case of ZrO_2, the conductivity increases with increasing dopant concentration. When a maximum is reached, defect association reduces the mobility of oxygen vacancies and as a result the total conductivity drops. Similar behavior is observed for all acceptor dopants in CeO_2. One of the problems with CeO_2 based electrolyte is the electronic conduction at lower oxygen partial pressures (pO_2) at high temperature. Tuller et al. studied 5 mol% Y_2O_3 doped CeO_2 to investigate the ionic transference number and developed an electrolytic domain (T vs pO_2), refers to the range of pO_2 and temperature at which t_i is greater than 0.99, reported that the Y_2O_3-CeO_2 composite behaved as an ionic conductor with transference number $t_i \geq 0.99$ [15]. The departure from stoichiometry creates vacancies as well as electronic defects in CeO_2 which in turn contribute to conductivity in the form of electronic conductivity. The reaction which leads to nonstoichiometry is expressed as:

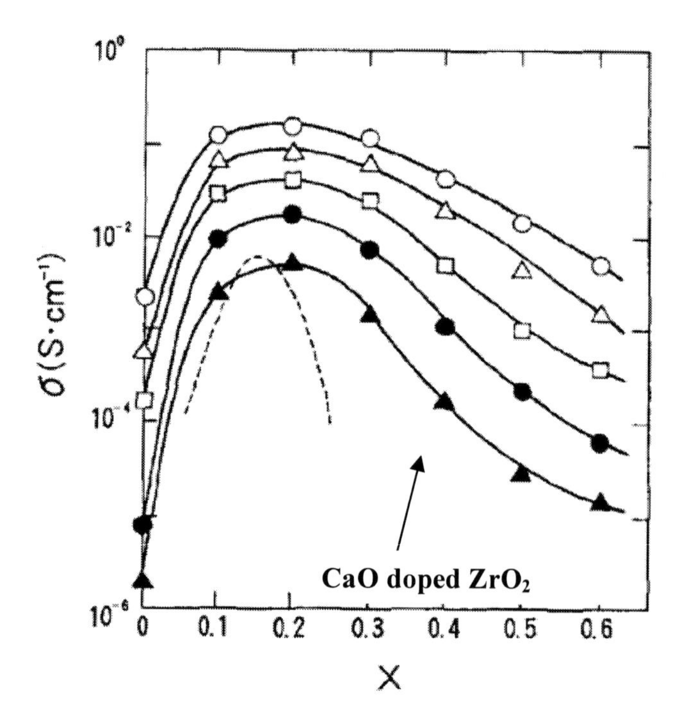

Figure 5. The variation of ionic conductivity (σ) with samarium content (x) in samaria doped ceria measured at 500-900 °C: (O) 900 °C; (Δ) 800 °C; (□) 700 °C; (•) 600 °C; (▲) 500 °C. The dotted line represents CaO doped ZrO_2 at 800 °C [9]. Reproduced with permission from Inaba, H. and Tagawa, H. *Solid State and Ionics,* 1996, 1-16. Copyright © 1996, Elsevier Science Ltd.

$$O_O = V_O^{\bullet\bullet} + 2e' + \frac{1}{2}O_2\,(g)\ldots \tag{3}$$

where e' = electronic defect (Ce'_{Ce}). The mass action equation for the reaction can be written as

$$n_V\, n_e^2\, pO_2^{\frac{1}{2}} = K_1(T) = K_{10}\exp\left(-\frac{\Delta H_i}{kT}\right) \tag{4}$$

where n_e = electron concentration; pO_2 = oxygen partial pressure; and ΔH_i = Change in enthalpy of the reaction. Also $n_e \ll n_M$ and $n_V \cong n_M$. The electronic conductivity is given by

$$\sigma_e = n_e\, e\, \mu_e \tag{5}$$

$$\mu_e = \left(\frac{b}{T}\right)\exp\left(\frac{-E_e}{kT}\right)\ldots \tag{6}$$

E_e = activation energy. The total conductivity can be written as

$$\sigma = \sigma_i + \sigma_e \tag{7}$$

where σ_i is independent of pO_2 ; σ_e is proportional to $pO_2^{\frac{-1}{4}}$ and this pressure dependence separates σ_i from σ_e. The ionic transference number is given by

$$t_i = \frac{\sigma_i}{\sigma_e} \tag{8}$$

Figure 6 shows the electrical conductivity of 5 mol% Y_2O_3 doped CeO_2 as a function of pO_2 in the temperature range of 635-1150 °C. One can clearly see that σ increases with decreasing pO_2. The tendency of increase of σ with decreasing pO_2 suggests the presence of non-stoichiometric carriers. The data in the graph is fitted with a slope of ¼. A similar power law has been reported in the case of CeO_2-La_2O_3 [16] and CeO_2-CaO [17].

Figure 7 shows the ionic transference number as a function of pO_2 in the temperature range 635-1150 °C, which are derived for the data in Figure 6 [15]. It is very well clear from Figure 7 that at lower temperatures, t_i remains close to 1 over a large range of pO_2. Among these materials CeO_2 is capable of forming a range of solid solutions with other materials due to its large open space in the crystal structure. Examples of solid solutions studied so far are given below. CeO_2- ZrO_2 [18], CeO_2-PrO_2 [19], CeO_2-TbO_2 [20], CeO_2-UO_2 [21], $GdCeO_2$-Al_2O_3[22] and $GdCeO_2$-(LiNa)CO_3 [23].

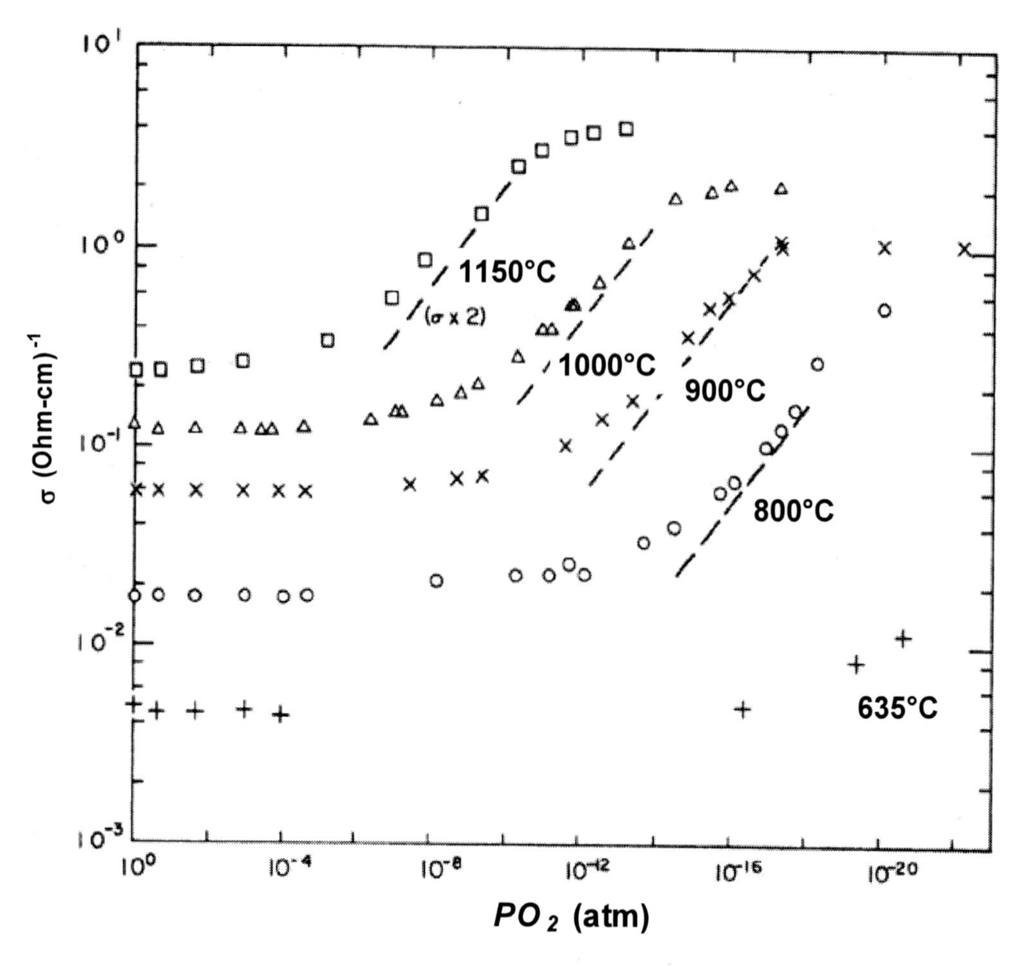

Figure 6. Conductivity as a function of pO_2 for 5% Y_2O_3 doped CeO_2 sample where the dashed lines represent the contribution of the electronic conductivity [15]. Reproduced with permission from Tuller, H. L. et al. *J. Electro. Chem. Soc.* 1975, *122,* 255-259. Copyright © 1975, The Electrochemical Society, USA.

5. CeO_2 –ZrO_2 Composites

Electrical properties of CeO_2-ZrO_2 composites have been investigated by many researchers. It was reported that the solid solutions of CGO_xYSZ_{1-x} exhibited lower conductivity than that of samples with CGO or YSZ alone due to the increase in activation energy caused by decrease in mobility of oxide ions. Butler et al., [24] studied this through theoretical calculations and reported that the activation energy (Ea) depended on elastic strain energy on the association enthalpy for the defect pair, which in turn was the effect of ionic radius of the dopants. Therefore, changes of Ea in CGO_xYSZ_{1-x} system may be due to contribution by the ionic radius difference between Ce^{4+} and Zr^{4+}.

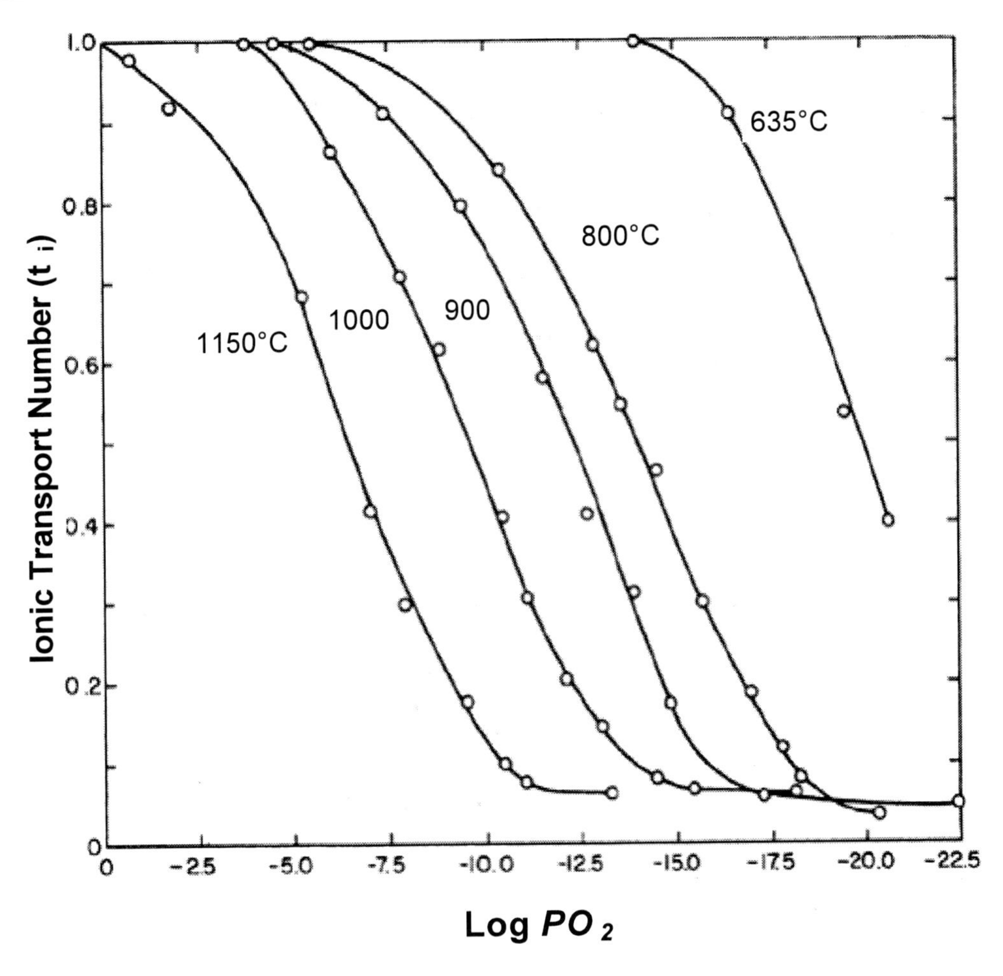

Figure 7. Ionic transference numbers as a function of pO_2 and temperature corresponding to the conductivity data measured from Figure 6 [15]. Reproduced with permission from Tuller, H. L. et al. *J. Electro. Chem. Soc.* 1975, *122,* 255-259. Copyright © 1975, The Electrochemical Society, USA.

Chiodelli et al. [18] studied the properties of CeO_2-ZrO_2 solid solutions by varying CeO_2 molar fractions between 16-100%. Their studies revealed that the concentrations of both electronic and ionic defects are constant in the temperature range between 500 and 1000 °C and in the 1-10^{-6} atm pO_2 range.

Therefore, the ionic conductivity is independent of CeO_2 content in the whole composition range. The ZrO_2-CeO_2 solid solution with less than 50 mol% of CeO_2 exhibited similar magnitude of ionic and electronic conductivity whereas, the electronic conductivity increased compared to ionic conductivity when the CeO_2 content is increased more than 50 mol%.

At the temperature range between 650 and 1000 °C, the electronic conductivities of ZrO_2-CeO_2 solid solutions showed different dependence on the oxygen partial pressure. The slope of logarithm of conductivity versus logarithm of oxygen partial pressure changed from -1/4 to -1/5 due to higher concentration of localized electrons with increase in temperature.

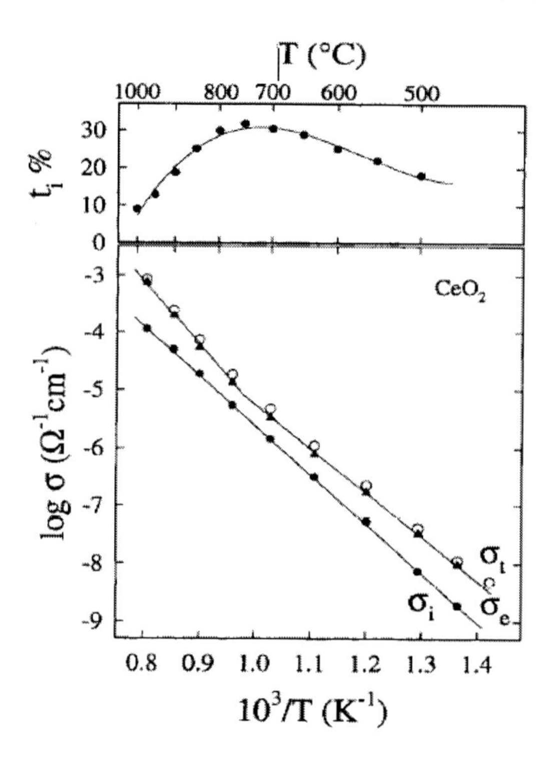

Figure 8. Arrhenius plots of total conductivity, ionic contribution and electronic conduction in pure CeO_2. Ionic and electronic contributions are calculated from the ionic transference numbers reported in the upper portion of the figure [18]. Reproduced with permission from Chiodelli, G. et al. *Solid State Ionics* 1996, *91,* 109-121. Copyright © 1996, Elsevier Science Ltd.

Figure 8 shows the Arrhenius plots of total, ionic and electronic conductivity of pure CeO_2 as a function of temperature. In the case of pure CeO_2, the Arrhenius plot which represents ionic contribution is a straight line. The activation energy is 1.7 eV. At the same time, the Arrhenius plots which represents electronic conductivity are two straight lines: one between 500-750 °C and the other between 750-1000 °C. The activation energy of first one is 1.5 eV and the second one is 2.3 eV. Between 500-750 °C the electronic conductivity is caused by movement of holes introduced by impurities. Between 750 °C and 1000 °C, the electronic conductivity is due to the movement of electrons localized in the cerium ions. Figure 9 show the Arrhenius plots of total, ionic and electronic conductivities of 80 mol% CeO_2-20 mol% ZrO_2 solid solutions. In this case, both the Arrhenius plots of ionic as well as electronic conductivities showed straight lines with a well defined slope. The activation energy of ionic conduction is 1.3 eV which is lower than pure CeO_2. The activation energy of electronic conduction is 1.5 eV. The slope of the line is -1/4 at 650 °C and -1/5 at 950 °C. The variation observed in the slope could be explained by Panhans model for pure CeO_2 [25].

A thin film of YSZ or doped CeO_2 can also be utilized to improve the electrolyte stability in solid oxide fuel cells. In one case, a thin layer of YSZ is placed between doped CeO_2 electrolyte and anode to prevent the reduction of CeO_2 [26]. Similarly, a thin layer of doped CeO_2 is placed between YSZ electrolyte and La containing cathode to prevent the reaction between ZrO_2 and La to form $La_2Zr_2O_7$ [27]. However, at higher temperatures the interlayer fails to prevent the reaction between the electrolyte and electrodes.

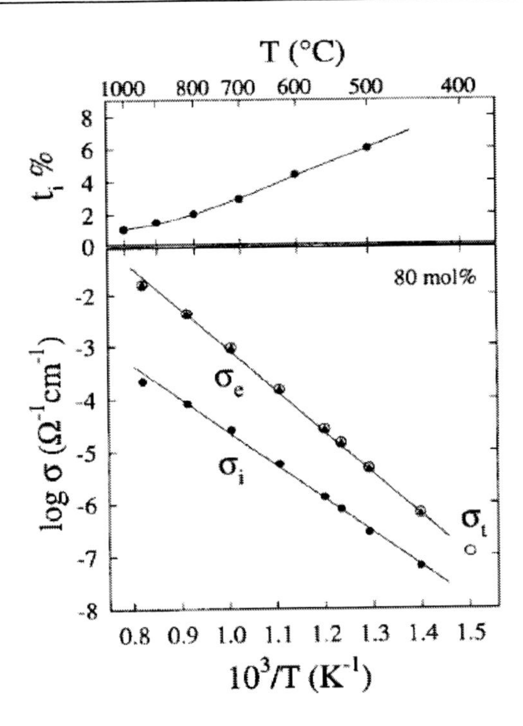

Figure 9. Arrhenius plots of total conductivity, ionic contribution and electronic contribution of conduction in the solid solution containing 80 mol% CeO_2. Ionic and electronic contributions are calculated from the ionic transference numbers reported in the upper portion of the figure [18]. Reproduced with permission from Chiodelli, G. et al. *Solid State Ionics* 1996, *91,* 109-121. Copyright © 1996, Elsevier Science Ltd.

However, the electronic conductivity values are two orders higher than pure CeO_2. The difference in electronic conductivity between pure CeO_2 and ZrO_2-CeO_2 solid solution is ascribed to a change in carrier mobility due to different transportation mechanisms. The main effect of ZrO_2 addition to CeO_2 is broadening of the conduction band. At the same time, no change occurs at the valence band and the hole mobility. The mobile carriers are quasi-free conduction electrons.

In the case of pure CeO_2, the mobile carriers are holes. Figure 10 shows the electronic conductivities of 80 mol% CeO_2, 20 mol% ZrO_2 solid solution as a function of oxygen partial pressure for three temperatures, 650 °C, 800 °C and 950 °C, respectively.

6. CeO_2-PrO_2 Composites

Praseodymium-Cerium oxide (PCO) solid solutions are known for its mixed electronic and ionic conductivity at lower partial pressures of oxygen and higher temperature to use as an electrode material for solid oxide fuel cells [19]. The mixed electronic and ionic conductor (MIEC) such as PCO also find application as an oxygen separation membrane to produce high purity oxygen simply by allowing oxygen to diffuse through a solid electrolyte under an applied oxygen activity gradient. In order for this process to occur, a compensating flow of electronic charge must counteract the flow of oxygen vacancies through a material. An MIEC allows this to happen through a counter flow of electrons. In a conventional SOFC electrode, the reaction between oxygen ions from the electrolyte, electrons from the external circuit and oxygen gas from the atmosphere occurs on "triple point boundaries" of electrode material. In contrast, a mixed ionic and electronic conducting electrode allows the reaction to occur uniformly across all surfaces of the material, rather than just at the triple point boundaries. As a consequence the efficiency of the electrode improves.

Praseodymium (Pr) exists in two valences state Pr^{4+} and Pr^{3+} and reduces more easily compared to Ce. Adding Pr to CeO_2 facilitates the reduction of Pr at much higher pO_2 values than those required to reduce pure CeO_2. A large non-stoichiometry is possible in the PCO system due to the wide range of solid solubility between CeO_2 and Pr_6O_{11} [29-31]. Moreover, PCO is an ideal material for investigating electrical properties of those systems in which a large concentration of multi-valent cation can be incorporated into a very stable, well studied host material without significantly changing the host material. Takasu et al., investigated the electrical properties of the PCO solid solution between 0 to 70% as a function of temperature [19]. It was found that PCO exhibits catalytic activity and mixed ionic electronic conductivity. Figure 11 shows result of X-ray diffraction (XRD) phase analysis of PCO as a function of Pr content. PCO up to 70% of Pr, was physically stable and lattice constant decreases linearly with increasing Pr addition and eventually phase separates into two different fluorite phases when the amount of Pr was higher than 80%. Figure 12 shows the results of total conductivity of PCO measured at 10 kHz frequency as a function of temperature. One can clearly see that the conductivity increased with increasing Pr content. It is also very interesting to note that the conductivity curves did not obey Arrhenius behavior.

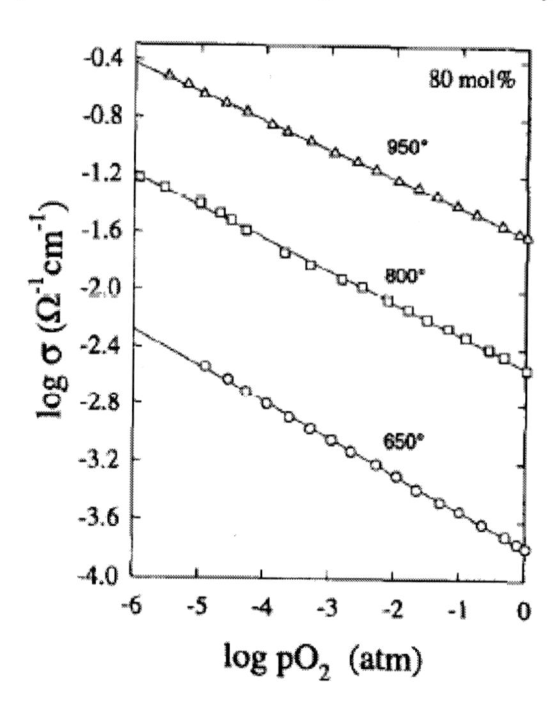

Figure 10. Electronic conductivity of 80 mol% CeO_2- 20 mol% ZrO_2 solid solution as a function of oxygen partial pressure for three temperatures, 650 °C, 800 °C and 950 °C [18]. Reproduced with permission from Chiodelli, G. et al. *Solid State Ionics* 1996, *91,* 109-121. Copyright © 1996, Elsevier Science Ltd.

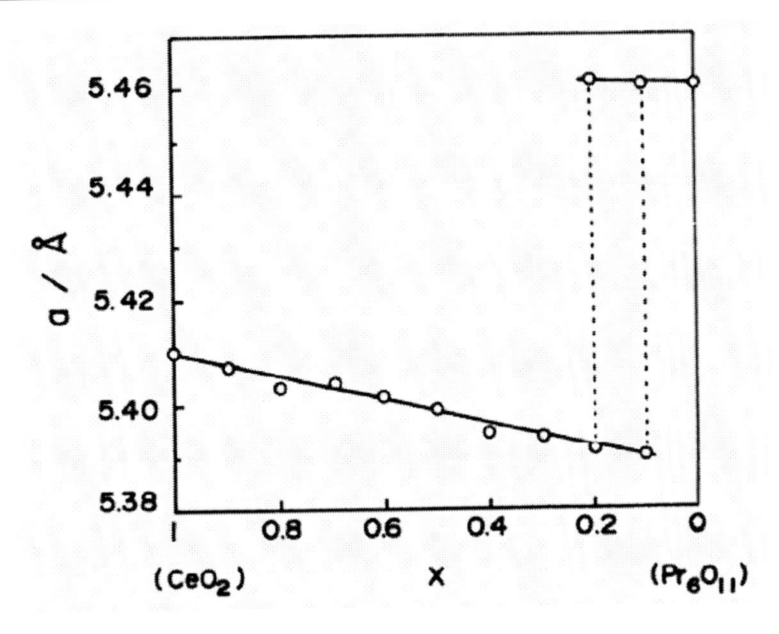

Figure 11. XRD result showing lattice constant as a function of Pr content in PCO [19]. Reproduced with permission from Takasu, Y. et al. *J. Appl. Electrochem.* 1984, *14,* 79-81. Copyright © 1984, Springer.

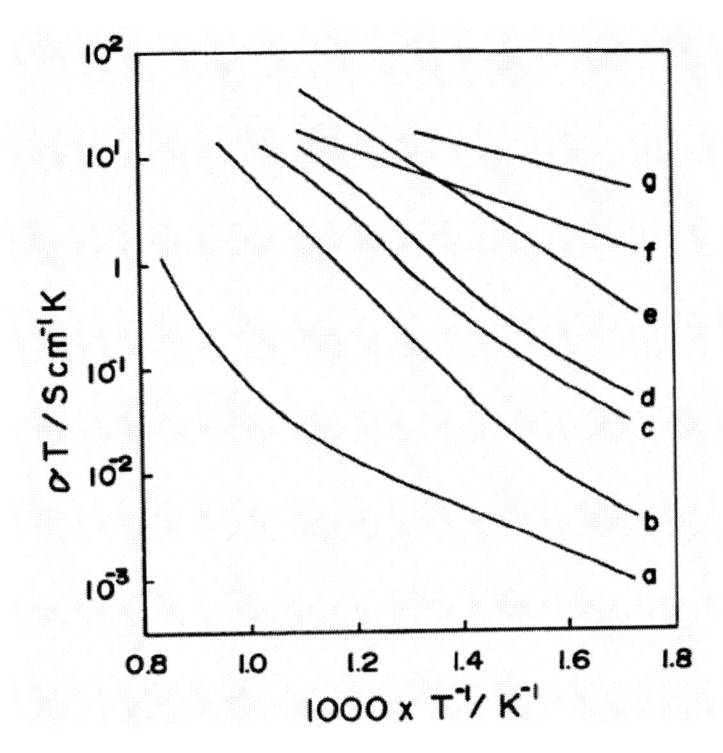

Figure 12. Conductivity of PCO as function of temperature for compositions range from 10% Pr (a) to 70% Pr (g) with an increment of 10% in each step [19]. Reproduced with permission from Takasu, Y. et al. *J. Appl. Electrochem.* 1984, *14,* 79-81. Copyright © 1984, Springer.

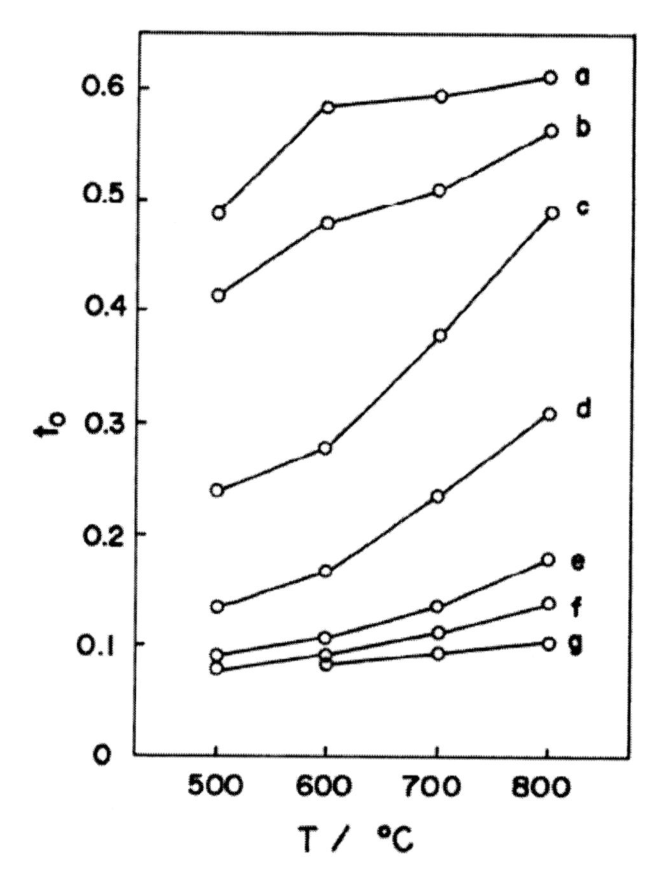

Figure 13. Ionic transference number of PCO as a function of temperature [19]. Reproduced with permission from Takasu, Y. et al. *J. Appl. Electrochem.* 1984, *14,* 79-81. Copyright © 1984, Springer.

Figure 13 shows the results of ionic transference number of PCO as a function of temperature. The transference number decreased with increasing the concentration of Pr due to electronic conduction. If the material is a perfect ionic conductor, then the transference number would be one.

Nauer et al. [30] performed similar experiment to investigate the transference number of PCO and found that the value they obtained was not in agreement with Takasu's value even though the values of lattice parameters were in good agreement. Nauer also reported the second phase formation with 40% Pr instead of 70% Pr and significantly higher conductivities than those reported by Takasu.

Shuk et al., [31] prepared PCO through hydrothermal route and found that the lattice parameter values up to 30% Pr solid solutions were in good agreement with the values reported by Takasu [19] and above 30% the value differ.

Figure 14 shows the results of ionic transference number measurements performed by Shuk et al. [31] indicate that the results were in agreement with those reported by Takasu except 10% PCO, which was considerably higher than that of Nauer and Takasu reported value. Porat et al. studied the electrical conductivity of 55% PCO as a function of pO_2 [32]. They reported a p-n transition in the material at low temperatures and relatively high pO_2 values due to the impurity band conductivity.

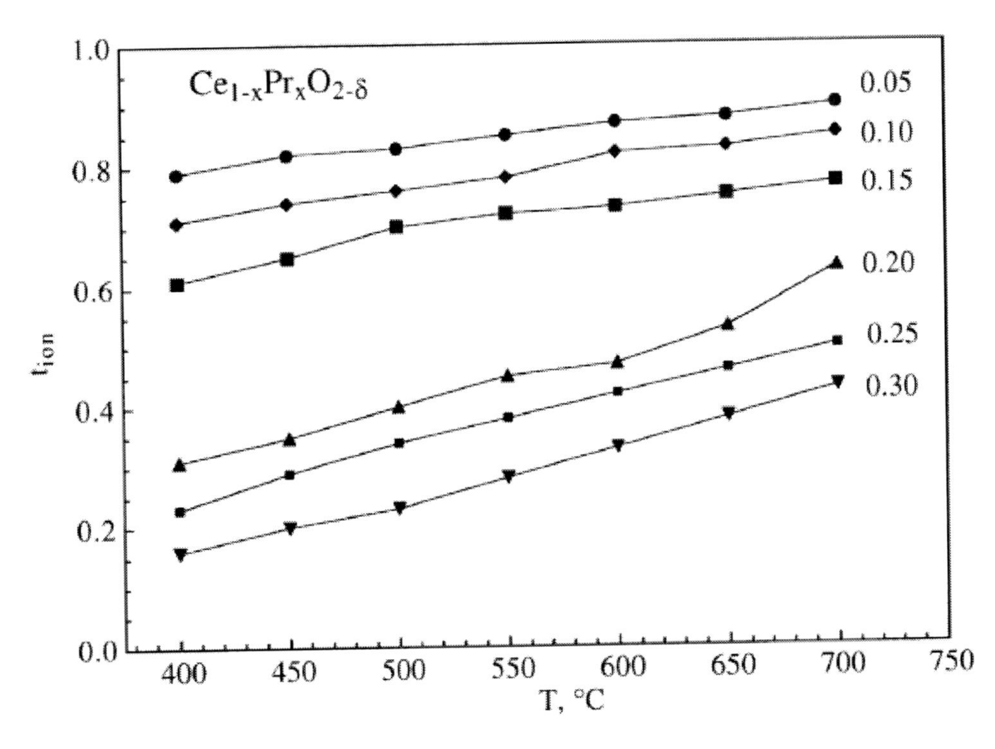

Figure 14. Ionic transference numbers in PCO as determined by Shuk et al. [31]. Reproduced with permission from Shuk, P. et al. *Solid State Ionics,* 1999, *116,* 217-223. Copyright © 1999, Elsevier Science Ltd.

7. CeO_2 -TbO_2 Composites

Shuk et al. [20] investigated the structural, thermal and electrical properties of CeO_2-TbO_2 nano-composites. $Ce_{1-x}Tb_xO_{2-\delta}$ (x = 0-0.30) nano-composite powder with particle size in the range of 33-40 nm was synthesized from appropriate quantities of ($Ce(NO_3)_3.6H_2O$) and ($Tb(NO_3)_3.5H_2O$). The precipitated gels were introduced to an autoclave and hydro thermally heated at 260 °C for 10 hours. Sintering was performed at a lower temperature of 1300-1400 °C. High temperature X-ray diffraction (HT-XRD) technique was used to study in situ the phase stability of the fluorite structure of the prepared CeO_2-TbO_2 composite from 25 °C to 800 °C. The HT-XRD results are shown in Figure 15.

The major phase observed was $Ce_{1-x}Tb_xO_{2-\delta}$. The intensities of the peaks remain constant throughout the temperature range. There was no shift observed in the XRD peaks when the temperature is increased from 25 °C to 800 °C.

The results confirmed the fluorite structure for CeO_2 and $TbO_{1.75}$ even after heat treatment at 800-1000 °C for 14 days and no phase separation occured. Figure 16. shows the lattice constants calculated from the XRD pattern. The lattice constant “a” increases linearly with temperature. At 50 °C, the lattice constant of $Ce_{1-x}Tb_xO_{2-\delta}$ composite for x = 0.3 is 0.5410 nm. When the temperature is increased to 400 °C, the lattice constant is increased to 0.5432 nm. At 800 °C, it shows the value of 0.5454 nm. It is interesting to note that the value of lattice constant decreases with increase in Tb content (x).

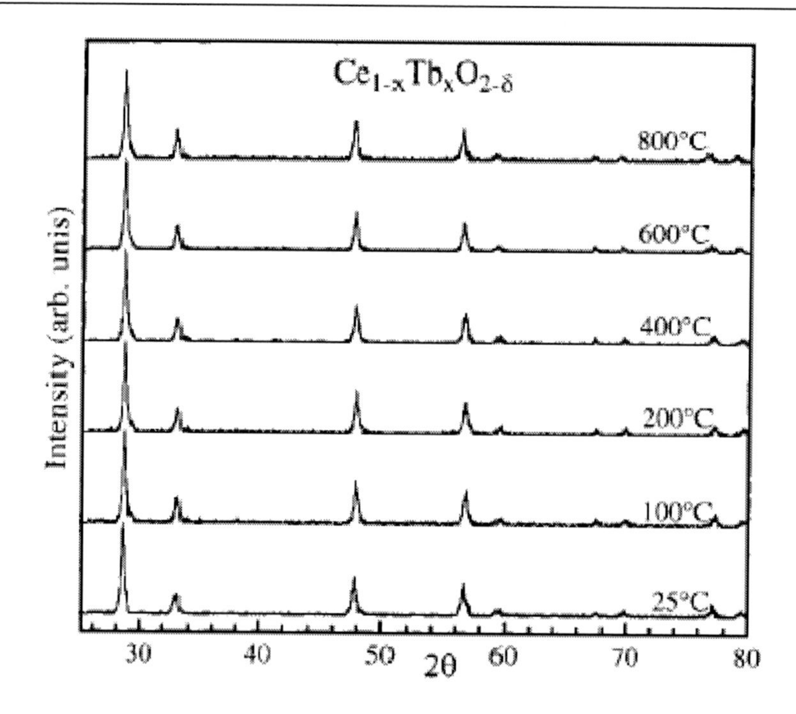

Figure 15. High temperature X-ray diffraction (HT-XRD) results of $Ce_{1-x}Tb_xO_{2-\delta}$ solid solution performed from 25°C to 800 °C [20]. Reproduced with permission from Shuk, P. et al. *Chem. Mater.* 1999, *11,* 473-479. Copyright © 1999, American Chemical Society.

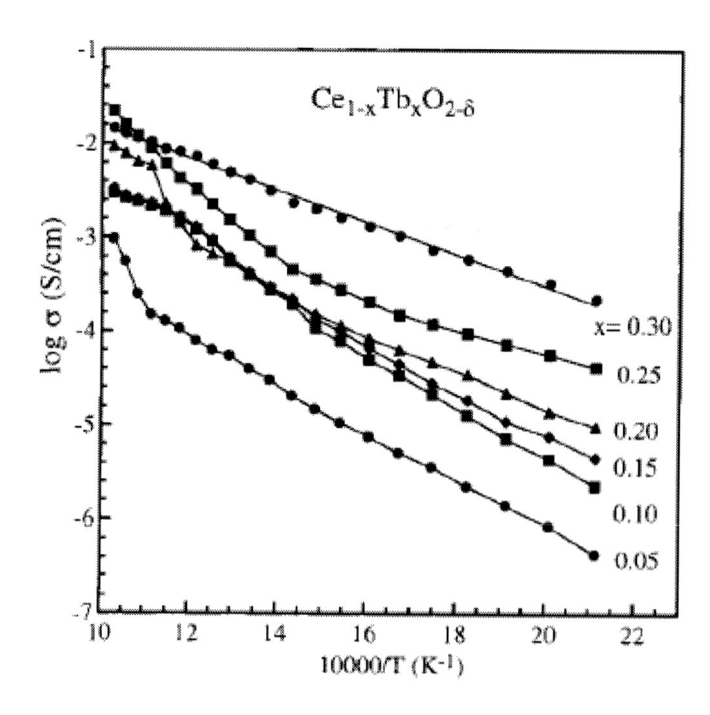

Figure 16. Arrhenius plots of the ionic conductivity of $Ce_{1-x}Tb_xO_{2-\delta}$ solid solutions [20]. Reproduced with permission from Shuk, P. et al. *Chem. Mater.* 1999, *11,* 473-479. Copyright © 1999, American Chemical Society.

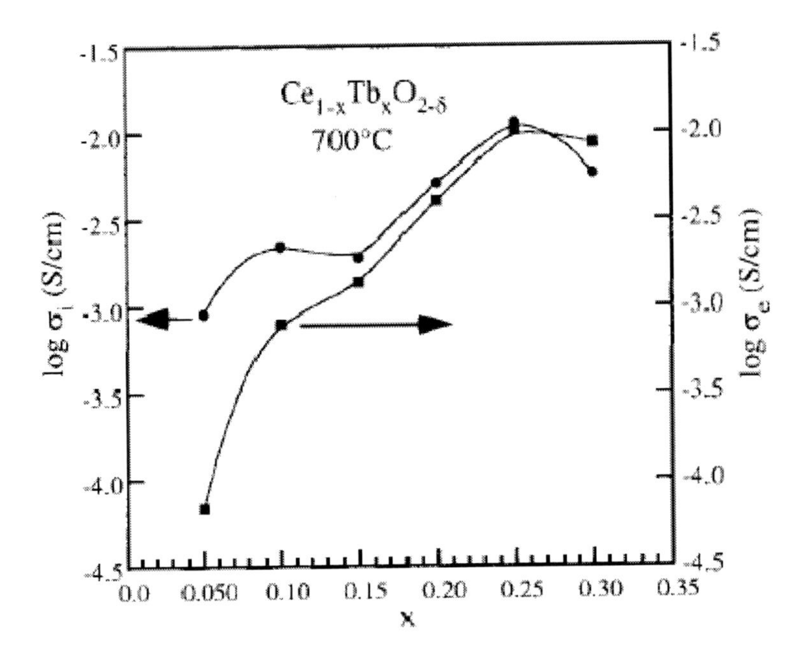

Figure 17. Concentration dependence of the ionic and electronic conductivities of $Ce_{1-x}Tb_xO_{2-\delta}$ solid solutions at 700 °C [20]. Reproduced with permission from Shuk, P. et al. *Chem. Mater.* 1999, *11,* 473-479. Copyright © 1999, American Chemical Society.

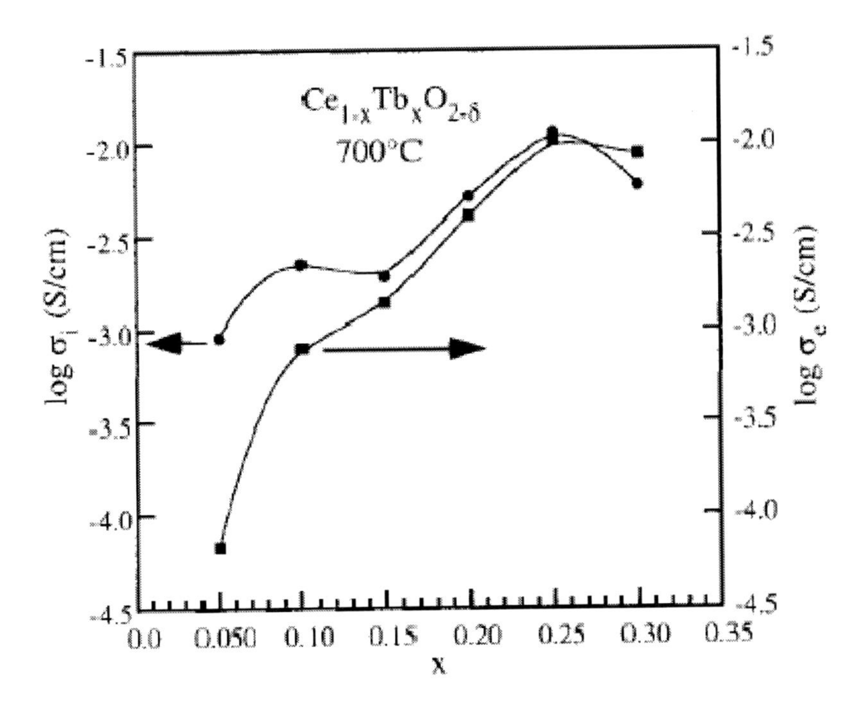

Figure 18. Concentration dependence of the ionic and electronic conductivities of $Ce_{1-x}Tb_xO_{2-\delta}$ solid solutions at 700 °C [20]. Reproduced with permission from Shuk, P. et al. *Chem. Mater.* 1999, *11,* 473-479. Copyright © 1999, American Chemical Society.

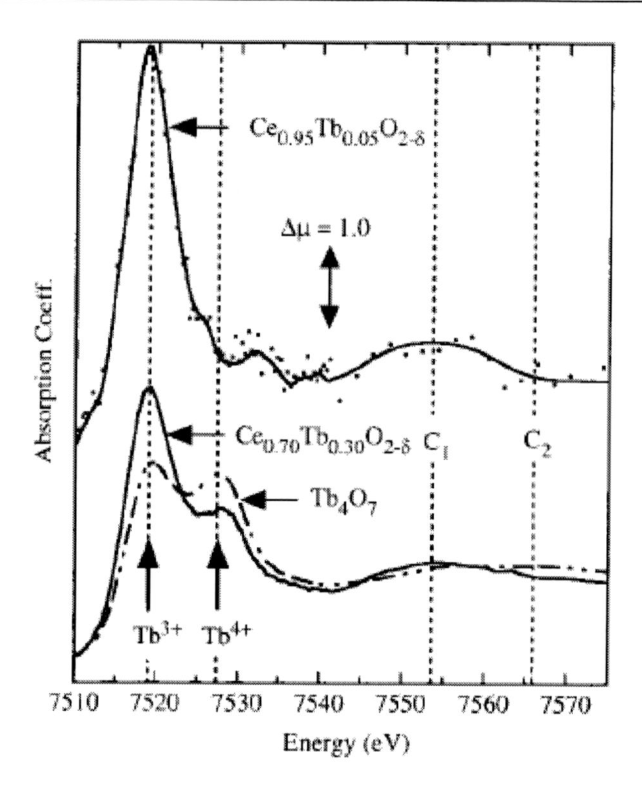

Figure 19. Tb L3 X-ray absorption edges of $Ce_{1-x}Tb_xO_{2-\delta}$ and Tb_4O_7 standard (with a formal valence of 3.5) with separate Tb^{3+} and Tb^{4+} features identified [20]. Reproduced with permission from Shuk, P. et al. *Chem. Mater.* 1999, *11,* 473-479. Copyright © 1999, American Chemical Society.

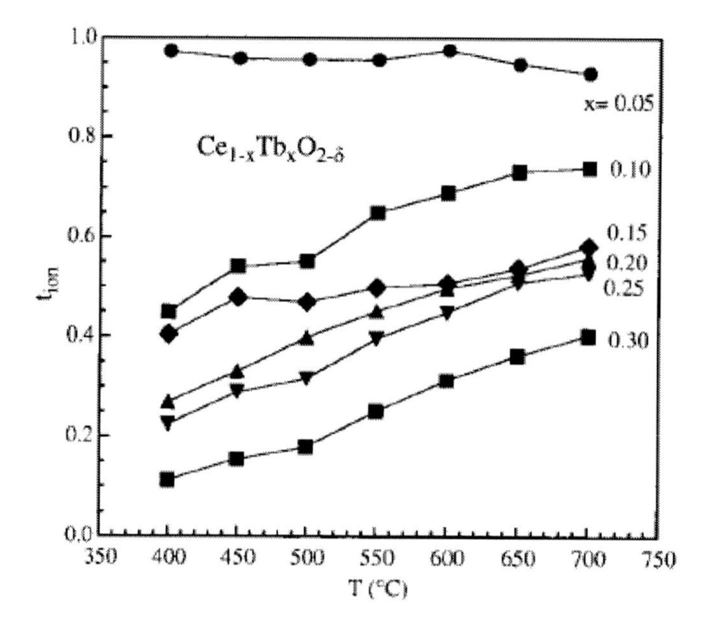

Figure 20. Ionic transference number of $Ce_{1-x}Tb_xO_{2-\delta}$ solid solutions at various temperatures [20]. Reproduced with permission from Shuk, P. et al. *Chem. Mater.* 1999, *11,* 473-479. Copyright © 1999, American Chemical Society.

The Kroger Vink notation which describes the addition of TbO_2 into CeO_2 can be written as

$$2CeO_2 \xrightarrow{2TbO_{2-\delta}} 2Tb_{Ce} + V_O^{\bullet\bullet} + \frac{1}{2}O_2 \tag{9}$$

Oxygen vacancies $\left(V_O^{\bullet\bullet}\right)$ are created by the substitution of Ce^{4+} by Tb^{3+}. The ionic conductivity increases with Tb addition up to x = 0.25 as shown in Figure 16. Moreover the oxide ion conductivity increases slowly with Tb addition compared to Sm and Ca addition in CeO_2. The composite exhibited 1-2 order of magnitude higher oxygen ion conductivity than that of the most commonly used solid electrolyte, stabilized zirconia at 650 °C. The existence of equilibrium between Tb^{3+} and Tb^{4+} at ambient conditions resulted reduction of oxide ion conductivity in $Ce_{1-x}Tb_xO_{2-\delta}$. When x is more than 0.25, the electronic conductivity exceeds ionic conductivity as shown in Figure 17 due to the conduction of polarons.

X-ray absorption near edge spectroscopy (XANES) was used to find the ratio of Tb^{3+} to Tb^{4+} in $Ce_{1-x}Tb_xO_{2-\delta}$. The results were shown in Figure 18.

For lower concentration of Tb (x = 0.05) Tb^{3+} state is dominant. When the concentration of Tb is increased (x = 0.3), Tb^{4+} line appeared stronger than the previous case. However, the Tb^{3+} line remain dominant. The transference number measurements were performed to separately study the ionic and electronic contribution to the total conductivity. It was observed that the electronic contribution increased when the concentration of Tb is increased due to the motion of polarons by a thermally activated hopping mechanism. Figure 19 shows the results of ionic transference number measurement as a function of temperature.

8. CeO_2-UO_2 Composites

Urania (UO_2) is known as a nuclear fuel and belongs to fluorite crystal structure [21]. UO_2 can also be used as an electrode material for fuel cells applications. The urania-ceria solid solution is a unique system for studying the defect structure and electronic transport due to its identical crystal structure [15].

In the case of CeO_2-UO_2 solid solution, UO_2 rich end, behaves like a p-type semiconductor, whereas at the CeO_2 rich end behave like an n-type semiconductor such a way that overall the material will behave like a highly nonstoichiometric i.e. at elevated temperatures, they both form a complete soild solution ideal to study defect chemistry of both an oxygen deficient composition and oxygen rich composition in a single crystal structure. Both UO_2 and CeO_2 exhibits fluorite crystal structure, which is a simple cubic anion lattice with every other interstitial position is filled with a cation, which results in a fairly open structure capable of supporting large concentrations of defects. Both CeO_2 and UO_2 exhibit large deviation from stoichiometry [33,34].

The relative ionization energy for uranium and cerium in the CeO_2-UO_2 solid solution has been investigated. A consistent model has been set up with various works done in this area. Their postulates are:

1. The highest filled band is UO_2 5f band, which is very close to the Fermi level.
2. An energy gap for conduction of ~2 eV is present between this band and lower edge of conduction band.
3. The oxidation of UO_2 results in an increased electron density of about 4 eV below the Fermi level which further clarifies that the oxygen interstitials are about 4 eV below Fermi level.

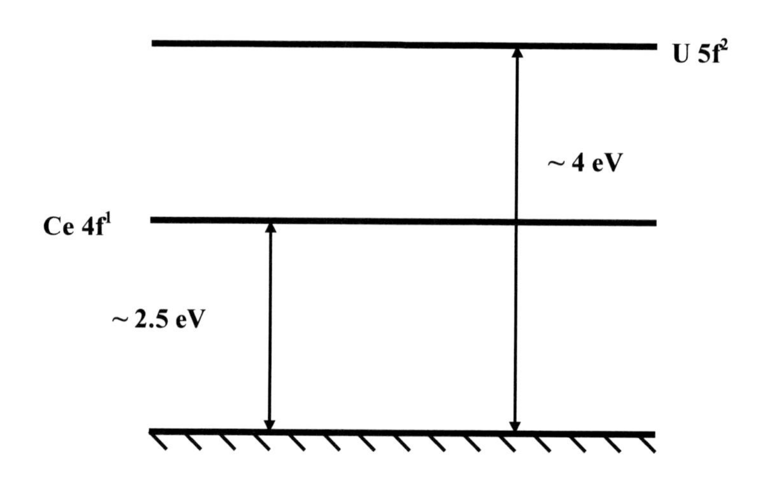

Figure 21. Relative ionization energies of cerium and uranium in cubic fluorite structure.

Studies conducted by Hass et al. [35] in order to measure optical absorption on CeO_2 thin films revealed a band gap of 5.2-5.5 eV even though an absorption takes place at 2.95 eV due to transition of Ce^{3+} to conduction band. This indicates that the band gap which is required for the conduction in CeO_2 is ~2.5 eV and the $V_O^{\bullet\bullet}$ defect is ~0.56 eV below the Ce 4f. The $V_O \rightarrow V_O^{\bullet} + e$ has small energy where $V_O^{\bullet}$ should occur about 0.6 eV below the Ce 4f level.

The alignment of the bands are such a way that the top of the O_2 p level is non bonding and hence shifted to little during compound formation. The Ce^{3+} lies about 1.5 eV below the U $5f^2$ level and about 2.5 eV above the O_2 p band edge. The O_i'' is an oxygen like state, occurs in CeO_2 at same energy as O_2 p and UO_2. The uncharged oxygen vacancy is missing oxygen plus two extra electrons positioned on the nearest neighbor metal ions. The energy of ionized oxygen vacancy is that energy needed to remove these electrons to the conduction band. Figure 21 below shows the relative ionization energies of cerium and uranium.

They concluded that the relative ionization energies of cerium and uranium indicate that cerium lies significantly below the uranium and would act as an ionized acceptor at a range of temperatures in UO_2 rich solid solutions. Similarly U^{5+}, lies significantly above $Ce4f^1$ level, should act as a simply ionized donor at all temperature range.

A Defect Model: The authors have studied the defect chemistry of U doped CeO_2 and Ce doped UO_2 based on mass action relations of point defects in an ideal solution. This type of analysis was first started by Wagner and Schottky [36]; their study was later known by Kroger Vink notation [37] that analyzes defect interrelationship between two species which depends on the following parameters.

Table 2. Quasi chemical defect reactions occur in UO_2 doped CeO_2

$$O_O^X = O_i'' + V_O^{\bullet\bullet}$$; **Frenkel pair production**

$$null = e' + h^\circ$$; **Electron-hole pair production**

$$O_O^X = V_O^{\bullet\bullet} + 2e' + \frac{1}{2} O_2(g)$$; **Reduction reaction**

$$\frac{1}{2} O_2 (g) = O_i'' + 2h^\circ$$; **Oxidation reaction**

$$V_O^\bullet = V_O^{\bullet\bullet} + e'$$; **Second ionization of oxygen vacancies**

$$U_{Ce} = U_{Ce}^\bullet + e'$$; **Ionization of uranium ions**

Equilibrium equations are: $\left[O_i''\right]\left[V_O^{\bullet\bullet}\right] = K_1$; $n\,p = K_2$

$$\left[V_O^{\bullet\bullet}\right] n^2\, pO_2^{\frac{1}{2}} = K_3 ; \quad \frac{\left[pO_2\right]^{\frac{1}{2}}}{\left[O_i''\right] p^2} = K_4$$

$$\frac{\left[V_O^{\bullet\bullet}\right] n^2\, pO_2^{\frac{1}{2}}}{n^2 p^2 \left[O_i''\right]\left[V_O^{\bullet\bullet}\right]} = \frac{pO_2^{\frac{1}{2}}}{p^2\left[O_i''\right]} = K_4 = \frac{K_3}{K_1 K_2^2} = \frac{\left[V_O^{\bullet\bullet}\right] n}{\left[V_O^{\bullet}\right]} K_5 ; \quad \frac{\left[U_{Ce}^{\bullet}\right] n}{\left[U_{Ce}\right]} = K_6$$

$$n + 2\left[O_i''\right] = p + 2\left[V_O^{\bullet\bullet}\right] + \left[V_O^{\bullet}\right] + \left[U_{Ce}^{\bullet}\right];$$

$$\left[U_{Ce}^{*}\right] + \left[U_{Ce}^{\bullet}\right] = \left[U_{Ce}\right]_{TOTAL} = U_T$$

1. The interrelationship between the defect concentrations of different species.
2. The pO_2 dependencies of the defect species.

The authors made the following assumptions that [37-39]:

1. The defects were present in low concentrations i.e. an ideal solution model where all but the most likely defects are ignored.
2. Defect activities are replaced by concentrations which are related through simple reaction constants.

The relative ionizations of uranium and cerium suggest that the uranium will remain ionized over the entire range of accessible oxygen partial pressures. Similarly, the oxygen interstitials lie well below uranium. All these facts point out that as long as uranium and oxygen interstitials are the prevalent ionic defects, then the oxygen interstitials will remain fully ionized. CeO_2 is an oxygen deficient material while UO_2 is an iso structural material with excess oxygen. The most important intrinsic ionic defects will be vacancies V_O and interstitial oxygen ions O_i. Uranium dopants ions on ceria sites (U_{Ce}) are assumed to be singly ionized. Electrons (e) and holes (h) are considered. In CeO_2 the following defects are considered: $O_i^{''}, V_O^{\bullet\bullet}, V_O^{\bullet}, e', h^{\circ}, U_{Ce}^{\bullet}$ by considering these defects the following quasi chemical defect reactions are written as given in Table 2.

The solution to the above equations specifies the defect concentration of all defects in terms of the equilibrium constants and pO_2. Figure 22 shows a schematic diagram of defect chemistry model of UO_2 doped CeO_2 assuming-fully ionized conditions [36-39]. Each curve is divided into regions I through IV corresponds to regions of different defect chemistry. At low pO_2, region I, $n \infty\ pO_2^{-\frac{1}{4}}$ and electro neutrality condition is $[U_{Ce}^{\bullet}] = 2\ [O_i'']$. Regions III predominate over a wide range of oxygen partial pressure at low temperature. This region is defined by the electroneutrality condition $n = [U_{Ce}^{\bullet}]$ and n is independent of pO_2. This way the controlling defect region can be identified. At higher pO_2, the concentrations of $O_i'' \propto pO_2^{\frac{1}{6}}$. If the controlling defect is O_i', then the proportionality will be $pO_2^{\frac{1}{4}}$. Real materials exhibit smooth transitions from one defect regime to another. Single charged species show a steep dependence at high pO_2. If the effective charge to excess oxygen ratio is ½, the proportionality would be $pO_2^{\frac{1}{3}}$. At lower pO_2, a pO_2 independent conductivity region is $\infty\ pO_2^{-\frac{1}{6}}$. The corresponding region II with an electro neutrality condition is controlled by electrons compensating from doubly ionized oxygen vacancies. According to Willis cluster, Kroger Vink notation can be written as $[2O_i''{}_{111}\ \ 1O_i''{}_{110}\ \ 2V_O^{\bullet\bullet}]''$. The subscripts denote the direction along which the interstitials were displaced. The net excess oxygen and net charge of this defect make it equivalent to a single O_i''. This defect has the same pO_2 dependence as O_i'', but the defect most likely trap a hole and its net charge is -1. The defect $[2O_i''{}_{111}\ \ 1O_i''{}_{110}\ \ 2V_O^{\bullet\bullet}]''$ would have pO_2 dependence like that of O_i'. The result of this singly charged cluster showed a steeper pO_2 dependence of defect species than fully ionized isolated defects. At high pO_2 concentration of O_i'' is proportional to $pO_2^{\frac{1}{6}}$, however, if controlling defect is O_i', the proportionality is $pO_2^{\frac{1}{4}}$. If we consider the presence of some partially ionized defect clusters, then the effective charge to excess oxygen ratio is ½ and the proportionality would be $pO_2^{\frac{1}{3}}$. The electrical conductivity (σ) of a material gives simple and reliable information about the concentration of carriers. The conductivity can be written as the sum of conductivities of each charge carrying species.

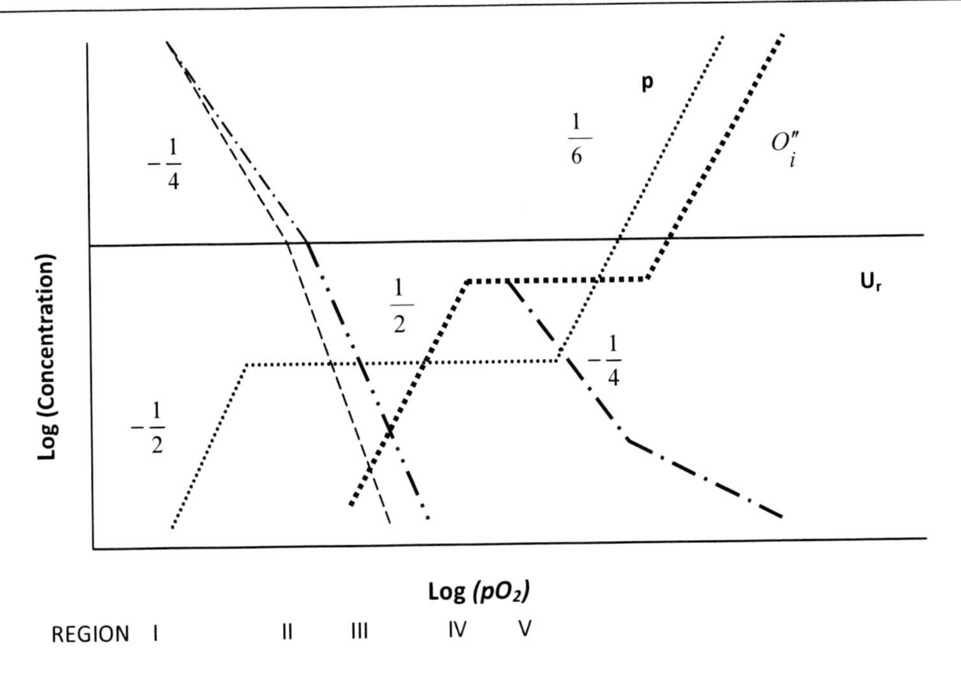

Figure 22. Schematic diagram of defect chemistry model of UO$_2$ doped CeO$_2$ assuming-fully ionized conditions [39].

$$\sigma_{Total} = \sigma_1 + \sigma_2 + \sigma_3 + . \tag{10}$$

where $\sigma_i = n_i \, q_i \, \mu_i$ (11)

Where, σ_i = conductivity of species i, q_i = charge of the carrier I, n_i = number of carriers per unit volume, and μi = mobility of species i. In a broad band semiconductor with a nearly empty conduction band, the mobility is dominated by impurity and phonon scattering. This mobility is a weak function of temperature. The mobility for hopping type condition is:

$$\mu_{Polaron} = \left(\frac{\mu_0}{T}\right) \exp\left(\frac{-E_M}{K_B T}\right) \tag{12}$$

where E_M is the activation energy for jump process, T is the temperature in Kelvin and K_B is Boltzmann constant.

$$\mu_0 = P e \bar{c} q a^2 \nu \tag{13}$$

Where P is the probability that a jump occurs during any given lattice vibration, $\bar{c}$ = concentration of available sites to which carrier may jump, e = charge of an electron, ν = jump frequency (longitudinal vibration frequency for the material) and a = jump distance. It is important to note that in fluorite crystals with carriers on the cation sublattice, $a = \frac{a_0}{1.414}$ where a_o is the lattice parameter. When CeO_2 is doped with UO_2 in region III as shown in figure, U in $Ce_{1-y}U_yO_2$ where y is the mole fraction. The $n = y\left(\frac{4}{a_0}\right)$ and $\bar{c} = A - Bx$, where A is the number of electrons per Ce site in the conduction band 1<A<2 and B is a constant B = 2. If U sites are not available as carrier sites, then according to band diagram, U_{Ce} will be at energy sufficiently above Ce_{Ce} and would not be available for jumps. Therefore, conductivity in terms of dopant concentration is:

$$\sigma = q\, y\left(\frac{4}{a_0}\right)\left(\frac{\left(P-(A-By)q\,a^2\,v\right)}{\frac{k}{T}}\right)\exp\left(\frac{-E_M}{k_B\,T}\right)\ldots \quad (14)$$

The mobility for ionic conduction takes the form of small polarons hopping. Applying Nerns't Einstein equation:

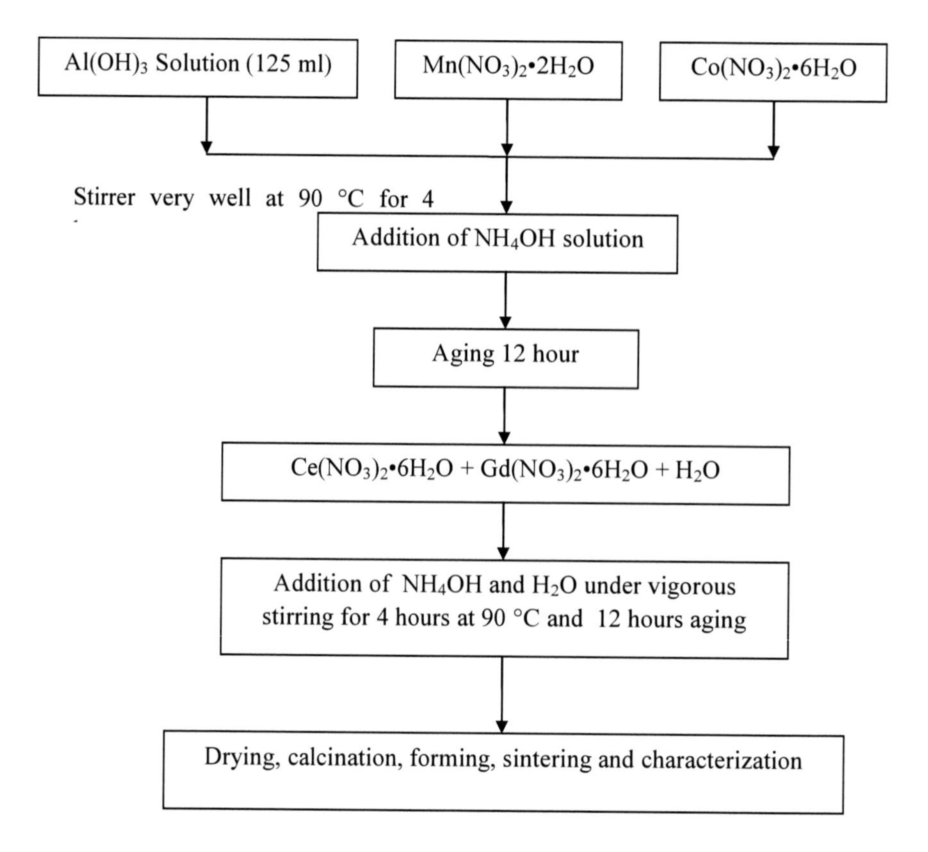

Figure 23. Procedure for preparation of Gd doped CeO_2 nano particles [22]. Reproduced with permission from Chockalingam, R. NYSCC, Alfred University. Copyright © 2007, Alfred University.

$$\mu_{Ion} = \left(\frac{\left(f\, D_{Ion}\, n_{Ion}\, q_{Ion}^{2} \right)}{\left(k_B\, T \right)} \right) \quad (15)$$

Where f is a function of the crystal structure related to number of equivalent sites for diffusing ions and D_{ion} = diffusion coefficient for the ion. If we consider charge carriers as vacancies, then $D_{Ion} = \frac{D_v\, D_{Tracer\ \ Lattice}}{n_{Vacancy}}$. The previous researches in this regard revealed that at concentrations greater than 10 mol% dopants the mobility of charge carriers is a function of dopant concentrations. As the composition of the material is changed, by adding more U on CeO_2, one or two electrons are added to the Ce 4f band.

9. CeO_2-Al_2O_3 Composites

It is known that the addition of an insulating phase to a conducting phase would decrease the conductivity of the resulting composite [22]. In contrast, addition of an insulating phase to an ionic conducting phase demonstrated increase in total conductivity [40]. The systems exhibiting such a phenomenon include ionic conductors such as LiI, AgI, AgCl and CaF_2 mixed with the insulating phase such as SiO_2, ZrO_2 and Al_2O_3. The increased ionic conductivity in these composite materials is attributed to interfacial effect or strain-induced defects at the grain boundary and disproportionate change in defects near the boundary based on Fermi level effects. In 1973, Liang [40] was the first to propose an enhanced conductivity effect in a system composed of 2Li•2LiOH•Al_2O_3. Zhu et al. [41] reported that thin films of composites of ceria with SiO_2, Al_2O_3 and ZrO_2 exhibited enhanced ionic conductivity compared with bulk ceria. Maier [42] reported enhanced ionic conductivity when a second phase is added to the ionic conductor. He argued that the enhanced conductivity is due to the defect formed either in the form of vacancies or interstitials. Qi [43] studied 15 mol% of Sm_2O_3 doped CeO_2 with an insulating second phase, demonstrated increased conductivity from 600 °C to 800 °C at low oxygen partial pressure. They also reported that the electronic conductivity was completely suppressed at 600 °C at low oxygen partial pressure. Additions of oxides of cobalt and manganese as electron acceptors at the grain boundary interface can enhance this effect further. For this purpose a complex Al_2O_3/CeO_2 composite structure was prepared by coating CoO and MnO doped Al_2O_3 nano particles with Gd-CeO_2.

Figure 23 illustrates the procedure for coating of cobalt and manganese doped Al_2O_3 with Gd-CeO_2 [44,45].

Sintering was performed at (a) 1300 °C (b) 1400 °C (c) 1450 °C (d) 1550 °C with a heating rate of 2 °C and 5 °C per minute for 5 hours hold at the sintering temperatures using conventional techniques. Microwave sintering was performed at 1350 °C with a heating rate of 10 °C per minute for 30 minutes hold at sintering temperature. Figure 24 shows the XRD patterns of Gd doped CeO_2 powder coated on 0.33 at% Co, Mn doped Al_2O_3 calcined at various temperatures. The major phase identified at room temperature without calcination was ammonium hydrogen oxalate hydrate and $Gd_{0.2}Ce_{0.8}O_{1.9}$ as indicated in the X-ray pattern. The presence of the ammonium hydrogen oxalate hydrate was not detected in the X-ray at 300 °C. This result is in agreement with the TGA-DTA analysis indicating complete decomposition of organic phase. Crystallization of Gd doped CeO_2 began at temperature as low as 300 °C. Diffraction peaks corresponding to $Gd_{0.2}Ce_{0.8}O_{1.9}$ phases have been identified for the sample calcined at 300 °C indicating the start of crystallization. The crystallization was completed at 900 °C for 4 hours of heat treatment. Thereafter, no change in the intensity of the peaks was observed in the sample annealed at 900°C for 6 hours. The X-ray result revealed that the crystallization of Gd-CeO_2 was completed at 900 °C for 4 hours of heat treatment. In the case of Gd-CeO_2 coated on 0.68 at% Co and Mn doped Al_2O_3 crystallization was completed at 900 °C for 4 hours as shown in Figure 25 no visible change in the intensity of XRD peaks were observed when the sample was heat treated at 900 °C for 6 hours.

Figure 26 shows the scanning electron microscopy (SEM) micrographs of conventionally and microwave sintered $Gd_{0.2}CeO_{0.8}$-0.33Mn-0.33Co-Al_2O_3 samples. Conventional samples were sintered at 1350 °C and 1450 °C with a heating rate of 5 °C per minute for 5 hours hold. Microwave sintered samples were sintered approximately 1250 °C and 1350 °C with a heating rate of 10 °C per minute for 45 minutes hold. Conventionally sintered sample at 1350 °C exhibited approximately 1μm average diameter and distribution of the grains were non-uniform. In contrast, microwave sintered samples at 1250 °C exhibited uniform grain size distribution with average grain diameter of 500 nm. Moreover, uniformly coated crystalline grains were observed in the case of microwave sintered samples at 1350 °C compared to conventionally sintered samples at 1450 °C. In the case of microwave sintered samples, decreasing sintering temperature to 1250 °C from 1350 °C decreased the average grain size from 1 μm to 500 nm. Based on the above results, the sintering temperature of the microwave sintered samples was optimized to 1250 °C.

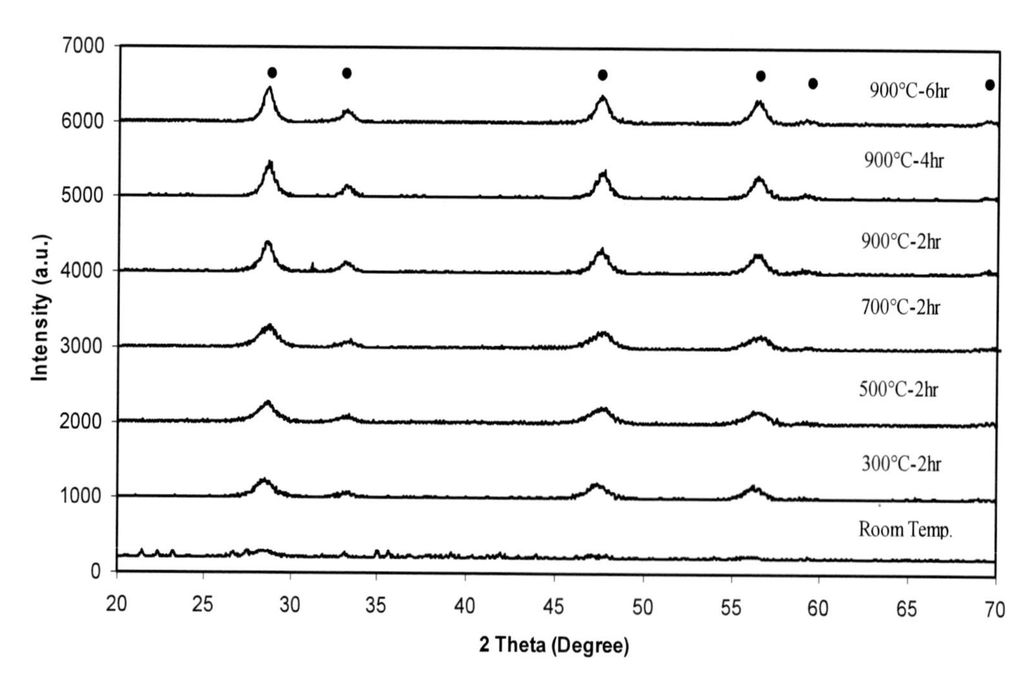

Figure 24. XRD analysis of $Gd_{0.2}CeO_{0.8}$-0.33Mn-0.33Co-Al_2O_3 coated powder calcined at various temperatures starting from room temperature through 900°C (● represents $Gd_{0.20}Ce_{0.80}O_{1.80}$ matches with PDF card 01-75-162) [46]. Reproduced with permission from Chockalingam, R. et al. *J. European Ceramic Soc.* 2008, *28*, 959–963. Copyright © 2008, Elsevier Science Ltd.

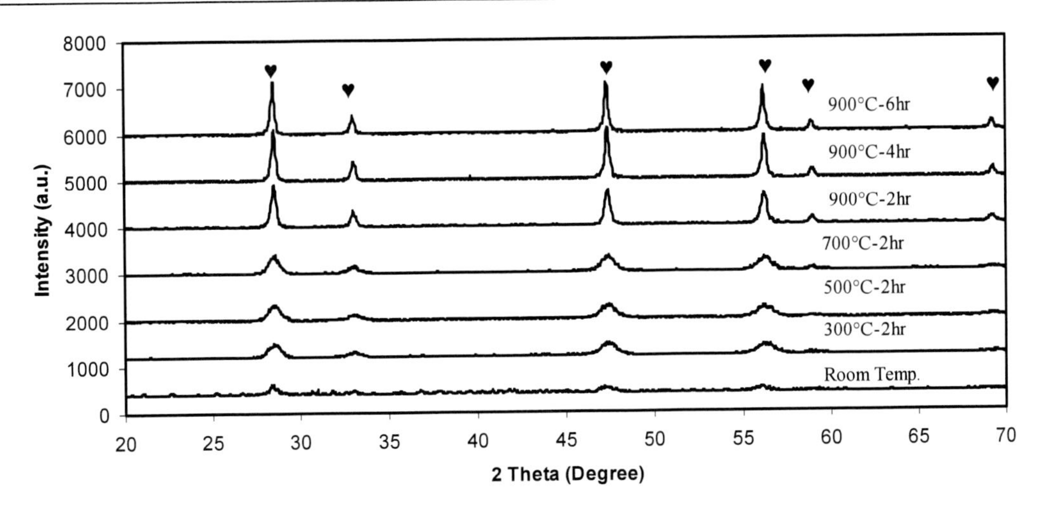

Figure 25. XRD analysis of $Gd_{0.20}Ce_{0.80}O_{1.80}$ -0.68Mn-0.68Co-Al_2O_3 coated powder calcined at different temperatures starting from room temperature through 900 °C (♥ represents $Gd_{0.20}Ce_{0.80}O_{1.80}$ matches with PDF card 01-075-0152) [46]. Reproduced with permission from Chockalingam, R. et al. *J. European Ceramic Soc.* 2008, *28*, 959–963. Copyright © 2008, Elsevier Science Ltd.

The observed uniform grain size distribution of microwave sintered samples is attributed to the volumetric nature of microwave heating. Moreover, the fast heating rate and the resulting enhanced reaction kinetics of microwave sintering favored the reduction of grain size. The chemical compositions of the sample analyzed include Ce, Gd, Al, O, Co, Mn, Au, Pd and C. The peaks Au and Pd originated from conductive coating of the sample and the peak C originated from carbon sample holder. Figure 27 shows the SEM micrographs of conventionally and microwave sintered $Gd_{0.2}Ce_{0.8}O_{1.9}$-0.68Mn-0.68Co-Al_2O_3 samples. Conventional sintering was performed with a heating rate of 5 °C per minute for 5 hours hold at 1450 °C and microwave sintering was performed at 1350 °C with a heating rate of 10 °C per minute for 45 minutes. Uniform grain size distribution was observed. The grain diameter approximately 500-700 nm were observed in the case of microwave sintered samples. In contrary, conventional sintered samples, the grain boundaries were not visible. For each composition, the activation energy was presented in Table 2. The calculated activation energies range from 0.77 to 1.02 eV and gadolinia doped ceria (GDC) sample exhibited lowest activation energy of 0.77 eV indicating highest conductivity. The sample which contains alumina shows increased activation energy, indicating decreased conductivity compared to Gd-CeO_2 (GDC).

Table 2. Summary of conductivity in air and activation energy [22]

Sample Specifications	Sintering Temperature(°C)	Atmosphere	σ_0 (Scm^{-1})	E_A (eV)	R^2
GDC	1450	Air	1.67×10^4	0.77	0.991
COAT-B	1450	Air	0.14×10^4	0.98	0.999
COAT-C	1450	Air	0.15×10^4	0.91	0.936
COAT-B-MW	1350	Air	0.19×10^4	0.99	0.968
COAT-C-MW	1350	Air	0.14×10^5	0.97	0.99

Reproduced with permission from Chockalingam, R. NYSCC, Alfred University, 2007. Copyright © 2007, Alfred University.

Figure 26. SEM micrograph of $Gd_{0.2}Ce_{0.8}O_{1.9}$-0.33Mn-0.33Co-Al_2O_3 sintered samples [46]. Reproduced with permission from Chockalingam, R. et al. *J. European Ceramic Soc.* 2008, *28,* 959–963. Copyright © 2008, Elsevier Science Ltd.

Figure 27. SEM micrograph of $Gd_{0.2}Ce_{0.8}O_{1.9}$-0.68Mn-0.68Co-Al_2O_3 sintered samples [46]. Reproduced with permission from Chockalingam, R. et al. *J. European Ceramic Soc.* 2008, *28,* 959–963. Copyright © 2008, Elsevier Science Ltd.

Table 3. Summary of conductivity in nitrogen and activation energy

Sample Specifications	Sintering Temperature(°C)	Atmosphere	σ_0 (Scm^{-1})	E_A (eV)	R^2
GDC	1450	N_2	2×10^4	0.78	0.987
COAT-C	1450	N_2	0.34×10^4	0.82	0.959
COAT-B-MW	1350	N_2	0.09×10^4	0.99	0.998
COAT-C-MW	1350	N_2	0.21×10^5	0.87	0.979

However, the addition of cobalt and manganese decreased the activation energy, which indicates increased ionic conductivity possibly due to the fact that electronic trapping mechanism is operating. It was also observed that the sintering mechanisms have neither significant influence on ionic conductivity nor in activation energy.

It is clearly observed in the Table 3 that GDC exhibited lowest activation energy in agreement with highest electronic conductivity. On the other hand microwave sintered samples COAT-B indicated highest activation energy and lowest electronic conductivity compared to conventional sintered sample with identical chemical compositions. The reduction in electronic conductivity in the case of 0.68 at% Mn, Co doped Al_2O_3/CeO_2 composite microwave sintered at 1350 °C can be attributed to the trapping of electrons at the interface between Al_2O_3 and CeO_2. In addition, the availability of more grain boundary area due to the reduced grain size of microwave sintered samples as reveled by the microstructure enhance electronic trapping Figure 26. The difference in conductivity among the compositions increased with decreasing temperature, indicating electron trapping mechanism which is more evident at low temperatures such as 400 °C. Higher activation energy of 1.02 eV in the case of microwave sintered sample (COAT-B-MW-N_2) further confirmed the possibility of enhanced electronic trapping compared to conventionally sintered sample COAT-B-CON-N_2 which has activation energy of 0.9 eV.

Even though ceria based electrolytes are known for their superior ionic conductivity at lower temperatures (650-800 °C) it becomes partially reduced at the fuel side of the cell, causing significant n-type electronic conductivity. This causes a lower efficiency due to a voltage drop due to the cell internal short circuit current. In general, the atmosphere controlled 4-point DC electrical conductivity measurements are used to evaluate the n-type electronic contributions as a function of oxygen partial pressure. Data can be fitted to the defect model where the ionic conductivity is assumed constant, and the electronic conductivity contribution is assumed to follow a 1/4 power to pO_2.

For ceria based solid solutions doped with trivalent cations, the dominant ionic defects are oxygen vacancies. In Gd-CeO_2, Gd^{3+} ions occupy substitutional positions with respect to Ce^{4+}. The concentrations of vacancies are determined by the dopant concentrations. However, when the cerium ions become reduced, the contribuition of oxygen vacancies increases. Two defect reactions based on Kroger Vink notations are required to describe these effects.

$$O_O = V_O^{\bullet\bullet} + 2e' + \frac{1}{2}O_2(g) \qquad (16)$$

$$e' = Ce'_{Ce} \qquad (17)$$

Equation (16) describes the formation of oxygen vacancies and electronic defects and the equation (17) describes the formation of electronic defects localized in ceria ions (Polarons).

The mass action equation for equation (16) can be written as:

$$n_V\, n_e^{\,2}\, P_{O_2}^{\frac{1}{2}} = K_1\,(T) = K_1 \exp\left(-\frac{\Delta H_1}{kT}\right) \tag{18}$$

n_e = The electron concentration; pO_2 = The oxygen partial pressure and ΔH_1 = The change in enthalpy of reaction and in doped samples.

$n_e << n_M$ and $n_V \approx n_M$

The electronic conductivity may be written as:

$$n_e = \left[\frac{K_1(T)}{n_M}\right]^{\frac{1}{2}} \left(P_{O_2}\right)^{-\frac{1}{4}} \tag{19}$$

In spite of the fact that $n_e << n_V$, however, the electronic conductivity

$$\sigma_e = n_e\, e\, \mu_e \tag{20}$$

The electronic conductivity is significant because, $\mu_e >> \mu_i$

Since electrons migrate by a hopping mechanism, μ_e is given by

$$\mu_e = \left(\frac{b}{T}\right) \exp\left(-\frac{E_e}{kT}\right) \tag{21}$$

where E_e is the activation energy. From equations [20] and [21],

$$\sigma_e \propto \left(P_{O_2}\right)^{-\frac{1}{4}} \tag{22}$$

The Total conductivity is given by

$$\sigma \equiv \sigma_i + \sigma_e \tag{23}$$

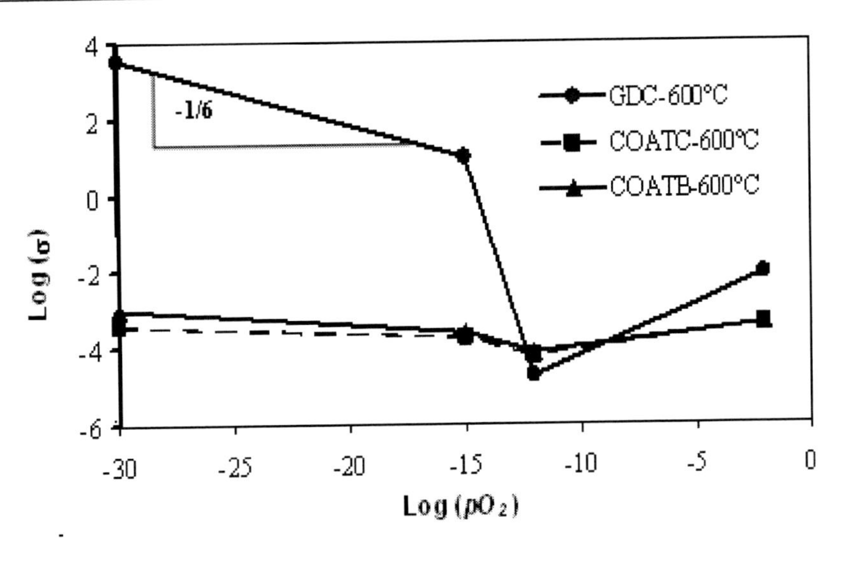

Figure 28. Dependence of electrical conductivity as a function of oxygen partial pressure for samples GDC, Coat C and Coat B at 600 °C [46]. Reproduced with permission from Chockalingam, R. NYSCC, Alfred University, 2007. Copyright © 2007, Alfred University.

In the above equation σ_e is depended on the partial pressure $pO_2^{-\frac{1}{4}}$ through a relation defined by equation (15) and σ_i is independent of partial pressure. The pressure dependence of σ_e separates σ_i. Therefore, the ionic transference number is defined as

$$t_i = \frac{\sigma_i}{\sigma} \tag{24}$$

The results of conductivity measurements were shown in Figure 28-30 and the oxygen partial pressure dependence of conductivity at 600 °C for the samples GDC, Coat-C and Coat-B are shown in Figure 28. In the case of GDC, at low oxygen partial pressure range (between10^{-15} to 10^{-30}) a slope of -1/6 was observed which is anticipated for Gd doped CeO_2. This implies that electronic conductivity is the dominant mechanism operating at this range of partial pressure. In contrary, a slope close to zero was observed in the case of Coat-B (4×10^{-6}) and Coat-C (7×10^{-7}). The magnitude of the slope is decreased drastically from values that would correspond to electronic conduction. However, more experiments need to conduct in order to confirm the result. Hence, the significant reduction observed in electronic conductivity in the case of Coat-B and Coat-C compared to GDC may be due to the trapping of electrons at the interface between Co and Mn doped Al_2O_3 and Gd-CeO_2 which suppresses the electronic contribution to the total conduction. Variations in the concentrations of dopants, Co and Mn levels (0.33 at%- 0.68 at%) had no significant influence on the electronic conductivity at low oxygen partial pressures. Similar trend was observed when the measurements were repeated at 800 °C and 1000 °C as indicated in Figures 29 and 30. The conductivity measurements of GDC sample as a function of oxygen partial pressure for three different temperatures 600 °C, 800 °C and 1000 °C indicates that at low oxygen partial pressure region (between 10^{-15} to 10^{-30}) conductivity decreases with increasing temperature. This point to the fact that at low oxygen partial pressure increases in ionic conduction overwhelms the increase in electronic conduction. The same trend is observed in the case of Coat-C and Coat-B samples as a function of oxygen partial pressure at 600 °C, 800 °C and 1000 °C respectively.

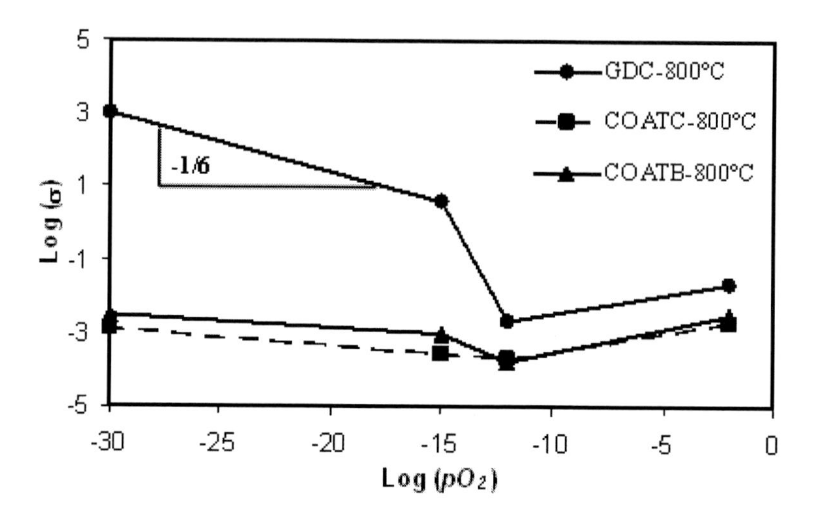

Figure 29. Dependence of electrical conductivity as a function of oxygen partial pressure for samples GDC, Coat C and Coat B at 800 °C. Reproduced with permission from Chockalingam, R. NYSCC, Alfred University, 2007. Copyright © 2007, Alfred University.

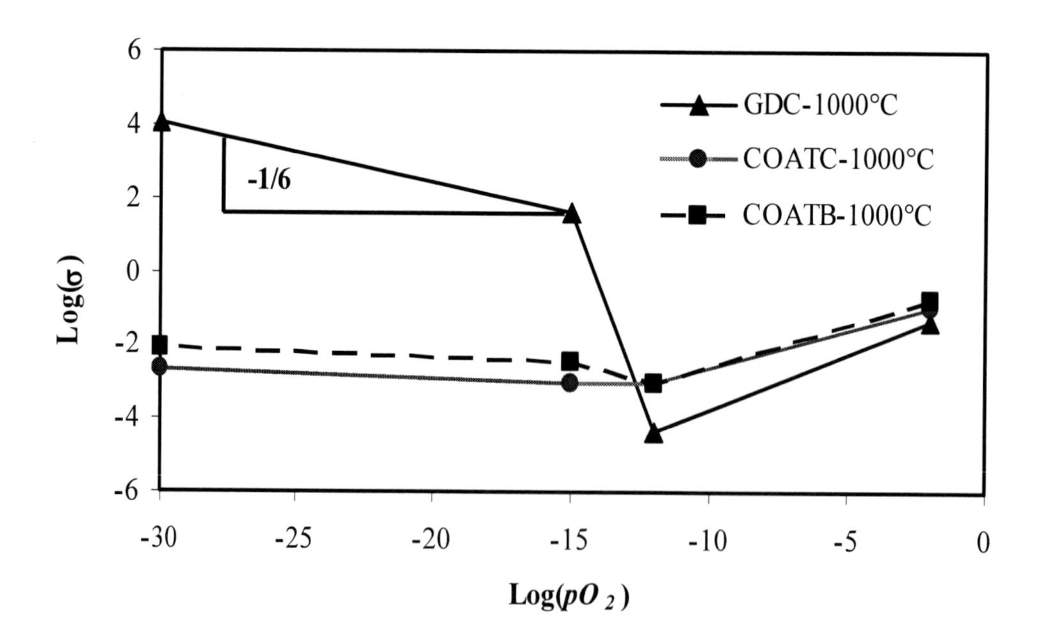

Figure 30. Dependence of electrical conductivity as a function of oxygen partial pressure for samples GDC, Coat C and Coat B at 1000 °C. Reproduced with permission from Chockalingam, R. NYSCC, Alfred University, 2007. Copyright © 2007, Alfred University.

10. $GdCeO_2$-$(LiNa)CO_3$ Composites

Reducing the operating temperature of solid oxide fuel cells is another important area of research in fuel cells. State-of-the-art zirconia solid oxide fuel cells (SOFC) suffer variety of challenges for commercialization due to their high operating temperatures of 800 °C. Recently, carbonate based ceria composite electrolytes attracted the attention of many researchers due to their excellent performance at a temperature of 300-600 °C [47-49]. A typical example is GDC-$(LiNa)CO_3$. These composite materials exhibit mixed H^+ and O^{2-} ionic conductivity and show better efficiency for direct operation of the hydrocarbon fuels, such as methanol [50]. There are many reports describing enhanced ionic conductivity, such as cationic doped ceria based composite materials such as gadolinia doped ceria (GDC)-NaCl, GDC-NaOH, GDC-MCO_3 (M = Ca, Ba, Sr), samaria doped ceria (SDC)-M_2CO_3 (M = Li, Na, K) [51]. The enhanced performance of the fuel cell was explained as an interfacial superionic conduction mechanism in the two phase region. The higher defect concentration in the amorphous Na_2CO_3 phase contribute significantly to Na^+ and O^{2-} interactions and facilitate the oxygen ion transportation through the interfacial mechanism [51]. The Na_2CO_3 also acted as a diffusion barrier layer to reduce the electronic conduction of CeO_2 by suppressing the reduction of Ce^{4+} to Ce^{3+} in hydrogen atmosphere above 450 °C [51].

GDC-$(LiNa)CO_3$ nano composite electrolytes are prepared by adding appropriate amount of Gd doped CeO_2 nano powder to a mixture of 53 mol% Li_2CO_3 and 47 mol% Na_2CO_3 [50]. The details of the compositions prepared are given in Table 4. The crystalline phase present in the Gd-CeO_2-$(LiNa)CO_3$ composites with varying amount of $(LiNa)CO_3$ sintered at 600 °C is shown in Figure 31. The major phase identified is Gd-CeO_2 and minor phase is $LiNaCO_3$ for all the compositions and no extra phase is formed during sintering at 600 °C.

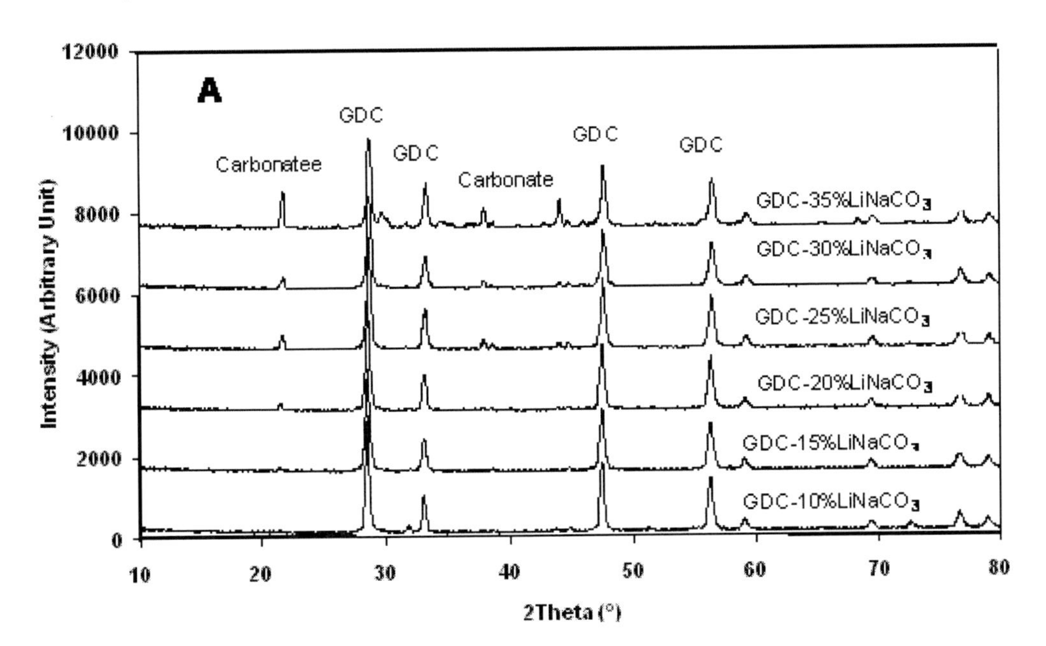

Figure 31. XRD results of the sample Gd-CeO_2-$(LiNa)CO_3$ sintered at 600 °C.

The SEM images designated as (A), (B), (C) in Figure 32 show the microstructures of $GdCeO_2$ with 10, 25 and 35 wt.% $(LiNa)CO_3$, respectively. The grain morphology of GDC carbonate composite with 10 wt% $(LiNa)CO_3$ showed small needle like carbonate crystals compared to the samples which contain 25 and 35 wt.% of carbonate. One can clearly see that porosity increases with increasing carbonate contents in the case of GDC-$(LiNa)CO_3$ samples. The sample which contains 10 wt% carbonate showed lowest porosity and 35 wt% carbonate showed highest porosity. Discontinuous voids and pores are clearly seen in all the cases except GDC carbonate composite 10 wt% $(LiNa)CO_3$, indicating that at 600 °C carbonate melted and covered the GDC agglomerated particles.

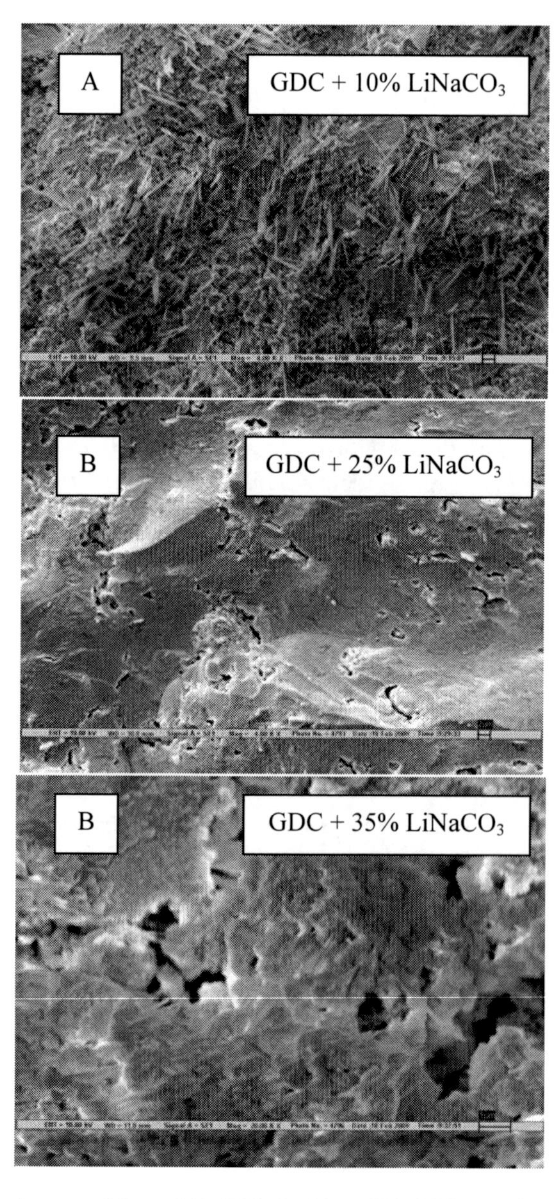

Figure 32. SEM images of compositions: (A) $GdCeO_2$-10 wt.%$(LiNa)CO_3$, (B) $GdCeO_2$-25 wt.% $(LiNa)CO_3$ and (C) $GdCeO_2$-35 wt.% $(LiNa)CO_3$.

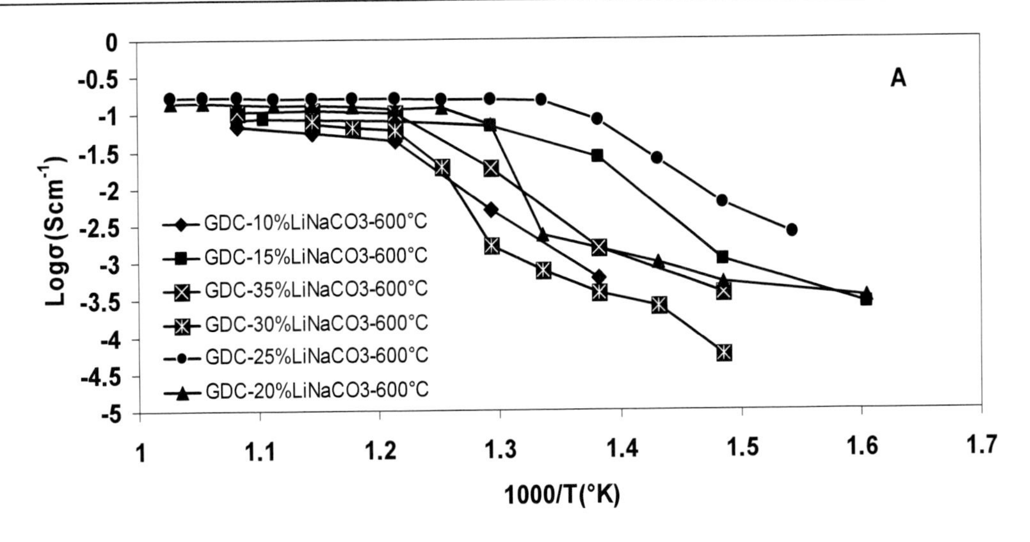

Figure 33. The ionic conductivities of $GdCeO_2(LiNa)CO_3$ samples with varying amount of $(LiNa)CO_3$ sintered at 600 °C.

Figure 33 shows the ionic conductivity of $GdCeO_2(LiNa)CO_3$ samples with varying amount of $(LiNa)CO_3$ sintered at 600 °C. Highest conductivity of 0.16 Scm^{-1} was obtained for $GdCeO_2$-25 wt.% $(LiNa)CO_3$. Conductivity reduced when the sintering temperature increased. The sample sintered at 650 °C showed a conductivity of 0.11 Scm^{-1} and at 700 °C showed a conductivity of 0.05 Scm^{-1}.

The highest conductivity observed in the case of $GdCeO_2$-25 wt.% $(LiNa)CO_3$ sample could be explained based on the melting transition of eutectic composition of 53 mol% Li_2CO_3 and 47 mol% Na_2CO_3 which in turn created more defects leading to enhanced ionic conductivity [52]. Similar observations have been reported previously by Maier [53]. He predicted the possibility of the existence of an equilibrium concentration of point defects in any ionic crystal due to the increase of configurational entropy.

He also argued that increase in number of defects leads to interaction between defects causing reduction in the total free enthalpy. Then a transition occurs from a low disordered state to a high disordered state. Such transitions may be superionic transition as observed in the case of AgI, where β-AgI transforms to α-AgI phase [53-55].

Based on the above collective result one can argue that the GDC-carbonate two phase regions create a space charge region and a local phase transition occurs at the boundary. Small activation enthalpies in this region create a liquid like surface conduction leading to higher conductivity [51,53,54]. An alternative explanation for higher ionic conduction may be caused by the reduction in electronic conductivity of ceria. Similar observation has previously reported in the case of CeO_2-Al_2O_3 composites. High conductivity could also may caused by hybrid conduction of proton and oxygen ion [55-58].

Table 5 shows the activation energies of $GdCeO_2$ $(LiNa)CO_3$ samples with varying amount of $(LiNa)CO_3$ sintered at 600 °C. One can see that the activation energy follows the similar trend observed in the ionic conductivity of the composite. The sample which contains 25 wt% carbonate showed lowest activation energy of (0.088 eV) at a temperature range of 500-650 °C.

Table 4. The details of the compositions prepared

Sample Number	Weight % of GDC	Weight % of $(LiNa)CO_3$
1	90	10
2	85	15
3	80	20
4	75	25
5	70	30
6	65	35

Table 5. The activation energy of $GdCeO_2$-$(LiNa)CO_3$ samples with varying amount of $(LiNa)CO_3$ sintered at 600 °C

Sample Number	Sample ID	Temperature (°C)	Slope	Activation Energy(eV)	R^2
1	GDC-10 wt.% $LiNaCO_3$	500-650	4.1855	0.3479	0.9995
2	GDC-15 wt.% $LiNaCO_3$	500-650	1.981	0.1647	0.9615
3	GDC-20 wt.% $LiNaCO_4$	500-650	1.9674	0.1636	0.974
4	GDC-25 wt.% $LiNaCO_5$	500-650	1.0635	0.0884	0.9817
5	GDC-30 wt.% $LiNaCO_6$	500-650	3.549	0.2951	0.9967
6	GDC-35 wt.% $LiNaCO_7$	500-650	8.642	0.7185	0.6976

SUMMARY

Although doped ZrO_2 electrolyte is known for its better electrical and mechanical properties, it requires high operating temperature, greater than 800 °C and costly refractory for thermal confinement. This led to an interest in researchers to look for alternative electrolyte materials capable of operating at a lower temperature. Doped CeO_2 is considered one such material which is capable of operating at intermediate temperature range of 500 to 700 °C. The disadvantage of CeO_2 is its high electronic conductivity at lower partial pressures of oxygen and poor mechanical strength. Addition of an insulating second phase to CeO_2 matrix is suggested as one method to reduce the electronic conductivity. The other advantage of adding suitable second phase such as Al_2O_3 and ZrO_2 to ceria is the improvement in mechanical strength. An alternative method to reduce to electronic conductivity of CeO_2 is to reduce the operating temperature below 500 °C. This is possible by mixing a suitable carbonate phase to CeO_2 matrix. The successful operation of SOFC also requires mixed electronic and ionic conducting electrodes with matching thermal expansion coefficient with that of electrolytes in order to reduce the thermal stresses during thermal cycling of the cells. Long term continuous operation of fuel cells also requires that the electrolyte, anode and cathode should be chemically and structurally stable under reduced conditions.

CeO_2-ZrO_2 composite has not shown any improvement in ionic conductivity, whereas the mechanical properties improved significantly. Addition of PrO_2 into CeO_2 has shown a positive trend in terms of ionic conductivity. The ionic conductivity of CeO_2-TbO_2 composite increased with Tb addition up to x = 0.25. The composite exhibited 1-2 order of magnitude higher oxygen ion conductivity than that of the most commonly used solid electrolyte, stabilized zirconia, at 650 °C. In the case of CeO_2-UO_2 solid solution, UO_2 rich end, behaves like a p-type semiconductor, whereas at the CeO_2 rich end behave like an n-type semiconductor such a way that overall the material will behave like a highly nonstoichiometric soild solution ideal to study defect chemistry of both an oxygen deficient composition and oxygen rich composition in a single crystal structure. CeO_2-Al_2O_3 composite has shown significant reduction in electronic conductivity due to the trapping of electrons at the interface between trace amounts of Co and Mn doped Al_2O_3 and Gd-CeO_2, which suppresses the electronic contribution to the total conduction. It is also important to reduce the operating temperature of fuel cell, which reduces the cost of interconnects and sealants. $GdCeO_2$ $(LiNa)CO_3$ composite exhibited reduced operating temperature of 550 °C and ionic conductivity of 0.16 Scm^{-1} for $GdCeO_2$-25 wt.% $(LiNa)CO_3$. Based on the collective information we concluded that addition of second phase to CeO_2 matrix is one of the methods to improve the electrical as well as mechanical properties of CeO_2 based intermediate temperature solid oxide fuel cells.

REFERENCES

[1] Basu, S.; *Recent Trends in Fuel Cell Science and Technology*; 233; Springer: New York, NY, 2007; 8-12.
[2] Knauth, P.; Tuller, H.L. *J. Am. Ceram. Soc.* 2002, *85,* 1654-1680.
[3] Tuller, H. L. *Solid State Ionics* 2000, *131,* 143-157.
[4] Basu, R. N. In *Recent Trends in Fuel Cell Science and Technology*; Basu, S.; Ed., 2007; India, Anamaya Publishers: New Delhi, ND, 286-330.
[5] Haile, S. M.; *Acta Mater.* 2003, 51, 5981-6000.
[6] Weber, W. J.; Tuller, H. L.; Mason, T. O.; Cormack, A. N. *Mater. Sci. and Eng. B* 1993, *18,* 52-71.
[7] Badwal, S. P. S; Drennan *J. Mater. Sci.* 1987, *22,* 3231-3239.
[8] Barsoum, M; *Fundamentals of Ceramics*; Mc GRAW-HILL Series in Materials Science and Engineering; The Mc Graw-Hill Companies Inc: USA. 1997.
[9] Inaba, H.; Tagawa, H. *Solid State Ionics* 1996, 1-16.
[10] Huang, K. In *Materials for fuel cells*; Gasik, M., Ed., Woodhead publishing Ltd, CRC Press: Boca Raton, FL 2008; 280-343.
[11] Badwal, S. P. S. *Solid State Ionics* 1992, *52,* 23-32.
[12] Badwal, S. P. S. *J. Mat. Sci.* 1983, *18,* 3230-3242.
[13] Fergus, J. W. *J. Power Sources* 2006, *162,* 30-40.
[14] Yahiro, H.; Eguchi, Y.; Eguchi, K.; Arai, H. *J. Appl. Electrochem.* 1988, *18,* 527-531.
[15] Tuller, H. L.; Nowick, A. S. *J. Electro. Chem. Soc.* 1975, *122,* 255-259.
[16] Bluminthal, R. N.; Brugner, F. S.; Garnier J. E. *J. Electro. Chem. Soc.* 1973, *120,* 1230-1237.
[17] Neuimin, A. D.; Pal'guev, S. F.; Chebotin, V. N. In *Electrochemistry of Molten and Solid Electrolytes*; Smirnov, M.V., Ed.; Consultant Bureau: New York, NY, 1963, Vol.2, 79-84.
[18] Chiodelli, G.; Flora, G.; Scagliottib, M. *Solid State Ionics* 1996, *91,* 109-121.
[19] Takasu, Y.; Sugino, T.; Matsuda. *J. Appl. Electrochem.* 1984, *14,* 79-81.

[20] Shuk, P.; Greenblatt., M.; Croft, M. *Chem. Mater.* 1999, *11,* 473-479.
[21] Markin, T. L.; Crouch, E. C. *J. Inorg. Nucl. Chem.* 1970, *32,* 77-85.
[22] Chockalingam, R.; *Synthesis and characterization of Gd-CeO_2/Al_2O_3 nano composite electrolyte for solid oxide fuel cells.* M.S. Thesis, NYCC, Alfred University, NY, 2007.
[23] Chockalingam, R.; Jain, S.; Basu, S. *Integrated Ferroelectrics* 2010, *115,* 1-12.
[24] Butler, V.; Catlow, C. R. A.; Fender, B. E. F.; Harding J. H. *Solid State Ionics* 1983, *8,* 109-113.
[25] Panhans, M. A.; Blumenthal, R. N. A. *Solid State Ionics* 1993, *60,* 279-298.
[26] Kim, S. G.; Yoon, S. P.; Nam, S. W.; Hyun, S. H.; Hong, S. A. *J. Power Sources* 2002, *110,* 222-229.
[27] Simner, S. P.; Bonnett, J. F.; Canfield, N. L.; Meinhardt, K. D.; Sprenkle, V. L.; Stevenson J. W. *Electrochemical Solid State Letters* 2002, *5,* 173-182.
[28] Nauer, M.; Ftikos, C.; Steele, B. C. H. *J. Euro. Ceram. Soc.* 1994, *14,* 493-499.
[29] Nauer, M.; Ftikos, C.; Steele, B. C. H. *J. Appl. Electrochem.* 1994, *14,* 79-81.
[30] Ftikos, C.; Nauer, M.; Steele, B. C. H. *J. European Ceram. Soc.* 1993, *12,* 267-270.
[31] Shuk, P.; Greenblatt. *Solid State Ionics* 1999, *116,* 217-223.
[32] Porat, O.; Tuller, H. L.; Shelef, M.; Logothetis, E. M. *Mat. Res. Soc.,* Davies P. K.; Jacobson A. J.; Torardi C. C. Ed., *Materials Research Society*, Pittsburg, PA, 1997, 453, 531-555.
[33] Aronson, S.; Belle, J. *J. Chem. Phy.* 1958, *29,*151-162.
[34] Saito, Y. *J. Nucl. Mater.* 1974, *51,* 112-119.
[35] Hass, G.; Ramsey, J. B.; Thun, R. *J. Opticals. Amer.* 1958, *48,* 324-330.
[36] Wagner, C.; Schottky, W. Z. *Z. Phys. Chem. Bill.* 1931, 163-172.
[37] Kroger, F. A.; Vink, H. J. *Solid State Physics* 1956, *3,* 307-435.
[38] Wagner, Jr. J. B. In *Transport in Nonstoichiometric Compounds*; Simkovich,G.; Stubican, V.S. Ed., 1985, Plenum Press: New York, NY, 3-16.
[39] Catlow, C. R. A. *Proc. R. Soc. Lond. A* 1977, *353,* 533-544.
[40] Liang, C. C. *Electrolyte materials for high voltage solid electrolyte battery systems.* 1973. U.S. Pat. 3713897.
[41] Zhu, B.; Liu, X.; Zhou, P.; Zhu, Z.; Zhu,W.; Zhou, S. *J. Mater. Sci. Lett.* 2001, *20,* 591-594.
[42] Maier, J. *J. Phys. Chem. Solids* 1985, *46,* 309-320.
[43] Qi. X. *Fluorite-type (ceria) mixed oxygen ion-electron conducting ceramic membranes.* Ph.D. Thesis, University of Cincinnati, Cincinnati, OH, 2000.
[44] Chockalingam, R.; Chockalingam, S.; Giesche, H.; Amarakoon, V. R. W. *Processing and Fabrication of Advanced Materials* XVI (PFAM XVI), Singapore, December 2007.
[45] Amarakoon, V. R. W.; Giesche, H.; Chockalingam, R.; Del Regno, G. E. *Composite electrolyte material having high ionic conductivity and depleted electronic conductivity and method for producing same.* 2008. US Patent # 20080248363.
[46] Chockalingam, R.; Amarakoon, V. R. W.; Giesche, H. *J. European Ceramic Soc.* 2008, *28,* 959–963.
[47] Chockalingam, R.; Giesche, H.; Amarakoon, V. R. W. *Ceramic Transactions*, Sosman Symposium, American Ceramic Society. DT USA. 2007. 456-500.
[48] Goodenough, J. B. *Nature* 2000, *404,* 821-823.
[49] Shao, Z. P.; Haile, S. M. *Nature*, 2004, *431,* 170-173.

[50] Huang, J. M. *Elect. Chem. Commns.* 2007, *9,* 2601-2605.

[51] Zhu, B. *Elect. Chem. Commns*. 2008, *10,* 302-305.

[52] Zhu, B. *Solid Oxide FCs-IV*; Singhal, S.C.; Ed., 1999; Electrochemical Society: Pennington, NY, 244-256.

[53] Maier J. *Prog. Solid St. Chem*. 1995, *23,* 171-263.

[54] Nabae, Y.; Pointon, K. D.; Irvine, J. T. S. *Energy Environmental Sci.* 2008, *1,* 148-155.

[55] Zhu, B. Next generation fuel cell R & D. *Int. J. Energy Res.* 2006, *30,* 895-903.

[56] Raza, R.; Wang, X.; Ma, Y.; Liu, X., Zhu, B. *Int. J. Hydrogen Energy*, 2010, *35,* 2684-2688.

[57] Xia, C.; Li, Y.; Tian, Y.; Liu, Q.; Wang, Z.; Jia, L.; Zhao, Y.; Li, Y. *J. Power Sources* 2010, 195, 3149-3154.

[58] Chockalingam, R.; Basu, S. *J. Int. Hydrogen Energy*, (accepted). 2010.

In: Advanced Organic-Inorganic Composites
Editor: Inamuddin

ISBN 978-1-61324-264-3

Chapter 11

COMPOSITE MATERIALS IN DENTAL APPLICATIONS

Geeta Rajput[*] and Sarwat Husain Hashmi
Department of Prosthodontics, Dr. Ziauddin Ahmad Dental College,
Aligarh Muslim University, Aligarh 202002, India

ABSTRACT

Composite materials are tooth coloured restorative materials used for filling in anterior teeth, posterior teeth especially in low stress areas and for cementation of crowns bridges inlays and onlays. It is one of the core build-up materials also used for cementation of acrylic and porcelain veneers and as pit and fissure sealants. It is a product which consists of at least two distinct phases normally formed by blending together components having different structures and properties. The purpose of this is to produce a material having properties which could not be achieved from any of the individual components alone. The two main components of composite filling materials are the resin phase and the reinforcing filler. The beneficial properties contributed by the resin are the ability to be moulded at ambient temperature coupled with setting by polymerization achieved in a conveniently short time. The beneficial properties contributed by the filler are rigidity, hardness, strength and a lower value for the coefficient of thermal expansion. The effect of filler depends on the type, shape, size and amount of filler incorporated and, often, the existence of efficient coupling between the filler and resin. The composite materials are supplied in two paste system set by chemical reaction or single paste system where the polymerization is initiated by visible blue light. The polymerization shrinkage and abrasive wear are two main disadvantages of these materials, which decrease its life span and causes discolouration of material.

Historically silicate cement was used as aesthetic restorative material followed by acrylic resins and then by composite resins. Acrylic resins are unfilled resins and their aesthetical qualities and insolubility in oral fluids make them superior to silicate cement but they had other weaknesses. With the advancement in the polymer science new resins reinforced by means of fillers have been developed called composites. Composites are the materials made up of two or more distinct phases and have bulk properties significantly different and superior from those of any of the constituents. Natural composites are tooth enamel and dentine.

[*] Corresponding author's e-mail: geetarajput70@gmail.com

ABBREVIATIONS

Bis-GMA	Bisphenol - A glycidylmethacrylate
BP	Benzoyl peroxide
BPA	Bisphenol - A
CMC	Ceramic matrix composite
CQ	Camphorquinone
DM	Dimethacrylate
DMAEMA	Dimethylaminoethyl methacrylate
DNA	Deoxyribonucleic acid
FRCs	Fibre reinforced thermoplastic composite
GIC	Glass ionomer cement
KHN	Knoop hardness number
LED	Light emitting diode
MMC	Metal matrix composite
MPP	Malleated polypropylene
PAC	Plasma arc curing
PMC	Polymer matrix composite
QTH	Quartz tungsten halogen
TEGDMA	Triethylene glycol dimethacrylate
UDMA	Urethane dimethacrylate

1. INTRODUCTION

1.1. Dental Composites

These are also known as tooth coloured restorative material or resin based composites. These are highly cross-linked polymeric materials reinforced by a dispersion of amorphous silica, glass, crystalline or organic resin filler particles and/or short fibres bounded to matrix by silane coupling agents.

1.2. Phases of Composites (Structural Components)

Following are the phases of composites.

1.2.1. Matrix Phase

It is primary phase, having a continuous character. This phase holds dispersed phase. In dental composites, a plastic resin material that forms a continuous phase and binds the filler particles. For example include bisphenol A- glycidyl methacrylate or urethane dimethacrylate (UDMA) or trithlene glycol dimethacrylate.

1.2.2. Dispersed (Reinforcing) Phase

This phase is embedded in matrix in a discontinuous form and stronger than matrix. In dental composites, fillers like quartz, colloidal silica, glasses/ceramics etc. are dispersed in matrix.

1.2.3. Coupling Agents

It is bonding agent between filler and resin matrix. Example includes organo silanes. The properties namely coefficient of thermal expansion, setting contraction and surface hardness depend almost on filler content. Strength and modulus of elasticity generally increase with addition of filler, as it increases abrasion resistance, probably as a result of increased surface hardness. If the added filler is translucent, the optical properties of the resin are improved and a more lifelike appearance produced. The resins used in composite materials are invariably based in methacrylate monomers.

Simple but most materials now utilize dimethacrylates. These monomers undergo a small contraction on setting and form a highly cross-linked three dimensional network with properties which are generally better than those of polymethylmelthacrylate.

2. Classification of Composites

Composite materials are classified as follows:

2.1. Based on Matrix Material

2.1.1. Metal Matrix Composites (MMC)

These are composed of a metallic matrix that are aluminium, magnesium, iron, cobalt, copper and a dispersed ceramic or metallic phases like oxides, carbides of lead, tungsten and molybdenum.

2.1.2. Ceramic Matrix Composites (CMC)

These are composed of a ceramic matrix and embedded fibres of other ceramic material as dispersed phase.

2.1.3. Polymer Matrix Composites (PMC)

These are composed of a matrix from thermoset unsaturated polyester, epoxy polyester or thermoplastic polycarbonate, polyvinyl chloride, nylon, polystyrene and embedded glass, carbon, steel or Kevlar fibres as dispersed phase.

2.2. Based on Reinforcing Material Structure

2.2.1. Particulate Composites

It consists of a matrix reinforced by a dispersed phase in the following form of particles.

- Composites with random orientation of particles

- Composites with preferred orientation of particles

2.2.2. Fibrous Composites

Fibrous composites are of two types.

2.2.2.1. Short Fibre Reinforced Composites

It consists of a matrix reinforced by a dispersed phase in form of discontinuous fibres (length <100 μm).

- Composites with random orientation of fibres
- Composites with preferred orientation of fibres

2.2.2.2. Long Fibre Reinforced Composites

It consists of matrix reinforced by a dispersed phase in form of continuous fibres.

- Unidirectional orientation of fibres
- Bidirectional orientation of fibres (woven)

2.2.3. Laminate Composites

When a fibre reinforced composite consists of several layers with different fibre orientations, it is called multi layer or angle ply composite.

2.3. Resin based Composites

These composites are commonly used as dental composites.

2.3.1. Based on Filler Particles

2.3.1.1. Conventional/Traditional Composites (Macro Filler)

Also referred to as conventional or macro filled composites. Filler used is finely ground amorphous silica and quartz. Average size of filler particles is 8 to 12 μm but particles as large as 50 μm may also present. Filler loading 70 to 80 wt% or 60 to 70 vole %

Properties

Compressive strength:	250–300 MPa
Tensile strength:	50–65 MPa
Elastic modulus:	8–15 GPa
Thermal expansion coefficient:	25–35 ppm/°C
Knoop hardness number (KHN):	55 KHN
Water sorption:	0.5–0.7 $\mu g/cm^2$

All properties are improved in comparison to unfilled acrylics. Compressive strength is improved by 300% to 500% because of stress transfer from matrix to filler particles. Composites that contain quartz or amorphous silica as filler are radiolucent.

Clinical Consideration

Used in class I, II, III, IV, V types of lesions on tooth surfaces and high stress areas. Finishing of restoration produces a roughened surface as does tooth brushing and masticatory wear overtime. Restorations prone to discoloration because of susceptibility of roughened surface to retain stain. Poor resistance to occlusal wear is major clinical problem. Therefore, inferiorly placed materials are used as posterior composites.

2.3.1.2. Small Particle filled Composites (SPFCs)

Amorphous silica and glasses containing heavy metals are use as filler which provide radiopacity to materials. Primary filler consists of silane coated ground particles. Colloidal silica is approximately 5 wt% to adjust viscosity of paste for packing into cavity. Matrix resin is bisphenol A - glycidyl methacrylate or urethane dimethacrylate.

Diluents monomer triethlene glycol dimethacrylate (TEGDMA) is added to make it clinically acceptable. For improving surface smoothness and various properties inorganic filler are added in a size range of ~0.5 to 3 μm, but with a fairly broad size range distribution. So inorganic filler load is 80-90 wt% and 65 to 77 vol%.

Properties

Compressive strength:	350–400 MPa
Tensile strength:	75–90 MPa
Elastic modulus:	15–20 GPa
Thermal expansion coefficient:	19-26 ppm/°C
Water sorption:	0.5-0.6 mg/cm^2
Knoop hardness number:	50-60 KHN

Such properties impart greater wear resistance and decrease in polymerization shrinkage. These materials are radio opaque since they have glass containing heavy metals. Radio-opacity is an important property for materials used for restoration of posterior teeth to facilitate diagnoses of secondary caries. With time this composite soften and become more prone to wear and deterioration, which reduces long term durability of restoration, because heavy metal glass fillers are softer and more prone to hydrolyze and leach in water.

Clinical Consideration

It is used for high stress and abrasion prone areas such as in class IV sites. So, these composites are used in areas requiring optimal polishability. Polished surface is not as smooth as microfilled composites.

2.3.1.3. Microfilled Composites

In microfilled composites use of colloidal silica particles as inorganic filler over come problems of surface roughening and low translucency associated with previous composites. Microfilled composites are further classified based on particle size as:

Homogenous microfill:	Particles size - 0.04 μm
Heterogeneous microfill:	Particle size - 0.04-0.4 μm

These composites have surface smoothness similar to unfilled direct filling acrylic resins. Colloidal silica have larger surface area that must be wetted by monomer but this phenomenon cause increase in viscosity and produce undue thickening. To solve this problem, methods employed are:

- To sinter colloidal silica so that particles several tenths of a micrometer size are obtained.
- Common method to increase filler loading is to make new filler particles by grinding a pre-polymerized composite that is highly loaded with colloidal silica particles.

Properties

Compressive strength:	250–350 MPa
Tensile strength:	30–50 MPa
Elastic modulus:	3-6 GPa
Thermal expansion coefficient:	50–60 ppm/°C
Knoop hardness number:	25–35 KHN
Water sorption:	1.4–1.7 mg/cm^2

Properties of microfilled composites are inferior to traditional composites but superior to unfilled acrylic resins. Microfilled remains wear resistant for several years as comparable with those of highly filler loaded composites. They provide smoothest surface finish available among aesthetic composites restorations. A major short coming is weak bonding between composites particles and clinically cured matrix, facilitating wear by chipping mechanism; so these composites are not use on stress bearing surfaces. If placed in proximal contact of anterior teeth "drifting" may occur. Hence, they are preferred for restoring teeth with carious lesion on smooth surfaces.

Clinical Consideration

Used for class III and IV restorations, diamond burs, rather than fluted tungsten carbide burs are recommended for trimming them to minimize risk of chipping. Because of their smooth surface they are resin of choice for aesthetic restoration of anterior tooth, particularly in non stress bearing areas and sub-gingival areas. Due to decreased physical properties, these composites are not used in stress bearing areas like class II and IV sites.

2.3.1.4. Hybrid Composites

As name implies, hybrid composites contains two kinds of filler particles. It consists of colloidal silica and ground particles of glasses containing heavy metals, constituting filler content of approximately 75 to 80 wt%. The glasses have an average particle size of about 0.4 to 1.0 μm. Colloidal silica represents 10–20 wt% of total filler content.

Properties

Compression strength:	300-350 MPa
Tensile strength:	40-50 MPa
Elastic modulus:	11-15 GPa
Thermal expansion coefficient:	30-40 ppm/°C

Knoop hardness number: 50-60 KHN
Water sorption: 0.5-0.7 mg/cm^2

These properties are superior to those of microfill composites. As the filler particles have heavy metal atoms, they show sufficient radiopacity for radiographic detection of secondary caries and various diagnostic tasks.

Clinical Consideration

Used for anterior restoration including class IV sites. Hybrid (large particles) use in high stress areas require improved polishability (classes I,II,III and IV). Hybrid (midifiller) use in high stress areas require improved polishability (classes III and IV). Hence, widely used for stress bearing posterior restorations. Hybrid composites are further divided into three classes:

- *Flowable Composites* have reduced filler level. They flow readily, spread uniformly and intimately adapt to a cavity form to produce a desired anatomy. They are more susceptible to wear. Use in class II posterior preparations and other situations in which access is difficult. Use in class I restorations in gingival areas and to prevent caries use in similar manner as pit and fissure sealants. Use in areas of poor accessibility and no exposure to wear.
- *Packable Composites* materials were introduced in late 1990s. They are more viscous to afford a “feel” on insertion similar to that of amalgam. Their development is an attempt to accomplish two goals:
  - Easier restoration of a proximal contact.
  - The handling properties are similiar to amalgam but time required is twice as needed for restoration.
  - They consist of elongated, fibrous filler particles of about 100 μm in length, and/or textured surfaces that tend to interlock and resist flow.
- *Nanofill Composites* are referred to as nanofill hybrids. They contain filler particles that are extremely small (0.005-0.01 μm). These particles can be easily agglomerated; a full range of filler sizes is possible. Consequently, high filler level can be generated in the restorative material resulting in good physical properties and aesthetics. It is highly polishable so these materials have become a popular composite restorative material of choice.

2.3.2. Based on Techniques of Insertion

2.3.2.1. Chemically Cured Composites

Chemically cured composites are synthesized by using aromatic tertiary amine activator e.g. N,N-dimethyl-p-toluidine in presence of benzoyl peroxide (BP) as initiator. This process is also referred as cold curing or self curing polymerization process. During mixing it is impossible to avoid incorporating air into mix, thereby forming pores that weaken structure and trap oxygen which inhibits polymerization during curing. Insert while it is still plastic for better adaptation to cavity walls. Working time is not in control after mixing of initiator to activator.

2.3.2.2. Light Activated Composites

They are supplied as single paste contained in a light proof syringe. The free radical initiating system, consists of

- *Photosensitizer* - Camphorquinone (CQ) that absorbs blue light with wavelength between 400-500 nm (0.2 wt% or less).
- An amine initiator such as dimethylaminoethyl methacrylate (DMAEMA) 0.15 wt%).

In all composites *inhibitors* like butylated hydroxytoluene in concentration of 0.01 wt% are added to extend storage lifetime for all resins and sufficient working time. This composite have greatly improved depth of cure, a controllable working time and other advantages over chemical cured composites. Exposure time of 40 sec or less is required to light cure a 2 mm thick layer as compared with several minutes for chemically activated.

They must be incrementally placed when bulk approximately 2 to 3 mm because of limited depth of light penetration. Cost of light curing unit is high.

There is no porosity as seen in chemically activated resins and also allow operator to complete insertion and contouring before curing. Light sources use are photo curing with visible (Blue) light and curing lamps like: light emitting diodes (LED) lamps, quartz tungsten halogen (QTH) lamps, plasma arc curing (PAC) lamp and argon laser lamps.

Dual Cure Resins

It consists of two light curable pastes, one containing benzoyl peroxide and other an aromatic tertiary amine. When these two pastes are mixed, light curing is promoted by amine/CQ combination and chemical curing is by amine/BP interaction. These resins are applied in areas where sufficient light penetration is not allowed for monomer conversion. Air inhibition and porosity are the common problems with dual cure resins.

4. Biocompatibility of Composites

It relates to the effects on the pulp from two aspects.

- The inherent chemical toxicity of the material.
- The marginal leakage of oral fluids.

Adequately polymerized composites are least toxic and relatively bio-compatible because they exhibit minimal solubility. Very rarely patients and dental personnel can develop an allergic response to these materials. Light activated materials are more susceptible to cause long term pulp inflammation because of inadequately cured composite materials at the floor of a cavity which serve as a reservoir of diffusible components. Such composites can release leachable constituents adjacent to the pulp. The second concern is associated with shrinkage of composites during polymerization and the subsequent marginal leakage. This causes bacterial ingrowths and lead to secondary caries or pulp reactions. Therefore, restorative procedure must be such that all such problems are minimized.

Bisphenol A (BPA), a precursor of bis-GMA has been shown to be a xenoestrogen or a synthetic compound that mimics effects of estrogen by having an affinity for estrogen receptors but its effect on humans are still unclear. BPA has recently also been shown to exhibit antiandrogenic activities which may prove to be detrimental in organ development. It is found that BPA and BPA– DM (Dimethyacrylate) applied to cancer cell significantly increase cell proliferation and DNA synthesis, similar to effect of estrogens. Controversy surrounds this issue because it is unclear how much BPA or BPA-DM is released to oral cavity and what dosage is enough to affect human health.

5. Properties of Composites

5.1. Polymerization Shrinkage

Microhybrid shows less shrinkage during setting than microfilled types. Two techniques have been proposed to overcome or minimize the effect of polymerization shrinkage.

(i) One method is to insert and polymerize the composites in layers, thus reducing shrinkage effects.
(ii) The second method is to prepare a laboratory (indirect) composite inlay on a die and than to cement inlay to tooth with a thin layer of low viscosity composite cement.

5.2. Thermal Conductivity

It closely matches that of enamel and dentin and lower than metallic restorations. Hence, provide good thermal insulation for dental pulp.

5.3. Thermal Expansion

Composites show greater change in dimensions with changes in oral temperatures than tooth structure will. The more resin matrix, the higher is the linear coefficient of thermal expansion, since polymer has a higher value than the filler. As a result, microfilled composites have higher values for thermal expansion than microhybrid composites.

5.4. Water Sorption

The microfilled composites have a greater potential for being dis-colored by water-soluble stains. The effect of water sorption on degradations of properties of composites is irreversible.

5.5. Radiopacity

Most microhybrid composites are radiopaque. One microfilled composite contain barium glass, ytterbium trifluoride, making it radiopaque. Most composites are radiopaque when compared to dentin but are radiolucent when compared to enamel. Radiopacity is an advantageous property for diagnosis of caries.

5.6. Compressive and Flexural Strengths

The compressive strength of microhybrid composites is higher than that of microfilled composites, with the strength generally increasing linearly with volume fraction of filler. Since composite restorations most likely fail in tension or bending, their tensile and flexural strengths are of special interest.

5.7. Elastic Modulus

The elastic modulus or stiffness of composites is dominated by the amount of filler and increases exponentially with the volume fraction of filler. The lower filler content of microfilled composites results in elastic modulus of one fourth to one half of more highly filled microhybrid composites. This stiffness is important in applications where high biting forces are involved and wear resistance is essential. However the rate of bond failure of class V cervical restorations was higher for microhybrid composites when compared with microfilled composites, the lower modulus of microfilled composites probably reduced the stress on bond of restoration to dentin.

5.8. Hardness and Wear

Knoop hardness of composites is exponentially related to volume fraction of filler and is less related to hardness of filler. The higher filler content of microhybrid is important in providing higher resistance to non-recoverable penetration and abrasive wear. Abrasive wear is, however, only one aspect of wear process.

5.9. Bond Strength

The bond strength of composites to acid etched enamel and dentin are about same (14-30 MPa) when a universal bonding agent is used. Etching of enamel and dentin are recommended to remove smear layer, resulting from cavity preparation, before application of bonding agent. Since polymerization stresses are of same magnitude as bond strengths to tooth structure marginal leakage may not be entirely prevented. It should be pointed out that bond strength of 20 MPa is estimated to be required to prevent marginal gaps as a result of polymerization shrinkage.

6. Indication of Composite Resins

(i) Class I, II, III, IV, V, VI restoration.
(ii) Foundation or core build-ups.
(iii) Sealants and preventive resin restoration.
(iv) Aesthetic enhancement procedures.
- Partial veneers
- Full veneers
- Tooth contour modifications
- Diastema closures

(v) Cements for indirect restoration.
(vi) (Temporary restoration.
(vii) Periodontal splinting.

7. Contra Indication of Composite Resins

The America Dental Association does not support the use of composite in teeth with heavy occlusal stress, sites that can not be isolated or patient who are allergic or sensitive to composite materials.

8. Disadvantage of Composites

8.1. Marginal Leakage

Resin is firmly anchored to the etched enamel at the other margins; the material tends to pull away from gingival margins during curing because of polymerization shrinkage. This is one of greatest problem of composites that leads to formation of a gap at that interface especially in class II and V restorations. It leads to marginal staining and secondary caries is enhanced.

8.2. Radiopacity

Resins are inherently radiolucent. Therefore, leaking margins, secondary caries, poor proximal contacts, wear of proximal surfaces and other problems cannot be detected unless adequate radiographic contrast can be achieved. However, all composites are not radiolucent; most demonstrate sufficient radiopacity to subside these problems. Some of flowable composites are still radiolucent, so precautions must be taken while using them for proximal boxes. Radiopacity is an important property for any posterior restorative material. A wide range of radiopacity values have been considered to be adequate, but exceeding radiopacity of enamel by a large degree will have the effect of obscuring radiolucent areas caused by gap formation or secondary caries.

8.3. Wear

Composites are superior materials for anterior restorations in which aesthetics and occlusal forces are low. Colour changes are minimal, marginal adaptation is good and recurrent decay is low. Wear of posterior composite restorations are observed at the contact area where stresses are the highest. Ditching at the margins within composite is also observed for posterior composites, probably resulting from inadequate bonding and stresses. The loss of surface contour in anterior composites results from combination of abrasive wear from tooth brushing and chewing and erosive wear from degradation of composites in the oral environment. Products use as posterior composites have better wear resistance than anterior or all- purpose composites.

8.4. Post Operative Sensitivity

It is believed to result from micro leakage of bacteria or induced internal stress. Incremental placement of the composites, excellent isolation during placement and use of bases to protect the pulp are recommended solutions.

8.5. Water Absorption

This property of material reduces its effectiveness as a restorative material of choice. Material with high filler content exhibits lower water absorption values.

9. Repair of Composites

It is repaired by placing new materials over old composite. When restoration is just placed and polymerized, it may still have an oxygen inhibited layer of resin on surface. Additions of new composites can be made directly to this layer. A recently cured and polished composite still have 50% of un-reacted methacrylate groups to co-polymerize with newly added material. As restoration ages, fewer and fewer un-reacted methacrylate groups remain and greater cross-linking reduces ability of fresh monomer to penetrate into the matrix. The polished surfaces expose filler surfaces that are free from silane, which does not chemically bond to the new composite layer. So, a silanate bonding agent is applied before placing the new composite but the strength of repaired composite is less than half strength of original material.

10. Fiber Reinforced Thermoplastic Composites (FRCs) in Dentistry

FRCs is used for treatment of misaligned teeth requiring a stable material under stress in a moist environment. FRCs is used as splints, retainers, space maintainers. The main advantage of FRC is aesthetics (translucency), direct bondability to tooth structure and easy forming of appliance in the dental office. The performance of the material is strongly dependent on its resistance to hydrolytic deterioration of components. Annealed polycarbonate and malleated polypropylene (MPP) reinforced with bare E-glass fibres have an appropriate combination of mechanical properties and environmental stability for potential orthodontic applications. FRCs can provide a continuous range of stiffness without changing the cross-sectional dimensions, simply by varying fibre volume fraction.

11. Compomers

These materials provide the combined benefits of composites and glass ionomers. These materials have two main constituents dimethacrylate monomer with two carboxylic groups presents in their structure and filler that is similar to ion-leachable glass present in glass ionomer cement (GIC). These materials set *via* a free radical polymerization reaction, do not have ability to bond to hard tooth tissues and have significantly lower levels of fluoride release than GICs. Their applicability as orthodontic adhesives and amalgam bonding systems has also been reported. Compomers are tooth coloured materials and so their aesthetics can immediately be seen as better than that of dental amalgams. Constant reformulations of these types of materials may eventually lead to them being comparable or even superior to existing composites, but as long as they do not set *via* an acid-base reaction and do not bond to hard tooth structures, they cannot and should not be classified with GICs. They are after all, just other dental composites.

Lower flexural modulus of elasticity, compressive strength, flexural strength, fracture toughness and hardness along with significantly higher wear rates compared to clinically proven hybrid composites, have been reported for these materials.

12. Clinical Technique

12.1. Initial Clinical Procedures

The patient is scheduled for operative appointment after complete examination, diagnosis and treatment plan is finalized. Also a brief review of chart, treatment plan and radiograph should precede each restorative procedure.

12.2. Local Anaesthesia

It is required in many operative procedures. Profound anaesthesia contributes to a more pleasant and uninterrupted procedure and usually results in a marked reduction in salivation. These effects of local anaesthesia contribute to better operative dentistry, especially when placing bonded restorations.

12.3. Preparation of the Operating Site

It is necessary to clear the operative site with slurry of pumice to remove plaque, pellicle and superficial stains. Calculus removal with appropriate instruments also may be needed. These steps create a site more receptive to bonding. Prophylactic pastes containing flouridating agents, glycerine or fluorides act as contaminants and should avoid a possible conflict with the acid etching technique.

12.4. Shade Selection

Special attention should be given to matching the colour of natural tooth with the composite material. The shade of tooth should be determined before the teeth are subjected to any prolonged duration because dehydrated teeth become lighter in shade as a result of a decrease in translucency. The colour varies with translucency, thickness and distribution of enamel and dentine and age of patient. Other factors such as fluorosis, tetracycline shining and endodontic treatment also affect tooth colour. Most manufacturers provide shade guides for their specific materials which usually are not interchangeable with materials from other manufacturers. Good lighting is necessary when colour selection is made. Natural light is preferred for selection of shades. If no windows are present in operatory site to provide natural daylight, colour-corrected operating lights or ceiling lights are available to facilitate accurate shade selection.

In choosing the appropriate shade, one holds the entire shade guide near the teeth to determine general colour. The selection should be made as rapidly as possible. If more time is needed, the eyes should be rested by looking at a blue or violet object for a few seconds.

Final shade selection can be verified by patient with use of a hand mirror. If additional shades are needed, they may be obtained by mixing two or more of the available shades together or by adding colour modifiers, which may be available from the manufacturers. To be more certain of proper shade selection, a small amount of material of selected shade can be placed directly on tooth, in close proximity to area to be restored and cured.

12.5. Isolation of Operating Site

Isolation of the area is imperative if desired bond is to be obtained. Contamination of etched enamel or dentine by saliva results in a significantly decreased bond, likewise contamination of composite material during insertion results in degradation of physical properties.

12.5.1. Rubber Dam

A heavy rubber dam is an excellent means of acquiring access, vision and moisture control. For proximal restorations, the dam should be attempted to isolate several teeth mesial and distal to operating site. If lingual approach is indicated, for an anterior tooth restoration, it is better to isolate all anterior teeth and include first premolars.

12.5.2. Cotton Rolls (With or without Retraction Cord)

An alternate method of obtaining a dry operating field is the use of cotton roll isolation. A cotton roll is placed in facial vestibule directly adjacent to tooth being restored. When the gingival extension of a tooth preparation is to be positioned sub-gingivally, or near the gingiva, a retraction cord can be used to retract temporarily and reduce seepage of tissue fluids into operating site. If haemorrhage control is needed, the cord can be saturated first with a liquid astringent material such as haemodent. A piece of cord approximately 0.5 to 1 mm in diameter and 8 to 10 mm long is usually sufficient, depending on dimension of involved gingival crevice. The cord is always inserted into crevice and not on top of gingiva. Any retraction cord placement must be done judiciously to avoid blunt dissection of gingival tissue or periodontal attachment. No loose strand should be left exposed.

When restoring posterior proximal surfaces, a wedge should be placed firmly into gingival embrasure preoperatively. Also, a preoperative assessment of occlusion should be made. Knowing the preoperative location of occlusal contacts is important in planning the restoration outline form and establishing proper occlusal contact on restoration.

12.6. Finishing of Composites

Optimum finishing and polishing of composite resins is a very important step in the completion of the restoration. Residual surface roughness can encourage bacterial growth, which can lead to a myriad of problems including secondary caries, gingival inflammation and surface staining. Several methods for the finishing and polishing of composite resins have been advocated. The best possible finish is produced by not polishing the surface at all, at least for surfaces that have polymerized next to a matrix strip. The smoothest surface on a restoration can be obtained by curing the composite against a smooth matrix strip. This minimizes porosities as well as the oxygen-inhibited layer. Use of a scalpel blade or any thin, sharp-edged instrument to remove flash on the proximal areas is recommended. Trimming forces should be applied either parallel to the margin or toward the gingival tissue. Coarse to ultrafine aluminium oxide discs, tungsten carbide bur, fine diamond tips with extra fine polishing paste, silicon based systems and silicon carbide-impregnated polishing brushes and points are used for finishing the composites. Application of surface sealer or a low viscosity resins are used to seal surface porosities and micro-cracks.

CONCLUSION

The clinical performance of dental restorations is judged on the basis of long term clinical trials, preferably those based on randomized, controlled experimental designs. However, a recent evidence-based review of the longevity of amalgam and composite restorations was based on a critical review of clinical data over 10 years. The most consistent survival levels are exhibited by amalgam restorations. The performance of posterior composites has greatly improved during the past decade relative to amalgams.

In summary, it is clear that the clinical use of amalgam has continued in many countries because of its ease of use, relatively low cost, wear resistance, freedom from excessive shrinkage during setting, and its high survival probabilities. In spite of all controversies surrounding its use, amalgam continues to exhibit superior clinical characteristics compared with several of the currently available tooth-coloured restorative materials.

REFERENCES

Journal Articles

[1] Academy of Dental Materials: International Congress on Dental Materials: in *Transaction International Congress on Dental Materials*, Houston, Tx, Academy of Dental Materials, Baylor College of Dentistry, 1990.

[2] Bowen R. L. US Patent 3, 006, 112, 1962.

[3] Lutz, F.; Philips, R. W. *J. Prosthet Dent.* 50, 1983, 480-488.

[4] Swartz, M. L.; Phillips, R. W.; Rhodes, R. *Visible light activated Resins: Depth of Cure, JADA 106*, 1983, 634-7.

[5] Farah, J. W; Dougherty, E. W. *Oper. Dent.* 1981, *6,* 95-101.

[6] Willems, G.; Lambrechts, P.; Braem, M.; Celis, J. P. *Dent. Mater.* 1992, *81,* 310-319.

Article from Books

[1] Alberts, H.; *Tooth Coloured Restoratives: Principles and Techniques* 9th Ed, Hamilton, Ontario, Canada, B C Decker Inc 2002.

[2] Baum, L.; Phillips, R. W.; Lund, M. R. *Textbook of Operative Dentistry* 3rd Ed., Philadelphia, WB, Saunders, Co., 1995.

[3] Theodore, M. R.; Harald, O. H. *Sturdevants Art and Science of Operative Dentistry*, 5th Ed., 2009, 502-512.

In: Advanced Organic-Inorganic Composites
Editor: Inamuddin

ISBN 978-1-61324-264-3

Chapter 12

ORGANIC/INORGANIC COMPOSITES FOR ENERGY STORAGE IN ELECTROCHEMICAL CAPACITORS

Sadaf Zaidi[a,*] ***and Inamuddin***[b]
[a]Department of Chemical Engineering, Faculty of Engineering and Technology, Aligarh Muslim University, Aligarh 202002, India
[b]Department of Applied Chemistry, Faculty of Engineering and Technology, Aligarh Muslim University, Aligarh 202002, India

ABSTRACT

Among the energy storage devices, different battery technologies have pervaded the day to day lives of the people. They are there from the small zinc-air button cells to the lead acid batteries used in automobiles. In sharp contrast, electrochemical capacitors, devices that too store energy, are less known to the public; even though they have several distinct advantages over batteries. Electrochemical capacitor technology has marched quietly ahead and has reached a stage where it is a full grown industry with sales of several hundred million dollars per annum. It is poised for rapid growth in the near future due to the expansion of power quality needs, emerging energy management and conservation applications and the increased reliance on renewable energy sources. The present status of electrochemical capacitor technology is the result of research being conducted globally to continually design and develop new materials for use as components for electrochemical capacitors, notably for the construction of electrodes. Efforts are being made to overcome the limited energy density of electrochemical capacitors.

This chapter, while briefly touching upon the fundamentals of electrochemical capacitors and their comparison with batteries, attempts to review the state-of-the-art research in materials for the construction of electrodes for electrochemical capacitors, albeit partially, owing to the voluminous amount of literature available in the field. Special emphasis is given to organic/inorganic composites. The work done using different methods for the synthesis and evaluation of organic/inorganic composite electrodes for use in electrochemical capacitors is reported.

* Corresponding author's e-mail:sadaf63in@yahoo.com

ABBREVIATIONS

AB	Acetylene black
ACFa	Activated carbons fabric
ACFis	Activated carbons fibers
ACs	Activated carbon(s)
APC	Activated porous carbon
BMPC	Bimodal porous carbon
CA	Carbon aerogel
CC	Carbon cloth
CCDC	Calcium carbide derived carbon
CNF(s)	Carbon nanofibers(s)
CNT(s)	Carbon nanotube(s)
CNTA	Carbon nanotube array
COP(s)	Conducting organic polymer(s)
CV	Cyclic valtammetry
DOE	US Department of Energy
DP	Diamond powders
DTA	Differential thermal analysis
ECs	Electrochemical capacitors
EDLC(s)	Electric double-layer capacitor(s)
EDS	Energy dispersive spectroscopy
EIS	Electrochemical impedance spectroscopy
EMITFSI	1-ethyl-3-methylimidazolium Bis((trifluoromethyl)sulfonyl)imide
ESR	Equivalent series resistance
FESEM	Field emission scanning electron microscopy
FTIR	Fourier transform infrared spectroscopy
GCD	Galvanostatic constant-current charge-discharge
GNS	Graphene nanosheet
HEC	Hybrid electrochemical capacitor
HNTs	Halloysite nanotubes
MPBCF	Mesoporous pitch based carbon foam
MWCNT(s)	Multi-walled carbon nanotube(s)
NEC	Nippon Electric Company
NMP	N-methyl pyrrolidone
OMC	Organic mesoporous carbon
PAni	Polyaniline
PANINWs	PAni nanowires
PC	Porous carbon
PEDOT	Poly (3, 4-ethylenedioxythiophene)
PNMA	Poly (n-methyl aniline)
PNR	Poly neutral red
POMs	Polyoxometalates
PPy	Polypyrrole
PS	Porous silicon

PSS	Poly(styrene sulfonic acid)
PSSMA	Poly(4-styrene sulfonic acid-co-maleic acid)
PTFE	Poly(tetrafluoroethylene)
PTh	Polythiophene
PTS	p-toluene sulfonic acid
PYR14TFSI	N-butyl-N-methyl pyrrolidinium bis (trifluoromethanesulfonyl) imide
RM	Red mud
RTILs	Room temperature ionic liquids
SCE	Standard calomel electrode
SEM	Scanning electron microscopy
SOC	State of charge
SOHIO	Standard Oil Company of Ohio
SS	Stainless steel
SWCNT(s)	Single-walled carbon nanotube(s)
SWCNTsf	Functionalized SWCNTs
$TEABF_4$	Tetraethyl ammonium tetrafluoroborate
TEM	Transmission electron microscopy
TGA	Thermo-gravimetric analysis
TOS	p-toluene sulphate
XPS	X-ray photoelectron spectroscopy
XRD	X-ray diffraction

1. INTRODUCTION

The consumption of energy in the world is increasing at a very fast rate with the ever increasing population and development. Coupled with this, is the growing realization about the need to save the environment in the wake of its reckless abuse by man; thus, the growing demand for low or even zero-emission sources of energy. Efficient, clean, and renewable energy sources seem to be the answer. Of particular interest is electricity that can be generated from renewable sources, such as solar energy and biomass. However, the efficient utilization of electricity generated from these and other renewable sources necessitates some form of electrical energy storage, given the fact that renewable energy sources often rely on weather or climate to work. Sometimes, their form is not suitable for direct use. Moreover, energy consumption varies significantly on diurnal and seasonal bases according to the demand by industrial, commercial and residential activities, especially in extremely hot and cold climate countries, where the major part of the variation is due to domestic space heating and air conditioning. Energy storage is a generic name for specific techniques for storing energy derived from some primary source in a convenient form for use at a later stage when a specific demand arises, most often at a different location. There are many methods of energy storage. Some of them are in current use, while others are in various stages of development. Figure 1 shows the Ragone plot for various electrical energy storage devices. The existing energy storage technologies are incapable of meeting the requirements for efficiently using electrical energy in applications such as transportation vehicles, commercial and residential requirements, and consumer devices. Energy storage is critical in enhancing the applicability, performance and reliability of a wide range of energy systems as the gap between energy supply and its demand can be eliminated by the use of proper energy storage options.

The need of the hour is to give due importance to energy storage at par with that being given to the development of new energy sources, in order to have a secure energy future. Advances in the performance of energy storage systems are intricately linked with the advances in materials science and in the make-up of materials for energy storage systems. The emphasis in this chapter is primarily on electrical energy storage in electrochemical capacitors (ECs). Electrochemical capacitors have attractive features such as a fairly broad operational range of power and energy densities, fast charging, reliability, large number of charge-discharge cycles and wide operating temperatures. Both ECs and batteries are based on electrochemistry, and the basic difference between electrochemical capacitors and batteries is that the former store energy directly as charge, whereas the latter store energy in chemical reactants capable of generating charge.

The electrode is the key part of the electrochemical capacitors, so the electrode materials are the most important in determining the properties of ECs [1]. In this chapter, the fundamentals of ECs, their comparison with batteries, the storage principles and characteristics of electrode materials for ECs such as, carbon-based materials, transition metal oxides and conductive polymers, are briefly discussed. Special emphasis is laid on an up-to-date, state-of-the-art review of the organic/inorganic composite materials being developed for use as electrodes in ECs.

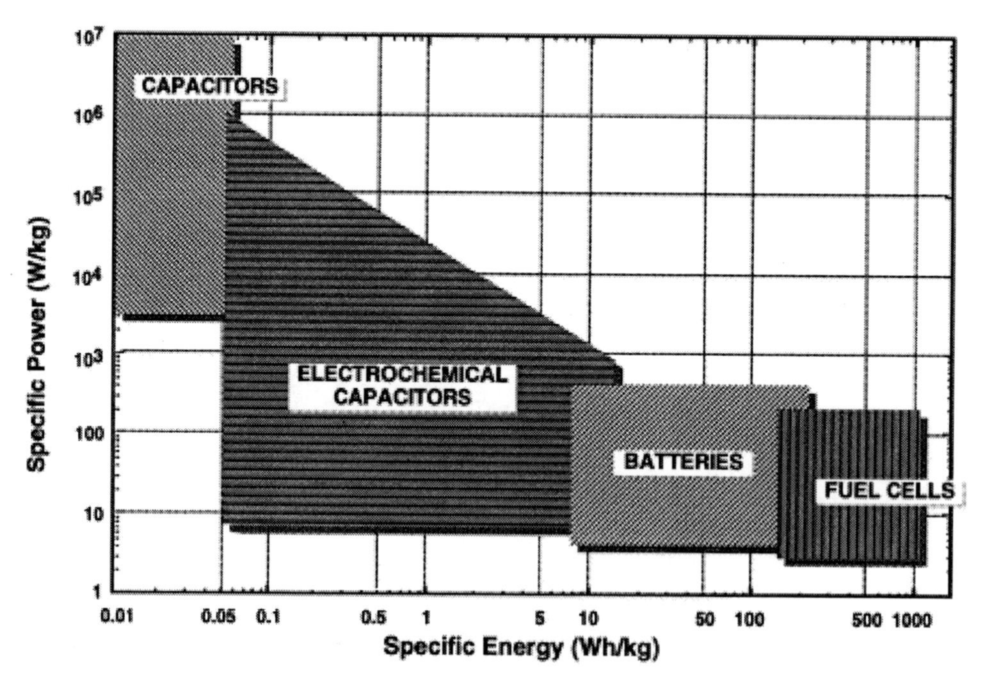

Figure 1. Sketch of Ragone plot for various energy storage and conversion devices. The indicated areas are rough guide lines. Reproduced with permission from Kotz, R. et al. *Electrochim Acta* 2000, *45,* 2483-2498. Copyright © 2000, Elsevier B.V.

2. Fundamentals of Electrochemical Capacitors

Capacitors which store energy within the electrochemical double layer at the electrode-electrolyte interface are variously referred to by manufacturers in promotional literature as 'supercapacitors' or 'ultracapacitors', 'power capacitors', 'gold capacitors' or 'power cache'. Consequently, they are also quite properly referred to as 'electric double layer capacitors'. Since, there are in general contributions to the capacitance other than by double layer effects (say, faradic processes); it is therefore apt to refer to all such devices as electrochemical capacitors. A German physicist, Hermann von Helmholtz, first gave the concept of double-layer capacitance in 1853 [2]. The EC was invented in 1957 by H. I. Becker of General Electric Company and was patented the same year [3]. However, Becker's device was not very practical as both the electrodes were required to be immersed in a container of electrolyte; therefore, it was never commercialized. It was Robert A. Rightmire, an electrochemist at the Standard Oil Company of Ohio (SOHIO), to whom can be attributed the invention of the device in the form now commonly used. Nippon Electric Company (NEC) of Japan licensed the technology from SOHIO and introduced the first EC products to the market place as memory back-up devices in computers in 1957 [4]. An ultracapacitor development program was initiated in 1989 by the U.S. Department of Energy (DOE), and short term as well as long term goals were defined for 1998-2003 and after 2003, respectively [5]. Capacitors in general, physically store electrical energy as equal quantities of positive and negative charge on opposite sides of an insulating material. When the two sides are connected by a conducting path, current flows until complete charge balance is achieved. The capacitor is returned to its charged state by applying a voltage across the electrodes. Because no chemical or phase changes take place, the process is highly reversible and the charge-discharge cycle can be repeated virtually without limit. In construction, the EC comprises of two electrodes (anode and cathode), an electrolyte and a separator that electrically isolates the two electrodes. Each electrode-electrolyte interface represents a capacitor; therefore, the complete cell comprises two capacitors in series. Depending on the charge storage mechanisms, ECs can be classified as electric double-layer capacitors (EDLCs) and faradic pseudocapacitors. A third class of ECs is the hybrid electrochemical capacitor (HEC), in which one of the electrodes is of a double-layer carbon and the other electrode is of a pseudocapacitance material or battery-like. In most of the hybrid electrochemical capacitors, the pseudocapacitance material has been used as the cathode. Pseudocapacitance materials in HECs, not only increase the specific capacitance, but also extend the working voltage. Thus, the energy density of such devices can be substantially higher than for EDLCs [1]. They can also be assembled using two non-similar mixed metal oxides or doped conducting polymer materials [6]. In the HEC using a battery-like electrode with a double-layer electrode, the capacity of the battery-like electrode is generally many times greater than the capacity of the double layer capacitor electrode. This asymmetric design provides exactly twice the capacitance of the symmetric design. Also, the operating voltage of the asymmetric design is larger, due to the two electrodes having different rest potentials. Both of these factors contribute to a higher energy density than can be achieved with a symmetric design. Several novel asymmetric electrochemical capacitor designs are under development [7,8] using a lithium ion intercalation electrode in an organic electrolyte at 3.8 V or a carbon electrode with a lead dioxide battery-like electrode using H_2SO_4 as the electrolyte, operating at 2.1 V with the potential of being very low cost [8]. Each of these designs can provide high cycle-life due to the electrode capacity asymmetry. Today, considerable research emphasis is being given to asymmetric electrochemical capacitors because of the very attractive performance features.

2.1. Mechanism of Energy Storage in an Electric Double Layer Capacitor

Energy is stored in an electric double layer capacitor (EDLC) as charge separation in the double-layer formed at the interface between the solid electrode material surface and the liquid electrolyte in the micropores of the electrodes. The amount of electrical charge accumulated due to the pure electrostatic forces that are typical for EDLC depends on the surface of the electrode-electrolyte interface and on the easy access of the charge carriers to this interface [9]. A schematic of an EDLC is given in Figure 2. The ions displaced in forming the double layers in the pores are transferred between the electrodes by diffusion through the electrolyte [6]. As an excess or deficiency of charge builds up on the electrode surface, ions of the opposite charge build up in the electrolyte near the electrolyte-electrode interface in order to provide electro-neutrality [1]. The double layer capacitance (C) at each electrode interface is given by Eq. (1), where, ε_r is the relative permittivity of the solution, and A is the surface area of the electrode material.

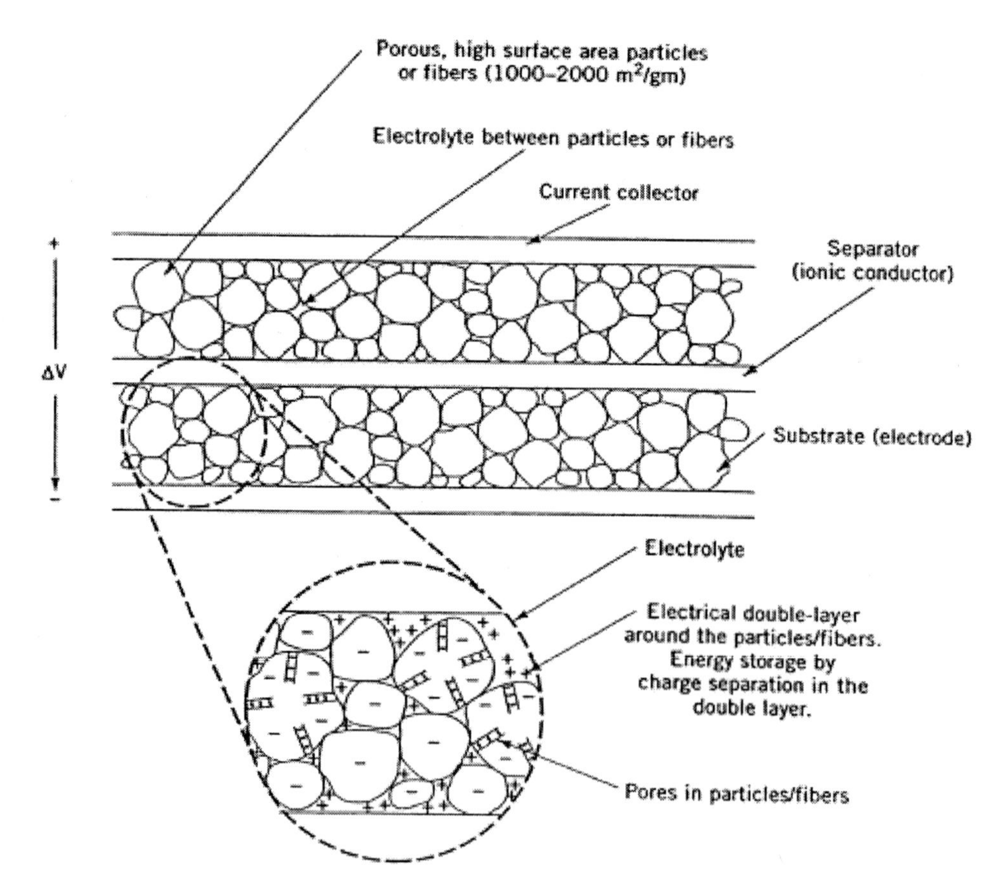

Figure 2. Schematic of a double-layer ultracapacitor [6]. Reproduced with permission from Burke, A. *J. Power Sources* 2000, *91,* 37-50. Copyright © 2000, Elsevier B.V.

$$C = \frac{\varepsilon_r A}{d} \tag{1}$$

The stored energy (E) and the maximum peak power (P) of an EC are given by equations (3) and (4), respectively, where C is the capacitance in Farads (F), V is the nominal voltage in volts, and R is the equivalent series resistance in ohms. Maximum peak power is achieved with a matched load.

$$E = \frac{CV^2}{2} \tag{2}$$

$$P = \frac{V^2}{4R} \tag{3}$$

From the above equations, it is clear that the capacitance of an EC is dependent on the characteristics of the electrode material and the concentration of the electrolyte. A very high value of surface area of porous electrodes recorded to be as large as 1000-2000 m^2/cm^3 [10] as well as a high concentration of electrolyte, impart a very high capacitance to ECs (150-300 F/g). However, the measured specific capacitances of electrode materials using carbon are far less (75-175 F/g for aqueous electrolytes and 40-100 F/g for organic electrolytes) because for most carbon materials, a relatively large fraction of the surface area is in pores that cannot be accessed by the ions in the electrolyte. This is especially true for organic electrolytes for which the size of the ions is much larger than in aqueous electrolytes [6]. Electrode materials for EDLCs should thus have a high fraction of their pore volume comprising large pores to allow easy access to electrolyte ions. Cell voltage is also an important parameter in an EC, as it affects both the stored energy as well as the power. The cell voltage is however dependent on electrolyte stability. Aqueous electrolytes, such as sulfuric acid and potassium hydroxide, are advantageous in that they have high ionic conductivity but they suffer from low breakdown voltage of ~1.2 V. On the other hand, organic electrolytes such as acetonitrile and propylene carbonate containing dissolved quaternary ammonium salts, allow higher operating voltages around 3 V without breakdown. On the flip side, their electrical resistance is at least an order of magnitude greater than that of aqueous electrolytes. Since the key factors influencing the performance of ECs are the properties of electrolytes and the specific surface area of electrode materials, therefore an understanding and modification of the surface properties are important in achieving high power and energy density [11,12].

2.2. Mechanism of Energy Storage in a Pseudocapacitor

Another type of ECs is called pseudocapacitor. This is a class of energy storage devices that undergoes electron transfer reactions yet behaves in a capacitive manner. Pseudocapacitance arises when, for thermodynamic reasons, the charge q required for the progression of an electrode process is a continuously changing function of the potential V [13]. In other words, pseudocapacitance on electrodes is due when the application of a potential induces faradaic current from reactions such as electrosorption or from the oxidation–reduction of electroactive materials (e.g., Co_3O_4, IrO_2, and RuO_2) [14-17] or the doping or undoping of active conducting polymer material [6]. Such systems exhibit many times higher values of maximum capacitance than carbon double-layer systems. For redox systems, these values could be 10-100 times than those obtainable from EDLCs [14]. Figure 3 shows the mechanism of pseudocapacitance in a conducting polymer. When the conducting polymer is being charged, it loses electrons and becomes polycations, causing the anions in the solution (Cl^- in this case) to intercalate into the conducting polymer in order to maintain electro-neutrality.

3. Comparison of Electrochemical Capacitors and Batteries

The fundamental difference between batteries and ECs is on the basis of their charge storage mechanisms. Batteries store energy in chemical reactants capable of generating charge, whereas the ECs store energy directly as charge. ECs have low specific energy or energy density, high power density (discharge at high current density) as shown in Figure 1, very high charge/discharge rates, a sloping discharge curve that directly provides state of charge (SOC), essentially unlimited cycle life, high tolerance to deep discharge, environmental friendliness and maintenance-free operation. They experience no volume or phase changes during charge/discharge cycling. In contrast, batteries have high specific energy or energy density (due to chemical reactions in the battery), limited charge/discharge rates (because chemical reactions in a battery are not fully reversible), relatively flat discharge curves, and limited cycle life and require periodic maintenance. The important distinguishing properties of secondary batteries and electrochemical capacitors are listed in Table 1.

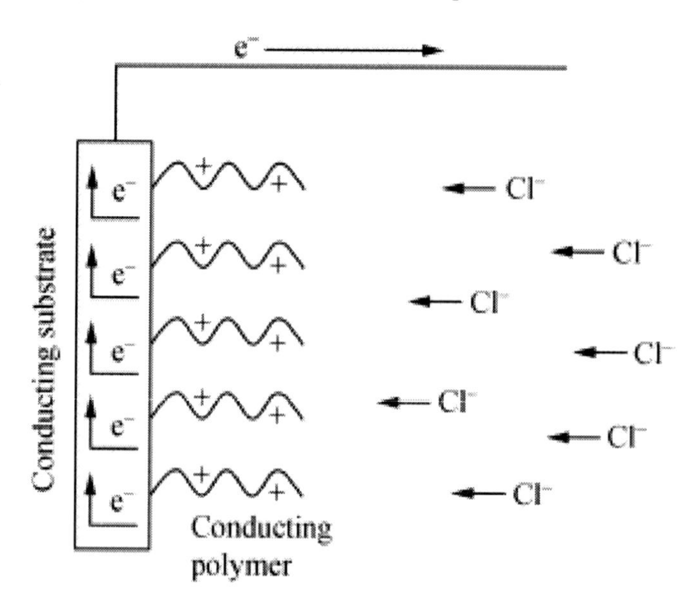

Figure 3. Illustration of pseudocapacitance in a conducting polymer [19]. Reproduced with permission from Peng, C. et al. *Progress in Natural Science* 2009, *18*, 777-788. Copyright © 2008, Elsevier B.V.

Thus, battery discharge rate, and therefore power performance, is limited by the reaction kinetics of the battery electrodes and the mass transport of the reactants, which means that charge and discharge rates are often different. The thermodynamics of the battery reactants dictate that the system operating voltage is relatively constant and, to the first order, independent SOC of the battery. Consequently, it is often difficult to measure the SOC of a battery precisely. In contrast, symmetric ECs have no reaction kinetics to limit charge and discharge rates and thus have exceptionally high power capability for both discharge and charge cycles. During charge or discharge, capacitor voltage is not constant but increases linearly (at constant charge current); thus the SOC of an EC is much more easily determined than that of a battery. Apart from relatively low energy density, the ECs are constrained by low cell voltage (<1V for aqueous and ~3V for organic electrolytes), and sensitivity to impurities [18].

Table 1. Comparison of properties of secondary batteries and electrochemical capacitors

Property	Battery	Electrochemical Capacitor
Storage Mechanism	Chemical	Physical
Power limitation	Electrochemical reaction kinetics, active materials conductivity, mass transport	Electrolyte conductivity in separator and electrode pores
Energy limitation	Electrode mass (bulk)	Electrode surface area
Output voltage	Approximate constant value	Sloping value-state of charge known precisely
Charge rate	Reaction kinetics, mass transport	Very high, same as discharge rate
Cycle life limitations	Mechanical stability, chemical reversibility	Side reactions
Life limitation	Thermodynamic stability	Side reactions

Reproduced with permission from Miller, J. R. and Simon, P. *The Electrochemical Society Interface* 2008, *17,* 31-32. Copyright © 2008, *Electrochemical Society.*

4. Materials for Electrochemical Capacitors

Depending on the material technology used for the manufacture of the electrodes, ECs can be categorized into EDLCs, pseudocapacitors and a new category of ECs called hybrid capacitors. These capacitors can be either symmetric or asymmetric. Symmetric ECs use the same electrode material for both the electrodes. Asymmetric ECs use two different materials for the anode and the cathode. The most recent EC designs are asymmetric and one of the electrodes is capacitor-like while the other is battery-like. The capacitor electrode is identical to those used in symmetric ECs, being made of a high-surface-area nanoporous carbon displaying double layer charge storage, while the battery-like electrode functions exactly like an electrode in a rechargeable battery. The capacity of the battery-like electrode is generally many times greater than the capacity of the porous carbon electrode. Due to the very small depth of discharge of the battery-like electrode during operation, a much higher cycle life is there than is normally possible for rechargeable batteries. Reaction kinetics generally does not limit charge or discharge rates of asymmetric designs because the battery-like electrode is usually oversized and therefore operates at substantially lower current density than normal in a battery. The development and taxonomy of ECs is shown in Figure 4. With respect to electrode materials, there are three main categories: carbon-based, transition metal oxides, and conducting organic polymers (COPs) or synthetic metals. The specific capacitance of selected materials is given in Table 2. In addition, new materials being developed are organic/inorganic composites made from combinations of COPs with carbon-based materials and transition metal oxides. In this section, a brief review of the developments in research on electrodes made from carbon-based materials, transition metal oxides and COPs has been presented.

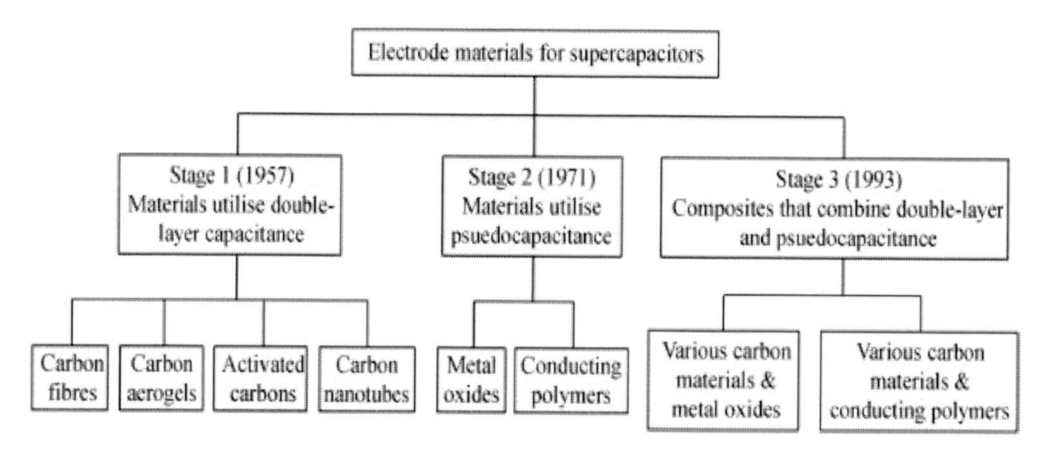

Figure 4. Taxonomy of the supercapacitor materials. Reproduced with permission from Peng, C. et al. *Progress in Natural Science* 2009, *18*, 777-788. Copyright © 2008, Elsevier B.V.

4.1. Carbon-based Materials

The most widely used commercial material for ECs is carbon, in its dispersed and conducting forms. The carbon electrodes in ECs comprise high-surface-area carbon in various forms, obtainable by an activation pre-treatment of existing carbon materials made initially from thermal carbonization of polymers, wood, coal, coconut shells or pitch [14]. Generally used activation pre-treatments are those with hot nitrogen [20], hydrogen [21], carbon dioxide or steam flux [22], that result in pore accessibility and high surface area[19].Carbon materials are endowed with high surface area (1000 to 2500 m^2/g), a tailored pore structure and high pore volume, good electrical conductivity, high purity, good corrosion resistance, relatively low cost, appropriate surface conditioning, and optimum structure; edge rather than basal planes [18]. These, combined with cycle stability, non-toxicity and wide temperature range [1], make them suitable for use as electrode materials for ECs. Though a large specific surface area is desirable for increased specific capacitance, but it is not necessary that increased specific surface area may always result in an increase in specific capacitance. This is because the pore size distribution may comprise of a substantial number of micropores (<2 nm wide) in which it is difficult to adsorb large solvated ions of the electrolyte at high charge-discharge rates. To obtain a high capacitance under such conditions, mesopores (2-50 nm wide) are desirable. However, under slow charge- discharge conditions, a microporous carbon with high specific surface area can give high capacitance. In recent reviews [1,23], several references are available that refer to studies conducted on carbon-based materials as electrode for ECs. Such materials are: activated carbons (ACs), activated carbons fibers (ACFis), exfoliated carbon fibers, templated porous carbon, poly(tetrfluoroethylene) (PTFE), carbide-derived carbons, carbon aerogels and xerogels,graphites, carbon nanotubes (CNTs), carbon nanofibers (CNFs), nano-sized carbons and carbons containing heteroatoms (oxygen, nitrogen, boron, metals like Co, Ni, Mn,Ru, Mo, etc).

Table 2. The specific capacitance of selected electrode materials [6]

Material	Density (g/cm^3)	Electrolyte	F/g	F/ cm^3
Carbon cloth	0.35	KOH Organic	200 100	70 35
Carbon black	1.0	KOH	95	95
Aerogel carbon	0.6	KOH	140	84
Particulate from SiC	0.7	KOH Organic	175 100	126 72
Particulate from TiC	0.5	KOH Organic	220 120	110 60
Anhydrous RuO_2	2.7	H_2SO_4	150	405
Hydrous RuO_2	2.0	H_2SO_4	650	1300
Doped conducting polymers	0.7	Organic	450	315

Where fundamental studies on the performance of single electrode have been done, the conventional three electrode electrolytic cells have been used, comprising of the working, counter and reference electrodes. For the evaluation of the cell performance similar to an actual capacitor, a two-electrode cells system has been the norm. This system is essential in estimating the energy density, power density and the cycle life of the cell. The future research direction of this material is directed toward higher specific surface area, rational pore distribution, smaller internal resistance and surface modification of carbon material [1].

4.2. Transition Metal Oxides

Transition metal oxides are considered to be prospective materials for electrodes in ECs due to their attractive properties. They are endowed with very high specific capacitance coupled with very low resistance, resulting in a high specific power [1]. Among the transition metal oxides, RuO_2 is the most promising material for electrode manufacture due to its high specific capacitance, long cycle life, high conductivity, and excellent electrochemical reversibility, as well as its high rate capability [24]. But its scarcity and high cost due it being a precious metal (Ru), low porosity and toxic nature [25] are the major disadvantages for commercial production of RuO_2. Alternatives to RuO_2 including $Ru_{1-y}Cr_yO_2/TiO_2$, NiO, MnO_2, $MnFe_2O_4$, Fe_3O_4, WC, V_2O_5, $VN_{1.08}O_{0.36}C_{10.1}$ and porous silicon are under investigation [26-29]. The research efforts are directed towards compounds providing high cyclability and capacitance [1]. In recent years, manganese oxide (MnO_2) has emerged as a very promising material for ECs because of its high specific capacitance, ability to charge-discharge rapidly, good cycle stability and low cost [30]. Added to these is that MnO_2 can be used in neutral aqueous electrolytes, unlike $RuO_2.xH_2O$ and NiOOH which can only be used in strong acidic or alkaline electrolytes, thus causing environmental problems [25]. However, while the specific capacitance values of hydrous RuO_2 are in the range of 720 to 1000 F/g, MnO_2 has lower specific capacitance values in the range 150-300 F/g that are far from its theoretical value of ca.1400 F/g [31-36]. Therefore, it is imperative to further improve the capacitive performance of this material. Transition metal oxides are the best electrode materials for redox pseudocapacitors. Most importantly, transition metals possess several oxidation states and are reasonably conductive. Due to cost consideration, work in this field is focused on the development of cheaper metal oxides as EC electrode materials by optimizing the particle size and distribution. Research needs to be directed towards: preparing transition metal oxides which have large specific surface area by different methods that reduce their cost; mixing transition metal oxides, like RuO_2 and IrO_2 with the other cheaper metal compounds in order to expand electrochemical windows and to enhance the capacitance of the electrode materials; finding able, low cost substitutes in order to reduce cost; and to assemble hybrid super-capacitors with proper electrode materials [1].

4.3. Conducting Organic Polymers

Until some years ago, polymers were known to us as insulators. They did not conduct electricity. But this conception was changed through the exciting discovery and development of conducting organic polymers by Alan J. Heeger, Alan G. MacDiarmid and Hideki Shirakawa, for which the trio were awarded the Nobel Prize in Chemistry for the year 2000. During the course of their study, it was discovered that while having a metallic appearance, polyacetylene was not a conductor but on oxidation with chlorine, bromine or iodine vapor, the polyacetylene films thus made, were 10^9 times more conductive than the original polyacetylene. This treatment with halogens was called “doping”. The doped polyacetylene had a conductivity of 10^5 Siemens per meter (S/m) which was higher than any of the previously known polymers. Conjugated double bonds form the backbone of a COP. These bonds between carbon atoms are alternately single and double. Every bond contains a localized sigma (σ) bond which forms a strong chemical bond. Every double bond also has a less strongly localized, weaker pi (л) bond. In addition to conjugation, it is the dopant through which the charge carriers in the form of electrons or “holes” are injected into the polymer that impart conductivity to it. Conducting polymers are pseudocapacitive materials in which the bulk of the material undergoes a fast redox reaction which acts like capacitance. They exhibit superior specific energies than carbon based EDLCs because of the fact that in them, the bulk of the material reacts and not just the surface layer as in the case of an EDLC. A conducting organic polymer EC can achieve a specific energy of 10 Wh/kg [37] as compared to carbon-carbon symmetric supercapacitors that can attain a specific energy of 3-5 Wh/kg [37-39]. On the other hand, the EDLCs exhibit faster kinetics as only the surface is involved in the charging and discharging. In comparison to metal oxides, COPs have low cost and have high charge density. Devices with low equivalent series resistance (ESR), high power and high energy can be developed. The charge density of polyaniline (PAni) is 140 mAh/g, which is slightly lower than that of $LiCoO_2$ [40,41] but much higher than carbon devices that often exhibit less than 15 mAh/g (perhaps ~40 mAh/g for the individual electrode) [42]. COPs can be p-doped with (counter) anions when oxidized and n-doped with (counter) cations when reduced [43]. Conducting organic polymers can be easily manufactured, particularly as thin films, because of their plastic properties. By increasing the doping level, the specific energy of COPs can be increased further. But they begin to degrade due to the changes in their volume (swelling and shrinking) that is caused by the enhanced doping/dedoping (intercalation/de intercalation) of ions. They exhibit a favorable morphology and are light weight (specific gravity of 1.0-1.2). COPs can be operated in nonaqueous or aqueous electrolytes. It was found that p-dopable polymers are more stable against degradation than n-dopable polymers [44]. ECs made exclusively from COPs can have three configurations [45].

Type I (symmetric) using the same p-dopable polymer for both electrodes.

Type II (asymmetric) using two different p-dopable polymers with a different range of electroactivity.

Type III (symmetric) using the same polymer for both electrodes with the p-doped form used as the positive electrode and the n-doped form used as the negative electrode.

Three main methods have been reported in literature for determining the capacitance of conducting organic polymer materials. These include galvanostatic constant-current charge-discharge (GCD), cyclic voltammetry (CV) and electrochemical impedance spectroscopy (EIS). They are illustrated in Figure 5.

Organic materials for ECs include: polythiophene and its derivatives, polypyrrole, polyaniline, poly-1,5-diaminoanthraquinone, polyquinoxaline, polyindole, cyclic indole trimer, polyacene, tetramethylpyridine derivatives (stabilized radical), and metal complex polymer [46]. Though a number of studies have been carried out on COPs as electrodes for ECs, but the main COPs studied are polyaniline (PAni), polypyrrole (PPy) and derivatives of polythiophene (PTh). Among the conducting polymers, polyaniline (PAni) has cornered much attention as an electrode material for ECs, due to its many desirable properties. It exhibits high electroactivity, a high doping level, a high specific capacitance in an acidic medium (400-500 F/g) [37], good environmental stability, controllable electrical conductivity, and can be easily processed [45].

PAni shows the most variance in specific capacitance values of all COPs. It was found that depending upon the method of synthesis; the specific capacitance is higher for PAni. Some other notable studies on PAni have also been conducted on various aspects of this COP for use in ECs. They have been cited by Snook et al. [43]. Polypyrrole (PPy) displays a great deal of flexibility in electrochemical processing than most conducting polymers [46].

Substantial research has been carried out to investigate its role as an electrode material for ECs.

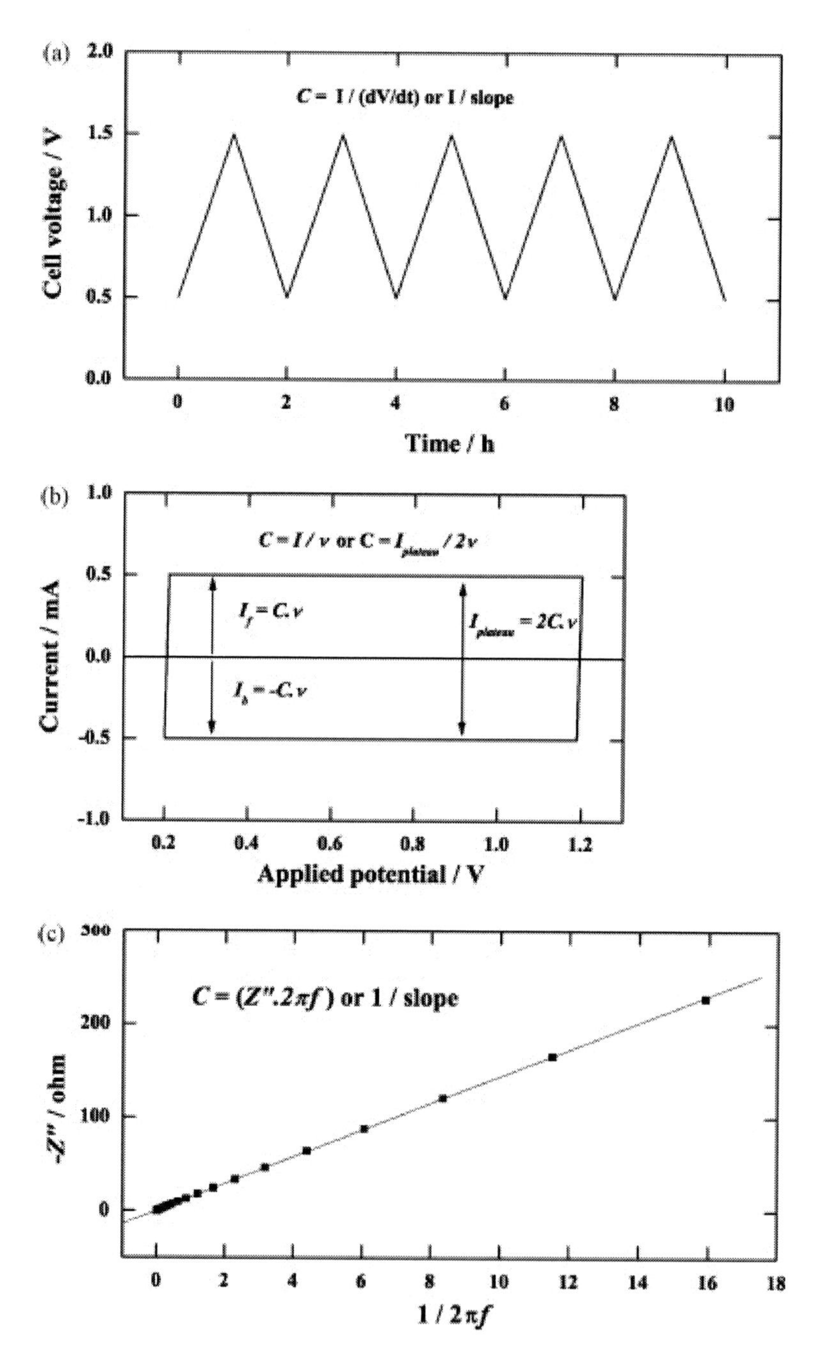

Figure 5. Idealized schematic representation showing measurement of capacitance of supercapacitor devices and electrodes from (a) constant-current charge–discharge curve; (b) cyclic voltammogram (rectangular component of response, ignoring redox peaks); (c) low-frequency component of imaginary impedance data from electrochemical impedance spectroscopy (EIS); where C is capacitance (F), I is current (A), v is scan rate (V s^{-1}), Z'' is imaginary impedance (ohm), f is frequency (Hz) [43]. Reproduced with permission from Snook, G. A. et al. *J. Power Sources* 2011, *196,* 1-12. Copyright © 2010, Elsevier B.V.

Since PPy cannot be n-doped, it is used only as a cathode. It has a large specific capacitance of 480 F/g [47]. Suematsu et al. [48], have shown that PPy when doped with multi-charged anions such as SO_4^{2-}, physical cross linking takes place, resulting in higher diffusivity and higher capacitance. Various derivatives of polythiophene (PTh) have been found to be suitable for use as electrode material for ECs; the most desirable properties being of poly (3, 4-ethylenedioxythiophene) (PEDOT). This polymer is electron rich and therefore exhibits a low oxidation potential [49]. It has a wide potential window over which the capacitance is high (1.2-1.5V). It has a low band-gap of 1-3 eV, a high conductivity in p-doped state (300-500 S/cm) [50], good thermal and chemical stability and high charge mobility that result in fast electrochemical kinetics [50-53]. It possesses good film forming properties and has a long cycle life [49]. Due to its large molecular weight and a doping level of 0.33 [50], it has a low specific capacitance of around 90 F/g. PEDOT has found application in both symmetric ECs (Type-1 device) and a hybrid device with a PEDOT cathode and an activated carbon anode [51].

5. Organic/Inorganic Composites as Electrode Materials for Electrochemical Capacitors

The properties of conducting polymers can be greatly enhanced by forming composites between the conducting polymer and other materials such as carbon (including carbon nanotubes) [19], inorganic oxides and hydroxides [54,55], and other metal compounds [56-58]. By anchoring polyoxometalates (POMs) within the network of COPs such as PAni or PPy, leads to the fabrication of hybrid materials in which the hybrid materials maintain their form and activity while taking advantage of the conducting properties and polymeric nature of the hybrid [59-62]. These electrodes can allow the device to be made symmetric (Type I or Type III) with the same positive and negative electrodes. The composite materials have improved conductivity (particularly at the more negative/reducing potentials) and better cyclability, mechanical stability, specific capacitance and processability [50].

To further improve the properties of electrodes for ECs, work is also being done on triple hybrid materials [63]. In Figure 6 a schematic of a triple hybrid is given. The methods reported for synthesizing COP based composites include chemical polymerization, electrochemical deposition of COPs on substrate electrodes and electro-co-deposition of COP/substrates. Owing to the tremendous amount of work done on organic/inorganic composites, it is beyond the scope of this chapter to review it in totality. Thus, an attempt has been made in this section to present some pertinent research done on the synthesis and analysis of organic/inorganic composites for use as electrodes in ECs, so as to highlight the great prospects of work in this field.

In order to take advantage of their desirable properties, a very attractive option is to have composites of COPs and carbon based materials, mainly carbon nanotubes (CNTs). Malinauskas [64] has carried out a review on the chemical polymerisation of conducting polymers and their derivatives on different materials, i.e. polymer, glass, fiber, textile, silica, and metal oxides. It was observed that almost all types of material can be coated with conducting polymers and the new composite materials yielded can be used in various fields. Wang et al.[65], have reviewed some of the current research on chemical synthesis of conducting polymers in the presence of fullerenes and CNTs, and the possible applications of these composites. In this section, the prime focus is on the review of recent literature about organic/inorganic composites for use as electrodes in ECs.

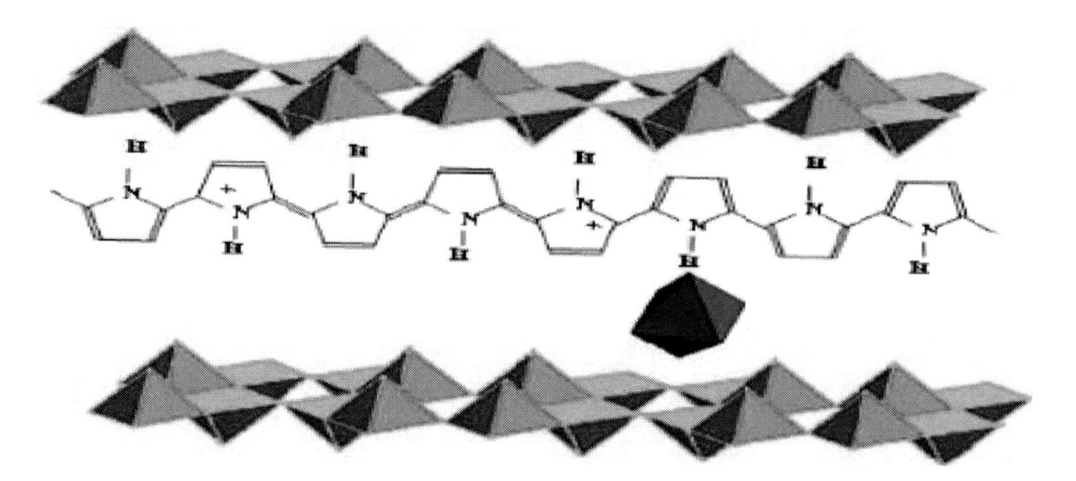

Figure 6. Schematic diagram of the desired triple hybrid formed by a hexacyanoferrate-doped conducting polymer (polypyrrole in this case) intercalated into layered V_2O_5 [63]. Reproduced with permission from Cuentas-Gallegos, A. K. et al. *J. Power Sources* 2006, *161,* 580-586. Copyright © 2006, Elsevier B.V.

5.1. Organic/Inorganic Composites by Chemical Polymerization

Chemical synthesis of COPs is achieved by oxidizing the corresponding monomers using an oxidizing agent. It is a cheap and efficient route for the bulk-production in industry. COPs produced by this method have high conductivity, negligible solubility in aqueous solutions and good stability. Chemical polymerization of aniline can be done in aqueous solution by a variety of oxidants, such as $FeCl_3$, $KBrO_3$, $KClO_3$, $K_2Cr_2O_7$, KIO_3, $KMnO_4$, and $(NH_4)S_2O_8$ [66]. Pyrrole can be oxidized by various transition metal ions from metallic salts, such as $CuCl_2$, $FeCl_3$, $K_3Fe(CN)_6$, $Fe(NO_3)_3$, and $Fe_2(SO_4)_3$, [67]. Thiophene and its derivatives require an organic solution to produce COPs by chemical oxidation. The good solubility of monomers of aniline and pyrrole in aqueous solutions results in cost effective and clean production of their polymers by chemical polymerization as compared with the synthesis of thiophene and its derivatives. The major disadvantage of chemical polymerization is that the product lacks the level of homogeneity and integrity as that produced by electrochemical polymerization. Investigations have been carried out[68-70], to study the effect of various parameters such as pH, relative concentration of reactants, polymerization temperature and time, different oxidizing agents and different protonic acids on chemical polymerization has been studied [68-70]. Research has shown that the application of ultrasound to the process of polymerization has resulted in rapid mixing [71], formation of free radicals and mechanical shocks [72,73]. Lota et al. [50] found from a study of the PEDOT/CNTs composites that, due to the presence of nanotubes, a synergistic effect of the two components was observed, resulting in an efficient energy extraction from PEDOT even though the CNTs and chemically polymerized PEDOT were only mixed mechanically by stirring. Here, the quantity of PEDOT in the composite was only 15 wt% and it showed a good capacitance value of 95 F/g in a three-electrode system, compared with the capacitance of pure CNTs, which was only 10–15 F/g, and the capacitance of pure PEDOT, which ranges from 80 to 100 F/g. Polymer coated carbon materials show larger charge/discharge rates and higher specific capacitance in application. Conducting polymers can facilitate the ions from the electrolyte in quickly reaching the internal surfaces of the carbon electrode; thereby increasing the accessibility of the carbon materials with high surface area during high rates of charge/discharge [74] resulting in high values of specific capacitance. Ryu et al. [75] observed that synthesized activated carbon/polyaniline composite gave a specific capacitance of 273 F/g at a scan rate of 50 mV/s. Kim et al. [76], found that carbon nano-fiber /polypyrrole composite showed a capacitance of 545 F/g at a scan rate of 200 mV/s. However, it was observed that increasing the polymer content in the polymer/carbon composites not only increased the capacitance, but it also increases the charging time constant. PAni/multi-walled carbon nanotubes (MWCNTs) composites were prepared [77,78] using in situ chemical polymerization and microwave-assisted polymerization [79]. For pure PAni nanofibers, the initial specific capacitance was high but rapidly decreased on continuous cycling. The use of MWCNTs with PAni resulted in a high initial value of specific capacitance as 606 F/g with good retention on cycling. In situ polymerization of aniline monomer was done in the presence of graphene oxide sheets to prepare a novel organic/inorganic composite [80]. The composite with a mass ratio of aniline and graphite oxide as 100:1exhibited a high conductivity of 10 S/cm at 22 °C. The GCD analysis of the said composite in the potential range of 0 to 0.45 V at 200 mA/g resulted in a high specific capacitance of 531 F/g as compared to 216 F/g for pure PAni. The morphology of the PAni nanofibers was fibrillar which was induced and greatly influenced by doping with graphene oxide sheets.

A graphene nanosheet (GNS)/PAni composite was prepared by using in situ polymerization [81]. Scanning electron microscopy (SEM), transmission electron microscopy (TEM), X-ray diffraction (XRD) and Raman spectroscopy were used to study the microstructure and morphology of the composite. The electrochemical properties were characterized by CV and GCD techniques. A high specific capacitance of 1046 F/g was obtained at a scan rate of 1 mV/s compared to 115 F/g for pure PAni. Energy density of the composite reached 39 Wh/kg at a power density of 70 kW/kg. A triple hybrid organic/inorganic composite of graphene nanosheet (GNS), carbon nanotubes (CNTs), and PAni was synthesized through in situ polymerization [82]. The GNS/CNTs/PAni composite exhibited a specific capacitance of 1035 F/g at 1 mV/s in 6 M KOH solution, which is a little lower than 1046 F/g reported for GNS/PAni composite. But, it is much higher than 115 F/g reported for pure PAni and 780 F/g reported for CNTs/PAni composite. The addition of 1 wt% CNTs into the triple hybrid seem to be responsible for providing a highly conducting path as well as mechanical strength to it. GNS acted as the support material for the deposition of PAni particles. The composite exhibited very good cyclic performance in comparison to the binary hybrids of PAni with GNS and CNTs. After 1000 charge/discharge cycles, the attenuation in the initial specific capacitance was only 6% for the GNS/CNTs/PAni composite whereas the corresponding values for the GNS/PAni and CNTs/PAni composites were 52% and 67%, respectively.

In an exhaustive review [83], an advance in the modification of graphene and the fabrication of graphene-based polymer nanocomposites has been discussed. The structure, preparation and properties of polymer/graphene nanocomposites have been described from literature. Most of the properties of polymer/graphene nanocomposites were superior to the base polymer matrix as well as other carbon fillers (CNTs, CNFs, and graphite) based composites. In situ chemical oxidation polymerization was used to prepare calcium carbide derived carbon (CCDC)/PAni composite [84]. Fourier transform infrared spectroscopy (FTIR), SEM, TEM and nitrogen sorption isotherms were used for the study of the structure and morphology of the composite. The electrochemical properties were assessed by CV, GCD and cycle life measurements. The specific capacitance of the CCDC/PAni composite was as high as 713.4 F/g measured by CV at 1 mV/s compared with 154.0 F/g for the CCDC electrode. The capacitance retention by the composite after 1000 cycles was 80.1%. Chemical polymerization of pyrrole by chemical dipping method on the surface of a porous graphite fiber matrix was carried out to prepare composite electrodes for supercapacitors [85]. The electrodes showed a specific capacitance of 400 F/g and a coulombic efficiency of 96.99 %.

In another study [86], organic/inorganic composites of PPy/carbon aerogel (CA) with different PPy contents were synthesized by chemical oxidation polymerization through ultrasonic irradiation. SEM and TEM were used to examine the morphology of the composites. It was observed that the specific capacitance values of all PPy/CA composites were higher than those of CA. The best value of specific capacitance was 433 F/g, obtained from 35 wt% of PPy/CA composite using 6 mol/l KOH electrolyte, while the specific capacitance for CA was reported to be 174 F/g. The 35 wt% PPy/CA composite exhibited small electrochemical resistance and good cyclic stability. After the initial deterioration due to the instability of PPy till 500 cycles, the specific capacitance stabilized at a fixed value. The composite, while enhancing the capacitance of the carbon material, improved the cycle life of PPy. An interesting novel microwave hydrothermal technique was used to grow thorn-like methyl orange (Mo)-Iron(III) chloride complex on CNTs [87]. This complex was successfully used to obtain a PPy/CNTs composite with controllable morphology. The morphological and structural studies of the nanocomposites were executed using TEM, energy dispersive spectroscopy (EDS), infrared spectroscopy and XRD. The composite had a specific capacitance of 304 F/g. A series of chemical characterizations revealed that the growth of granular-like PPy was closely related to the shape and size of the oxidative complex of Mo-$FeCl_3$ on the CNTs. It was predicted that if the thorns of the Mo-$FeCl_3$ complex could be developed into long fibers, highly oriented PPy nanofibers or nanotubes covered on the CNTs would be obtained, leading to a significant improvement in the electrochemical performance. In a very recent investigation [88], nano organic/inorganic composites were synthesized as electrodes for electrochemical capacitors. Chemically modified ordered mesoporous carbon was prepared by wet oxidative method in HNO_3 with different concentrations leading to acidic surface groups and good surface chemical properties. Nano-thin PPy layers were coated on the modified ordered mesoporous carbon by chemical oxidation polymerization process to synthesize modified ordered mesoporous carbon/PPy nanocomposites. Field emission scanning electron microscopy was used to carry out the structural and morphological characterizations of the modified mesoporous carbon/PPy composites. Pseudocapacitive behavior of deposited PPy on modified ordered mesoporous carbon was studied by CV, GCD, and EIS. It was observed that the composite with 82 wt% PPy attained a maximum specific capacitance of 427 F/g or 478 F/g after correcting for the weight % of the PPy phase at a current density of 5 mA/cm^2. This value was much higher than that reported for pure PPy electrodes. The excellent characteristics of modified ordered mesoporous carbon are responsible for the improvement in the performance of the composite. The composite when compared with pure PPy electrodes was reported to have more active sites for the Faradic reaction, larger specific capacitance, and better cyclic stability. Organic/inorganic composites of CNFs/poly(3,4-ethylenedioxythiophene) (PEDOT)/poly(styrene sulfonic acid) (PSS) were fabricated [89] for symmetric and asymmetric electrochemical capacitors. A small amount of PEDOT-PSS acted as a dispersing agent for CNFs and as an enhancer for electrical conductivity.

Zhou et al. [90], studied MnO_2/PAni composite thin films and the composite formed after chemically depositing this film on a porous carbon (PC) electrode. The structure and electrochemical properties of the composite films were investigated by X-ray photoelectron spectroscopy (XPS), XRD and CV techniques. At a scanning rate of 50 mV/s, the PC/MnO_2/PAni composite electrode produced a specific capacitance of 250 F/g while the corresponding value was 500 F/g for the MnO_2/PAni composite film in a 0.1 M Na_2SO_4 solution. However, the major drawback was that only 60% specific capacitance could be maintained after 5000 cycles, which was due to the slow dissolution of MnO_2 in the form of Mn^{2+} ions in the electrolyte. In another study [91], PAni and polypyrrole (PPy) were used with MoO_3 to prepare the respective composites *via* oxidative polymerization. The oxidant used was ammonium vanadate with H_2SO_4 in the presence of aqueous suspension of MoO_3. Complete analyses of the MoO_3/ PAni and MoO_3/PPy were done by FTIR, thermo-gravimetric analyses (TGA), and differential thermal analysis (DTA). The conductivity of MoO_3/PAni and MoO_3/PPy composites were of the order of 10^{-2} and 10^{-3} S/cm respectively. Song et al. [92], first prepared PAni chemically and used it in 1m H_2SO_4 aqueous electrolyte and to improve its cyclability composite was prepared with Nafion. This composite exhibited a high degree of cyclic stability and higher specific capacitance compared with pure PAni electrode. When subjected to 104 cycles in the voltage range of 0.2 to 0.6 V at 500 mV/s, the capacitance retention of PAni/Nafion composite was 84% while it was only 62% for pure PAni electrode. The composite was found to be a compatible matrix for the dispersion of hydrous RuO_2.

A new inorganic substrate in the form of cheap and waste red mud (RM) was used [93] to synthesize organic/inorganic composites with different amounts of PAni *via* chemical oxidation polymerization of aniline in an acidic medium using ammonium peroxydisulfate as the oxidant. The conductivities were in the range of 0.42 to 5.2 S/cm. Characterization was done by IR-VIS spectroscopy, SEM and XRD. Thermal stability was tested by TGA analysis. The thermal conductivity was measured in situ by ageing at 125 °C. The RM/PAni composites were made by coating 13, 24 and 31 wt% of aniline on red mud in 1 M HCl solution. It was concluded that RM/PAni composites made using cheap waste substrate can be used further for the synthesis of conductive composites with polymeric matrices. Yuan et al. [94], synthesized MnO_2/MWCNTS composites by chemically depositing MnO_2 onto the surface of MWCNTs wrapped with poly(sodium-p-styrene sulfonate). Further, PAni was coated on to the MnO_2/MWCNTs composite to form a triple hybrid. A mixed Na_2SO_4-H_2SO_4 was used to evaluate the electrochemical performance of the composite by CV and chronopotentiometry. The PAni served the dual purpose of acting as physical barrier in restraining the MnO_2/MWCNTs composite from reductive dissolution as well as contributing in energy storage in the mixed acidic electrolyte. A specific capacitance of 384 F/g delivered at 2.5 mA/cm^2, and a specific retention of 79.7% over 1000 continuous charge/discharge cycles were reported for the hybrid, which were much higher than those obtained for the

MnO_2/MWCNTS composites in 0.5 M Na_2SO_4-0.5M H_2SO_4 acidic electrolytes. The importance of the addition of PAni to the MnO_2/MWCNTs composite was clearly emphasized in this study. A one-step in situ oxidative polymerization of aniline was done [95] in the presence of nano particles of TiO_2 to form fibriform PAni/TiO_2 nanocomposite having a well ordered structure and containing 80% conducting PAni by mass with a conductivity of 2.45 S/cm at 25°C. The specific capacitance of the composite was 33 F/g at a constant current density of 1.5 A/g. It can be subjected to 1000 charge/discharge cycles in the voltage range of 0.05 to 0.55V. A PAni/SnO_2 composite was fabricated [96] using nanostructured SnO_2 prepared by sol-gel method and aniline polymerized in the suspension of nanocrystalline SnO_2. XRD and field emission scanning electron microscopy (FESEM) were used to characterize the structure and morphology of SnO_2 and PAni/SnO_2 composite. The electrochemical evaluation was done by CV, GCD and EIS. A high specific capacitance of 305.3 F/g with a specific energy of 42.4 Wh/kg and coulombic efficiency of 96% were reported. The energy density of the composite was three times as compared with pure SnO_2. Excellent cyclic stability was obtained by the composite. Attenuation in specific capacitance was only 4.5% after 500 cycles. Chemical oxidative polymerization was done to synthesize a PAni/poly neutral red (PNR)/TiO_2 composite [97]. XRD and SEM studies showed a uniform dispersion of TiO_2 particles in the net-like structure of PAni/PNR matrix without aggregation. The PAni/TiO_2 structure also improved the conductivity for electron transport during the electrochemical reaction. CV, EIS, and GCD measurements in 1 M H_2SO_4 solution were made. The specific capacitance of the PAni/PNR/TiO_2 composite, obtained by GCD, was 335 F/g at a constant current of 5 mA compared with 260 F/g for the PAni/PNR matrix. Cyclic performance, in the form of attenuation of 20% in specific capacitance after 1000 charge/discharge cycles, was also observed. The study showed that PAni/PNR/TiO_2 composite exhibited good electrochemical performance in terms of high specific capacitance and cyclic stability.

In another study [98], chemical polymerization of PPy was carried out in aqueous Fe_2O_3 colloidal solution using $FeCl_3$ as oxidant and tosylate anions as doping agent to synthesize PPy/Fe_2O_3 nanocomposite. The composite was analyzed by XRD, FTIR spectroscopy and TGA. CV and GCD techniques were used to assess the charge storage properties of the composite. Room temperature ionic liquids (RTILs) namely, 1-ethyl-3-methylimidazolium bis((trifluoromethyl) sulfonyl) imide (EMITFSI) and N-butyl-N-methyl pyrrolidinium bis (trifluoromethanesulfonyl) imide (PYR14TFSI) were used as electrolytes, giving 72 mAh/g and 62 mAh/g at 1 mA/cm^2 discharge current, respectively. These values were more than twice the values obtained with pure PPy. The specific capacitance values for the nanocomposite electrodes were 210 F/g and 190 F/g in EMITFSI and PYR14TFSI respectively, and their cyclability showed only a 3-5% capacitance attenuation after one thousand cycles for both ionic liquid electrolytes. The improved performance of this composite via-a-vis the pure components was due mainly to the modification of the morphological properties of the polymer in terms of increased specific surface area and a suitable pore size distribution. In situ chemical oxidative polymerization method was used [99] to synthesize PPy/halloysite nanotubes (HNTs) nanocomposites with coaxial morphology based on self-assembled monolayer amine functionalized HNTs. The characterization was done by TEM, XRD, TGA, electrical conductivity measurements at different temperatures, CV and GCD measurements. The highest conductivity of 40 S/cm was exhibited at room temperature and a weak dependence of conductivity on temperature was observed from 298K to 423K. The maximum capacitance was 522 F/g after correcting for the weight of the PPy phase at a current density of 5 mA/cm^2 in a 0.5M Na_2SO_4 electrolyte. Chin et al. [100], synthesized tetrapropyl ammonium-manganese oxide/PPy (TPA/MnO_2/PPy) nanocomposites by in situ polymerization and sol-gel process. Optimization of nanocomposite films showed higher charge capacities which were due to enhanced material utilization as a result of optimized microstructural parameters namely, their porosity, pore size distribution and specific area. The study revealed that the nanocomposite was a prospective material for thin film electrochemical capacitors. In another study [101], poly 3,4-ethylenedioxdythiophene (PEDOT) based $NiFe_2O_4$ conducting nanocomposites were prepared for evaluation as electrode material for electrochemical capacitors. Sol-gel technique was used to synthesize nanocrystalline nickel ferrites of size 5-20 nm. The bulk PEDOT was prepared by aqueous micellar dispersion polymerization and the PEDOT nanotubes/$NiFe_2O_4$ composite was prepared by reverse microemulsion polymerization in n-hexane medium. XRD, SEM, TEM and IR spectroscopy were used for studying the structural morphology. CV and GCD were employed for evaluating the electrochemical properties of the electrode materials in acetonitrile solvent containing 1 M $LiClO_4$ electrolyte. The specific capacitance of the nanocomposite material was found to be 250 F/g, which was higher in comparison to its constituents namely $NiFe_2O_4$ (127 F/g) and PEDOT (156 F/g). In the case of PEDOT/$NiFe_2O_4$ composite, the formation of mesoporous nanotubes over the surface of nickel ferrite nanoparticles takes place, and it plays a vital role in the accumulation of highest internal charge compared to total surface area, leading to highest specific capacitance. $NiFe_2O_4$, though having a low specific capacitance, imparts a synergistic effect to the PEDOT/$NiFe_2O_4$ nanocomposite. EIS data revealed that nano nickel ferrite in the composite not only helps to reduce intrinsic resistance through the synthesis of more number of mesoporous structures but also enhances the kinetics of electron transfer *via* redox process leading to increase in the pseudocapacitance as compared to electrical double capacitance in the PEDOT/$NiFe_2O_4$ nanocomposite. Good cyclic stability of the PEDOT/$NiFe_2O_4$ nanocomposite was observed. There was only a loss of only 16% in specific capacitance over 500 charge/discharge cycles as compared to the initial value. The study concluded that the PEDOT/$NiFe_2O_4$ nanocomposite was a bright contender for use as electrode material in the development of safe and cost effective ECs.

In situ chemical polymerization was carried out to prepare variable mass composite powders of PPy doped with 4 mg $RuCl_3.xH_2O$ [102]. CV, GCD and EIS in 1 M $NaNO_3$ were used to study the electrochemical properties of the composite. The morphology and crystalline structure as examined by TEM showed a small particle size, uniformly dispersed and well crystallized microstructure. The composite had positive attributes of a higher density of active centers, lower charge-transfer resistance, larger specific capacitance, and better rate performance than pure PPy under similar conditions. As such, it seemed a prospective material for use in ECs.

5.2. Organic/Inorganic Composites by Electrochemical Polymerization

Electrochemical deposition of conducting organic polymers on suitable substrate preforms is another method to obtain organic/inorganic composites for ECs. Downs et al. [103] may be probably credited for making the first attempt to electrochemically deposit conducting polymers on carbon nanotubes, where multi-walled nanotube (MWNT) electrodes were used for the deposition of PANI films; a higher current density and more effective polymerisation resulted as compared with those deposited on Pt electrode. Thereafter, a lot of work has been reported on the preparation of COP/CNT composites via electrochemical polymerisation and the evaluation of the obtained composites. The substrates for electro-deposition of COPs have been in various forms namely as, well aligned CNT arrays [104], pellets pressed with binder [105-107], and CNT buckypaper [108,109]. The methods adopted for electrochemical deposition include potentiodynamic polymerisation, galvanostatic polymerisation or potentiostatic polymerisation. Studies by Gupta and Miura [106,107], were conducted to investigate the effect of microstructure on specific capacitance of PAni/SWCNTs composites. In situ potentiostatic deposition of PAni on SWCNTs at a potential of 0.75 V versus standard calomel electrode (SCE) was carried out to synthesize PAni/SWCNTs. The specific capacitance of the composite was strongly influenced by its microstructure, which was correlated to the wt% of the PAni deposited onto the SWCNTs. 73 wt% PAni, deposited onto SWCNTs at 10 mA/cm^2 gave the highest specific capacitance of 463 F/g, specific power of 228 Wh/kg and specific energy of 2250 W/kg. Excellent cyclic stability was observed. Only 5% attenuation in original specific capacitance was there during the first 500 cycles and just 1% during the next 1000 cycles. The composites prepared by electrochemical deposition also exhibited much higher specific capacitance, specific energy and power than CNT or COPs alone [106]. Nuria et al. [108] observed that electrochemically grown conducting polymers decreased the contact resistance between CNTs thereby producing composites with improved electrical conductivity. The fact that COPs were deposited on CNT preforms rather than individual tubes, made it impossible to obtain homogeneous composites of COPs and CNTs. The electro- deposition only takes place on CNTs that have a good contact with the monomer solution while the CNTs inside have little or no polymer coatings, resulting in a heterogeneous structure. Therefore, as the polymer grows thicker, the material will behave like pure PAni and the effects of the CNTs will be diminished. However, it was pointed by Chen et al. [104] that electro-deposition of COPs on aligned CNT arrays gave a more homogeneous structure and a higher utilization of CNT surface for deposition. Efforts have been made to deposit COPs on porous carbon electrochemically [110,111], with encouraging results. Electrochemical polymerization of PAni was carried out on the surface of activated porous carbon (APC) [112]. SEM and X-ray photoelectron spectroscopy (XPS) were used to characterize the surface morphology and chemical composition of the electrodes. The electrochemical properties were studied by CV, EIS and GCD within a potential range of 0-0.6 V. The 60% increase in the specific capacitance of the composite electrode vis a vis the bare activated porous carbon was attributed to the contribution of faradic pseudocapacitance by the conducting PAni. Hu et al. [113] deposited PAni on activated carbon fabric (ACFa) by means of a potentiostatic method. A sandwich-type supercapacitor consisting of two such similar electrodes was evaluated. A high reversibility and good stability was reported. The ACFa/PAni composite with a polymerization charge density of 9 C/cm^2 showed encouraging results for applicability to electrochemical capacitors between 0.2 to 0.6 V in 1M $NaNO_3$. The specific capacitance was 300 F/g at this polymerization density. Good stability upto 500 cycles was reported. In another study [114], the electrochemical polymerization was used to fabricate composite electrodes of PAni and activated carbon powders (ACP). First, the ACP electrodes were synthesized by coating a slurry containing 90% ACP having specific area of 3000 m^2/g along with a solvent N-methyl pyrrolidone (NMP) onto platinum (Pt) followed by evaporation of the solvent in a vacuum oven, and pressed by a roll press. PAni was then electrochemically loaded onto the Pt/ACP electrode to produce the Pt/ACP/PAni composite. Another electrode of Pt/PAni was also fabricated. The aqueous 1 M H_2SO_4 electrolyte was prepared in three states viz. liquid, gel, and solid-like gel using 3.85, 10.8 and 13.7% SiO_2, respectively. The performance of all the three electrodes was tested in the three forms of the electrolyte. The Pt/ACP/PAni electrode gave the highest specific capacitance of 270 F/g in I M H_2SO_4 solid-like gel, while the Pt/ACP and Pt/Ani electrodes recorded 144 F/g and 209 F/g, respectively. It was inferred that the electrochemical performance of ACP can be improved by compositing it with PAni and employing a solid-like gel electrolyte. Three composites were prepared with PAni using platinum (Pt), stainless steel (SS) and porous carbon substrates through electrochemical deposition [111]. The variables of the experiment included different concentrations of aniline monomer and H_2SO_4 electrolyte. The results were compared to identify the best composite and the optimum conditions for use as electrode in ECs. Low concentrations of aniline and H_2SO_4 produced PAni at slow rates but the network possessed morphology of nanofibrils, which was desirable. Specific capacitance of this composite was reported as 1600 F/g and charge/discharge current densities as high as 45 mA/cm^2 were attained. The composite exhibited fair cyclic stability over a long cycle life.

Zhang et al. [115], reported a study in which electrochemical deposition led to a tube-covering-tube nanostructured PAni/vertically aligned carbon nanotube array (CNTA) composite. The composite had a high specific capacitance of 1030 F/g, superior rate capability of 95% capacity retention at 118 A/g and high stability, giving on a 5.5% capacity loss after 5000 charge/discharge cycles. The work highlights the fact that CNTA is an ideal substrate for depositing conducting organic polymers to synthesize advanced functional materials for many applications. The influence of conducting PAni coating on the properties of mesoporous pitch based carbon foam (MPBCF) has been studied [116]. The PAni/MPBCF composite was obtained by electropolymerization of aniline on MPBCF at various scan rates; the better capacitive behavior resulting at 100 and 150 mV/s. Under these fast rates, thinner films of PAni coatings were combined with more porous structure of carbon foam. EIS data of PAni revealed that under high scan rates, the PAni film developed, showed ~90° phase angle; thereby exhibiting ideal capacitive behavior. The conductivity of the composite was also higher than that of carbon foam. The composites of PAni and activated carbon (AC) were studied for their possible use as electrodes for electrochemical capacitors [117]. Aniline was polymerized over AC using CV. The morphological properties of AC and PAni/AC composites were obtained by SEM. EIS and GCD measurements were made to assess the electrochemical composites and their constituents. The PAni/Ac composite had a specific capacitance of 587 F/g as against 140 F/g for pristine AC, owing to the faradic reaction of PAni with the electrolyte. The PAni thin film was uniformly deposited on the surface of the AC, forming an interlinked porous network. The PAni/Ac electrodes showed better cyclic stability in comparison to the PAni electrodes. It was shown that adding PAni to Ac can greatly improve the performance of the electrodes. In another study [118], the effect of microstructure on the capacitive behavior of PAni/carbon nano tube array (CNTA) composites was studied. CV was used to electrodeposit PAni onto CNTA electrodes in the range 100-5000 cycles. It was observed that the morphology of the composite varied with the CV cycles of electrodeposition. The optimum condition was obtained for the composite synthesized by 100 cycles, corresponding to best rate performance, highest specific capacitance and longest cycle life in comparison with carbon fiber cloth, PAni deposited on stainless steel substrate and CNTA electrode. The best performance of PAni/CNTA composite was attributed to high ionic conductivity, high electronic conductivity, high utilization of electrode materials and best accommodation of the volume changes during charge/discharge. It was deduced from this study that the microstructure and capacitive performance of the PAni/CNTA composites can be simply controlled by varying the electrodeposition CV cycles. PAni/MWCNTs composites were synthesized using in situ electrochemical polymerization of an aniline solution with different quantities of MWCNTs [119]. The electrochemical behavior of these composites was tested by CV, GCD and EIS. The composite films showed much better performance than pure PAni film electrode. The highest specific capacitance obtained was 500 F/g for the composite film containing 0.8 wt% MWCNTs. The PAni/MWCNTs film had a highly porous structure with a higher doping density and lower defect density. More active sites are available for faradic reaction. These composites exhibited lower resistance, better cyclic stability and better charge-transfer than pure PAni owing to the presence of MWCNTs. In an investigation [120], PAni was electropolymerized on undoped 100 nm diamond powders (DP) in H_2SO_4 solution to form a composite. A potential range of 0.2 to 0.9 V was used for CV to grow the PAni on the DP electrode. Field emission scanning microscopy (FESEM) data showed that the diamond particles were coated by PAni films with globular and fibroid surface morphology, forming rough and porous surfaces. CV and EIS analyses reveal that the introduction of PAni reduces the resistance of the composite and improves the capacitance of the diamond powders.

A study was carried out on a PAni/bimodal porous carbon (BMPC) composite for assessing its use as an electrode for electrochemical capacitors. Three dimensionally ordered macroporous carbons having walls of mesosized spherical pores were synthesized by a colloidal crystal method [121]. This bimodal porous carbon was used to prepare the PAni/BMPC composite by electrochemical polymerization of aniline in the macropores of the BMPC. It was found that deposition of PAni decreased both the specific surface area as well as the porosity of the composite. Ethylene carbonate and diethyl carbonate containing 1 mol/dm $LiPF_6$ mixed solution was used as the electrolyte for assessing the electrochemical properties of the composite. The discharge capacity of this composite electrode was reported as 111 mAh/g in the potential range of 2.0 to 4.0 V vs Li/Li^+, which includes contributions from double layer capacity (30 mA/g) and the redox capacity (81 mA/g). A cost effective and simple method for the fabrication of flexible PAni nanowires (PANINWs)/carbon cloth (CC) composite *via* electrochemical deposition has been reported [122]. The composite exhibits a high gravimetric capacitance of 1079 F/g at a specific energy of 100.9 Wh/kg and a specific power of 12.1 kW/kg coupled with an exceptionally high area –normalized capacitance of 1.8 F/cm^2. EIS analysis showed that the diffusion length of protons within PANINWs is 60 nm, which indicated that the electrochemical performance of the electrode was not limited by the thickness of the PANINWs. Also, the electrochemical behavior of the composite remains intact, even if it is bent under high cuvature. The attractive properties of the composite make it a potential material for large scale application as electrode for flexible and electrochemically stable ECs. Electrochemical polymerization was carried out to synthesize composite films of PPy with SWCNTs and functionalized SWCNTs (SWCNTsf) for their possible use in ECs [123]. The functionalization of SWNTs was done by suspending in concentrated H_2SO_4/HNO_3 solution and sonicating in a water bath. The morphology of the composites was studied by FESEM. The electrochemical properties of the composite films were assessed by CV and EIS in 1 M KCl aqueous solutions. The performance of the PPy/SWCNTsf composites was the best, followed by PPY/SWCNTs and pure PPy films. The PPy/SWCNTsf had a specific capacitance of 200 F/g at a scanning rate of 200 mV/s. The charge transfer resistance was very small. The high conductivity of the CNTs and their mesoporous structure contributed to their very low resistance and almost ideal capacitance behavior even in deep discharge state. It was concluded that the PPy/SWCNTsf composites had a great potential as electrode materials for supercapacitors followed by PPy/SWCNTs composite films. Wang et al. [124], doped PPy films with p-toluene sulphate (TOS^-), ClO_4^- and Cl^- electrochemically. CV, GCD and EIS were used to study the electrochemical properties of the electrodes. The morphological and structural analyses were done by SEM and XRD. PPy/Cl and PPy/TOS possessed a highly porous structure. They exhibited a rectangular shape of voltammetry curve even at a scan rate of 50 mV/s, a linear variation of voltage with respect to time in the charge/discharge process, and almost ideal capacitance in low frequency, even in the deeply charged/discharged state in 1 mol/l /KCl solution. It was observed that PPy/TOS had the fastest charging/discharging rate and a specific power of 10 W/g even though its specific capacitance was low at 141.6 F/g as compared with PPy/Cl with 270 F/g. The PPy/Cl gave the highest energy density of 35.3 mWh/g. The charge transfer resistance of PPy/TOS and PPy/Cl was low. Both of these composites gave stable cyclic performance. Lota et al. [50] studied the organic/inorganic composites prepared from PEDOT and MWCNTs or acetylene black (AB) for supercapacitor application. Three methods were used to prepare the composites viz chemical polymerization, electrochemical polymerization and from a homogeneous mixture of PEDOT and CNTs. The electrolytes used were acidic (1 M H_2SO_4), alkaline (6 M KOH), and organic (1 M ($TEABF_4$)). Among these, the organic electrolyte showed high energy storage. The composites from carbon and PEDOT allowed for the formation of three dimensional volumetric capacitor with quick charge propagation. The optimum composition of the composite was: 20-30 wt% CNTs and 70-80 wt% PEDOT. CNTs when replaced with AB gave a moderate loss in capacitance and a weaker support to the composite. The specific capacitance range was from 60 to 130 F/g in symmetric ECs and 160 F/g in the asymmetric configuration. PEDOT/CNTs composite had a better cyclic stability. The electrochemical method of synthesis was found to give the best results but was the most complicated of all the three. The chemical deposition was very promising but there was a loss of 20 F/g in specific capacitance. The composite obtained by simple mixing of carbon and PEDOT gives the lowest specific capacitance but was still attractive due to its simplicity.

PAni was deposited by electrochemical polymerization on substrates of nickel [125] and stainless steel [126] in the presence of p-toluene sulfonic acid (PTS). The electrochemical properties were determined by CV, GCD and EIS techniques. A specific capacitance of 405 F/g was achieved for PAni/nickel substrate composite. FTIR, XRD and SEM were used for morphological and structural analyses. In yet anther study [127] the effect of TritonX-100 in enhancing the capacitance of PAni/nickel electrodes was studied using the same techniques for assessing the electrochemical behavior as well as the morphological and structural analyses, as done in previous studies [125,126]. It was observed that the cyclic voltammogram (CV) for PANI electrode prepared in the presence of TX 100 had a markedly higher area under the curve, which implies an enhancement in the charge storage. A new nanocomposite containing manganese oxide (MnO_2) nanoparticles and PAni doped poly(4-styrene sulfonic acid-co-maleic acid) (PSSMA) was synthesized by electrochemical doping-deposition technique [128]. The morphological and structural characterizations were done by SEM, XRD and XPS. The SEM images showed that the MnO_2 particles were uniformly dispersed in the porous structure of PAni-PSSMA. The XRD analysis revealed the distortion of the crystal structure of β-MnO_2 after incorporation into the PAni-PSSMA matrix and the predominance of PAni. The electrochemical properties were determined using CV and chronopotentiometry in 0.5 M Na_2SO_4. Significant improvement in the values of specific capacitance of the composite electrode was observed compared to that of PAni. The specific capacitance of PANI–PSSMA–MnO_2 was found as 50.4 F/g while that of PANI was 18.5 F/g. The composite had a specific capacitance of 556 F/g, when only the mass of MnO_2 was considered.

Three forms of PAni namely, leucoemeraldine, emeraldine, and pernigraniline were electrosynthesized on stainless steel (Type-304) substrates, and their supercapacitive behavior was studied [129]. It was found by using CV that emeraldine had the highest specific capacitance of 258 F/g for a 0.7 μm thickness. Lee et al. [130], synthesized PPy/ruthenium oxide (PPy/RuO_x) nanocomposites by electrochemical polymerization of PPy in the pores of an anodized aluminium oxide template followed by the deposition of RuO_x on the surface of the resultant PPy nanorod array by an electrochemical method. The maximum specific capacitance of the PPy/RuO_x composite obtained by GCD test as a function of the amount of RuO_x was 419 mF/cm^2 (681 F/g) at 1 mA/ cm^2. Cyclic stability was well retained even after 100 redox cycles. This composite provided greater contact area with the electrolyte ions. Overall, its properties were attractive for its use as an electrode material for ECs. A three dimensional PEDOT-poly (styrene sulfonic acid) (PSS) matrix was loaded with hydrous RuO_2 particles using electrochemical polymerization to form PEDOT-PSS-$RuO_2.xH_2O$ composite electrode [131]. By varying the number of potential cycles in CV, the quantity of hydrous RuO_2 particles in the composite was controlled. Raman spectroscopy confirmed the incorporation of hydrous RuO_2 particles in the PEDOT-PSS matrix and SEM analysis revealed the uniform dispersion of hydrous RuO_2 particles in the same. Electrochemical properties were assessed in 0.5M H_2SO_4 using chronopotentiometry and CV. A maximum specific capacitance of 653 F/g was obtained. This composite too showed great promise as a supercapacitor material. In a study [132], hexacyanoferrate analogues of iron (Fehcf), cobalt (Cohcf) and nickel (Nihcf) were used for modifying PEDOT using multicyclic polarization of PEDOT/$Fe(CN)_6^{3-/4-}$, films in aqueous electrolytes containing nickel, cobalt and iron salts. Good cyclic stability was obtained over hundreds of cycles. Specific capacitance of 70,70 and 50 F/cm^3 was achieved for PEDOT/Cohcf, PEDOT/Nihcf and PEDOT/Fehcf electrodes, respectively. The films of PEDOT containing Fehcf, Cohcf and Nihcf gave higher specific capacitance than the polymer PEDOT without an inorganic multinuclear redox network. The three dimensional matrix of PEDOT-PSS-PAni was synthesized by interfacial polymerization of aniline into PEDOT-PSS [133]. A proton migration passage was provided by the pending sulfonic acid groups on PSS. The inclusion of PAni into the PEDOT-PSS improved the conductivity due to the decrease in the distance for electron transport along the conjugated polymeric chain. Manganese oxide (MnO_2) was electrodeposited in the PEDOT-PSS-PAni three dimensional matrix. The characterization of the composites was done using FESEM, X-ray photoelectron spectroscopy (XPS), and CV techniques. Significant improvement in specific capacitance of PEDOT-PSS-PAni and PEDOT-PSS-PAni-MnO_2 composites was achieved. The latter showing the higher value of 372 F/g when only considering MnO_2 mass. While providing an improvement in conductivity for electron transport during the electrochemical reaction, the conductive matrix of PEDOT-PSS-PAni also facilitated the dispersion of MnO_2 particles without aggregation, as proved by SEM analysis. PEDOT/MnO_2 composites were synthesized by a sequential synthetic method and a one-step co-electrodeposition method [134]. In the sequential synthetic method, free standing MnO_2 rods (10 μm long and 1.5 μm in diameter) were synthesized without a template through anodic deposition from a dilute solution of manganese acetate on Au coated Si substrates and then PEDOT was electrodeposited on MnO_2 to produce coaxial rods. Agglomerated PEDOT/MnO_2 particles were produced by the one-step co-electrodeposition method. The composite produced by the sequential method exhibited better specific capacitance and redox performance in comparison to manganese oxide and co-electrodeposited composite. Sequentially produced PEDOT/MnO_2 rods in which MnO_2 rods were synthesized at a current density of 5 mA/cm^2 and then coated with PEDOT at 1 V for 45s, achieved a specific capacitance of 285 F/g with a 92% retention of capacitance after 250 cycles in 0.5 M Na_2SO_4 at 20 mV/s. The co-electrodeposited PEDOT/MnO_2 composite achieved a capacitance of 195 F/g with retention of 87% of the original value. In a recent study [135], PEDOT nanowires/MnO_2 nanoparticles composites were synthesized by simply soaking the PEDOT nanowires in $KMnO_4$ solution. Characterization of these nanowires was done by SEM and TEM. It was revealed that the MnO_2 nanoparticles had uniform sizes and fine dispersion in PEDOT matrix. The reduction of $KMnO_4$ *via* redox exchange of permanganate ions with the functional group on PEDOT was forwarded as the likely cause for the formation and dispersion of MnO_2 nanoparticles into the nanopores of the PEDOT nanowires. The loading amount and size of the MnO_2 nanoparticles was controlled by varying the concentration of $KMnO_4$ and the reaction time. CV and GCD were used to estimate the electrochemical properties of the composite. The MnO_2 nanoparticles, with their very high exposed surface area, imparted the composite with a high specific capacitance of 410 F/g for electrochemical capacitor application as well as high Li storage capacity of 300 mAh/g for use as cathode for lithium ion battery. The highly conductive and porous PEDOT matrix enabled fast charge/discharge of MnO_2 nanoparticles, without agglomeration.

5.3. Organic/Inorganic Composites by Electrochemical-Co-Deposition

Organic/inorganic composites for electrodes in ECs can also be produced by electro-co-deposition of COP with the substrate. The substrates comprise of carbon-based materials and transition metal oxides. Chen et al.[136], first reported the synthesis of COP-CNT composite in a stable aqueous solution containing anionic CNTs and pyrrole monomers.The CNT performed the dual functions of a charge carrier during electrodeposition and a strong conducting dopant in the PPy-CNT composite. Composites by electro-co-deposition have been made using CNTs with three important COPs, namely, PPy [136,137], PAni [138], and PEDOT [139]. The organic/inorganic composites by electro-co-deposition exhibited enhanced specific capacitance, improved thermal stability, good conductivity, and improved electrode kinetics [19]. Such composites displayed a homogeneous network structure that facilitates both electron and ion movement. Sun et al. [140] obtained composites of PAni/MnO_2 through electro-co-deposition of PAni and MnO_2 on carbon cloth by potential cycling from 0.2 to1.45 V in aqueous solutions of aniline and $MnSO_4$. The morphologies of the composite were greatly improved by the co-deposition of PAni and MnO_2, as fibrous structures with large effective areas instead of granular structures for PAni, were obtained. A specific capacitance of 532 F/g at 2.4 mA/cm^2 discharging current was achieved for hybrid films in the presence of Mn^{2+}. This value was 26% higher than that of similarly synthesized PAni. A coulombic efficiency of 97.5% over 1200 cycles with 24% loss in specific capacitance was reported, which showed good cyclic stability. In a recent study [141], the electrochemical performance of a composite of PAni and MnO_2 in an organic electrolyte has been reported. The PAni/MnO_2 film was electro-co-deposited on high surface area and conductive active carbon (AC) electrode. The electrochemical properties were assessed using 1 M $LiClO_4$ in acetonitrile as the electrolyte. The PAni/MnO_2 composite exhibited a high specific capacitance of 1292 F/g and around 82% capacitance retention after 1500 charge/discharge, at a current density of 4.0 mA/cm^2 with a coulombic efficiency of 95%. Further, an asymmetric capacitor was fabricated using PAni/MnO_2/AC as positive and pure AC as negative electrodes. It delivered a specific energy of 61 Wh/kg at a specific power of 172 W/kg in the range 0 to 2 V. It was deduced that the organic electrolyte was very compatible with PAni/MnO_2 composite, for application in modern electrochemical capacitors.

CONCLUSION

The importance of ECs as effective energy storage devices has been highlighted. The mechanisms of energy storage through electric double layer and pseudocapacitance have been discussed, along with the comparison of ECs with secondary batteries. Research work pertaining to a variety of materials for the synthesis of electrodes for ECs, including carbon-based materials, transition metal oxides, conducting organic polymers, and most importantly organic/inorganic composites, has been reviewed. It was found that the methods being used by researchers for the preparation of organic/inorganic composite electrodes include chemical oxidation polymerization, electrochemical polymerization and electro-co-deposition. For studying the morphology and structure of the composites a number of advanced techniques such as SEM, TEM, FTIR, FESEM, etc., have been employed. For characterization of the electrochemical properties, most workers have used GCD, EIS and CV techniques.

The race is on to develop ECs that are robust, efficient, cheap and reliable, with the capability to function as energy storage devices of the future. Technological challenges exist for improving the performance of ECs. New materials and processes are needed to improve their charge storage capabilities in order to enhance both their power and energy densities. It is necessary to develop a fundamental understanding of the physical and chemical processes occurring in the ECs because without it, advanced design concepts would not materialize. The advent of nanotechnology has opened new vistas to design nanocomposites with exceptional features for use in ECs. Not only is the field of research very fertile and unexplored with respect to electrodes in ECs, but there is equal, if not less, scope for research in the development of electrolytes that have high ionic conductivity combined with a broad range of electrochemical, chemical and thermal stability with long term sustainability. Ample avenues exist for modeling to understand and predict new configurations of ECs.

ACKNOWLEDGMENTS

The authors wish to gratefully acknowledge and thank the researchers from all over the globe in the field of electrochemical capacitors, whose research has been cited by us for the purpose of this review chapter as well as the publishers of such work.

REFERENCES

[1] Zhang, Y.; Feng, H.; Wu, X.; Wang, L.; Zhang, A.; Xia, T.; Dong, H.; Li, X.; Zhang, L. *Intl. J. Hydrogen Energy* 2009, *34*, 4889-4899.
[2] Conway, B. E. Electrochemistry Encyclopedia. http://electrochem.cwru.edu/ed/encycl/art-c03-elchem-cap.htm (accessed Oct 10, 2010).
[3] Becker, H. I. US Patent 2800616, July 23, 1957.
[4] NEC-Tokin web site.http://www.nectokin (accessed Nov 19, 2010).
[5] Murphy, T. C.; Wright, R. B.; Sutula, R. A.; in: Electrochemical Capacitors II, Proceedings, vols. 96–25, Delnick, F.M.; Ingersoll, D.; Andrieu, X., Naoi, K. (Eds.), The Electrochemical Society, Pennington, NJ, 1997, p. 258.
[6] A. Burke, *J. Power Sources* 2000, *91,* 37-30.
[7] Balducci, A.; Henderson, W. A.; Mastragostino, M.; Passerini, S.; Simon, P.; Soavi, F. *Electrochim. Acta* 2005, *50*, 2233-2237.
[8] Kazaryan, S. A.; Razumov, S. N.; Litvinenko, S. V.; Kharisov, G. G.; Kogan, V. I. *J. Electrochem. Soc.* 2006, *153*, A 1655-1671.
[9] Frackowiak, E.; Béguin, F. *Carbon* 2001, *39,* 937-950.
[10] Bard, J.; Faulkner, L. R. *Electrochemical Methods, Fundamentals And Applications*; 2nd ed. John Wiley & Sons, Inc. 2001; 12–13.
[11] Katakabe, T.; Kaneko, T.; Watanabe, M.; Fukushima, T.; Aida, T. *J. Electrochem. Soc.* 2005, *152*, A 1913–1916.
[12] Barbieri, O.; Hahn, M.; Herzog, A.; Kotz, R. *Carbon* 2005, *43*, 1303-1310.
[13] Conway, B. E. *J. Electrochem. Soc.* 1991, *138*, 1539–1548.
[14] Conway, B. E. *Electrochemical Supercapacitors: Scientific Fundamentals And Technological Applications;* Kluwer Academic Publishers/Plenum Press, Dordrecht/New York,1999; pp
[15] Wu, M. S.; Chiang, C. J. *Electrochem. Solid State Lett.* 2004, *7*, A 123–126.
[16] Sugimoto, W.; Iwata, H.; Murakami, Y.; Takasu, Y. *J. Electrochem. Soc.* 2004, *151*, A 1181–1187.
[17] Dong, X.; Shen, W.; Gu, J.; Xiong, L.; Zhu, Y.; Li, H.; Shi, J. *J. Phys. Chem. B* 2006, *110*, 6015–6019.
[18] *Technology and Applied R&D Needs for Electrical Energy Storage- Resource Document for Workshop on Basic Research Needs for Electrical Energy Storage*, Office of Science, U.S. Department of Energy, March 2007.
[19] Peng, C.; Zhang, S.; Jewell, D.; Chen, G. Z. *Prog. Nat. Sci.* 2008, *18*, 777-788.
[20] Phillips, J.; Xia, B.; Menendez, J. A. *Thermochim. Acta* 1998, *312*, 87–93.
[21] Oliveira, L. C. A.; Silva, C. N.; Yoshida, M. I.; Lago, A. M. *Carbon* 2004, *42,* 2279–2284.

[22] Pastor-Villegas, J.; Duran-Valle, C. J. *Carbon* 2002, *40*, 397–402.
[23] Inagaki, M.; Konno, H.; Tanaike, O. *J. Power Sources* 2010, *195*, 7880-7905.
[24] Kim, I. H.; Kim, K. B. *J. Electrochem. Soc.* 2006, *153*, A 383–389.
[25] Li, G.; Feng, Z.; Ou, Y.; Wu, D.; Fu, R.; Tong, Y. *Langmuir* 2010, *26*, 2209-2213.
[26] Wu, M. S.; Huang, Y. A.; Yang, C. H.; Jow, J. J. *Int. J. Hydrogen Energy* 2007, *32*, 4153–4159.
[27] Choi, D.; Blomgren, G. E.; Kumta, P. N. *Adv. Mater.* 2006, *18*, 1178–1182.
[28] Castro, E. B.; Real, S. G.; Pinheiro D. L. F. *Int. J. Hydrogen Energy* 2004, *29*, 255–61.
[29] Morishita, T.; Soneda, Y.; Hatori, H.; Inagaki, M. *Electrochim. Acta* 2007, *52,* 2478-2484.
[30] Wang, Y.; Zhitomirsky, I. *Langmuir* 2009, *25,* 9684-9689.
[31] Athouel, L.; Moser, F.; Dugas, R.; Crosnier, O.; Belanger, D.; Brousse, T. *J. Phys. Chem. C* 2008, *112*, 7270–7277.
[32] Devaraj, S.; Munichandraiah, N. *J. Phys. Chem. C* 2008, *112*, 4406–4417.
[33] Xu, M.; Kong, L.; Zhou, W.; Li, H. *J. Phys. Chem. C* 2007, *111*, 19141–19147.
[34] Liu, R.; Lee, S. B. *J. Am. Chem. Soc.* 2008, *130*, 2942–2943.
[35] Fischer, A. E.; Pettigrew, K. A.; Rolison, D. R.; Stroud, R. M.; Long, J. W. *Nano Lett.* 2007, *7*, 281–286.
[36] Subramanian, V.; Zhu, H.; Vajtai, R.; Ajayan, P. M.; Wei, B. *J. Phys. Chem. B* 2005, *109*, 20207–20214.
[37] Talbi, H.; Just, P. E.; Dao, L. H. *J. Appl. Electrochem.* 2003, *33*, 465-473.
[38] Pasquier, A. D.; Laforgue, A.; Simon, P.; Amatucci, G. G.; Fauvarque, J. F. *J. Electrochem. Soc.* 2002, *149*, A 302-306.
[39] Laforgue, A.; Simon, P.; Fauvarque, J. F.; Mastragostino, M.; Soavi, F.; Sarrau, J. F.; Lailler, P.; Conte, M.; Rossi, E.; Saguatti, S. *J. Electrochem. Soc.* 2003, *150*, A 645-651.
[40] Nohma, T.; Kurokawa, H.; Uehara, M.; Takahashi, M.; Nishio, K.; Saito, T. *J. Power Sources* 1995, *54*, 522-524.
[41] Peng, Z. S.; Wan, C. R., Jiang, C. Y. *J. Power Sources* 1998, *72*, 215-220.
[42] Rudge, A.; Raistrick, I.; Gottesfeld, S.; Ferraris, J. P. *Electrochim. Acta* 1994, *39*, 273-287.
[43] Snook, G. A.; Kao, P.; Best, A. S. *J. Power Sources* 2011, *196*, 1-12.
[44] Ryu, K. S.; Wu, X.; Lee, Y. G.; Chang, S. H. *J. Appl. Polym. Sci.* 2003, *89*, 1300-1304.
[45] Ryu, K. S.; Kim, K. M.; Park, Y. J.; Park, N. G.; Kang, M. G.; Chang, S. H. *Solid Ionics* 2002, *152*, 861-866.
[46] Naoi, K.; Morita, M. *The Electrochemical Society Interface* 2008, *17,* 44-48.
[47] Fan, L. Z.; Maier, J. *Electrochem. Commun.* 2006, *8*, 937–940.
[48] Suematsu, S.; Oura, Y.; Tsujimoto, H.; Kanno, H.; Naoi, K. *Electrochim. Acta* 2000, *45*, 3813-3821.
[49] Stenger, J. D.; Smith, J. D.; Webber, C. K.; Anderson, N.; Chafin, A. P.; Zong, K. K.; Reynolds, J. R. *J. Electrochem. Soc.* 2002, *149*, A 973-977.
[50] Lota, K.; Khomenko, V.; Frackowiak, E. *J. Phys. Chem. Solids* 2004, *65*, 295-301.
[51] Ryu, K. S.; Lee, Y. G.; Hong, Y. S.; Park, Y. J.; Wu, X. L.; Kim, K. M.; Kang, M. G.; Park, N. G. ; Chang, S. H. *Electrochim. Acta* 2004, *50*, 843-847.
[52] Hong, J. I., Yeo, I. H., Paik, W. K. *J. Electrochem. Soc.* 2001, *148*, A 156-163.
[53] Ghosh, S.; Inganas, O. *Electrochem. Solid State Lett.* 2000, *3*, 213-215.

[19] Peng, C.; Zhang, S. W.; Jewell, D.; Chen, G. Z. *Prog. Nat. Sci.* 2008, *18*, 777-788.

[54] Huang, L. M.; Wen, T. C.; Gopalan, A. *Electrochim. Acta* 2006, *51*, 3469-3476.

[55] Mallouki, M.; Tran-Van, F.; Sarrazin, C.; Simon, P.; Daffos, B.; De, A.; Chevrot, C.; Fauvarque, J. *J. Solid State Electrochem.* 2007, *11*,398-406.

[56] Gomez-Romero, P.; Chojak, M.; Cuentas-Gallegos, K.; Asensio, P. J. Kulesza, J. A.; Casan-Pastor, N.; Lira-Cantu, M. *Electrochem. Commun.* 2003, *5*,149-153.

[57] Gomez-Romero, P.; Cuentas-Gallegos, K.; Lira-Cantu, M.; Casan-Pastor, N. *J. Mater. Sci.* 2005, *40*, 1423-1428.

[58] Kulesza, P. J.; Skunik, M.; Baranowska, B., Miecznikowski, K.; Chojak, M.; Karnicka, K.; Frackowiak, E.; Beguin, F.; Kuhn, A.; Delville, M. H.; Starobrzynska, B.; Ernst, A. *Electrochim. Acta* 2006, *51*, 2373-2379.

[59] Gomez-Romero, P.; Lira-Cantu, M. *Adv. Mater.* 1997, *9*, 144-147.

[60] Kulesza, P. J.; Chojak, M.; Miecznikowski, K.; Lewera, A., Malik, M. A.; Kuhn, A. *Electrochem. Commun.* 2002, *4*, 510-515.

[61] Boyle, A.; Genies, E.; Lapkowski, M. *Synth. Met.* 1989, *28*, 769-774.

[62] Barth, M.; Lapkowski, M.; Lefrant, S. *Electrochim. Acta* 1999, *44*, 2117-2123.

[63] Cuentas -Gallegos, A. K.; Romero, P. G. *J. Power Sources* 2006, *161*, 580-586.

[64] Malinauskas, A. *Polymer* 2001, *42*, 3957–72.

[65] Wang, C.; Guo, Z. X.; Fu, S.; Wu, W.; Zhu, D. *Prog. Polym. Sci.* 2004, *29*, 1079–141.

[66] Feast W. J. *Synthesis of Conducting Polymers*. In: Handbook of Conducting Polymers. New York: Marcel Dekker Inc.; 1986. p. 1–43.

[67] Nalawa, H. S.; Dalton, L. R.; Schmidt, W. F.; Rabe, J. G. *Polym. Commun.* 1985, *27*, 240–242.

[68] Cao, Y.; Andreatta, A.; Heeger, A. J.; Smith, P. *Polymer* 1989, *30*, 2305–2311.

[69] Armes, S. P.; Miller, J. F. *Synth. Met.* 1988, *22*, 385–393.

[70] Ayad, M. M.; Salahuddin, N.; Shenashin, M. A. *Synth. Met.* 2004, *142*, 101–106.

[71] Bradley, M.; Grieser, F. *J. Colloid Interface Sci.* 2002, *251*, 78–84.

[72] Li, W. K.; Chen, J.; Zhao, J. J.; Zhang, J.; Zhu, J. *Mater. Lett.* 2005, *59*, 800–803.

[73] Xia, H.; Wang, Q. *Chem. Mater.* 2002, *14*, 2158–65.

[74] Ingram, M. D.; Pappin, A. J.; Delalande, F.; Poupard, D.; Terzulli, G. *Electrochim. Acta* 1998, *43*, 1601–1605.

[75] Ryu, K. S.; Lee, Y. G.; Kim, K. M.; Park, Y. J.; Hong, Y. S.; Wu, X.; Kang, M. G.; Park, N. G.; Song, R. Y.; Ko, J. M. *Synth. Met.* 2005, *153*, 89–92.

[76] Kim, J. H.; Sharma, A. K.; Lee, Y. S. *Mater. Lett.* 2006, *60*, 1697–1701.

[77] Dong, B.; He, B.; Xu, C.; Li, H. *Mater. Sci. Eng. B* 2007, *143*, 7–13.

[78] Sivakkumar, S. R.; Kim, W. J.; Choi, J.; MacFarlane, D. R.; Forsyth, M.; Kim, D. *J. Power Sources* 2007, *171*, 1062–1068.

[79] Mi, H.; Zhang, X.; An, S.; Ye, X.; Yang, S. *Electrochem. Commun.* 2007, *9*, 2859–2862.

[80] Wang, H.; Hao, Q.; Yang, X.; Lu, L.; Wang, X. *Electrochem. Commun.* 2009, *11*, 1158–1161.

[81] Yan, J.; Wei, T.; Shao, B.; Fan, Z.; Qian, W.; Zhang, M.; Wei, F. *Carbon* 2010, *48*, 487-493.

[82] Yan, J.; Wei, T.; Fan, Z.; Qian, W.; Zhang, M.; Shen, X.; Wei, F. *J. Power Sources* 2010, *195*, 3041–3045.

[83] Kuilla, T.; Bhadra, S.; Yao, D.; Kim, N. H.; Bose, S.; Lee, J. H. *Prog. Polym. Sci.* 2010, *35*, 1350-1375.
[84] Zheng, L.; Wang, Y.; Wang, X.; Li, N.; An, H.; Chen, H.; Guo, J. *J. Power Sources* 2010, *195*, 1747–1752.
[85] Park, J. H.; Ko, J. M.; Park, O. O.; Kim, D. *J. Power Sources* 2002, *105*, 20–25.
[86] An, H.; Wang, Y.; Wang, X.; Zheng, L.; Wang, X.; Yi, L.; Bai, L.; Zhang, X. *J. Power Sources* 2010, *195*, 6964–6969.
[87] Mi, H.; Zhang, X.; Xu, Y.; Xiao, F. *Appl. Surf. Sci.* 2010, *256*, 2284–2288.
[88] Zhang, J.; Kong, L.; Cai, J.; Luo, Y.; Kang, L. *Electrochim. Acta* 2010, *55*, 8067–8073.
[89] Cuentas-Gallegos, A. K.; Rinćon, M. E. *J. Power Sources* 2006, *162*, 743–747.
[90] Zhou, Z.; Cai, N.; Zhou, Y. *Mater. Chem. Phys.* 2005, *94*, 371–375.
[91] Ballav, N.; Biswas, M. *Mater. Lett.* 2006, *60*, 514–51.
[92] Song, R. Y.; Park, J. H.; Sivakkumar, S. R.; Kim, S. H.; Ko, J. M.; Park, D.; Jo, S. M.; Kim, D. Y. *J. Power Sources* 2007, *166*, 297–301.
[93] Gök, A.; Omastová, M.; Prokes, J. *Eur. Polym. J.* 2007, *43*, 2471–2480.
[94] Yuan, C.; Su, L.; Gao, B.; Zhang, X. *Electrochim. Acta* 2008, *53*, 7039–7047.
[95] Bian, C.; Yu, A.; Wu, H. *Electrochem. Commun.* 2009, *11*, 266–269.
[96] Hu, Z.; Xie, Y.; Wang, Y.; Mo, L.; Yang, Y.; Zhang, Z. *Mater. Chem. Phys.* 2009, *114*, 990–995.
[97] Xu, H.; Cao, Q.; Wang, X.; Li, W.; Li, X.; Deng, H. *Mater. Sci. Eng. B* 2010, 104-108.
[98] Mallouki, M.; Tran-Van, S. C.; Chevrot, C.; Fauvarque, J. F. *Electrochim. Acta* 2009, *54*, 2992–2997.
[99] Yang, C.; Liu, P.; Zhao, Y. *Electrochim. Acta* 2010, *55*, 6857–6864.
[100] Chin, S. F.; Pang, S. C. *Mater. Chem. Phys.* 2010, *124*, 29–32.
[101] Sen, P.; De, A. *Electrochim. Acta* 2010, *55*, 4677–4684.
[102] He, B.; Zhou, Y.; Zhou, W.; Dong, B.; Li, H. *Mater. Sci. Eng. B* 2004, *374*, 322–326.
[103] Downs, C.; Nugent, J.; Ajayan, P. M.; Duquette, D. J.; Santhanam, K. S. V. *Adv Mater.* 1999, *11*, 1028–1031.
[104] Chen, J. H.; Huang, Z. P.; Wang, D. Z.; Yang, S. X.; Li, W. Z.; Wen, J. G.; Ren, Z. F. *Synth. Met.* 2002, *125*, 289–94.
[105] Frackowiak, E.; Jurewicz, K.; Szostak, K.; Delpeux, S.; Béguin, F. *Fuel Process Technol.* 2002, *77–78*, 213–219.
[106] Gupta, V.; Miura, N. *Electrochim. Acta* 2006, *52,* 1721–1726.
[107] Gupta, V.; Miura, N. *J. Power Sources* 2006, *157,* 616–620.
[108] Nuria, F. A.; Martti, K.; Viera, S.; Dettlaf-Weglikowska, U.; Roth, S. *Diamond Relat. Mater.* 2004, *13*, 256–260.
[109] Baibarac, M.; Baltog, I.; Godon, C.; Lefrant, S.; Chauvet, O. *Carbon* 2004, *42,* 3143–3152.
[110] Chen, W. C.; Wen, T. C.; Teng, H. *Electrochim. Acta* 2003, *48*, 641–649.
[111] Mondal, S. K.; Barai, K.; Munichandraiah, N. *Electrochim. Acta* 2007, *52,* 3258–3264.
[112] Wei-Chih, Chen, Ten-Chin, W. *J. Power Sources* 2003, *117*, 273–282.
[110] Chen, W.; Wen, T.; Teng, H. *Electrochim. Acta* 2003, *48*, 641-649.
[113] Hu, C.; Li, W.; Lin, J. *J. Power Sources* 2004, *137*, 152–157.
[114] Ko, J. M.; Song, R. Y.; Yu, H. J.; Yoon, J. W.; Min, B. G.; Kim, D. W. *Electrochim. Acta* 2004, *50*, 873–876.

[115] Zhang, H.; Cao, G.; Wang, Z.; Yang, Y.; Shi, Z; Gu, Z. *Electrochem. Commun.* 2008, *10*, 1056–1059.
[116] Sipahi, M.; Parlak, E. A.; Gul, A.; Ekinci, E.; Yardim, M. F.; Sarac, A. S. *Prog. Org. Coat.* 2008, *62*, 96–104.
[117] Wang, Q.; Li, J.; Gao F.; Li, W.; Wu, K.; Wang, X. *New Carbon Materials* 2008, *23*, 275–280.
[118] Zhang, H.; Cao, G.; Wang, W.; Yuan, K.; Xu, B.; Zhang, W.; Cheng, J.; Yang, Y. *Electrochim. Acta* 2009, *54*, 1153–1159.
[119] Zhang, J.; Kong, L.; Wang, B.; Luo, Y.; Kang, L. *Synth. Met.* 2009, *159*, 260–266.
[120] Zhao, X. Y.; Zang, J. B.; Wang, Y. H.; Bian, L. Y.; Yu, J. K. *Electrochem. Commun.* 2009, *11*, 1297–1300.
[121] Woo, S.; Dokko, K.; Nakano, H.; Kanamura, K. *J. Power Sources* 2009, *190*, 596–600.
[122] Horng, Y.; Lu, Y.; Hsu, Y.; Chen, C.; Chen, L.; Chen, K. *J. Power Sources* 2010, *195*, 4418–4422.
[123] Wang, J.; Xu, Y.; Chen, X.; Sun, X. *Compos. Sci. Technol.* 2007, *67*, 2981–2985.
[124] Wang, J.; Xu, Y.; Chen, X.; Du, X.; Li, X. *Acta Phys. Chim. Sin.* 2007, *23*, 299-304.
[125] Girija, T. C.; Sangaranarayanan, M. V. *J. Power Sources* 2006, *156*, 705–711.
[126] Girija, T. C.; Sangaranarayanan, M. V. *Synth. Met.* 2006, *156*, 244–250.
[127] Girija, T. C.; Sangaranarayanan, M. V. *J. Power Sources* 2006, *159*, 1519–1526.
[128] Liu, F.; Hsu, T.; Yang, C. *J. Power Sources* 2009, *191*, 678–683.
[129] Jamadade, V. S.; Dhawale, D. S.; Lokhande, C. D. *Synth. Met.* 2010, *160*, 955–960.
[130] Lee, H.; Cho, M. S.; Kim, I. H.; Nam, J. D.; Lee, Y. *Synth. Met.* 2010, *160*, 1055–1059.
[131] Huang, L.; Lin, H.; Wen, T.; Gopalan, A. *Electrochim. Acta* 2006, *52*, 1058–1063.
[132] Andrzej, L.; Nowak, P. *J. Power Sources* 2007, *173*, 829–836.
[133] F. Liu, *J. Power Sources* 2008, *182,* 383–388.
[134] Babakhani, Douglas, G. I. *Electrochim. Acta* 2010, *55,* 4014–4024.
[135] Liu, R.; Duay, J.; Lee, S. B. *ACS Nano* 2010, *4*, 4299-4307.
[136] Chen, G. Z.; Shaffer, M. S. P.; Coleby, D.; Dixon, G.; Zhou, W.; Fray, D. J.; Windle, A. *Adv. Mater.* 2000, *12*, 522–526.
[137] Hughes, M.; Chen, G. Z.; Shaffer, M. S. P.; Fray, D. J.; Windle, A. H. *Compos. Sci. Technol.* 2004, *64*, 2325–2331.
[138] Wu, M.; Snook, G. A.; Gupta. V.; Shaffer, M.; Fray, D. J.; Chen, G. Z. *J. Mater. Chem.* 2005, *15*, 2297–303.
[139] Peng, C.; Snook, G. A.; Fray, D. J.; Shaffer. M. S. P.; Chen, G. Z. *Chem. Commun.* 2006, *4*, 629–631.
[140] Sun, L.; Liu, X. *Eur. Polym. J.* 2008, *44*, 219–224.
[141] Zou, W.; Wang, W.; He, B.; Sun, M.; Yin, Y. *J. Power Sources* 2010, *195*, 7489–7493.

In: Advanced Organic-Inorganic Composites
Editor: Inamuddin
ISBN 978-1-61324-264-3

Chapter 13

NANOCOMPOSITE COATINGS FOR CORROSION CONTROL

Rana Sardar[*]
Department of Applied Chemistry, Aligarh Muslim University, Aligarh 202002, India

ABSTRACT

Corrosion usually takes place and degrades material surfaces based on environmental chemistry. There are several popular ways of decreasing corrosion rates to improve the lifetime of materials and devices. In the last two decades, nanotechnology has been playing an increasing important role in supporting innovative technological advances to manage the corrosion. Nanostructured coatings have recently attracted increasing interest because of the possibilities of synthesizing materials with unique physical–chemical properties. Highly sophisticated surface related properties, such as optical, magnetic, electronic, catalytic, mechanical, chemical and tribological properties can be obtained by advanced nanostructured coatings, making them attractive for industrial applications in high speed machining, tooling optical applications and magnetic storage devices because of their special mechanical, electronic, magnetic and optical properties due to size effect. There are many types of design models for nanostructured coatings, such as nanocomposite coatings, nano-scale multilayer coating, superlattice coating, nano-graded coatings, etc. Designing of nanostructured coatings needs consideration of many factors, e.g. the interface volume, crystallite size, single layer thickness, surface and interfacial energy, texture, epitaxial stress and strain, etc., all of which depend significantly on materials selection, deposition methods and process parameters. A nanocomposite coating comprises of a nanocrystalline phase and an amorphous phase. Nanocomposite coatings exhibit hardness significantly exceeding that given by the rule of mixture. Nanocomposite coatings can be hard, superhard or even ultra-hard, depending on coating design and application. Extensive theoretical and experimental efforts have been made to synthesize and study these nanocomposite coatings with superhardness and high toughness.

---

[*] Corresponding author's e-mail: ranasardar_alig@rediffmail.com

In this article a brief introduction to corrosion and application of nanotechnology that is nanocomposite coatings in the field of corrosion prevention of metals is reviewed. Recent research and developments in this area are discussed in designing efficient coating materials which provide superior resistance to corrosion.

ABBREVIATIONS

AC	Alternating current
Al_2O_3	Aluminum oxide
AMBIO	Advanced nanostructured surfaces for the control of biofouling
AP	Atmospheric pressure
a-Si:N	Amorphous silicon nitride
BMA	Butyl methacrylate
CdO	Cadmium oxide
CF	Corrosion fatigue
CP	Cathodic protection
CRS	Cold-rolled steel
CS	Carbon Steel
CVD	Chemical vapour deposition
DC	Direct current
EIS	Electrochemical impedance spectroscopy
FBR	Fluidized bed reactor
HIC	Hydrogen-induced cracking
HVOF	High velocity oxy-fuel thermal spraying
HVSFS	High-velocity suspension flame spraying
I_a	Anodic current
I_c	Cathodic current
ICP	Inherently conducting polymer
KPS	Potassium persulphate
LME	Liquid metal embrittlement
MAO	Micro arc oxidation
MMA	Methyl methacrylate
MMT	Montmorillonite
MoS_2	Molybdenum sulfide
MS	Mild steel
Na^+-MMT	Na^+-montmorillonite
$NaAlO_2$	Sodium aluminate
NACE	National association of corrosion engineers
Nc	Nano crystalline
NH_3	Ammonia
PANI	Polyaniline
PCN	Polymer clay nanocomposite
PEO	Plasma electrolytic oxidation
POA	Poly(o-anisidine)
POEA	Poly(*o*-ethoxyaniline)

POT	Poly(*o*-toluidine)
PPy	Polypyrrole
PS	Polystyrene
PU	Polyurethane
PVD	Physical vapour deposition
SCC	Stress corrosion cracking
SCE	Saturated calomel electrode
SDS	Sodium dodecyl sulphate
SEM	Scanning electron microscopy
SNAP	Self-assembled nanophase particle
SS	Stainless steel
SSC	Sulfide stress cracking
St	Styrene
TEM	Transmission electron spectroscopy
Tg	Glass transition temperature
Ti–C:H	Titanium-containing-amorphous hydrocarbon
TiN	Titanium nitride
TiO_2	Titanium oxide
WPU	Waterborne polyurethane
XPS	X-ray photoelectron spectroscopy
XRD	X-ray diffraction

1. INTRODUCTION

Corrosion is a naturally occurring phenomenon commonly defined as the deterioration of a substance (usually a metal) or its properties because of a reaction with its environment [1]. Like other natural hazards such as earthquakes or severe weather disturbances, corrosion can cause dangerous and expensive damage to everything from vehicles, home appliances, and water and wastewater systems to pipelines, bridges, and public buildings. Corrosion has a major impact on the economy of a nation. The loss to the world economy due to corrosion is in the billions per year. Thus from national economic point of view, it is necessary for scientists and engineers to adopt various ways and means to minimize the losses due to corrosion. With technological and industrial growth, the use of metals and their alloys is increasing very rapidly and any step in the direction of understanding the nature of corrosion, its mechanism and the way to control it, would be of great help to nation's economy. The science of corrosion prevention and control is highly complex, exacerbated by the fact that corrosion takes many different forms and is affected by numerous outside factors. Corrosion professionals must understand the effects of environmental conditions such as soil resistivity, humidity, and exposure to salt water on various types of materials; the type of product to be processed, handled, or transported; required lifetime of the structure or component; proximity to corrosion-causing phenomena such as stray current from rail systems; appropriate mitigation methods; and other considerations before determining the specific corrosion problem and specifying an effective solution. The first step in effective corrosion control, however, is to have a thorough knowledge of the various forms of corrosion, the mechanisms involved, how to detect them, and how and why they occur [2].

2. Cost of Corrosion

Corrosion Costs and Preventive Strategies in the United States [3], was initiated by National Association of Corrosion Engineers (NACE) International and conducted by Corrosion Costs Technologies, Inc. and determined the annual direct cost of corrosion to be a staggering $276 billion or 3.1% of the gross domestic product. This finding was based on an analysis of direct costs by industry sector; when based on the cost of corrosion prevention and control, the study found that $121 billion is attributed to corrosion control methods, services, research and development, and education and training. Corrosion is so prevalent and takes so many forms that its occurrence and associated costs never will be completely eliminated; however, the study estimates that 25 to 30% of annual corrosion costs could be saved if optimum corrosion management practices were employed.

3. The Consequences of Corrosion

The consequences of corrosion are many and varied and the effects of these on the safe, reliable and efficient operation of equipment or structures are often more serious than the simple loss of a mass of metal. Failures of various kinds and the need for expensive replacements may occur even though the amount of metal destroyed is quite small. Some of the major harmful effects of corrosion can be summarized as follows:

- Reduction of metal thickness leading to loss of mechanical strength and structural failure or breakdown. When the metal is lost in localized zones so as to give a crack-like structure, very considerable weakening may result from quite a small amount of metal loss.
- Hazards or injuries to people arising from structural failure or breakdown (e.g. bridges, cars, aircraft).
- Loss of time in availability of profile-making industrial equipment.
- Reduced value of goods due to deterioration of appearance.
- Contamination of fluids in vessels and pipes e.g. beer goes cloudy when small quantities of heavy metals are released by corrosion.
- Perforation of vessels and pipes allowing escape of their contents and possible harm to the surroundings. For example, a leaky domestic radiator can cause expensive damage to carpets and decorations, while corrosive sea water may enter the boilers of a power station if the condenser tubes perforate.
- Loss of technically important surface properties of a metallic component. These could include frictional and bearing properties, ease of fluid flow over a pipe surface, electrical conductivity of contacts, surface reflectivity or heat transfer across a surface.

- Mechanical damage to valves, pumps, etc., or blockage of pipes by solid corrosion products.
- Added complexity and expense of equipment which needs to be designed to withstand a certain amount of corrosion, and to allow corroded components to be conveniently replaced.

4. Chemistry of Corrosion

In most of the cases metallic state represents the state of high energy. Therefore, metals have a natural tendency to react with other substances and go back to lower energy state with subsequent release of energy. All metals show decrease in free energy by undergoing reaction with the environment, (except noble metals, which are found in native state in nature). Since most metallic compounds, and especially corrosion products, have little mechanical strength a severely corroded piece of metal is quite useless for its original purpose.

Virtually all corrosion reactions are electrochemical in nature, at anodic sites on the surface the iron goes into solution as ferrous ions, this constituting the anodic reaction. As iron atoms undergo oxidation to ions they release electrons whose negative charge would quickly build up in the metal and prevent further anodic reaction, or corrosion. Thus, this dissolution will only continue if the electrons released can pass to a site on the metal surface where a cathodic reaction is possible. At a cathodic site the electrons react with some reducible component of the electrolyte and are themselves removed from the metal. The rates of the anodic and cathodic reactions must be equivalent according to Faraday's Laws, being determined by the total flow of electrons from anodes to cathodes which is called the "corrosion current".

Since the corrosion current must also flow through the electrolyte by ionic conduction, the conductivity of the electrolyte will influence the way in which corrosion cells operate. The corroding piece of metal is described as a "mixed electrode" since simultaneous anodic and cathodic reactions are proceeding on its surface. The mixed electrode is a complete electrochemical cell on one metal surface.

The most common and important electrochemical reactions in the corrosion of iron are as follows:

Anodic reaction (corrosion)

$$Fe \rightarrow Fe^{2+} + 2e^- \qquad (1)$$

Cathodic reactions (simplified)

$$2H^+ + 2e^- \rightarrow H_2 \qquad (2a)$$

or

$$H_2O + \frac{1}{2} O_2 + 2e^- \rightarrow 2OH^- \qquad (2b)$$

Reaction 2 a is most common in acids and in the pH range 6.5–8.5 the most important reaction is oxygen reduction 2 b. In this latter case corrosion is usually accompanied by the formation of solid corrosion debris from the reaction between the anodic and cathodic products.

$$Fe^{2+} + 2OH^- \rightarrow \underset{\text{Iron(II) hydroxide}}{Fe(OH)_2} \quad (3)$$

Pure iron(II) hydroxide is white but material initially produced by corrosion is normally greenish color due to partial oxidation in air.

$$2Fe(OH)_2 + H_2O + \tfrac{1}{2} O_2 \rightarrow \underset{\text{Hydrated iron(III) oxide}}{Fe(OH)_3} \quad (4)$$

Further hydration and oxidation reactions can occur and the reddish rust that eventually forms is a complex mixture whose exact constitution will depend on other trace elements which are present. Because the rust is precipitated as a result of secondary reactions it is porous and absorbent and tends to act as a sort of harmful poultice which encourages further corrosion. For other metals or different environments different types of anodic and cathodic reactions may occur. If solid corrosion products are produced directly on the surface as the first result of anodic oxidation, these may provide a highly protective surface film which retards further corrosion, the surface is then said to be "passive". An example of such a process would be the production of an oxide film on iron in water, a reaction which is encouraged by oxidising conditions or elevated temperatures.

$$2Fe + 3H_2O \rightarrow Fe_2O_3 + 6H^+ + 6e^- \quad (5)$$

5. Factors Influencing Corrosion

The nature and the extent of corrosion depend largely on the metal and the environment. Thus, factors like structural features of the metal, nature of the environment and the type of reactions that occur at the metal/environment interface have to be considered for the understanding of the corrosion phenomenon.

The important factors, which influence the corrosion process are as follows:

- Nature of the metal
- Nature of the environment
- Electrode potential
- Temperature
- Solution concentration
- Aeration
- Agitation
- pH of the solution
- Nature of the corrosion products and hydrogen over potential

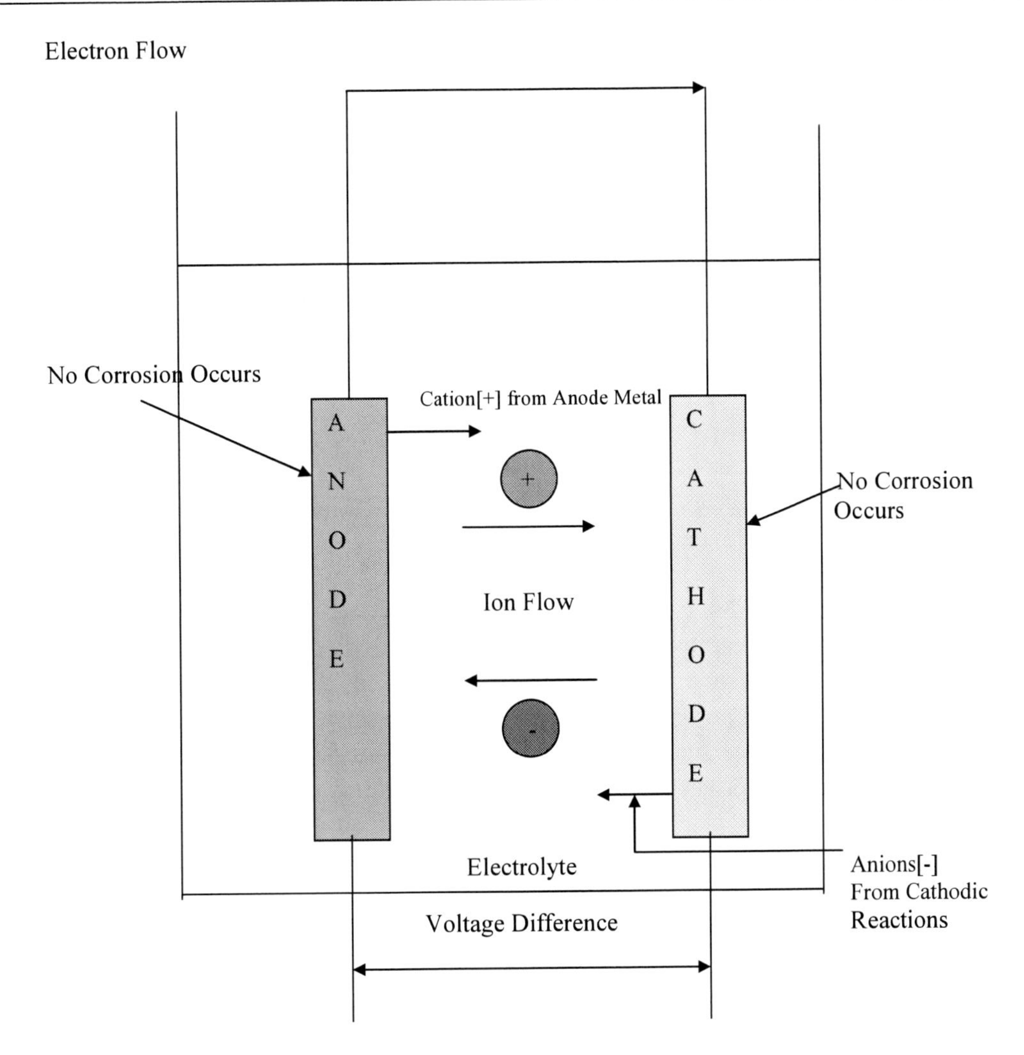

Figure 1. Electrochemical Cell.

6. Classification of Corrosion

There are 10 primary forms of corrosion:

All forms of corrosion, with the exception of some types of high-temperature corrosion, occur through the action of the electrochemical cell (Figure 1). The elements that are common to all corrosion cells are an anode where oxidation and metal loss occur, a cathode where reduction and protective effects occur, metallic and electrolytic paths between the anode and cathode through which electronic and ionic current flows, and a potential difference that drives the cell. The driving potential may be the result of differences between the characteristics of dissimilar metals, surface conditions, and the environment, including chemical concentrations.

6.1. Uniform Attack Corrosion

Uniform attack is the most common form of corrosion and is characterized by a chemical or electrochemical reaction which proceeds uniformly over the entire exposed metal surface. The metal becomes thinner and eventually fails. It is recognized by a roughening of the surface and usually by the presence of corrosion products.

Uniform corrosion can be prevented or reduced by using proper materials including protective coatings, inhibitors or cathodic protection (CP) an electrochemical technique used for corrosion control.

6.2. Localized Corrosion

Localized corrosion occurs at discrete sites on a metal surface. Types of localized corrosion include pitting, crevice, and filiform corrosion.

6.2.1. Pitting

It is one of the most destructive insidious forms of corrosion. It is an example of highly localized corrosion. Pitting is a deep, narrow attack that can cause rapid penetration of the substrate (metal) wall thickness. It is characterized by corrosive attack in a localized region surrounded by non-corroded surfaces or surfaces that are attacked to a much lesser extent. Pits initiate at defects in a protective or passive film. The corrosion is caused by the potential difference between the anodic area inside the pit which often contains acidic, hydrolyzed salts and the surrounding cathodic area. Pitting corrosion can be controlled by protective coatings, or CP, or by modifying the environment.

6.2.2. Crevice Corrosion

Intense localized corrosion in shallow holes, occurs at localized sites where free access to the surrounding environment is restricted, such as crevices where materials meet either metal-to-metal or metal-to-nonmetal. Crevices also can form under deposits of debris or corrosion products. It is also known as concentration cell corrosion, Crevice corrosion can be controlled by the use of coatings to seal crevices, and CP.

6.2.3. Filiform Corrosion

Filiform corrosion is a special form of oxygen cell corrosion occurring beneath of organic or metallic coatings on materials. This form is recognizable by the appearance of a fine network of random threads of corrosion product beneath the coating material. Filiform corrosion is associated with mild surface contamination of solid particles deposited from the atmosphere or residue on the metal surface after processing. When exposed to humid conditions (relative humidity greater than 60%), these surfaces often will suffer filiform corrosion. Corrosion proceeds because of the potential difference between the head of the advancing filament, which becomes anodic, with a low pH and lack of oxygen compared to the cathodic area immediately behind the head. This type of corrosion, particularly on painted surfaces, can be prevented by proper cleaning and preparation of the metallic surface and then applying a protective coating to a clean and dry surface.

Table 1. Galvanic series for metals in seawater

Active (More negative) end	Magnesium
	Zinc
	Aluminum Alloys
	Carbon Steel
	Cast Iron
	13% Cr (Type 410 Stainless Steel (SS) (Active)
	18-8 (Type 304) SS (Active)
	Naval Brass
	Copper
	70-30 Copper-Nickel Alloy
	13% Cr (Type 410) SS (Passive)
	Titanium
	18-8 (Type 304) SS (Passive)
	Graphite
	Gold
Noble (More positive) end	Platinum

6.3. Galvanic Corrosion

Galvanic corrosion occurs when a potential difference exists between two dissimilar metals immersed in a corrosive solution. The potential difference produces a flow of electrons between the metals. The less corrosion resistant metal becomes the anode and the more corrosion resistant one the cathode. The mechanism is a classic electrochemical cell electrons flow through a metallic path from sites where anodic reactions are occurring to sites where the electrons cause cathodic reactions to occur. At the same time, there is a migration of ions (charged particles) in the electrolyte. The positive ions are charged metal particles dissolved at the anode; loss of these ions causes the corrosion of the anode. The negative ions come from the cathodic reactions. Thus, the anode corrodes and the cathode does not. In fact, the cathode is protected against corrosion (Figure 1).

A galvanic series can be used to determine the likely interactions between adjoining metals. Table 1 shows an example of a galvanic series for metals in seawater. When two metals are connected together, the more active one becomes the anode (corroding) and the less active one the cathode.

Galvanic corrosion can be controlled with modification of the environment, materials selection, CP, barrier coatings, electrical isolation (the connection between dissimilar metals is insulated to break the electrical continuity), and design. Rivets, bolts, and other fasteners should be of a more noble metal than the material to be fastened.

6.4. Environmental Cracking

Unlike many other forms of corrosion where corrosion occurs over long periods of time, environmental cracking can occur very rapidly. Because it is unanticipated, it can be catastrophic. Environmental cracking is the brittle failure of an otherwise ductile material caused by the combined action of corrosion and tensile stress. It can be identified by tight cracks that are at right angles to the direction of maximum tensile stress. Types of environmental cracking include stress corrosion cracking (SCC), hydrogen-induced cracking (HIC), liquid metal embrittlement (LME), and corrosion fatigue (CF).

6.4.1. Stress Corrosion Cracking (SSC)

SCC refers to cracking caused by the simultaneous presence of tensile stress and a specific corrosive medium. During SCC the metal may be virtually unattacked over most of the surface, but fine cracks progress through the metal.

Examples of media that promote SCC of specific alloys include caustic with carbon steel (CS), chlorides with stainless steel (SS), and ammonia (NH_3) with copper alloys. Usually there is incubation period during which cracking initiates on a microscopic level, followed by propagation. Typically, there is little metal loss or general corrosion associated with SCC.

6.4.2. Hydrogen-Induced Cracking (HIC)

HIC results in the brittle failure of materials when exposed to an environment where hydrogen can enter the metal. Caused by the combined action of tensile stress and hydrogen. A cathodic phenomenon, HIC occurs when the normal evolution of hydrogen at cathodic sites is inhibited and the atomic hydrogen in the cathodic reaction enters the metal. Higher-strength alloys (those with a tensile strength of 1,034 MPa or greater) are more susceptible to HIC than lower-strength alloys. Sulfide stress cracking (SSC) is a specific form of HIC wherein the presence of sulfides suppresses the evolution of hydrogen.

6.4.3. Liquid Metal Embrittlement (LME)

LME refers as the decrease in strength or ductility of a metal or alloy as a result of contact with a liquid metal. A normally ductile material under tensile stress while in contact with a liquid metal may exhibit brittle fracture at low stress levels. Unlike fracture by SCC, LME is not time dependent cracking and can begin immediately upon application of stress. Embrittling agents cause different reactions depending on the alloy; for example, SS are quite resistant to degradation when contacted by liquid metal whereas mild CS and copper-based alloys can become severely embrittled.

6.4.4. Corrosion Fatigue (CF)

CF occurred by the combined action of a cyclic tensile stress and a corrosive environment. It is characterized by a premature failure of a cyclically loaded part and is a serious cause of failure that necessitates major expenditures for repair. The petroleum industry encounters CF problems in the production of oil, the exposure of drill pipes and sucker rods to brines and sour crudes causes costly failures and loss of production.

Environmental cracking can be controlled by using protective coatings, modification of the environment, materials selection, CP, reduction in residual surface stress, and by changing the design to lower tensile stress levels.

6.5. Flow-Assisted Corrosion

Flow-assisted corrosion is caused by the combined action of corrosion and fluid flow. This includes erosion-corrosion, impingement, and cavitation.

6.5.1. Erosion-Corrosion

Erosion-corrosion is the increase in the rate of attack of a metal because of relative movement between a corrosive medium and the metal surface. Generally, this movement is rapid and mechanical wear is involved. Metal is removed from the surface either in the form of dissolved ions or in the form of solid corrosion products which are mechanically swept from the metal surface. This attacked is characterized by the appearance of grooves, gullies, waves, rounded holes. An erosion corrosion type of attack can be observed in piping systems. Prevention of erosion corrosion damage can be achieved by selection of materials with better resistance to erosion corrosion, proper design, alteration of the aggressiveness of the environment, coatings and CP.

6.5.2. Impingement

Impingement is caused by turbulence or impinging flow (directed at roughly right angles to the materials). Entrained air bubbles tend to accelerate this action, as do suspended solids. This type of corrosion occurs in pumps, valves, and orifices; on heat exchanger tubes; and at elbows and tees in pipelines. It usually produces a pattern of localized attack with directional features. When a liquid is flowing over a surface, there usually is a critical velocity below which impingement does not occur and above which it rapidly increases. Impingement first received attention because of the poor behavior of some copper alloys in seawater.

6.5.3. Cavitation

Cavitation damage occurs in surfaces where high velocity liquid flow and pressure changes are experienced. This mechanical damage process is caused by collapsing bubbles in a flowing liquid, usually forming deep aligned pits in areas of turbulent flow. Cavitation occurs when protective films are removed from a metal surface by high pressures generated by the collapse of gas or vapor bubbles. In general, higher-strength alloys are more resistant to this type of corrosion than lower-strength alloys. When cavitation damage is caused primarily by corrosion following the removal of protective films, the corrosion portion of the damage may predominate. Under extreme cavitation conditions, the cavitation itself is capable of removing the metal directly and corrosion effects are insignificant.

Flow-assisted corrosion can be controlled by modification of the environment, proper materials selection, protective coatings, CP, and controlling flow velocity and patterns through design.

6.6. Intergranular Corrosion

Intergranular corrosion consists of localized attack at, and adjacent to, the grain boundaries with relatively little corrosion of the grains and results in disintegration of the alloy and loss of strength. Intergranular corrosion can be caused by impurities at the grain boundaries, enrichment of one of the alloying elements or depletion of one of these element in the grain boundary areas.

Intergranular corrosion can be controlled using proper material selection, design, fabrication, and weld procedures; modification of the environment; and heat treatment, which dissolves the undesirable constituents at the grain boundaries.

6.7. Dealloying

Dealloying is a corrosion process in which one constituent of an alloy is removed preferentially, leaving an altered residual structure. Many alloys consist of mixtures of elements (i.e., zinc and copper are alloyed to produce brass), and one element can be anodic with respect to the other element(s) and can selectively corrode by galvanic action. This phenomenon is commonly detectable as a color change or drastic change in mechanical strength. For example, brasses will turn from yellow to red and cast irons will turn from silvery gray to dark gray. It can be controlled through materials selection, modification of the environment, protective coatings, CP, and design (e.g., controlling temperature to minimize hot-wall effects in heat exchangers).

6.8. Fretting Corrosion

Fretting corrosion is defined as metal deterioration caused by repetitive slip at the interface between two surfaces in contact that were not intended to move in that fashion. The motion between surfaces either removes protective films or combined with the abrasive action of corrosive products, mechanically removes material from surfaces in relative motion. Fretting results in the metal loss in the area of contact; production of oxide and metal debris; galling, seizing, fatiguing, or cracking of the metal; loss of dimensional tolerances; loosening of bolted or riveted parts; and destruction of bearing surfaces. It can be controlled through materials selection, designing to avoid motion between surfaces, and the use of lubricants such as molybdenum sulfide (MoS_2).

6.9. High Temperature Corrosion

Direct chemical reactions, rather than the reactions of the electrochemical cell, are responsible for the deterioration of metals by high-temperature corrosion. The actual temperature at which corrosion occurs depends upon the material and the environment, but corrosion usually starts when the temperature is within approximately 30 to 40% of the alloys melting point. High-temperature corrosion usually is associated with the formation of thick oxide or sulfide scales, with reactions that cause internal swelling of the metal. It is dependent on reactions that include oxygen effects, sulfidation, carburization, decarburization (hydrogen effects), halide effects, and molten-phase formation. Methods to control this type of corrosion are largely confined to materials selection and design, although limited modification of the environment can be achieved and protective coatings can be effective.

6.10. Microbial Corrosion

This general class covers the degradation of materials by bacteria, moulds and fungi or their byproducts. It can occur by a range of actions such as attack of the metal or protective coating by acid by-products, sulphur, hydrogen sulphide or ammonia, direct interaction between the microbes and metal which sustains attack. Prevention can be achieved by selection of resistant materials, frequent cleaning, control of chemistry of surrounding media and removal of nutrients, use of biocides, cathodic protection.

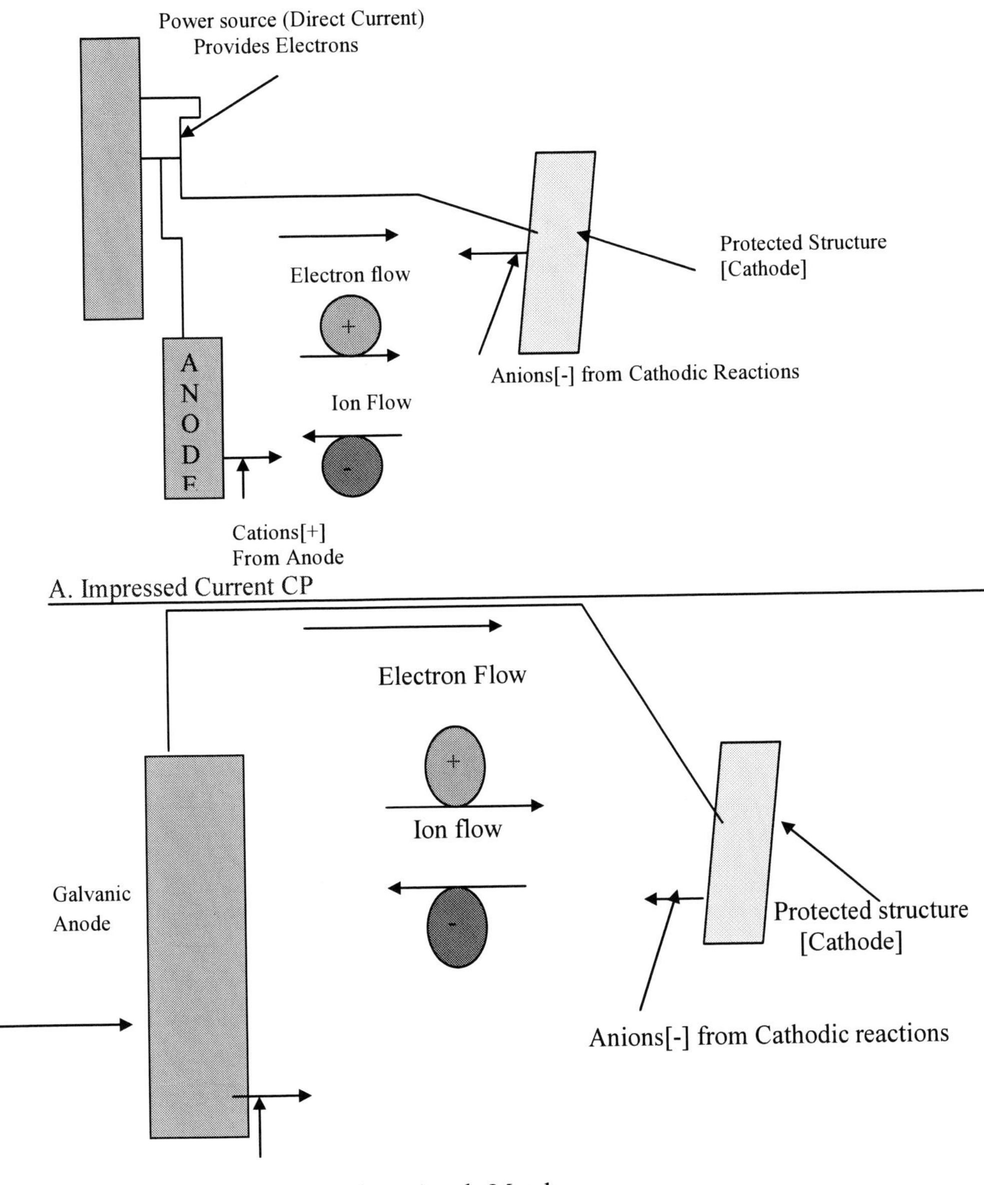

Figure 2. Cathodic Protection Electrochemical Technique.

7. Methods of Corrosion Control

There are four common methods used in controlling corrosion. They include protective coatings and linings, cathodic protection, materials selection, and corrosion inhibitors.

- *Coatings and linings* are principal tools for defending against corrosion. Putting a barrier between a corrosive environment and the material to be protected is a fundamental method of corrosion control. These substances are often applied in conjunction with cathodic protection systems to provide the most cost-effective protection for a structure. There are many organic and metallic coating systems.
- *Cathodic protection (CP)* is a technology which uses direct electrical current to counteract the normal external corrosion of a structure that contains metal, such as an underground petroleum storage tank or natural gas pipeline. On new structures, CP can help to prevent corrosion from starting; on existing structures; CP can help to stop existing corrosion from getting worse (Figure 2). CP is widely used in several environments including water and soil. It is often used in combination with coatings that reduce the exposed surface area to receive protective current.
- *Materials selection* refers to the selection and use of corrosion resistant materials such as stainless steels, plastics, and special alloys to enhance the lifespan of a structure. Some of the most common materials used in constructing a variety of facilities, such as steel and steel reinforced concrete, can be severely affected by corrosion.
- *Corrosion inhibitors* are substances which, when added to a particular environment, decrease the rate of attack of that environment on a material such as metal. They can help to extend the life of equipment, prevent system shutdowns and failures, avoid product contamination, prevent loss of heat transfer, and preserve an attractive appearance of structures. They control corrosion by forming thin films that modify the environment at the metal surface, some retard corrosion by adsorption to form a thin invisible film, others form bulky precipitates that coat the metal and protect it from attack.

8. Nanoscience and Protective Coatings

Among the available methods of corrosion prevention, protective coatings are probably the most widely accepted system due to ease of application and cost effectiveness. Coatings provide a durable corrosion resistant layer. Nanoscience is currently enabling evolutionary changes in several technology areas but new paradigms will eventually have a much wider and revolutionary impact. In the area of coatings, new approaches utilizing nanoscale effects can be used to create coatings with significantly optimized or enhanced properties. The ultimate impact of nanoscience and nanotechnology in the area of coatings, and other potential application areas, will depend on ability to direct the assembly of hierarchical systems that include nanostructures. Nanostuctured materials (1–100 nm) are known for their outstanding mechanical and physical properties due to their extremely fine grain size and high grain boundary volume fraction [4]. In this article a brief introduction to corrosion and application of nanotechnology that is nanocomposite coatings in the field of corrosion prevention of metals is reviewed. Recent research and developments in this area are discussed in designing efficient coating materials, which provide superior resistance to corrosion. Significant progress has been made in various aspects of synthesis of nano-scale materials. The focus is now shifting from synthesis to manufacture of useful structures and coatings having greater wear and corrosion resistance. Existing physical vapour deposition (PVD) and chemical vapour deposition (CVD) processes for preparing microcrystalline coatings can be used to produce nano-structured coatings by modifying the process parameters or using feedstock powders having nano-grained structure. Deposition of coatings by thermally activated processes like high velocity oxy-fuel thermal spraying (HVOF) has been successfully used for producing nanocrystalline coatings [5,6]. Nanostructures promote selective oxidation, forming a protective oxide scale with superior adhesion to the substrate. A polymer nanocomposite coating can effectively combine the benefits of organic polymers, such as elasticity and water resistance to that of advanced inorganic materials, such as hardness and permeability [7]. Improvements in environmental impact can be achieved utilizing nanostructure particulates in coating and eliminating the requirement of toxic solvents. Nano-sized silica has proved to be an alternative to phosphate–chromate pretreatment that is hazardous due to toxic hexavalent chromium [8]. Nano cobalt–phosphorus is compatible with most existing electroplating equipment and positioned as an effective replacement for the hexavalent chromium [9]. Medical benefits of nanostructured diamond, hydroxyapatite and metalloceramics coatings are well reviewed [10]. Nanocomposite coatings based on hydroxyapatite nanoparticles can provide better corrosion protection of titanium that can be utilized for fabrication of advanced biomedical implants [11]. Development of paints and finishing materials with self-cleaning properties, discoloration resistance, graffiti protection and high scratch and wear resistance caused a major revolution in the concrete world [12]. Work is in progress for developing nanoparticle based organic corrosion inhibitors, biocidal coatings and non-skid coatings [13].

9. Polymer Nanocomposite Coatings

Conducting polymers are rapidly gaining traction in new applications with increasingly processable materials with better electrical and physical properties and lower costs. The new nanostructured forms of conducting polymers particularly provide fresh air to this field with their higher surface area and better dispersability [14]. They have been used as host matrices in various composite films. Organic or inorganic particles can be mixed with or incorporated in the conducting polymers to modify their morphology, conductivity and different physical properties depending upon the application, such as corrosion protection. Polycrystalline nanocomposites that consist of conductive polymers possess excellent properties. Nanoparticulate dispersions of organic metal polyanilines in various paints at low concentrations can cause tremendous effects in corrosion protection [15]. Melt dispersion of polyaniline leads to fine particles, which self-organize into complex ultra fine networks. Some specific nanoconducting polymers which enhance corrosion resistance are polyaniline, polythiophene and polypyrrole. To enhance the oxidizing power of the polymers, incorporation of strong oxidizing species in the polymer has been envisaged. Polypyrrole nanocomposites with oxides, especially with Fe_3O_4 have prospects for use in corrosion protection of iron [16]. Polypyrrole nanocomposites with montmorillonite (MMT) clay showed better corrosion protection compared to undoped polypyrrole [17].

Yeh et al. [18] synthesized a series of polymer-clay nanocomposite (PCN) materials consisting of polystyrene (PS) and layered montmorillonite clay by effectively dispersing the inorganic nanolayers of MMT clay in the organic PS matrix *via* thermal polymerization. PCN coatings with low clay loading (1 wt.%) on cold-rolled steel (CRS) were found to be superior in anticorrosion to those of bulk PS.

A series of polymer-clay nanocomposite materials that consist of siloxane-modified epoxy resin and inorganic nano layers of montmorillonite clay has been prepared through a thermal ring opening polymerization using 1,3-bis(3-aminopropyl)-1,1,3,3-tetramethyl disiloxane as a curing agent. These materials at low clay concentration in the form of coating on cold-rolled steel were found to be much superior in corrosion protection over those of pure epoxy resin [19]. The epoxy resin-clay nanocomposite materials have significant advantages over standard epoxy resins such as lower water absorption, lower cure shrinkage, moderate glass transition temperature (Tg), and higher tensile strength.

Electrochemical synthesis of polypyrrole (PPy) film was achieved on mild steel (MS), in monomer containing 0.1 M phenylphosphonic acid solution. It has been found that the electrode surface could only become completely passive, after a few successive cycles in solution of 0.1 M pyrrole + 0.1 M phenylphosphonic acid. Results showed that the coating exhibited important anodic protection behavior on mild steel. The percent protection efficiency value (E%) was found to be 98.4% and the percent total porosity value (P%) was determined to be 0.752%, after 96 h exposure time to corrosive solution [20].

Strongly adherent poly(aniline-*co*-*o*-toluidine) coatings were synthesized on low-carbon-steel substrates by the electrochemical copolymerization of aniline with *o*-toluidine and sodium tartrate as the supporting electrolyte. The results showed that the poly(aniline-*co*-*o*-toluidine) coatings provided more effective corrosion protection to low-carbon steel than the respective homopolymers. The corrosion rate depended on the feed ratio of *o*-toluidine used for the synthesis of the copolymer coatings [21].

Poly(2,5-dimethylaniline) coatings were synthesized on copper (Cu) by electrochemical polymerization of 2,5-dimethylaniline in aqueous salicylate solution by using cyclic voltammetry. Data showed that the poly(2,5-dimethylaniline) coating has excellent ability to protect the Cu against corrosion [22].

Poly(o-anisidine) (POA) coatings were electrosynthesized on copper from an aqueous solution containing o-anisidine and sodium salicylate by using cyclic voltammetry, galvanostatic and potentiostatic modes. Potentiodynamic polarization and electrochemical impedance spectroscopy studies revealed that the POA act as a protective layer on Cu against corrosion in 3% NaCl solution. The positive shift in the corrosion potential for the POA-coated Cu indicated the protection of the Cu surface by the POA. The POA coating synthesized by using cyclic voltammtery, galvanostatic and potentiostatic modes reduced the corrosion rate of Cu almost by a factor of 100, 100 and 7, respectively. The cyclic voltammetry was proved to be the better mode to adopt for the synthesis of more compact and strongly adherent POA coatings on Cu. It was found that the electrochemical polymerization of o-anisidine on Cu takes place after the passivation of its surface *via* the formation of Cu_2O and/or copper salicylate complex and resulted in the generation of shiny, uniform and strongly adherent POA coatings. The optical absorption spectroscopy revealed the formation of the emeraldine salt form of POA. The results showed that the POA has outstanding potential to protect Cu against corrosion in a chloride environment [23].

The poly(*o*-ethylaniline) coatings were electrochemically synthesized on 304-stainless steel by using cyclic voltammetry from an aqueous salicylate medium. The results demonstrated that the poly(*o*-ethylaniline) coating provides excellent protection to both localized and general corrosion of 304-stainless steel [24].

Poly(*o*-ethoxyaniline) (POEA) coatings were synthesized on copper (Cu) by electrochemical polymerization of *o*-ethoxyaniline in aqueous salicylate solution by using cyclic voltammetry. The performance of POEA as protective coating against corrosion of Cu in aqueous 3% sodium chloride (NaCl) was assessed by the potentiodynamic polarization technique and electrochemical impedance spectroscopy (EIS). The results of the potentiodynamic polarization and EIS studies demonstrated that the POEA coating has ability to protect the Cu against corrosion. The corrosion potential was about 0.330 V versus saturated calomel electrode (SCE) more positive in aqueous 3% NaCl for the POEA coated Cu than that of uncoated Cu and reduces the corrosion rate of Cu almost by a factor of 140 [25].

A series of novel advanced environmentally friendly anticorrosive materials have been successfully prepared by effectively dispersing nanolayers of Na^+-montmorillonite (Na^+-MMT) clay into water-based polyacrylate latex (i.e., vinyl acrylic terpolymers). First of all, a polyacrylate latex was synthesized through co-polymerizing organic monomers of methyl methacrylate (MMA), butyl methacrylate (BMA) and styrene (St) using conventional emulsion polymerization technique with sodium dodecyl sulphate (SDS), 1-pentanol and potassium persulphate (KPS) as surfactant, co-surfactant and initiator, respectively. Subsequently, the commercial purified hydrophilic Na^+-MMT was effectively dispersing into the polyacrylate latex through the direct solution dispersion technique. The water-based Na^+-PCN materials loaded with low content of Na^+-MMT when in the form of coating on the CRS coupons was found to be remarkably superior in anticorrosion efficiency over those of neat polyacrylate based on a series of electrochemical measurements of corrosion potential, polarization resistance, corrosion current, and impedance spectroscopy in saline [26].

A series of waterborne polyurethane (WPU)/Na^+-montmorillonite clay nanocomposite materials have been prepared by effectively dispersing the inorganic nanolayers of commercially purified Na^+-MMT clay in WPU matrix through direct aqueous solution dispersion technique. PCN materials in the form of coating at low Na^+-MMT clay loading up to 3 wt% coated on the cold-rolled steel coupons were found to exhibit superior corrosion protection effect over those of neat WPU. A series of polyurethane (PU)/organo-MMT nanocomposite (denoted by organo-PCN) materials were also prepared which showed good corrosion protection [27].

Poly(*o*-toluidine)/cadmium oxide (POT-CdO) nanoparticle composite coating for corrosion protection of mild steel in chloride environment has been studied. The POT-CdO nanoparticle composite coating was synthesized on mild steel from aqueous tartrate solution containing CdO-nanoparticles (size ~18 nm) by using cyclic voltammetry. These coatings were characterized by cyclic voltammetry, UV–Visible absorption spectroscopy, fourier transform infrared spectroscopy, scanning electron microscopy, and X-ray diffraction measurements. The corrosion protection aspects of the resulting POT-CdO nanocomposite structure were investigated in aqueous 3% NaCl solution by potentiodynamic polarization technique and electrochemical impedance spectroscopy. The results revealed that the POT-

CdO nanoparticle composite acts as a protective coating on mild steel and reduces the corrosion rate of mild steel almost by factor of 70 [28].

Efforts are presently underway to develop methods of corrosion protection of different metals that are more effective and also more environmentally friendly than the present techniques. Recently, there has been an upsurge of interest in the prevention of metallic corrosion using coating strategies based on inherently conducting polymers (ICP), thus, creating, an important research field mainly due to restrictions on the use of heavy metals, which have been considered environmentally unacceptable and toxic. Polyaniline (PANI) and its derivatives are among the most frequently studied ICPs used for corrosion protection. In addition, the use of PANI for corrosion protection of metals has been of wide interest. Nanostructured materials engineering extends the possibility of engineering 'smart' coatings that can release corrosion inhibitors on demand when the coating is breached, stressed or an electrical or mechanical control signal is applied to the coating [29]. Ideally, a 'smart' corrosion inhibiting coating will generate or release an inhibitor only when demanded by the initiation of corrosion. Inherently conducting polymer (ICP) films containing inhibiting anions as the dopant anions can release them when the film is coupled to a breach in the coating. Research has developed chromate free corrosion inhibiting additives in which organic corrosion inhibitors are anchored to nanoparticles with high surface areas that can be released on demand [30].

10. Ceramic Nanocomposite Coatings

Ceramic coatings offer an interesting alternative to produce a protective layer over a steel structure due to the excellent chemical, corrosion and thermal resistance of the ceramic materials. Nanoparticles of diamond and chemical compounds used for hard coatings (Al_2O_3, SiC and ZrO_2,) are commercially available [31], with particle sizes in the range 4-300 nm. Within tribology, a new development has been considered to deposit nanocoatings from colloids, e.g. of graphite. Nano-sized silica has proved to be an alternative to toxic chromate conversion coating [8]. Metal precoat based on the combination of a nanostructured metallic oxide of ceramic-type, with metals like Ti and Zr produces nanometre-range conversion coating, while the conventional phosphate layers are within micron range [32].

Amalgamation of suitable nanoparticles in paints is well commercialized in order to improve the properties. During the painting process, e.g. of automobiles, the ceramic nanoparticles float around freely in the liquid paint. At higher temperature, the ceramic nanoparticles crosslink into a dense network instead of the long molecular chains found in conventional paint. This allows the lacquer to provide a much more effective scratch protection against normal wear and tear and allows the paint to retain its gloss. Shen et al. [33] have enhanced the importance of nano-TiO_2 in the development of high corrosion resistance and hydrophobic coatings. Hydrophobic coatings with low wet ability are possible to effectively prevent the water on the substrate surface, and provide excellent corrosion resistance in wet environments. Hydrophobicity of the porous coatings is attributed to air trapped in the nanopores that limits water accessibility and concentration of corrosive species in the stainless steel holes, and hence retards the anodic dissolution process. The corrosion potential of the nano-TiO_2 and fluoro alkyl silanes/nano-TiO_2-coated electrodes is found much nobler than that of the 316 l stainless steel substrate in Ringer's solution.

Deposition of ceramic nano-scaled particles during the electroplating process brings improvements in properties at reasonable cost. Deterioration is observed in the corrosion resistance when the particles were co-deposited. Euler et al. [34] produced a series of nickel nano-ceramic composites, with co-deposition of particles of Al_2O_3 and TiO_2 as a single primary particle in the nanometre range (10–30 nm) at one end of the scale and as agglomerates up to a size of a micrometre at the other. Successful incorporation of particles up to 2 volume% has been established despite the problem of possible agglomeration. The decrease in corrosion resistance is explained by an accelerated diffusion of chloride ions along the interface between nickel and the incorporated particles. The high surface energy and agglomeration tendency of the nanoparticles in highly conductive metal electrolytes will tend to impede uniform distribution of the particles.

Friction and wear characteristics of titanium-containing amorphous hydrocarbon (Ti–C:H) coatings were measured during unlubricated sliding against WC–Co. These Ti–C:H coatings consist of nanocrystalline TiC clusters embedded in an amorphous hydrocarbon (**a**-C:H) matrix, i.e., they were TiC/**a** C:H nanocomposites. The elastic modulus and hardness of the coatings exhibit smooth variations with increasing Ti composition. An abrupt transition occured in the friction coefficient and wear rate of the coatings over a relatively narrow (20–30 atomic %) Ti composition range [35].

Meng et al. [36] have synthesized a series of Ti–Si–N coatings with 0–20 at.% Si by high-density plasma-assisted vapor phase deposition. Data showed that the series of Ti–Si–N coatings were nanocomposites, consisted of a nm-scale mixture of crystalline titanium nitride (TiN) and amorphous silicon nitride (a-Si:N). The hardness of the series of Ti–Si–N coatings was found to be less than 32 GPa.

Ceramic coatings with 1100–1600 HK_{50g} hardness were deposited on Ti–6Al–4V alloy substrates using a micro arc oxidation (MAO) technique, based on a dielectric barrier discharge created during anodic oxidation in an aqueous electrolyte. The influences of electrolyte concentration, deposition time and the cathodic to anodic current ratio (I_c/I_a) on phase composition and mechanical properties of the coatings have been studied. The results showed α-Al_2O_3 (aluminium oxide) phase, which greatly improved the hardness of the coatings, can be obtained at high concentration of sodium aluminate ($NaAlO_2$) and its relative content increases with decreasing I_c/I_a ratio. The direct pull-off test and impact test results indicated that the films have the excellent adhesion with the substrate [37].

TiSiN nanocomposite coatings were deposited on stainless steel by chemical vapor deposition in a fluidized bed reactor (FBR) at atmospheric pressure (AP) (AP/FBR-CVD) by reaction of $TiCl_4$ and $SiCl_4$ with NH_3 at 850 °C. TiSiN coatings with a Si content of 9 at.% showed a hardness of 28 GPa (the hardness of TiN and SiN_x coatings was around 21 GPa) and a lower oxidation rate under dry air at 600 °C. The results showed that AP/FBR-CVD could be tuned for the deposition of nanocomposite ceramic coatings [38].

High velocity oxy-fuel thermal spraying (HVOF) sprayed ceramic coatings can be considered as potential candidates for applications, where good chemical and corrosion resistance of ceramics is needed. Due to the less porous structure of the HVOF sprayed coating as compared to the plasma sprayed one, the protection capability of the coating is increased. It has been shown that HVOF spraying of ceramics was an effective method to produce dense, well-adhered ceramic coatings with good environmental protection capability [5].

Ceramic coatings were prepared on aluminum alloy by plasma electrolytic oxidation method in an environmental-friendly solution. Evolvement of surface morphology, composition and thickness of ceramic coatings on aluminum alloy were studied during plasma electrolytic processes by scanning electron microscopy and X-ray diffractometer. Data has shown that the two-layer structure ceramic coatings consist of different states of Al_2O_3 such as α-Al_2O_3, γ-Al_2O_3 and θ-Al_2O_3. For short treatment times, the formation processes of ceramic coatings followed a linear kinetics, while for longer times the kinetics departs from the linear behavior [39].

Nanocomposite coatings of CrN/Si_3N_4 and $CrAlN/Si_3N_4$ with varying silicon contents were synthesized using a reactive direct current (DC) unbalanced magnetron sputtering system. Study revealed that both CrN/Si_3N_4 and $CrAlN/Si_3N_4$ nanocomposite coatings exhibited cubic B1 NaCl structure in the XRD data, at low silicon contents ($<9\%$ by weight). A maximum hardness and elastic modulus of 29 and 305 GPa, respectively were obtained from the nanoindentation data for CrN/Si_3N_4 nanocomposite coatings, at a silicon content of 7.5% by weight (24 and 285 GPa, respectively for CrN). The hardness and elastic modulus decreased significantly with further increase in silicon content. $CrAlN/Si_3N_4$ nanocomposite coatings exhibited hardness and elastic modulus of 32 and 305 GPa, respectively at a silicon content of 7.5% by weight (31 and 298 GPa, respectively for CrAlN). The thermal stability of the coatings was studied by heating the coatings in air for 30 min in the temperature range of 400–900 °C. The microstructural changes as a result of heating were studied using micro-Raman spectroscopy. The Raman data of the heat-treated coatings in air indicated that CrN/Si_3N_4 and $CrAlN/Si_3N_4$ nanocomposite coatings, with a silicon content of approximately 7.5% by weight were thermally stable up to 700 and 900 °C [40].

Compound ceramic coatings were prepared on Ti-6Al-4V alloy by pulsed single-polar plasma electrolytic oxidation (PEO) in K_2ZrF_6 electrolyte. The corrosion resistance of the coated samples was examined through the potentiodynamic anodic curves in 3.5% NaCl solution. The coatings prepared for short PEO time was composed of *m*-ZrO_2, *t*-ZrO_2 and $ZrTiO_4$ and a little ZrP_2O_7; while increasing PEO time, the content of ZrP_2O_7 was increased and became the main crystalline. The Ti content in the coating near the substrate was decreased sharply while the content of Zr was increased greatly. The thickness of the coating was increased with the PEO time, but the coatings turned rougher and more porous [41].

A novel process named high-velocity suspension flame spraying (HVSFS), has been developed to omit the granulation step and to perform direct spraying of liquid nanoparticle dispersions in a high-velocity oxygen fuel spraying torch with robot controlled kinematics. This process has been well suited to produce dense and desired, very thin coatings of various oxide-based nanocomposites and cermets for tribological and further applications in automotive, aerospace and mechanical engineering [42].

Ni–TiO_2 nanocomposite coatings were prepared by codeposition of Ni and nanoparticles of TiO_2 powder onto surface of Al alloy. The influences of the TiO_2 nano-particulates concentration and current density on the composition of nanocomposite coatings were investigated. Results have shown that Ni–TiO_2 composite coatings exhibited higher corrosion resistances in comparison with pure Ni coating. The morphologies of the coated, eroded and corroded surfaces of pure Ni and Ni–TiO_2 composite coatings were observed using scanning electron microscopy (SEM). The results also showed that TiO_2 nanparticles were uniformly distributed into Ni matrix and significantly improved the microhardness and erosion wear properties of the Ni coating [43].

Ceramic coatings have been produced on 6061 Al alloy substrates in a weak alkaline-silicon electrolyte under different alternating current (AC) voltage at 60 Hz combined with a 200 V, DC voltage for 5 minutes by a plasma electrolytic oxidation process. Analysis showed that a two-layer structure coating consist of different states of α-Al_2O_3, β-Al_2O_3 and γ-Al_2O_3, where Si concentrates in the hard outer layer, and Cl predominates in the soft internal layer. The influence of deposition AC voltages on the kinetics of coating formation, coating microhardness, the number and size of the discharging channels was investigated [44].

Recent research to develop eco-friendly coatings based on salts like silica, ceria, vanadia, or molybdate conversion coatings for aluminium alloys and Al-composite as alternatives to toxic chromate-based systems has been discussed. Newly developed Ni-P nanoparticle alumina and Ni-P-W were studied. Results confirmed the vital role of the surface modification prior to applying the coatings on the corrosion resistance and adhesion performance of chemical conversion coatings. A new coating systems based on etching the substrate surface with dilute alkaline solution followed by oxide thickening in boiled water prior to applying the coatings was found promising to improve the adhesion as well as the corrosion resistance [45].

An industrial scale cathodic arc assisted middle frequency magnetron sputtering system was introduced and Cr–Si–N nanocomposite coatings with various amounts of Si content have been synthesized. Energy dispersive spectroscope, X-ray diffraction, transmission electron microscopy, nanoindentation technique and friction measurement were employed to characterize the microstructure and mechanical properties of the coatings. The microstructure and hardness of the coatings were greatly influenced by the Si content. The nano crystalline (nc) CrN/a-Si_3N_4 structure was formed in the coatings and exhibited a maximum hardness of 39.8 GPa. The surfaces of the samples were smooth and the adhesion between the coating and substrate was higher than 50 N. These properties make it possible for industrial applications [46].

Ti–Si–N nanocomposite coatings were synthesized by using a cathodic arc assisted middle-frequency magnetron sputtering system in an industrial scale. X-ray photoelectron spectroscopy, X-ray diffraction, and transmission electron microscopy were employed to investigate the chemical bonding and microstructure of the coatings. Atomic force microscope and scanning electron microscope were used to characterize the surface and cross-sectional morphologies of the samples. The coating was found to be nc-TiN/a-Si_3N_4 structure and exhibit a high hardness of 40 GPa when the Si content was 6.3 at.%. [47].

TiN-containing amorphous Ti–Si–N (nc-TiN/a-Si_3N_4) nanocomposite coatings were deposited by using a modified closed field unbalanced middle frequency magnetron sputtering system which is arc assisted and consists of two circles of targets, at a substrate temperature of 400 °C. The coatings exhibit good mechanical properties that are greatly influenced by the total gas pressure and N_2/Ar ratios. For coatings prepared at a N_2/Ar ratio of 3:1, the hardness increases from 24 GPa at a total gas pressure of 0.2 Pa–58 GPa at 0.4 Pa, and then, the hardness decreases gradually when the total gas pressure was further increased. On the other hand, the friction coefficient decreases monotonously with increasing total gas pressure. X-ray diffraction (XRD), X-ray photoelectron spectroscopy (XPS) and high resolution transmission electron spectroscopy (TEM) experiments showed that the coatings contain TiN nanocrystals embedded in the amorphous Si_3N_4 matrix. The coating deposited under optimum conditions exhibits excellent tribological performance with a low friction coefficient of 0.42 and a high hardness of 58 GPa. These properties make it possible for industrial applications [48].

11. Self-Assembled Nanophase Coating

Generally, sol–gel method includes the hydrolysis–condensation process followed by condensation polymerization upon film application. However, the evaporation process results in voids and channels throughout the solid gel and cannot provide subsequent corrosion protection due to the high cracking potential. Studies have shown that incorporation of nanoparticles to the sol can increase the coating thickness, without increasing the sintering temperature [49]. Electrophoretic deposition of commercial SiO_2 nanoparticles suspended in an acid-catalysed SiO_2 sol on AISI 304 stainless steel substrates results in coatings as thick as 5 mm with good corrosion resistance [49].

Mixing together of nanoparticles in the hybrid sol-gel systems leads to an increase in the corrosion protection properties due to lower cracking potential and lower porosity as well [50]. Incorporation of inorganic nanoparticles is a way to insert corrosion inhibitors, preparing inhibitor nanoreservoirs for self-repairing pre-treatments with controlled release properties [50,51].

Studies have shown that sol–gel films containing zirconia nanoparticles improve the barrier properties. Doping this hybrid nanostructured sol–gel coating with cerium nitrate brings additional improvement to corrosion protection. In the sol–gel matrix zirconia particles present act as nanoreservoirs resulting a prolonged release of the cerium ions [52].

The recent discovery of a method of forming functionalized silica nanoparticles *in situ* in an aqueous sol–gel process, and then crosslinking the nanoparticles to form a thin film, is an excellent example of a nanoscience approach to coatings. This self-assembled nanophase particle (SNAP) surface treatment based on hydrolyzed silanes, containing a cross linking agent substantially free of organic solvents and Cr-containing compounds promotes adhesion of overcoat layers more effectively.

Unlike chromate based treatments, SNAP coatings provide barrier-type corrosion resistance but do not have the ability to leach corrosion inhibitors upon coating damage and minimize corrosion of the unprotected area [50]. The SNAP surface coating could replace the currently used chromate containing surface treatment and can provide the basis of long lived coating for aluminum alloys [8,53]. The ability to design coating components from the molecular level upward offers potential for creating multifunctional coatings.

Molecular simulation approaches have been used to enhance the understanding of complex chemical interactions in coatings related processes [54]. Soucek and coworkers [55] have studied polyurea and polyurethane inorganic/organic films using different sol-gel precursors such as organofunctional alkoxysilanes or non-functional organoalkoxysilane. The ceramer films exhibited enhanced adhesion and corrosion resistance properties *via* a self-assembly phase separation mechanism. In addition, the new inorganic/organic hybrid materials based on the plant oils were also investigated. The hybrid films showed the higher tensile strength and modulus [55].

12. Self-Cleaning Paints and Biocidal Coatings

The design and development of surfaces is of great interest that not only provides biocidal activity but are also easy to clean and even self-cleaning as well. Most of these types of coatings attain their biocidal capacity by specific nanoparticle incorporation example silver and titanium oxide (TiO_2) [56,57]. Nano TiO_2 provide anti-UV, anti-bacterial and self-cleaning paints properties that is hydrophobic properties, which causes water droplets to bead-off of a fully cured surface picking up dirt and other surface contaminants along the way. This self-cleaning hydrophobic property clean and maintain important surfaces and to accelerate drying, leaving the surface with minimal spotting. Cai et al. [58] has utilized corona treatment technique, inert sol–gel coating and anatase TiO_2 layer. With the corona treatment, an organic surface was activated to allow a uniform TiO_2 sol–gel coating. Nanoparticles treated with Al_2O_3 molecules enhance hydrophobicity and increase scratch resistance as well.

Microbial evolution may cause corrosion, dirt, bad odour and even serious hygiene and health hazards on a wide variety of surfaces. Advanced Nanostructured Surfaces for the Control of Biofouling (AMBIO), a European Union research project [59] is working keenly how to prevent the growth of organisms on surfaces under marine conditions to avoid biofouling. Main aim of the project is to use nanostructuring to significantly reduce the adhesion of organisms to surfaces in marine/aquatic environments, and thus controlling the fouling process inspite the use of toxic biocides like copper and organotin compounds which resist fouling by killing organisms. Nanostructuring of the surface alters the wetting properties and is meant to signal that the site is not suitable for the organisms to settle. The project aims to synthesize new nanostructured polymers that are stable under marine conditions. Even though no alternatives to the use of biocides are available at present, creation of nanostructured surfaces could offer an innovative and environment friendly solution to the problem of biofouling [59]. Research has developed new biocidal coating systems that prolong biocidal activity by immobilizing such additives on nanoparticles; the embedded biocides are designed to be released into the environment only when needed, thus extending the lifetime of the biocidal activity [13].

Conclusion

Corrosion is a phenomenon of universal interest. Corrosion can cause dangerous and expensive damage to everything from vehicles, home appliances, water and waste watersystem to pipelines, bridges and public buildings. The seriousness of the problem has made the corrosion scientists all over the world very conscious and active. Now a day's prevention of corrosion has become a part of their struggle. In the last two decades nanotechnology has been playing an increasing important role in supporting innovate technological advances to manage the corrosion. Nanostructured coatings have recently attracted increasing interest because of the possibilities of synthesizing materials with unique physical and chemical properties due to their extremely fine grain size and high grain boundary volume fraction. Numerous methods of prevention have been suggested and among them corrosion control through the application of nanocomposite coatings have received the attention of the scientist to a very great extent because they exhibit very high resistance to corrosion attack, long term stability in aggressive environment, and economical.

REFERENCES

[1] Delinder, L. S.; *Corrosion Basics: An Introduction*, (Houston, TX: NACE, 1984.
[2] NACE International Basic Corrosion Course Handbook (Houston, TX: NACE, 2000.
[3] Koch, G.H.; Brongers, M. P. H.; Thompson, N. G.; Virmani, Y. P; Payer, J. H. *Corrosion Costs and Preventive Strategies in the United States,* (Washington D.C.: FHWA, 2001).
[4] Nalwa, H. S. (ed.), *Handbook of Nanostructured Materials and Nanotechnology*, Academic Press, San Diego, 2000; Vol. 1.
[5] Turunen, E.; Hirvonen, A.; Varis, T.; Falt, T.; Hannula, S. M.; Sekino, T.; Niihara, K., *Azojomo* 2007, *3*, 1-8.
[6] Honggang, J.; Maggy, L.; Victoria, L. T.; Enrique, J. L. Synthesis of nanostructured coatings by high velocity oxygen fuel thermal spraying; Nalwa, H. S.; Ed.; In *Handbook of Nanostructured Materials and Nanotechnology*, Academic Press, San Diego, 2000; Vol. 1, pp 159–209.
[7] Wang, Y.; Limb, S.; Luob, J. L.; Xub, Z. H. *Wear.* 2006, *260*, 976–983.
[8] Voevodin, N.; Balbyshev, V. N.; Khobaib, M.; Donley, M. S. *Prog. Org. Coat.* 2003, *47*, 416–423.
[9] Bjerklie, S. *Met. Finish.* 2005, *103*, 46–47.
[10] Catledge, S. A.; Fries, M.; Vohra, Y. K. In *Encyclopedia of Nanoscience and Nanotechnology* Nalwa, H. S. Ed.; Nanostructured surface modifications for biomedical implants; American Scientific Publishers: California, 2004; Vol. 7, pp 741–762.
[11] Pang, X.; Zhitomirsky, I. *Int. J. Nanosci.* 2005, *4*, 409–418.
[12] Sobolov, K.; Gutierrez, M. F. *Am. Ceram. Soc. Bull.* 2005, *84*, 14–17.
[13] www.tda.com/Library/docs/Nanomaterials%20for%20Coatings% 205-17-04.pdf
[14] Rout, T. K.; Jha, G.; Singh, A. K.; Bandyopadhyay, N.; Mohanty, O. N. *Surf. Coat. Technol.* 2003, *167*, 16–24.
[15] Wessling, B.; Posdorfer, *J. Synth. Met.* 1999, *102*, 1400–1401.
[16] Garcia, B.; Lamzoudi, A.; Pillier, F.; Le, H. N. T., Deslouis, C. *J Electrochem. Soc.* 2002, *149*, 52–60.
[17] Yeh, J. M.; Chin, C. P. *J. Appl. Polym. Sci.* 2003, *88*, 1072–1078.
[18] Yeh, J. M.; Joe, L. S; Guang, L. C; Chang, Y. P; Hsiang Y. Y; Cheng, F. C. C. *J. Appl. Polym. Sci.* 2004, *92,* 1970-1976.
[19] Yeh, J. M.; Yin, H. H.; Lun, C. C.; Fen, S. W.; Hsiang, Y. Y. *Surf. Coat. Technol.* 2006, *200*, 2753-2763.
[20] Tuken, T.; Yazıcı, B.; Erbil, M. *Appl. Surf. Sci.* 2006, *252*, 62311-2318.
[21] Pritee, P.; Sainkar, S. R.; Patil, P. P. *J. Appl. Polym. Sci.* 2006, *103,* 1868–1878.
[22] Shinde, V.; Gaikwad, A. B.; Patil, P. P. *Appl. Surf. Sci.* 2006, *253,* 1037-1045.
[23] Chaudhari, S.; Sainkar, S. R.; Patil, P. P. *Electrochim Acta* 2007, *53,* 927-933.
[24] Chaudhari, S.; Sainkar, S. R.; Patil, P. P. *Prog. Org. Coat.* 2007, *58,* 54-63.

[25] Chaudhari, S.; Patil, P. P. *Electrochim Acta* 2007, *53,* 927-933.
[26] Lai, M. C.; Chang, K. C.; Yeh, J. M.; Liou, S. J.; Hsieh, M. F.; Chang, H. S. *Eur. Polym. J.* 2007, *43*, 4219-4228.
[27] Yeh, J. M.; Yao, C. T.; Hsieh, C. F.; Lin, L. H.; Chen, P. L.; Wu, J. C.; Yang, H. C.; Chi-Phi, W. *Eur. Polym. J.* 2008, *44,* 3046-3056.
[28] Chaudhari, S.; Gaikwad, A. B.; Patil, P. P. *J. Coating. Technol. Res.* 2010, *7,* 119-129.
[29] Kendig, M.; Warren, L. *Prog. Org. Coat.* 2003, *47*, 183–189.
[30] De Souza, S. *Surf. Coat. Technol.* 2007, *201,* 7574-7581.
[31] Jensen, H.; Sorensen, G. *Surf. Coat. Technol.* 1996, *84*, 500–505.
[32] Droniou, P.; Fristad, W. E. *Coatings* 2005, *38*, 237–239.
[33] Shen, G. X.; Chen, Y. C.; Lin, L.; Lin, C. J.; Scantlebury, D. *Electrochim Acta* 2005, *50*, 5083–5089.
[34] Euler, F.; Jakob, C.; Romanus, H.; Spiess, L.; Wielage, B.; Lampke, T.; Steinha, S. *Electrochim Acta* 2003, *48*, 3063–3070.
[35] Cao, D. M.; Feng, B.; Meng, W. J.; Rehn, L. E.; Khonsari, M. M. *Appl. Phys. Lett.* 2001, *79*, 329-331.
[36] Meng, W. J.; Zhang, X. D.; Shi, B.; Jiang, J. C.; Rehn, L. E.; Baldo, P. M. *Surf. Coat. Technol.* 2003, *163,* 251-259.
[37] Sun, X.; Jiang, Z.; Xin, S.; Yao, Z. *Thin Solid Films* 2005, *147,* 194-199.
[38] Perez-Mariano, J.; Lau, K. H.; Sanjurjo, A.; Caro, J.; Casellas, D.; Colominas, C. *Surf. Coat. Technol.* 2006, *201,* 2217-2225.
[39] Chao, G. W.; Guo-Hua, L.; Chen, H.; Guang-Liang, C.; Wen-Ran, F.; Yang, S. Z. *Mat. Sci. Eng. A* 2007, *447*, 158-162.
[40] Barshilia, H. C.; Deepthi, B.; Rajam, K. S. *Surf. Coat. Technol.* 2007, 201, 9468-9475.
[41] Zhongping, Y.; Yanli, J.; Zhaohua, J.; Fuping, W.; Zhendong, W. *J. Mater. Process Tech.* 2008, *205,* 303-307.
[42] Gadow, R.; Kern, F.; Killinger *Mat. Sci. Eng.* B 2008, *148,* 58-64.
[43] Abdel, A. A.; *Mat. Sci. Eng. A* 2008, *474*, 181-187.
[44] Wang, K.; Kim, Y. J.; Hayashi, Y.; Lee, G. C.; Koo, B. H. *J. Ceram. Process Res.* 2009, *10*, 562-566.
[45] Hamdy, A. S. *Int. J. Nanomanu.* 2009, *4*, 235-241.
[46] Zou, C. W.; Wang, H. J.; Li, M.; Yu, Y. F.; Liu, C. S.; Guo, L. P.; Fu, D. J. *J. Alloys Comp.* 2009, *485,* 236-240.
[47] Yang, Z. T.; Yang, B.; Guo, L. P.; Fu, D. J. *Appl. Surf. Sci.* 2009, *255,* 4720-4724.
[48] Zou, C. W.; Wang, H. J.; Li, M.; Yu, Y. F.; Liu, C. S.; Guo, L. P.; Fu, D. J. *Vacuum* 2010, *84,* 817-822.
[49] Castro, Y., Ferrari, B., Moreno, R.; Duran, A. *Surf. Coat. Technol.* 2004, *182*, 199–203.
[50] Zheludkevich, M. L.; Miranda, S. I. M.; Ferreira, M. G. S. *J. Mater. Chem.* 2005, *15*, 5099–5111.
[51] Zheludkevich, M. L.; Serra, R.; Montemor, M. F.; Miranda, S. I. M.; Ferreira, M. G. S. *Surf. Coat. Technol.* 2006, *200*, 3084–3094.
[52] Zheludkevich, M. L.; Serra, R.; Montemor, M. F.; Yasakau, K. A.; Miranda, S. I. M.; Ferreira, M. G. S. *Electrochim Acta* 2005, *51*, 208–217.
[53] Voevodin, N. N.; Kurdziel, J. W.; Mantz, R. *Surf. Coat. Technol.* 2006, *201*, 1080–1084.

[54] Balbyshev, V. N.; Anderson, K. L.; Sinsawat, A.; Farmer, B. L.; Donley, M. S. *Prog. Org. Coat.* 2003, *47*, 337–341.
[55] Soucek, M.; Zong, Z.; Johnson, A. *J. Coat. Technol. Res.* 2006, *3*. 133-140.
[56] Li, R.; Chen, L. *Chinese Patent*, CN 10027622, 2005.
[57] Morrow, W. H.; McLean, L. J. U.S. Patent, US 2003059549, 2003.
[58] Cai, R.; Van, G. M.; Aw, P. K.; Itoh, K.; *C. R. Chim.* 2006, *9*, 829–835.
[59] www.ambio.bham.ac.uk

In: Advanced Organic-Inorganic Composites
Editor: Inamuddin
ISBN 978-1-61324-264-3

Chapter 14

COMPOSITE MATERIALS FOR ORTHOPAEDIC AIDS

***Geeta Rajput*[a,*] *and Aftab Ahmad Iraqi*[b]**
[a]Department of Prosthodontics,
Dr. Ziauddin Ahmad Dental College,
Aligarh Muslim University,
Aligarh 202002, India
[b]Department of orthopaedic Surgery,
Aligarh Muslim University,
Aligarh 202002, India

ABSTRACT

Nowadays, the use of composite materials becomes common in orthopaedic surgery. These are the materials which mainly consist of polymer materials and fibres as a reinforcement phase. This combination improves physical mechanical and biological properties. These materials commonly used for making artificial limbs and bones. Materials may be bioabsorbable or non bioabsorbable which have been utilized for the fabrication of fractures as well as for soft tissue fixations. Such kind of fixation will gradually transfer the load to the healing tissue, reduced need or hardware removable and radiolucency, which facilitates radiographic evaluation. The ultimate aim is to strengthen the bone and become a part of normal human body functions.

ABBREVIATIONS

PEEK	Polyether ether ketone polymer
PES	Polyethersulphone
PMMA	Polymethylmethacrylate

* Corresponding author's e-mail: geetarajput70@gmail.com

1. INTRODUCTION

Composite materials consist of two or even more different material components or phases, which are combined with the aim to improve physical, mechanical and/or biological properties. Such structures are designed to fulfill very specific requirements with respect to a selected device application making full use of their higher weight-specific strength and/or stiffness. Furthermore, these materials offer an opportunity for constructing radiolucent devices. In medical technology, composite materials mainly consist of a polymer matrix and fibers as a reinforcement phase. An important development has been the usage of carbon-fibre reinforced polymer-matrix for composite limb. The matrix includes polysulfone or polyetherketones. There is a wide array of matrix materials available like polymers, metals and ceramics. The use of these materials and ongoing efforts to make artificial limbs and braces provide great benefits to patients. Bioabsorbable materials have been utilized for the fixation of fractures as well as for soft-tissue fixation. These implants offer the advantages of gradual load transfer to the healing tissue, reduced need or hardware removal and radiolucency, which facilitates postoperative radiographic evaluation.

1.1. Types of Composite Materials

1.1.1. Biodegradable

Biodegradable (or absorbable), self-reinforced polymeric composites fulfill the demands of secure orthopaedic fixation materials because of their high strength, appropriate stiffness and strength retention which can be tailored according to the healing rate of damaged tissues. Hydroxyapatite particles and poly (L-lactide) composites for internal fixation of bone fractures have been developed based on the hypothesis that incorporation of hydroxyapatite particles in a poly(L-lactide) matrix might enhance bone bonding (Figure 1). Composites improved the strength of the interface between bone and plate. This improved interfacial strength lead to a substantial decrease in the frequency of implant loosening in the treatment of fractured bones by internal fixation.

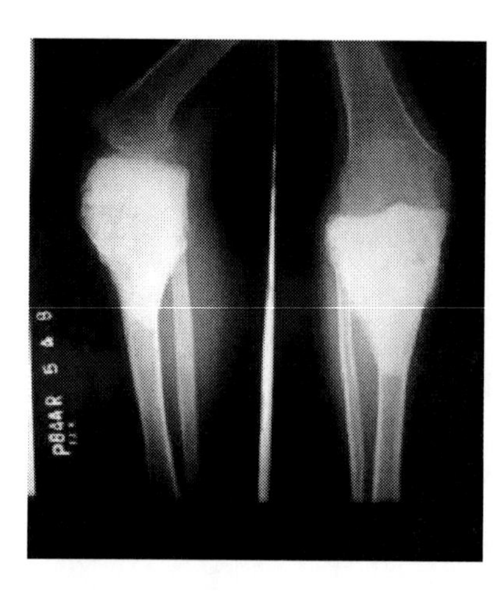

Figure 1. Bone cement used as composite.

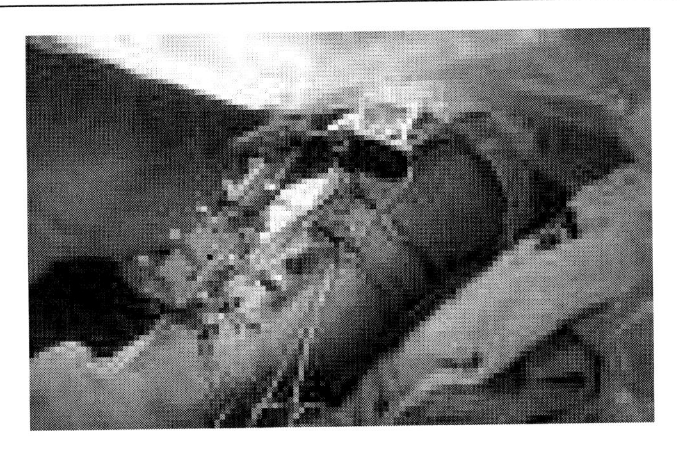

Figure 2. Carbon fibre composite based external fixator.

Figure 2 illustrated the Carbon fibers/polyether ether ketone polymer (C/PEEK) composite materials are being developed for use as orthopaedic implant materials.

Wear is an issue of increasing importance in orthopaedic implants; particulate debris generated by the wearing of biomaterials may be a causal factor leading to osteolysis and implant loosening. Therefore, numerical and experimental studies were completed to characterize the wear of C/PEEK composite materials (Figure 3) in comparison to current orthopaedic implant materials. The composite implant exhibited 10-40% lower contact stresses in the distal region compared to a titanium-alloy implant of identical design [1].

1.1.2. Non-biodegradable

Nowadays these materials are not common in use.

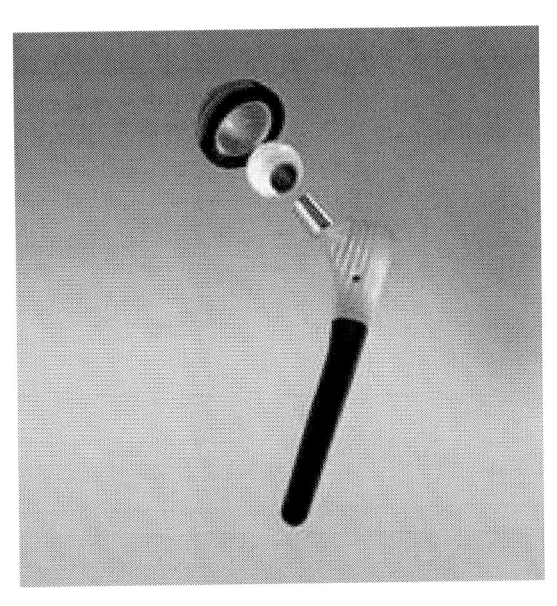

Figure 3. PEEK-OPTIMA carbon fiber composite in combination with a metal insert has been used to create a unique elastically tailored prosthetic hip capable of withstanding high loads for many millions of cycles.

1.2. Carbon Fibre Composites in Orthopaedics

Carbon fibre has negative coefficient of expansion in axial direction, which makes them resistant to thermal shock. They also possess low coefficient of friction and very good fatigue characteristics. The dimensional stability and very low coefficient for thermal expansion are properties instrumental for its use in tooling. The reduced coefficient of friction than glass makes carbon fibre more effective against steel.

The ability of carbon fibre composites to resist chemical environment depends principally on the matrix systems. With the suitable selection of the matrix resin, it is possible to manufacture composites that exhibit excellent resistance to chemicals. Therefore, using polyether sulphone as the matrix for carbon fibres makes an excellent combination of high chemical resistance.

In addition to heat deformation resistance, the polymer is also resistant to chemical change on heating. It is therefore, capable of absorbing a high degree of thermal and ionizing radiation without cross-linking.

The principal features of the polyhether sulphone are their exceptional resistance to creep, good high temperature resistance, rigidity and above all transparency and self extinguishing characteristics. The transparency characteristic of polyethersulphone (PES) enables the composites to be radiolucent.

These excellent properties of carbon fibre and polyether sulphone and moreover, the compatibility of the fibre with PES matrix makes them suitable for such orthopaedic applications (Figure 4 and 5).

The stainless steel external fixators are found heavy for the patients. Composite fibre ring is not only lighter but also due to its radiolucency, it enables the surgeons read the X-rays with better accuracy. Moreover, the high strength to weight ratio of composite made up of carbon fibre and engineering polymers makes it suitable for high impact strength, as needed for the surgical procedures. Other examples of orthopaedic appliances are:

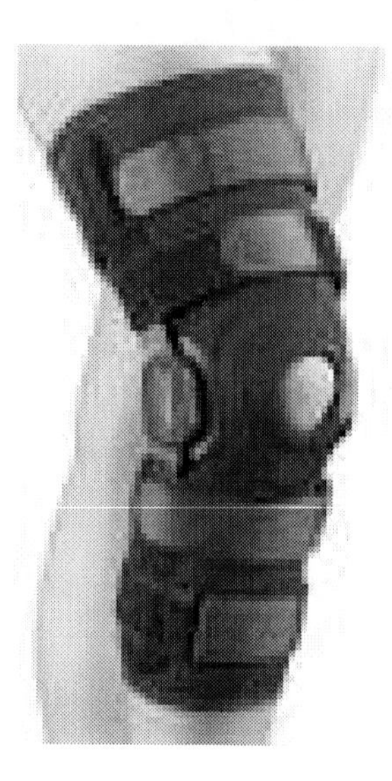

Figure 4. Composite knee braces.

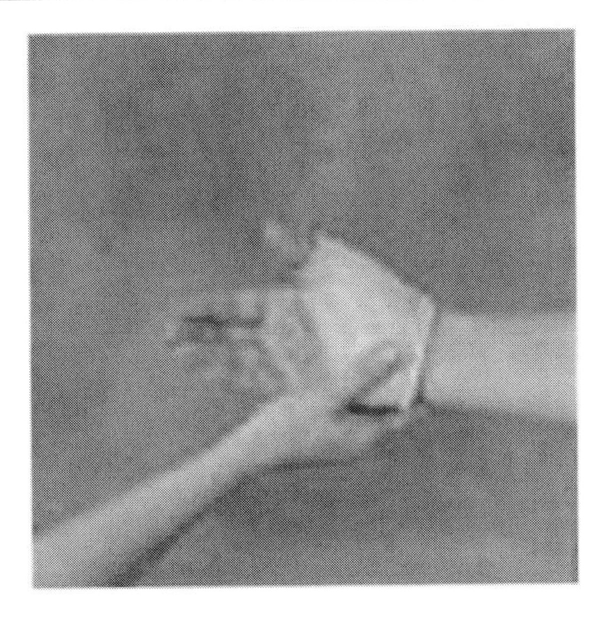

Figure 5. Composite wrist braces.

- Ring fixator system like Italian femoral arches,
- Long connection plate
- Foot rings
- Carbon fibre half rings
- Carbon fibre rods
- Limb reconstructive systems as external fixator

Further, composites have found a good place in various orthopaedic instruments like screwdrivers, taps, osteotomes gauges, etc. for cutting and drilling various sections of bones.

1.3. Injectable Composite Material Suitable for Use as a Bone Substitute

In orthopaedic surgery, biomaterials can be applied in number of diseases of the skeletal system where it is necessary to replace or supplement bone tissue, from the most common ones, related to age, such as osteoporosis, bone arthrosis, arthritis, to the most seriouss ones, such as sarcomas and bone cysts. The need for the use of a bone substitute may arise subsequent to fracture of the bone tissue, in cases where the normal processes of repair and re-growth do not take place in physiological time or do not take place at all, or following surgery for the removal of a tumoral mass or bone cyst, in which case there is a need to fill the cavity produced by surgery.

In addition to this, the increasingly wider use of the arthroscopy technique resulted in an increased interest for a search for injectable materials capable of being conveniently inoculated into bone cavities, allowing optimum filling without the need to know their shape and dimensions in advance, or even, when possible, eliminating the need for surgery. At the present time, the most widely used injectable material in orthopaedic surgery is polymethylmethacrylate (PMMA), which however develops appreciable quantities of heat on application and may result in the necrosis of the tissues with which it comes into contact.

2. Applications of Composite Materials

2.1. Below Knee Composite Artificial Limbs

In India, commonly used exoskeleton type artificial legs are made up of high density polyethylene (Figure 6 and 7). Though the imported endoskeleton type of limbs is available in India, they are very expensive. As physical deformities aggravate the economic woes of the victims in our country, it calls for an indigenous development to restore the functional normalcy of physically challenged people at an affordable price.

2.2. Carbon Fibre Composite Based External Fixator Rings

Instruments like osteotomes, chisel have been developed using PES matrix reinforced with carbon fibre to make the handles. This makes the product lighter but of adequate strength to take beating with a hammer during surgical procedure.

- It is also radiolucent and does not interfere with image intensifier for observing procedure on screen.
- Polyethersulphone shows good resistance to chemical attack. As the surgical chisels require high temperature sterilization, PES is found suitable for the purpose.
- Appliances like limb reconstruction system, ring fixator system (half rings, short connection plates, foot rings, Italian femoral arches, long connection plates, and carbon fibre rods) were developed based on carbon fibre composites. This new material has been developed as a substitute for stainless steel currently being used for the external ring fixator for the bone-healing system.
- The conventional stainless steel external fixators are heavy for the patients. The external fixators made up of composite are not only lighter but also due to their radiolucency, they enable the surgeons read the X-rays with better accuracy.
- The importance of advanced composites for future lightweight materials is derived from the high stiffness combined with high strength and low density of fibre backbone. In this respect, carbon fibre surpasses the more economic glass fibre (Figure 8).

2.3. Total Hip Replacement

In general, the femoral head is replaced with a cast alloy, usually an austenitic stainless steel or Co-Cr alloy, secured, usually by polymethylmelthacrylate bone cement, in the marrow cavity and seated in a high density polyethylene acetabular cup. The alloys used have proved satisfactory, but less than ideal, in such applications; as a result, attention has recently been devoted to the development of even more "bioinert" prosthetic materials with comparable strength properties. Figure 9 shows titanium alloys and ceramics, principally alumina. The brittle nature of ceramics, however, presents new problems for prosthesis. All such synthetic prosthetic materials hitherto used, however, suffer from one major defect: the prosthetic material eventually becomes detached from the bone to which it was originally affixed. That is, hitherto, major surgical prosthetic operations have been intrinsically impermanent.

The present invention seeks to provide composite materials suitable for use in orthopaedics, especially as endoprosthetic materials, which, in vivo, do not become detached from bone to which they are affixed. The particulate inorganic solid component of the composite is present both to reinforce the composite and to enhance its stiffness. Such types of suitable inorganic solids are usually non-metallic and include ceramics, preferably calcium salts. Other particulate inorganic solids include chalk, fly ash and silica. Bone can be considered as a biphasic composite material, mineral as one phase, and collagen and ground substance as the other. In accordance with a further aspect of this invention there is provided a composite of a homo- or co-polyolefin with a particulate inorganic solid for use in surgery as a prosthesis, especially where the composite is as hereinabove defined; and a prosthesis prepared from such a composite, preferably an endoprosthesis, particularly for the direct engagement of bone, which may be a fracture fixation device, a jaw prosthesis or a prosthesis for the simple substitution of a local section of bone; especially, however, the endoprosthesis is a bone joint device, particularly for partial or total replacement of the hip or knee joints. In particular, the composite may be used to fabricate either or both of the femoral head and stem and the acetabular cup (Figure 10) into which the head seats in vivo, although it may be used in the prosthesis of any joint affected by arthrosis.

CONCLUSION

Composite inserts are much better than previously used metallic inserts in limited cases. These are especially used in the areas where gradual load transmission to the healing areas of the bone is required.

REFERENCES

[1] Albert, K.; Schledjewski, R.; Harbaugh, M.; Bleser, S.; Jamison, R.; Friedrich, K. *Biomed. Mater. Eng.* 1994, 4, 199-211.

[2] Herich, L. L. *J. Amer. Ceram.* 1991, *74*, 1487-1510.

[3] Tuan, R. S., *Overview of cells and 67th Annual Academy of Orthopaedic Surgery* 1998, *8*, 119-123.

[4] Peter, S. J.; Miller, M. J.; Yasko, A. W.; Yaszemski, M. J.; Milos, A. G.; *J. Biomed. Mater. Res.* 1998, *43,* 422-427.

In: Advanced Organic-Inorganic Composites
Editor: Inamuddin
ISBN 978-1-61324-264-3

Chapter 15

BIPHASIC LAYER MATERIALS IN THIN LAYER CHROMATOGRAPHIC ANALYSIS OF INORGANIC AND ORGANIC MATRICES

Ali Mohammad[a,*] ***and Abdul Moheman***[b]
[a]Department of Applied Chemistry, Faculty of Engineering & Technology, Aligarh Muslim University, Aligarh 202002, India
[b]Department of Chemistry, Faculty of Science, Aligarh Muslim University, Aligarh 202002, India

ABSTRACT

This chapter covers the literature published on the use of biphasic layer materials during last thirty-one years (1980—2010) in the analysis of organic and inorganic species by thin layer chromatography (TLC). It presents the types of materials used in biphasic layer, mobile phase and technique involved in the separation, identification and determination of inorganic/organic substances present either single, or as components of closely related mixtures in a variety of matrices. Examples of the use of biphasic layer materials in TLC/HPTLC analyses of inorganic and organics are described. In some cases, TLC in combination with other sophisticated analytical techniques has been used for quantitative analysis.

The main features of using biphasic layer materials include the better differential migration, increase separation efficiency and improved resolution of analytes. Until now, the biphasic handmade TLC plates prepared from different types of bulk materials have been commonly used widespread. Because of the poor quality and lack of reproducibility, the handmade plates are now being replaced by biphasic precoated plates. The most frequently used layer material as one of the components of biphasic layers has been silica gel compared to other layer materials.

* Corresponding author's e-mail: alimohammad08@gmail.com

ABBREVIATIONS

$(NH_4)_2SO_4$	Ammonium sulfate
2D TLC	Two dimensional thin layer chromatography
$AcONH_4$	Ammonium acetate
BuOAc	Butyl acetate
BuOH	Butanol
CA	Cholic acid
C_6H_6	Benzene
C_6H_{14}	Hexane
$CaCl_2$	Calcium chloride
CCl_4	Carbon tetrachloride
CDC	Chenodeoxycholic acid
CH_2Cl_2	Dichloromethane
$CHCl_3$	Chloroform (Trichloromethane)
CrO_4^{2-}	Chromate ion
DCA	Deoxycholic acid
Dil HNO_3	Dilute nitric acid
DMSO	Dimethyl sulfoxide
Et_2O	Diethyl ether
EtOAc	Ethyl acetate
EtOH	Ethanol
FAAS	Flameless atomic absorption spectrometry
FCDE	Modified natural diatomaceous earth
FTIR	Fourier transform infra-red
GC	Gas chromatography
GCA	Glycocholic acid
GDCA	Glycodeoxycholic acid
GLCA	Glycolithocholic acid
$HClO_4$	Perchloric acid
H_2O_2	Hydrogen peroxide
HOAc	Acetic acid
HPLC	High performance liquid chromatography
HPTLC	High performance thin layer chromatography
hR_F	R_F x 100
KSCN	Potassium thiocyanate
LCA	Lithocholic acid
Me_2CO	Dimethyl ketone
MeCOEt	Ethyl methyl ketone
MeOH	Methanol
$MgCl_2$	Magnesium chloride
MoO_4^{2-}	Molybdate ion
MS	Mass spectroscopy
MSS	Microspherical silica gel
NaAc	Sodium acetate

$NaCl$	Sodium chloride
$NaClO_4$	Sodium perchlorate
$NaHCO_3$	Sodium bicarbonate
$NaNO_3$	Sodium nitrate
ND	Not detected
NH_3	Ammonia
NH_4Cl	Ammonium chloride
NH_4NO_3	Ammonium nitrate
NH_4OH	Ammonium hydroxide
NO_2^-	Nitrite ion
n-PrOH	n-Propanol
pH	Potentia hydrogenii
Pr_2O	Dipropyl ether
REEs	Rare-earth elements
R_F	Retardation factor
RPTLC	Reversed phase thin layer chromatography
SAIE	Stannic arsenate ion-exchange gel
SCN^-	Thiocyanate ion
TBP	Tributyl phosphate
TGBA	Thiophene-2-glyoxal-p-bromo-anil
TGDEA	Thiophene-2-glyoxal-p-diethylamino-anil
TGDMA	Thiophene-2-glyoxal-p-dimethylamino-anil
THF	Tetrahydrofuran
TLC	Thin layer chromatography
UV	Ultraviolet light
v/v	Volume/volume
w/w	Weight/weight
WO_4^{2-}	Tungstate ion

1. INTRODUCTION

1.1. Thin Layer Chromatography

Thin layer chromatography (TLC) is an open-bed chromatographic technique with extensive versatility, much already utilized, but still with great potential for future development into different areas of research. It was first proposed by Izmailov Schraiber in 1938 [1]. TLC is useful to determine the components in a mixture and to check the purity of a compound. By observing the appearance of a product or the disappearance of a reactant, it can also be used to monitor the progress of a reaction. In all chromatographic methods including TLC it is necessary to have two phases to achieve a successful separation. These phases are designated as the stationary phase and the mobile phase. Although TLC is still being used primarily as an adsorption technique, it has been found useful for separations based on ion exchange, partition and other modes used in chromatography. First the sample to be analyzed is dissolved in an appropriate solvent and applied as spots or bands on TLC plate approximately 1 cm from the edge. An eluent (single solvent or mixture of solvents) is allowed to flow through the layer starting at a point just below the applied samples. Most commonly this is performed in a glass rectangular tank in which the eluent is poured to give a depth of about 5 mm. The plate is placed in the tank of chromatography chamber and the chamber is covered with a lid to get saturated with solvent vapors. As the eluent front migrates through the layer, the components of the sample also migrate. When the solvent front reached to at point near the top of the sorbent layer, the plate or sheet is removed and dried. The spots or bands on the developed layer are visualized, if required, under UV light or by chemical treatment or derivatisation. For quantitative determination zones can be removed, analyte is eluted from the layer and determined using suitable analytical technique. In modern TLC, sample application, development and recording of the chromatograms is realized by fast automated procedures.

Although thin layer chromatography is an analytical method in its own, it is also complementary to other chromatographic and spectroscopic techniques. It has been successfully hyphenated with high performance liquid chromatography (HPLC), gas chromatography (GC), flameless atomic absorption spectrometry (FAAS), mass spectroscopy (MS), photoaccuoustic spectroscopy, Fourier transform infra-red (FTIR), and Raman spectroscopy to provide more useful analytical data.

The ratio of the distance that a compound moves to the distance that the eluent front moves is called the retention factor, R_F (as shown in Figure 1). The value of which is always lesser than 1.0.

In order to obtain a reproducible R_F values, the following basic factors should be considered:

Quality of adsorbent
Thickness of the layer
Activity of the layer
Relative humidity
Amount of sample applied
Solvent quality
Temperature
Equilibrium between liquid and vapor in tanks

1.1.1. TLC Problems and Remedial Steps

Problem	*Cause*	*Remedy*
Over migration	Developer too polar	Reduce polarity
Under migration	Developer too non-polar	Increase polarity
Distorted solvent front	Developer not equilibrated	Proper equilibration
Distorted spots	Wrong adsorbent or high concentration of analyte	Change plates / reduce concentration
No separation	Wrong developer or adsorbent	Change developer or plate adsorbent
Tailing	Spot overloading, basic/acidic component	Reduce concentration, increase acidity / basicity

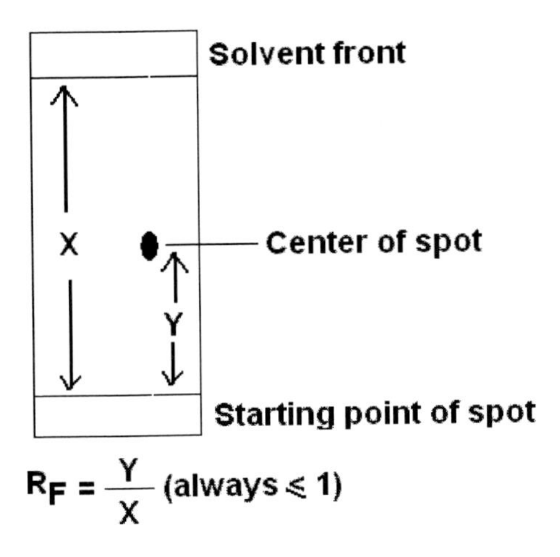

Figure 1. A chromatogram showing measurements for R_F calculations.

1.2. Layer Materials

Various layer materials are available, some of which have been more widely used than others [2-4]. More important ones will be reviewed in this chapter. For optimum separation, it is important that correct material is chosen. Before selecting sorbent material consideration must be given to the compounds to be separated. Parameters such as the polarity, solubility, ionisability, molecular weight, shape and size of the analytes are all important to influence the separation mechanism. The more commonly used layer materials in TLC analyses of inorganic and organics have been the following:

1.2.1. Silica Gel

Silica gel, also called silicic acid is the most versatile among all stationary phases used in adsorption as well as in partition modes of TLC. It is a white amorphous porous material, usually made by precipitation from silicate solutions by addition of acid. It consists of silanol groups, Si-OH and siloxane structures, Si-O-Si. The silanol group represents adsorption-active surface centers that are able to interact with solvent and solute molecules during the separation process. An increase in surface activity of silica results in lowering of R_F values due to change in silica porosity and humidity. For TLC, silica gel with 5 to 40 μm particle size and specific pore volume of 0.5-2.0 ml/g is generally found useful. Water content plays an important role in the retention of analytes on the chromatographic layer. The water is held in the structure either as physically adsorbed or hydrogen-bonded water. Activation of TLC plate by heating upto 105 ^{o}C for 30 min is necessary to removes the physically bound water.

1.2.2. Cellulose

Cellulose is another layer material that has a polymeric structure consisting of glycopyranose units joined together by oxygen bridges. The chemical formula is $(C_6H_{10}O_5)_x$. The specific surface area is about 2 m^2/g. The presence of hydroxyl groups in cellulose is available readily for hydrogen bonding. Micro-crystalline form of cellulose commonly called '*Avicel*' a fine powder used widely in both TLC and HPTLC. '*Avicel*' is formed by dissolving the amorphous part of native cellulose by hydrolysis with hydrochloric acid. The resolution of sample spots is generally diffused as that obtained on silica gel layer. The diffusion of the chromatographic zones is greatly reduced with HPTLC cellulose. Many separations have been based on cellulose and it is still widely in use research laboratories [5-8].

1.2.3. Alumina

Like silica gel, alumina or aluminum oxide, is a synthetic adsorbent. It is of three types:

The first type is acidic in character (pH range of 4.0 to 4.5) and has the following structure,

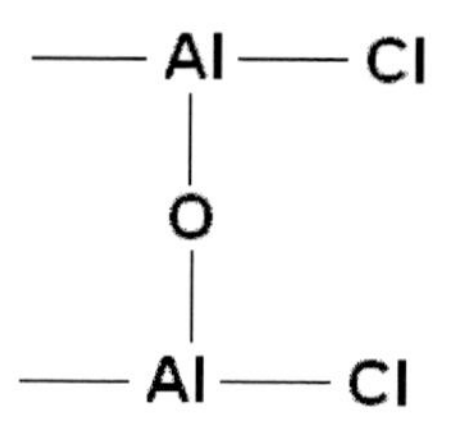

Acidic alumina

The second type of alumina is defined as neutral (pH between 7.0 and 8.0) and is thought to have the following structure,

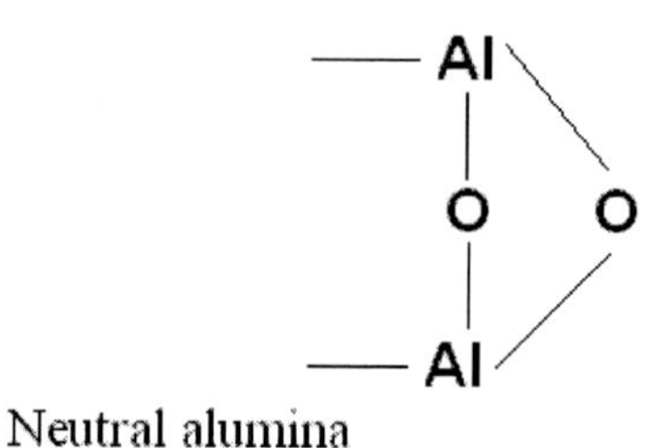

Neutral alumina

Finally, the third type of alumina is defined as basic (pH approximately 9.0 to 10.0) and thought to have the following structure,

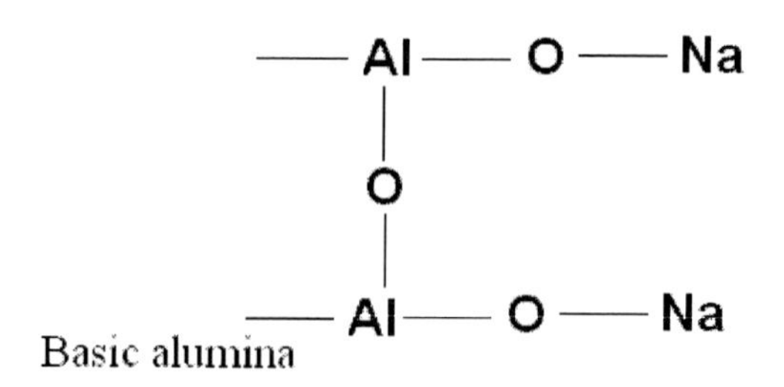

Basic alumina

Of the three types, basic alumina is the most widely used. The surface area and a particle size of alumina normally employed in TLC have been 100-250 m^2/g, 20 μm respectively. Owing to the high density of hydroxyl groups (about 13μmol/m^2), alumina tends to adsorb water and become deactivated. For this reason, it is necessary to activate aluminum oxide layers, before use by heating at 105—110°C. Many separations of inorganic and organics have been reported on alumina layer by adsorption and partition chromatography [9-13].

1.2.4. Kieselguhr

Kieselguhr is a purified and thermally treated diatomaceous earth. It is a neutral adsorbent having a lower resolving power. It is mixed with silica gel to increase its adsorption capacity. It is mainly used as a support for mobile phase in partition TLC. Kieselguhr has very large pores and, as a consequence, has a relatively small surface area (1-5 m^2/g). The variability of pore size and surface area limits the use of kieselguhr for high quality precision TLC.

1.2.5. Ion Exchangers

The ion exchange materials commonly used in TLC are inorganic ion exchangers, ion exchange resins, ion exchange cellulose and ion exchange bonded phases. The inorganic ion exchange materials are zeolites (aliuminosilicates), apatite and hydroxy apatite. However, some of the phosphates, molybdates, tungstates and vanadates of the tetravalent metals have also been found useful for ion exchange separations. However, the most popular ion exchange substrates used in TLC are the ion exchange resins and the ion exchange celluloses. The major difference between an ion exchange resin and ion exchange cellulose is that, in the former, the ion exchange groups are attached to a largely dispersive matrix whereas with ion exchange cellulose, the groups are attached to a largely polar matrix. Thus, in separations carried out on ion exchange resins, the dominant interactive forces that control retention are ionic and dispersive. In contrast, the dominant forces that will control retention on ion exchange cellulose may be ionic and polar.

Ion exchange bonded phases which are widely used in TLC as layer material are usually silica based. Carboxylic acids and amino groups are frequently used to provide ionogenic properties to bonded phases. Unfortunately, silica based bonded phases are unstable at extremes of pH due to the silica matrix. Consequently, ion exchange bonded phases have a limited pH range over which they can be employed effectively. As a result, ion exchange resins are preferentially chosen as the stationary phase, both in LC (liquid chromatography) and TLC, for separations where ionic interactions dominate and control the level of retention.

1.2.6. Polyamide

Polyamide phases in TLC are produced from polycaprolactam (nylon 6), polyhexamethyldiaminoadipate (nylon 6, 6), or polyaminoundecanoic acid (nylon 11). They can be represented by the following formula:

$$-[NH-(CH_2)_6-NH-CO-(CH_2)_x-C)]_n$$

If x = 4, then the product is nylon 6, 6, if x = 8 then the product is nylon 6, 10 etc.

A range of precoated sheets with aluminum or plastic backing are commercially available. The chromatographic separation of inorganic and organics on polyamide depends on the hydrogen bonding capabilities of its amide and carbonyl groups. The relative retention of the analytes depends on the nature of eluting solvent. As the solvent migrates through the sorbent, the analytes separate in accordance with their ease of displacement. Many separations have been reported on polyamide layer material [14-16].

1.2.7. Miscellaneous

Other less commonly used layer materials include magnesium silicate, chitin, starch, sephadex^{TM} and talc. Magnesium silicate is a white, hard powder often known under the name of florisil^{TM}. The manufacturer, floridin, (Pittsburgh, USA) gives the surface area of the TLC grade florisil 298 m^2/g and the pH as 8.5. The chromatographic characteristics of magnesia (florisil^{TM}) are similar to those of silica gel, but the material is basic in nature in contrast to silica gel which is acidic. It has been reported as suitable for the separation of closely related compounds and positional isomers [17,18]. Chitin is a polysaccharide composed mainly of 2-acetamide-2-deoxy-D-glucan molecules linked *via* oxygen linkages in a similar type of structure to cellulose but it is basic in nature. Typical specific surface area is low, only 6 m^2/g. It has been used for the separations of inorganic and organics [19]. Starch is a carbohydrate consisting of a large number of glucose units joined together by glycosidic bonds. This polysaccharide is produced by all green plants. Starch molecules arrange themselves in the plant in the form of semi-crystalline granules. Each plant species has a unique starch granular size: rice starch is relatively small (about 2 μm) while potato starches have larger granules (up to 100 μm). Cassava, Guinea corn and Irish potato starches are extracted from the tubers of *Manihot utilissima* and *Solanum tuberosum* and seeds of *Sorghum vulgare.* As shown earlier, this material has excellent separation characteristics for some other classes of organic compounds [20]. Sephadex are polymeric, cross-linked dextran gels, hydrophilic and neutral in nature. Sephadex is a trade name of pharmacia fine chemicals, which are available in coarse (100-300 μm), medium (50-150 μm), fine (20-80 μm) and superfine (10-40 μm) particle size distribution. The use of these materials in TLC as layer material has been limited because they require pre-swelling for many hours before use. These materials have been found suitable for the separation of peptides and nucleic acids [21]. Talc, a major component of baby powder, is a magnesium silicate mineral with the composition $Mg_3[Si_4O_{10}](OH)_2$ and has layered lattice structure. Its available surface for adsorption markedly depends on particle size. Talc has been used as a TLC adsorbent to separate fatty acids, lanatosides, amino acids, the sugars and the flavonoids [22-24]. Thus, baby powders based talc may be safely used as TLC adsorbents. A number of other materials have also been used as TLC layer materials such as chitosan [25], soil [26-32], egg shell [33-36], calcium sulfate [37] activated bentonite [38], kaolinite [39], china clay [40] etc.

2. Biphasic Layer Materials

Biphasic layer materials can be obtained by mixing two adsorbents bearing different functional groups. By varying the fractions of the constituents in a 'mixture' one can alter the affinity of adsorbent towards inorganic or organics within the limits defined by the properties of pure individual adsorbents. Generally, the retention of inorganic or organics on biphasic layer material is intermediate between the values for individual stationary phases (Figure 2 a). Figure 2 a, provides a very simple way to regulate the separation ability and to create adsorbents for special tasks.

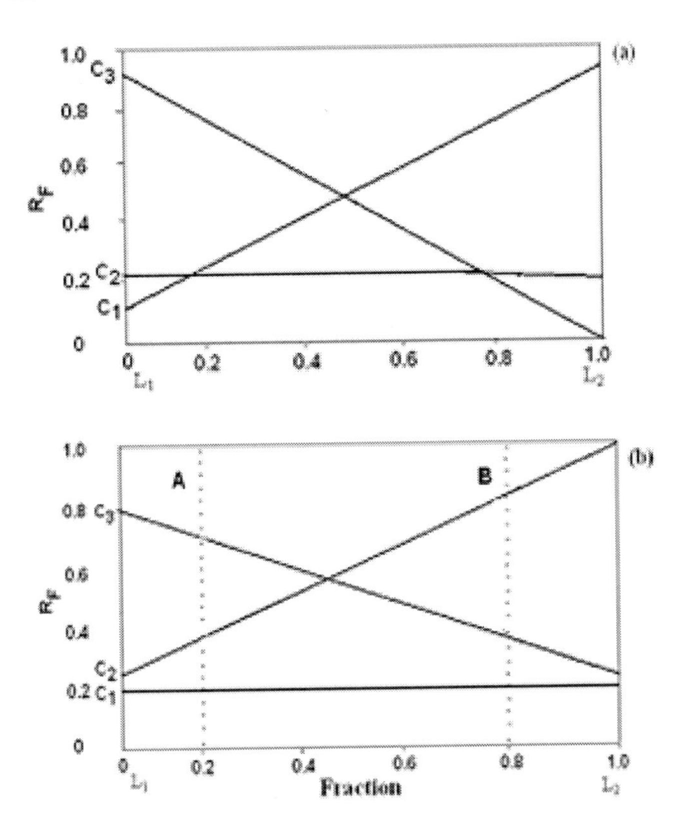

Figure 2. Use of biphasic layer materials (a) Change of the retention order of components: $C_1 > C_2 > C_3$ (as for L_1 alone); $C_3 > C_2 > C_1$ (as for L_2 alone); $C_2 > C_1$, C_3 (as for $L_1 + L_2$, w/w) C_1, C_2, C_3 are components (analytes); L_1 and L_2 denotes layer material respectively. (b) Improvement of separation, the use of individual phases results in poor separation; "mixtures" A, B show good separation.

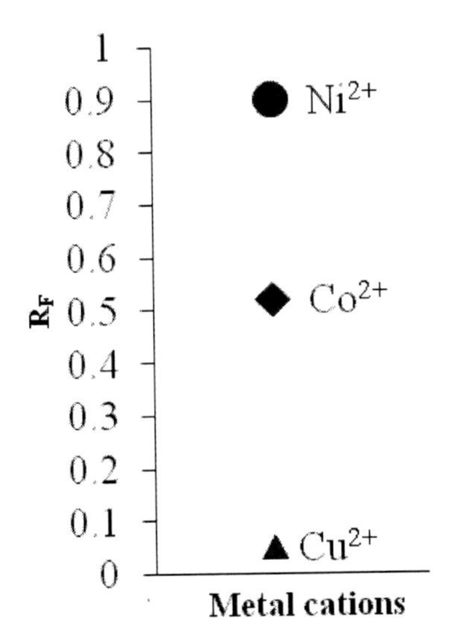

Figure 3. Separation of coexisting Cu^{2+}, Co^{2+} and Ni^{2+} on 0.2 M TBP impregnated biphasic layer material (SAIE—silica gel, 10:1, w/w).

Somewhat less interesting is the potential ability to use biphasic layer materials to enhance the separation of multi-components mixtures of inorganic or organic (Figure 2 b). Of the three component mixture ($C_1 + C_2 + C_3$) shown in the Figure 2 b, one pair is not separated on the L_1-adsorbent and another is not separated on the L_2-phase. The mixtures corresponding to lines A and B provide a reasonable separation of all components. This is evidently important for a chromatographic analysis.

The aim of this chapter is to encapsulate the scattered information on the application of biphasic (or mixed layers) such as silica gel—cellulose, silica gel—alumina, cellulose—alumina, silica gel—kieselguhr, alumina—kieselguhr, silica gel—cation or anion exchange resin, silica—polyamide, silica—florisil, silica—chitin, silica—chitosan and silica—starch etc. in TLC analysis of inorganic and organic compounds.

2.1. Applications

2.1.1. Thin Layer Chromatographic Analysis of Inorganics

Biphasic layer materials in TLC have been proved useful for the analysis of inorganic substances. The chromatographic effectiveness of biphasic layer materials is found to be better than single layer materials [41]. The most interesting point is that the biphasic layer materials may have the greatest affinity to any inorganic ions which is not bound most strongly on any parent individual adsorbent. Sharma et al. [42] has described the measurement of mobility of twenty metal ions on single (starch, alumina) as well as on biphasic (starch—alumina, starch—talc, silica gel—talc and alumina—talc in 1:1 ratio) layer materials using aqueous carboxylic acids as mobile phase. Authors suggested that admixture of talc-with starch proved very useful in effecting a number of analytically important separations of metal ion. Mohammad and Jabeen [43] studied the separation and mobility of heavy metal cations on single (soil or silica gel) or biphasic (mixture of soil-silica gel) layer materials with surfactant mediated eluent systems. The study suggests that, the mobility of all cations was insignificant on pure soil layers, irrespective of the nature of mobile phase used. Addition of silica gel into soil leads to the increase of mobility and facilitates the separation of metal cations. In order to compare the chromatographic performance of biphasic layer materials, the silica gel was replaced by alumina, cellulose, kieselguhr and fly ash in the mixture of soil and silica gel (1:9) and the TLC plates coated with these layer materials were used to determine the mobility of metal ions. In order to investigate the effect of silica gel 'G' concentration in binary mixture, Mohammad and Hena [44] developed a method for mutual separation of Al^{3+}, Fe^{3+} and Ti^{4+} on single as well as on biphasic layer materials. Separation of coexisting Al^{3+}, Fe^{3+} and Ti^{4+} is possible only if silica gel 'G' is kept $\geq$ 50% (w/w) in mixed layer materials. Low weight ratios of silica gel 'G' with alumina, cellulose and kieselguhr G causes tailing in the spot of Fe^{3+}. Thin layer of pure cellulose, alumina and kieselguhr 'G' fails to resolve the three-component mixture of Al^{3+}, Fe^{3+} and Ti^{4+}. A rapid and sensitive TLC method comprising polychrome A (a copolymer of mixed 1, 4-and 1, 5-di (methacryloyloxymethyl) naphthalene and styrene with starch, dextrin of a polyvinyl alcohol containing 10-14% acetate groups as binder) as stationary phase and 0.5 M HCl-EtOH, 9:1 as mobile phase for measurement of mobility of some inorganic (Re, Mo, V and W) was developed by Gawdzik and coworkers [45]. A new biphasic layer material consisting of stannic arsenate ion exchange gel and silica gel has been utilized for the separation and identification of coexisting Co^{2+}, Ni^{2+} and Cu^{2+} ions from their mixture [46]. Tributyl phosphate (TBP) was used as an eluent as well as an impregnate. The TLC system comprising of stannic arsenate ion exchange (SAIE) gel mixed with silica gel in 10:1 ratio (w/w) impregnated with 0.2 M TBP as stationary phase and 0.1 M potassium thiocyanate as mobile phase was identified to be the best for providing well resolved spots of Cu^{2+}, Co^{2+} and Ni^{2+} ions from their mixture (Figure 3).

Mohammad and coworkers [47] have reported a simple and accurate reversed phase TLC system comprising of 0.2 M of tributyl-phosphate-impregnated stannic arsenate-silica gel biphasic layer material as stationary phase and 1.0 M KSCN + 5.0 M HCI + 1.0 M NaCI (8:1:1) as mobile phase for determination of Fe^{3+} in rock sample. The retention behavior of heavy metals on layer comprising mixtures of silica gel and inorganic ion exchange gels has been examined [48]. The layers of practical utility for the separation of cations have been identified. Mohammad et al. have determined mobility of some cations and anions on biphasic sorbent layers using aqueous methanol containing tributylphosphate and formic acid as mobile phase [49]. Mohammad and Agrawal [50] have reported a new mixed layer material for the isolation of nickel from industrial wastes. The mobility of various anions on plain adsorbent layers (silica gel, alumina or cellulose) as well as on layers containing different combinations of silica gel and alumina or cellulose has been examined [51]. The mobility (R_F value) was found to depend on the composition of sorbent materials used (Figure 4). To achieve difference in migration of anions, cellulose and kieselguhr were mixed together in 4:1, 3:2 and 1:1 (w/w) to get a set of biphasic adsorbents on which the anions were chromatographed using mixture of 0.1 M NH_4OH and Me_2CO in 1:9 (v/v) as mobile phase [52]. Out of three layer materials, cellulose-kieselguhr, 3:2 is identified the best combination in term of spot compactness and sufficient difference in hR_F values of SCN^-, NO_2^-, CrO_4^{2-}, MoO_4^{2-} (Figure 5). Some application of biphasic layer materials in inorganic TLC that have not been covered in the text, are summarized in Table 1.

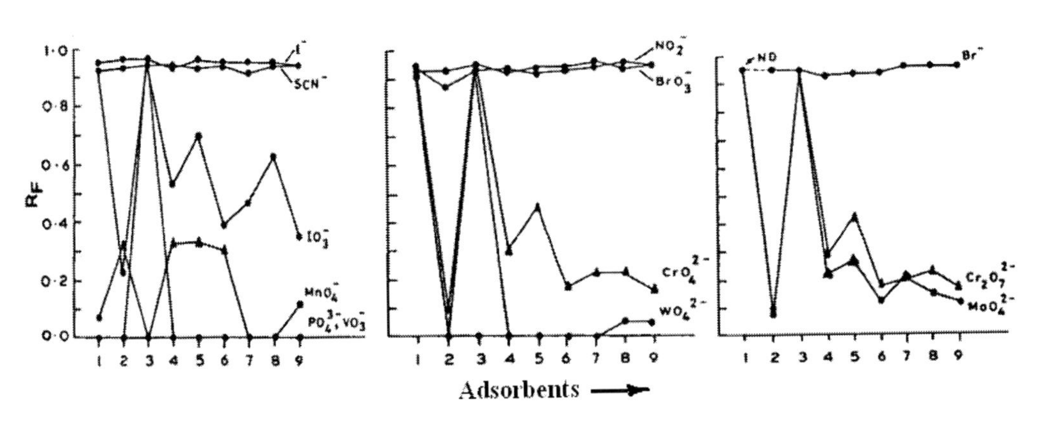

Figure 4. Effect of nature of adsorbents on R_F values of anions. Adsorbents: 1, silica gel (SG); 2, alumina (AL); 3, cellulose (CL); 4, AL−SG (1:1); 5, AL−SG (1:2); AL−SG (2:1);6, AL−CL (1:1); 7, AL−CL (1:2); AL−CL (2:1). ●, Compact spot; ▲, tailed spots; ND, not detected. Reproduced with permission from Mohammad, A. et al. *Microchem. J.* 1993, *47,* 379-385. Copyright © 1993, Elsevier Science Ltd.

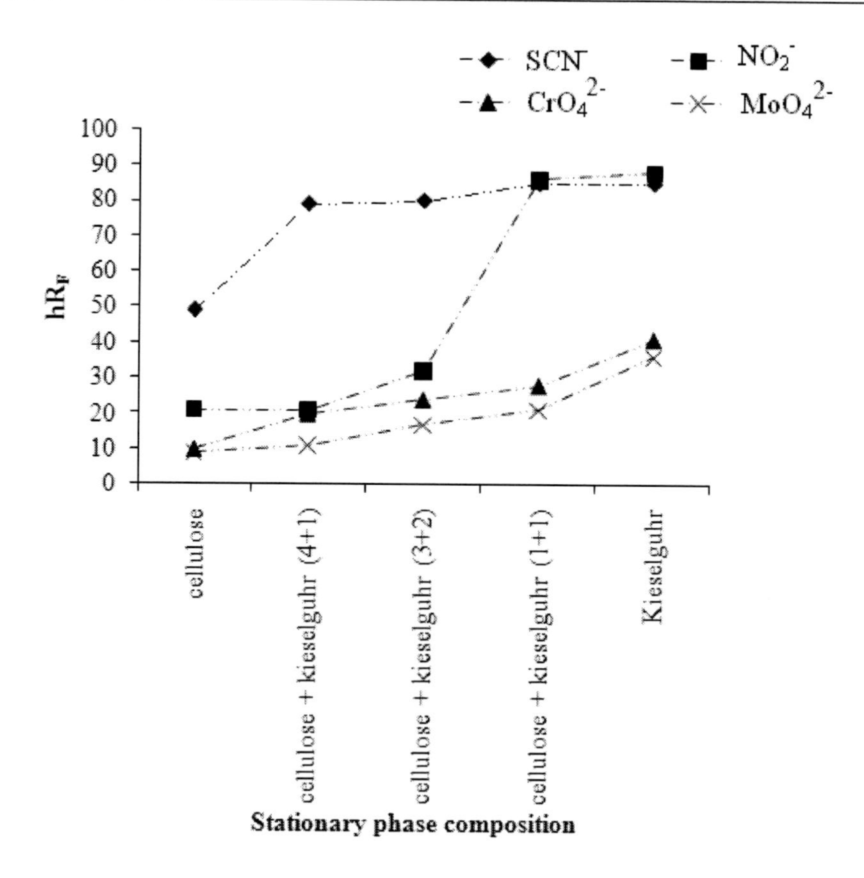

Figure 5. Effect of stationary phase composition on the mobility of inorganic anions using mobile phase 0.1 M NH_4OH + Me_2CO in 1:9.

Table 1. Biphasic layer materials used in thin layer chromatographic analysis of inorganic

Analyte	Layer Material	Mobile Phase	Remarks	Ref.
Forty-nine ions including W, Ha, Tl, Mo, Re, Sn, Pb, Cr, Ni, Cd, Zr, Co, Te, Sc, Sb, Cu, Ag, As, Be, Fe, Mg, Th, Y, Ge	Cellulose phosphate–microcrystalline cellulose (3:1)	Aqueous acetic acid (0.1-3.0 M) and mixtures of acetic acid and ammonium acetate solutions	Development time 35-45 min; layer thickness 0.25 mm; run 17 cm; qualitative separations.	[53]
α- and β- naphthylanils of methyleneglyoxal of Ti, UO_2, Au and Hg	Silica gel–starch (19:1)	MeOH, EtOH, BuOH, AcOH, BuOH–AcOH (1:1, 1:2, 2:1), aqueous BuOH–AcOH–$CHCl_3$ (5:5:1, 5:5:2, 5:5:3) and BuOH–C_6H_{14}–$CHCl_3$ (1:2:1, 2:1:1)	Layer thickness 1 mm; temperature 35 °C; BuOH-C_6H_{14}-$CHCl_3$ (1:2:1) was most suitable mobile phase for the separation.	[54]
Rh(III), Ir(III), Ir(IV)	Acteola–cellulose	2.0 M HCI with different concentrations of H_2O_2	Separation by anion exchange TLC.	[55]

Analyte	Layer Material	Mobile Phase	Remarks	Ref.
Ni, Mn, Co, Cu	Silica gel–vionit CS-32 ion exchange resin	Mixtures of Me_2CO, HCI and water in various ratios	Separation on silica gel was poor; addition of ion exchange resin markedly improved the resolution.	[56]
Au, Ir(III), Ir(IV), Pt, Pd, Ru, Rh	Acteola–cellulose	Different concentrations of HCI and aqueous chloride solutions of Li, Mg, Na, Ca, Sr and Ac	Qualitative analysis.	[57]
Forty metal ions	Dowex 1 or 50–plain cellulose (Avicel SF)	HOAc–HCl	Qualitative analysis.	[58]
Cis and trans isomers of Ti, tetrahedral and trigonal planar isomers of Hg, square planar and tetrahedral isomers of Zn with p-diethylaminoanil of anthraceneglyoxal and thiourea	Silica gel–starch (19:1)	n-Butanol–benzene (1:1, v/v), n-butanol and acetic acid	Separation of various isomers.	[59]
Ti, Mn, Fe, Co, Cu, Zn complexed with the p-diethylaminoanil of anthracene glyoxal alone and alongwith thiourea	Silica gel–starch (19:1)	Acetone, n-butanol–benzene	Cations of complexes were identified by means of migration rates and spectroscopic correlations.	[60]
Rh(III), Ir(III), Ir(IV)	Acteola–cellulose	3.0 or 5.0 M HCI containing $NaClO_4$	Anion exchange separations.	[61]
Rare-earth elements	Mixture of silica gel, starch and NH_4NO_3	TBP-MeCOEt, EtOAc and HNO_3	Separation and sepectrophotometric determination of rare-earth elements; application to the determination of La, Ce, Pr and Nd in monazite.	[62]
Ba, Y, La, Ce	Silica gel–hydrous manganese dioxide	Dil. HNO_3	Quantitative separation from Sr.	[63]
Sn, Mn, Fe, Co, Ni, Cu, Zr, Y, La, Pr, Nd, Sm, Gd, Dy, as their ions and as their complexes with the p-dimethylaminoanil of 3-benzoylmethyl glyoxal (DMABC)	Silica gel–starch (24:1, w/w)	CH_2Cl_2, $CHCl_3$, CCl_4, C_6H_6, BuOH, C_6H_6–BuOH (1:1, 2:1, 1:2)	Ascending technique; layer thickness 1mm, temperature 30-62 °C; qualitative separations.	[64]
Au, Ru, Rh, Pd, Os, Ir, Pt, Cu, Fe, Co and Ni	Semicrystalline Sn phosphate ion exchanger–silica layers	Various solvent systems	Studies of the chromatographic behavior of noble metal ions on a semicrystalline inorganic ion exchanger.	[65]

Table 1. Continued

Analyte	Layer Material	Mobile Phase	Remarks	Ref.
Light rare-earth metals	Silica gel–NH_4NO_3–carboxymethyl cellulose (5:0.64:0.16, w/w)	EtOAc–dioxane–P_{204}–HNO_3 (2.4:2:1.1:0.4)	Effective separation of light rare-earths in the presence of heavy rare-earths.	[66]
Al, Ni, Cr, Mn, Co, Zn, Fe	Silica gel–chitosan	n-BuOH–conc HCl (2:1, 3:1, 4:1, 9:1, 17:3)	The best solvent for the separation of a cation mixture is n-BuOH–conc HCI (17:3).	[67]
Light rare-earth metals	Silica gel H–NH_4NO_3–CM-cellulose (5.0:0.64:0.16, w/w)	TBP–THF–Et_2O–HNO_3 (1:9:9:1.5)	Separation of light rare-earth metals in monazite.	[68]
Ru, Au, Pd, Pt, Rh, Os, Ir, Ag	Tin pyrophosphate and silica gel containing sodium carboxymethyl cellulose as binder	Acids (nitric, tartaric, citric, perchloric, formic), bases (aqueous NH_3, trimethylamine), neutral compounds (NH_4Cl, NH_4NO_3, $AcONH_4$ or mixtures of these with organic solvents (EtOH, MeOH, n-PrOH, Me_2CO)	Qualitative analysis.	[69]
Heavy rare-earth metals	Silica gel–sodium carboxymethyl cellulose–NH_4NO_3	Et_2O–THF–bis-(2-ethylhexyl) phosphate –HNO_3	Separation of Cd, Tb, Dy and Ho.	[70]
Rare-earth elements	Silica gel–NH_4NO_3 –sodium carboxymethyl cellulose (33:4:1, w/w)	2-ethylhexyl (2-ethylhexyl) phosphoric acid (P_{507})—methylisobutylketone—iso-Pr_2O—HNO_3 (1:12:6:2.4, v/v)	Qualitative separations of rare-earth metals.	[71]
Rare-earth elements	Silica gel H–microcrystalline cellulose–NH_4NO_3	Bis-(2-ethylhexyl) phosphate (P_{204})–P_{507}–THF–HNO_3–iso-PrOH, 3:2:90:19:280, 17:2:110:18:20, v/v)	Separation of lanthanides.	[72]
Platinum complexes with ketoanils	Silica gel G–starch (95:5, w/w)	Methyl chloride, dioxane, AcOH, MeOH, EtOH, aqueous BuOH, BuOH–AcOH (1:1, 1:2, 2:1), aqueous BuOH–$CHCl_3$ (5:1, 5:2), aqueous BuOH–AcOH–$CHCl_3$ (5:5:1, 5:5:2, 5:5:3)	Quantitative separation of platinum metal (Ru, Pd, Ir, Pt) complexes with TGDMA, TGDEA, TGBA; layer thickness 1 mm; ascending technique; run 10 cm.	[73]
Rare-earth elements	Silica gel–starch–ammonium rhodantate, (2.8:0.15:0.5, w/w)	Tri-n-octylamine or tri-iso-octylamine–4-methyl-2-pentanone–isopropylether–isopropylalcohol–HNO_3 (1:8:8:6:0.75, v/v)	R_F were governed by the amount of each component in the eluent; discussion of the mechanisms of REE; application of the method to the separation of REEs in monazite.	[74]

Analyte	Layer Material	Mobile Phase	Remarks	Ref.
Rare-earth elements	Silica gel–starch –ammonium rhodantate (2.8:0.15:0.5, w/w)	Trimethylammonium chloride–n-octylalcohol–petroleum ether –HCI (2:10:30:1, 2:6:30:1.2 and 2:7:30:1, v/v)	Qualitative separations; detection limits 0.01 mg; increase in R_F value with increasing volume of trimethylammonium chloride in mobile phase.	[75]
Heavy rare-earth metals	Silica gel–NH_4NO_3–CM -cellulose (5.0:0.64:0.16, w/w)	Bis(2-ethylhexyl) phosphate–diisopropylether–diethylether–nitric acid (1:10:6:1.1) Mono(2-ethylhexyl) phosphate–isopropyl ether–ethyl ether–nitric acid (1:8:8:1.1)	La, Sm, Eu, Gd, Tb, Dy, Ho, Er, Tm, Yb, Lu and Y are separated from each other; developer showed better separation efficiency.	[76]
28 metal ions including Ru, Pd, W, Pt, Au, Mo and UO_2^{2+}	Zr(IV) antimonate– silica gel G (1:1)	HNO_3 (10^{24}-1.0 M), DMSO–0.1 M HNO_3 (1:0, 4:1, 3:2, 1:4), dioxane–0.1 M HNO_3 (1:0, 4:1, 3:2, 2:3, 1:4)	Quantitative separation of Ru^{3+} from other metal ions.	[77]
La, Ce, Pr, Nd, Sm, Gd, Dy, Er, Y, Pb	Silica gel–NH_4NO_3–CM-cellulose (5.0 :0.64 :0.16, w/w)	Mixed solvent systems consisting of mono(2-ethylhexyl phosphate, 4-methylpentanone, nitric acid, isopropyl ether and / or THF in different ratio	Qualitative separations and applications of the method to the analysis of rare-earth ores.	[78]
Cu, Cd, Co, Pb	Silica gel H–sodium carboxymethyl cellulose	0.2 M acetic acid–0.2 M sodium acetate systems	Quantification by densitometry, detection limits in the nanogram range; the method was applied to the determination of Cu and Pb in ground water and electroplating waste water.	[79]
Au, Pt, Pd, Cr, Mn, Fe, Co, Ni, Cu, Ba, Al, Bi, Pb, Zn, Ag	Acteola-–cellulose	Mixtures of 2.5 M HCI, 2.5 M NaCI and 0.6% H_2O_2 in different ratios	Application of method for the analysis of platinum powder and two kinds of Au alloys.	[80]
18 anions	Alumina and alumina–silica gel (1:1, 1:2, 2:1)	Mixed acidic organic solvent systems containing formic acid	Discussion of the chromatographic behavior of anions on single or biphasic layer materials.	[81]
Metal ions	Silica–inorganic ion exchange gel	Sixteen different solvent systems	Qualitative separations.	[82]
Some anions	Alumina–silica gel G (1:1, 1:2, 2:1)	Distilled water	Qualitative separations, effect of $CaCI_2$, $MgCI_2$ and $NaHCO_3$ on the separation of anions; identification of NO_2^- in artificial sea water.	[83]

Table 1. Continued

Analyte	Layer Material	Mobile Phase	Remarks	Ref.
Cations: Ni^{2+}, Co^{2+}, Cd^{2+}, Cu^{2+}, UO_2^{2+}, VO^{2+}, $Fe^{2+/3+}$, Al^{3+}, Th^{4+}, Mo^{6+}, W^{6+}, Pb^{2+}, $Hg^{+/2+}$, Bi^{3+}, Ag^{+} and Tl^{+} Anions: CrO_4^{2-}, $Cr_2O_7^{2-}$, IO_3^{-}, IO_4^{-}, BrO_3^{-}, SCN^{-}, $Fe(CN)_6^{3-}$, NO_2^{-}	1:9 mixtures of a synthetic inorganic ion-exchanger with silica gel, alumina, or cellulose	1.0% Methanolic TBP−1.0 M aqueous formic acid (1:4)	Several binary cation separations of analytical interest were achieved. The mutual ions were detected by use of conventional spot-test reagents.	[84]
Rare-earth benzoates (La, Ce, Nd, Sm, Dy and Yb)	Various combination of silica−alumina	Pure single or two-component solvent systems	Some mutual separations of benzoates were realized on silica and mixed adsorbent layers developed with some organic and nonacidic solvent systems. The best separation conditions were obtained with non-impregnated silica-alumina, 9:1 with 0.5M NH_3.	[85]
Pr(IV), W(VI), Au(III), Fe(III), TI(I), As(III), Se(IV), Sb(III)	silica mixed with cerium (III) nitrate and sodium silicate	0.6 M NH_3, 0.01 M $HClO_4$, 1.0 M AcOH, 0.5 M $NaNO_3$ solution	Determination of mobility of metal ions on cerium (III) silicate. A new ion-exchanger.	[86]
Mn, Fe, Ni and Cu or Zn	Silica gel−cellulose (2:1)	Double distilled water	Identification of suitable chromatographic systems for separation of some transition metal chlorosulfate on mixed adsorbent layers.	[87]
Cu(II), Co(II) and Ni(II)	Silica gel G F_{254}-−NaAc (40:1) in 0.2% carboxymethyl cellulose sodium solution	Single and mixed developing agent systems	In the benzene/CCl_4 and toluene/CCl_4 mixed developing agent systems, the R_F values of complexes of Co(II), Ni(II) and Cu(II) were different than in single developing agents. Trace (Cu(II) and Ni(II) in environmental reference material and in water sample were determined successfully and the results were satisfactory.	[88]
Al^{3+}, Fe^{2+}, Si^{4+} and Ti^{4+} in bauxite samples	Silica gel H—silica gel G, silica gel H—cellulose, silica gel G—cellulose, silica gel G 60—cellulose (in 1:1, 1:2, 2:1, 3:7, 7:3, 1:9, 9:1 ratios)	Several single and mixed solvent systems	Thirty bauxite samples of different geological origin have been studied. Binary and ternary separations of Al^{3+}, Fe^{2+}, Ti^{4+} and Si^{4+} in bauxite were achieved. Quantitative analysis by scanning densitometry.	[89]

2.1.2. Thin Layer Chromatographic Analysis of Organics

Many investigations have confirmed the usefulness of biphasic layer materials in the analysis of organic compounds. According to Wu [90], a biphasic layer consisting of silica and microcrystalline cellulose was most suitable for TLC separation and quantitative analysis of amino acids, nucleosides, organic acids and saccharides etc. Mohammad et al. [29] reported that the presence of silica gel in soil layer activates the movement of pesticides resulting in enhanced migration of pesticides. Mohammad and coworkers [91] developed a method to investigate the effect of additives on the migration of pesticides through soil modified with silica gel, alumina, or cellulose. Authors suggested that, the presence of silica gel, alumina, or cellulose in soil influences the mobility of pesticides. A few studies [92-94] have been reported on mixed inorganic ion-exchange layers. It has been found that the presence of ion-exchange material in the adsorbent enhances the separation potential of the layer. Two-dimensional (2D) TLC is a very popular method for separation of multicomponent mixtures. In 2D TLC the success of a separation depends on the difference between the selectivity of the two chromatographic systems used. Nyiredy [95] described the technique of joining two different adsorbent layers to form a single plate. Preliminary experiments have shown that it is possible to separate mixture of pesticides on bilayer adsorbents by 2D TLC [96-101]. To optimize layer materials for direct TLC-FTIR coupling, Bauer et al. [102] compared chromatograms of phenazene, caffeine or paracetamol obtained on 1:1 mixed silica gel 6o—magnesium tungstate with that obtained on silica gel (Figure 6).

Thin layer chromatography is one of the most suitable methods for the separation and quantification of bile acids in biological samples. Pyka and Dolowy [103] investigated the separation of seven selected bile acids, such as cholic acid (C), glycocholic acid (GCA), glycolithocholic acid (GLCA), deoxycholic acid (DCA), chenodeoxycholic acid (CDCA), glycodeoxycholic acid (GDCA) and lithocholic acid (LCA) by adsorption TLC using silica gel and of silica gel plus kieselguhr as layer materials. They observed certain differences in R_F values of the studied bile acids on mixed layer prepared from silica gel 60 and kieselguhr F_{254}. Pyka et al. [104] studied the influence of temperature on retention and separation of selected bile acids (cholic acid, glycocholic acid, glycolithocholic acid and lithocholic acid) using adsorption TLC on aluminium plates precoated with the mixture of silica gel 60 F_{254} (# 1.05554) and kieselguhr F_{254} (# 1.05567). Mixtures consisting of n-hexane, ethylacetate and acetic acid in different volume ratios were used as mobile phase. They suggested that the right choice of temperature can improve the separation of some bile acids. Thin layer chromatography has been used for the identification and determination of lipophilic vitamins, steroid hormones and their metabolities in a variety of samples, such as biological samples, plants and pharmaceutical formulations. Pyka et al. [105] has examined the mobility (in term of R_F values) of lipophilic vitamins (α-tocopherol acetate, α-tocopherol, cholecalciferol) and steroid hormones (estradiol, testosterone, hydrocortisone) on silica gel 60 (#105553) and mixture of silica gel 60 and kieselguhr F_{254} (#105567) layers by normal phase thin layer chromatography (NPTLC). The results presented in Figure 7 indicate that, the R_F values of studied vitamins and steroid hormones were lower on silica gel 60 as compared to their values on mixture of silica gel 60 and kieselguhr F_{254}, but the order of retention was the same.

Various chitosan derivatives fixed on silica gel through adsorption or chemical reactions are effectively used as chiral stationary phases [106,107]. Using TLC method, Kabulov et al. [108] examined the feasibility of using the chitosan-silica nanocomposite sorbent as a stationary phase for separating cytisine alkaloid and some of its derivatives. The chromatographic effectiveness of microspherical silica gel is enhanced by blending it with chitosan. The results presented in Figures 8 and 9 suggests that the chitosan–silica sorbent is more effective stationary phase for separating cytisine alkaloid and its derivatives using a 6:1 (v/v) chloroform—methanol mixture as a mobile phase. On the chitosan–silica sorbent, selectivity indices are higher than on micro spherical silica gel (Figure 9).

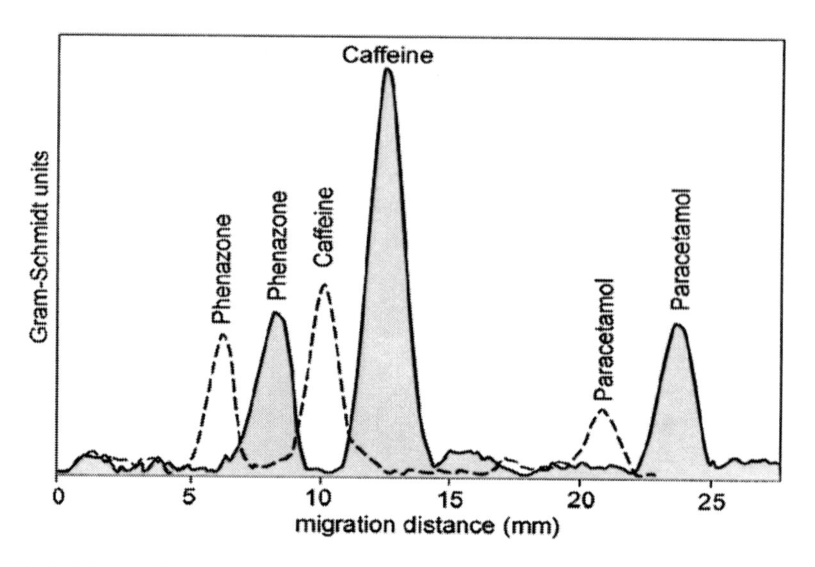

Figure 6. HPTLC-FTIR on-line coupling. Chromatogram of test substances on 1:1 mixed silica gel 60-magnesium tungstate (---) and on silica gel (—). Reproduced with permission from Bauer, G. K. et al. *J. Planar Chromatogr.* 1998, 2, 84–89. Copyright © 1998, Springer.

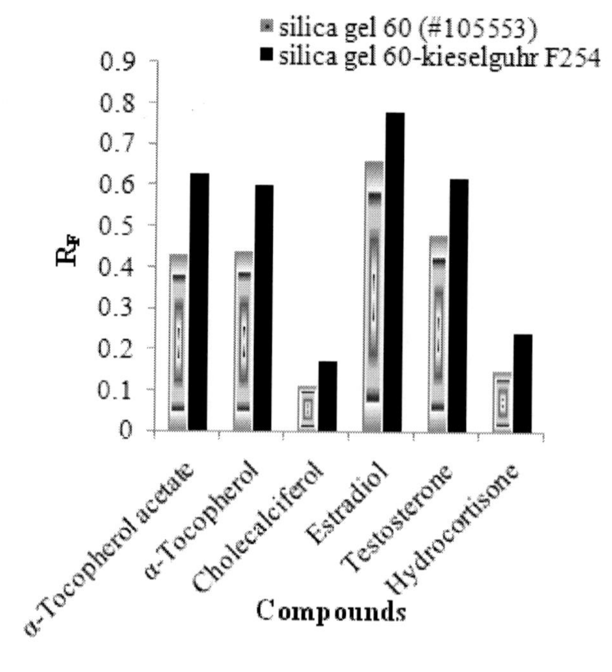

Figure 7. R_F values of selected lipophilic vitamins and steroid hormones investigated on silica gel 60 (#105553) and mixture of silica gel 60 and kieselguhr F_{254} (#105567).

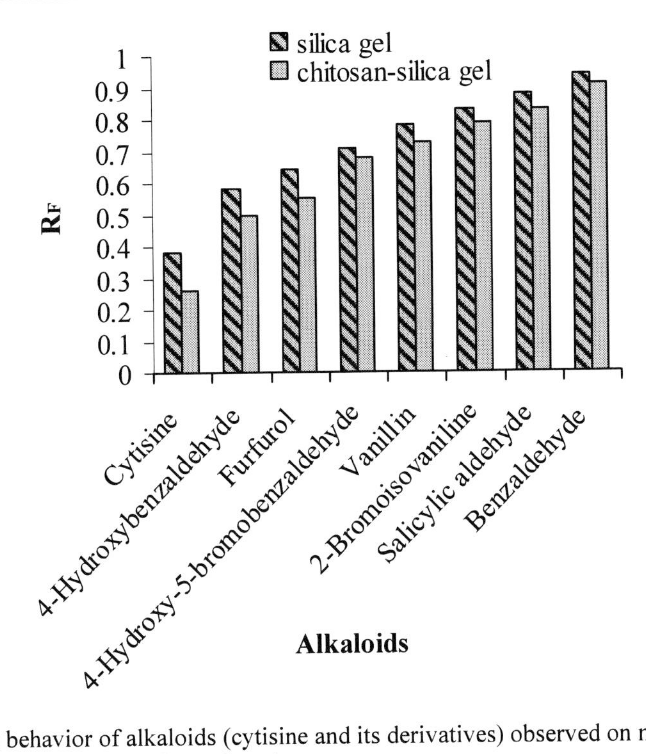

Figure 8. Retention behavior of alkaloids (cytisine and its derivatives) observed on microspherical silica gel and chitosan-silica nanocomposite sorbent layer.

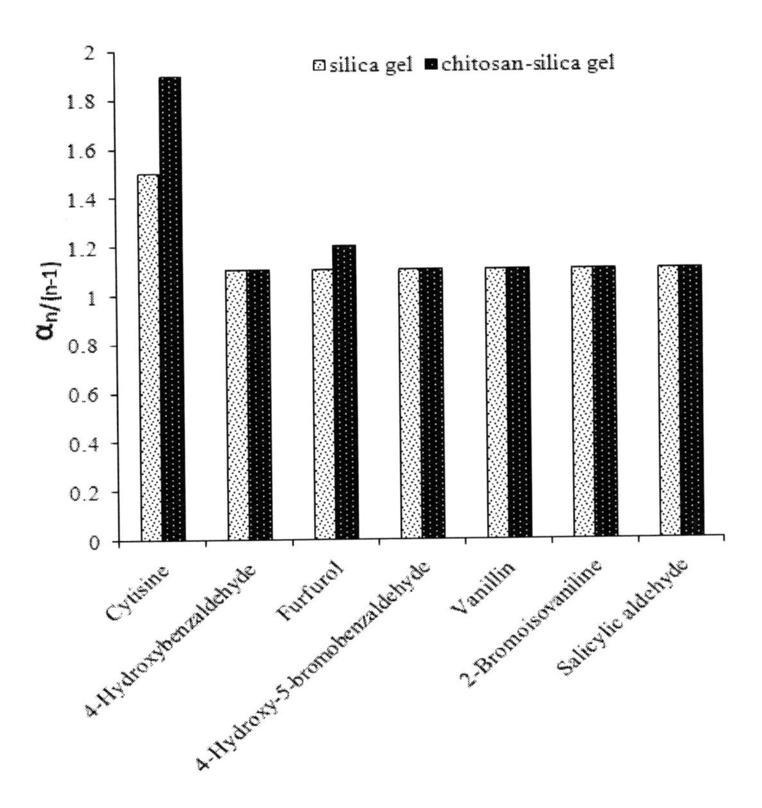

Figure 9. (Continued).

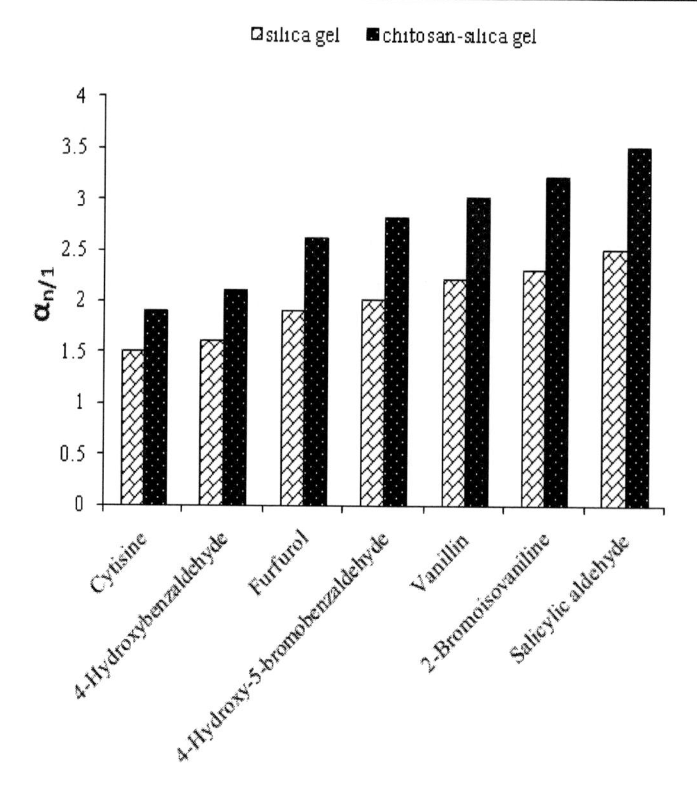

Figure 9. Selectivity indices of separation of cytisine and its derivatives on both single and biphasic layer materials.

Biphasic silica-magnesia adsorbents have been proved to be useful for the separation of various organic compounds during last decades. The retention behavior of positional isomers (phenol, aniline derivatives, quinoline bases and polyaromatic hydrocarbons on mixed silica-magnesia layers using various binary eluents have been examined by Waksmundzka-Hajnos and Wrońska [109]. The influence of magnesia content in mixed adsorbents on retention was investigated by comparison of the retention of solutes on magnesia using the same eluents. Analysis of the structural effects of solutes and the position of points on $R_{M\ magnesia}$ Vs. $R_{M\ mix}$ enables comparison of chemical character and distribution of active centers on magnesia and silica-magnesia surfaces.

Waksmundzka-Hajnos [110] has compared the retention behavior of ortho- and para-isomers of phenol, aniline derivatives and quinoline bases on mixed silica-magnesia adsorbent with the retention behavior of solutes on alumina and silica. Modified natural diatomaceous earth (FCDE-1) and commercial silica gel 60 G F254 (Si-60GF254) layer materials have been very useful in TLC on using as stationary phases individually or in combination. Commercially available red and blue ink samples were run on layers of Si60GF254 and FCDE-1 individually and on various FCDE-1 and Si-60GF254 mixtures. Butanol-ethanol-2 M ammonia (3:1:1, v/v) and butanol-acetic acid-water (12:3:5, v/v) mixtures were used as mobile phases. It was observed [111] that the R_F values of ink components were increased when the FCDE-1 content in the stationary phase is increased. The important examples showing the utility of biphasic layer materials in TLC of organic compounds are presented in Table 2.

Table 2. Biphasic layer materials used in thin layer chromatographic analysis of organic

Analyte	Layer Material	Mobile Phase	Remarks	Ref.
20 Essential amino acids	Silica — polyamide (5:1)	MeOH—BuOAc—AcOH—pyridine (20:20:10:5)	TLC of amino acids. Detection with 0.1% ninhydrin in acetone and heating at 60°C for 60 min.	[112]
21 Amino acids	Silica gel, alumina, cellulose, kieselguhr and their mixtures	Water	Comparison of retention behavior of different supports-determination of lipophilicity of amino acids by RPTLC.	[113]
Amino acids	Cellulose—silica gel (5:2)	Two solvent systems	Description of a developing chamber for horizontal TLC.	[114]
Amino acids	Cellulose—silica gel	-	Two-dimensional TLC for determination of amino acids. Visualization by spraying with cadmium-ninhydrin. Quantification by densitometry.	[115]
Amino acids	Microcrystalline cellulose—silica gel (5:2 or 6:4)	EtOAc—Me_2CO—MeOH—parapentanol—26% NH_3—water 9:3:3:1:1:3 for the 1st dimension and BuOH—Me_2CO—isopropanol–formic acid– water 18:8:8:3:6 for the 2nd dimension	Separation of amino acids by two-dimensional TLC. Detection by spraying with cadmium-ninhydrin reagent.	[116]
14 Amino compounds	Silica gel, alumina, cellulose and their mixture	Mixed organic eluents containing benzene	Development of rapid and reliable TLC method for separation and identification of amines. The results obtained on mixed layer materials have been compared with on plain layer materials.	[117]
Acid, alkaline and water soluble food dyestuffs and water soluble vitamins B_2, B_6, B_{12}	Silica gel—GDX	Hexadecyltrimethylammonium bromide	Study of the use of surfactants in TLC for the separation of acid, alkaline, water soluble food dyestuffs and water soluble vitamins B_2, B_6, B_{12}. Comparison of the results with those from alkyl-bonded silica layers.	[118]
Amino acids	Zeolite—cellulose (1:1)	Water—saturated phenol—EtOH—water—AcOH(12:4:4:1)	TLC of amino acids on laboratory prepared layers. Detection by spraying with 1% ninhydrin solution in 1-butanol and heating at 80°C for 20 min. Determination by visible reflectance spectrometry image analysis and densitometry. Investigations of the precision, detection limit and limit of quantification for each method.	[119]

Table 2. Continued

Analyte	Layer Material	Mobile Phase	Remarks	Ref.
21 Amino acids	Silica—tin (IV) selenium arsenate (1:1) and silica-calcium sulfate	Dimethyl sulfoxide	TLC of α-amino acids. Detection by spraying with ninhydrin. The R_F values under different separation conditions are reported.	[120]
17 Amino acids	Cellulose—silica gel (5:2)	Iso-PrOH—EtOAc—Me_2CO—MeOH—isopentanol—NH_3—water (9:3:3:1:1:3:3) and BuOH—Me_2CO—isopropanol—formic acid—water (18:8:8:3:6)	Two-dimensional TLC of amino acids in Bos grunniens Linnaeus horn hydrolysate. Detection by spraying with 0.5% ninhydrin in acetone.	[121]
30 Amino acids, 40 organic acids and 20 nucleosides and related compounds	Cellulose—silica gel	Isopropanol—EtOAc—Me_2CO—MeOH—para-amylalcohol—NH_3—water (9:3:3:1:1:3:3) BuOH—isopropanol—Me_2CO—formic acid—water (18:8:8:3:6) Tert-BuOH—EtOH-NH_3—water (5:6:1:3), isopentanol—petroleumether—EtOH—formic acid—propionic acid—water (45:100:25:20:1:1) Isopropanol—EtOAc—MeOH—NH_3—water (5:3:1:3:1), isopropanol—saturated $(NH_4)_2SO_4$—water (4:77:19)	Simultaneous separation of the metabolins of amino acids, organic acids and nucleosides by two-dimensional thin layer chromatography. Detection under UV 254 nm, by spraying with 0.1% acridine in ethanol, by spraying with ninhydrin-cadmium acetate reagent.	[122]
Phenolic Compounds such as nitrophenol, halophenols, cresols and aminophenols etc.	Stannic arsenate—silica gel (1:4)	Several single and mixed mobile phases	TLC of phenolic pesticide residues in banana fruits and plant tissues. Mobile phase ethanol—1.0 M citric acid, 1:3 (V/V) was identified best for mutual separation of m-niltrophenol, p-nitrophenol, picric acid and p-aminophenol.	[123]
Antihypertensive	Aminoplast—silica gel G	Benzene—cyclohexane—ethylmethyl ketone (15:10:X, where X = 5-15)	New rapid and simple TLC of lisinopril, cilazapril, captopril, quinapril and ramipril on aminoplast layers prepared from a suspension of aminoplast (a carbamide formaldehyde polymer) and silica gel G in a mixture of water and ethanol. Detection by exposure to iodine vapor.	[124]

Analyte	Layer Material	Mobile Phase	Remarks	Ref.
Selected bile acids	Silica gel—kieselguhr	n-Hexane—ethyl acetate—acetic acid in various volume compositions	Separation was achieved with the compositions 4:4:1 and 22:21:5. Detection by 10% aqueous sulfuric acid followed by heating at 120 °C for 20 min. Densitometric measurement at 250 nm.	[125]
Penicillins	Stannic arsenate—cellulose (1:4)	Acetonitrile—acetic acid—chloroform (1:1.5:4.5) Acetone—acetic acid—chloroform (5:5:6)	Amoxycillin and ampicillin were selectively separated from binary and ternary synthetic mixtures. Amoxycillin and ampicillin in commercial drugs were also separated and quantitatively determined.	[126]
Nicotinic acids and its derivatives	HPTLC plates precoated with mixture of silica gel 60 and kieselguhr F254	Mixture of an acetone—n—hexane in different volume compositions	Chromatographic separations were achieved on HPTLC non impregnated and impregnated with 2.5% and 5.0% aqueous solutions of $CuSO_4$. Improvement of chromatographic separation on impregnated plates is observed only in the case of esters nicotinic acids.	[127]

CONCLUSION

Biphasic layer materials have proven useful in thin layer chromatographic analyses of inorganic and organic substances in the past and are currently considered being more promising. Silica gel has been the most commonly used component of mixed layer materials. The work performed on analysis of inorganic and organic substances during 1980—2010 using biphasic layer materials has been collected from all available sources such as research papers, review articles, chemical abstracts and camag bibliography service (cbs). Care has been taken to provide as much information as possible in a condensed form without omissions. In general, biphasic handmade TLC plates have been more frequently used in comparison to commercially available precoated plates. In few cases precoated biphasic layers have been used in combination with other spectroscopic methods. Due to poor quality and lack of reproducibility of handmade TLC plates, the development of modern biphasic precoated TLC plates, becoming increasingly more important.

ACKNOWLEDGEMENT

The authors are grateful to the publishers and editors who granted permission to reproduce figures from published articles in their journals. Dr. Abdul Moheman is thankful to CSIR, New Delhi (India) for providing financial assistance.

REFERENCES

[1] Izmailov, N. A.; Schraiber, M. S. *Farmatsiya* 1938, *3,* 1.

[2] Hauck, H. E.; Mark, M. In *Handbook of Thin Layer Chromatography,* Sherma, J.; Fried, B. eds., Marcel Dekker, Inc., New York, 1996, 101-128.

[3] Simion, G. *J. Chromatogr. Sci.* 2002, *40,* 538-549.

[4] Lepri, L; Cincinelli, A. In *Encyclopedia of chromatography*; Cazes, J.; Third Ed.; Taylor & Francis: London, 2005, 854-858.

[5] Mohammad, A.; Iraqi, E.; Khan, I. A. *J. Surfact. Deterg.* 1992, *2,* 523-529.

[6] Flieger, J.; Tatarczak, M. *J. Chromatogr. Sci.* 2008, *46,* 565-573.

[7] Mohammad, A.; Laeeq, S. *Der Pharma Chemica* 2010, *2,* 281-286.

[8] Mohammad, A.; Laeeq, S.; Moheman, A. *J. Bioanal. Biomed.* 2010, *2,* 055-059.

[9] Maki, Y. *J. Radioanal. Chem.* 1975, *27,* 33-45.

[10] EL-Imam, Y. M. A.; Evans, W. C.; Grout, R. *J. Phytochemistr.* 1988, *27,* 2181-2184.

[11] Vuckovic, N.; Juranic, N.; Radnovic, D. J.; Celap, M. B. *J. Chromatogr.* 1989, *466,* 227-232.

[12] Lei, Q.; Li, Y. *Fenxi Ceshi Tongbao* 1990, *9,* 22-25.

[13] Mohammad, A.; Iraqi, E. *J. Surfact. Deterg.* 1999, *2,* 85-90.

[14] Ge, Z.; Lin, H. *Fenxi Huaxua* 1992, *22,* 1369-1372.

[15] Mŭcaji, P.; Nagy, M.; Liptaj, T.; Pronayonga, N.; Svajdlenka, E. *J. Planar Chromatogr.* 2009, *22,* 301-304.

[16] Wang, K.; Liu, Z.; Huang, J.-A.; Fu, D.; Lu, F.; Gong, Y.; Wu, X. *J. Planar Chromatogr.* 2009, *22,* 97-100.

[17] Rössler, H. In *Thin Layer Chromatography A Laboratory Handbook*, Stahl, E. (Ed.), Springer-Verlag, Berlin, Germany, 1969, 29.

[18] Waksmundzka-Hajnos, M.; Wrońska, B. *Chromatographia* 1996, *43,* 405-412.

[19] Rozylo, J. K.; Chomicz, D. G.; Malinowska, I. In *Instrumental High Performance Thin Layer Chromatography* (Würzburg, 1985) Kaiser, R.E. (Ed.), Bad Dürkheim, Germany, 1985, 173–187.

[20] Abere, T. A.; Okeri, H. A.; Okafor, L. O. *Tropical J. Pharmaceut. Res.* 2005, *4,* 331-339.

[21] Rössler, H. In *Thin Layer Chromatography A Laboratory Handbook,* Stahl, E. (Ed.), Springer-Verlag, Berlin, Germany, 1969, 40–41.

[22] Walsch, B. J. M. *J. Chem. Educ.* 1967, *44,* 294-296.

[23] Sthal, E. *Thin-Layer Chromatography,* Springer, New York, 1969.

[24] Ergül, S. *Chemistry* 2009, *18,* 36-48.

[25] Lepri, L.; Desideri, P. G.; Tonturi, G. *J. Chromatogr.* 1978, *147,* 375-381.

[26] Sanchez-Camazano, M.; Arienzo, M.; Sanchez-Nartin, M. J.; Crisanto, T. *Chemosphere* 1995, *31,* 3793-3801.

[27] Ravanel, P.; Liegeois, M. H.; Chevallier, D.; Tissut, M. *J. Chromatogr. A* 1999, *864,* 145-154.

[28] Khan, S. U.; Moheman, A. *J. Environ. Sci. Engg.* 2005, *47,* 310–315.

[29] Mohammad, A.; Moheman, A; Seema, *Acta Universitatis Cibiensis, Seria F, Chemia* 2008, *11,* 15-27.

[30] Mohammad, A.; Moheman, A; Seema, *J. Planat Chromatogr.* 2008, *21,* 453-460.

[31] Mohammad, A.; Moheman, A. *J. Planat Chromatogr.* 2010, *23,* 28-34.
[32] Mohammad, A.; Moheman, A. *Austr. J. Basic Appl. Sci.* 2010, *4,* 3635-3642.
[33] Mohammad, A.; Iraqi, E. *J. Planar Chromatogr.* 1999, *12,* 288-292.
[34] Misra, A. K.; Saxena, P. S.; Gupta, U. *Indian J. Chem. Technol.* 2002, *9,* 432-437.
[35] Misra, A. K. *Indian J. Chem. Technol.* 2003, *10,* 367-369.
[36] Misra, A. K. *Indian J. Chem. Technol.* 2007, *14,* 407-411.
[37] Khan, H. A. *Chromatographia,* 2006, *64,* 423-427.
[38] Popov, A.; Stefanov, K.; Compt. R. *Acad. Bulg. Sci.* 1968, *21,* 673-675.
[39] Fayez, M. B. E.; Gad, G.; Nasr, I.; Radvan, A. S. *J. Chem. UAR* 1967, *10,* 49-54.
[40] Sheen, B. *J. Chromatogr.* 1971, *60,* 363-370.
[41] Mohammad, A.; Nasim, K.T.; Ahmad, J.; Najar, M. P. A. *Analusis* 1995, *23,* 243-247.
[42] Sharma, S. D.; Sharma, S. C.; Sharma, C. *J. Planar Chromatogr.* 1999, *12,* 440-445.
[43] Mohammad, A.; Jabeen, N. *Indian J. Chem. Technol.* 2003, *10,* 79-86.
[44] Mohammad, A.; Hena, S. *Sep. Sci. Technol.* 2004, *39,* 2731-2750.
[45] Gawdzik, B.; Gajbakyan, D. S.; Matynia, T.; Sarkisyan, A. R. *J. Planar Chromatogr.* 1990, *3,* 280-282.
[46] Mohammad, A.; Iraqi, E.; Khan, I. A. *Chromatography* 2000, *21,* 29-36.
[47] Mohammad, A.; Iraqi, E.; Sirwal, Y. H. *Sep. Sci. Technol.* 2003, *38,* 2255-2278.
[48] Mohammad, A.; Ajmal, M.; Fatima, N.; Khan, M. A. M. *J. Planar Chromatogr.* 1992, *5,* 368-375.
[49] Mohammad, A.; Ajmal, M.; Fatima, N.; Yousuf, R. *J. Planar Chromatogr.* 1994, *7,* 444-449.
[50] Mohammad, A.; Agrawal, V. *Acta Universitatis Cibiniensis Seria F, Chemia*, 2002, *5,* 5-17.
[51] Mohammad, A.; Tiwari, S. *Microchem. J.* 1993, *47,* 379-385.
[52] Mohammad, A.; Chahar, J. P. S. *J. Chromatogr. A* 1997, *774,* 373-377.
[53] Shimiza, T.; Miyazaki, A.; Saitoh, I. *Chromatographia 1980, 13,* 119-121.
[54] Upadhyay, R. K.; Bajpai, U.; Bajpai, A. K. *J. Liq. Chromatogr. Relat. Technol.* 1980, *3,* 1913-1919.
[55] Morita, T.; Hamada, T.; Ishida, K. *Freseniu's Z. Anal. Chem.* 1981, *305,* 377-379.
[56] Gocan, S.; Cecilia, C. *Babes-Bolyai (Ser.) Chem.* 1981, *26,* 31-34.
[57] Ishita, K.; Morita, T.; Hamad, T. *Fresenius'Z. Anal. Chem.* 1981, *305,* 257-261.
[58] Kuroda, R.; Hasoi, N. *Chromatographia* 1981, *14,* 359-362.
[59] Upadhyay, R. K.; Sharma, V. K.; Singh, V. P. *J. Liq. Chromatogr. Relat. Technol.* 1982, *6,* 1141-1153.
[60] Kumari, V.; Upadhyay, R. K.; Singh, V. P. *J. Liq. Chromatogr. Relat. Technol.* 1983, *6,* 155-164.
[61] Hamada, T.; Morita, T.; Ishida, K. *Fresenius'Z. Anal. Chem.* 1983, *316,* 23-25.
[62] Chem, Y.; Zheng, K.; Luo, H.; Cheng, J. *Fenxi Huazue* 1983, *11,* 101-106.
[63] Le Van So, *Radiochem. Radioanal. Lett.* 1983, *59,* 53-57.
[64] Upadhyay, R. K.; Sharma, M. R.; Rastogi, R. K. *J. Liq. Chromatogr. Relat. Technol.* 1984, *7,* 2813-2820.
[65] Yin, B. H.; Lin, J. L. *Chinese J. Chem. (Huaxue Tongbao)* 1985, *7,* 25-27.
[66] Lin, M.; Zong, W.; Jia, X.; Hu, Zh.; Zheng, J. *Chinese J. Chem. World (Huaxue Shijie)* 1986, *27,* 261-264.
[67] Moslowska, J.; Mlodzikowski, Z. *Chemica Analityczna* 1986, *31,* 193-199.

[68] Ren, S.; Zong, W.; Zia, X.; Hu, Zh.; Zheng, J. *Chinese J. Chem. World* (*Huaxue Shijie* 1986, *27,* 403-406.
[69] Liu, L.; Liu, J.; Hung, Z.; Cheng, J. *Wuhan Daxue Xuebao, Ziran Kexueban* 1987, 64-70.
[70] Lin, H.; Yi, S. *Chinese J. Chromatogr.* (*Sepu*) 1987, *5,* 53-55.
[71] Wang, J.; Hu, C. *Huaxue Tongbao* 1987, *6,* 1041-1045.
[72] Cheng, G.; Hu, Z.; Jia, Xi.; Zheng, J. *Chinese J. Chromatogr. (Sepu)* 1988, *6,* 108-110.
[73] Upadhyay, R. K. *Chromatographia* 1988, *25,* 324-326.
[74] Wang, J.; Hu, Zh. *Chinese J. Chem. World (Huaxue Shijie)* 1988, *10,* 136-138.
[75] Luo, H.; Mo, J. *Fenxi Shyanshi* 1988, *7,* 27-30.
[76] Wang, J.; Hu, Zh.; Zheng, J. *Yankunag Ceshi* 1988, *7,* 10-14.
[77] Rajput, R. P. S.; Misra, A. K.; Agrawal, S. *J. Planar Chromatogr.* 1988, *1,* 349-350.
[78] Wang, J.; Hu, Zh. *Fenxi Huaxue* 1988, *16,* 740-742.
[79] Cheng, G.; Hu, Zh.; Jia, X. *Fenxi Cashi Tongbao* 1989, *8,* 55-58.
[80] Panesar, K. S.; Singh, O. V.; Tandon, S. N. *Anal. Lett.* 1990, *23,* 125-133.
[81] Mohammad, A.; Ajmal, M.; Fatima, N.; Ahmad, J. *J. Liq.Chromatogr. Relat. Technol.* 1991, *14,* 3283-3300.
[82] Yoshioka, M.; Araki, H.; Seki, M.; Miyazaki, T.; Utsuki, T.; Yaginuma, T.; Nakano, M. *J. Chromatogr.* 1992, *603,* 223-229.
[83] Mohammad, A.; Tiwari, S. *Microchem. J.* 1993, *47,* 379-385.
[84] Mohammad, A.; Ajmal, M.; Fatima N.; Yousuf, R. *J. Planar Chromatogr.* 1994, *7,* 444-449.
[85] Mohammad, A.; Nasim, K. T.; Najar, M. P. A. *J. Planar Chromatogr.* 1996, *9,* 445-449.
[86] Husain, S. W.; Avanes, A.; Ghoulipour, V. *J. Planar Chromatogr.* 1996, *9,* 67-69.
[87] Mohammad, A.; Najar, M. P. A. *Acta Chrom*atogr. 2001, *11,* 154-170.
[88] Zhu, X.; Zhang, H.; Wang, B. *Sep. Sci. Technol.* 2005, *40,* 3289-3298.
[89] Najar, M. P. A.; Janbandhu, K. R.; Bhukte, P. G.; Misra, R. S. *Indian J. Chem. Technol.* 2009, *16,* 65-73.
[90] Wu, Y. *Chinese J. Chromatogr.* 1989, *7,* 112-114.
[91] Mohammad, A.; Khan, I. A.; Jabeen, N. *J. Planar. Chromatogr.* 2001, *14,* 283-290.
[92] Varshney, K.G.; Khan, A.A.; Maheshwari, S.M. Bull. Chem. Soc. Jpn 1992, *65,* 2773
[93] Nabi, S.A.; Gupta, A.; Khan, M.A.; Islam, A. Acta Chrom. 2002, *12,* 201-210.
[94] Nabi, S.A.; Khan, M.A. Acta Chrom. 2003, *13,* 161-171.
[95] Nyiredy, Sz. In: Planar Chromatography, Sz. Nyiredy (Ed.) A Retrospective View for the Third Millennium, Springer, Budapest, 2001, 103.
[96] Tuzimski, T.; Soczewiński, E. *J. Chromatogr. A* 2002, *961,* 277–283.
[97] Tuzimski, T.; Soczewiński, E. *Chromatographia* 2002, *56,* 219–223.
[98] Tuzimski, T.; Bartosiewicz, A. *Chromatographia* 2003, *58,* 781–788.
[99] Tuzimski, T.; Soczewiński, E. *J. Planar Chromatogr.* 2003, *16,* 263–267.
[100] Tuzimski, T.; Wojtowicz, J. *J. Liq. Chromatogr. Relat. Technol.* 2005, *28,* 277–287.
[101] Tuzimski, T. *J. Planar Chromatogr.* 2007, *20,* 13–18.
[102] Bauer, G. K.; Pfeifer, A. M.; Hauk, H. E.; Kovar, K. A. *J. Planar Chromatogr.* 1998, *2,* 84—89.
[103] Pyka, A.; Dolowy, M. *J. Liq. Chromatogr. Relat. Technol.* 2004, *27,* 2987-2995.

[104] Pyka, A.; Dolowy, M.; Gurak, D. *J. Liq. Chromatogr. Relat. Technol.* 2005, *28,* 631-640.

[105] Pyka, A.; Babuska, M.; Dziadek, A.; Gurak, D. *J. Liq. Chromatogr. Relat. Technol.* 2007, *30,* 1385-1400.

[106] Cass, Q. B.; Bassi, A. L.; Matlin, S. A. *Chirality* 1996, *8,* 131-135.

[107] Franco, P.; Senso, A.; Minguillon, C.; Oliveros, L. *J. Chromatogr. A* 1998, *796,* 265-272.

[108] Kabulov, B. D.; Shakarova, D. Sh.; Shpigun, O. A.; Negmatov, S. S. *Russian J. Phys. Chem. A* 2008, *82,* 924-927.

[109] Waksmundzka-Hajnos, M.; Wronska, B. *Chromatographia* 1996, *43,* 405-412.

[110] Waksmundzka-Hajnos, M. *Chromatographia* 1996, *43,* 640-646.

[111] Ergül, S.; Kadan, I.; Savasci, S.; Ergül, S. *J. Chromatogr. Sci.* 2005, *43,* 394-400.

[112] Srivastava, S.; Bushan, R.; Chauhan, R. *J. Liq. Chromatogr. Relat. Technol.* 1984, *7,* 1359-1365.

[113] Gullner, G.; Cserhati, T.; Bordas, B.; Szogyi, M. *J. Liq. Chromatogr. Relat. Technol.* 1986, *9,* 1919-1931.

[114] Wu, Y. *Chinese Anal. Chem.* 1987, *15,* 945-949.

[115] Wu, Y.; Xiong, L.; Nan, G. *Chinese J. Chromatogr. (Sepu)* 1987, *5,* 256-259.

[116] Wu, Y.; Gong, Sh. *Chinese J. Adv. In Biochem. & Biophys.* 1988, *15,* 452-453.

[117] Ajmal, M.; Mohammad, A.; Anwar, S. *Microchem. J.* 1990, *42,* 206-217.

[118] Yin, P.; Yan, Ch. *Chinese J. Chromatogr. (Sepu)* 1994, *12,* 35-36.

[119] Petrovic, M.; Kastelan-Macon, M. *J. Chromatogr. A* 1995, *704,* 173-178.

[120] Siddiqi, Z. M.; Rani, S. *J. Planar Chromatogr.* 1995, *8,* 141-143.

[121] Lei, G.; Cao, Y. *Chinese J. Chromatogr. (Sepu)* 1996, *14,* 158-160.

[122] Wu, Y.; Kuang, B.; Long, F. *Chinese J. Chromatogr. (Sepu)* 1996, *14,* 259-263.

[123] Nabi, S. A.; Sikarwar, A. *Acta Chrom.*1999, *9,* 123-132.

[124] Perisic-Janic, N. U.; Agbaba, D. *J. Planar Chromatogr.* 2002, *15,* 210-213.

[125] Pyka, A.; Dolowy M. *J. Liq. Chrom. Relat. Technol.* 2004, *27,* 2613-2623.

[126] Nabi, S. A.; Khan, M. A.; Khowaja, S. N.; Alimuddin, Acta Chrom. 2006, *16,* 164-172.

[127] Pyka, A.; Klimczok, W. *J. Liq. Chrom. Relat. Technol.* 2008, *31,* 526-542.

In: Advanced Organic-Inorganic Composites
Editor: Inamuddin
ISBN 978-1-61324-264-3

Chapter 16

SPECTROSCOPIC, MORPHOLOGICAL AND ELECTROCHEMICAL PROPERTIES OF POLYANILINE: FLY ASH NANO COMPOSITES

Sipho Mavundla[a], Amir-Al-Ahmed[b], Leslie Petrik[c], Priscilla G. L. Baker[a] and Emmanuel I. Iwuoha[a,*]

[a]SensorLab, Department of Chemistry, University of the Western Cape, Bellville 7553, Cape Town, South Africa

[b]Center of Research Excellences in Renewable Energy, King Fahd University of Petroleum and Minerals, Dhahran 31261, Saudi Arabia

[c]Environmental and Nanotechnology Research Group, Department of Chemistry, University of the Western Cape, Bellville 7553, Cape Town, South Africa

ABSTRACT

Polyaniline (PANI)/fly ash (FA) (PANI-FA) nanocomposites were prepared, in which polyaniline was doped with the metal oxides present in the coal combustion fly ash. SEM image confirmed the formation of nanostructured composite materials. Nanotube clusters were formed when the reaction mixture containing 0.1 M aniline and 40% (w/w) FA in 2 M HCl was aged before polymerization. On the other hand the polymerisation of an acid solution (2 M HCl) of 0.1 M aniline, 40% (w/w) FA and poly(styrene sulphonic acid) (PSSA) produced PANI-FA-PSSA nanorod composites. UV-Vis and FTIR (Fourier transform infrared) spectroscopy data showed that the nanotubes and nanorods are basically polyemeraldine salts containing some conducting partially protonated pernigraniline salts, which ensure that PANI retains its conductivity when doped with FA or PSSA. This was also confirmed by the electrochemical response of the composite. Thermogravimetric analysis (TGA) showed that the metal oxide in FA stabilizes PANI-FA (nanotube) or PANI-PSSA-FA (nanorods) composites up to 800 °C.

* Corresponding author's e-mail: eiwuoha@uwc.ac.za

ABBREVIATIONS

APS	Ammonium persulphate
BAS	Bioanalytical System
CSIR	Centre for Science and Industrial Research
CV	Cyclic voltammetry
DMF	Dimethylformamide
EDX	Energy dispersive X-ray
FA	Fly ash
FTIR	Fourier transform infrared
PANI	Polyaniline
PSSA	Poly(styrene sulphonic acid)
NRF	National Research Foundation
SEM	Scanning electron microscopy
TGA	Thermogravimetric analysis
UV-Vis	Ultra violate Visible spectra
XRD	X-ray diffraction

1. INTRODUCTION

Fly Ash (FA) is a by-product of coal combustion plants, which mainly generates electricity. It is collected with electrostatic precipitators and bag filters or allowed to escape into the atmosphere. It is an alkaline grey powder with pH ranging from 9 to 9.9 [1]. There are a large number of coal based power plants around the world, which produces a huge quantity of FA, causing serious environmental problems. Less than half of this ash is used as a raw material for concrete manufacturing and construction and the remaining is directly dumped in land fill or open space. In South Africa about 90% of electricity is produced by burning of coal in power plants that dispose nearly 25 million metric tons of FA per year [1,2]. The main components of South African FA are; silica (SiO_2) ~50%, aluminium (Al_2O_3) ~30%, calcium oxide (CaO) ~6%, magnesium oxide (MgO) ~6%, iron oxide (Fe_2O_3) ~4% and titanium oxide (TiO_2) ~3% [2]. These metal oxides have been used by other laboratories [3,4] in the preparation of nanocomposites. Al_2O_3 and Fe_2O_3 have been used as catalyst support for the production of nanotubes. Huang [5] has reported that carbon nanotubes can be grown on the silica spheres by pyrolysis reaction involving iron (II) phthalocynanine, silica and carbon nanotubes. Electrically conducting polymers, especially, polyaniline (PANI) has been studied extensively. However, due to its poor processibility the application of PANI is limited [2,6-9]. Many schemes have been tried to overcome this problem, such as doping with bulky organic acids, blending with other polymers, preparing composites and nanocomposites etc. Nanocomposites prepared with transitional metals are thought to have application in manufacture of electronic or photonic devices [3,10-13]. The aim of this work is to use the FA (an industrial waste) to produce composite materials with polyaniline. This may help in both the ways, use of industrial waste can help the environment related issues and at the same time trying to make PANI more applicable.

2. EXPERIMENTAL

2.1. Materials

Dimethylformamide (DMF), acetone, ammonium persulphate, aniline, (25% v/v in water), hydrochloric acid and poly(styrene sulphonic acid) (PSSA) were purchased from Sigma-Aldrich. Aniline was redistilled at low pressure before use. Fly ash was obtained from Matla Power Station, South Africa. Only freshly produced FA was used in this study. The main constituents of the Matla FA as determined by X-ray diffraction (XRD) analysis are Al_2O_3 ~30%, CaO ~5%, Fe_2O_3 ~4%, MgO ~2%, TiO_2 ~2% and SiO_2 ~50% [7].

2.2. Synthesis of Polyaniline-Fly Ash Composites

Polyaniline (PANI) was synthesized using reported method [2,6,9]. Four different composite samples of PANI and FA were prepared: (i) first sample, 100 ml of 0.1 M aniline solution in 2 M HCl and 40 % (w/w) FA were stirred for 30 min in a beaker kept over an ice bath. This was followed by a slow addition of 0.1 M ammonium persulphate (APS) with continuous stirring for 8 h. Resultant green-colored mixture was filtered and the solid product was washed with water and acetone and dried at room temperature. (ii) For the synthesis of aged PANI-FA composite, 100 ml solution of 2 M HCl containing 0.1 M aniline and 40% FA was placed in a furnace for 48 h at 100 °C and then cooled to room temperature and placed in an ice bath at 0-5 °C. Ammonium persulphate was added to the mixture as described before. Subsequently the vessel was removed from the ice bath and left at room temperature for 12 h. (iii) Polyaniline composites containing PSSA was prepared by dissolving 10.93 ml of PSSA in a solution consisting of 555 ml of water, 9.3 ml of 2 M HCl and 10.68 ml of aniline which was placed in a water bath. Then 80 ml aqueous solution containing 13.84 g of APS was added slowly to the aniline-PSSA mixture with constant stirring for 12 h. The reaction mixture was filtered and the PANI-PSSA solid was washed and dried as previously described. (iv) The preparation of PANI-PSSA-FA composite material is similar to PANI-PSSA process, but in this case 40% FA (w/w) in aniline solution was used.

2.3. Characterization of Composites

Electronic properties of the polymeric composite were studied by UV-Vis absorbance experiments performed with UV/Vis 920 Spectrometer (GBC Scientific Instruments, Australia) at room temperature. Samples containing 0.005 g composite per 10 ml DMF were used in all UV-Vis measurements which were carried out at wavelength range of 300-800 nm using quartz cuvettes. Energy dispersive X-ray (EDX) analysis and scanning electron microscopy (SEM) data were obtained with a Hitachi X-650 Scanning Electron Microanalyser (5-40 kV). Fourier transform infrared (FTIR) spectra of the composite materials were obtained from a Perkin Elmer, Paragon 1000PC FTIR at room temperature and in the range of 4000 to 600 cm^{-1} and a Perkin Elmer Thermogravimetric Analyzer (TGA) - 7 was used for thermogravimetric experiments covering 45 °C to 900 °C temperature range at a heating rate of 20 °C min^{-1} in nitrogen atmosphere.

A Bioanalytical System (BAS) 50W electrochemical workstation (Bioanalytical Systems, USA) and a 20 ml conventional three-electrode cell setup (from BAS) were used for all the electrochemical experiments. Platinum disk working electrode (1.77×10^{-2} cm^2), Ag/AgCl reference electrode and a Pt mesh auxiliary electrode was placed in an acid paste of PANI-FA composites (slurry of 0.05 g composite materials in 5 ml 1 M HCl). After the paste was degassed for 15 min, voltammograms were recorded for scan rates of 100-1000 mV/s from an initial potential, E_i of -200 mV to a switch potential, E_λ of +1200 mV at 20 °C.

3. Results and Discussion

3.1. Spectroscopic Studies

UV-Vis spectra of various polyaniline composites are shown in Figure 1. PANI-PSSA-FA, PANI-PSSA-FA-aged and PANI-FA show two distinct absorption peaks around 340 nm and 635 nm, which are the characteristic peaks for emeraldine base [3,14,15]. It is noteworthy that only PANI composites containing FA give emeraldine salt, that's why in PANI-PSSA composites there is no emeraldine base peak. This is not unexpected since FA contains metal oxides that stabilize PANI in the emeraldine salt form. A small peak is also visible at 450 nm for PANI-PSSA-FA indicating the presence of some partially protonated pernigraniline salt, which is the conducting form of PANI. This is further confirmed by the peaks at 360-380 nm due to the π-π^* transition of benzoid rings [3,14,15] that are retained in PANI (a), PANI-FA (b) and PANI-PSSA(d). The spectra of PANI (a), PANI-FA (b) and PANI-PSSA (d) show absorbance peak at 450 nm.

This peak is associated with the presence of conducting partially protonated pernigraniline salt and conveys the fact that PANI retains its conductivity when doped with FA or PSSA. The peak at 340 nm is characteristic of all four composites containing FA, although the intensity of the absorbance is higher for composites for which their synthesis involves the aging of the reaction mixture.

In the FTIR spectra Figure 2, all the composites showed peaks at and around 1558 cm^{-1} and 1469 cm^{-1} because of the quinoid and benzoid rings of PANI, respectively [1,6,7,11,16]. Position of the peaks agrees with UV-Vis results as they are due to emeraldine salt form of PANI. The PANI-PSSA composite has a very sharp peak at 1393 cm^{-1} due to the C-N stretching [1,17] and this peak is also observed as a weak signal in PANI-PSSA-FA composite spectrum and are assignable to the presence of various metal oxides.

The peaks around 1490, 1018, 796, 1100 cm^{-1} are also attributed to the different metal oxides present in the composites [1,18]. FTIR peaks at 1250 and 1300 cm^{-1} are retained in both PANI and PANI composites, although the intensity of absorption is higher for the composites containing FA. This confirms the UV/Vis result which shows the presence of pernigraniline salts in PANI composites that contain FA.

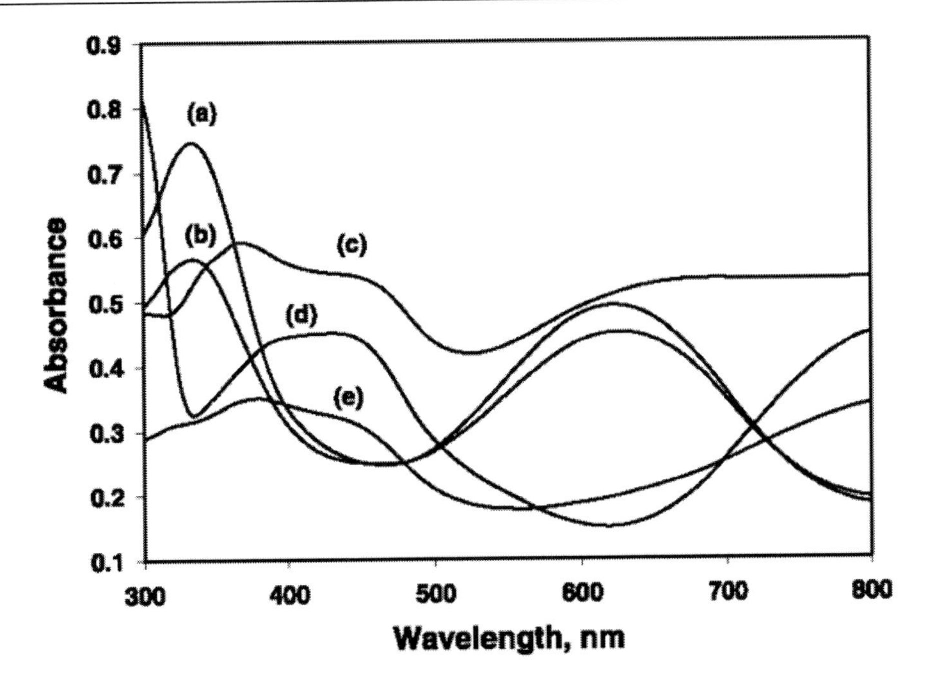

Figure 1. UV-Vis spectra of polyaniline-fly ash composites with PANI–FA (40%) aged (a), PANI–PSSA–FA (40%) (b), PANI (c), PANI–PSSA (d) and PANI–FA (40%) (e). Samples were prepared in dimethylformamide.

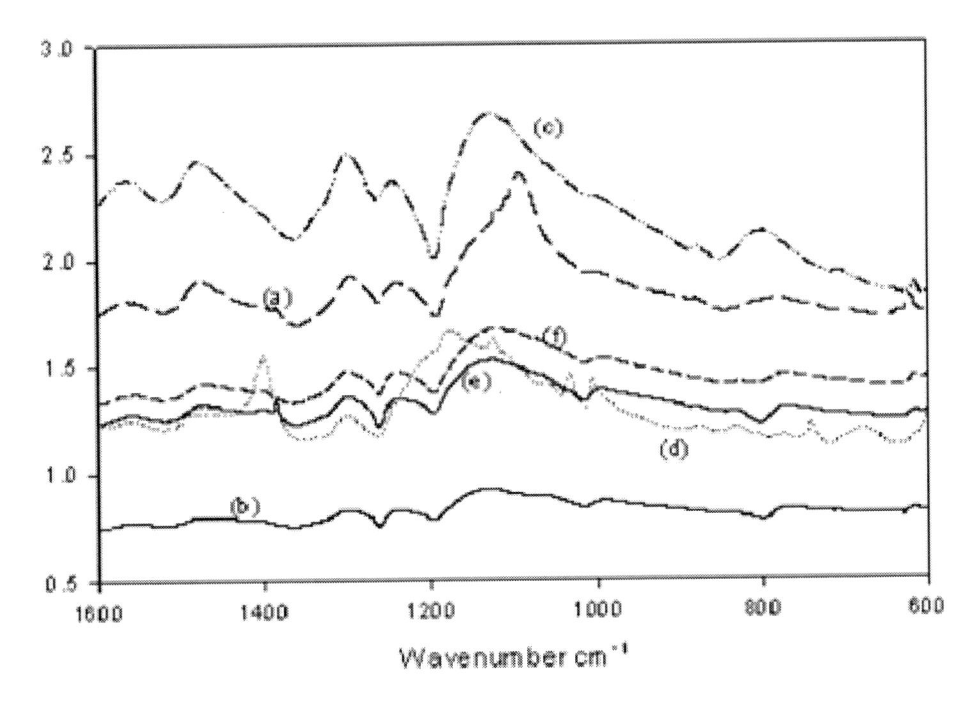

Figure 2. FTIR spectra of PANI-FA composites, PANI (a), PANI-FA (b), PANI-FA aged (c), PANI-PSSA (d), PANI-PSSA-FA (e) and PANI-PSSA-FA aged (f).

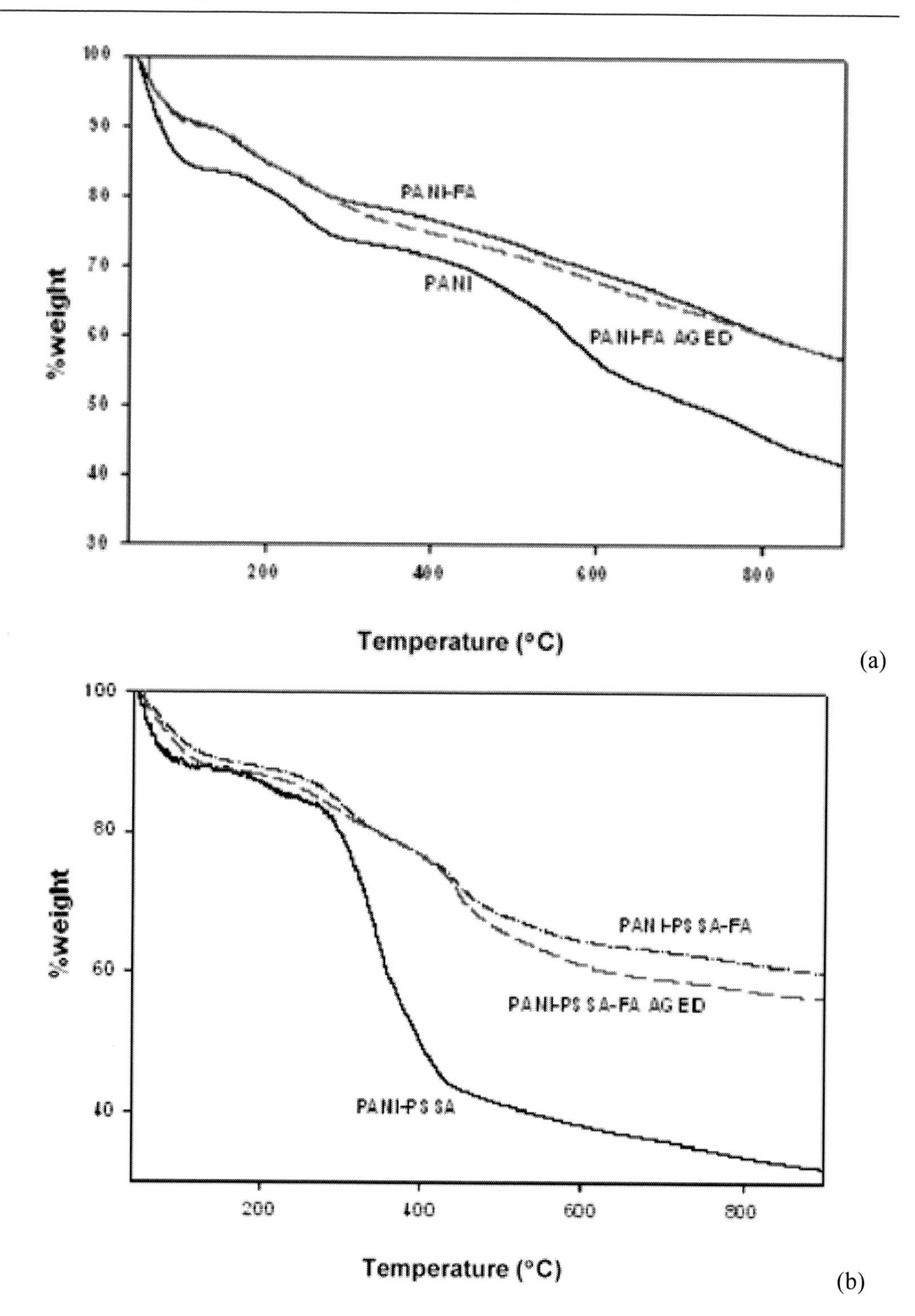

Figure 3. TGA of PANI, PANI-FA and PANI-FA aged (a) and PANI-PSSA, PANI-PSSA-FA and PANI-PSSA-FA aged (b).

Figure 4. SEM images of polyaniline composites: PANI (a), FA (b), PANI-FA (c), PANI-FA aged (d), PANI-PSSA (e), PANI-PSSA-FA (f) and PANI-PSSA-FA aged (g).

3.2. Morphological Studies

TGA thermograms of PANI and PANI-FA composites without PSSA are shown in Figure 3 (a). The initial decrease in mass was observed at about 120 °C and it is due to the removal of moisture from the polymer. A small decrease in mass occurs between 160-240 °C suggesting that the polymer still contains some volatile materials like HCl. The onset temperature for the degradation of PANI is 270 °C which is accompanied by decrease in mass with the increasing temperature. Both aged and un-aged PANI-FA composites exhibit a three-step degradation mechanism. Initial mass loss occurs at temperature of about 100 °C because of removal of moisture and further decrease in mass occurs at temperatures of up to 200 °C which is associated with the removal of volatile material (such as HCl) trapped within polymer composites. There is also a slow decrease in mass from 220-350 °C temperature range which is due to the slow degradation of the polyaniline backbone of the composites. As temperature approaches 800 °C, the composites undergo slow degradation losing only 40% of its initial mass. This is mainly due to the presence of different metal oxides which are stable at high temperatures and can also form complex with the degraded polymer [5,11,16]. Figure 3 (b) contains the TGA thermograms of composites of PANI and PANI-FA composites containing PSSA. These composites are also exhibit three-step degradation mechanism. There is an initial decrease in mass at temperatures up to 100 °C due to the removal of the moisture from the materials followed by a gradual decrease in mass between 100 °C and ~300 °C associated with the removal of volatile materials, mainly HCl. Then a large lowering of mass occurs at temperatures 300 °C up to 420 °C corresponding to the degradation of PANI and PSSA. In the case of PANI-PSSA-FA composite the first two thermogravic steps are due to the same reasons described above for PANI-PSSA, which indicates that the material is quite stable up to 320 °C. The decrease in mass observed between ~320 °C and 450 °C is mainly due to the degradation of polyaniline and PSSA. At temperatures higher than 450 °C, the degradation process is slow up to 800 °C mainly due to the presence of different stable oxides from FA as previously explained. It is clear from the thermograms that the PANI-PSSA composite which does not have any FA lost about 80% of its initial mass, whereas PANI-PSSA-FA composites retained 40% of their mass up to 800 °C [5,11,16].

Scanning electron microscopy (SEM) images of the polyaniline composites are shown in Figure 4. Figures 4 (a) and 4 (b) are the micrographs of PANI and FA, respectively. PANI-FA composite, Figure 4 (c), shows morphological differences from those of the individual constituents. However, the aging of PANI-FA before polymerisation results in the formation of nanotube clusters (60 to 120 nm cross-sectional diameters) as shown in Figure 4 (d). UV-Vis studies demonstrated that the aging of PANI-FA stabilizes PANI in the emeraldine salt state, it means that the nanotubular clusters of PANI-FA are in the emeraldine salt state. Other researchers [5] implicated SiO_2, a major constituent of FA, as having large surface curvature that causes PANI to form composites with nanotubular morphology. Figure 4 (e) shows the SEM of PANI-PSSA. Analysis of the figure indicates that the inclusion of PSSA in the composite leads to the formation of a gel-like PANI composite. Figure 4 (f) shows that PANI-PSSA-FA crystallizes out as rods (95 to 450 nm cross-sectional diameter). Since PANI-FA did not form nanorods, it indicates that PSSA in the polymeric composite plays a role in making PANI to crystallize as nanorods. Figure 4 (g) micrograph shows that aging the reaction mixture during the preparation of PANI-PSA-FA results in the composites forming gel like morphology. It is essential to point out that the aging of the reactants before oxidative polymerisation leads to the formation of polymeric nanotube clusters and PSSA causes the formation of gel-like polymer. The mechanism for this process is still unclear. EDX data's are presented in Table 1 shows that the formation of nanostructured composite with FA and PANI generally requires relatively high content of Si, Al, Cu, Ti and other metal ions in the resultant material.

3.3. Electrochemical Studies

Cyclic voltammogram of PANI-PSSA-FA is plotted in Figure 5. The redox peaks in the figure have been assigned using Pekmez formalism [19,20]. The first redox peak is leucoemeraldine/leucoemeraldine radical cation (A/A′), the second one is the emeraldine radical cation/emeraldine (B/B′) and the third redox peak is the pernigraniline radical cation/pernigraniline (C/C′). From the experimental design the voltammograms represent the electrochemical behavior of the PANI composite at high potential scan rate. Only electrode processes (electron transfer reaction at the electrode and any accompanying physical or chemical process) that have rate constants comparable to the fast potential scan rates (200–

1000 mVs^{-1}) will be discernible in the voltammograms. The 200 mVs^{-1} cyclic voltammogram of PANI-PSSA-FA have formal potentials, $E^{\circ\prime}$ (200 mVs^{-1}), values of 105 mV (A/A′), 455 mV (B/B′) and 670 mV (C/C′). The corresponding cathodic to anodic peak current ratio, I_{pc}/I_{pa} (200 mVs^{-1}), is 'unity' for all the three redox couples. However at higher scan rates (500–1000 mVs^{-1}) the $E^{0\prime}$ value of the A/A′ redox couple vary with scan rate while those B/B′ and C/C′ are independent of scan rate within limits of experimental error of ± 10 mV. For example the $E^{\circ\prime}$ (1000 mVs^{-1}) values are 157 mV (A/A′), 465 mV (B/B′) and 660 mV (C/C′). The I_{pc}/I_{pa} (1000 mVs^{-1}) values are 0.7 (A/A′), 1.0 (B/B′) and 0.8 (C/C′). These results indicate that the redox species exhibit quasi-reversible electrochemistry, at scan rates lower than 200 mVs^{-1}. But at higher scan rates the I_{pc}/I_{pa} (1000 mVs^{-1}) values of 0.7 for the A/A′ redox couple means that the anodic process represented by A′ consist of electron transfer reaction at the electrode coupled to a diffusion process that is detectable only at fast scan rates, with the result that the anodic currents are higher than the cathodic currents and $E^{0\prime}$ values shifts anodically with scan rate. The diffusion processes being detected at high scan rates could be intra-paste charge transfer within the electroactive polymer paste or intra-molecular charge transportation along the polymer backbone. The later event is more likely since ΔE° {= $E^{\circ\prime}{}_{(1000\ mVs^{-1})}$ - $E^{\circ\prime}{}_{(200\ mVs^{-1})}$} value is only 50 mV which is less than 65 mV normally associated with surface-bound electroactive polymers undergoing charge propagation along polymer chain. Figure 5 (b) shows that the electrochemistry of PANI–PSSA is similar to those of PANI–PSSA–FA nanorods. The three redox couples of the 200 mVs^{-1} voltammogram display fairly reversible electrochemistry as shown by the I_{pc}/I_{pa} (200 mVs^{-1}) value which is approximately unity for each of the redox couples. But at 1000 mVs^{-1} when the contribution of other fast processes become evident in the voltammograms, the I_{pc}/I_{pa} (1000 mVs^{-1}) value drops to 0.4 for A/A′ and increased to 1.2 for C/C′. This shows that charge propagation is more effective when the polymer is in the form of leucoemeraldine radical cation or pernigraniline radical cation, and for the emeraldine or its radical cation, charge transportation does not make significant contribution to the overall current. Changing from PANI–PSSA to PANI–PSSA–FA decreases both $E^{\circ\prime}$ (200 mVs^{-1}) and $E^{\circ\prime}$ (1000 mVs^{-1}) values. The effect of this is that FA causes a cathodic shift in the $E^{\circ\prime}$ values of the polymer irrespective of the redox state of the material.

3.4. Conductance Measurement

Normal conductivity measurements using a four pin probe or conductivity cell gives the net conductance of all the components of the material. However, the conductivity or resistance of individual oxidization state cannot be obtained in this way. This section demonstrates the determination of the conductance profile of all the six redox states of the PANI–PSSA–FA by applying the classical Ohm's relation to multiple potential scan rate cyclic voltammetry (CV). In this approach the conductance, C, is given by $C = \Delta I_v/\Delta E_v$ where $\Delta I_v = I_p$ (1000 mVs^{-1}) – I_p (200 mVs^{-1}) and $\Delta E_v = E_p$ (1000 mVs^{-1}) – E_p (200 mVs-1). The conductance values of all the redox states of PANI–PSSA–FA in comparison with those of PANI–PSSA are given in Table 2. The average value for all the six redox states is the conductance of the material, which was calculated as 1.21×10^{-2} S for PANI–PSSA–FA and 1.7×10^{-2} S for PANI–PSSA.

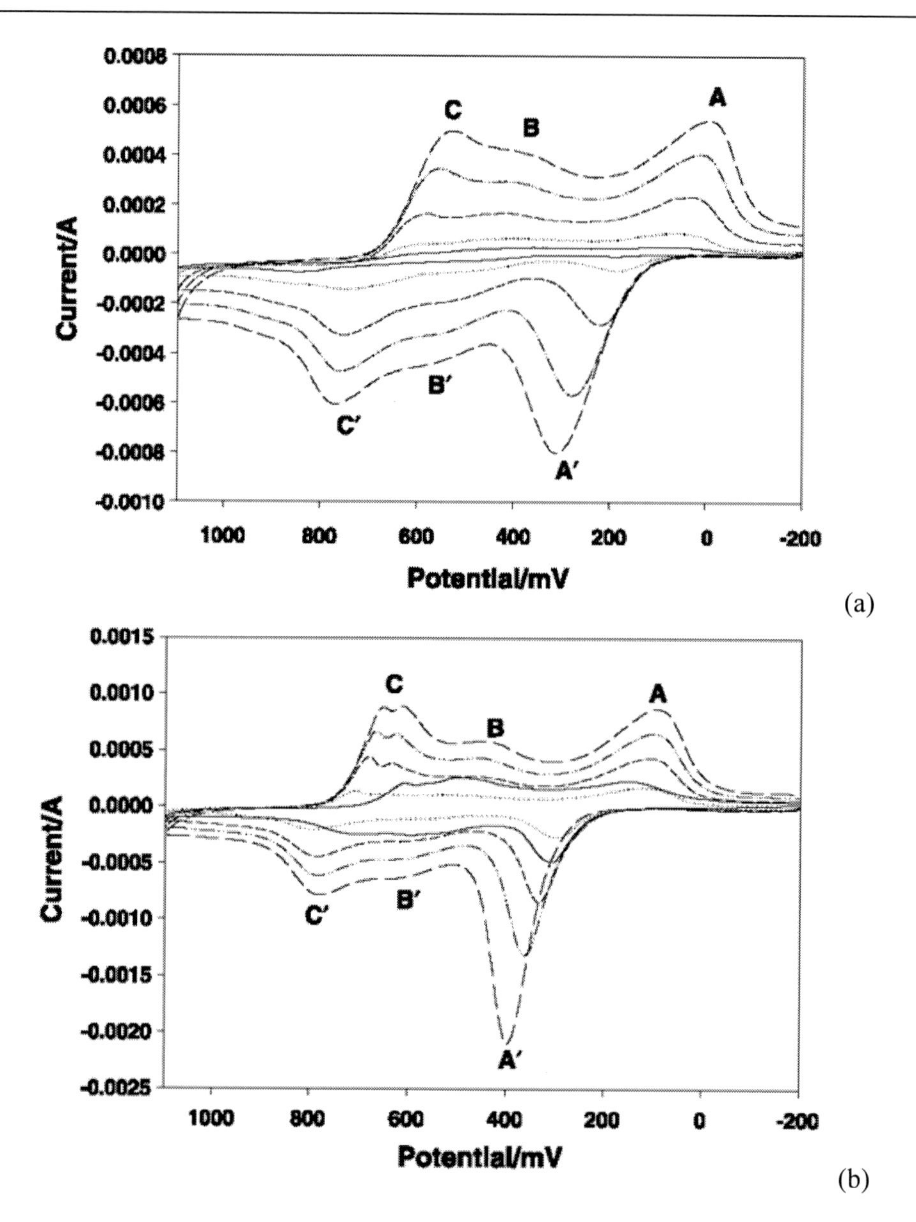

Figure 5. Multi-scan rate cyclic voltammograms of PANI–PSSA–FA (40%) (a) and PANI–PSSA (b) paste in 1M HCl performed at of 100, 200, 500, 750 and 100 mVs^{-1} (from inside outwards) . (A/A′) leucoemeraldine/leucoemeraldine radical cation; (B/B′) emeraldine radical cation/emeraldine; (C/C′) pernigraniline radical cation/pernigraniline.

There is also a general decrease in the conductance of the redox states of the composite on changing from the nanotubular PANI–PSSA to nanorodular PANI–PSSA–FA. These conductance values calculated for the nanocomposites agree with what is reported for similar substances [21,22]. The data in Table 2 shows that as the nanorod composite is oxidized, A′→ B′→ C′, the conductance of the polymer composite increases. This trend is related to the formation of pernigraniline redox state which contains polarons that increased the conductivity of the polymeric nanorod. However, the emeraldine redox state (B′) exhibited low conductance value in both PANI–PSSA–FA (0.7×10^{-2} S) and PANI–PSSA (0.53×10^{-2}

S) compared to the two other redox states. This may be related to the occurrence of π–π* transition of benzoid rings and the formation of charge transfer excitation of the quinoid structure characteristic of emeraldine which combine to reduce the conductance of the polymer.

CONCLUSION

SEM and EDX analyses show that the formation of nanostructured composite with FA and PANI generally requires relatively high content of Si, Al, Cu, Ti and other metal ions in the resultant material. This could be achieved either by aging a mixture of the reactants (aniline and FA) before polymerisation or by the inclusion of PSSA in the reaction mixture. Polyaniline composites that did not contain FA or PSSA didn't form any nanostructures. However, PSSA caused PANI-FA to form nanorod composites by introducing high sulphur content (22% S) and incorporating large amount of Ca into the composite. As shown in Table 1, PANI-PSSA-FA has 5 times the Ca content of aged PANI-FA (which is nanotubular) and there is a tremendous decrease in the C, Si, Al and Ti contents of PANI-PSSA-FA nanorods compared to those of aged PANI-FA (nanotube clusters). Aged PANI-PSS-FA (gel-like) contains half the amount of S and Ca found in un-aged PANI- A-PSSA (nanorod), and between two to four times decrease in the amount of Si, Al and Ti associated with aged PANI-FA (nanotube cluster). However, the presence of 11% S in aged PANI-FA-PSSA, which was lacking in aged PANI-FA led to the formation of gel like PANI composite. This study showed that FA can be used to control the morphology of polyaniline composite materials, and the method of preparation can be used to determine whether nanorods, nanotubes or gels are formed. Thermogravimetric analysis data indicate an improvement in thermal stability of PANI composites containing FA up to 800 °C at which temperature the compound retains 80% of its mass compared to 40% for composites without FA.

ACKNOWLEDGMENTS

This study was funded by National Research Foundation (NRF) of South Africa. The authors are also grateful for postgraduate bursary granted to Sipho Mavundla by Centre for Science and Industrial Research (CSIR) of South Africa.

REFERENCES

[1] Raghavendra, S. C.; Syed, K.; Revanasiddappa, M.; Ambika Prasad, M. V. N.; Kulkarni A. B. *Bull. Mater. Sci.* 2003, *26,* 733-739.

[2] Petrik L. F.; White, R.; Klink, M.; Somerset, V.; Key, D.; Iwuoha, E.; Burgers, C.; Fey, M. V. *Water Research Commission (WRC) of South Africa,* final report no. K5/1242.

[3] Zhang, Z.; Wei, Z.; Zhang, L.; Wan, M. *Acta Mater.* 2005, *53*, 1373-1379.

[4] Ahmed, S.; Shan, J.; Petrik, L.; Linkov, V. M. *Anal. Sci.* 2004, *20*, 1283-1287.

[5] Huang, S. *Carbon* 2003, *41*, 2347-2352.

[6] Puda, A.; Ogurtsova, N.; Korzhenkob, A.; Galina, Shapoval, G. *Prog. Polym. Sci.* 2003, *28*, 1701–1753.

[7] Dimitriev, O. P. *Macromol.* 2003, *37*, 3386-3395.

[8] MacDiarmid, A. G. *Synth. Met.* 2002, *125*, 11-22.

[9] Heeger, A. J. *Synth. Met.* 2002, *125*, 23-42.

[10] Bae, W.J.; Kim, K. H.; Jo, W. H. *Macromol.* 2004, *37*, 9850-9854.

[11] Schnitzler, D. C.; Meruvia, M. S.; Mmelgen, I. A. H.; Zarbin, A. J. G. *Chem. Mater.* 2003, *15*, 4658-4665.

[12] Cho, M. S.; Park, Y.; Yeong, J.; Choi, H. *J. Mater. Sci. Engg.* 2004, *24*, 15-18.

[13] Ramanathan, K.; Bangar, M. A.; Yun, M.; Chen, W.; Mulchandani, A.; Mnyung, M. V. *Nano Lett.* 2004, *4*, 1237-1239.

[14] Zhang, Z.; Wei, Z.; Wan, M. *Macromol.* 2002, *35*, 5937-5942.

[15] Chandrakanthi, N.; Careem, M. A. *Polym. Bullet.* 2000, *45*, 113-120.

[16] Park, M.; Onishi, K.; Locklin, J.; Caruso, F.; Advincula, R. C. *Langmuir* 2003, *19*, 8550-8554.

[17] Somania, R. P.; Marimuthua, R.; Mandaleb, A. B. *Polymer* 2001, *42*, 2991-3001.

[18] Mollah, M. Y. A.; Promreuk, S.; Schennach, R.; Cocke, D. L. *Fuel* 1999, *78*, 1277-1282.

[19] Pekmez, N.; Pekmez, K.; Yildiz, A. *J. Electroanal. Chem.* 1993, *348*, 389-398.

[20] Iwuoha, E. I.; de Villaverde, D. S.; Garcia, N. P.; Smyth, M. R.; Pingarron, J. M. *Biosens. Bioelectro.* 1997, *12*, 749-761.

[21] Lua, X.; Xub, J.; Wong, L. *Synth. Met.* 2006, *156*, 117-123.

[22] Long, Y.; Chen, Z.; Duval, J. L.; Zhang, Z.; Wan, M. *Physica B* 2005, 370, 121-130.

In: Advanced Organic-Inorganic Composites
Editor: Inamuddin

ISBN 978-1-61324-264-3

Chapter 17

SYNTHESIS AND SPECTRO-CHEMICAL INTERROGATION OF METAL OXIDE MODIFIED POLYPYRROLE HYBRID COMPOSITES

Richard Akinyeye*[a,b*], *Priscilla G. L. Baker*[a] *and Emmanuel Iwuoha

[a]SensorLab, Department of Chemistry, University of the Western Cape,
Bellville, Cape Town, 7535, South Africa

[b]Department of Chemistry, University of Ado Ekiti,
Ado Ekiti, Ekiti State, Nigeria

ABSTRACT

The chemical synthesis and characterisation of metal oxide modified polypyrrole using nano powder of tungsten oxide and zirconium oxide as polymerisation template under different concentration of oxidant is presented. The hybrid materials which were based on the combination of the reduced forms of the metal oxides, i.e. WO_{3-x} and ZrO_{2-y}; with polypyrrole were prepared from the in-situ oxidation of pyrrole in aqueous acidic solution of the individual metal oxides. Variation of the concentration of oxidant used i.e. ammonium persulphate (APS) was made by using two mole concentration ratios of 0.2 and 1 for both $PPyWO_3$ and $PPYZrO_2$ respectively to produce dry metal oxide modified polypyrrole. The scanning electron micrograph (SEM) of the dry, granular powder of the polymers revealed agglomerated nano-bundles with diameters of about 75-300 nm. The tungsten oxide modified polypyrrole ($PPyWO_3$) gave circular nano structures but zirconium oxide modified polypyrrole ($PPyZrO_2$) gave islands of globular nano-bundles. Electron diffraction X-ray (EDX) analysis of the micrographs provided information on the degree of incorporation of the metal oxides in the respective metal oxide modified polypyrrole as well as to the observed yield pattern. While 11.55% incorporation of W was found in $PPyWO_3$, 7.55% of Zr was found in $PPyZrO_2$. Spectroscopic investigation provided features of possible over oxidation in the nanohybrids resulting from the combination of the oxidising nature of the metal oxide coupled with that of the oxidant used. A red shift in absorption maximum for PPy from the normal π to π^* absorption at about 295 nm was seen at 325 nm in both $PPyWO_3$ and $PPyZrO_2$. This might be due to the inclusion of the solvated tungsten oxide and zirconium oxide along the polymers matrix. Characteristic fourier transform infra red spectroscopy (FTIR) spectra band at 822

cm^{-1} ($PPyWO_3$) and 558 cm^{-1} ($PPyZrO_2$) and an extra sharp peak due to the C = C/C-C vibrational modes at 1636-1720 cm^{-1} were observed as further evidence of the metal oxide loading into their respective polypyrrole matrix. The reduced forms of WO_{3-x} and ZrO_{2-y} were speculated as the form of inclusion of the metal oxides in the polymer matrix, where x and y are integers showing the extent of reduction in each case.

ABBREVIATIONS

APS	Ammonium per sulphate
d/m	Mole concentration ratio of metal oxide to monomer (pyrrole)
DMSO	Dimethyl sulfoxide
FTIR	Fourier transform infra-red spectroscopy
ITO	Tin doped indium oxide
MOS	Metal oxide sensors
MOSFET	Metal oxide semiconductor field effect transistors
NPs	Nanoparticles
NRs	Nanorods
o/m	Mole concentration ratio of oxidant (APS) to monomer (pyrrole)
PPy/Fe_2O_3	Polypyrrole doped with iron oxide
PPyDW	Polypyrrole prepared in distilled water without use of dopant of NSA
PPyNQS	Polypyrrole doped with naphthaquinone sulphonic acid
PPyNSA	Polypyrrole doped with naphthalene sulphonic acid
$PPyWO_3$	Tungsten oxide modified polypyrrole
$PPyZrO_2$	Zirconium (IV) oxide modified polypyrrole
Py	Pyrrole
SEM	Scanning electron microscopy
SEM-EDX	Scanning electron microscopy with electron diffraction X-ray (EDX) analysis
UV-Vis	Ultra violet-Visible spectrocopy
WO_3	Tungsten (VI) oxide nanopowder
ZrO_2	Zirconium (IV) oxide nanopowder

1. INTRODUCTION

Metal oxide modified polypyrrole have been used as organic/inorganic hybrid electrodes for increased charge storage in electrochemical storage devices such as supercapacitors and batteries [1-3]. While transition metal oxides incorporated into the matrix of double layer activated carbons are used as double layer capacitors, based on their high specific capacitance and relatively low electric conductivity. The metal oxide nanohybrid-polypyrrole offers room for increased charge storage capacity and conductivity because of the advantage of being both electroactive and conductive. The effective energy storage generated from the nanohybrid of a fairly conducting transition metal oxide with an electroactive and conducting organic polymer like polypyrrole offers room for improved technological possibilities such as electrochemical sensors and supercapacitors. In this study, in-situ reaction between pyrrole and tungsten (VI) oxide (WO_3) and zirconium (IV) oxide (ZrO_2) in acidic solutions are used to produce insoluble metal oxide-polypyrrole composites *via* oxidation with aqueous solution of ammonium persulfate. While the pyrrole is oxidized to polypyrrole, the metal oxide in solution also gets reduced to form insoluble particles in the process. The simultaneous reactions coupled with stirring allows for the incorporation of the insoluble metal oxide particles into the interstitial pores of the polymer.

The main hindrance to polypyrrole processability is its insolubility in most organic solvents. Amongst the approaches that have been used to improve its processability is the production of nano-structurised sulphonated polypyrrole with inherent intrinsic electro-activity [4,5]. We have self-assembled conducting polypyrrole (PPy) micro/nanotubes and films using β-naphthalene sulphonic acid (NSA) both chemically and electrochemically prepared *via* this approach [6,7]. Micellic clusters of the dopant (NSA) and its complex with pyrrole (PPyNSA) served as template for this tubular growth.

In this work, we attempt to improve the processability of polypyrrole composite using metal oxide template. The ability of transition metals to exhibit multiple oxidation states makes metal-oxide doped conducting polymers as suitable intermediate material for the catalytic exchange of electrons in many heterogeneous electrochemical systems [8]. This also is aimed at synergistically improving selectivity and stability. Metal oxide doped polypyrrole is the hybrid polymer obtained from the chemical or electrochemical coupling of electrically conducting metal oxides with polypyrrole. These polymers which have semiconductor properties find potential applications in electrochemical systems such as sensors, batteries and fuel cells [9]. The basic components of a metal oxide semiconductor (MOS) are shown in Figure 1.

Various electrochromic applications have been documented for modified polypyrroles. Rocco et al. fabricated an electrochromic device combining PPy and WO_3 which consisted of tin doped indium oxide (ITO) coated with PPy/dodecylsulfate, an ITO electrode coated with WO_3 and a liquid junction [10]. The light filtering capacity and stability of the solid-state device was dependent on the thickness of the PPy film. The chromatic contrast was stable after 15,000 double potential chromatographic steps [11]. Polypyrrole (PPy) nanocomposites reinforced with tungsten oxide (WO_3) nanoparticles (NPs) and nanorods (NRs) were fabricated by a surface-initiated polymerization method recently by Zhu et al. [2]. The electrical conductivity was observed to depend strongly on the particle loadings, molar ratio of oxidant to pyrrole monomer, and the filler morphology. WO_3 NRs were observed to be more efficient in improving the electrical conductivity, dielectric permittivity, and thermal stability of the resulting nanocomposites as compared to those with WO_3 NPs. Many metal oxide semiconductor sensors (MOS) from materials such as TiO_2, WO_3, In_2O_3 and other oxides have been used in the assembly of the different chemical sensor systems, metal oxide semiconductor field effect transistors (MOSFET) and as gas sensing elements in E-noses [9]. While conducting polymer based chemical sensors could be used at ambient temperature, MOS sensors are used at elevated temperatures. Doping the metal oxide with noble catalytic metals can be used to modify the selectivity of the MOS devices through the changing of working temperature of the sensing element (250-400 °C), or by modifying the grain size [11]. Figure 2 shows conducting polymer sensor used to measure changes in conductance/resistance at different operating conditions. It does not require the use of a reference electrode.

The preparation of tungsten and zirconium oxide modified polypyrrole and assessment of the yield, morphological and spectroscopic properties of the hybrid polymer formed under different synthesis conditions have been investigated. This provided information on the effects of pH, concentration of oxidant, concentration and the type of metal oxide used on the morphology and spectroscopic properties of the polypyrrole metal oxide composites.

2. EXPERIMENTAL

2.1. Chemicals

All chemicals used for this study were purchased from Sigma-Aldrich Pvt. Ltd. South Africa. The pyrrole (98%) used was distilled at reduced pressure, saturated with argon gas, and kept in 1 ml ampoules in the dark at 4 °C prior to use. Tungsten oxide (nanapowder), zirconium oxide (nanopowder), acetone (99.8%), ammonium persulphate, hydrochloric acid (32%) and methanol-anhydrous (99.8%) were used without further treatment. Distilled water (specific resistance 18 MΩ, Milli-Q, Millipore) used was generated through a Milli-Q water purification apparatus (Millipore).

2.2. Chemical Synthesis of Tungsten Oxide Modified Polypyrrole ($PPyWO_3$) and Zirconium Oxide Modified Polypyrrole ($PPyZrO_2$)

Similar procedure as used for PPyNSA [7] was employed in preparing polypyrrole composites using two separate metal oxides, namely tungsten (VI) oxide (WO_3) and zirconium (IV) oxide (ZrO_2). The optimum synthesis temperature condition of 0 °C and a dopant to monomer mole concentration ratio of 0.8 was employed. Variation of the concentration of oxidant used for polymerisation was made by using two mole concentration ratios of 0.2 and 1 for both $PPyWO_3$ and $PPYZrO_2$ respectively.

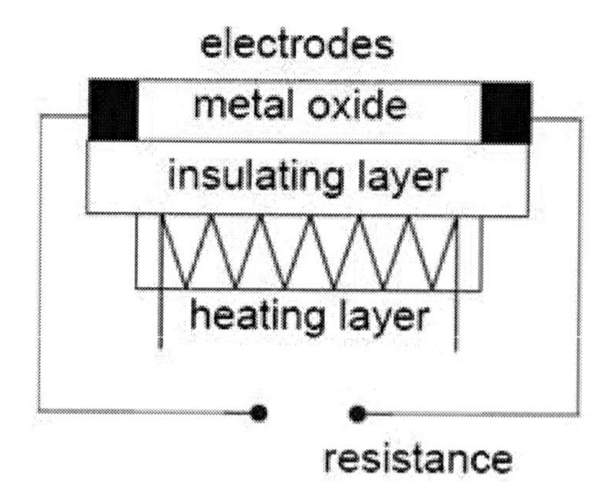

Figure 1. Scheme for a metal oxide semiconductor sensor.

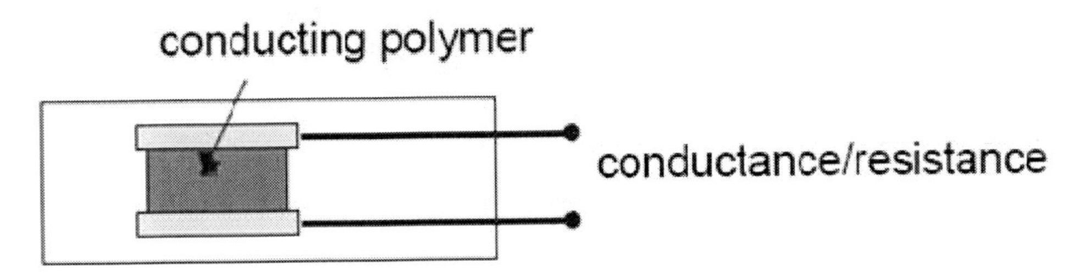

Figure 2. Scheme of a conductometric based sensor using a conducting polymer material.

In a typical procedure for the preparation of tungsten oxide modified polypyrrole (PPyWO_3) from a dopant (WO_3) to monomer (Py) mole concentration ratio (d/m) of 0.8 and an oxidant ammonium persulphate (APS) to monomer concentration ratio (o/m) of 0.2; 0.529 ml (0.0075 mol) of pyrrole was dissolved and stirred in 20 ml aqueous solutions of 1.391 g (0.0060 mol) of WO_3 at 0 °C while stirring for 15 min. The light green liquor solution was cooled to 0 °C after which a 10 ml aqueous solution of oxidant (APS) containing 0.3484 g (0.0015 mol) was added drop wisely to give a black colloidal solution. The beakers were rinsed with distilled water to make up the total volume of liquor to 50 ml and kept stirred for 24 hr at 0 °C. Similarly a more highly oxidised PPyWO_3 was synthesized using the d/m mole concentration ratio of 0.8 but o/m mole concentration ratio of 1 (i.e. 1.745 g of APS (0.0075 mol) at 0 °C). Polymerization process was terminated in each case by the addition of excess methanol and the liquor allowed to age. The resulting polypyrrole precipitate was vacuum filtered and washed sequentially with excess distilled water, methanol and acetone until a colourless filtrate was obtained. The precipitates obtained were dried in a vacuum oven at 25 °C for 12 hours. The apparent yield (wt/wt) was determined from the weight (g) of dry powder from the respective mass of metal oxide used. Table 1 presents the summary of concentrations of the reacting materials used for the synthesis and %yield of PPyWO_3 and PPyZrO_2, respectively from different synthesis conditions. The morphological and spectroscopic properties of the dry product of tungsten oxide modified polypyrrole (PPyWO_3) and zirconium oxide modified polypyrrole (PPyZrO_2) obtained using the experimental procedure described above are discussed underneath.

2.3. Morphological Investigation using Scanning Electron Microscopy (SEM)

The synthesized polypyrrole-metal oxide hybrid samples were examined under a scanning electron microscope (Hitachi X650 Micro-analyzer) with interchangeable accelerating voltages of 10 kV to 30 kV for optimum sensitivity. In the SEM experiment, about 0.01g of the polypyrrole sample was spiked onto a carbon coated sample holder charged with gold to improve surface electrical conductivity. The charged sample is subsequently transferred into the micro-analyzer where it is transversed by electron beam. The signals produced are collected by an appropriate detector, amplified and displayed on the cathode ray tube at different magnifications. The magnification of the image is the relationship between the length of the scan line on the specimen and that on the cathode ray tube. Energy dispersive X-ray (EDX) spectra and the elemental analysis for carbon, sulphur, oxygen, tungsten and zirconium were captured on the polymer's micrograph by stigmation of the relevant area on the micrograph where the nanoparticles were assayed.

2.4. Spectroscopic Investigations Using Ultra Violet-Visible and Fourier Transform Infra Red Spectroscopy

Two broad spectroscopic techniques, namely: Ultra Violet-Visible spectroscopy (UV-Vis) and Fourier transform infra red spectroscopy (FTIR) were employed in this study. The UV-Vis absorption spectra of the $PPyWO_3$ and $PPyZrO_2$ were recorded at room temperature on a GBC UV/Vis 920 spectrophotometer (GBC Scientific Instruments, Australia) between 200 and 900 nm using a 1 cm path length quartz cuvette and 99.6% dimethyl sulfoxide (DMSO) was used as reference solvent. UV-Vis measurements were made with the filtrate obtained from dispersions of the polymer materials in the solvent. Spectra obtained in each case were processed and investigated for characteristic absorptions of the materials. The band gap peculiarities of the materials were explored using the wavelengths of maximum absorption.

The Fourier transform infra red spectra of the polymers were recorded on a Perkins Elmer FT-IR Spectrometer, Paragon 1000PC. In each case, less than 0.0010 g of each polymer was ground in a medium of 0.4 g of dried KBr salt, and placed in the pallet to obtain a fairly transparent pellet. The spectra were recorded in the wavenumber region of 400 to 4000 cm^{-1}. The characteristic set of absorption bands in the spectrum were used to identify various functional groups predominating in the various polymeric states. The bands observed were compared with that from a metal oxide free polypyrrole (PPy) that was prepared from distilled water (DW) under similar synthesis conditions. This was called (PPyDW).

3. Results and Discussion

3.1. Synthesis and Yield Pattern of the Metal Oxide Modified Polypyrrole

After the addition of the tungsten oxide solution to pyrrole solution a light greenish colloidal solution was formed whereas the mixture with zirconium oxide formed a white milky colloidal solution. Upon the drop wise addition of the oxidant solution of ammonium persulphate to either of this solution a gradual black colouration was observed which increases with addition and stirring. Actually, the polymerization of pyrrole into PPy and the reduction of the metal oxide to the reduced form occur instantaneously and simultaneously. The mixture was stirred overnight at room temperature for complete reaction. The yield of the modified polypyrrole prepared under different chemical synthesis conditions is shown in Table 1. From previous studies, we have shown that the optimum conditions for the chemical synthesis of nanostructured polypyrrole with good product yield are low temperature (0 °C), an acidic medium of pH ~1.5, an oxidant to monomer concentration ratio of 0.2 and dopant to monomer ratio of 0.8 [6,7]. While polymerization is impaired at higher temperature, use of higher o/m concentration ratio (1.0), leads to the formation of some degradation products which gets co-precipitated with the polymer composites.

Table 1. Yield of metal-oxide modified polypyrrole per weight of metal oxide used under different synthesis conditions

Code	Pyrrole (m) used (ml)	Metal-oxide (d) used (g)	APS (o) used (g)	Ratios [d/m and o/m]	pH of medium	Colour change	%Polymer yield (wt/wt)
$PPyWO_3$ A	0.529 mL (0.0075 mol)	1.391 g (0.0060 mol)	0.3484 g (0.0015 mol)	d/m (0.80) o/m (0.20)	1.68	Light green to black after oxidation	1.138 g (82 %)
$PPyWO_3$ B	0.529 mL (0.0075 mol)	1.391 g (0.0060 mol)	1.745 g (0.0075 mol)	d/m (0.80) o/m (1.00)	1.36	Light green to black after oxidation	1.889 g (136 %)
$PPyZrO_2$ A	0.529 mL (0.0075 mol)	0.740 g (0.0060 mol)	0.3484 g (0.0015 mol)	d/m (0.80) o/m (0.20)	1.77	White milky solution to black	0.753 g (102 %)
$PPyZrO_2$ B	0.529 mL (0.0075 mol)	0.740 g (0.0060 mol)	1.745 g (0.0075 mol)	d/m (0.80) o/m (1.00)	1.37	White milky solution to black	1.215 (169 %)

* defined as follows: 'd/m' is mole concentration ratio of metal oxide to monomer (pyrrole); 'o/m' is mole concentration ratio of oxidant (APS) to monomer (pyrrole).

The pH indicated on the table was determined for the liquor after the polymerization reaction in the 0.1 M HCl (pH = 1.2) was completed. The %yields observed represent the %of the dry polymer (g) obtained from the equal volume of pyrrole used (ml) per different mass (g) of metal oxide used. The polymerization yield of the hybrid polymers prepared from a more acidic medium of 'o/m' mole concentration ratio of 1.0 (pH ~1.36) is higher than those from the lower 'o/m' mole concentration ratio of 0.2 at a fixed 'd/m' mole concentration ratio of 0.8 (pH ~1.7).

Table 2. Comparative trend of elemental composition (C, S, O, W, Zr, others) in different modified polypyrroles prepared from a d/m mole concentration ratio of 0.8 and o/m concentration ratio of 0.2 by EDX spectroscopic analysis

Samples	C (%)	S (%)	O (%)	W (%)	Zr (%)	Others (%)	Total (%)
PPyNSA	91.84	5.24	2.66	-	-	0.26	100.00
PPyNQS	89.37	5.46	4.58	-	-	0.67	100.08
$PPyWO_3$	82.55	1.47	4.05	11.77	-	0.22	100.06
$PPyZrO_2$	87.23	1.15	3.92	-	7.55	0.15	100.00

With the use of o/m concentration ratio of 0.2, the yields were 82% and ~100% for $PPyWO_3$ and $PPyZrO_2$, respectively. This is indicative of efficient and optimized polymerization of the metal oxide with the pyrrole monomer from the mother liquor. At higher o/m concentration ratio of 1.0, the yield was 136% and 169% for $PPyWO_3$ and

$PPyZrO_2$, respectively, a sign of significant admixture of co-precipitates with the PPy-metal oxides. More polymer hybrid was obtained from the zirconium oxide modified polypyrrole than that from tungsten oxide.

This could be attributed to the lower mole equivalent of the zirconium oxide that is required to react with a given quantity of the monomer relative to tungsten oxide (indicating that Zr is more susceptible to oxidation). Optimisation of the chemical synthesis conditions to determine desirable maximum loading of metal oxide (d/m) into the polymer matrix for the production of a highly conducting polymer composite state is still being investigated.

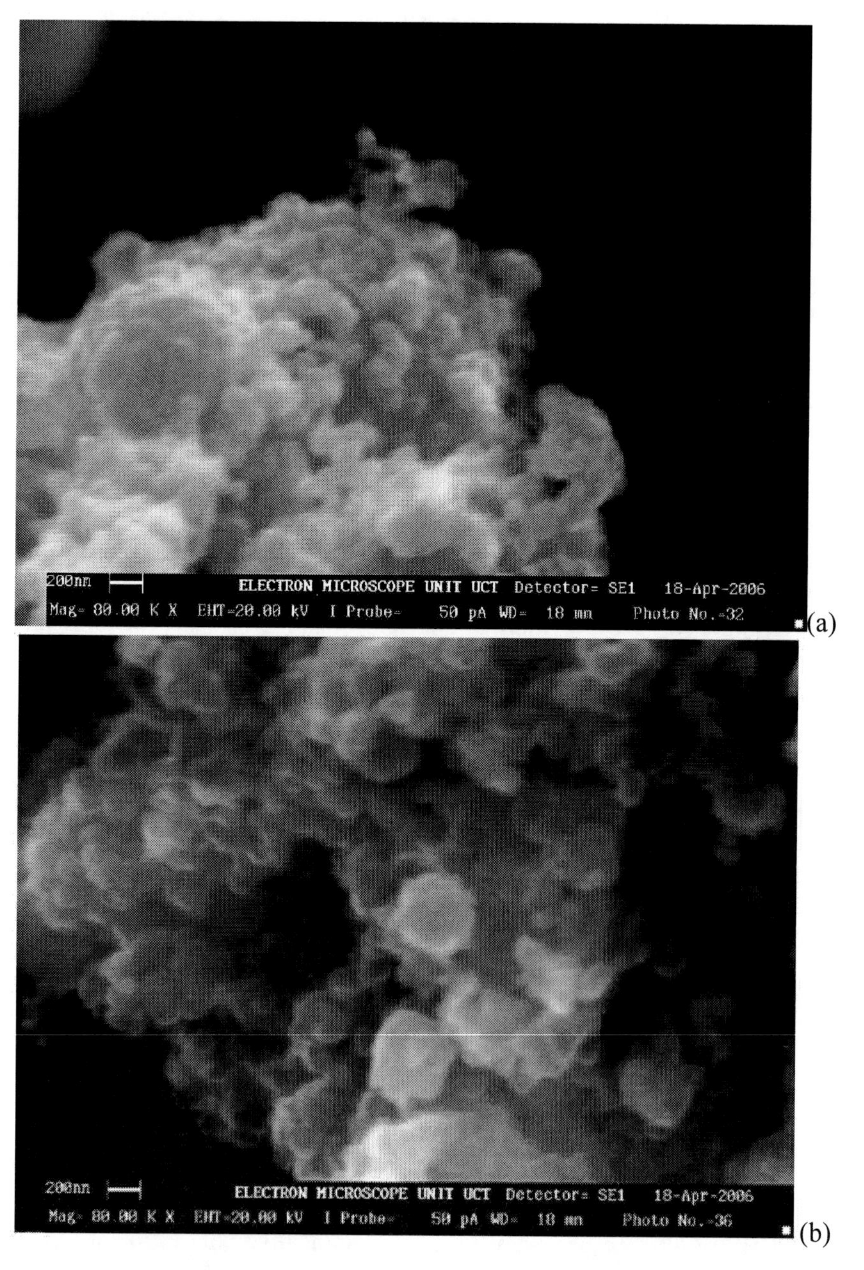

Figure 3. SEM micrographs of dry powder of metal oxide modified polypyrrole showing the typical fibrous nanostructures from $PPyWO_3$ [d/m 0.8; o/m 0.2] (a) and $PPyZrO_2$ [d/m 0.8; o/m 0.2] (b).

3.2. Morphological and Electron Diffraction X-ray (EDX) Examination

The SEM micrograph (Fig. 3) on the polypyrrole from the o/m mole concentration ratio of 0.2 were taken to see if nanohybrids were formed and to determine relative metal oxide loading in the polymer matrix. Figure 3 show the fibrillar morphology of the nanofibres forms of the polypyrrole hybrids. The micrograph of the dry, granular powder of the tungsten oxide modified polypyrrole (Figure 3 a) gave agglomerated nanobundles with circular diameters of about 75-300 nm while that from zirconium oxide modified polypyrrole (Figure 3 b) gave islands of globular nanobundles of similar diameter as $PPyWO_3$. Similar morphology with spherical particles of particle sizes of 400-500 nm was reported for nanohybrid of PPy/Fe_2O_3 nanocomposite [3].

Table 2 presents the proximate elemental composition obtained from the EDX spectra of different polypyrroles. Evidence of the incorporation of the metal oxides was revealed in the respective metal oxide modified polypyrrole. While 11.55% of W was found in $PPyWO_3$, 7.55% of Zr was in $PPyZrO_2$. The higher product yield observed from the $PPyZrO_2$ was also corroborated by the higher carbon content of 87% for $PPyZrO_2$ as against about 83% in $PPyWO_3$ (Table 3). Furthermore, the data shows lower % carbon from the metal oxide modified polymers ($PPyWO_3$ and $PPyZrO_2$) compared to the organic acid modified polymers (PPyNSA and PPyNQS).

The order of decreasing % carbon is PPyNSA > PPyNQS > $PPyZrO_2$ > $PPyWO_3$. The relatively lower % sulphur in the metal oxide modified polypyrrole is attributed to the absence of surfactants whereas the sulfonated polypyrroles have higher % sulphur. The observed 1.5% sulphur in the metal oxide modified polypyrrole must have originated from the oxidant (APS) used in the polymers preparation. Higher values of % S obtained for the surfactant modified PPy is therefore not surprising since there is S contribution from the surfactant used as dopant.

3.3. UV-Vis Spectroscopic Studies

The spectroscopic properties of metal oxide modified polypyrrole are presented and discussed below. Figures 4 present the UV-Vis spectra for the metal oxide modified polypyrrole. Similar trend of spectra were obtained on the polymer composites of $PPyWO_3$ (Figure 4 a) and $PPyZrO_2$ (Figure 4 b) using different o/m mole concentration ratios of 0.2 and 1.

A small absorption maxima at 325 nm was observed on the spectra of $PPyWO_3$ and $PPyZrO_2$ respectively for the conjugated double bonds (i.e. π to π^* transition). The red shift from the normal π to π^* absorption at about 295 nm might be due to the inclusion of the solvated tungsten oxide and zirconium oxide along the polymers matrix [13]. The polaronic absorption is poorly formed at about 480 nm (Figure 4 a). Similar trend of spectra were seen for the $PPyZrO_2$ (Figure 4 b). The absence of sharp polaronic and/or bipolaronic absorptions at higher wavelengths may be an indication of insignificant charge carriers in the system and thus further optimisation may be required to improve the polymers conductivity.

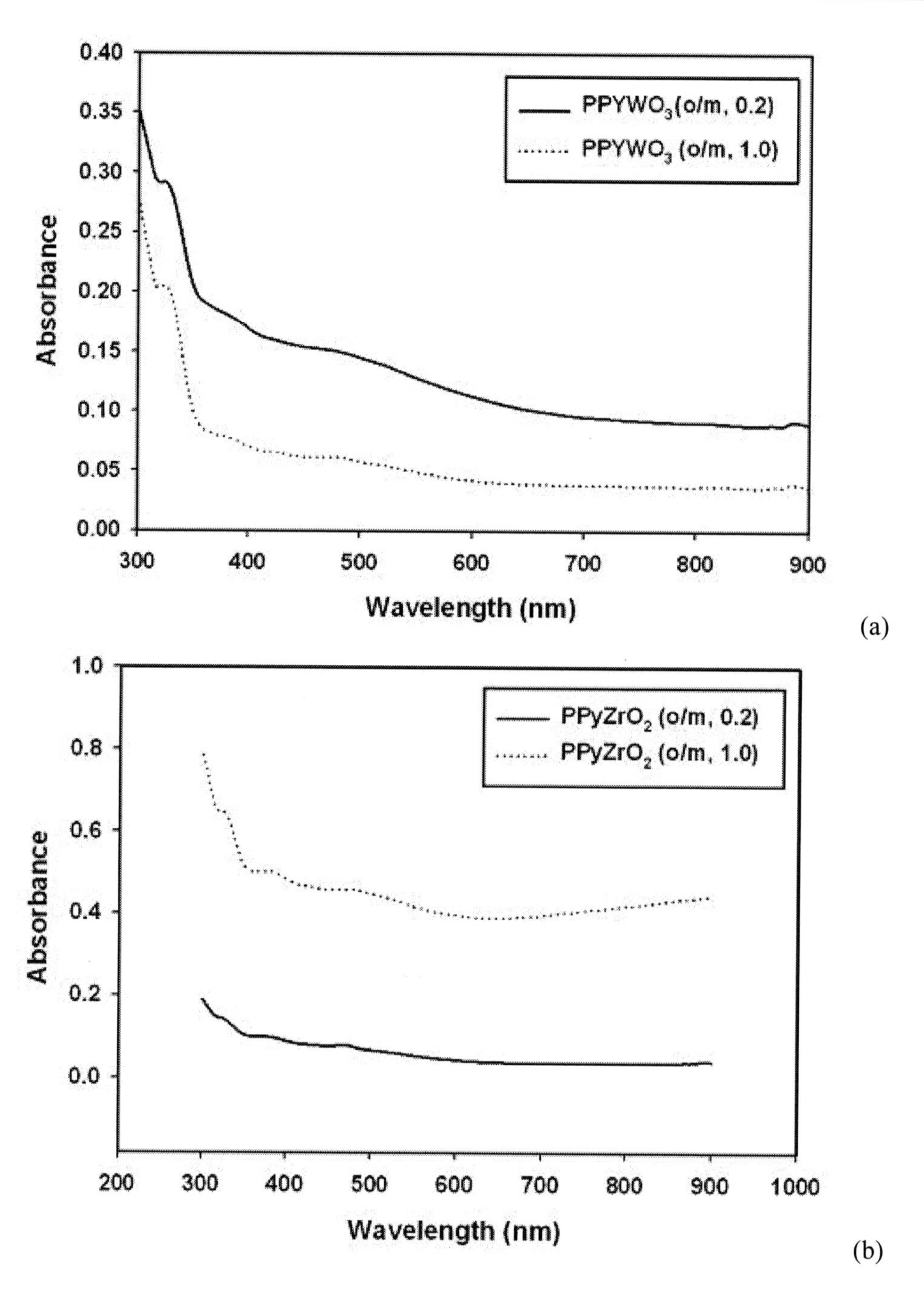

Figure 4. UV-Vis results for metal oxide modified polypyrrole prepared under different synthesis conditions: (a) $PPyWO_3$ from o/m 0.2 and 1.0; and (b) $PPyZrO_2$ from o/m 0.2 and 1.0.

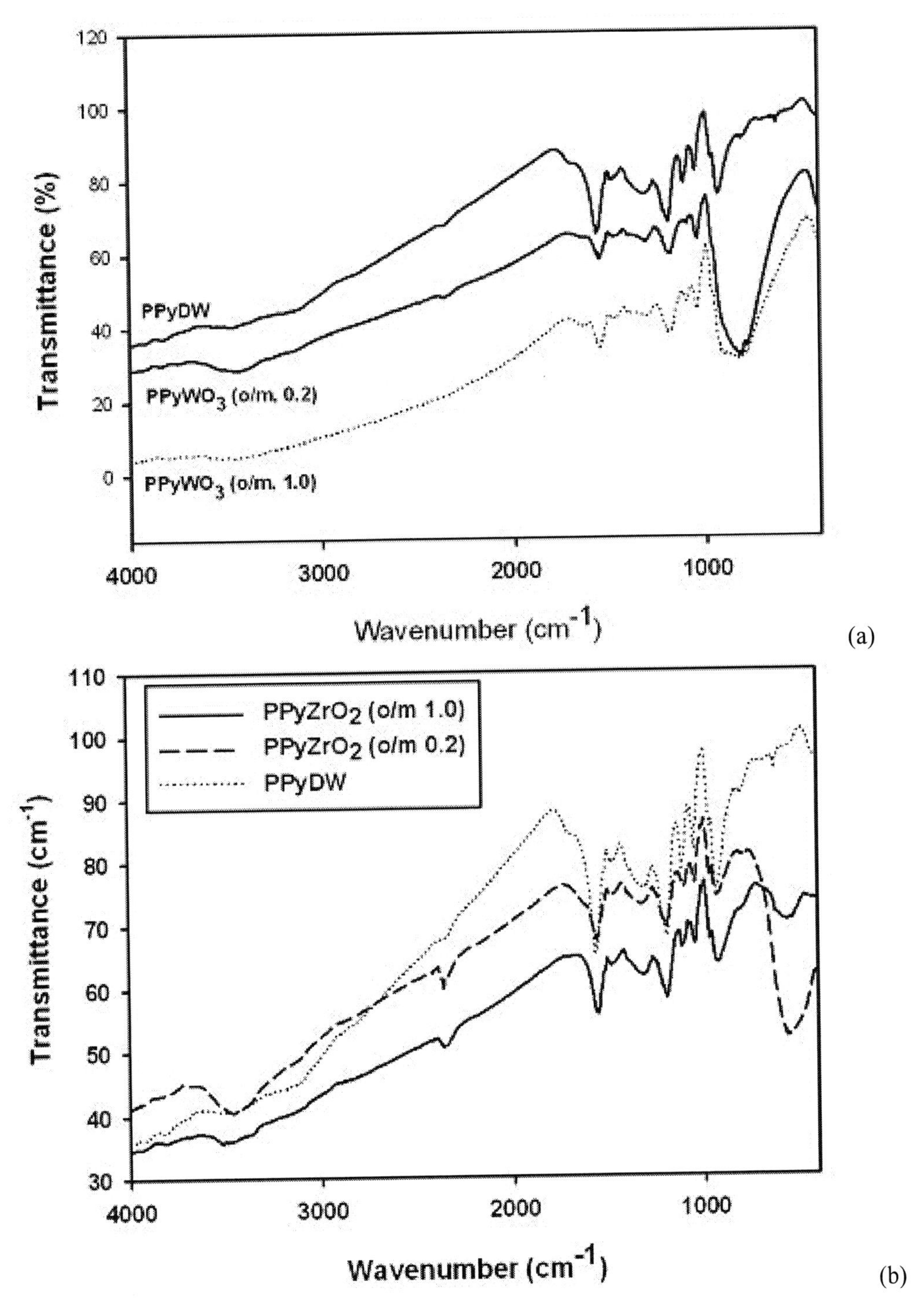

Figure 5. FTIR spectra of metal oxide modified polypyrroles in KBr medium, Fig. 5a: PPyDW, $PPyWO_3$ (d/m 0.8; o/m 0.2) and $PPyWO_3$ (d/m 0.8; o/m 1.0); and Fig. 5b: PPyDW, $PPyZrO_2$ (d/m 0.8; o/m 0.2) and $PPyZrO_2$ (d/m 0.8; o/m 1.0).

Table 3. Major shifts of bands (cm^{-1}) in FTIR spectra of PPyDW, $PPyWO_3$ (d/m 0.8, o/m 0.2) and $PPyWO_3$ (d/m 0.8, o/m 1.0) from undoped polypyrrole major bands [13]

Major Bands (cm^{-1})	PPyDW	$PPyWO_3$ (o/m 0.2)	$PPyWO_2$ (o/m 1.0)
3421 (N-H Str)	Absent	3420	3413
3100 (C-H Str)	Absent	Absent	Absent
1535 (C=C & C-C Str)	1554	1550	1547
1450 (N-H Str)	Absent	1461	1458
1295 (C-H & N-H Def)	1308	1305	1305
1050 (C-H Def)	1047	1040	1040

Table 4. Major shifts of bands (cm^{-1}) in FTIR spectra of PPyDW, $PPyZrO_2$ (d/m 0.8, o/m 0.2) and $PPyZrO_2$ (d/m 0.8, o/m 1.0) from undoped polypyrrole major bands [13]

Major Bands (cm^{-1})	PPyDW	$PPyZrO_2$ (o/m 0.2)	$PPyZrO_2$ (o/m 1.0)
3421 (N-H str)	Absent	3461	3468
3100 (C-H str)	Absent	Absent	Absent
1535 (C=C & C-C str)	1554	1556	1558
1450 (N-H str)	Absent	Absent	Absent
1295 (C-H & N-H def)	1308	1311	1307
1050 (C-H def)	1047	1045	1045

3.4. Fourier Transform Infra Red (FTIR) Spectral Studies

Figure 5 shows the FTIR spectra of the metal oxide modified polypyrrole. Figure 5 a, gives spectra for $PPyWO_3$ and metal oxide free polypyrrole (PPyDW) while Figure 5 b gives the spectra for $PPyZrO_2$ and metal oxide free polypyrrole (PPyDW) respectively. The Tables 3 and 4 present's data on some characteristic bands for the polymers and compared with those reported for polypyrrole by Geetha and Trivedi [13]. All the characteristic absorption bands for the metal oxide modified polypyrrole were observed especially at the fingerprint region of 1000-1700 cm^{-1} with slight variation in the absolute values (Tables 3 and 4). The usual N-H stretching at 3400 cm^{-1} in neutral polypyrrole is present in both $PPyWO_3$ and $PPyZrO_2$ samples, though with slight variations, indicating that the polymers are not in the doped state as was observed in the earlier investigation on PPyDW, PPyNSA (d/m 0.8, o/m 0.2) and PPyNSA (d/m 0.8, o/m 1.0) [6].

The most interesting feature in the FTIR spectra is the lower wavelength absorption band at 822 cm^{-1} for $PPyWO_3$ and 558 cm^{-1} for $PPyZrO_2$ (Figure 4). Similar low band absorption at around 500 cm^{-1} was reported for polypyrrole-manganese (IV) oxide nanocomposite hybrids [1]. Furthermore, an extra sharp peak due to the C=C/C-C vibrational modes at 1636-1720 cm^{-1}, depending on the metal ion and the oxidant concentration used, was seen in the spectra of the metal oxide modified polypyrroles. This peak is assigned to the shifted peak from 1540 cm^{-1} due to significant overoxidation of the polypyrrole matrix [6,13,14]. The oxidising nature of the metal oxide coupled with that of the oxidant might have caused this over-oxidation. This will invariably reduce the electronic conductivity and ultimately loss or reduced electrochemical activity. These characters in the metal oxide modified polypyrrole attest to the incorporation of the metal oxide in the polypyrrole. The included inorganic component in the hybrid is a reduced metal ion resulting from the oxidative polymerisation. Thus, the reduced form in which the metal oxide precipitate could be represented as WO_{3-x} and ZrO_{2-y}, respectively where x and y are integers showing the extent of reduction in each case [1].

CONCLUSION

In this paper, we described a direct and very simple procedure for the chemical synthesis of $PPyWO_3$ and $PPyZrO_2$ and investigated the morphological and spectroscopic properties of the nanohybrids that were formed. This preliminary investigation on metal oxide modified polypyrrole has shown that significant modification of the polypyrrole matrix is generated by the in-situ chemical polymerisation with the nano powder of the metal oxides. The metal oxide nanoparticles served as a support to the polymerisation process of pyrrole and this led to a more porous structure with a higher specific surface area. Nanohybrids are readily formed via the in-situ polymerisation of pyrrole in acidic aqueous medium. Two simultaneous processes have been hypothesised; namely the oxidation of pyrrole to polypyrrole and the reduction of the WO_3 to WO_{3-x} (or ZrO_2 to ZrO_{2-y}) where x and y are the respective degree of reduction after polymerisation. With efficient and optimised reduction of the metal oxide and oxidation of the metal oxide (where $x = 3$ and $y = 2$), the polymer hybrid ends up with the elemental nano particles of the metals on the polymer chain. The investigated polymers do not show significant features indicative of reasonable concentration of charge carriers for electrocatalytic application. This is evidenced by the absence of sharp polaronic and/or bipolaronic absorptions which presuppose insignificant charge carriers in the system and thus further optimisation may be required to improve the polymers conductivity. Further optimisation of synthesis conditions to come up with appropriate metal oxide loading for specific application is on course. Evidence of incorporation of the metal oxide into the polymers matrix was provided by both SEM-EDX and FTIR spectroscopic analysis. One of the characteristic features in the FTIR spectra is the lower wavelength absorption band at 822 cm^{-1} for $PPyWO_3$ and 558 cm^{-1} for $PPyZrO_2$. The electrochemical synthesis, characterisation and test application of the nanohybrids is still being investigated. The electrochemical interrogation for the oxidation and reduction propertiess of the hybrid polymers is also under investigation. These hybrid polymers may find applications as electrochromic devices and metal oxide sensors (MOS) that could be operated at elevated temperatures.

ACKNOWLEDGMENTS

The authors gratefully acknowledge the financial support from National Research Foundation of South Africa.

References

[1] Cuentas-Gallegos, A. K.; Gómez-Romero, P. *J. New. Mat. Electrochem. Systems* 2005, *8, 181-188.*

[2] Zhu, J.; Wei, S.; Zhang, L.; Mao, Y.; Ryu, J.; Mavinakuli, P.; Karki, A. B.; Young, D. P.; Guo, Z., *J. Phys. Chem. C* 2010, *114*, 16335–16342.

[3] Mallouki, M.; Tran-Van, F.; Sarrzin, C.; Simon, P.; Daffos, B. De A.; Chevrot, C.; Fauvarque, J. *J. Solid State Electrochem.* 2007, *11,* 398-406.

[4] Yang, Y. S.; Liu, J.; Wan, M. X.; *Nanotechnology* 2002, *13,* 771-773.

[5] Yang, Y. S.; Wan, M. X.; *J. Mater, Chem.* 2001, *11,* 2022-2027.

[6] Akinyeye, R. O.; Sekota, M.; Baker, P.; Iwuoha, E. *Fullerenes, Nanotubes and Carbon Nanostructures* 2006, *14,* 49-55.

[7] Akinyeye, R. O;, Michira, I.; Sekota, M.; Al-Ahmed A.; Baker, P.; Iwuoha, I. *Electroanalysis* 2006, *18,* 2441-2450.

[8] Martínez-Millian, W.; Toledano, T. T.; Smit, M. A. *Int. J. Electrochem. Sci.* 2010, *5,* 931–943.

[9] James, D.; Scott, S. M.; Ali., S. Z.; O'Hare, W. T. *Microchimica Acta* 2005, *149,* 1-17.

[10] Rocco, A. M.; De Paoli, M. A.; Zanelli, A.; Mastragostino, M. *Electrochimica Acta* 1996, *41,* 2805-2816.

[11] De Paoli, M. A., Zanelli, A.; Mastragostino, M.; Rocco, A. M. *J. Electroanal. Chem.* 1997, *435,* 217-224.

[12] Kemp, W. (1988). *Organic Spectroscopy*, ELBS/Macmillan, 2nd ed., 1988; 204–206.

[13] Geetha, S.; Trivedi, D. C. *Mater. Chem. Phys.* 2004, *88,* 388–397.

[14] Rodríguez, I.; Scharifker, B. R.; Mostany, J. *J. Electroanal. Chem.* 2000, *491,* 117-125.

In: Advanced Organic-Inorganic Composites
Editor: Inamuddin
ISBN 978-1-61324-264-3

Chapter 18

POLYMER COMPOSITES: TOXICITY, ENVIRONMENT AND HUMAN HEALTH

***Abhinav Srivastava*[a]*, Mayank Pandey*[a]*, Pradeep K. Mathur*[b] *and Vinod P. Sharma*[a,*]**

[a]Indian Institute of Toxicology Research, Lucknow 226001, India
[b]Department of Chemistry, University of Lucknow,
Lucknow 226001, India

ABSTRACT

We are living in plastic era. Polymers and plastics are indispensible part of the daily life. Plastics and their composites are being used on a large scale for packaging of food stuffs, drinking water and other ingredients. Polymer composites are becoming popular as they are durable, resistant to environmental conditions and wide applicability despite of their high cost. Polymer composites are heterogeneous substances consisting of two or more materials which have their individual characteristics. The combination of varied compatible materials brings about new physical, chemical and mechanical desirable properties. The composite engineering is an interesting field of science and requires multi-disciplinary knowledge for the understanding of chemical interaction of ingredients, structural changes during processing, selection of appropriate method and resultant consequent product.

Above all, suitability assessment is must for the risk free use of polymer composites and they must be fabricated as per intended usage and regulatory ISO or BIS guidelines. The efforts are being made worldwide for developing environmental friendly and cost effective composites extending the utilization of polymers from renewable resource into new value-added bio-based composites.

---

* Corresponding author's e-mail: vpsitrc1@rediffmail.com

ABBREVIATIONS

ASTM	American Society for Testing and Materials
BIS	Bureau of Indian Standards
CFCs	Chloro Fluoro Carbons
FDCA	Furan Di Carboxylic Acid
FRP	Fiber Reinforced Material
GFRP	Glass Ffiber Reinforced Plastic
HBCDs	Hexa Bromo Cyclo Decanes
HMF	5- Hydroxy Methyl Furfural
IS	Indian Standard
ISO	International Organization for Standardization
MEKP	Methyl Ethyl Ketone Peroxide
PBDEs	Poly Brominated Diphenyl Ethers
SBR	Styrene Butadiene-latex Rubber
UV	Ultra Violet
VOCs	Volatile Organic Compounds

1. INTRODUCTION

The application of plastic and polymeric composites for application in spacecrafts, defense products, alternate building and construction materials is rapidly increasing. Greatest advantage of composite materials is the strength, stiffness and design flexibility combined with lightness. The polymer composites are materials made from polymers or matrix along with addition of fillers or raw natural ingredients as reinforcement. The different materials with different attributes may have varied matrices and reinforcement and thus their properties may differ.

The reinforcement material imparts its special mechanical and/ or physical properties to enhance the matrix properties. Thus the synergism produces materials with properties unavailable or nonfeasible from the individual constituent materials. This allows the designer of the product to select the optimum combination e.g. strengthening, reinforcement characteristics, solidification, colour, texture, shape etc.

Engineered composite materials are evolving with new innovations and market demand of the 21st Century. The end product requirements of the moulded materials may have variable constituents or features.

Most commercially produced composites are fabricated using the polymeric matrix material often called a resin solution. The different polymers available in the market depend upon the starting raw ingredients viz. polyesters, vinyl ester, epoxy, phenolic, polypropylene, poly ether- ether ketone, natural and partially treated raw materials. The strength of the product is dependent on the ratio of the resin, fiber and additives.

2. HISTORICAL PERSPECTIVES AND INNOVATIONS

The history of composites dates back to the biblical times as mentioned in the biblical Book of Exodus, Moses's mother built the ark from rushes, pitch and slime—a kind of fiber-reinforced composite, according to the modern classification of material. Since ancient times fibers have been used as reinforcement materials.

The concept of using fibers of different natural products along with mortar and straw in mud- bricks is historically well known. The need was felt for replacement of asbestos in concrete and other building materials once the health risks associated with the substances were discovered. The innovations are in progress for the utilization of steel, glass and synthetic fibers in concrete and other matrices for desired tailor made specifications. The newly developed engineered composites are 500 times more resistant to cracking and 40% lighter than traditional concrete.

The recent studies on fiber reinforced composite materials have revealed that addition of fibers provides residual strength and controlled cracking. Few studies using waste carpet fibers in concrete are environmental friendly used of recycled carpet waste. The application of Styrene Butadiene-latex Rubber (SBR) is increasing rapidly. The carbon nano tubes are used as polymer anti- oxidants [1]. The oxidation of polystyrene, poly ethylene, poly propylene, poly (vinylidene fluoride) is retarded by carbon nano tube. Incorporation of boron into the nano tubes enhances the electron affinity of the tubes and leads to a small entries in anti-oxidant efficiencies.

2.1. Properties

Composite material have gain popularity (despite their high cost) in high performance products that need to be light weight yet strong enough harsh loading conditions such as arrow space components (tails, wings, propellers), antennas, reflectors, boat and scull hulls, bicycle frames, racing car bodies, storage tanks, fishing rods etc. the composite material are also becoming more common in orthopedic surgery, dentistry, launch vehicles, The physical properties of composite materials are generally anisotropic i.e. independent of direction of applied force. But is typically orthotropic i.e. different depending upon the different directional based. The stiffness of the composite is dependent on the design, type and orientation of fiber axis. The FRP materials may be categorized into short fiber reinforced material and continues fiber reinforced material. The continuous reinforced materials often constitute a layered or laminated structure. In few cases the pre- impregnation is done with suitable chemicals or monomers.

The techniques that take advantage of anisotropic properties of the materials include mortise and tendon joints and Pi joints in synthetic composites. The material properties can be characterized by young's modulus, the shear modulus and the Poisson's ratio. Polyester resin mixed with fiber glass e.g. glass fiber reinforced plastic (GFRP) have yellowish tint and is suitable for multiple projects and construction works. The hardness is provided by use of a catalyst known as methyl ethyl ketone peroxide (MEKP) the epoxy resin is used as structural materials or as glue. Its weaknesses are it is UV Sensitive and may tend to degrade with passage of time. Hybrid composite metal hip resurfacing [2] had the potential to reduce stress shielding, preserve bone stock, and prevent from bone fracture compared to conventional metallic hip resurfacing implants.

3. ENVIRONMENTAL AND HEALTH IMPLICATIONS OF POLYMER COMPOSITES

The polyvinyl based composites may release dioxin and furans on combustion. Fiber glass has increased in popularity since the discovery due to its low toxicity. Most synthetic vitreous fibers are not crystalline like asbestos; they do not split longitudinally to form thinner fibers. They also generally have less bio- persistence (0.04-10%) in biological tissues than asbestos fibers as they can undergo dissolution and transverse breakage. Most of the composites are resistant to common solvents, acid, alkalis and cutting fluids. There advantages include flexibility, less assembly time for incorporating multiple components into one casting. The conventional wood composites which have a high degree of customer acceptance and market demand include particle board, fiber board and flake board. Most of these composites are made from recycled wood. A heat curing adhesive holds the wood components together. There are reports indicating that poly brominated diphenyl ethers (PBDEs) and hexa bromo cyclo decanes (HBCDs) [3] may be present as bio accumulative products in the environment.

They are used as additives in few plastics, foams, electronics, fabrics, building insulation materials, furniture, electrical equipments etc. the level of concentration may vary and thus there is a need of complete understanding of mechanism of toxicity of such chemicals and the potential for additives and synergistic effects (EHP, 2010). The Research and Development organizations around the globe are experimenting with plant based plastics with different properties as alternatives in a bid to lower CO_2 and reduce the use of petroleum as oil stocks decline. 5-Hydroxy methyl furfural (HMF) is being converted into furan di carboxylic acid (FDCA) to serve as petroleum based precursor for the fabrication of plastic bottles.

Figure 1. Plastic Debris and its Environmental Implications.

The building industries are contributing to sufficient extent in depletion of resources, generation of waste and energy conservation [4]. The promotion of utilization of renewable resources as CO_2 neutral building materials can only be considered sustainable when it does not result in faster deforestation. (Emmanuel, 2004). Polymer composites made from polystyrene may not biodegrade for hundreds of years and may be resistant to photolysis. Moreover, degradation of materials creates potentially harmful liquid and gaseous by- product that could contaminate ground water and other environmental components (Figure 1). The efforts are being made for development of composites which are degradable to a large extent. The recycling of polystyrene products and conversion into useful value added products is the demand of the day. However, when polystyrene is burnt at temp of 800- 900^{0}C, the product of combustion consist of complex mixture of polycyclic aromatic hydrocarbons (PAHs). Beside carbon black, carbon mono oxide and release styrene monomers under uncontrolled conditions.

3.1. Relevance of Suitability Assessment

The high molecular mass polymer itself does not pose a toxic hazard being inert. There is, however, a likelihood that some transfer/migration of polymer additives, adventitious impurities, such as monomers, catalyst remnants, residual polymerization solvents of low molecular mass fractions may occur from the plastics in to the packaged materials with consequent toxic hazard to the consumers of products packed in plastics. The occurrence of acute toxicity due to plastic materials in contact with food stuffs is most unlikely; since only trace quantities are potentially toxic materials are likely to migrate. However, the accumulation of these materials with time may lead to hazards which may be serious (BIS 10141:2001). The fabricated products are to be evaluated before and after construction for predicting and preventing failures of composites. The safety and suitability parameters are dependent on the regulatory guidelines and intended usage conditions (Figure 2).

Figure 2. Suitability Assessment of Polymer Composites

It is preferred that the composite materials are tested through non- destructive methods including ultrasonic, thermography, shearography, X-ray radiography, finite elemental analysis, chromatography and state of art techniques.

3.1.1. Salient Standards

Few useful standards with respect to polymer composites are IS 9833, 9845, 10141, 10142, 10146, 10149, 10151, 10171, 10909, 11434, 12229, 12252, 12709, 13449, 13601, 14399, 13360, BIS 5736 [5]; ISO 10993, ASTM 978-08, ASTM A20-06; ASTM F 2068-09, ASTM C 1018-07, ASTM 3678 [6]. The new methods are being devised for polymerizing conventional lipophilic using super critical CO_2 along with added stabilizer. It is anticipated to have potential applications with elimination of the toxic by- products generated in conventional polymer manufacturing. Super critical CO_2 is a highly pressurized form of chemically benign gas that has properties of a liquid. In future chemical companies may be able to utilize this technology for plastic production and curb the use of environmentally hazardous chlorofluorocarbons (CFCs) and volatile organic compunds (VOCs) viz. toluene, methylene chloride, which presents risks to human health. The efforts are in progress for the development of green composites for minimizing the environmental impact of polymer composite production. The life cycle assessment is of paramount in importance at every stage of a product's life from initial synthesis to final disposal for a sustainable environment. This includes clean processing, applications, recycling bio-degradation and reprocessing.

3.1.1.1. Bio-based Polymer Composites

Modern composites are extending the utilization of polymers from renewable resource into new value-added products. The study and utilization of natural polymers is the need of today due to environmental issues and the rapidly depleting conventional energy resources. Bio composites are generally produced by embedding natural fibres (e.g. flax, hemp, ramie) into a bio polymeric matrix made of derivatives from cellulose, starch, lactic acid, etc. [7]. Natural polymers (like starch) are generally moisture sensitive which limits their applications. Although to overcome this problem multilayer composites are being developed having thermoplastic starches with laminated water-resistant, biodegradable polymers. Composites based on biodegradable polymers viz. poly (3-hydroxybutyrate), poly(4-hydroxybutyrate) and poly(3- hydroxybutyrate-co-4-hydroxybutyrate) have been suggested suitable for heart valve tissue engineering or direct implantation [8]. Natural fibres are of basic interest since they not only have the functional capability to substitute the widely used glass fibers, but are also advantageous in view of weight and fiber matrix adhesion, specifically with polar matrix materials. They have good potential for use in waste management hence promoting the eco friendly concept of Green Chemistry due to their biodegradability and their much lower production of ash during incineration.

4. Polymer Based Nano-Composites

The combination of polymers and nanoparticles is opening a new way for engineering flexible composites that exhibit the demanded electrical, optical, or mechanical properties [9]. Metal nano particles and complex oxide nano particles fabricated after doping may be used for biocide applications. Silver based nano- particles can be prepared in a form suitable for effective biocide water born polymeric coatings for applications in bathroom fixtures, wound dressings, biological sensors, controlled drug delivery mechanisms, surface desired physical characteristic. In composites, the varied reinforcing levels may affect the mechanical properties depending on the chemistry of nano particles, its aspect ratio, dissemination and interfacial interactions with polymer matrix [10]. The production of polymer based nano-particles can be achieved through solid state methods, vapor methods, chemical synthesis and gas phase synthesis methods. This may lead to improved mechanical, electrical, optical, barrier and flame retarded behavior using melt compounding, plasma spraying techniques or polymerization. Many nano particles are stabilized using organic surfactants, with a compatibilization role similar to that of sizing for fiber glass for traditional composites. The optimization of interfacial compatibility in case of nano- composites is vital from interaction of the polymer with nano particles point of view and product performance. There is a going trend of polymer nano-composites market throughout the globe. Widespread commercial usage of nanocomposites is limited due to lack of structure-property relationships and effective processing techniques effective at both the nanoscale as well as at the microscopic levels. There is a need of hierarchically structured composites in which each sublayer contributes a distinct function to yield a mechanically integrated, multifunctional material.

CONCLUSION

The integrated waste management of plastic products is desirable for managing the waste in the environmental and economically sustainable nature. The practices are to be adopted for efficient disposal or collection of solid waste to avoid adverse effect on public health. The plastic products should be manufactured as per the regulatory guidelines, evaluated for their characteristics and ensure to minimize the adverse effect on human health and biodiversity. There is significant need for research and development in the field of polymer blends from renewable resources having both the eco friendly aspect as well as the economic viability.

ACKNOWLEDGMENT

We are thankful to the Director, Indian Institute of Toxicology Research (Council for Scientific and Industrial Research), Lucknow for providing basic infrastructural support and encouragement.

REFERENCES

[1] Watts, P. C. P.; Fearon, P. K.; Hsu, W. K.; Billingham, N. C.; Kroto, H. W.; Walton, D. R. M. *Mater Chem.* 2003, *13,* 491-495.

[2] Bougherara, H.; Martin, N. *Mat. Sci.* 2008, *368985,* 1-4.

[3] Schecter, A.; Haffner, D.; Colacino, J.; Patel, K.; Papke, O.; Opel, M.; Birnbaum, L. *Envir. Health Persp.* 2010, *118,* 357-362.

[4] Emmanuel, *Buil. and Envir.* 2004, *39,* 1253-1261.
[5] Bureau of Indian Standards (BIS) guidelines on Plastics (1981-2009).
[6] American Society for Testing and Materials (ASTM).
[7] Yu, L.; Dean, K.; Li, L. *Prog. Polym. Sci.* 2006, *31,* 576–602.
[8] Stamm, C.; Khosravi, A.; Grabow, N.; Schmohl, K.; Treckmann, N.; Drechsel, A.; M.; Schmitz, K.P.; Haubold, A.; Steinhoff, G. *Ann. Thorac Surg.* 2004, *78,* 2084-2093.
[9] Balazs, A.C.; Emrick, T.; Russell, T.P. *Science* 2006, *314,* 1107-1110.
[10] The Project on emerging nano- technologies inventories, Wood Row Wilson International Center for Scholars, Washington DC. 2009.

In: Advanced Organic-Inorganic Composites
Editor: Inamuddin

ISBN 978-1-61324-264-3

Chapter 19

SYNTHESIS, CHARACTERISATION OF POLYANILINE-CO-POLY(2,2'-DITHIODIANILINE) AND ITS SENSOR APPLICATION FOR TRACE METAL DETERMINATION

***Vernon Somerset*[a,*], *Emmanuel I. Iwuoha*[b] *and Lucas Hernandez*[c]**

[a]NRE, Council for Scientific and Industrial Research (CSIR), Stellenbosch, 7599, South Africa
[b]SensorLab, Chemistry Department, University of the Western Cape, Bellville, South Africa
[c]Facultad de Ciencias, Universidad Autónoma de Madrid, Campus de Cantoblanco, Cantoblanco, 28049, Madrid, Spain

ABSTRACT

This chapter reports the preparation of the co-polymer of aniline (ANI) with 2,2′-dithiodianiline (DTDA) in acidic solution. Cyclic voltammetry (CV), Fourier Transformed Infrared Spectroscopy (FTIR) and UV-Vis spectroscopy were used to identify the differences in optical properties and structures between the co-polymer and the homopolymer of PANI and/or poly(2,2′-dithiodianiline) (PDTDA).

The prepared PANI-co-PDTDA co-polymer exhibited the electrochemical properties of both PANI and PDTDA. The incorporation of the S–S links displaying their distinct redox behaviour in the cyclic voltammetric results obtained. A modified carbon paste electrode (CPE) containing the PANI-co-PDTDA co-polymer was constructed and the CPE/PANI-PDTDA sensor prepared in situ with an antimony film at the carbon paste substrate was applied successfully for lead ion determination. For this sensor various parameters were optimised with reference to square wave anodic stripping voltammetric (SWASV) signals.

[*] Corresponding author's e-mail: vsomerset@csir.co.za

ABBREVIATIONS

Abbreviation used	Full form
ANI	Aniline
APS	Ammonium persulfate
BASi	Bioanalytical Systems Incorporated
CPE	Carbon paste electrode
CPEs	Carbon paste electrodes
CV	Cyclic voltammetry
DMF	*N,N*-Dimethylformamide
DPASV	Differential pulse anodic stripping voltammetry
DTDA	2,2′-Dithiodianiline
FE-SEM	Field emission scanning electron microscope
FTIR	Fourier transform infrared spectroscopy
GCE	Glassy carbon electrode
MCPE	Modified carbon paste electrode
PANI	Polyaniline
PANI-co-PDTDA	Polyaniline-co-poly(2,2′-dithiodianiline)
PDTDA	Poly(2,2′-dithiodianiline)
PMT	Poly(3-methylthiophene)
PPY	Polypyrrole
SEM	Scanning electron microscopy
SPCEs	Screen-printed carbon electrodes
SWASV	Square wave anodic stripping voltammetry

1. INTRODUCTION

For the modifications of the transducer surface in electrochemical sensor applications a variety of methods have been applied to achieve better sensitivity and selectivity. In amperometric systems, the use of glassy carbon, platinum and gold electrodes are common and it is the modification of these transducer surfaces that has receive such extensive attention. The use of conducting polymers for the modification of the transducer surface has also been growing, with the use of polyaniline (PANI), poly(3-methylthiophene) (PMT) and polypyrrole (PPY). These polymers are electronically conducting organic polymers that are easily electro-synthesised onto the transducer surfaces by means of electro-oxidation of their monomers [1-4].

Among the family of conducting polymers, PANI has seen the widest use and application due to its unique properties that include: (i) ease of preparation in aqueous medium, (ii) good stability in air, (iii) ease of doping with various surfactants; (iv) excellent electronic properties; (v) excellent catalytic properties; (vi) display of electrochromic properties; (vii) and improved conductivity in the doped form [5-10]. Recent years have seen the investigation and synthesis of co-polymers in the search of newer polymers, which exhibit improved mechanical strength and processability [5].

The co-polymerisation of aniline with substituted anilines provides another route to improve the processability of the resulting polymers. The use of 2,2′-dithiodianiline (DTDA) in the co-polymerisation with aniline is showing increased interest due to the special properties of this organic disulfide compound. Some of the important properties include low cost, low toxicity, low operating temperature and high theoretical energy density. Furthermore, PDTDA also exhibit properties such as good stability, redox behaviour and kinetic reversibility that are also required in materials for energy storage devices [5,11].

The polymerisation of PDTDA can be done using electro- or chemical oxidative polymerisation and the resulting polymer (PDTDA) show the redox activity of the –S-S– bond and exhibit the electronic conductivity of PANI [12].

In the search for new conducting polymers, the latest trend is to combine existing ones and co-polymerise them to form new compounds that exhibit the combined and/or improved properties of the individual homopolymers. It has been demonstrated that PANI can act as an effective catalyst to accelerate the slow redox process of polymers containing disulfide bonds. Furthermore, there is also interest to synthesize a conducting polyaniline derivative that contains –S-S– links in its structure [13,14].

This chapter reports the use of 2,2′-dithiodianiline (DTDA) as a co-monomer with aniline (ANI) in the co-polymerisation process to form polyaniline-co-poly(2,2′-dithiodianiline), abbreviated as PANI-co-PDTDA. The influence of synthesis conditions, such as reaction time, the concentration of reactants, reaction temperature, and the molar ratio of reactants were studied and reported.

2. Materials and Methods

2.1. Materials and Reagents

The reagents aniline (99%), *N,N*-dimethylformamide (98%), graphite powder (<20 micron) and mineral oil were obtained from Aldrich, Germany. Atomic absorption standard solutions (1000 μgl^{-1} of antimony(III), bismuth(III), mercury(II), cadmium(II), lead(II), and nickel(II) were also purchased from Aldrich, Germany. Potassium chloride, methanol (99.8%), ethanol (99%) sulphuric acid (95%) and hydrochloric acid (32%) were purchased from Merck, South Africa. The 2,2′-dithiodianiline (95%), diethyl ether (99.8%), dimethyl sulfoxide (99.6%) and ammonium persulfate (APS) were purchased from Fluka (Germany) and used as received. All other chemicals were of analytical grade and were used as received. All solutions were always prepared using Milli-Q (Millipore) water [9,15].

2.2. Apparatus

Cyclic voltammetry was performed with the use of a conventional three-electrode cell using a BASi Epsilon Electrochemical Analyzer and Workstation. Screen-printed carbon electrodes (SPCEs) and carbon paste electrodes (CPEs) were used as the working electrodes. An Ag/AgCl electrode (saturated KCl) and a platinum wire (diam. 1 mm) were used as the reference and auxiliary electrode, respectively. All electrochemical experiments were carried out in a single compartment electrochemical cell at room temperature of 22 ± 1°C [15].

Differential pulse anodic stripping voltammetry (DPASV) experiments were performed with 2 ml aliquot samples. DPASV involved the following steps: (a) the pre-concentration step at – 1.2 V for 120 s; (b) the differential pulse anodic stripping voltammograms were recorded when swept from - 1.2 V to + 0.4 V after 2 s quiescence. The experimental parameters used were: deposition potential, - 1.2 V; accumulation time, 120 s; differential pulse amplitude (peak to peak), 50 mV; step amplitude, 5 mV; pulse width, 50 ms; pulse period, 200 ms [16].

Scanning electron microscopy (SEM) measurements for morphology studies were performed using a LEO 1525. Field emission scanning electron microscope (FE-SEM) with interchangeable accelerating voltages (maximum of 15.00 kV) for optimal sensitivity were used. Samples were mounted on aluminium stubs using conductive glue and were then coated with a thin layer of carbon [17].

2.3. Electrode Preparation

SPCEs were obtained from the Sensors and Separations Group, Department of Chemical Sciences, Dublin City University, Dublin 9, Ireland. The fabrication of the SPCE is described by Grennan et al. [15,18-19].

The carbon paste electrode is prepared by hand mixing of 1 g of spectroscopic graphite powder with 1.0 g of the PANI-co-PDTDA polymer sample and 0.6 ml of mineral oil. This mixture was homogenised and packed into 0.6 mm diameter piston-driven polyethylene tube. After the collection of each stripping voltammogram for the different concentrations evaluated, the surface of the carbon paste electrode (CPE) was mechanically renewed by cutting away 0.5 mm of the CPE surface and then smoothing it on filter paper [20].

2.4. Preparation of Polymer Films

Polyaniline was also electropolymerised as a monopolymer on a glassy carbon electrode (GCE) surface, using a 10 ml solution of 0.2 M aniline and aqueous 1 M HCl, followed by cycling the potetntial repetitively at 50 mVs^{-1} from – 200 to + 1100 mV for 20 cycles at 25 °C. The polymer collected on the GCE surface was then dissolved in DMF and the solution prepared for spectroscopic analysis [15].

Chemical preparation of the PANI-co-PDTDA co-polymer included the following. A total of one hundred and ninety milliliters (ml) of Milli-Q water was used, to which 3.1 ml of 0.2 M aniline and 56 ml of concentrated H_2SO_4 (5 M) was added in a beaker and stirred for approximately 5 min with a magnetic stirrer. This was followed by the addition of 7.6 g of ammonium persulfate (APS), before hand dissolved in 10 ml of Milli-Q water and then added slowly to the first solution. The solution was heated to approximately 70 °C during stirring (using a magnetic stirrer bar), which continued for 30 min until a thick green polyaniline solution started forming. To this solution 4.97 g of 2,2′-dithiodianiline (0.2 M) crystals was added with continued stirring and the solution was monitored until all the crystals had dissolved. Stirring continued for 2 hours at room temperature. After stirring, the solution was filtered, washed with Milli-Q water, methanol and diethyl ether, followed by drying in a desiccator at room temperature [21,22].

Electrochemical preparations of the conducting co-polymer were done as follows. A 10 mL solution consisting of 0.2 M aniline, 0.02 M 2,2′-dithiodianiline and aqueous 5 M H_2SO_4 was prepared and heated in a water bath to 70°C to dissolve the 2,2′-dithiodianiline crystals. The co-polymer film of PANI and PDTDA was grown electrochemically on the surface of a SPCE by repetitive cyclic voltammetric scanning at 50 mV/s from − 200 to + 1100 mV, for 10 cycles at 25 °C. Each SPCE was then rinsed with Milli-Q water and immersed in the water until use [15,18].

2.5. UV-Vis Spectroscopic Characterisation

UV-Vis spectra were recorded between 200 and 1000 nm using a 1 cm path quartz cuvette and *N,N*-dimethylformamide (DMF) as the reference solvent, on a ThermoFisher Spectronic™ Helios™ range UV-Vis spectrometer with VISION PC software. Solutions for subsequent spectroscopic studies were obtained by dissolving PANI-co-PDTDA polymer powder samples in DMF [5,15,23].

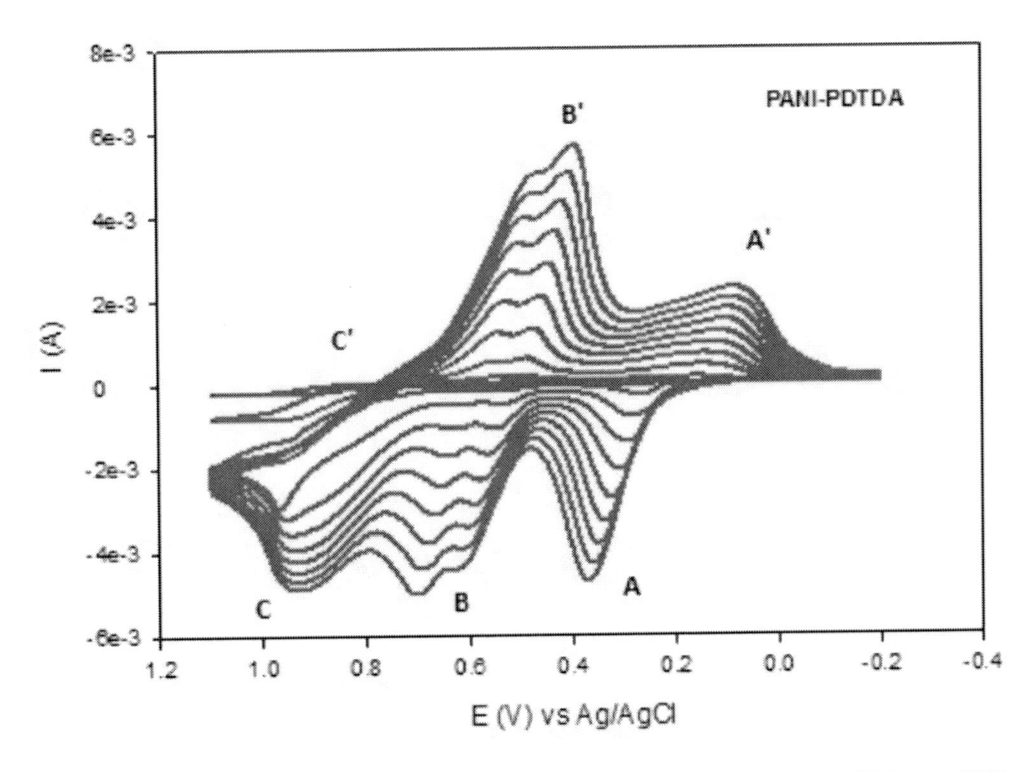

Figure 1. Cyclic voltammogram displaying the electropolymerisation of PANI-co-PDTDA on a GCE surface. The potential was cycled between − 200 and + 1100 mV at a scan rate of 100 mV/s for 10 cycles.

2.6. The Fourier Transform Infrared (FTIR) Spectroscopic Characterization

FTIR spectra were recorded at wavenumber range of 400–4000 cm^{-1} for the PANI-co-PDTDA polymer powder samples dissolved in DMF and recorded with a Bruker ALPHA-T, FTIR spectrometer fitted with a Bruker Optics aligned RockSolid™ interferometer [3,15,24].

2.7. Stripping Voltammetric Procedure

Square-wave anodic stripping voltammetry (SWASV) was performed under the following conditions. The preconcentration step was done at a deposition potential of – 1300 mV, applied to the CPE/PANI-co-PDTDA copolymer sensor for 240 s with the solution being constantly stirred. During this step an alloy mixture of antimony and the lead metal ions (from a 0.01 M HCl solution) was deposited on the transducer surface using SWASV. After deposition, the anodic stripping voltammogram was recorded with the potential step, amplitude and frequency at 4 mV, 25 mV and 15 Hz respectively. The potential was scanned from – 100 mV to + 400 mV and the stripping voltammograms were recorded automatically [15,25-26].

3. Discussion

The results presented for the co-polymerisation of PANI-co-PDTDA in this study was performed on a GCE. Aniline is very soluble in HCl and this is the preferred supporting electrolyte very often used in the electropolymerisation of PANI on a transducer surface. The 2,2′-dithiodianiline was not so soluble in HCl at room temperature and after using different concentrations of H_2SO_4, it was found to be soluble in 5 M H_2SO_4 that was then used as the supporting electrolyte for the co-polymerisation of aniline with 2,2′-dithiodianiline. The acidic solution containing the monomers of aniline and 2,2′-dithiodianiline was first dissolved in a water bath at 60 °C, prior to performing electropolymerisation on the GCE surface [15].

Figure 1 shows the cyclic voltammogram for the electropolymerisation results obtained for the co-polymemrisation of PANI-co-PDTDA on a GCE surface at a scan rate of 50 MV/s, cycling the potential between – 200 and + 1100 mV.

The electropolymerisation results shown in Figure 1 for the PANI-co-PDTDA co-polymer, exhibits good redox activity with 3 redox couples obtained. These results indicate that the co-polymer has a strong PANI backbone as the redox couples corresponds to the results obtained for PANI electropolymerisation in previous studies, with the formation of the redox couples at approximately the same potentials as for PANI [15,27-28].

Furthermore, on closer inspection of the PANI-co-PDTDA co-polymerisation results, it is observed that redox couple (B/B′) represents two separate redox processes occurring during electrosynthesis. The results indicate that during benzoquinone and hydroquinone formation in the PANI backbone, another process of self-doping/undoping of thiolate anions (-S) formed by the reductive cleavage of S–S bonds in the PDTDA backbone, of the PANI-co-PDTDA co-polymer is also occurring. These novel co-polymerisation results for PANI-co-PDTDA show the influence of the S–S links in the electropolymerisation process [13-15].

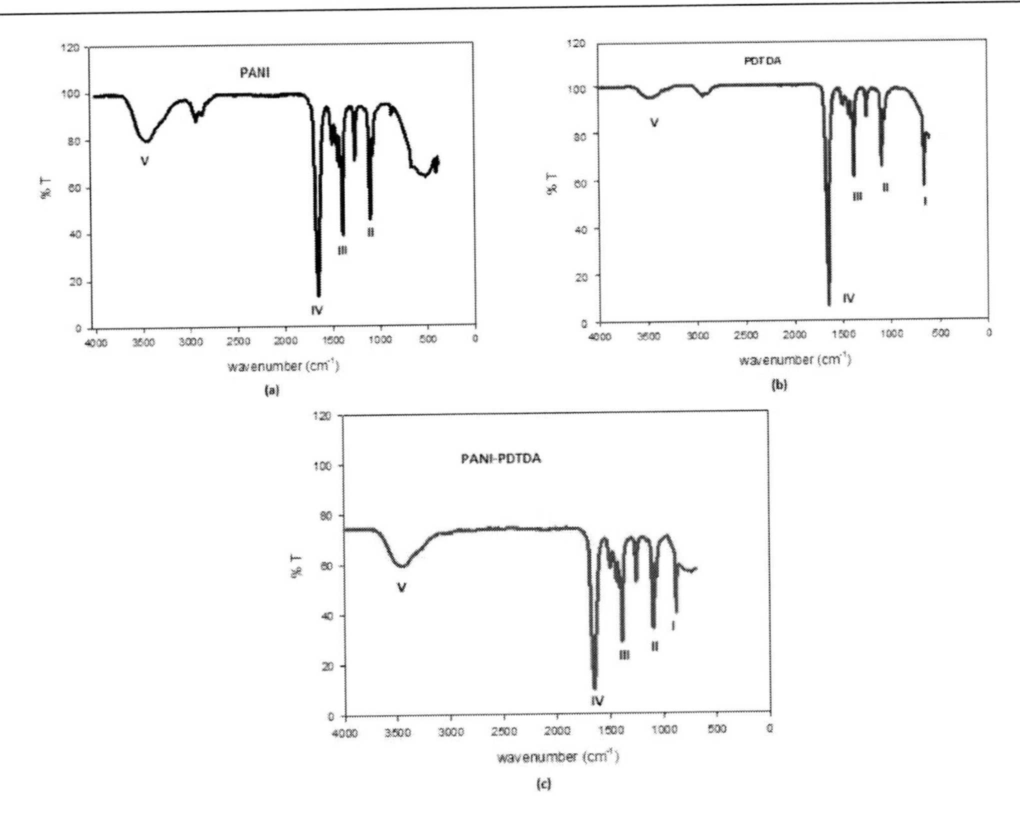

Figure 2. The FTIR results of the different polymers investigated with PANI (a), PDTDA (b) and the co-polymer of PANI-co-PDTDA (c).

The FTIR results obtained for the characterisation of the individual polymers of PANI and PDTDA, which will be compared to the PANI-co-PDTDA co-polymer, are shown in Figure 2. The FTIR results displayed in Figure 2 (a) show that four characteristic stretching frequencies were obtained for the PANI polymer, while five stretching frequencies were obtained for the PANI-co-PDTDA polymer. For band I at ca. 720 cm^{-1} in the co-polymer, the results of the C–S stretching in the DTDA units are displayed, indicating that DTDA has been incorporated in the co-polymer structure. The results for band I can only be seen in Figure 2 (b) and (c) with the S-atoms found in PDTDA. The results for band II at ca. 1120 cm^{-1} can be assigned to the C–N stretching in all three polymer structures. This is followed by the results for band III at ca. 1300–1400 cm^{-1} that are assigned to the C–C stretching of the benzenoid aniline units of the polymers. Band IV at ca. 1568–587 cm^{-1} can be attributed to C–N quinoneimine stretching in the polymer structures. The stretching frequency of band V at ca. 3250–3500 cm^{-1} is attributed to the N–H absorption bands in all three polymer structures [11,13,15,21,29].

From these results for the structural analysis of the different polymers, it can be concluded that the PANI-co-PDTDA co-polymer contains sulphur units and its structure is composed of quinoid and benzenoid rings linked by amine-imine centers, with the polyaniline backbone strongly intact.

In Figure 3 the results of the UV-Vis characterisation of the homo-polymer of PANI and the PANI-co-PDTDA co-polymer are shown.

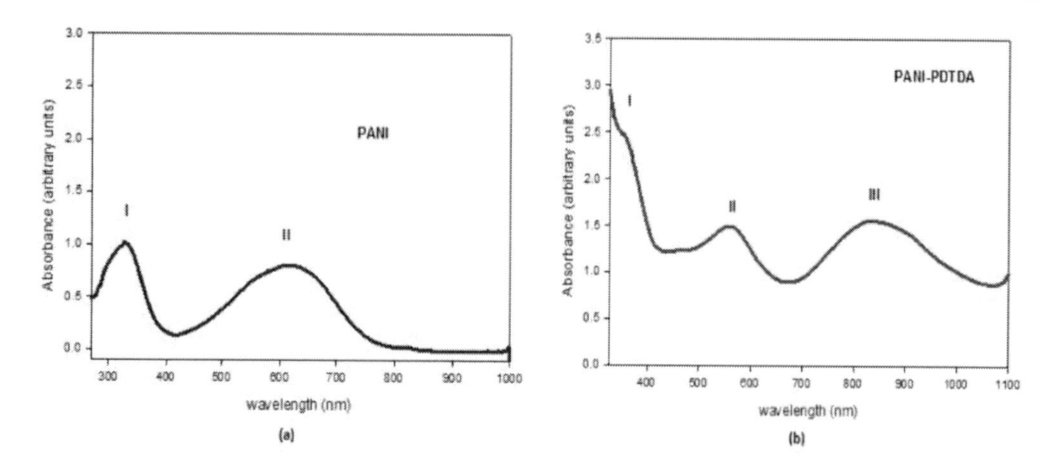

Figure 3. Displays the results for the UV-Vis spectroscopic spectra of the polymer PANI (a) and the co-polymer of PANI-co-PDTDA (b).

The UV-Vis results shown in Figure 3 (a) for PANI indicates that 2 characteristic peaks are observed at approximately 320 nm (I) and 610 nm (II) in the spectrum of the polymer. The results for band I at 320 nm can be attributed to the π–π^* transition, as the N-atoms in the benzenoid rings of PANI are excited. The results for band II at 610 nm (n–π^* transition) can be assigned to absorption in the visible range that is ascribed to exciton formation in the quinoid rings [9,15,30,31].

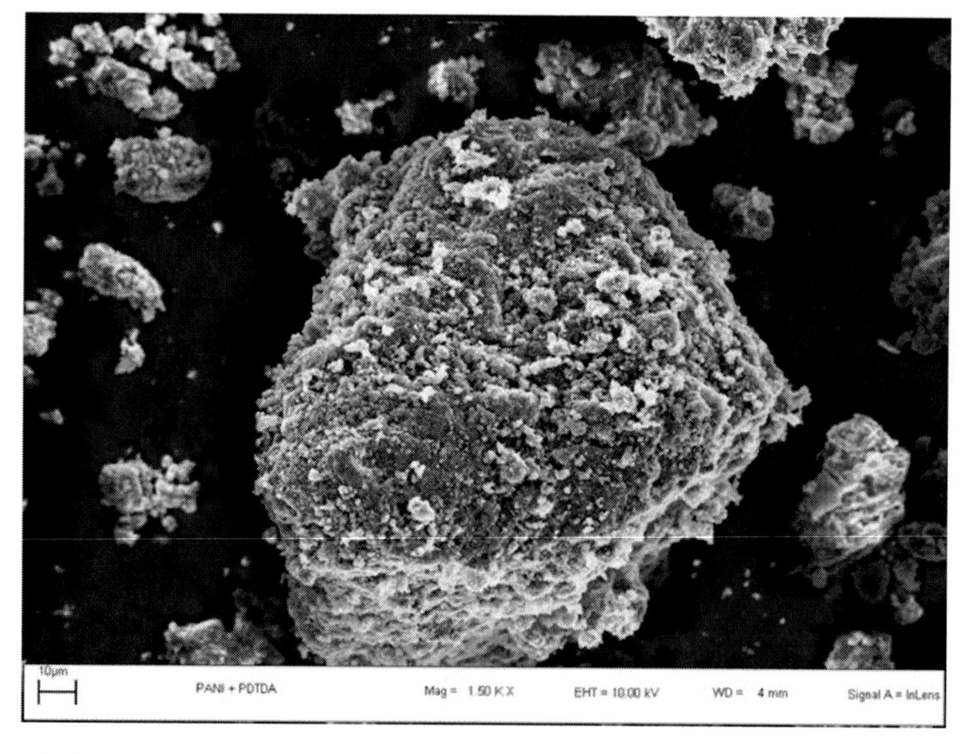

Figure 4. The SEM microscopy for the co-polymer of PANI-co-PDTDA.

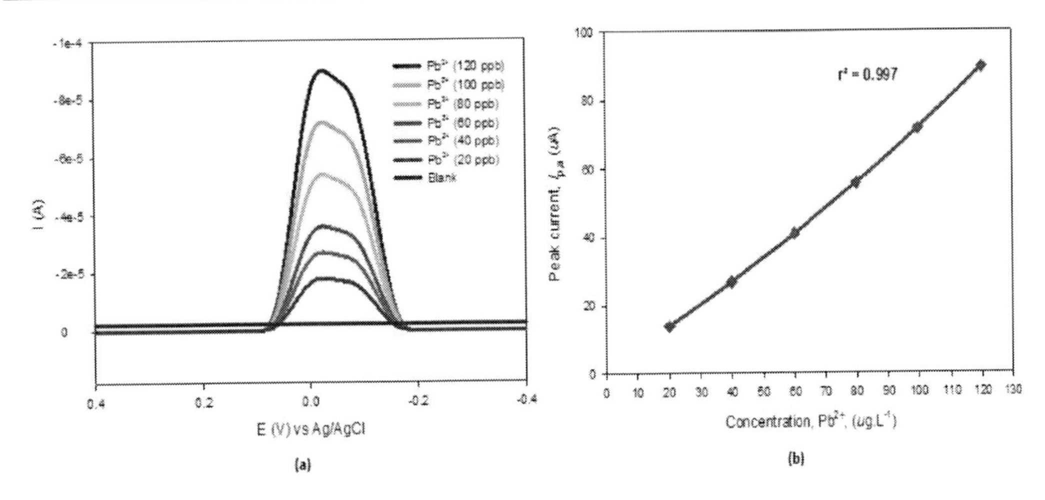

Figure 5. SWASV determination of Pb^{2+} concentrations varying between 20 and 120 ug/l. The solution contained a 0.01 M HCl solution, with 1 mg/l Sb(III), the deposition potential was – 1.2 V, deposition time was 240 s and the equilibration time was 2 s.

Analysis of the spectroscopic results for the PANI-PDTDA co-polymer, displays three maxima with the first peak not so strongly defined. The results for band I at approximately 360 nm can be attributed to the N-atoms in the benzenoid rings of the PANI backbone being excited. The results for band II shows a hypsochromic shift from 620 to 560 nm, for the PANI-PDTDA co-polymer in comparison with that of band II of PANI. These results can be attributed to the S–S– bonds substituted on the benzene rings of PDTDA in the PANI-PDTDA co-polymer structure. The presence of the S–S bonds has caused an increase in the band gap due to their steric strain on the oligomer structure, increasing the torsional angle between adjacent rings. The presence of band III in Figure 3 (b) can be attributed to the protonation of the imino site and the formation of polaron bands in the PANI-PDTDA co-polymer structure [12-13,15,32]. Figure 4 displays the results for the SEM analysis of the PANI-co-PDTDA co-polymer.

The SEM micrograph (Figure 4) for the co-polymer of PANI-co-PDTDA exhibits a well-defined structure, with micro morphological dimensions. The micrograph also shows a big crystal with smaller crystals forming on top of it. This can be attributed to the two amino groups in the PDTDA structure that act as nucleating sites for the growth of the polymer chain [13,33].

After successful synthesis and characterisation of the PANI-co-PDTDA co-polymer, this polymer was used in the construction of a modified carbon paste electrode (MCPE) for stripping voltammetric analysis of lead (Pb^{2+}) ions in aqueous solutions. Optimisation of the supporting electrolyte was performed and a 0.01 M HCl solution in the presence of 1 mgl^{-1} Sb(III) for stripping voltammetric analysis was used. The results obtained are displayed in Figure 5.

The results for the square-wave anodic stripping voltammetric determination of a series of standard solutions of Pb^{2+}, performed after optimisation of the deposition potential and deposition time, are displayed in Figure 5. The results also show that the stripping peak current obtained with the CPE/PANI-co-PDTDA sensor has a linear relationship for the concentrations ranging between 20 and 120 μgl^{-1}. The calibration curve also exhibit excellent linearity with a correlation coefficient of 0.997, whereas the peak current increased linearly with metal concentration yielding a slope of 0.74 $\mu A \mu g^{-1}$ for the lead ion solutions analysed [15-16,26,34].

CONCLUSION

The results in this chapter have shown that a PANI based co-polymer containing S–S links can be prepared by incorporating PDTDA in the co-polymer structure. The prepared PANI-co-PDTDA co-polymer exhibited the electrochemical properties of both PANI and PDTDA with the incorporation of the S–S links displaying their distinct redox behaviour in the cyclic voltammetric results obtained. Furthermore, FTIR and UV-Vis spectroscopic analysis of the co-polymer has shown the presence of the C–S bonds in the PANI-co-PDTDA structure. This chapter further reported the preparation, optimisation and practical application of the CPE/PANI-PDTDA sensor, which was prepared in situ with an antimony film at the carbon paste substrate for lead ion determination. The use of the chemically modified carbon paste electrode exhibited an interesting alternative in the anodic stripping voltammetic determination of metal ions. The modified carbon paste electrode (MCPE) further revealed attractive electro-analytical performance in acidic medium of pH≈ 2 and compares favourably with other antimony and bismuth analogues for lead determination. Further studies with the application of this sensor will focus on real sample analysis and the detection of other metal ions such as zinc, cobalt and cadmium.

ACKNOWLEDGMENT

This study was supported by funding received from the CSIR's parliamentary grant (PG) process and the South Africa/Spain Bilateral Programme of the National Research Foundation (NRF) of South Africa. The collaboration with the research laboratory of the SensorLab, Chemistry Department, University of the Western Cape (UWC) in Bellville, South Africa is greatly acknowledged.

REFERENCES

[1] Erdogdu, G.; Karagozler, A. E. *Talanta* 1997, *44,* 2011-2018.

[2] Somerset, V. *Mercaptobenzothiazole-On-Gold Biosensor Systems for Organophosphate and Carbamate Pesticide Compounds. Un-published Ph.D Thesis.* University of the Western Cape, Bellville, South Africa, 2007; Vol. 1, pp 1-541.

[3] Akinyeye, R.; Michira, I.; Sekota, M.; Al-Ahmed, A.; Baker, P.; Iwuoha, E. *Electroanal.* 2006, *18*, 2441-2450.

[4] Kinyanjui, J. M.; Wijeratne, N. R.; Hanks, J.; Hatchett, D. W. *Electrochim Acta* 2006, *51,* 2825-2835.

[5] Rajendran, V.; Gopalan, A.; Buvaneswari, R.; Vasudevan, T.; Wen, T. C. *J. Appl. Pol. Sci.* 2003, *88,* 389-397.

[6] Morrin, A.; Wilbeer, F.; Ngamna, O.; Moulton, S. E.; Killard, A. J.; Wallace, G. G.; Smyth, M. R. *Electrochem. Comm.* 2005, *7,* 317-322.

[7] Guimard, N. K.; Gomez, N.; Schmidt, C. E. *Prog. Pol. Sci.* 2007, *32,* 876-921.

[8] Michira, I.; Akinyeye, R.; Somerset, V.; Klink, M. J.; Sekota, M.; Al-Ahmed, A.; Baker, P. G. L.; Iwuoha, E. *Macromol. Symp.* 2007, *255,* 57-69.

[9] Somerset, V.; Klink, M.; Akinyeye, R.; Michira, I.; Sekota, M.; Al-Ahmed, A.; Baker, P.; Iwuoha, E. *Macromolecular. Symp.* 2007, *255,* 36-49.

[10] Jianming, J.; Wei, P.; Shenglin, Y.; Guang, Li. *Synth. Met.* 2005, *149,* 181-186.

[11] Cotarelo, A. A.; Huerta, F.; Quijada, C.; Pérez, J. M.; del Valle, M. A.; Vázquez, J. L. *J Electrochem. Soc.* 2006, *153,* A2071-A2076.

[12] Cho, J. S.; Sato, S.; Takeoka, S.; Tsuchida, E. *Macromol.* 2001, *34,* 2751-2756.

[13] Chen, S. S.; Wen, T. C.; Gopalan, A. *Synth. Met.* 2003, *132,* 133-143.

[14] Huang, L. M.; Wen, T. C.; Yang, C. H. *Mat. Chem. Phys.* 2002, *77,* 434-441.

[15] Somerset, V.; Leaner, J.; Mason, R.; Iwuoha, E.; Morrin, A *Electrochim. Acta* 2010, *55,* 4240-4246.

[16] Li, Y.; Liu, X.; Zeng, X.; Liu, Y.; Liu, X.; Wei, W.; Luo, S. *Sens. Actuat. B,* 2009, *139,* 604-610.

[17] Somerset, V.; Leaner, J.; Mason, R.; Iwuoha, E.; Morrin, A. *J. Environ. Anal. Chem.* 2010, *90,* 671-685.

[18] Grennan, K.; Killard, A. J.; Smyth, M. R. *Electroanal.* 2001, *13,* 745-750.

[19] Grennan, K.; Killard, A. J.; Smyth, M. R. *Electroanal.* 2005, *17,* 1360-1369.

[20] Tesarova, E.; Baldrianova, L.; Hocevar, S. B.; Svancara, I.; Vytras, K.; Ogorevc, B. *Electrochim. Acta* 2009, *54,* 1506-1510.

[21] Bhadra, S.; Singha, N. K.; Khastgir, D. *J. Appl. Pol. Sci.* 2007, *104,* 1900-1904.

[22] Lee, Y. G.; Ryu, K. S.; Chang, S. H. *J. Power Sources* 2003, *119,* 321-325.

[23] Wang, Y.; Jing, X. *Pol. Test.* 2005, *24,* 153-156.

[24] Somerset, V. S.; Petrik, L. F.; White, R. A.; Klink, M. J.; Key, D.; Iwuoha, E. *Talanta* 2004, *64,* 109-114.

[25] Honeychurch, K. C.; Hart, J. P.; Cowell, D. C. *Electroanal.* 2000, *12,* 171-177.

[26] Wang, Z. M.; Guo, H. W.; Liu, E.; Yang, G. C.; Khun, N. W. *Electroanal.* 2010, *22,* 209-215.

[27] Mathebe, N. G. R.; Morrin, A.; Iwuoha, E. I. *Talanta* 2004, *64,* 115-120.

[28] Somerset, V. S.; Klink M. J.; Baker, P. G. L.; Iwuoha, E. I. *J. Environ. Sci. Health* 2007, *B42,* 297-304.

[29] Kinyanjui, J. M.; Wijeratne, N. R.; Hanks, J.; Hatchett, D. W. *Electrochim Acta* 2006, *51,* 2825-2835.

[30] de Albuquerque, J. E.; Mattoso, L. H. C.; Fariac, R. M.; Masters, J. G.; MacDiarmid, A. G. *Synth. Met.* 2004, *146,* 1-10.

[31] Zheng, L.; Li, J. *J Electroanal. Chem.* 2005, *577,* 137-144.

[32] Lee, Y. G.; Ryu, K. S.; Lee, S.; Park, J. K.; Chang, S. H. *Pol. Bull.* 2002, *48,* 415-421.

[33] Chang, T. K.; Wen, T. C. *Synth. Met.* 2008, *158,* 364-368.

[34] Hwang,G. W.; Han, W. K.; Park, J. S.; Kang, S. G. *Talanta* 2008, *76,* 301-308.

In: Advanced Organic-Inorganic Composites
Editor: Inamuddin

ISBN 978-1-61324-264-3

Chapter 20

BASIC PROPERTIES OF HIGH TEMPERATURE SUPERCONDUCTORS AND ITS APPLICATIONS

Intikhab Aalam Ansari *

Department of Physics and Astronomy, College of Science,
King Saud University, Riyadh, Kingdom of Saudi Arabia

ABSTRACT

Superconductivity is the quantum mechanical phenomenon occurring on a macroscopic scale in certain materials as electrical resistance drops to zero at transition temperature. It was discovered by Kamerlingh Onnes in 1911. It is not the property of individual atoms but the collective effect of whole sample which is characterized by Meissner effect. This caused the expulsion of any sufficiently weak magnetic field from the interior of the sample that directs the superconducting state. The electron-phonon interaction in superconductors is responsible for the electron attraction, leading to the electron coupling. Many scientists had taken interest to unravel the mystery of superconducting state because only 0.01% electrons are accountable for conduction. The electrical resistivity of the material drops abruptly to zero when the sample is cooled below its transition temperature. In this chapter, the basic properties of high temperature superconductors and its practical applications are discussed in brief. Indeed, the electronics, sensors, electromagnets, power applications, transportations and cryocoolers are very promising fields of engineering applications of superconductors.

ABBREVIATIONS

ASTRA	Advanced Space Technologies for Robotics and Automation
A	Amplitude
BCS	Bardeen, Cooper and Schrieffer
E	Charge on the electron
j_c	Critical current density

* Corresponding author's e-mail: intikhabansari@yahoo.com

Dc	Direct current
ξ_N	Effective coherence length
CERN	European Organization for Nuclear Research
E_F	Fermi level
V_F	Fermi velocity
I_0	Finite current
HTSC	High temperature superconductors
A	Isotope coefficient
JEs	Josephson effects
LHC	Large hadron collider
LN_2	Liquid nitrogen
Λ	London penetration depth
H	Magnetic field
B	Magnetic induction
Maglev	Magnetic-levitation
M	Magnetization
M	Mass of the superconducting element
MRI	Medical resonance imaging
K	Momentum
N	Normal metal
Θ	Phase difference
H	Planck's constant
X	Position vector
PE	Proximity effect
SQUID	Superconducting quantum interference device
S	Superconductor
SI	Superconductor-insulator
S-I-S	Superconductor–insulator–superconductor
SN	Superconductor-normal metal
SNS	Superconductor-normal metal-superconductor
T_c	Transition temperature

1. Introduction

Superconductivity was first discovered by Kamerlingh Onnes [1] in 1911. He observed that the resistivity suddenly drops to zero below 4.2 K in the element mercury. Superconductors are the perfect conductors of electricity, in which resistivity abruptly drops to zero at a certain temperature called transition temperature (T_c). Later it was found that the origin of the zero resistance is nothing but a perfect diamagnetism, also called the *Meissner effect*, which states that the superconductor expelled the magnetic flux from the interior of the sample when it cooled below the T_c in presence of weak external magnetic fields. As far as high temperature superconductors (HTSC) are concern, these are the materials which have T_c above 30 K. In 1986 Bednorz and Müller [2] discovered the first HTSC in LaBaCuO. After that, enormous copper-based superconductors have been investigated including YBaCuO,

BiSrCaCuO and HgTlBaCaCuO having T_c 92 K, 110 K and 138 K, respectively. HTSC always need liquid nitrogen (77 K) as a coolant rather than liquid helium (4.2 K). From the application point of views HTSC are significant because liquid nitrogen is relatively inexpensive (refrigeration cost is 1000 times less rather than liquid helium) and easily handled coolant. In order to find the room temperature superconductivity, researchers around the globe are engaged in search of this noble material. Recently, some new class of iron-based superconductors have been discovered including LaOFFeAs, PrOFFeAs and SmOFFeAs encompasses T_c 26 K, 52 K and 55 K, respectively.

The Bardeen, Cooper and Schrieffer (BCS) theory was emerged in 1957 which broadly explain the phenomenon of superconductivity in metals but on the contrary, it did not provide a generalized law for predicting the occurrence of superconductivity in different compounds.

2. History of High Temperature Superconductors

After superconductivity was discovered by Kamerlingh Onnes in 1911, many superconductors were discovered and several steps had been taken for the enhancement of transition temperature. When the BCS theory in 1957 provided an efficient explanation for superconducting phenomena, many scientists believed that higher transition temperatures could not be reached without finding new superconducting phenomena. In 1954, Fröhlich [3] had proposed a model of high-temperature superconductivity in charge density wave systems, and in 1964 Little [4] proposed a so-called excitonic superconductivity model. However, no such superconductors have actually been found.

Sleight [5] reported in 1975 that superconductivity was seen in $BaPb_{(1-x)}Bi_xO_3$ and that the transition temperature of this material was 13 K which changes with the Bi/Pb ratio. This was the opening of research on the oxide superconductors.

The year of 1980 was very crucial; scientist all over the globe began looking for new types of superconductivity. A new superconductor, $PbMo_6S_8$ was found by Chevrel [6] in Switzerland. It had high upper critical field, but its transition temperature was still only ~16 K. The Japanese government started a project entitled "New Superconducting Material" in 1984.

Bednorz and Müller [2] in 1986 discovered a new Ba-doped $LaCuO_3$ superconductor. Takagi et al. [7] reported that the transition temperature for superconductivity in Ba-doped La_2CuO_4 was 30 K. In addition, in early 1987 Tanaka [8] reveal the possibility of two-dimensional superconductivity due to the layer structure of this material. At that time, Anderson [9] proposed the new model of high-temperature superconductivity based on the two-dimensional resonating valence band model. At that moment, many scientists began to think that the superconductivity of Ba-doped La_2CuO_4 could not be explained by the BCS theory.

In 1987 Chu et al. [10] found a new superconductor, $YBa_2Cu_3O_7$, with a transition temperature above 90 K. Many scientists all over the world then took interest to search for new superconducting materials with higher transition temperatures. American Physical Society organized a special conference on the topic high-temperature superconductivity on March 15 of 1987. Remarkable advances were made in experimental research, and new materials were found in quick succession. As a result the critical temperature reached 112 K in a Bi compound, 126 K in a Tl compound and 135 K in an Hg compound. The physical properties of these compounds were also investigated very seriously, and it was confirmed from different techniques that in all cuprate superconductors the superconductivity occurred in very thin layers including CuO_2 planes.

The physical properties in the normal state of the high temperature superconductors are so complex that any theory cannot so far explain them in a reliable way. Deeper perceptive on "the strongly correlated system" may be necessary in order to explain the origin of the high temperature superconductivity. The superconducting state is a state of matter. Therefore, it has its own specific characteristics and basic properties. Furthermore, we will discuss the basic properties of HTSCs.

3. Basic Properties of High Temperature Superconductors

In 1957 Bardeen, Cooper and Schrieffer developed a theory called BCS theory that correctly predicts the Meissner effect and reproduces the isotope effect as well. According to this theory, the electron-phonon interaction in conductors is responsible for the electron attraction at low temperature formed a Cooper pair. The single electron has –1/2 spin, which are *fermions* but a Cooper pair is composite *boson* because of integer spin. The basic properties of high temperature superconductors are discussed below.

3.1. Zero Resistance

This is one of most important basic property of superconductor discovered by Kamerlingh Onnes in 1911. He found that the electric resistance in mercury dropped to zero at temperatures below 4.2 K. From the application point of view, zero resistance is beneficial in high field electromagnets because of negligible power dissipation. This zero resistance property was demonstrated by closed ring of superconductor. Small direct current (dc) amplitude of current was induced over two and half years into closed superconducting ring, no change in the persistent current was measured. This implies that decay time of supercurrent in a superconducting ring was not less than 10^5 years, which refer the resistivity 10^{-24} Ωcm in this superconducting ring. This peculiar property of superconducting state is the most intriguing one and was used in real time applications from microscopic scale to macroscopic scale. Prior, it was believed that superconductivity is destroyed by heating of the sample but later experiment shows that it also vanished by applying magnetic field. Soon after the discovery of type II superconductors, the issue of presence of magnetic field in superconductors has become the subject of further study.

3.2. The Meissner Effect

The Meissner effect is a very important characteristic of superconductors due to the perfect diamagnetism and plays a vital role in the applications where the magnetic field is applied concurrent with superconductivity.

When the superconductor is placed under the weak external magnetic field and cooled through T_c, it expels the magnetic field from its interior. This phenomenon is called the Meissner effect and it was discovered by Meissner [11] in 1933.

In any superconducting sample, the relation among magnetic induction B, magnetic field H and magnetization M is given by [12]

$$B = \mu_0 (H + M) \tag{1}$$

In a perfect diamagnet, $B = 0$ for $M = -H$. This condition is shown in the Figure 1.

Levitated magnet is the most interesting demonstration of the Meissner effect. If we place a small magnet having $T < T_c$ above the superconductor, the magnet will levitate above the superconductor. The gravitational force exerted on the magnet is poised by the magnetic pressure in presence of inhomogeneous magnetic flux. If we rotate the magnet, it continues to rotate for a long time without any friction. This is the crucial example of non-resistive magnetic bearing using the Meissner effect.

London equation gives one of the theoretical descriptions of the Meissner effect in terms of London penetration depth, λ. This equation shows that when magnetic field is applied parallel to surface of the superconductor it falls off exponentially inside the superconductor over a distance of 50 nm (as shown in Figure 2) and described by the following expression

$$B(x) = B_0 \exp(-x/\lambda) \ldots \tag{2}$$

3.3. Flux Quantization

After the advent of BCS theory, it has been confirmed that the superconductivity is a quantum phenomena taking place on a macroscopic scale and its corresponding physical quantities are also quantized i.e. have discrete values.

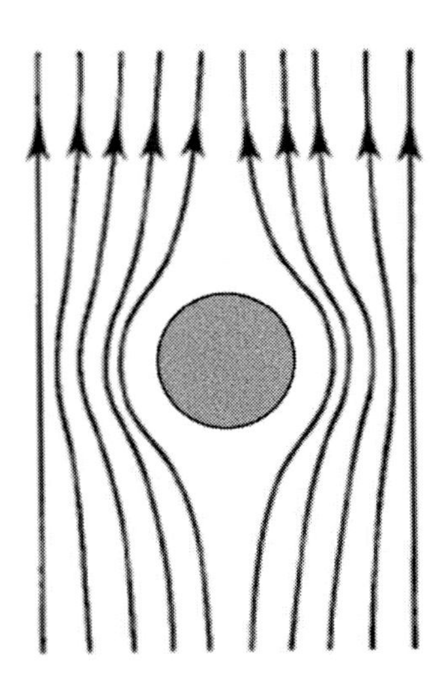

Figure 1. The Meissner effect in a superconductor expels a weak magnetic field.

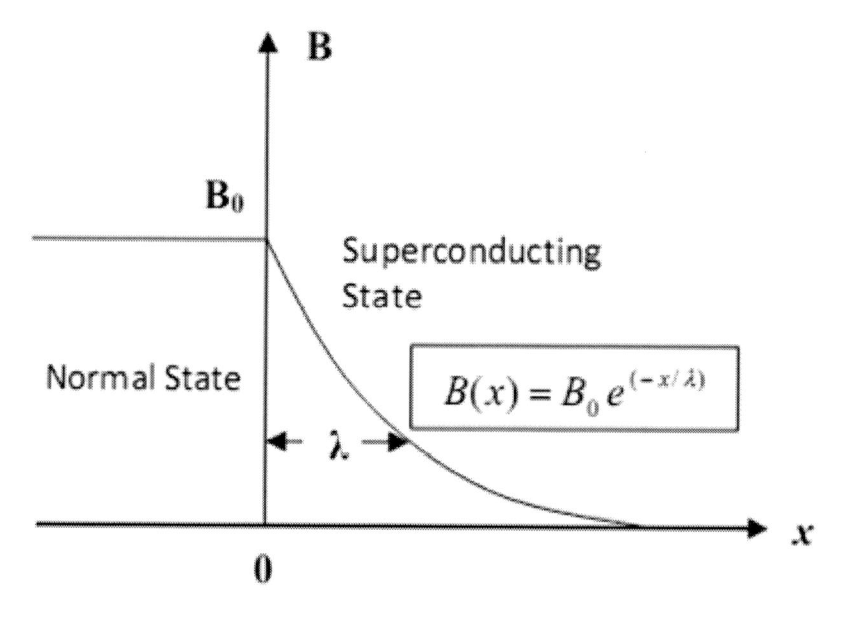

Figure 2. Decay of magnetic field inside a superconductor.

It is evident from several experiments that type II superconductors as well as HTSCs exhibit the flux quantization but type I superconductors have not shown any discrete values. Flux quantization is the main property that separates the type I superconductor from the type II.

One of the demonstrations that shows the flux quantization is superconducting ring experiment possesses phase of a wave function called the flux quantum or fluxoid.

$$\phi_0 = h/2e = 2.0678 \times 10^{-15}\, Webers \quad (3)$$

Where h is Planck's constant (6.626×10^{-34} J.sec) and e^- is the charge on the electron (1.6×10^{-19} Coulombs).

Flux quantization gives a considerable result that any applied magnetic flux enclosed by the ring superconductor have discrete value whenever it penetrate a superconductor. This experiment is based on the Onsager's hypothesis [13] which concluded that particle circulating on the superconducting ring has a charge $2e^-$ in agreement with BCS theory of Bose-Einstein condensate composed of Cooper pair of charge $2e^-$. Flux quantization was experimentally reported by Deaver-Fairbank and Doll-Näbauer [14,15] in 1961 that gives the agreement with BCS theory.

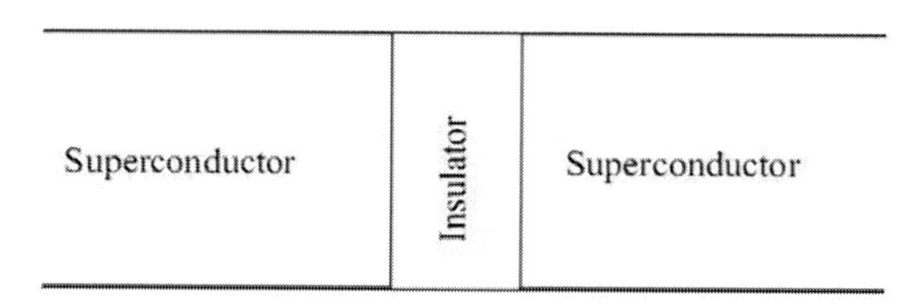

Figure 3. Schematic diagram of Josephson junction.

3.4. The Josephson Effects

In 1962, Josephson [16,17] predicted that supercurrent can flow through thin oxide insulating layer sandwiched between two superconductors called the Josephson effects (JEs) as shown in the Figure 3. Two superconductors are prepared of the same material. Furthermore, Anderson and Rowell [18] experimentally confirmed this effect. JEs are quantum statistical effects prominences on a macroscopic scale. Above T_c the Superconductor–Insulator–Superconductor (*S-I-S*) system shows the potential drops but when T reaches below T_c the potential drops in both superconductors vanishes because of zero resistance. This is the case of flowing supercurrent into insulator I with no energy loss.

The *V-I* curves shows the finite current I_0 (Order ~mA) even at zero potential. When a weak magnetic field is applied, the current I_0 goes down as shown in Figure 4. When voltage (~mV) is elevated high enough, the normal tunneling current appears.

It is well known that when a phase difference appears between the order parameters of superconductors separated by insulating thin layer, a superconducting tunneling current proportional to that phase difference flows across the insulating barrier. Schematic diagram of Josephson junction is shown in Figure 3. If we presume that the order parameter is constant and the gradient of the phase is uniform in the insulating region, this fact leads the equation.

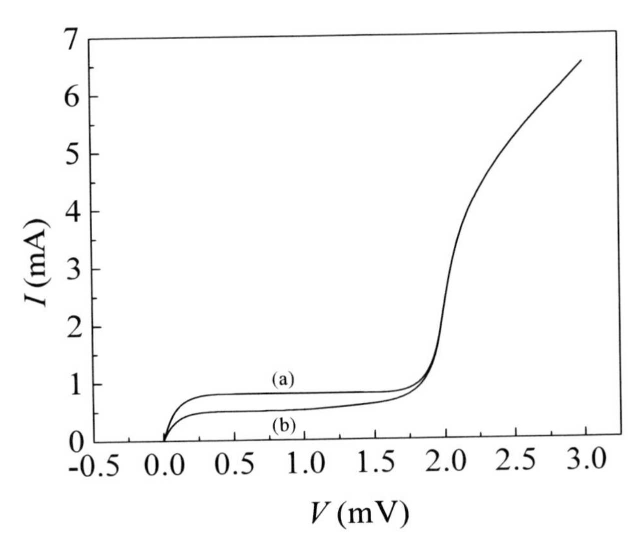

Figure 4. *I-V* characteristic curve indicates a Josephson tunneling current. (a) $B = 0$, (b) $B = 0.4$ Gauss.

$$j = j_c\theta \tag{4}$$

This equation is correct when the phase difference θ is very small but when θ becomes large enough then above equation starts to deviate and makes a relationship

$$j = j_c \sin\theta \tag{5}$$

Let us now discuss summarily how j_c is varied by applying magnetic field and size of the junction. Result shows that by applying a magnetic field to a tunneling junction, j_c is found to be a non-monotonic function of field strength, as shown in Figure 5. This occurs due to quantum interference effect caused by the phases of the order parameters. The critical current density of Josephson junction, J_c i.e. maximum value of Eq. (4) varies with magnetic field as

$$J_c = j_c \left| \frac{\sin(\pi \Phi / \Phi_0)}{\pi \Phi / \Phi_0} \right|, \tag{6}$$

where Φ is the magnetic flux embedded inside the junction and Φ_0 is the flux quantum. This form of equation makes an analogy to interference pattern due to Fraunhofer diffraction by a single slit. When the total magnetic flux is an integer multiple of flux quantum, i.e. $\Phi = n\ \Phi_0$, $n = 1, 2, 3\ldots.$, the critical current density of the junction becomes zero. In this state the phase inside the junction diverge over 2π and zero critical current density turns out from the interferences of the positive and negative currents of the same magnitude. This is the phenomenon of dc Josephson effect. The superconducting quantum interference device (SQUID) works on this property in which a very weak magnetic flux density can be measured.

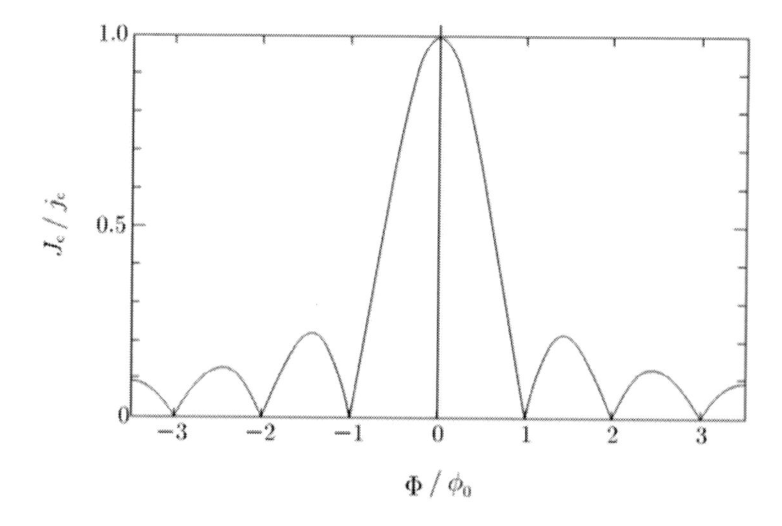

Figure 5. Critical current density vs. magnetic flux inside the junction.

In order to detect the extremely weak magnetic field, a SQUID is used. The actual resolution of such a device can much better than a single flux quantum, ~$10^{-5}\ \Phi_0$. The SQUID acts as a flux to voltage convertor to directly measure the change in flux, variations of magnetic field or its gradient as the sample moves through a superconducting detection coil attached to the SQUID.

Let us now take an example to understand the mechanism of SQUID. We make a circuit of closed ring superconductor having two parallel Josephson junctions. The super-current I, breaks up into I_1 and I_2 below the transition temperature. When a magnetic field is applied perpendicular to the ring downwards on the paper. The magnetic flux changed continuously and passes through the junctions. The total current I rejoin and describes by the following equation.

$$I = I_0 \cos(\pi \Phi / \Phi_0) \tag{7}$$

Here I_0 is constant, $\Phi_0 = \pi\hbar/e^-$ and Φ is flux embedded into the ring. This phenomenon shows that two supercurrent I_1 and I_2 macroscopically dispersed by 1 mm and can interfere just as two coherent beams from the same source. This phenomenon is called Josephson interference. The block diagram of SQUID is shown in Figure 6.

Now, we will discuss the AC Josephson effect. In this event, when a constant voltage V is applied across the barrier, then an AC supercurrent flows with characteristic frequency, $\omega = 2eV/\hbar$ across the junction. Whenever Cooper pairs crosses the barrier, it emits a photon of energy $\hbar\omega = 2eV$. When we place the tunneling junction in a microwave cavity, one can perceive the equidistant steps called the Shapiro steps in $I(V)$ characteristics. One can measure the ratio of e/h by using the relation $\hbar\omega = 2eV$, which is the most precise determination of this ratio so far.

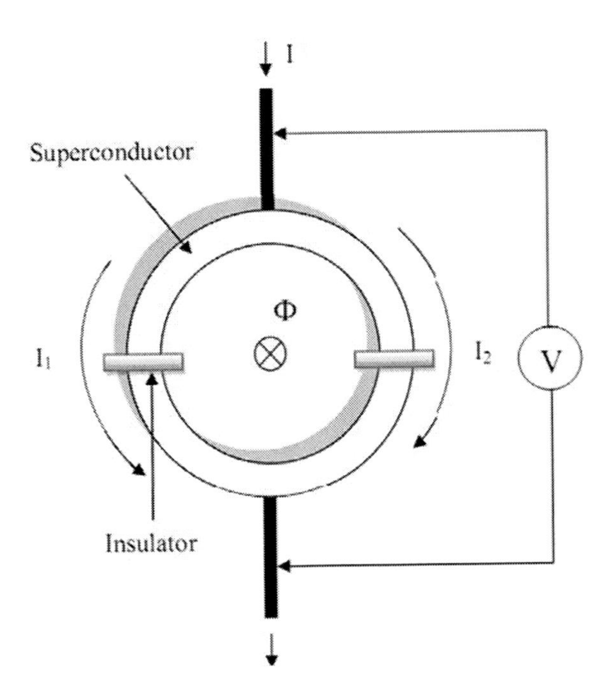

Figure 6. Sketch of superconducting quantum interface device (SQUID).

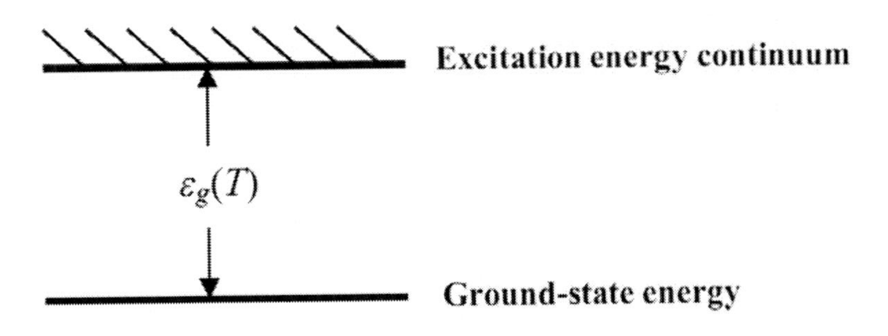

Figure 7. Excitation energy spectrum with a gap.

3.5. Energy Gap

The finite gap between the continuous band of excitation energy and ground state energy level is called the energy gap, $\varepsilon_g(T)$ as shown in Figure 7. There are several experiments which can measure the energy gap, $\varepsilon_g(T)$ viz. photo-absorption [19,20], quantum tunneling [21], heat capacity [22] and tunneling experiments.

Since the formulation of quantum mechanics, the phenomenon of tunneling came into existence. An electron has a finite probability of entering a classically forbidden region; this is the most important consequence of quantum mechanics. Therefore, the electron separates two classically allowed regions after tunnel through a potential barrier. The tunneling probability has exponential dependence on the potential barrier width; this is the main cause that experimental observation of tunneling experiment is considerable only for small barriers.

The first tunneling measurements between normal metal and superconductor were carried out in 1960. The energy gap in a superconductor is rather unlike that of the semiconductor. According to the band theory, energy bands are the result of the static lattice structure. The energy gap in the superconductor is extreme minor and occurred due to the attractive force between the electrons in the lattice. In a superconductor, the gap comes out on either side of the Fermi level, E_F, as shown in the Figure 8.

The obvious observation of the energy gap in the superconductor is the key for the strong confirmation of the BCS theory. The excited states of the superconductors are stained from the normal state. In the normal state, the minimum energy is zero whereas in the superconducting state, it is instead the smallest value of ε_g. Thus, the lattice structure in the superconducting state cannot take up arbitrary small amounts of energy. At zero temperature all electrons are stay underneath the energy gap and minimum energy $2\varepsilon_g(0)$ is required to excite the electrons across the gap, as shown in Figure 8. Phonon mediated electrons conquer its usual Coulomb repulsion and endure a net attraction to one another; consequently, they stay in lower energy state when paired.

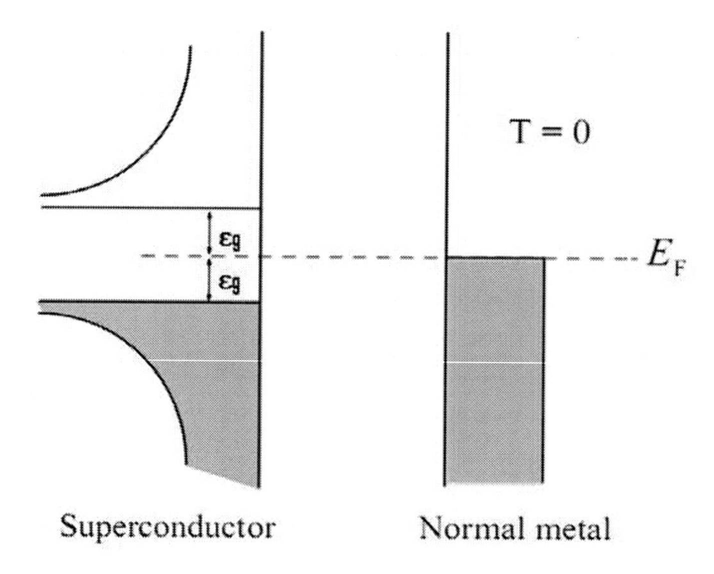

Figure 8. The density of state in the vicinity of Fermi level (E_F) in a superconductor and normal metal.

No electron can break that pair until they have acquired the sufficient energy to cross the gap. The temperature dependence of energy gap as per the BCS theory of superconductivity is depicted in reference [23]. As the temperature increases, the thermal energy ultimately overcomes the pairing interaction, and so superconductivity destroys as the energy gap goes to zero. In order to understand the superconducting behavior in lattice of superconducting material, understanding the energy gap is not enough. It is very essential to elucidate how tunneling occurs. In tunneling measurement $I\,(V)$ characteristics is used. Now, we will discuss the notion of this measurement.

As determined by the tunneling experiment, the energy gap depends on the temperature. Tunneling measurement is the most through method to determine the energy gap in the superconductors. The tunneling tests are of great interest evidently because they determine the density of state nearby to the energy gap. The energy gap becomes zero at the transition temperature, T_c and attains a maximum value when temperature approaches toward absolute zero [23].

As we know tunneling is rigorously a quantum-mechanical event, because in classical physics, there is no concept for the particle to penetrate a barrier. However there is a finite probability of finding an electron beyond a barrier. We can describe the moving particle by the help of wave function, ψ as shown below:

$$\psi(x,t) = A(x,t)\exp[i(k.x)]\ \exp(i\omega t) \tag{8}$$

Here, x, A and k denote the position vector, amplitude and momentum, respectively. In terms of classical physics, the particle cannot penetrate the barrier because the potential energy inside the barrier surpasses the total energy outside the barrier. In quantum-mechanics, however, there is finite probability of the penetrating a particle into a barrier. The probability of finding a particle at any specific point is $|A|^2$, and whenever the particle crosses through a barrier that would absolutely stop it, classically. This is the realism that current can flow from metal to metal through an insulating layer. This is the phenomenon on which scanning tunneling microscope and transmission electron microscope works.

The Fermi energy, in a normal metal, lies at the mid-point of a partially filled energy band; therefore the density of states in the region of Fermi level has no remarkable properties. Tunneling experiment exhibit an approximate flat density of state due to linear dependence of current with voltage. On the contrary, in a superconductor, the existence of an energy gap consign at centre on the Fermi level.

3.6. Proximity Effect

Proximity effect (PE) takes place when a superconductor (S) is placed in closed contact with normal metal (N). Generally the transition temperature T_c of the superconductor is reduced and we observed the precursor of weak superconductivity in the normal metal. If the contact between the superconductor and normal metal is adequate, the order parameter of the superconductor will changed near the interface. However, the superconductor S does not react with normal metal N but it induces the superconductivity into the N close to the interface layer. The wholesome distances calculated from the NS interface layer are the order of the coherence length, approximately 10^4 Å.

It is remarkable that when the contact between S and N is of superior quality, the Cooper pair diffuses from superconductors to normal metal and remains there for a moment. This phenomenon explicit the decline of Cooper pair density within the superconducting material. In this way, the normal metal induces the superconductivity. Thus, the proximity effect offers ascend to induce the superconductivity and becomes strapping at $T \leq T_c$. As a contact effect, PE is strongly associated to thermoelectric phenomena *viz.* the Peltier's effect or the fabrication of pn junction in the semiconductors. The PE development of T_c is largest when N has a large diffusivity rather than an insulator. The PE suppression of T_c in a superconductor is largest when the N is ferromagnetic because in presence of internal magnetic field Cooper pair breaks and superconductivity vanishes.

Now, we will discuss the interface layer between S and N. The shape of the interface layer is assumed to be flat and coincides with the plane x = 0, as depict in reference [24]. The semi-space is exposed by the superconductor and the metal for x > 0 and x < 0, respectively. Due to the induction, the order parameter penetrates the normal metal up to finite length called the effective coherence length, ξ_N and exponentially decay followed by the relation, $\psi_n \propto \exp(-|x|/\xi_N)$. Inside the normal metal N, when the electron mean free path is extremely large than effective coherence length, $l_n >> \xi_N$, the effective coherence length is describes by

$$\xi_N = \frac{h v_F}{2\pi k_B T}, \tag{9}$$

Here, v_F is Fermi velocity in normal metal. This expression shows that when $T \to 0$, $\xi_N \to \infty$. The decay of order parameter in N region is quite slower than exponential part. When $l_n << \xi_N$, the effective coherence length is expresses by

$$\xi_N = \left(\frac{h v_F l_n}{6\pi k_B T} \right)^{1/2} \tag{10}$$

Estimating Eqs (9) and (10) yields ξ_N in the order of 10^3–10^4 Å.

Assume that d is the thickness of a superconductor and this thickness make contact with normal metal, is adequately large, i.e. $d >> \xi_{GL}$, the transition temperature of the superconductor is almost unaltered. Conversely, when a superconducting thin film of small thickness $d << \xi_{GL}$ is deposited on the surface of a normal metal, the transition temperature of the whole system decreases drastically and becomes zero in some cases.

It is noticeable that PE will also observe in any Bose-Einstein condensate. The study of Superconductor-Normal metal (SN) and Superconductor-Insulator (SI) bilayers and multilayers is the live arena of PE. In superconductor-normal metal-superconductor (SNS) Josephson junctions, when phase coherence between the superconducting electrodes is forms through a quite thick normal layer, PE is observed.

The behavior of compound structure is anisotropic in nature both parallel and perpendicular to the interface. In high temperature superconductors exposed to magnetic field parallel to the interface, vortex disorder will primarily nucleate in the N or I layers and instability in nature is explicit when the field enhances and penetrates into S layers. Similarly, in conventional superconductors magnetic flux first penetrates the N layers. Alike qualitative changes do not take place when an applied field is perpendicular to the SI or SN interface. In SN and SI multilayers at low temperatures, the large penetration depths and coherence lengths will permit the S layers to retain a 3D quantum state. On enhancing the temperature the contact between the S layers is smashed that turns into 2D behavior. Interestingly, while an electron moves from normal metal to superconductors and came across the NS interface, if the entire electron energy is less than the energy gap of the superconductor, the electron come back from the interface. The anisotropic nature of SN and SI bilayers and multilayers are fruitful for understanding the composite critical field phenomena exhibited in highly anisotropic cuprate high temperature superconductors. The movement of a negative charge in the normal metal from the interface is analogous to movement of a positive charge in the superconductors in the opposite direction. Consequently, the process of the electron reflection produces an electrical current from the normal metal to the superconductors. This process was first proposed by Andreev and is called the *Andreev reflection* [25].

3.7. Isotope Effect

After the World War II, the isotopes of the various metals became available and soon isotope effect [26,27] has discovered. According to this effect, different superconducting metal have different transition temperatures and related by the following equation.

$$T_c m^{\alpha} = \text{Constant} \quad (11)$$

Here, m is the mass of the superconducting element and α is the isotope coefficient having value near to ½ for most of the conventional superconductors. However, there are a lot of conventional materials which possesses the quite different value of the coefficient; α. The isotope effect offered a vital role to the development of BCS theory of superconductivity for conventional superconductors. High temperature superconductors also reveal an unconventional behavior of the isotope effect in association with doping. The list of some conventional superconductors along with its different coefficient is shown in Table 1.

Table 1. Coefficients, α with different conventional elements

Element	α
Mg	0.55
Re	0.40
Os	0.21
Zr	0.00

Table 1 shows that the isotope effect is not universally true and even conventional superconductors not exhibit this effect. In unconventional superconductors the situation becomes more complicated. Let us take an example of CuO, varying the doping concentration; the isotope effect is almost gone in the optimally-doped region ($\alpha \approx 0.03$) but in underdoped region, α is near about 1.

We know that mercury has purely electrical conduction in nature; there should be no relation upon the atomic masses. Instead of this, isotope effect shows the dependence of transition temperature of the superconducting materials upon isotope mass explicit the evidence for interaction between the electrons and the vibrating lattice [26,27]. This statement supports the BCS theory of lattice coupling for electron pairs.

Enormous factors (not pertained to the pairing mechanism) widely change the value of the isotope coefficient while leaving the phonon spectrum unaffected. These notions include the magnetic impurities, non-adiabatic charge transfer or the proximity effect [28]. Such classes of isotope effects instantly sway both conventional and unconventional superconductors.

The isotope effect does not illustrate explicit results on the pairing mechanism responsible for the superconducting state. Especially, a small value of isotope coefficient does not entail that phonons are extraneous for the pairing mechanism. In fact, the Coulomb interaction, magnetic impurities, proximity effect can produce very different values of the isotope coefficient.

Biondi [20] depicts the result acquired for oxygen doping (^{16}O and ^{18}O) in $YBa_2(Cu_{1-x}Zn_x)_3O_{7-\partial}$ for the isotopic shift of T_c in various high temperature superconductors. We can easily see the magnetic impurities, Zn in the whole system.

Vladimir et al. [29] demonstrates the doping of Pr on the isotope coefficient. This affects the layers of the superconductor and introduces magnetic impurities [30]. Each contribution is shown by the separate lines. The lower line covers the non-adiabatic channel in absence of magnetic impurities. The central line exhibit the influence of adding magnetic impurities to the non-adiabatic channel. The solid, upper line is the best fit of the data.

It should be noted that the isotope effect in presence of magnetic impurities or the proximity effect deviate when T_c approaches to absolute zero. This attribute supports that one is approaching the phase transition from superconducting to normal state.

The new isotope effect is based on the shift of the magnetic field penetration depth upon isotope doping [31]. This effect has been observed recently by torque magnetometry [32]. The interesting feature of the penetration depth is that the isotope coefficient has dependence with temperature [33,34]. A similar dependence is also observed for the isotope effect in systems displaying the proximity effect. This effect has not been verified experimentally up to date.

4. Applications of Superconductors

High temperature superconductors explored the variety of practical application in different disciplines of engineering. The most common application of HTSC incorporate; magnetic shielding devices, medical resonance imaging (MRI), SQUIDs, sensors and signal processing devices. Properties of the superconducting materials are enhancing by day to day research and applications such as; power transmission, energy storage devices, particle accelerators, levitated transports and magnetic separators will become more realistic. Due to the ability of superconductors to conduct the electricity with zero resistance, its use can be exploited in the area of wire industries. These superconducting wires can be used in generators in order to produce the same amount of electricity with miniature equipment and less energy. The generated electricity could be stored in superconducting coils for a long span of time without any energy loss.

The practical application of HTSCs drops into two major areas - electromagnets and electronics. The conventional conductors such as copper or aluminum dissipate the large amount of heat because of this a significant fraction of electricity is lost. In electromagnets, on the contrary, superconducting winding have much lower power consumption and are best suited for high-field applications. High temperature superconducting wires retain its superconductivity while they possess the substantial current densities. HTSCs are also productive as the coatings for radio-frequency cavities. The field of electronics is a promising area for practical application of HTSCs with thin-film interconnections, superconducting field-effect transistors and Josephson junction integrated circuits. Superconductors are relevant for logic devices and memory cells due to quick toggle speeds and compact wiring delays. Logic delay of 13 picoseconds and switching times of 9 picoseconds had been successfully demonstrated. In order to fabricate the integrated devices, superconductors are pooled with semiconductors at the temperature of liquid nitrogen (77 K). SQUIDs are the most sensitive detectors based on Josephson junction technology for sensing small variation in magnetic field up to the order of 10^{-15} Tesla that are similar to produced by human brain and heart. SQUID based gradiometer is indeed an excellent instrument for constructive evaluation of inorganic materials. The enhanced energy gap in HTSCs permits the production of superconducting electromagnetic radiation detectors employed from x-ray to infrared region [35].

In early 2001, the recent discovery of superconductivity at 40 K in MgB_2 reveals the scientific interest due to its simple electronic structure and low fabricating cost as well. Extensive work on MgB_2 has now shown that it is a conventional s-wave superconductor but with the unconventional property that it contains two superconducting gaps. This two-gap property is very important to its high-field performance potential. A particularly rapid use of MgB_2 wire has been for low thermal conductivity current leads in the Advanced Space Technologies for Robotics and Automation (ASTRA)-2 satellite.

Magnetic-levitation (Maglev) is another example of the practical use of the superconductors using liquid helium as a refrigerant. Maglev trains can be made to drift on strong enough superconducting magnets by eradicating friction between the train and its tracks. Conventional electromagnets dissipate much of the electrical energy as heat as well as its size is also much larger then superconducting magnets. The Maglev trains used in Japan have an incredible speed of more than 500 km/hour.

Superconducting magnets are critical components of enormous technologies. MRI is playing an excellent role in diagnostic the cure of human-being. The strong enough magnetic fields used in these instruments is a practical application of superconductors. Likewise, particle accelerators employed in high-energy physics are very dependent on high-field superconducting magnets. Recently, the large hadron collider (LHC), the particle accelerator at European Organization for Nuclear Research (CERN), Geneva is the largest and perhaps more crucial scientific instrument ever built. High-energy particle research pivots on being able to accelerate sub-atomic particles to nearly the speed of light that required the superconductor magnets. Almost 10,000 superconducting magnets cooled by 130 tonnes of helium at 1.9 and 4.2 K that contains overall stored magnetic energy of about 15 000 MJ.

High quality Nd-Ti superconducting cables (weight ~1200 tonnes) have a concatenation of severe challenges in terms of critical currents, magnetization and inter-strand resistance [36].

New applications of the superconductors will become available whenever we find the enhanced transition temperature superconductors. Liquid nitrogen (LN_2) based superconductors provide the new way as compared to liquid helium superconductors. The possible finding of room temperature superconductors has the control to get superconducting devices into our daily lives.

Conclusion

High temperature superconductors are recently discovered by the physicist. New viable innovations initiate with the existing technology produced by the research scientists for the development of new products. The quick development in the field of high temperature superconductivity attends one to believe that practical applications of superconductors is restricted only by one's thoughts and time. Scientists are expecting the upcoming crucial technologies based on superconducting thin-films, wires and tapes. The development of superconductors in electronics, sensors, magnets, power applications, transportations and cryocoolers are very promising by increasing the transition temperature, upper critical fields and critical current densities of the superconducting materials.

References

[1] Onnes, H. K. *Akad. V. Wetenschappen* (Amsterdam) 1911, *14,* 113–115.
[2] Bednorz, J. G.; Müller, K. A. *Z. Phys.* B 1986, *64,* 189–193.
[3] Fröhlich, H. *Proc. Roy. Soc.* A 1954, *223,* 296–305.
[4] Little, W. A. *Phys. Rev.* A 1964, *134,* 1416–1424.
[5] Sleight, A. W.; Gillson, J. L.; Bierstedt, P. E. *Solid State Commun.* 1975, *17,* 27–28.
[6] Chevrel R.; Sergent, M.; Prigent, J. *Solid State Chem.* 1971, *3,* 515–519.
[7] Takagi, H.; Uchida, S.; Kitazawa, K.; Tanaka, S. *Jpn. J. Appl. Phys.* 1987, *26,* L123–L124.
[8] Tanaka, S.; Takagi H.; Uchida, S.; Kitazawa, K. *Jpn. J. Appl. Phys.* 1987, *26,* L218–L219.
[9] Anderson, P. W. *Science* 1987, *235,* 1196–1198.
[10] Wu, M. K.; Ashburn, J. R.; Torng, C; Hor, P. H.; Meng, R. L.; Gao, L.; Huang, Z. J.; Wang, Y. Q.; Chu, C. W. *J. Phys. Rev. Lett.* 1987, *58,* 908–910.
[11] Meissner, W.; Ochsenfeld, R. *Naturwiss* 1933, *21,* 787–788.
[12] Jackson, J. D. *Classical Electrodynamics*; Wiley: New York, 1962.
[13] Onsager, L. *Phyl. Mag.* 1952, *43,* 1006–1008.
[14] Deaver, B. S.; Fairbank, W. M. *Phys. Rev. Lett.* 1961, *7,* 43–47.
[15] Doll, R.; Näbauer, M. *Phys. Rev. Lett.* 1961, *7,* 51–52.
[16] Josephson, B. D. *Phys. Lett.* 1962, *1,* 251–253.
[17] Josephson, B. D. *Rev. Mod. Phys.* 1964, *36,* 216–220.
[18] Anderson, P. W.; Rowell, J. M. *Phys. Rev. Lett.* 1963, *10,* 486–489.

[19] Glover, R. E.; Tinkham, M. *Phys. Rev.* 1957, *108,* 243–256.
[20] Biondi, M. A.; Garfunkelm, M. *Phys. Rev.* 1959, *116,* 853–861.
[21] Giaever, I. *Phys. Rev. Lett.* 1960, *147,* 464–466.
[22] Phillips, N. E. *Phys. Rev.* 1959, *114,* 676–685.
[23] Giaever I., Megerle, K. *Phys. Rev.* 1961, *122,* 1101–1111.
[24] Mourachkine, A. *Room temperature superconductivity*; ISBN: 1-904602-27-4, Cambridge International Science Publishing: Cambridge, 2004, pp 60.
[25] Andreev, A. F. *Sov. Phys. JETP* 1964, *19,* 1228–1231.
[26] Maxwell, E. *Phys. Rev.* 1950, *78,* 477–477.
[27] Reyolds, C. A.; Serin, B.; Wright, W. H.; Nesbitt, L. B. P*hys. Rev.* 1950, 78, 487–487.
[28] Kresin, V. V.; Knight, W. *Pair Correlations in Many-Fermion Systems*, Plenum Press: New York, 1998, pp 25.
[29] Kresin, V. Z.; Bill, A.; Wolf, S. A.; Ovchinnikov, Y. N. *Phys. Rev. B.* 1997, *56,* 107–110.
[30] Bill, A.; Kresin, V. Z.; Wolf, S. A. *Z. Phys. B* 1997, *104,* 759–763.
[31] Bill, A.; Kresin, V. Z.; Wolf, S. A. *Phys. Rev. B* 1998, *57,* 10814–10824.
[32] Hofer, J.; Conder, K.; Sasagawa, T.; Zhao, G. M.; Willemin, M.; Keller, H.; Kishio, K. *Phys. Rev. Lett.* 2000, *84,* 4192–4195.
[33] Bill, A.; Kresin, V. Z.; Wolf, S. A. *Z. Phys. Chem.* 1997, *201,* 271–276.
[34] Bennemann, K. H.; Ketterson, J. B.; Joachim, A. K. *Handbook of Superconductivity*; Springer-Verlag, 2003.
[35] Ford, P.; Saunders, G. *The Rise of Superconductors.* Taylor & Francis, 2003.
[36] Rossi, L. *Supercond. Sci. Technol.* 2010, *23,* 034001 (1–17).

In: Advanced Organic-Inorganic Composites
Editor: Inamuddin
ISBN 978-1-61324-264-3

Chapter 21

POLYMERS AND COMPOSITES

Zeid A. AL-Othman and Mu. Naushad [*]
Department of Chemistry, College of Science, Building 5,
King Saud University, Riyadh, Saudi Arabia

ABSTRACT

The conversion of inorganic ion exchange materials into polymer composite ion exchange materials is the latest development. Sol-gel derived polymer composite materials have found numerous applications in the areas of chemistry, biochemistry, engineering, and material science. The organic-inorganic polymer composite materials prepared via the sol-gel technique have attracted significant attention in the literature. The combination of organic polymer and inorganic precursors yields composite materials that have mechanical properties not present in the pure materials. Organic inorganic polymer composite ion exchange materials show the improvement in its granulometric properties that makes more suitable for the application in column operations. The binding of organic polymer also introduces the better mechanical properties in the end product, i.e. composite ion exchange materials. In the present study, a review on the history, introduction and applications of polymer composite ion exchange materials are studied.

ABBREVIATIONS

CCFA	Composites, characterization, fabrication and application
CFRPs	Carbon-fiber-reinforced polymer composites
CMC	Carbon matrix composite
CNTs	Carbon nanotubes
ISE	Ion-selective electrode
PMC	Polymer matrix composite
PVC	Poly(vinyl chloride)
SWNT	Single-walled nanotubes

[*] Corresponding author's e-mail: shad81@rediffmail.com

1. History of Polymer and Composites

The word polymer is derived from the Greek words -poly- meaning "many", and -meros meaning "part". Jons Jacob Berzelius coined this term in 1833. Although his definition of a polymer was quite different from the modern definition. Perhaps the earliest important work in polymer science was done by Henri Braconnot in 1811 on derivative cellulose compounds. Leo Baekeland (1907) created the first synthetic polymer bakelite by reacting phenol and formaldehyde at precisely controlled temperature and pressure. Bakelite was then publicly introduced in 1909. In 1922, Hermann Staudinger proposed that polymers consisted of long chains of atoms held together by covalent bonds and he was awarded by the Nobel Prize for this in 1953. Synthetic polymer materials such as nylon, polyethylene, Teflon, and silicone have formed the basis for a growing polymer industry. These materials were used to synthesize polymer composite materials. Composite materials have became known as a major class of structural elements and are either used as substitutions for metals/traditional material in aerospace, automotive and other industries. Composite Materials have come to existence as long as the history of human being is known. Wood is a natural composite of cellulose fibres in a matrix of lignin [1]. The most primitive man-made composite materials were straw and mud used to form bricks for building construction. The ancient brick-making process can still be seen on Egyptian tomb paintings in the Metropolitan Museum of Art. Mummification may be one of the very first human composite made which is documented in the history of ancient Egypt. The oldest known mummy found is dated 3000 BC [2]. The world's largest composite structure manifesting the civilization in human history belonged to the city of Arg-e-Bam with area of approximately 180,000 square meters. The city of Arg-e-bam was made of bricks consist of straw and mud with a history as old as 2000 BC [2]. The history of modern composites probably began in 1937 when the Owens corning Fiberglass Company began to sell fiberglass around the United States. The speed of composite development, already fast, was accelerated during World War II. In the hysterical days of the war, among the last parts on an aircraft to be designed were the ducts. Since all the other systems were already fixed, the ducts were required to go around the other systems, often resulting in ducts that were twisting, turning, and placed in the most difficult to access locations. Metal ducts just could not easily be made in these "horrible" shapes. Composites seemed to be the answer. Some of the war-oriented applications were converted directly to commercial applications such as fiberglass reinforced polyester boats. By 1948 several thousand commercial boats had been made. In 1961 a patent was issued to A. Shindo for experimentally producing the first carbon (graphite) fiber but several years later Courtalds Limited of the United Kingdom was the first to produce commercially feasible carbon fibers. Nowadays, composites have found their place in the world. Composites are made up of individual materials referred to as constituent materials. There are two categories of constituent materials: matrix and reinforcement. At least one portion of each type is required. The matrix material surrounds and supports the reinforcement materials by maintaining their relative positions. The reinforcements impart their special mechanical and physical properties to enhance the matrix properties. A synergism produces material properties unavailable from the individual constituent materials, while the wide variety of matrix and strengthening materials allows the designer of the product or structure to choose an optimum combination. As a matter of fact, the modern development of polymeric materials and high modulus fibres introduced a new generation of composites. Polymeric composites were mainly developed for aerospace applications where the reduction of the weight was the principal objective, irrespective of the cost. The need of exploring new markets in the field of polymeric composites has recently driven the research in Europe towards the development of new products and technologies. In particular, activities on thermoplastic based composites and composites based on natural occurring materials (environmentally friendly, biodegradable systems) have been of relevant interest in many European countries.

2. Introduction

It is well-accepted fact that the development of mankind today is directly or indirectly dependent on advanced technology materials (high performance materials) that perform better and open new dimensions in research and development. Among the major dependents in materials in recent years are polymer composite materials. In fact, polymer composites are now one of the most important classes of engineered materials, as they offer several outstanding properties as compared to conventional materials. These materials have found increasingly wider utilities in the general areas of chemical sensors [3-9], toxic metal ion separation [10-13], chromatography, fabrication of selective materials [3-9], and electrical and optical applications [14,15]. The disciplines like analytical chemistry, electro-analytical, medical, agriculture, potable water, power generation, textile and environmental have been using these materials. The term "composite materials" may, perhaps, be simply defined on the classical definition of a composite material as given in Longman's dictionary "something combining the typical or essential characteristics of individuals making up a group". Thus composite materials are those, which are formed by the combination of one or more other materials with a polymer matrix to produce a material with a combination of desirable properties from individual components. Thus, in addition, the constituent phases must be chemically dissimilar and separated by a distinct interface. In designing composite materials, scientists and engineers have ingeniously combined various metals, ceramics, and polymers to produce a new generation of extraordinary materials. Most composites have been created to improve combinations of mechanical characteristics such as stiffness, toughness, and ambient and high temperature strength. The constituents retain their identities in the composite; that is, they do not dissolve or otherwise merge completely into each other although they act in concert. Normally, the components can be physically identified and exhibit an interface between one another. Polymer composites are increasingly gaining importance as substitute materials for metals in applications within the aerospace, automotive, marine, sporting goods and electronic industries. Their light weight and superior mechanical properties make them especially suited for transportation applications. Polymer based composite materials or fiber reinforced polymer composites constitutes a major category of composites materials with a wide range of applications. They offer very attractive properties, which can be tailored to the specific requirements by careful selection of the fiber, matrix, fiber configuration (short, long, strength, woven, braided, laminated, etc.) and fiber surface treatment. Glass-fiber-reinforced polymer composites have found extensive applications including fiber-glass boats, pressure vessels, and airplane panels. Glass fibers are the earliest known fibers in advanced composites compatible with polyester and epoxy matrices. Boron-fiber- reinforced composites were used as structural components. Three dimensional carbon-carbon (c-c) composites are extensively used in aerospace and defense structures. Mechanical properties of these composites are strongly dependent on the architecture and interface behavior. Carbon-fiber-reinforced polymer composites (CFRPs) are known for their strength, durability and light-weight. Application of CFRP includes sport goods and components in aerospace industry. Current active research areas in composite materials provide some insight into technologies that will mature to practical application in both the short and long terms. There are several worldwide ongoing research themes which are developed based on today's emerging technological challenges in characterization, fabrication and application plotting the future directions for research activities in composite materials. To discuss the latest research challenges in composites as well as future research themes and directions, the first international conference on composites: characterization, fabrication and application (CCFA-1) held in Kish Island on December 2008. Composite materials were also used as ion exchange materials. Ion exchange technique is the well-known analytical technique for the treatment of water for a very long time. Inorganic ion exchange materials were used for this purpose but they had some limitations. For instance, these materials, in general were reported to be not very much reproducible in behavior and fabrication of the inorganic adsorbents into rigid beads type media suitable for column operation was quite difficult. Organic ion exchangers were also used but they also have limitations. One of the severest limitations of the organic ion exchanger was its poor thermal stability for instance, the mechanical strength and removal capacity of ordinary organic ion-exchange ion exchangers tend to decrease under high temperature conditions which are frequently encountered in processing liquid radioactive waste stream [16, 17]. In order to obtain a combination of these advantages associated with organic and inorganic materials as ion exchangers, attempt have been made to develop ion exchangers by combination of multivalent metal acid salts and organic polymers providing a new class of organic-inorganic polymer composite ion exchangers. The conversion of inorganic ion exchange materials has been taking place into composite ion exchange materials is the latest development in this discipline. Sol-gel derived composite materials have found numerous applications in the areas of chemistry, biochemistry, engineering, and material science [3-15]. The organic-inorganic composite materials prepared *via* the sol-gel technique have attracted significant attention in the literature [10-14]. The combination of organic and inorganic precursors yields composite materials that have mechanical properties not present in the pure materials. Organic inorganic composite ion exchange materials show the improvement in its granulometric properties that makes more suitable for the application in column operations. The binding of organic polymer also introduces the better mechanical properties in the end product, i.e. composite ion exchange materials. More recently, some organic-inorganic composite ion exchange materials have been developed. Khan et al. have reported polypyrrole Th(IV) phosphate [3] polyaniline Sn(IV) arsenophosphate [18], polystyrene Zr(IV) tungstophosphate [19] used for the selective separation of Pb^{2+}, Cd^{2+} and Hg^{2+} respectively, and ion exchange kinetics of M^{2+}-H^{+} exchange and adsorption of pesticide [20] have also carried out on these composite materials. Beena Pandit et al. have synthesized such type of composite ion exchange materials, i.e. O-chlorophenol Zr(IV) tungstate and p-chlorophenol Zr(IV) tungstate [21]. Styrene supported Zr(IV) phosphate composite material [22], and fibrous composite ion exchange materials such as polymethyl methacrylate, polyacrylonitrile, styrene and pectin based Ce(IV) phosphate, Th(IV) phosphate and Zr(IV) phosphate [23] having a great analytical applications, have been investigated by Varshney et al. Polyaniline Zr(IV) tungstophosphate has been synthesized by Gupta et al. [24] which was used for the selective separation of La^{3+} and UO_2^{2+}. Chanda et al. reported polyacrylic acid coated SiO_2 as a new ion exchange material. These composite ion exchangers have good ion exchange capacity, high stability, reproducibility and selectivity for specific heavy metal ions indicating its useful environmental applications. Mu. Naushad et al. have also synthesized and characterized new composite cation exchange materials viz- acrylonitrile stannic(IV) tungstate [11], EDTA-Zr(IV) phosphate [12], EDTA-Zr(IV) iodate [13] and cellulose acetate Zr(IV) molybdophosphate [25]. Cellulose acetate Zr(IV) molybdophosphate have offered a variety of technological opportunity for quantitative determination and separation of Cr(III) from a synthetic mixture of metal ions and Ca(II) from commercially available vitamin and mineral formulation namely Recovit. The practical applicability of acrylonitrile stannic(IV) tungstate was demonstrated in the quantitative separation of Fe^{3+} and Zn^{2+} contents of a commercially available pharmaceutical formulation Fefol-Z. EDTA–zirconium phosphate composite ion exchanger was found to have a good ion exchange capacity, high chemical stability and selectivity towards toxic metal ions. As a consequence, this material may be suitably utilized for the removal and isolation of toxic heavy metal ions for environmental monitoring. EDTA–zirconium iodate has been used for the qualitative separation of Pb(II) ions from a synthetic mixture and Mg(II) and Al(III) ions from a commercially available pharmaceutical formulation digene.

A crystalline organic–inorganic composite exchanger, acrylamide stannic silicomolybdate was synthesized by Amjad et al [26] and was used several binary separations of metal ions viz. Cd^{2+}–Pb^{2+}, Cd^{2+}–Cu^{2+}, Al^{3+}–Pb^{2+}, Al^{3+}–Cu^{2+}, Zn^{2+}–Pb^{2+}, Zn^{2+}–Cu^{2+}. The quantitative separation of Cu^{2+} and Zn^{2+} were also achieved on commercially available pharmaceutical formulation I-Vit by using the packed column of acrylamide stannic silicomolybdate. Acrylamide zirconium (IV) arsenate [27] has been used for the selective separation of Pb^{2+} from a synthetic mixture containing Mg^{2+}, Ca^{2+}, Sr^{2+}, Zn^{2+} and Cu^{2+}, Al^{3+}, Ni^{2+}, Fe^{3+}. Poly-o-methoxy aniline Zr(IV) molybdate [28] composite was found efficient ion-exchanger for the adsorption of heavy metal ions from aqueous waste stream. This composite showed the electrical conductivity in the semi-conducting range. Conducting properties of this composite is under consideration for the development of electrochemically switchable ion-exchanger. A polymeric-inorganic composite cation-exchanger, Nylon-6,6 Sn(IV) phosphate [29] was synthesized via mixing of polymer Nylon-6,6 into the matrices of inorganic precipitate of Sn(IV) phosphate. Using this electroactive composite material, a new heterogeneous precipitate based selective membrane electrode was fabricated for the determination of Hg(II) ions in solutions. So, we can say that polymer composite materials have the applications in diverse fields.

3. Applications of Polymer Composite Materials

3.1. Polymer Composite Materials as Ion-exchange Materials

As described above polymer composite materials are used as ion exchange materials. 'Organic-inorganic' composite ion-exchangers developed by incorporation of organic polymers into the inorganic matrices possessed good ion-exchange capacity, high chemical, thermal and mechanical stabilities, improved reproducibility and granulometric properties [30-33]. The introduction of various organic polymers, chelating or intercalating agents enhance the selectivity towards particular ionic moiety. Thus these ion-exchangers are specific in selectivity and can solve diverse problems of analytical chemistry and other fields, such as metal ion separations, materials for kidney dialysis, conduction, ion transport devices such as fuel cells, thermoelectric batteries etc. Ion exchange fibers, ion exchange membranes, ion selective electrodes, have been used in heavy metal separations, preconcentrations as well as metal recovery from various systems to decrease the pollution load in the environment.

3.2. Polymer Composite Materials as Ion Selective Electrodes

Increasing concern with environmental and personal protection together with widespread requirements for accurate process control has created a need for new and improved sensors for measuring both physical and chemical parameters. Thus, increasing demand for chemical surveillance in environmental protection, medicine and many industrial processes has created the need for sensors with features such as high selectivity, sensitivity, reliability and sturdiness. These demands can often be satisfied by ion-selective electrodes (ISEs), which are commonly used owing to their simplicity, lower cost and fast provision for analytical results and high selectivity for heavy toxic metal ions. Ion selective membrane electrodes are mainly membrane based devices, consist of permselective ion conducting materials. The composition of the membrane is designed to yield to potential that is primarily due to the ion interest (via selective binding process, *e.g.*, ion exchange which occurs at the membrane-solution interface). The purpose is to find membrane that will selectively bind the analyte ions leaving co-ions behind, possessing different ion recognition properties, have thus been developed to impart high selectivity.

Precipitate based ion-selective membrane electrodes are well known as they are successfully employed for the determination of several anions and caiton. The ion-exchange membrane obtained by embedding suitable precipitate ion-exchange are elecroactive materials in a polymer binder *i.e.* poly(vinyl chloride)(PVC) or epoxy resins (Araldite), have been extensively studied as potenitometric sensors. The use of 'organic-inorganic' composite ion-exchange materials formed by the combination of inorganic precipitate and organic polymer as electroactive compound in membrane electrodes has generated wide spread interest in developing new ISEs especially for heavy toxic metals [3-9, 34].

3.3. Application of Polymer Composites Comprising Carbon Nanotubes

Carbon nanotubes are being used in place of carbon fibers in making composites due to their high aspect-ratio, high strength and excellent thermal and electrical conductivity. These outstanding properties of carbon nanotubes (CNTs) make them an attractive candidate for making advanced composite materials with multifunctional features. Various CNT-polymer composites have been fabricated to improve the material properties such as electrical conductivity, mechanical strength, radiation detection, optical activity etc. Diverse applications in electronics, medicine, defense and aerospace have been envisioned due to the remarkable improvement on the material properties. A technique was developed by Ounaies *et al.* [35] for making actuating composite materials with polarizable moieties (eg. polyimide) and CNTs. With the aid of an *in-situ* polymerization under sonication and stirring, they achieved an increase of dielectric constant from 4.0 to 31 with a 0.1% volume fraction of SWNTs. Sensing gas molecules is significant to environmental monitoring, space missions and industrial, agricultural and medical applications. The detection of NO_2 and CO is important to monitor environmental pollution. Detection of NH_3 is required in industrial and medical environments. Kong *et al.* [36] demonstrated chemical sensors based on individual SWNTs. They found that the electrical resistance of a semi-conducting SWNT changed dramatically upon exposure to gas molecules such as NO_2 or NH_3. The existing electrical sensor materials including carbon black polymer composites operate at high temperatures for substantial sensitivity whereas the sensors based on SWNT exhibited a fast response and higher sensitivity at room temperature.

3.4. Application of Composite Materials for Lightweight and Smart Structures Design

The introduction of composite materials in new industrial market will surely bring great advances in term of dynamic performances and will represent a new challenge for design engineers and huge opportunities for the composite manufacturers. Conventional materials for building machine tools are cast iron, welded steel and, in some cases, aluminium-alloys. Although these materials represent a much consolidated technology for machine tools engineers, the need to ensure such high performances requires the investigation of a new class of materials with improved inherent properties in terms of specific stiffness and structural damping [37]. Besides the above applications, the composite materials have applications in following disciplines

- Separation and preconcentration of metal ions
- Nuclear separations
- Catalysis
- Redox systems
- Electrodialysis
- Hydrometallurgy
- Effluent treatment
- Ion exchange fibers

CONCLUSION

Polymer composites are now one of the most important classes of engineered materials, as they offer several outstanding properties as compared to conventional materials. These materials have found increasingly wider utilities in the general areas of chemical sensors, toxic metal ion separation, chromatography, fabrication of selective materials, and electrical and optical applications. So the synthesis of new composite materials is worthwhile for various applications.

Acknowledgment

The authors would like to acknowledge King Saud University, Riyadh, Saudi Arabia for the support to carry out this research work.

References

[1] Hon, D.; Shiraishi, N. *Wood and cellulose chemistry;* 2nd ed., Marcel Dekker, New York, 2001.
[2] Farahani, A. V. *Appl. Compos. Mater.* 2010, *17,* 63–67.
[3] Khan, A. A.; Alam, M. M. *React. Funct. Polym.* 2003, *55,* 277–290.
[4] Khan, A. A.; Inamuddin, Alam, M. M. *Mat. Research Bull.* 2005, *40,* 289–305.
[5] Varshney, K. G.; Tayal, N.; Gupta, U. *Coll. Surfaces A: Phys. and Engg. Aspects* 1998, *145,* 71–81.
[6] Khan, A. A.; Khan, A.; Inamuddin *Talanta* 2007, *72,* 699–710.
[7] Naushad, M. *Sensors & Trans. J.* 2008, *95,* 86-96.
[8] Naushad, M. *Bull. Mater. Sci.* 2008, *7,* 1–9.
[9] Gupta, V. K.; Singh, A. K.; Khayat, M. A.; Gupta, B. *Anal. Chim. Acta* 2007, *590,* 81–90.
[10] Nabi, S. A.; Shalla, A. H. *J. Hazard. Mater.*2009, *163,* 657–664.
[11] Nabi, S. A.; Naushad, M.; Bushra, R. *Chem. Engg. J.* 2009, *152,* 80-87.
[12] Nabi, S. A.; Naushad, M.; Bushra, R. *Adsor. Sci. Tech.* 2009, *27,* 423-435.
[13] Nabi, S. A.; Ganai, S. A.; Naushad, M. *Adsor. Sci. Tech.* 2008, *26,* 463-476.
[14] Khan, A. A.; Alam, M. M.; Mohammad, F. *Electrochimica Acta* 2003, *48,* 2463-2472.
[15] Inamuddin, Ismail, Y. A. *Desalination* 2010, *250,* 523–529.
[16] Macauly, K. W. C. *Radioactive Waste, Advanced Management methods for Medium Active Liquid Waste;* Harwood Academic, 1981.
[17] De, A. K.; Sen, A. K. *Sep. Sci. Technol.* 1978, *13,* 517.
[18] Khan, A. A.; Niwas, R.; Vershney, K. G.; *Coll. Surf. A. Phys. Engg. Aspects* 1999, *150,* 7-14.
[19] Khan, A. A.; Niwas, R.; Vershney, K. G. *Ind. J. Chem.* 1998, *37A,* 469-472.
[20] Khan, A. A.; Niwas, R.; Alam, M. M.; *Indian J. Chem. Tech.* 2002, *9,* 256-264.
[21] Pandit, B.; Chudasama, U. *Bull. Mater. Sci.* 2001, *24,* 265-272.
[22] Varshney, K. G.; Pandith, A. H. *Chem. Environ. Res.*1996, *5,* 1-8.
[23] Varshney, K. G.; Gupta, P.; Agrawal, A. *22nd National Conference in Chemistry, Indian Council of Chemists*; I.I.T, Roorkee, 2003.
[24] Gupta, A. P.; Varshney, P. K. *React. Funct. Polym.* 1996, *31,* 111.
[25] Nabi, S. A.; Naushad, M. *Coll. Surf. A: Physic. Engg. Aspects* 2008, *316,* 217–225.
[26] Khan, A. M.; Ganai, S. A.; Nabi, S. A. *Coll. Surf. A: Physicochem. Engg. Aspects* 2009, *337,* 141–145.
[27] Nabi, S. A.; Shalla, A. H. *J. Hazard. Mater.* 2009, *163,* 657–664.
[28] Inamuddin, Ismail, Y. A. *Desalination* 2010, *250,* 523–529.
[29] Khan, A. A.; Akhtar, T. *Electrochim. Acta* 2009, *54,* 3320–3329.

[30] AL-Othman, Z.A.; Inamuddin, Naushad, Mu. *Chem. Engg. Journal* 2011, *166*, 639-645.
[31] Nabi, S.A.; Bushra, R.; Naushad, Mu.; Khan, A.M. *Chem. Engg. Journal* 2010, *165*, 529-536.
[32] Nabi, S.A.; Bushra, R.; Al-Othman, Z.A.; Naushad, Mu. *Sepn. Scien. & Technology* 2011, *46*: In Press.DOI: 10.1080/01496395.2010.534759.
[33] AL-Othman, Z.A.; Inamuddin, Naushad, Mu. *Chem. Engg. Journal* 2011, DOI:10.1016/j.cej.2011.02.046.
[34] Pungor, E.; Toth, K. *Ion Selective Electrodes in Analytical Chemistry*, Plenum Press, New york, 1978, pp. 143.
[35] Ounaies, Z.; Park, C.; Harrison, J.S.; Holloway, N. M.; Draughon, G. K. *(US200608475, 2006).*
[36] Kong, J.; Franklin, N. R.; Zhou, C.; Chapline, M. G.; Peng, S.; Cho, K.; Dai, H. *Science* 2000, *287,* 622-625.
[37] Ashby, M. F. *Materials Selection in Mechanical Design;* Elsevier, 2005.

ACKNOWLEDGMENTS

I am most indebted to the grace of the Almighty 'One Universal Being', who inspires entire Humanity to knowledge, and who has given me the required favor to complete this work.

This book is the outcome of remarkable contributions of experts of interdisciplinary fields of science, with comprehensive, in-depth and up-to-date research works and reviews. I am thankful to all the contributing authors and their co-authors for their esteemed contribution to this book. I would also like to thanks to all publishers, authors and others who granted us permission to use their figures/tables. Although sincere efforts has been made to obtain copyright permissions from the respective owners and to include citation with the reproduced materials, I would still like to offer my sincere apologies to any copy right holder whose rights may have been unknowingly infringed.

I would like to express my deep sense of gratitude to my mentors, Mr. Farooq Ali (High school math's teacher), Late Mr. Sanaullah Khan, and Dr. H.S. Yaday (Intermediate and graduation chemistry teachers, respectively), who always inspired me to have lifelong interest in chemistry and Prof. Syed Ashfaq Nabi (Chairman, Department of Chemistry, Aligarh Muslim University (A.M.U.), India), Prof. Ishtiyaq Ahmad and Prof. Rakesh Kumar Mahajan (Department of Chemistry, Guru Nanak Dev University, Amritsar, India), Dr. B.D. Malhotra (Scientist F, NPL, New Delhi, India), Dr. Raju Khan (Scientist-C, NEIST, Assam, India), Prof. Seon Jeon Kim (Hanyang University, South Korea), Prof. Kenneth I. Ozoemena (University of Pretoria, South Africa), Prof. Saleem-ur-Rahman and Prof. S.M.J. Zaidi (King Fahd University of Petroleum and Minerals, Saudi Arabia), Prof. Sheikh Raisuddin (Jamia Hamdard University, New Delhi, India) and Prof. A.I. Yahya (Nizwa University, Oman) for their valuable suggestions, guidance and constant inspiration.

I especially thanks to Prof. Faiz Mohammad (Department of Applied Chemistry, A.M.U., India), who always encouraged me to step forward as a beginner in this journey. My sincere thanks are also to Prof. Ali Mohammad (Chairman, Department of Applied Chemistry, A.M.U. Aligarh), for his amicable nature, time to time support, and constant motivation. It is with immense gratitude that I thank my departmental colleagues Prof. M. Mobin, Prof. Asif Ali Khan, Prof. R.A.K. Rao, Dr. M.Z.A. Rafiqui, Dr. Abu Nasar, Dr. Rais Ahmad and Dr. Yasser Azim, without whose continuous encouragement, this book would have not been brought to its final form. I am feeling lack of words to express thanks to my friends and colleagues Dr. M.M. Alam (U.S.A.), Dr. Amir-Al-Ahmad (KFUPM, Saudi Arabia), Dr. Zafar Alam, Dr. Mu. Naushad, Dr. Mohammad Luqman, Dr. Salabh Jain, Dr. Hemendra Kumar

Tiwari, Dr. Adesh Bhadana, Dr. Shakeel Ahmad Khan, Satish Singh and others, for their timely help, good wishes, encouragement and affections.

I also thanks to my childhood friends, Mohammad Dilshad, Noor Mohammad, Sagheer Ahmad, Fajlur Rehman, Nadeem, whose memories are always surrounding me and inspiring to achieve a high status in the field of education as they fails to do so by personnel reason indeed of hard work.

I would like to make special thanks to my nearest cousins Mohammad Tariq, Mohammad Akhtar, Mr. & Mrs. Ghayasuddin, Mr and Mrs. Shahabuddin, Mrs. Chandni and Mr. and Mrs. Shakir, who encouraged me at every stage of my life. I am very lucky to have the blessings and love of my elder sisters Mrs. Salma Bano, Mrs. Nazma Bano, Mrs. Reshma, their husbands also and younger Abida Khanam, Nilofar and Gulbahar. They have always extended their support and wishes at every phase of my life and study. I also thank to my elder brother Mr. Ghayasuddin and Mrs. Mahjabeen and younger Hashmuddin, Md. Javeed and loving son Mohammad Uzair Inam, whose love pulled me out of all tensions and has inspired to move ahead. I feel short of words and full of emotions in thanking my parents for the unconditional love, constant inspiration and gracious support they have always provided.

Last but not least, I am thankful to my loving wife Mrs. Khushbu Jahan, to provide gracious supports and constant inspirations during the sleepless and tiring editing exercise of this book edition in night timings. Thanks.

INDEX

A

absorption spectra, 149, 151, 152, 482
absorption spectroscopy, 418, 419
access, 82, 359, 366, 374, 375, 410, 530
accessibility, 93, 111, 359, 378, 385, 420
accommodation, 392
acetic acid, 108, 112, 448, 449, 451, 453, 456, 459
acetone, 457, 458, 459, 467, 480, 481
acetonitrile, 375, 389, 396
acidity, 15, 16, 18, 19, 440
acrylate, 181, 185
acrylic acid, 209
acrylonitrile, 108, 533
activated carbon, 94, 95, 104, 291, 292, 379, 383, 385, 390, 391, 478
activation energy, 206, 207, 319, 320, 322, 335, 339, 341, 342, 347, 348
active centers, 389, 456
active site, 36, 100, 103, 104, 216, 228, 387, 392
adaptation, 359, 364
additives, 6, 8, 9, 18, 27, 142, 171, 316, 420, 425, 453, 492, 494, 495
adhesion, 45, 83, 116, 169, 172, 177, 179, 181, 183, 186, 191, 194, 196, 200, 212, 417, 421, 423, 424, 425, 496
adhesive strength, 254
adhesives, 83, 365
adlayers, 37
adsorption, 35, 38, 39, 42, 44, 51, 59, 82, 83, 87, 88, 92, 93, 94, 95, 96, 97, 98, 99, 100, 101, 102, 107, 108, 109, 110, 111, 113, 115, 116, 117, 118, 119, 162, 192, 194, 212, 234, 292, 416, 439, 441, 443, 444, 453, 532, 533
aerogels, 379
aerospace, 252, 254, 422, 530, 531, 534
AFM, 33, 40, 52, 55, 132, 149, 150, 151, 158, 159
age, 74, 366, 433, 481
aggregation, 24, 25, 108, 177, 220, 388, 395
aggressiveness, 413
agriculture, 76, 531
Air Force, 106
Al2O3 particles, 96
alkaline media, 183
alkaloids, 455
alkenes, 38, 41
allergic reaction, 80
allylamine, 70, 73, 90
alopecia, 78
alters, 307, 425
aluminium, 33, 41, 44, 74, 110, 143, 192, 196, 223, 252, 293, 355, 367, 394, 421, 423, 442, 453, 466, 502, 535
amalgam, 359, 365, 367
ambient air, 142
amine(s), 22, 71, 72, 90, 92, 97, 102, 103, 115, 195, 210, 359, 360, 388, 457, 505
amino, 49, 54, 99, 101, 102, 104, 443, 444, 453, 457, 458, 507
ammonia, 58, 95, 195, 198, 412, 415, 456
ammonium, 59, 73, 107, 115, 118, 337, 371, 387, 389, 448, 450, 451, 467, 477, 479, 480, 481, 482, 501, 502
amplitude, 51, 502, 504, 514, 521
anatase, 9, 425
anatomy, 359
anchoring, 38, 43, 383
anemia, 78, 79, 81
aniline, 370, 384, 387, 391, 392, 394, 395, 418, 456, 465, 467, 475, 499, 501, 502, 503, 504, 505, 533
annealing, 142, 145, 162, 221, 225
anodization, 222, 223
anorexia, 78, 80
antacids, 78
anther, 393
antimony, 74, 81, 499, 501, 504, 508
antioxidant, 78

apathy, 78, 80
APC, 370, 390
aquatic systems, 118
aqueous solution, x, 9, 82, 90, 94, 100, 101, 108, 109, 111, 112, 115, 131, 291, 293, 305, 308, 384, 387, 393, 395, 418, 419, 459, 467, 479, 481, 507
argon, 195, 196, 199, 213, 233, 264, 360, 480
aromatic hydrocarbons, 51, 119
aromatics, 292
arsenic poisoning, 74, 78
arthritis, 78, 433
arthroscopy, 433
asbestos, 249, 493, 494
assessment, 367, 480, 491, 496
asthma, 79
astringent, 367
asymmetry, 374
atmosphere, 39, 77, 94, 143, 172, 178, 181, 186, 221, 233, 247, 324, 341, 345, 410, 466, 468
atmospheric pressure, 197, 421
atomic force, 40, 150, 159
atomic orbitals, 141
atoms, 17, 19, 21, 94, 95, 143, 212, 305, 359, 407, 505, 506, 507, 511, 530
attachment, 53, 57, 367
Au substrate, 45, 229
Auger electron spectroscopy, 33
automobile parts, 243, 245, 247, 251
automobiles, ix, x, 246, 249, 250, 252, 253, 369, 420
automotive applications, 4, 252, 253, 254

B

bacteria, 172, 183, 186, 292, 364, 415
band gap, 136, 141, 156, 163, 222, 269, 274, 275, 284, 332, 482, 507
barium, 110, 246, 362
barriers, 292, 520
base, 43, 96, 105, 106, 108, 178, 196, 253, 365, 386, 468
basicity, 440
batteries, x, 76, 219, 369, 372, 376, 377, 378, 396, 478, 479, 534
bauxite, 452
BCS theory, 513, 514, 515, 516, 520, 521, 523, 524
benefits, 365, 417, 430
benign, 3, 196, 291, 496
benzene, 42, 73, 85, 87, 449, 452, 457, 507
benzoyl peroxide, 359, 360
beryllium, 74, 75, 78
bias, 192, 206, 209, 212, 214, 231, 259, 261, 277, 280, 281, 282
bicarbonate, 113, 117, 439
bile, 453, 459
bile acids, 453, 459
bimetallic metal composites, x
binding energy, 52, 134, 135, 137
biochemistry, 69, 88, 529, 532
biocompatibility, 101, 116, 117, 180, 360
biodegradability, 177, 496
biodiversity, 497
biological activities, 81
biological samples, 453
biological systems, 79
biomass, 371
biomaterials, 431, 433
biomolecules, 54, 85
biopolymer, 115
bioseparation, 97, 101
Bis-GMA, 354
bismuth, 114, 313, 501, 508
bisphenol, 182, 354, 357, 361
blend films, 150
blends, 131
blindness, 78
blood, 79, 80
blood pressure, 80
BMA, 404, 419
body composition, 79
body weight, 79
Boltzmann constant, 132, 139, 335
bonding, ix, 23, 47, 50, 54, 82, 89, 104, 109, 141, 177, 179, 183, 185, 332, 355, 358, 362, 364, 365, 366, 423, 430, 441, 443
bonds, 35, 38, 41, 82, 84, 141, 174, 305, 380, 444, 501, 504, 507, 508
bone, 79, 80, 117, 172, 180, 429, 430, 433, 434, 435, 494
boson, 514
brain functioning, 80
branching, 265
brass, 77, 414
breakdown, 197, 375, 406
breathing, 78, 79
brittle hair, 81
brittle nature, 435
brittleness, 250
bromine, 305, 380
bronchitis, 80, 81
BTC, 33, 44
building blocks, 34, 37, 46, 47, 114
bulk materials, 437
butadiene, 201, 244
butyl methacrylate, 419
by-products, 82, 415

C

Ca^{2+}, 117, 533
cables, 178, 526
cadmium, 54, 74, 76, 78, 79, 81, 112, 419, 457, 458, 501, 508
caffeine, 453
calcination temperature, 9
calcium, 18, 80, 169, 186, 246, 386, 435, 444, 458, 466
calibration, 507
calixarenes, 54
cancer, 71, 76, 78, 79, 80, 249, 361
carbides, 178, 355
carbohydrate(s), 79, 80, 444
carbon atoms, 380
carbon dioxide, 58, 69, 83, 105, 378
carbon materials, 375, 378, 385
carbon monoxide, 3, 58
carbon nanotubes, 95, 96, 170, 192, 194, 220, 379, 383, 385, 390, 466, 534
carbon tetrachloride, 293
carbonization, 93, 94, 378
carbonyl groups, 443
carboxyl, 119, 209
carboxylic acid, 54, 163, 209, 446, 494
carboxylic groups, 365
carboxymethyl cellulose, 115, 450, 451, 452
carcinogen, 75
cardiomyopathy, 81
cardiovascular disease, 79
caries, 357, 359, 360, 362, 363, 367
casting, 8, 20, 38, 39, 117, 134, 144, 248, 251, 494
castor oil, 183
catalytic activity, 220, 221, 222, 223, 227, 228, 229, 230, 233, 324
catalytic effect, 215, 230
catalytic properties, 500
category a, 278
cathode materials, 263
cation, 19, 91, 107, 108, 111, 112, 113, 198, 201, 324, 331, 336, 446, 450, 452, 472, 474, 533
cavity preparation, 362
C-C, 478, 488
CDC, 438
CEC, 243, 248
cell membranes, 9
cell phones, 270, 284
cellulose, 70, 105, 115, 116, 182, 186, 441, 443, 444, 446, 447, 448, 449, 450, 451, 452, 453, 457, 459, 496, 530, 533, 536
central nervous system, 58
ceramic, x, 117, 174, 180, 243, 246, 247, 248, 249, 250, 251, 252, 254, 255, 313, 314, 349, 350, 355, 420, 421, 422
cerium, 312, 322, 331, 332, 333, 341, 424, 452
CERN, 512, 525
cesium, 2, 6, 143
chalcogenides, 113
challenges, 36, 43, 69, 193, 210, 248, 345, 396, 526, 532
charge density, 381, 390, 513
charge trapping, 259
chelates, 74
chemisorption, 39
chitin, 71, 116, 118, 444, 446
chitosan, 70, 71, 73, 94, 113, 115, 116, 117, 118, 182, 444, 446, 450, 453, 455
chloride anion, 117
chlorinated aliphatics, 292
chlorine, 292, 305, 380
chloroform, 293, 294, 295, 299, 300, 301, 302, 305, 454, 459
cholecalciferol, 453
cholesterol, 81
cholic acid, 453
chromatograms, 440, 453
chromatographic technique, 439
chromatography, x, 83, 293, 437, 438, 439, 440, 443, 453, 458, 460, 496, 531, 535
chromium, 74, 76, 79, 224, 417
chromosome, 80
cigarette smoke, 74
civilization, ix, 530
classes, ix, 35, 82, 96, 359, 444, 524, 531, 535
classification, 77, 83, 173, 493
clay minerals, 275
cleaning, 34, 40, 41, 169, 185, 263, 292, 410, 415, 417, 425
cleavage, 87, 274, 504
clinical trials, 367
clusters, 4, 7, 53, 131, 146, 147, 148, 149, 150, 151, 152, 156, 162, 163, 164, 197, 215, 218, 334, 421, 465, 472, 475, 479
CMC, 246, 247, 248, 250, 252, 354, 355, 529
C-N, 247, 250, 468
CO2, 41, 56, 57, 85, 92, 96, 132, 133, 200, 494, 495, 496
coal, 73, 75, 378, 465, 466
cobalt, 74, 76, 79, 109, 337, 341, 355, 394, 417, 508
coherence, 37, 146, 512, 522, 523
collaboration, 508
collagen, 435
collisions, 52, 197, 205

color, 76, 152, 156, 172, 257, 259, 261, 262, 266, 268, 270, 277, 278, 279, 280, 283, 408, 414
combined effect, 27
combustion, 73, 76, 96, 178, 465, 466, 494, 495
combustion processes, 96
commercial, 8, 12, 16, 76, 77, 88, 96, 100, 106, 136, 162, 210, 218, 221, 226, 227, 284, 371, 378, 380, 419, 424, 456, 459, 497, 530
community, 4, 38, 75, 83, 269
compatibility, 18, 179, 203, 270, 271, 432, 497
compensation, 204, 315
competition, 272
competitive conditions, 118
complexity, 4, 16, 110, 115, 205, 407
composite resin, 96, 353, 367
composition, 14, 17, 49, 50, 83, 88, 112, 120, 205, 228, 233, 253, 272, 312, 321, 331, 337, 339, 347, 349, 390, 393, 421, 422, 444, 447, 448, 483, 485, 534
compounds, x, 6, 13, 34, 39, 56, 60, 74, 75, 76, 77, 78, 79, 80, 84, 105, 114, 115, 117, 119, 224, 233, 275, 282, 292, 307, 308, 380, 383, 407, 420, 424, 441, 444, 450, 457, 458, 501, 513, 514, 530
concentration ratios, 477, 480, 485
condensation, 14, 83, 84, 87, 92, 174, 176, 180, 181, 182, 424
conditioning, 371, 378
conductance, 473, 474, 479
conducting polymer, ix, x, 169, 276, 280, 373, 376, 379, 381, 383, 384, 390, 404, 417, 420, 466, 479, 481, 484, 500, 501
conduction, 3, 7, 20, 23, 24, 27, 156, 160, 195, 200, 206, 207, 224, 312, 314, 318, 322, 323, 326, 331, 332, 335, 336, 343, 345, 347, 349, 511, 524, 534
configuration, 3, 203, 206, 215, 231, 261, 313, 393, 531
confinement, 36, 211, 212, 311, 348
conjugation, 380
constituent materials, 172, 245, 492, 530
constituents, 172, 174, 183, 353, 360, 365, 389, 391, 414, 444, 467, 472, 492, 531
construction, 41, 47, 369, 373, 466, 492, 493, 495, 507, 530
consumers, 495
consumption, 57, 279, 371, 525
contaminant, 71, 72, 74, 291, 292
contaminated water, 85, 104, 117, 119
contamination, 76, 119, 200, 220, 308, 366, 410, 416
contour, 363, 364
controversies, 368
convergence, 3
COOH, 18, 42, 45, 47, 89, 114
coordination, 8, 34, 35, 37, 41, 42, 47, 48, 49, 78, 82, 90, 91, 119
copolymer(s), 7, 22, 25, 72, 84, 92, 93, 99, 102, 111, 418, 446, 504
copolymerization, 200, 418
copper, 39, 41, 74, 76, 78, 79, 85, 109, 178, 180, 197, 214, 233, 279, 293, 355, 412, 413, 414, 418, 419, 425, 512, 525
correlation, 181, 449, 507
corrosion, ix, x, 145, 169, 171, 172, 175, 177, 178, 179, 182, 183, 193, 212, 224, 250, 251, 253, 300, 304, 378, 403, 404, 405, 406, 407, 408, 409, 410, 411, 412, 413, 414, 416, 417, 418, 419, 420, 421, 422, 423, 424, 425
corrosivity, 171
cost effectiveness, 252, 416
cotton, 115, 367
covalent bond, 13, 23, 25, 35, 82, 83, 95, 141, 530
covering, 154, 155, 391, 468
cracks, 44, 46, 227, 246, 367, 412
creep, 246, 250, 276, 432
creosote, 77
cross-linked polymers, 26
crown(s), 53, 54, 55, 353
crust, 76, 77
crystal growth, 274
crystal structure, 59, 312, 314, 315, 316, 317, 319, 331, 337, 349, 394
crystalline, 8, 17, 18, 20, 36, 39, 42, 45, 47, 49, 82, 98, 134, 142, 151, 160, 178, 186, 225, 279, 338, 345, 354, 389, 404, 421, 422, 423, 441, 444, 494, 533
crystallinity, 15, 18, 111, 142, 160, 172, 212, 222, 233
crystallites, 142, 147
crystallization, 18, 42, 46, 253, 338
crystals, 44, 45, 46, 96, 259, 267, 274, 336, 346, 502, 503, 507
cure, 181, 182, 186, 360, 418, 525
curing process, 181, 182
curing unit, 360
current ratio, 160, 421, 473
cutting fluids, 494
CVD, 192, 219, 228, 244, 251, 404, 417, 421
cycles, 49, 59, 100, 110, 117, 372, 377, 385, 386, 387, 388, 390, 391, 394, 396, 418, 431, 502, 503
cycling, 311, 348, 376, 385, 395, 502, 504
cyclodextrins, 54
cyst, 433
cytochrome, 79

D

damping, 251, 535
DCA, 438, 453, 494
deacetylation, 116
decay, 135, 259, 261, 262, 264, 265, 269, 271, 273, 274, 364, 514, 522
decay times, 273
decolonization, 110
decomposition, 108, 120, 292, 294, 296, 302, 337
defect formation, 223, 316
defects, 179, 203, 206, 272, 274, 318, 321, 331, 333, 334, 337, 341, 342, 347, 410
deficiencies, 316
deficiency, 13, 16, 374
deforestation, 495
deformation, 432
degradation, 15, 25, 117, 142, 143, 186, 205, 208, 209, 234, 253, 259, 291, 292, 293, 294, 295, 296, 297, 298, 299, 300, 301, 302, 303, 304, 305, 306, 307, 364, 366, 381, 412, 415, 471, 482, 495, 496
degradation process, 143, 472
dehydration, 5, 16, 103
delusions, 81
dementia, 78
demonstrations, 516
density values, 9
dental restorations, 367
dentin, 361, 362
Department of Energy, 134, 255, 370, 373, 397
deposition rate, 204, 212, 264
deposits, 73, 410
depression, 80, 81, 96
depth, ix, 50, 216, 221, 227, 360, 378, 440, 512, 515, 524, 539
deregulation, 80
derivatives, 39, 40, 92, 131, 141, 259, 381, 383, 384, 420, 453, 455, 456, 459, 496
desorption, 100, 101, 103, 105, 107, 109, 110, 192, 194, 212, 234, 292
destruction, 293, 296, 297, 299, 302, 414
detachment, 97
detectable, 414, 473
detection, 52, 54, 55, 56, 57, 58, 59, 274, 305, 359, 451, 457, 508, 518, 534
detergents, 220
developing countries, 193
deviation, 151, 331
dialysis, 101
diamines, 102
diarrhea, 79, 80
dibenzo-p-dioxins, 77
diffraction, 160, 477, 518
diffusion, 60, 109, 120, 132, 135, 137, 139, 143, 144, 163, 171, 177, 192, 195, 196, 219, 227, 232, 337, 345, 374, 392, 421, 442, 473
diffusion process, 132, 139, 473
diffusivity, 19, 383, 522
digital cameras, 193
diluent, 181
dimethacrylate, 70, 354, 357, 365
dimethylformamide, 183, 469, 501, 503
dimethylsulfoxide, 112
diodes, 258, 260, 270, 276, 285
discs, 249, 250, 367
diseases, 433
disorder, 315, 523
dispersion, 10, 19, 23, 51, 83, 97, 179, 195, 216, 220, 221, 228, 234, 251, 354, 387, 388, 389, 394, 417, 419
displacement, 52, 249, 443
disposition, 222
dissociation, 25, 26, 131, 137, 139, 144, 145, 146, 156, 162, 163, 197, 305
dissolved oxygen, 97, 301
distilled water, 452, 478, 481, 482
distortions, 59
distribution, ix, 12, 27, 91, 97, 162, 172, 179, 194, 200, 205, 219, 266, 298, 338, 339, 357, 366, 378, 379, 380, 388, 421, 444, 456
dizziness, 79, 81
DMF, 16, 21, 70, 466, 467, 500, 502, 503, 504
DNA, 78, 79, 354, 361
DOI, 188, 537
donors, 15, 91, 141, 163
dopants, 269, 313, 315, 316, 317, 318, 320, 334, 337, 343
doping, 26, 48, 93, 95, 117, 119, 233, 262, 269, 279, 281, 315, 316, 317, 376, 380, 381, 383, 385, 388, 392, 393, 466, 496, 500, 504, 523, 524
dosage, 23, 177, 199, 361
double bonds, 380, 485
dressings, 497
drinking water, 74, 76, 77, 78, 94, 292, 491
drug delivery, 97, 497
drying, 98, 142, 172, 183, 185, 425, 503
DSC, 170
ductility, 412
durability, 3, 25, 60, 83, 178, 182, 193, 222, 271, 274, 357, 532
dyes, 47, 88, 110, 115, 116, 117, 260, 268

E

earthquakes, 405
edema, 78

education, 406, 540
effluents, 76, 112
emission, x, 51, 55, 58, 132, 147, 154, 155, 219, 220, 257, 261, 262, 264, 265, 266, 267, 271, 272, 273, 274, 275, 277, 278, 281, 282, 283, 284, 370, 371, 386, 392, 500, 502
emitters, 262, 277, 279
emphysema, 78
enamel, 353, 361, 362, 363, 366
encapsulation, 85, 142, 143, 264
encouragement, 497, 539
endoskeleton, 434
engineering, ix, 60, 69, 88, 178, 191, 194, 254, 281, 420, 422, 432, 491, 496, 511, 524, 529, 532
enlargement, 156
entropy, 347
enzyme, 47, 78, 79, 88, 97
epoxy resins, 418, 534
equilibrium, 53, 95, 109, 110, 271, 331, 334, 347
equipment, 58, 76, 92, 260, 406, 407, 416, 417, 525
erosion, 73, 413, 423
ESR, 370, 381
ester, 39, 40, 131, 133, 136, 140, 163, 492
estrogen, 361
etching, 143, 221, 366, 423
ethanol, 14, 47, 59, 85, 93, 105, 229, 232, 456, 458, 501
ethers, 53, 54, 494
ethyl acetate, 459
ethylene, 72, 84, 87, 92, 93, 98, 178, 198, 277, 305, 493
evaporation, 39, 57, 144, 222, 263, 264, 391, 424
excitation, 137, 154, 205, 259, 264, 271, 272, 273, 281, 475, 520
exciton, 131, 134, 135, 137, 139, 144, 145, 146, 156, 162, 163, 261, 269, 272, 274, 506
exoskeleton, 434
expenditures, 412
experimental condition, 208, 234, 295, 301
experimental design, 367, 472
exploitation, 1, 119, 193, 269
exposure, 55, 57, 58, 74, 75, 76, 77, 78, 79, 80, 181, 194, 359, 405, 412, 418, 458, 535
expulsion, 511
external magnetic fields, 512
extinction, 51
extraction, 75, 81, 85, 91, 266, 267, 268, 384
extracts, 204, 275

F

fabrication, x, 35, 36, 38, 43, 44, 47, 48, 49, 53, 60, 77, 131, 142, 144, 145, 163, 194, 195, 212, 228, 229, 250, 251, 252, 257, 259, 263, 264, 267, 268, 280, 350, 383, 385, 392, 414, 417, 429, 494, 502, 522, 529, 531, 535
families, 115
fast processes, 473
fat, 79
fatty acids, 444
FDA, 70, 76, 77
feedstock, 417
Fermi level, 232, 332, 337, 512, 520, 521
fermions, 514
ferrite, 389
ferrous ion, 407
fertilizers, 76, 119
fetus, 80
fever, 81
filament, 410
filiform, 410
filler, 7, 9, 16, 23, 24, 25, 179, 180, 181, 353, 354, 355, 356, 357, 358, 359, 364, 386, 492
filling materials, 353
film formation, 60
film thickness, 179, 226
filters, 147, 154, 157, 232, 466
filtration, 46, 81, 251, 292
first generation, 85
fishing, 493
fission, 111
fissure sealants, 353, 359
fixation, 429, 430, 435
flame, 404, 422, 497
flatulence, 78
flavonoids, 444
flexibility, 26, 60, 109, 177, 179, 180, 181, 183, 211, 275, 281, 284, 381, 492, 494
flow system, x, 291, 293, 294, 295, 296, 297, 298, 299, 300, 301, 302, 303, 304, 305, 306, 307, 308
fluid, 81, 406, 413, 421
fluorescence, 51, 262, 264, 265, 271, 272
fluoride ions, 113
fluorine, 23, 25, 185, 205
fluorophores, 55
foams, 70, 119, 494
food, 76, 171, 457, 491, 495
force, 33, 52, 108, 132, 249, 423, 493, 520
formaldehyde, 72, 73, 92, 93, 94, 96, 109, 180, 244, 458, 530
formula, 18, 19, 84, 87, 111, 141, 441, 443
fouling, 191, 194, 425
fractures, 44, 247, 365, 429, 430
fragments, 50, 95, 194, 195, 197, 205
free energy, 407
free radicals, 25, 183, 195, 205, 384

free volume, 19
freedom, 367
freezing, 3
friction, 180, 212, 243, 249, 250, 421, 423, 432, 515, 525
fruits, 458
fullerene, 57, 131, 147, 150, 152, 163
functionalization, 13, 16, 18, 40, 176, 220, 392
funding, 508
fungi, 415
furan, 494

G

gallium, 58
gamma rays, 272, 273, 274
gas diffusion, 194, 211, 218
gas sensors, 56, 222
gas sorption, 42
gastrointestinal tract, 74
GCE, 500, 502, 503, 504
gene expression, 80, 81
genomic instability, 80
geometry, 172, 195, 221, 267
gingivitis, 80
glass transition, 24, 172, 175, 253, 263, 418
glasses, 40, 179, 185, 230, 275, 355, 357, 358
global warming, 133
glow discharge, 197, 198, 200, 201, 202, 203, 205, 206, 209, 210, 212
glucose, 79, 186, 444
glue, 493, 502
glycol, 55, 70, 72, 185, 354, 357
gold nanoparticles, 178
granules, 71, 117, 444
granulomas, 78
graphite, 44, 60, 94, 178, 227, 233, 385, 386, 420, 501, 502, 530
gravimetric analysis, 170, 371
gravitational force, 515
grids, 206, 209, 267
gross domestic product, 406
groundwater, 76, 77, 117, 292
growth, 36, 40, 42, 43, 44, 45, 46, 47, 48, 49, 60, 78, 83, 95, 115, 174, 193, 216, 219, 279, 367, 369, 386, 405, 425, 433, 479, 507
guidance, 539
guidelines, 74, 491, 495, 497, 498

H

half-life, 111
halogen(s), 35, 307, 354, 360, 380
hardness, 76, 82, 178, 185, 353, 354, 355, 356, 357, 358, 359, 362, 365, 403, 417, 421, 422, 423, 493
harmful effects, 406
harvesting, 135, 144, 145, 264
hazardous waste, 75, 134
hazards, 425, 495
headache, 78, 81
healing, 429, 430, 434, 435
health, 72, 73, 74, 76, 77, 79, 80, 181, 271, 284, 425, 493
heart rate, 80
heartburn, 78
heat capacity, 520
heat transfer, 406, 416
heating rate, 337, 338, 339, 468
heavy metals, 74, 82, 127, 128, 250, 357, 358, 406, 420, 447
helium, 16, 195, 513, 525, 526
hemoglobin, 78
hemolytic anemia, 81
hemp, 180, 496
heterogeneous catalysis, 219
hexachlorobenzene, 77
hexafluoropropylene, 2, 20
hexane, 199, 389, 453, 459
HFP, 2, 20, 21
high density polyethylene, 434
high strength, 246, 250, 252, 253, 430, 432, 434, 534
high temperature superconductors, x, 511, 512, 514, 522, 524
histogram, 222
history, x, 9, 268, 493, 529, 530
homogeneity, 23, 25, 83, 278, 280, 384
homopolymers, 418, 501
host, 47, 53, 54, 91, 105, 176, 269, 272, 273, 274, 324, 417
hot pressing, 96, 248
HRTEM, 192, 222
HSCT, 244, 248
humidity, 2, 9, 12, 15, 16, 18, 21, 26, 27, 54, 171, 172, 209, 249, 405, 410, 440, 441
hybridization, 5, 110
hydrocarbons, 292, 307, 456
hydrocortisone, 453
hydrolysis, 14, 83, 84, 87, 91, 113, 174, 181, 424, 442
hydrophilicity, 51, 101
hydrophobicity, 50, 51, 110, 175, 425
hydroquinone, 504
hydroxide, 70, 71, 95, 106, 117, 178, 375, 408, 439
hydroxyapatite, 117, 180, 417, 430
hydroxyethyl methacrylate, 72

hydroxyl, 9, 18, 22, 25, 41, 88, 94, 99, 114, 174, 182, 209, 441, 442
hygiene, 425
hypertension, 80
hypothesis, 430, 489, 516

I

ideal, 47, 138, 139, 144, 145, 274, 312, 324, 331, 332, 333, 349, 391, 393, 434
identification, 437, 447, 451, 453, 457
illumination, 147, 151, 154, 156, 160, 271, 279
image analysis, 457
immersion, 38, 46, 49, 60, 186, 210
immobilization, 38, 46, 53, 54, 60, 72, 85, 88, 100, 109, 115
impact strength, 247, 253, 432
implants, 186, 417, 426, 430, 431, 494
impotence, 81
impregnation, 27, 493
imprinting, 91
improvements, 24, 160, 230, 421
impurities, 3, 220, 272, 274, 322, 377, 413, 495, 524
in vivo, 186, 435
incidence, 51
incubation period, 412
indium, 142, 143, 230, 263, 267, 281, 478, 479
individual character, 491
individual characteristics, 491
individuals, 531
induction, 512, 515, 522
industrial wastes, 77, 447
industrialization, 74
industries, 74, 77, 171, 243, 246, 248, 252, 253, 495, 525, 530, 531
industry, 20, 58, 77, 180, 243, 244, 246, 251, 252, 268, 270, 273, 284, 369, 384, 406, 412, 532
inertia, 219
inflammation, 360, 367
information technology, 260
infrared spectroscopy, 370, 386, 419, 500
ingestion, 74
ingredients, 96, 491, 492
inhibition, 9, 78, 171, 181, 228, 360
inhibitor, 420, 424
initiation, 420
inorganic fillers, 7, 9, 16, 23, 24, 25, 27
insertion, 131, 151, 160, 161, 216, 220, 359, 360, 366
insomnia, 81
insulation, 361, 494
insulators, 380
insulin, 79
insulin sensitivity, 79
integrated circuits, 270, 271, 525
integration, 52, 195, 253, 280
integrity, ix, 116, 253, 384
interaction effect, 228
interdependence, 53
interference, 115, 117, 512, 518, 519
internal fixation, 430
internalization, 79
interphase, 8
iodine, 380, 458
ion bombardment, 212
ion exchangers, 71, 83, 96, 97, 108, 111, 113, 114, 119, 127, 443, 532
ion transport, 317, 345, 534
ion-exchange, 18, 99, 106, 112, 119, 211, 439, 452, 453, 532, 533, 534
ionic conduction, 314, 322, 336, 344, 347, 407
ionic polymers, 8
ionization, 135, 163, 197, 205, 274, 331, 332, 333
ionization potentials, 135, 163
ionizing radiation, 273, 274, 432
IR spectroscopy, 389
IRC, 89, 98
Ireland, 502
iridium, 93, 278
iron, 39, 41, 57, 58, 70, 74, 94, 95, 97, 102, 104, 105, 106, 110, 116, 117, 119, 179, 183, 250, 292, 355, 394, 407, 408, 418, 466, 478, 513, 535
irradiation, 44, 148, 153, 157, 181, 186, 386
irritability, 80, 81
isolation, 364, 367, 411, 447, 533
isomers, 444, 449, 456
isotope, 514, 523, 524
issues, x, 27, 36, 43, 55, 120, 145, 193, 226, 466

J

joints, 435, 493
Josephson junction, 516, 517, 518, 522, 525

K

K^+, 18, 54, 55, 56
KBr, 482, 487
kidney dialysis, 534
kidney failure, 78
kinetic model, 109
kinetics, 3, 4, 18, 44, 48, 51, 53, 60, 89, 96, 100, 108, 109, 144, 192, 194, 212, 214, 232, 234, 272, 339, 377, 378, 381, 383, 389, 395, 422, 423, 532
KOH, 379, 385, 386, 393

L

lactic acid, 244, 253, 496
lamella, 15, 17
lamination, 249
landfills, 77
lanthanum, 314
lattice parameters, 326
lattices, 43
layer-by-layer growth, 43
leaching, 26, 78
lead, 4, 7, 18, 23, 25, 74, 76, 77, 80, 81, 110, 119, 144, 146, 162, 180, 195, 199, 222, 226, 253, 267, 275, 280, 300, 304, 355, 360, 365, 367, 369, 373, 430, 495, 497, 499, 501, 504, 507, 508
leakage, 131, 139, 140, 146, 156, 159, 160, 162, 228, 360, 362, 364
learning, 78
LED, 261, 270, 275, 277, 279, 280, 287, 354, 360
lifetime, 21, 142, 191, 194, 234, 257, 360, 403, 405, 425
ligand, 35, 46, 47, 56, 58, 88, 89, 90, 91, 102, 104, 107, 114
light emitting diode, 258, 260, 261, 270, 277, 280, 283, 284, 287, 360
light scattering, 179
light-emitting diodes, 260
lignin, 116, 530
linear dependence, 521
linear function, 53
liquid chromatography, 438, 440, 443
liquid interfaces, 51
liquid phase, 39, 49
liquids, 179, 371, 388
lithium, 269, 373, 395
lithography, 37, 40
localization, 216, 227
longevity, 54, 367
loss of libido, 81
low temperatures, 275, 326, 341, 523
low-cost ZnO nanostructure/organic composite solar cells, ix, 162
LTD, 65
lubricants, 74, 119, 250, 414
luminescence, 57, 155, 259, 264, 271, 272, 274, 275, 276, 277, 281, 284, 285, 287
luminescence efficiency, 259
Luo, 400, 401, 461, 462, 509
lying, 17, 272

M

macromolecules, 115
macropores, 392
magnesium, 70, 94, 108, 251, 293, 355, 444, 453, 454, 466
magnesium alloys, 251
magnetic field, 196, 212, 263, 511, 514, 515, 516, 517, 518, 522, 524, 525
magnetic properties, 96, 97, 98, 104
magnetization, 515, 526
magnitude, 9, 59, 193, 202, 262, 279, 312, 317, 321, 331, 343, 348, 362, 375, 518
majority, 60, 268, 270
man, 171, 172, 260, 275, 371, 530
management, 3, 4, 195, 200, 224, 369, 406
manganese, 74, 76, 80, 107, 111, 114, 127, 128, 337, 341, 380, 389, 393, 395, 449, 488
manufacturing, 76, 77, 116, 212, 246, 252, 254, 259, 270, 271, 280, 284, 292, 466, 496
marine environment, 171
marrow, 434
mass, 8, 9, 34, 50, 51, 52, 105, 108, 132, 162, 198, 216, 218, 221, 224, 226, 317, 319, 332, 342, 377, 385, 388, 389, 394, 406, 433, 440, 471, 475, 481, 483, 495, 523, 524
material sciences, 33
material surface, 374, 403
materials science, 33, 255, 372
maximum sorption, 90, 111, 113, 115, 117
measurement, 27, 51, 52, 147, 154, 157, 228, 331, 382, 423, 446, 459, 521
measurements, 10, 27, 51, 54, 57, 157, 186, 209, 221, 227, 296, 326, 331, 341, 343, 386, 388, 391, 419, 441, 467, 473, 482, 502, 520
mechanical properties, 19, 23, 69, 111, 172, 176, 180, 182, 219, 251, 253, 311, 312, 348, 365, 421, 423, 496, 529, 531
media, 4, 55, 84, 94, 102, 115, 128, 174, 180, 183, 232, 412, 415, 532
medical, 76, 83, 430, 524, 531, 535
medicine, ix, 77, 171, 273, 534
melt, 176, 248, 251, 253, 277, 497
melting, 26, 178, 247, 249, 250, 251, 274, 316, 347, 414
melting temperature, 249
membranes, ix, x, 1, 3, 4, 5, 6, 7, 8, 9, 12, 14, 16, 18, 19, 20, 21, 22, 23, 24, 25, 26, 27, 43, 46, 48, 116, 118, 191, 193, 194, 196, 197, 198, 199, 200, 201, 202, 203, 204, 205, 206, 207, 208, 209, 210, 226, 230, 233, 350, 534
memory, 60, 78, 80, 373, 525
mental impairment, 78

mercury, 39, 74, 76, 80, 87, 95, 501, 512, 514, 524
metabolism, 79, 81
metabolized, 78
meter, 380
methacrylates, 181
methacrylic acid, 182
methanol, 1, 2, 3, 4, 5, 6, 7, 10, 12, 13, 14, 16, 19, 20, 22, 24, 26, 27, 59, 192, 194, 199, 200, 201, 206, 207, 212, 214, 220, 223, 224, 228, 229, 230, 232, 234, 345, 447, 454, 480, 481, 501, 503
methodology, 5
methyl methacrylate, 419
methylation, 78, 210
methylene blue, 110
methylene chloride, 496
Mg^{2+}, 94, 533
microbial community, 308
microcavity, 267
microcrystalline, 218, 417, 448, 450, 453
microemulsion, 389
microhardness, 181, 423
micro-lens, 267
micrometer, 84, 358
microscope, 33, 34, 44, 193, 210, 244, 258, 423, 481, 500, 502, 521
microscopy, 40, 132, 150, 159, 392, 506
microstructure, 43, 51, 194, 212, 341, 346, 385, 389, 390, 391, 423
microwave heating, 339
microwaves, 44, 197, 199, 233
migraines, 78
migration, 208, 220, 221, 394, 411, 437, 440, 447, 449, 453, 495
miniaturization, 191, 194, 200, 271
missions, 535
mixing, 17, 82, 109, 111, 112, 113, 173, 185, 270, 311, 348, 359, 366, 380, 384, 393, 444, 502, 533
modifications, 13, 115, 131, 146, 162, 220, 244, 363, 500
modified polymers, 485
modulus, 24, 178, 180, 245, 247, 250, 253, 355, 356, 357, 358, 362, 365, 421, 422, 424, 493, 530
moisture, 143, 171, 254, 271, 274, 366, 471, 496
mole, 336, 477, 480, 481, 483, 484, 485
molybdenum, 19, 79, 101, 144, 355, 414
momentum, 521
monolayer, 36, 38, 41, 43, 52, 53, 54, 55, 56, 57, 58, 225, 230, 388
monomers, 5, 194, 195, 197, 198, 199, 200, 205, 206, 208, 209, 234, 355, 384, 395, 419, 493, 495, 500, 504
morphology, 49, 131, 142, 144, 174, 176, 183, 187, 194, 222, 226, 346, 381, 385, 386, 388, 389, 390, 391, 392, 396, 417, 422, 472, 475, 479, 480, 485, 502
MRI, 512, 524, 525
multilayer films, 47, 49, 52
mutagenesis, 79

N

nafion, 185
naphthalene, 446, 478, 479
nausea, 78, 79, 80, 81
necrosis, 434
negative effects, 78
nervous system, 76, 80
neurotoxicity, 80
neutral, 50, 53, 54, 84, 105, 195, 197, 370, 380, 388, 442, 443, 444, 450, 488, 495
neutrons, 273
next generation, 219
NH2, 102, 114
nickel, 74, 77, 80, 93, 109, 144, 178, 180, 251, 299, 304, 314, 389, 393, 394, 421, 447, 501
nicotinic acid, 459
niobium, 70, 117
nitric oxide, 58
nitrides, 193, 232, 251
nitrobenzene, 94
nitrogen, 59, 93, 95, 193, 195, 200, 209, 220, 228, 233, 263, 293, 341, 378, 386, 468, 512, 513, 525, 526
Nobel Prize, 380, 530
noble metals, 116, 216, 407
nonionic surfactants, 87
nonlinear optics, 114
nucleating agent, 18
nucleation, 40, 44, 45, 48
nuclei, 44
nucleic acid, 97, 444
nutrients, 415

O

obstacles, 3, 20
obstructive lung disease, 79
occlusion, 367
oil, 76, 182, 183, 185, 200, 251, 412, 494, 501, 502
oligomers, 82, 174, 181, 268
one dimension, 36
opacity, 357
operations, 37, 77, 435, 529, 532
opportunities, 535
optical activity, 534

optical fiber, 57
optical properties, 55, 87, 175, 277, 355, 403, 499
optimal performance, 231
optimization, 21, 60, 208, 210, 234, 497
optoelectronic properties, 142
oral cavity, 361
orbit, 270
ores, 76, 451
originality, 203
osmium, 49, 55, 57
osmosis, 81, 292
osteoporosis, 78, 433
overtime, 357
oxalate, 107, 337
ozonation, 292
ozone, 34, 41, 263, 268

P

pain, 78, 81
paints, 77, 83, 172, 180, 183, 417, 420, 425
pairing, 521, 524
palladium, 48, 49, 232
PAN, 72, 93, 96, 108, 109, 111
pancreas, 81
parallel, 51, 142, 197, 203, 209, 212, 367, 515, 518, 522
partition, 439, 441, 443
passivation, 279, 294, 296, 302, 418
pathways, 7, 9, 13, 20, 137, 142, 151, 178, 222
PCM, 2, 20, 61
pellicle, 366
peptides, 47, 54, 444
perchlorate, 95, 106, 107, 120, 439
percolation, 142, 145, 222
perforation, 78
periodicity, 50
permeability, 1, 5, 10, 13, 14, 16, 19, 22, 24, 26, 108, 172, 179, 193, 199, 200, 207, 208, 210, 233, 277, 284, 417
permeation, 7, 9, 198, 221
permittivity, 374
peroxide, 354, 438, 493
PES, 429, 432, 434
pesticide, 77, 120, 458, 532
PET, 244
petroleum, 77, 119, 412, 416, 451, 465, 494, 539
pH, 9, 54, 55, 74, 75, 89, 90, 91, 95, 97, 98, 99, 101, 104, 105, 107, 108, 109, 112, 115, 116, 117, 118, 119, 172, 176, 305, 306, 307, 384, 407, 408, 410, 439, 442, 443, 444, 466, 480, 482, 483, 508
phase boundaries, 199
phase transformation, 274, 316
phenol, 56, 85, 92, 109, 456, 457, 530
phenolic resins, 93, 253
phonons, 272, 524
phosphate, 3, 6, 16, 17, 18, 21, 24, 25, 73, 76, 104, 106, 107, 111, 113, 114, 116, 117, 169, 186, 195, 230, 417, 420, 439, 443, 447, 448, 449, 450, 451, 532, 533
photocatalysts, 222
photoconductivity, 275, 284
photodegradation, 117
photoelectron spectroscopy, 34, 52, 193, 371, 405, 423
photoemission, 133, 163, 279
photographs, 217
photoluminescence, 257, 259, 262, 280, 283
photolysis, 495
photonics, 132, 162
photons, 50, 137, 139, 195, 264, 272, 273, 274
photooxidation, 81
photosensitivity, 145
photovoltaic cells, 222
photovoltaic devices, 131, 147, 148, 153, 157, 164
pitch, 93, 370, 378, 391, 493
pituitary gland, 80
plant diseases, 76
plants, 76, 172, 292, 444, 453, 466
plaque, 366
plasma membrane, 203, 234
plasma method, x, 220
plastic, 74, 76, 77, 252, 253, 256, 416, 491, 494, 495, 497
platelets, 176, 246, 251, 277
platform, 36, 54, 55, 56, 57
platinum, 39, 203, 211, 214, 218, 219, 221, 226, 227, 231, 232, 233, 391, 411, 450, 451, 468, 500, 501
PMDA, 72, 109
PMMA, 72, 93, 99, 170, 172, 185, 258, 278, 429, 434
pneumonia, 79, 80
point defects, 316, 332, 347
polar, 5, 10, 20, 21, 46, 171, 177, 422, 443, 496
polar groups, 177
polarity, 39, 139, 172, 440, 441
polarizability, 55
polarization, 15, 51, 206, 215, 221, 394, 418, 419
pollutants, 77, 81, 96, 104, 110, 119, 120, 292
pollution, 134, 220, 245, 534, 535
poly(2-hydroxyethyl methacrylate), 101
poly(methyl methacrylate), 72, 93, 97
poly(vinyl chloride), 534
polyacrylamide, 98, 115
polyamine, 73, 88, 89, 90
polyamines, 88

polybutadiene, 84
polycarbonate, 181, 355, 365
polychlorinated biphenyl, 77
polycondensation, 109
polycyclic aromatic hydrocarbon, 110, 495
polyesters, 171, 172, 181, 492
polyether, 23, 181, 431, 432
polyimide, 253, 254, 277, 535
polymer blends, 179, 497
polymer chain, 177, 473, 489, 507
polymer electrolytes, 9
polymer films, 47
polymer industry, 530
polymer materials, 275, 373, 381, 429, 482, 530
polymer matrix, 7, 19, 24, 25, 174, 178, 179, 181, 183, 243, 252, 253, 258, 278, 281, 386, 430, 478, 484, 485, 497, 531
polymer nanocomposites, 385
polymer structure, 173, 505, 507, 508
polymeric composites, 259, 430, 492, 531
polymeric materials, 96, 97, 252, 253, 354, 530
polymeric matrices, 387
polymerization process, 116, 205, 359, 386
polymerization temperature, 384
polymethylmethacrylate, 98, 172, 434
polypropylene, 253, 354, 365, 492
polysaccharide, 115, 444
polystyrene, 2, 22, 73, 84, 92, 96, 97, 107, 108, 111, 113, 133, 142, 172, 185, 203, 206, 253, 281, 355, 418, 493, 495, 532
polyvinyl acetate, 72, 99, 102
polyvinyl alcohol, 72, 99, 102, 446
polyvinyl chloride, 355
population, 80, 193, 371
porosity, 8, 44, 114, 115, 116, 179, 215, 221, 222, 227, 280, 346, 360, 380, 389, 392, 418, 424, 441
porous materials, 91, 104
porphyrins, 53, 57
potassium, 100, 108, 109, 375, 419, 447
power generation, 531
power plants, 466
precipitation, 81, 99, 105, 106, 110, 441
premolars, 366
prevention, x, 171, 404, 405, 406, 416, 420, 425
priming, 204
principles, x, 257, 259, 317, 372
probability, 212, 261, 262, 335, 520, 521
probe, 50, 52, 57, 196, 473
process control, 53, 57, 534
process gas, 204
product performance, 497
project, 248, 252, 425, 513
proliferation, 186, 361
propagation, 222, 265, 393, 412, 473
proportionality, 272, 334
propylene, 72, 84, 92, 375, 493
prosthesis, 435
prosthetic materials, 435
protection, 53, 69, 70, 119, 171, 172, 178, 179, 182, 183, 404, 410, 415, 416, 417, 418, 419, 420, 421, 424, 534
protective coating, 410, 412, 413, 414, 415, 416, 419, 420
proteins, 47, 78, 81, 116, 220
proton, 1, 3, 4, 5, 16, 21, 24, 194, 198, 207, 209, 230, 392
prototype, 3
proximal boxes, 363
psychosis, 80
PTFE, 2, 6, 7, 20, 193, 204, 205, 227, 292, 293, 295, 299, 302, 303, 371, 379
public health, 497
publishing, 349
pulmonary edema, 81
pulp, 77, 360, 361, 364
pumps, 407, 413
purification, 117, 119, 120, 293, 299, 302, 480
purity, 134, 270, 299, 323, 378, 439
PVA, 2, 6, 24, 26, 111, 244
PVC, 529, 534
pyrophosphate, 450

Q

quality of life, ix
quanta, 264
quantification, 453, 457
quantization, 516
quantum dot, 276
quantum mechanics, 520
quantum phenomena, 515
quantum state, 523
quartz, 34, 38, 49, 50, 51, 197, 354, 355, 356, 360, 467, 482, 503
quaternary ammonium, 22, 97, 107, 375

R

radiation, 22, 76, 79, 81, 96, 108, 113, 135, 147, 160, 169, 175, 259, 263, 271, 273, 274, 296, 525, 534
radical polymerization, 2, 22, 365
radicals, 25, 194, 195, 220
radio, 196, 209, 233, 254, 279, 357, 525
radioactive waste, 111, 532
radiography, 496

radioisotope, 108
radiopaque, 362
radius, 90, 317, 320
Raman spectroscopy, 385, 394, 422, 440
rare earth elements, 118
raw materials, 83, 492
reactant, 4, 221, 439
reaction temperature, 174, 303, 501
reaction time, 46, 186, 293, 296, 297, 303, 395, 501
reactivity, 41, 120, 181, 314
reagents, 452, 501
real time, 514
receptor, 36, 41, 53, 54, 55, 361
recognition, 53, 54, 55, 57, 58, 60, 534
recombination, 137, 139, 142, 155, 156, 194, 195, 261, 262, 265, 267, 270, 272, 274, 275, 281, 282
recombination processes, 137
recommendations, iv, 308
reconstruction, 434
recovery, 82, 88, 94, 106, 110, 116, 120
rectification, 160
recycling, 495, 496
red mud, 387
red shift, 477, 485
redistribution, 161
reduced lung function, 80
reflectance spectra, 161
reflectivity, 34, 52, 406
refractive index, 51, 82, 179, 266, 267
refractive indices, 179
regeneration, 58, 100, 105, 106, 107, 108, 110
regulations, 76, 77, 171, 292
reinforcement, 3, 172, 173, 180, 246, 247, 248, 250, 429, 430, 492, 493, 530
relaxation, 202, 258, 270, 272, 273, 274
reliability, 250, 372, 534
remediation, 83, 88, 109, 120, 291, 292, 308
remote sensing, 57
renewable energy, 131, 134, 193, 222, 369, 371
repair, 78, 412, 433
replication, 81
reprocessing, 496
reproduction, 284
repulsion, 520
requirements, 1, 7, 23, 188, 212, 246, 249, 255, 270, 280, 371, 430, 492, 531, 534
researchers, 45, 83, 171, 181, 257, 259, 311, 314, 320, 345, 348, 396, 397, 472, 513
residues, 458
resins, 88, 96, 100, 101, 102, 103, 106, 108, 171, 175, 181, 182, 253, 353, 355, 358, 360, 367, 443
resolution, 52, 192, 222, 271, 272, 279, 423, 437, 442, 449, 518
resorcinol, 93, 94, 180
resources, 182, 186, 193, 495, 496, 497
response, 36, 55, 56, 57, 58, 59, 60, 160, 198, 280, 360, 382, 465, 535
response time, 36, 55, 58, 60
restoration, 357, 358, 359, 362, 363, 364, 366, 367
restorative material, 353, 354, 359, 363, 364, 368
restrictions, 420
restructuring, 18
rhodium, 58
rings, 87, 252, 433, 434, 468, 475, 505, 506, 507
rods, 93, 144, 197, 221, 251, 395, 412, 433, 434, 472, 493
Romanus, 427
room temperature, 3, 8, 19, 21, 23, 25, 45, 57, 82, 137, 154, 157, 160, 181, 182, 183, 218, 221, 251, 274, 275, 299, 302, 316, 337, 338, 339, 388, 467, 482, 502, 503, 504, 513, 526, 535
root, 52, 150, 159, 232
root-mean-square, 52, 150, 159
roughness, 48, 50, 132, 150, 159, 199, 212, 367
Rouleau, 235
routes, 39, 47, 95, 275
Royal Society, 42, 44
rubber, 40, 77, 366
rules, 196, 278
rural areas, 74
ruthenium, 57, 58, 394
rutile, 9, 152, 156

S

safety, 72, 179, 181, 193, 244, 245, 252, 495
salinity, 172
saliva, 366
salts, 74, 171, 375, 384, 394, 410, 423, 435, 465, 468, 532
samarium, 317, 318
sapphire, 43, 49
saturation, 96, 139
scaling, 82
scanning calorimetry, 170
scanning electron microscopy, 132, 147, 183, 338, 370, 386, 419, 422, 467
scanning tunneling microscope, 521
scanning tunnelling microscopy, 40
scarcity, 380
scattering, 18, 34, 52, 162, 266, 274, 335
science, ix, 4, 35, 36, 56, 69, 88, 273, 353, 405, 491, 529, 530, 532, 539
scope, 38, 54, 187, 383, 396
scull, 493
seafood, 77

security, 53, 193
sediments, 108
segregation, 142, 144, 278
selected area electron diffraction, 222, 229
selectivity, 43, 55, 56, 57, 58, 60, 83, 90, 91, 95, 96, 102, 105, 106, 108, 110, 112, 113, 114, 146, 191, 194, 196, 198, 199, 453, 454, 479, 500, 533, 534
selenium, 74, 77, 81, 95, 458
self-assembly, 39, 93, 95, 424
self-destruction, 120
self-organization, 141, 142, 150
self-repair, 424
SEM micrographs, 218, 223, 224, 339, 484
semiconductor, 53, 57, 60, 134, 137, 146, 192, 220, 257, 264, 270, 271, 274, 275, 276, 281, 284, 312, 331, 335, 349, 478, 479, 480, 520, 522, 525
sensing, 36, 47, 53, 54, 55, 57, 59, 60, 85, 91, 96, 479, 525
sensitivity, 16, 49, 54, 56, 57, 58, 60, 377, 481, 500, 502, 534, 535
sensor functionalities, ix, 53
sensors, ix, 33, 35, 36, 45, 46, 53, 54, 58, 59, 60, 115, 183, 186, 276, 478, 479, 489, 497, 511, 524, 526, 531, 534, 535
septum, 78
shape, 43, 51, 91, 114, 115, 172, 173, 180, 221, 248, 250, 253, 272, 353, 386, 393, 433, 441, 492, 522
shear, 180, 493
shear strength, 180
shelter, 171
shock, 246, 250, 274, 432
shortness of breath, 79
showing, 20, 140, 156, 158, 160, 162, 196, 207, 210, 224, 262, 325, 382, 394, 441, 456, 478, 484, 489, 501
shyness, 80
Si3N4, 246, 248, 251, 422, 423
side chain, 7, 22, 134, 142
signal transduction, 81
signals, 481, 499
signs, 78, 271
silane, 46, 48, 49, 53, 85, 169, 175, 182, 354, 357, 364
silanol groups, 13, 88, 441
silver, 39, 41, 76, 115, 152, 156, 163, 183, 263, 299, 300, 302, 304, 305, 306, 307, 425
simulation, 424
sintering, 248, 337, 338, 339, 341, 345, 347, 424
SiO2, 6, 8, 13, 14, 20, 24, 25, 26, 27, 37, 38, 43, 46, 49, 58, 92, 176, 178, 185, 251, 337, 391, 424, 466, 467, 472, 533
skeleton, 172
sludge, 81, 110, 292
smart materials, 35
smoothing, 502
smoothness, 357, 358
SNS, 512, 522
SO42-, 105, 113, 117, 383
sodium, 84, 99, 107, 108, 112, 387, 418, 419, 421, 450, 451, 452
software, 503
solar cells, ix, 131, 133, 134, 135, 139, 141, 142, 143, 145, 146, 147, 148, 149, 151, 162, 165
sol-gel, 8, 9, 12, 14, 37, 38, 82, 83, 91, 109, 112, 173, 174, 181, 182, 185, 248, 275, 277, 280, 388, 389, 424, 529, 532
solid oxide fuel cells, 313, 314, 322, 323, 345, 349, 350
solid solutions, 316, 319, 320, 321, 322, 323, 326, 328, 329, 330, 332, 341
solid state, 14, 19, 83, 260, 283, 315, 497
solid surfaces, 36
solid waste, 497
solidification, 150, 492
solubility, 5, 9, 24, 74, 108, 119, 141, 324, 360, 384, 441
solvents, 20, 77, 93, 112, 142, 171, 196, 220, 292, 417, 440, 494, 495
sorption, 69, 82, 83, 85, 89, 90, 92, 95, 96, 97, 98, 100, 101, 102, 104, 105, 106, 107, 108, 109, 111, 112, 113, 114, 115, 117, 119, 120, 124, 356, 357, 358, 359, 386
space shuttle, 178
spatial location, 272
speciation, 117
species, 9, 20, 46, 48, 50, 51, 53, 55, 74, 81, 82, 84, 93, 104, 105, 174, 196, 197, 205, 223, 272, 274, 277, 332, 333, 334, 335, 417, 420, 437, 444, 473
spectral techniques, 48
spectro-chemical interrogation, x
spectroscopic, x, 51, 272, 279, 440, 449, 459, 480, 481, 482, 483, 485, 489, 502, 503, 506, 507, 508
spectroscopy, 34, 51, 52, 53, 55, 57, 65, 132, 133, 147, 163, 170, 206, 312, 331, 370, 381, 382, 386, 387, 404, 405, 418, 419, 423, 438, 440, 465, 477, 478, 482, 499
speed of light, 139, 268, 525
spin, 39, 134, 146, 152, 156, 270, 276, 278, 280, 281, 514
splinting, 363
standard deviation, 151
starch, 182, 186, 444, 446, 448, 449, 450, 451, 496
steel, 76, 77, 200, 203, 205, 246, 249, 250, 355, 371, 391, 392, 393, 394, 404, 405, 412, 416, 418, 419, 420, 421, 424, 432, 434, 493, 535
STM, 34, 40, 52

stoichiometry, 318, 324, 331
storage, x, 3, 42, 47, 97, 253, 279, 360, 369, 371, 372, 373, 375, 376, 377, 387, 388, 393, 395, 396, 403, 416, 478, 493, 501, 524
stress, 59, 180, 199, 246, 353, 356, 357, 358, 359, 362, 363, 364, 403, 405, 412, 493
stretching, 468, 488, 505
strong interaction, 5, 38, 177, 195, 219, 234
strontium, 74
structural changes, 491
structural characteristics, 183
styrene, 2, 16, 22, 113, 181, 203, 206, 208, 209, 371, 387, 393, 394, 419, 446, 465, 466, 467, 495, 532
substitutes, 117, 380
substitution, 41, 114, 331, 435, 530
sulfate, 2, 6, 73, 107, 110, 117, 185, 438, 444, 458
sulfur, 74, 205, 375, 459
sulphur, 185, 415, 475, 481, 485, 505
superconducting gap, 525
superconducting materials, 513, 524, 526
superconductivity, 219, 513, 514, 515, 521, 522, 523, 525, 526, 527
superconductor, 512, 513, 514, 515, 516, 518, 520, 521, 522, 523, 524, 525
superlattice, 403
suppression, 199, 522
surveillance, 534
survival, 367, 368
susceptibility, 357
sustainability, 193, 396
swelling, 18, 23, 26, 97, 116, 209, 381, 414, 444
SWNTs, 392, 535
symmetry, 163, 197
symptoms, 78, 80, 81
synergistic effect, 195, 228, 234, 384, 389, 494
synthetic fiber, 493

T

talc, 444, 446
tantalum, 179, 232
target, 36, 53, 75, 204, 205, 211, 212, 213
taxonomy, 378
TBP, 439, 445, 447, 449, 450, 452
technological advancement, 171
technological advances, 403, 425
technologies, 69, 88, 117, 252, 271, 281, 284, 369, 371, 498, 526, 532
teeth, 172, 353, 357, 358, 363, 364, 366
TEM, 170, 183, 193, 220, 221, 222, 226, 227, 228, 229, 371, 385, 386, 388, 389, 395, 396, 405, 423
temperature dependence, 206, 207, 521
template molecules, 85, 91
tendon, 78, 493
tensile strength, 10, 26, 27, 245, 250, 252, 253, 412, 418, 424
TEOS, 73, 84, 92, 93, 170, 176, 181, 185
testing, 145, 164, 230, 250
testosterone, 453
tetraethoxysilane, 83, 176
texture, 226, 403, 492
TGA, 170, 337, 371, 387, 388, 465, 466, 467, 470, 471
thermodynamics, 377
thermograms, 471
thermoplastics, 253
time use, 253
tissue, 367, 429, 430, 433, 496
titanate, 16, 108, 111, 258, 280
titania, 6, 16, 179, 219
titanium, 9, 41, 93, 110, 111, 144, 185, 186, 219, 221, 222, 246, 251, 252, 417, 421, 425, 431, 435, 466
toluene, 13, 85, 371, 393, 452, 496
tooth, 353, 354, 357, 358, 361, 362, 364, 365, 366, 367, 368
topology, 36, 186
total energy, 521
trace elements, 408
trade, 279, 444
transcription, 81
transducer, 60, 500, 504
transference, 318, 319, 321, 322, 323, 326, 327, 330, 331, 343
transformation, 80, 85, 99, 186, 293
transistor, 281
transparency, 82, 169, 175, 177, 179, 181, 277, 432
transport, 7, 9, 79, 131, 132, 133, 140, 141, 144, 145, 147, 151, 156, 163, 179, 193, 209, 216, 221, 255, 257, 258, 259, 260, 261, 262, 263, 269, 275, 278, 280, 282, 314, 331, 377, 388, 394
transportation, 3, 27, 76, 156, 163, 226, 323, 371, 473, 531
tribology, 420
tricarboxylic acid, 33
trihalomethanes, x, 291, 292, 305, 306, 307
tungsten, 19, 144, 178, 229, 354, 355, 358, 360, 367, 477, 479, 480, 481, 482, 484, 485
tungsten carbide, 178, 358, 367
tunneling, 34, 517, 519, 520, 521
turbulence, 413

U

ultrasound, 384
unconventional superconductors, 523, 524

uranium, 6, 107, 111, 331, 332, 333
urea, 176
urethane, 354, 357
urine, 78

V

vacancies, 315, 316, 317, 318, 323, 331, 333, 334, 337, 341, 342
vacuum, 58, 132, 134, 139, 144, 156, 162, 178, 197, 200, 205, 220, 227, 259, 263, 268, 271, 276, 391, 481
valence, 52, 141, 160, 272, 274, 316, 323, 330, 513
valve, 496
vanadium, 19, 74, 114, 144
vapor, 5, 38, 39, 51, 57, 192, 198, 218, 220, 244, 258, 259, 263, 380, 413, 421, 440, 458, 497
variables, 161, 391
variations, 144, 421, 488, 518
varieties, 36
vector, 512, 521
vegetable oil, x, 169, 182
vehicles, 3, 58, 191, 244, 254, 371, 405, 425, 493
velocity, 404, 413, 417, 421, 422, 426, 512, 522
versatility, 46, 439
vertebrates, 172
vessels, 200, 406, 531
vibration, 209, 335
vinyl chloride, 529
vinylidene fluoride, 2, 22, 493
viscose, 77
viscosity, 179, 181, 357, 358, 361, 367
vision, 33, 80, 256, 366
vitamin B1, 79
vitamin B12, 79
vitamins, 453, 454, 457
volatile organic compounds, 56
volatility, 39, 181
vomiting, 78, 79, 80, 81

W

Waterborne, 170, 181, 405
wavelengths, 482, 485
weak interaction, 38, 54
wear, 178, 180, 212, 224, 249, 250, 252, 353, 357, 358, 359, 362, 363, 364, 365, 367, 413, 417, 420, 421, 423, 431
welding, 81
wettability, 50, 196
wetting, 425
wheezing, 79
wind turbines, 172
wires, 227, 267, 525, 526
wood, 76, 77, 179, 181, 188, 253, 378, 494
workers, 42, 177, 178, 182, 183, 201, 215, 396
World Health Organization (WHO), 73, 74, 75, 76, 77
World War I, 523, 530
worldwide, 180, 491, 532

X

XPS, 34, 52, 55, 58, 133, 163, 193, 206, 228, 371, 394, 405

Y

yarn, 77
yield, 79, 174, 250, 262, 271, 272, 284, 294, 296, 297, 298, 300, 302, 305, 307, 308, 477, 480, 481, 482, 483, 485, 497, 534
ytterbium, 362

Z

zeolites, 6, 8, 42, 82, 108, 115, 443